# Lecture Notes in Artificial Intelligence     8733

## Subseries of Lecture Notes in Computer Science

T0215685

Dosam Hwang   Jason J. Jung
Ngoc Thanh Nguyen (Eds.)

# Computational
# Collective Intelligence

## Technologies and Applications

6th International Conference, ICCCI 2014
Seoul, Korea, September 24-26, 2014
Proceedings

 Springer

Volume Editors

Dosam Hwang
Jason J. Jung
Yeungnam University
Department of Computer Engineering
Dae-Dong Gyeungsan, Korea, 712-749
E-mail: {dosamhwang, j2jung}@gmail.com

Ngoc Thanh Nguyen
Wroclaw University of Technology
Institute of Informatics
Wybrzeże Wyspiańskiego 27, 50-370 Wroclaw, Poland
E-mail: ngoc-thanh.nguyen@pwr.edu.pl

ISSN 0302-9743                         e-ISSN 1611-3349
ISBN 978-3-319-11288-6                 e-ISBN 978-3-319-11289-3
DOI 10.1007/978-3-319-11289-3
Springer Cham Heidelberg New York Dordrecht London

Library of Congress Control Number: 2014947827

LNCS Sublibrary: SL 7 – Artificial Intelligence

*Typesetting:* Camera-ready by author, data conversion by Scientific Publishing Services, Chennai, India

Printed on acid-free paper

Springer is part of Springer Science+Business Media (www.springer.com)

# Computational Collective Intelligence – Technologies and Applications
## 6$^{th}$ International Conference ICCCI 2014

## Preface

This volume contains the proceedings of the 6$^{th}$ International Conference on Computational Collective Intelligence (ICCCI 2014) held in Seoul, Korea, September 24-26, 2014. The conference was co-organized by Wrocław University of Technology (Poland), Yeungnam University, Hanyang University and Dankook University (Korea). The conference was run under the patronage of the IEEE SMC Technical Committee on Computational Collective Intelligence.

Following the successes of the 1$^{st}$ ICCCI (2009) held in Wrocław, Poland, the 2$^{nd}$ ICCCI (2010) in Kaohsiung, Taiwan, the 3$^{rd}$ ICCCI (2011) in Gdynia, Poland, the 4$^{th}$ ICCCI (2012) in Ho Chi Minh city, Vietnam, and the 5$^{th}$ ICCCI (2013) in Craiova, Romania, this conference continued to provide an internationally respected forum for scientific research in the computer-based methods of collective intelligence and their applications.

Computational Collective Intelligence (CCI) is most often understood as a sub-field of Artificial Intelligence (AI) dealing with soft computing methods that enable making group decisions or processing knowledge among autonomous units acting in distributed environments. Methodological, theoretical, and practical aspects of computational collective intelligence are considered as the form of intelligence that emerges from the collaboration and competition of many individuals (artificial and/or natural). The application of multiple computational intelligence technologies such as fuzzy systems, evolutionary computation, neural systems, consensus theory, etc., can support human and other collective intelligence, and create new forms of CCI in natural and/or artificial systems. Three subfields of application of computational intelligence technologies to support various forms of collective intelligence are of special attention but are not exclusive: semantic web (as an advanced tool increasing collective intelligence), social network analysis (as the field targeted to the emergence of new forms of CCI), and multiagent systems (as a computational and modeling paradigm especially tailored to capture the nature of CCI emergence in populations of autonomous individuals).

The ICCCI 2014 conference featured a number of keynote talks, oral presentations and invited sessions, closely aligned to the theme of the conference. The conference attracted a substantial number of researchers and practitioners from all over the world, who submitted their papers for the main track subdivided into 10 thematic streams and 3 special sessions.

The main track streams, covering the methodology and applications of computational collective intelligence, included: knowledge integration, data mining for collective processing, fuzzy, modal and collective systems, nature inspired systems, language processing systems, social networks and semantic web, agent and multi-agent systems, classification and clustering methods, multi-dimensional data processing, web systems, intelligent decision making, methods for scheduling, image and video processing.

The special sessions, covering some specific topics of particular interest, included: collective intelligence in web systems, computational swarm intelligence, and cooperation and collective knowledge.

We received in total 205 submissions from 23 countries. Each paper was reviewed by 2-4 members of the International Program Committee and International Reviewer Board. Only 70 best papers have been selected for oral presentation and publication in the volume of the Lecture Notes in Artificial Intelligence series.

We would like to express our sincere thanks to the honorary chairs, Philip S. Yu, Pierre Lévy, Jin Hyung Kim, and Tadeusz Więckowski for their support.

We also would like to express our thanks to the keynote speakers - Francis Heylighen (Belgium), Il-Hong Suh (Korea), and Mirjana Ivanović (Serbia), for their world-class plenary speeches.

Special thanks go to the organizing chair, Dosam Hwang for his efforts in the organizational work. Thanks are due to the program co-chairs, Program Committee and the Board of Reviewers, essential for reviewing the papers to ensure the high quality of accepted papers. We thank the publicity chairs, special sessions chairs and the members of the Local Organizing Committee.

Finally, we cordially thank all the authors, presenters and delegates for their valuable contributions to this successful event. The conference would not have been possible without their supports.

It is our pleasure to announce that the conferences of ICCCI series are closely cooperating with the Springer journal *Transactions on Computational Collective Intelligence*, and the IEEE SMC Technical Committee on *Transactions on Computational Collective Intelligence*.

We hope and intend that ICCCI 2014 significantly contributes to fulfillment of the academic excellence and leads to even greater successes of ICCCI events in the future.


September 2014                                                    Dosam Hwang
                                                                 Jason J. Jung
                                                             Ngoc Thanh Nguyen

# ICCCI 2014 Conference Organization

## Honorary Chairs

Tadeusz Więckowski — Rector of Wrocław University of Technology, Poland

| | |
|---|---|
| Tadeusz Więckowski | Rector of Wrocław University of Technology, Poland |
| Pierre Lévy | University of Ottawa, Canada |
| Jin Hyung Kim | KAIST, Korea |
| Philip S. Yu | University of Illinois at Chicago, USA |

## General Chairs

| | |
|---|---|
| Ngoc Thanh Nguyen | Wrocław University of Technology, Poland |
| Sang-Wook Kim | Hanyang University, Korea |

## Steering Committee

| | |
|---|---|
| Ngoc Thanh Nguyen (Chair) | Wrocław University of Technology, Poland |
| Piotr Jędrzejowicz (Co-Chair) | Gdynia Maritime University, Poland |
| Shyi-Ming Chen | National Taiwan University of Science and Technology, Taiwan |
| Adam Grzech | Wrocław University of Technology, Poland |
| Kiem Hoang | University of Information Technology, VNU-HCM, Vietnam |
| Lakhmi C. Jain | University of South Australia, Australia |
| Geun-Sik Jo | Inha University, Korea |
| Janusz Kacprzyk | Polish Academy of Sciences, Poland |
| Ryszard Kowalczyk | Swinburne University of Technology, Australia |
| Ryszard Tadeusiewicz | AGH University of Science and Technology, Poland |
| Toyoaki Nishida | Kyoto University, Japan |

## Program Chairs

| | |
|---|---|
| Jason J. Jung | Yeungnam University, Korea |
| Piotr Jędrzejowicz | Gdynia Maritime University, Poland |
| Kazumi Nakamatsu | University of Hyogo, Japan |
| Edward Szczerbicki | University of Newcastle, Australia |

## Organizing Chair

| | |
|---|---|
| Dosam Hwang | Yeungnam University and KAIST, Korea |

## Liaison Chairs

Costin Badica                      University of Craiova, Romania
Sai Peck Lee                       University of Malaya, Malaysia

## Local and Web Chairs

Seungbo Park                       Dankook University, Korea
Duc Trung Nguyen                   Yeungnam University, Korea

## Special Session Chairs

David Camacho                      Universidad Autónoma de Madrid, Spain
Bogdan Trawinski                   Wrocław University of Technology, Poland

## Publicity Chair

Marcin Maleszka                    Wroclaw University of Technology, Poland
Tutut Herawan                      University of Malaya, Malysia

## Finance Chair

Pankoo Kim                         Chosun University, Korea

## Keynote Speakers

Prof. Francis Heylighen            Free University of Brussels, Belgium
Prof. Il-Hong Suh                  Hanyang University, Korea
Prof. Mirjana Ivanović             University of Novi Sad, Serbia
Prof. Key-Sun Choi                 KAIST, Korea

## Special Sessions

CI4BC 2014: 2nd International Workshop on Computational Intelligence for
Business Collaboration
Organizers: Jason J. Jung and Huu-Hanh Hoang

WebSys'2014: Collective Intelligence in Web Systems - Web Systems Analysis
Organizers: Kazimierz Choroś and Maria Trocan

CompStory14: The First International Symposium on Computational Story
Organizers: Seung-Bo Park, Eun-Soon Yoo and Jason J. Jung

FMADIS 2014: The 1st International Workshop on "Frontier Management and Intelligent Decision Support for Highly Ill-Structured Decision Problems
Organizers: Kun Chang Lee

CSI 2014: Computational Swarm Intelligence - High Effectiveness and Pick Efficiency in Optimization
Organizers: Urszula Boryczka

## International Program Committee

| | |
|---|---|
| Muhammad Abulaish | King Saud University, Saudi Arabia |
| Cesar Andres | Universidad Complutense de Madrid, Spain |
| Amelia Badica | University of Craiova, Romania |
| Amar Balla | Ecole Supérieure d'Informatique (ESI), Algeria |
| Dariusz Barbucha | Gdynia Maritime University, Poland |
| Nick Bassiliades | Aristotle University of Thessaloniki, Greece |
| Maria Bielikova | Slovak University of Technology in Bratislava, Slovakia |
| Olivier Boissier | ENS Mines Saint-Etienne, France |
| Urszula Boryczka | University of Silesia, Poland |
| Aleksander Byrski | AGH University of Science and Technology, Poland |
| José Luís Calvo-Rolle | Universidad de La Coruña, Spain |
| David Camacho | Universidad Autonoma de Madrid, Spain |
| Tru Cao | Ho Chi Minh City University of Technology, Vietnam |
| Frantisek Capkovic | Slovak Academy of Sciences, Slovakia |
| Dariusz Ceglarek | Poznan School of Banking, Poland |
| Krzysztof Cetnarowicz | AGH University of Science and Technology, Poland |
| Tzu-Fu Chiu | Aletheia University, Taiwan |
| Amine Chohra | Paris-East University, France |
| Kazimierz Choros | Wroclaw University of Technology, Poland |
| Dorian Cojocaru | University of Craiova, Romania |
| Mihaela Colhon | University of Craiova, Romania |
| Tina Comes | Centre for Integrated Emergency Management, University of Agder, Norway |
| Phan Cong-Vinh | NTT University, Vietnam |
| Irek Czarnowski | Gdynia Maritime University, Poland |
| Paul Davidsson | Malmo University, Sweden |
| Phuc Do | University of Information Technology, Vietnam |
| Tien Van Do | Budapest University of Technology and Economics, Hungary |
| Trong Hai Duong | Inha University, Korea |
| Atilla Elçi | Süleyman Demirel University, Turkey |

| | |
|---|---|
| Bogdan Trawinski | Wroclaw University of Technology, Poland |
| Jan Treur | Vrije University, The Netherlands |
| Olgierd Unold | Wroclaw University of Technology, Poland |
| Roberto De Virgilio | Università degli Studi Roma Tre, Italia |
| Iza Wierzbowska | Gdynia Maritime University, Poland |
| Drago Zagar | University of Osijek, Croatia |
| Danuta Zakrzewska | Lodz University of Technology, Poland |
| Constantin-Bala Zamfirescu | University of Sibiu, Romania |
| Katerina Zdravkova | University Sts Cyril and Methodius, Macedonia |

# Table of Contents

## Social Networks

## E-learning Systems

## Pattern Recognition

## Expert Systems and Applications

## GIS Applications

## Computational Intelligence

## Ontologies, Graphs and Networks

# Machine Learning

# Data Mining

# Cooperation and Collective Knowledge

## Computational Swarm Intelligence

## Collective Intelligence in Web Systems - Web Systems Analysis

# Agreement Technologies – Towards Sophisticated Software Agents [*]

Mirjana Ivanović and Zoran Budimac

Department of Mathematics and Informatics,
Faculty of Sciences, University of Novi Sad, Serbia
{mira,zjb}@dmi.uns.ac.rs

**Abstract.** Nowadays, agreements and all the processes and mechanisms implicated in reaching agreements between different kinds of agents are a subject of perspective interdisciplinary scientific research. Newest trend in Agent Technology is to enhance agents with "social" abilities. Agreement Technologies brings new flavor in implementation of more sophisticated autonomous software agents that negotiate to achieve acceptable agreements.

The paper presents key concepts in this area and highlights influence of Agreement Technologies on development of more sophisticated multi-agent systems.

## 1    Introduction

It is impossible to imagine contemporary world without agreements. Human ability to reach agreements is present in all their interactions and without them there is no cooperation in social systems. Human social skills represent an intriguing challenge for researchers and have led to the emergence of a new research field, Agreement Technologies (AT) [30]. AT refer to computer systems in which autonomous software agents negotiate with one another, typically on behalf of humans, in order to come to mutually acceptable agreements.

One among most important initiatives in the area of AT was big project COST Action IC0801. The Action was funded for 4 years (2008–2012), comprised about 200 researchers from 25 European countries and 8 institutions from other continents working on topics related to AT [20]. The overall mission of the project was to support and promote the harmonization of high-quality research towards a new paradigm for next generation distributed systems based on the notion of agreement between computational agents and support technology transfer to industry.

The rest of the paper is organized as follows. In Section 2, basic concepts of AT are briefly presented. Section 3 brings wider view on basic concepts of AT and their role in multi-agent environments. In section 4 several AT research projects are briefly presented. Last section concludes the paper.

---

[*] The work is partially supported by Ministry of Education and Science of the Republic of Serbia, through project no. OI174023: "Intelligent techniques and their integration into wide-spectrum decision support"

D. Hwang et al. (Eds.): ICCCI 2014, LNAI 8733, pp. 1–10, 2014.

## 2    Agreement Technologies in Brief

Software agents are essential concept appearing and supporting people in different working environments. Future open distributed systems are supposed to support interactions between software *agents* based on the concept of *agreements* where two key elements are needed: a normative context that defines the "space" of agreements that the agents can possibly reach; an interaction mechanism by means of which agreements are first established, and then enacted [21]. Interactions between sophisticated software agents can be abstracted to the establishment of *agreements for execution*, and a subsequent *execution of agreements*.

**Table 1.** Key dimensions of Agreement Technologies

| Dimension | Challenge |
|---|---|
| Semantics | Support of application-dependent ontology: domain-specific objects and language interpretation. |
| | Explicit and exploitable representation of environment aspects and semantics. |
| | Combining knowledge in large-scale open settings & reconciling subjective views. |
| | Learning the semantics of everything, out of cases of inspecting and exploiting the interactions of others with the environment (within specific contexts of interaction). |
| | Inventing commonly agreed languages for interaction. |
| Norms | To define a standard way for representing the events and actions that happen in an environment. |
| | To define in a standard way how to represent the context of the interactions in terms of properties of resources and their value. |
| | To define general mechanisms for contextualizing abstract norms defined at design-time into norms situated in specific spaces. |
| | To easily extend the functionalities/services provided by the environment for adding those required for norms management. |
| Organizations | To provide facilities to enter or exit a given organization to allow run-time recruitment of new members as well as voluntary desertion and/or expulsion of members. |
| | To support on-demand creation, deletion and modification of organizations. |
| | To give support to the institutional components of an organization, i.e. norms, powers, agreements. |
| | Agents must be able to make use of the elements of the environment, such artifacts, that provide all these previous functionalities and facilities. |
| Argumentation | Scaling up existing work, which typically considers single interactions between a small number of agents. |
| | Management of libraries and database of ontologies, protocols, agreements. |
| | Participating agents in a system then need the ability to reason about ontologies and interaction protocols and to invoke them as required. |
| Trust | To exploit the existing strong link between trust/reputation and the environment in electronic environments. |
| | To give the trust and reputation system the capacity to influence the actions of the agent to modify the environment so it becomes more trust and reputation "friendly". |

Accordingly agreements have to be changed *dynamically* at run-time, and there must be mechanisms for re-assessing and *revising* them during the execution. It introduces "interaction-awareness" term where software components explicitly represent and reason about agreements and associated processes. There are several key dimensions where new solutions for the establishment of agreements need to be developed [2]: *Semantics, Norms, Organizations, Argumentation*, and *Trust (Tab. 1)*.

Currently characteristic areas of applications of AT are *E-Commerce, Transportation Management and E-Governance but* in *near future seems that* AT will inevitable support *smart energy grids and virtual power plants* [25].

# 3     Key Dimensions of Agreement Technologies

In this section essential dimensions of AT will be presented separately.

**Semantics in Agreement Technologies.** Semantic Web consists of several standards [23]: XML, RDF, Ontologies [17], RIF, XQuery, SPARQL. In AT Semantic Web standards include additional elements: Policies, Norms and the Semantic Web "Trust Layer"; Evolution of Norms and Organizations; Implicit Versus Explicit Norms.

Semantic Web standards serve for representing the knowledge of local agents, in order to achieve a goal making agreement with other agents. In distributed, open and heterogeneous systems that use AT, formalisms of Semantic Web have limitations. Recently researchers have been proposing a number of formalisms for handling the situations in which pieces of knowledge are defined independently in various contexts. They are known as *contextual logics* or *distributed logics* or *modular ontology languages* and usually extend classical logics or the logics of Semantic Web.

**Norms in Agreement Technologies.** Norms recently have been an issue of growing interest in agent oriented research. They deal with coordination and security and started to be important mechanisms to regulate electronic institutions. Study of norms includes different views: cognitive science, behavioral and evolutionary economics, computational and simulation-based social science [12]. Accordingly an innovative understanding of norms and their dynamics (on individual and social level) emerged. Deontic logic is highly connected to norms. It is a formal system that attempts to capture the essential logical features of these concepts but still there are several key research questions and dilemmas connected to deontic logic that have to be resolved in order to fully use them in real environments.

Some specific architecture that incorporates interaction between beliefs, obligations, intentions and desires in the formation of agent goals is proposed in [4]. In 'BOID' architecture essential issue discussed is the interaction between 'internal' and 'external' motivations (deriving from norms of the agent's social context). Constitutive norms are extremely important mechanism [3] for normative reasoning in dynamic and uncertain environments (agent communication in e-contracting).

Works on model agents interactions (based on cooperation or coordination) [28] have been studying how these norms emerge. An interesting approach is presented in [27] where authors propose a data-mining for the identification of norms.

Also promising research area for the study of norms could be inclusion of humans where agents can learn from humans, software agents can recommend norms to humans that are most applicable in a given context.

**Organizations and Institutions in Agreement Technologies.** Open multi-agent systems and AT are promising technologies for organizations and institutions. Complex task or problem in organizations can be solved by appropriate declarative specifications to a number of agents. Agents can work together as teams in order to solve delegated task that helps reach the global goals of the organization. Besides, the notion of institution has been used within the agent community to model and implement a variety of socio-technical systems. Agents have to know how to access the services of the infrastructure and to make requests according to the available organizational specification. Such *"organization aware"* agents possess skills to contemplate the organization and decide whether or not to enter such a structure and whether or not to comply with the different rights and duties promoted by the organization.

In modern complex socio-technical systems it is not possible to possess and keep updated all the information about the environment. Agent-oriented modeling [29] presents a holistic approach for analyzing and designing organizations consisting of humans and technical components (both are *agents*). Recently several different organizational models have been developed: Moise [15], AGR [9], TAEMS [19], OperA [7], AGRE [9], MOISEInst [10], ODML [14], TEAM [31], AUML [22], MAS-ML [6].

**Augmentation and Negotiation in Agreement Technologies.** In last decade Argumentation has been researched extensively in computing especially for inference, decision making and support, dialogue, and negotiation in order to reach agreement. Agreement also benefits from negotiation, especially when autonomous agents have conflicting interests/desires but may benefit from cooperation. Formal logic provides a promising paradigm for modeling reasoning in the presence of conflict and uncertainty, and for communication between reasoning entities. The nature of argumentation is predominantly modular and most formal models adopt that: (1) arguments are constructed in some underlying logic that manipulates statements about the world; (2) interactions between arguments are defined; (3) given the network of interacting arguments, the winning arguments are evaluated. Recent work in computer science community has illustrated the potential for implementations of logical models of argumentation, and the wide range of their application in different software systems.

Nowadays the challenging area of research is Online negotiations.

**Trust and Reputation in Agreement Technologies.** Computational trust and reputation mechanisms at the moment have reached certain level of maturity.

Trust always denotes an agent (trustee) behavior that may interfere with the truster own goals. Equipping intelligent agents with ability to estimate the trustworthiness of interacting partners is crucial in improving their social interactions [26]. This means that agents use *computational trust models* based on trust theories to assist their trust-based decisions. *Degree of trust is another important element connected to trust.* The *strength* of trust is some kind of measure of the degree of trust. In [8] authors introduce situational trust by defining trust as a measurable belief that the truster has on the competence of the trustee in behaving in a dependably way, in a given period of time, within a given context and relative to a specific task. So to construct robust computational trust models, it is necessary to understand how trust forms and evolves. This will allow intelligent agents to promote their own trustworthiness, and to allow them to correctly predict others' trustworthiness even in case of new partnerships.

Reputation is a social concept as complex as trust. Interrelation between trust and reputation is rather ambiguous: 1) reputation is an antecedent of trust, and it may or may not influence the trust, 2) the process of reputation building is subject to specific social influences that are not present in the process of building trust.

Although computational *reputation is a field that* has its own set of research questions different researchers have proposed models of computational trust and reputation that integrate both social concepts, assuming the perspective of reputation as an antecedent of trust [16], [26].

# 4    Real World Applications

In last decade a lot of authors implemented different frameworks that demonstrate the use of AT in a variety of real-world scenarios [13], [1], [32]. This section briefly discusses three examples of such frameworks.

## 4.1    Augmentation and Negotiation in Agreement Technologies

In [13] author proposed a system that allows the technicians of a call centre to provide a high quality customer support employing case-based argumentation.

Nowadays, a lot of companies offer very similar products, prices and quality. They try to be better than their competitors by offering focused customer care supported by high quality and fast service within call centre.

On the other hand less experienced technicians are cheaper and it is interesting to provide them with a means for arguing, contrasting their views with other technicians and reaching agreements to solve (collaboratively) as many requests as possible [13].

In the system the technicians are representing as software agents. Agents are engaged in an argumentation process to try to find the best solution to be applied to each new incidence that the call centre receives. This hybrid system integrates an argumentation framework where agents are provided with argumentation capabilities and individual knowledge resources. In such Multi-Agent Systems software agents are capable to manage and exchange arguments taking into account the agents' *social context* (roles, dependency relations and preferences). This virtual call center is based on a society of agents that act on behalf of a group of technicians that must solve problems in a Technology Management Centre (TMC). Therefore, virtual call centre technicians support similar real agent's roles: operator, expert and administrator. Also, each agent can have its own values that it wants to promote or demote (adjust the reasons that an agent has to give preference to certain decisions) and that represent its motivation to act in a specific way.

When a new request is received appropriate *ticket* is generated representing the problem to be solved. In the system are possible also complex cases where a ticket must be solved by a group of agents (as technicians). They must argue to reach an agreement over the best solution to apply. Each agent also has its own knowledge resources to generate a solution for the ticket. This process is represented in Fig. 1.

The system has been implemented and tested in a real call centre. The Magentix2 agent platform (http://users.dsic.upv.es/grupos/ia/sma/tools/magentix2/index.php) has been used for implementation of this MAS. Magentix2 is a platform that provides new tools that allow the secure and optimized management of open MAS.

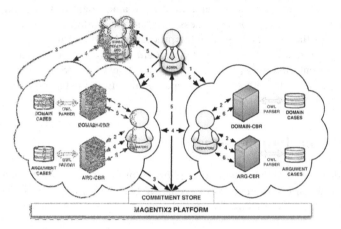

**Fig. 1.** Data-flow for the argumentation process of the call centre application

It has been integrated as an argumentation module that agents can use to persuade other agents to accept their proposed solutions as the best way to solve the problem.

## 4.2    ANTE: Agreement Negotiation in Normative and Trust-Enabled Environments

The ANTE framework is based on three main agreement technology concepts: negotiation, normative environments and computational trust [5]. ANTE is targeting B2B electronic contracting but could be seen as a more general. It addresses the issue of multi-agent collective work including negotiation for finding mutually acceptable agreements, the enactment of agreements and the evaluation of the enactment phase, in order to improve future negotiations. Computational trust may therefore be used to appropriately capture the trustworthiness of negotiation participants, both in terms of the quality of their proposals when building the solution. Computational trust adopted in ANTE framework is based on the ability to compute adequate estimations of trustworthiness in several different environments, including those of high dynamicity.

ANTE has been applying in different scenarios and its application in disruption management in Airline Operations Control Centre (AOCC) is presented here. The AOCC is the organization responsible for monitoring and solving operational problems that might occur during the execution of the airline operational plan. Disruption Management includes teams of experts specialized in solving problems related to aircrafts and flights, crewmembers and passengers.

MASDIMA (Fig. 2) is an agent-based application that represents AOCC of an airline company. AOCCs have a process to monitor the events and solve the problems, so that flight delays minimize costs: Crew Costs, Flight Costs, Passenger Costs, the cost of delaying or cancelling a flight from the passenger point of view. When a disruption appears, the AOCC needs to find the best solution that minimizes costs and get back to the previous operational plan. The crucial negotiation happened between the *Supervisor* and the *A/C, Crew and Pax* manager agents. The Supervisor acts as the organizer agent and the managers as respondents. Each manager does not possess the full expertise to be able to propose a solution to the supervisor. So manager needs to start an inter-manager negotiation to be able to complete their proposal and participate in the main negotiation. To minimize activities it is enough to have at least one manager for each part of the problem. Nevertheless, in the environment more than one agent with the same expertise in the same dimension of the problem can exist.

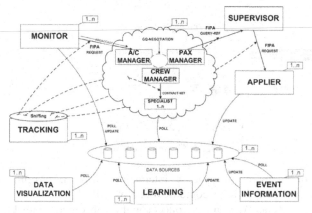

**Fig. 2.** MASDIMA Architecture

In this scenario, trust is used when the supervisor is evaluating negotiation proposals from the managers. The trust information is built from the application of the winner solution on the environment through the *Applier* agent. Applier agent is applying the winning solution to the environment, which checks the successful execution of the solution. Supervisor agent enables connection of monitoring facility with the trust engine to use evidence regarding the quality of solutions.

To present a proposal manager agents first need to find a candidate solution. To find these candidate solutions, each manager might have a team of problem solving agents that are able to find and propose those solutions.

## 4.3   The Augmentation Web

The plethora of argument visualization and mapping tools [18] testifies to the enabling function of argumentation-based models for human clarification and understanding, and for promoting rational reasoning and debate. The development of such tools is a consequence of existence of pile of discussion forums on the web, and the lack of support for checking the relevance and rationality of online discussion and debate. Such tools offer possibility of reuse of *readymade* arguments authored online (i.e. mining of arguments from online resources).

To facilitate the development of such tools and the reuse of authored arguments new systems and standards on the Internet are needed. This leads to the new concept *Argument Web* [24] (Fig. 3). It can serve as a common platform that brings together applications in different domains (e.g. broadcasting, mediation, education and healthcare) and interaction styles (e.g. real-time online debate, blogging).

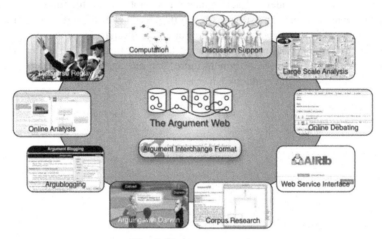

**Fig. 3.** The argument web

A number of examples of specific interactions with the Argument Web that illustrate usage of prototype tools could be found in [20].

Further in the Social Web, users connect with each other, share knowledge and experiences of all types, and debate with exchange of arguments (e.g. in comments on blogs). However, the argumentative structure is implicit, arguments need to be inferred, and debates are unstructured, often chaotic.

Most recent work on online systems and argumentation focuses on extracting argumentation frameworks using argument schemes and Semantic Web technology for editing and querying arguments [24].

Recently the way people access the Web is changing: they change their status in Facebook or Twitter. A possible reason of this in the Internauts' life style is that your Facebook friends will actually give you better information and decrease useless spam. Indeed, search engines are interested in opinion mining and sentiment analysis [11].

It seem that future use of web will be based on sentiment-aware search engine that mines large online discussion boards, include advanced clustering of result based on user agreement/sentiment, and fully integrates argumentation in a Social Web context.

## 5    Conclusions

The paper brings key concepts, dilemmas and possible usage of AT in open distributed environments predominantly based on multi-agent systems. These define environments that are based on norms, argumentations and trust within which agents interact. AT are obviously contemporary, interesting and promising research area. Its multidisciplinary and interdisciplinary character offers great future possibilities for applications in more intelligent and sophisticated artificial societies.

# References

1. Alberola, J.M., Such, J.M., Botti, V., Espinosa, A., Garcia-Fornes, A.: A Scalable Multiagent Platform for Large Systems. Computer Science and Information Systems 10(1), 51–77 (2013)
2. Argente, E., Boissier, O., Carrascosa, C., Fornara, N., McBurney, P., Noriega, P., Ricci, A., Sabater-Mir, J., Ignaz Schumacher, M., Tampitsikas, C., Taveter, K., Vizzari, G., Vouros, G.A.: Environment and Agreement Technologies. In: AT 2012, pp. 260–261 (2012)
3. Boella, G., van der Torre, L.W.N.: Constitutive norms in the design of normative multiagent systems. In: Toni, F., Torroni, P. (eds.) CLIMA 2005. LNCS (LNAI), vol. 3900, pp. 303–319. Springer, Heidelberg (2006)
4. Broersen, J., Dastani, M., van der Torre, L.: Beliefs, obligations, intentions and desires as components in an agent architecture. International Journal of Intelligent Systems 20(9), 893–920 (2005)
5. Cardoso, H.L., Urbano, J., Brandão, P., Rocha, A.P., Oliveira, E.C.: ANTE: Agreement Negotiation in Normative and Trust-Enabled Environments. In: PAAMS 2012, pp. 261-264 (2012)
6. da Silva, V.T., Choren, R., de Lucena, C.J.P.: A UML based approach for modeling and implementing multi-agent systems. In: International Joint Conference on Proceedings of the Autonomous Agents and Multi-Agent Systems, vol. 2, pp. 914–921. IEEE Computer Society, Los Alamitos (2004)
7. Dignum, V.: A model for organizational interaction: Based on agents, founded in logic. Ph.D. thesis, Universiteit Utrecht (2004)
8. Dimitrakos, T.: System models, e-risks and e-trust. In: Proceedings of the IFIP Conference on Towards the E-society: E-Commerce, E-Business, E-Government, I3E 2001, pp. 45–58. Kluwer, Deventer (2001)
9. Ferber, J., Gutknecht, O., Michel, F.: From agents to organizations: An organizational view of multi-agent systems. In: Giorgini, P., Müller, J.P., Odell, J.J. (eds.) AOSE 2003. LNCS, vol. 2935, pp. 214–230. Springer, Heidelberg (2004)
10. Gâteau, B., Boissier, O., Khadraoui, D., Dubois, E.: Moiseinst: An organizational model for specifying rights and duties of autonomous agents. In: Third European workshop on multi-agent systems (EUMAS 2005), Brussels, pp. 484–485 (2005)
11. Godbole, N., Srinivasaiah, M., Skiena, S.: Large-scale sentiment analysis for news and blogs. In: Proceedings of the international Conference on weblogs and social media (ICWSM), Salt Lake City (2007)
12. Gostojić, S., Milosavljević, B., Konjović, Z.: Ontological Model of Legal Norms for Creating and Using Legislation. Computer Science and Information Systems 10(1), 151–171 (2013)
13. Heras, S.: Case-based argumentation framework for agent societies. Ph.D. thesis, Departamento de Sistemas Informáticos y Computación. Universitat Politècnica de València (2011), http://hdl.handle.net/10251/12497,
14. Horling, B., Lesser, V.: A survey of multi-agent organizational paradigms. The Knowledge Engineering Review 19(4), 281–316 (2005)
15. Hübner, J.F., Sichman, J., Boissier, O.: A model for the structural, functional, and deontic specification of organizations in multi-agent systems. In: Bittencourt, G., Ramalho, G.L. (eds.) SBIA 2002. LNCS (LNAI), vol. 2507, pp. 118–128. Springer, Heidelberg (2002)

16. Huynh, T.D., Jennings, N.R., Shadbolt, N.R.: An integrated trust and reputation model for open multi-agent systems. Autonomous Agents and Multi-Agent Systems 13, 119–154 (2006)
17. Ivanovic, M., Budimac, Z.: An overview of ontologies and data resources in medical domains. Expert Syst. Appl. 41(11), 5158–5166 (2014)
18. Kirschner, P.A., Buckingham Shum, S.J., Carr, C.S.: Visualizing argumentation: Software tools for collaborative and educational sense-making. Springer, London (2003), http://oro.open.ac.uk/12107/
19. Lesser, V., Decker, K., Wagner, T., Carver, N., Garvey, A., Horling, B., Neiman, D., Podorozhny, R., NagendraPrasad, M., Raja, A., Vincent, R., Xuan, P., Zhang, X.: Evolution of the gpgp/taems domain-independent coordination framework. In: Autonomous Agents and Multi-Agent Systems, vol. 9(1), pp. 87–143. Kluwer Academic Publishers (2004)
20. Ossowski, S.: Agreement Technologies. Springer Series: Law, Governance and Technology Series, vol. 8, XXXV, p. 645 (2013)
21. Ossowski, S., Sierra, C., Botti, V.: Agreement Technologies: A Computing perspective. In: Ossowski, S. (ed.) Springer Series: Law, Governance and Technology Series, vol. 8, pp. 3–16 (2013)
22. Van Dyke Parunak, H., Odell, J.J.: Representing social structures in UML. In: Wooldridge, M.J., Weiß, G., Ciancarini, P. (eds.) AOSE 2001. LNCS, vol. 2222, pp. 1–16. Springer, Heidelberg (2002)
23. Polleres, A., Huynh, D.: Special issue: The web of data. Journal of Web Semantics 7(3), 135 (2009)
24. Rahwan, I., Zablith, F., Reed, C.: Laying the foundations for a world wide argument web. Artificial Intelligence 171, 897–921 (2007)
25. Ramchurn, S., Vytelingum, P., Rogers, A., Jennings, N.: Putting the "Smarts" into the smart grid: A grand challenge for artificial intelligence. Communications of the ACM 55(4), 86–97 (2012)
26. Sabater-Mir, J., Paolucci, M.: On Representation and aggregation of social evaluations in computational trust and reputation models. International Journal of Approximate Reasoning 46(3), 458–483 (2007)
27. Savarimuthu, B.T.R., Cranefield, S., Purvis, M.A., Purvis, M.K.: Obligation norm identification in agent societies. Journal of Artificial Societies and Social Simulation 13(4) (2010), http://jasss.soc.surrey.ac.uk/13/4/3.html
28. Sen, S., Airiau, S.: Emergence of norms through social learning. In: Proceedings of the Twentieth International Joint Conference on Artificial Intelligence (IJCAI), pp. 1507–1512. AAAI Press, Menlo Park (2007)
29. Sterling, L., Taveter, K.: The art of agent-oriented modeling. MIT, Cambridge/London (2009)
30. Shaheen, F.S., Wooldridge, M., Jennings, N.R.: Optimal Negotiation of Multiple Issues in Incomplete Information Settings. In: AAMAS 2004, pp. 1080–1087 (2004)
31. Tambe, M., Adibi, J., Alonaizon, Y., Erdem, A., Kaminka, G.A., Marsella, S., Muslea, I.: Building agent teams using an explicit teamwork model and learning. Artificial Intelligence 110(2), 215–239 (1999)
32. Vrdoljak, L., Podobnik, V., Jezic, G.: Forecasting the Acceptance of New Information Services by using the Semantic-aware Prediction Model. Computer Science and Information Systems 10(3), 1025–1052 (2013)

# False Positives Reduction on Segmented Multiple Sclerosis Lesions Using Fuzzy Inference System by Incorporating Atlas Prior Anatomical Knowledge: A Conceptual Model

Hassan Khastavaneh and Habibollah Haron

Faculty of Computing, Universiti Teknologi Malaysia,
81310 UTM Skudai, Johor, Malaysia
khassan3@live.utm.my, habib@utm.my

**Abstract.** Detecting abnormalities in medical images is an important application of medical imaging. MRI as an imaging technique sensitive to soft tissues shows Multiple Sclerosis (MS) lesions as hyper-intense or hypo-intense signals. As manual segmentation of these lesions is a laborious and time consuming task, many methods for automatic MS lesion segmentation have been proposed. Because of inherent complexities of MS lesions together with acquisition noises and inaccurate pre-processing algorithms, automatic segmentation methods come up with some False Positives (FP). To reduce these FPs a model based on fuzzy inference system by incorporating atlas prior anatomical knowledge have been proposed. The inputs of proposed model are MRI slices, initial lesion mask, and atlas information. In order to mimic experts inferencing, proper linguistic variable are derived from inputs for better description of FPs. The experts knowledge is stored into knowledge-base in if-then like statement. This model can be developed and attached as a module to MS lesion segmentation methods for reducing FPs.

**Keywords:** multiple sclerosis lesion, segmentation, false positive reduction, fuzzy inference system, atlas anatomical knowledge, MRI, MS.

## 1 Introduction

The goal of many medical image segmentation methods is to delineate abnormal structures such as tumors, lesions, masses, pathologies, and so forth. Alongside with the complexity of abnormal structures, acquisition noises are inherent characteristics of medical images that make the segmentation of abnormal structures of interest a complex task. Sensitivity or True Positive (TP) rate and specificity or True Negative (TN) rate both are two important performance measure metrics that all segmentation methods attempts to keep them in their highest possible values. Any attempt for increasing TP and TN rates leads the False Positive (FP) and False Negative (FN) rates to increase as well. In medical imaging

D. Hwang et al. (Eds.): ICCCI 2014, LNAI 8733, pp. 11–19, 2014.

applications it is common to segment all suspected structures to keep the sensitivity and specificity values in an acceptable range which leads the segmentation methods to come up with high number of FPs. These FPs should be reduced to make the segmentation methods robust and reliable enough for clinical trials. FP reduction process can be done either manually by a human operator or automatic. In automatic cases, usually a module is attached as a post-processing phase to segmentation methods.

Multiple Sclerosis (MS) is one of the most common diseases of Central Nervous System (CNS) including brain and spinal cord. In MS, the myelin sheets that shield the nerves are damaged which causes problems in message transmission. Loss of myelin goes along with an interruption in the ability of the nerves to conduct electrical impulses in the CNS that causes various symptoms. The areas where myelin is lost lesions appear as scar areas. Measuring the lesion load (size and number of lesions) is an important metric for diagnoses of MS; it is necessary for follow-up sessions and also observing the effect of drugs and research purposes. As Magnetic Resonance Imaging (MRI) is an imaging technique sensitive to soft tissues like brain, MS lesions appear as hyper-intense or hypo-intense signals depending on MRI modality. Figure 1 illustrate a FLAIR slice of axial view of brain with its corresponding lesion mask that masks hyper-intense signals together with a lesion mask that has some FPs.

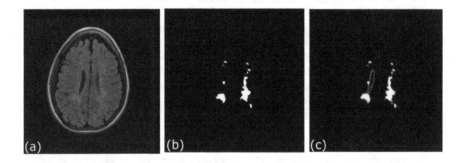

**Fig. 1.** (a) MRI slice. (b) Ground truth. (c) Lesion mask with potential regions (red areas) for FP.

Usually physicians delineate lesions by looking at the MRI slices manually. Manual segmentation of MS lesions is a cumbersome, time consuming and error prone task. So, automatic segmentation of MS lesions is a need. Many methods [1] [2] [3] and models [4] have been proposed for automatic segmentation of MS lesions based on Support Vector Machines (SVM), K-Means, Fuzzy c- Means (FCM), K-Nearest Neighborhood (KNN), Bayesian network, Markov Random Fields (MRF), and so forth. The accuracy of MS lesion segmentation methods is not satisfactory enough to be used in clinical trials because of the inherent complexity of lesions such as their diversity in shape, location, and size, their fuzzy borders, and partial volume effect. As a result of these complexities, mentioned segmentation methods come up with some FPs which removing them leads toward more accurate and reliable segmentation methods.

Regardless of lesion complexities, there are many reasons for occurring FPs in MS lesion segmentation methods; one of them is image artifacts that appear in some MRI modalities. Another reason is inaccuracy of pre-processing steps of segmentation methods like registration and skull-stripping. Existence of brain vessels also causes FPs. In some situations a voxel/pixel has properties of both lesion and normal tissue which increase the potential of occurring FPs.

Brain atlases are used to simplify and facilitate segmentation or classification of brain tissue by providing spatial context (prior anatomical knowledge). There are two types of atlases, probabilistic and topological. Probabilistic atlases give information that in some areas of brain MS lesions are more probable or less probable; in other words the probability of belonging an individual Voxel/pixel to a particular tissue type at a specified location is obtained from probabilistic atlases. Topological atlases are used for preserving topology as well as lowering the influence of competing intensity clusters in regions that are spatially disconnected. To obtain information from atlases for segmenting brain slices, the slices should be registered on the atlas.

Fuzzy Inference System (FIS) is a way of mimicking experts common sense by mapping an input space to an output space non-linearly using fuzzy set theory. It has many applications such as Control Systems, Robot Navigation, Voice Recognition, Machine Vision, and so forth. A FIS has four components namely, Fuzzifier, Defuzzifier, Inference Engine, and Fuzzy Knowledge Base (FKB). Fuzzifier converts the crisp inputs (sets) to fuzzy sets; the inference engine together with FKB which stores experts knowledge carries inferencing task; defuzzifier converts fuzzy output of inferencing component to a crisp set [5]. In the proposed model FIS is used to mimic human inferenceing for FP reduction.

In this article a voxel/pixel based model for reducing FPs on segmented MS lesions is proposed; the proposed model is based on FIS which reduce FPs with aid of atlas prior anatomical knowledge and definition of proper linguistic variables. This model can be developed, implemented, and attached to any MS lesion segmentation method for removing FPs.

The rest of this paper is organized as follow: the related works are discussed in section 2. The proposed conceptual model will be explained in section 3. Finally, the proposed conceptual model will be discussed and concluded in section 4.

## 2   Related Works

As mentioned previously, FP reduction could be manual or automatic. In manual cases a human operator detects and eliminates FPs from the segmentation results; in automatic cases usually machine learning algorithms are employed to detect and eliminate FPs without human intervention. Automatic reduction of FPs could be region based or voxel/pixel based. In region based methods features of segmented regions such as size and shape are used to judge whether the region is an actual lesion or FP; in voxel/pixel based methods usually the location and differences in statistics of individual voxel/pixel is used for FP detection and reduction.

In [6] FIS was used to reduce the FP pixel/voxels around the boundary of segmented lesion regions by defining three linguistic variables. The first linguistic variable is derived from the Euclidean distance between the pixel and boundary of the lesion region; the second linguistic variable is derived from the difference between the gray-scale of the pixel and the mean gray-scale of the lesion region; and the third linguistic variable is derived from the segmentation score of the pixel which is result of the initial segmentation step. The membership functions and the threshold values are calculated based on the analysis of the initial segmentation of the training subject dataset. Thirteen fuzzy rules have been defined which 8 of them keep the pixels as lesion and the rest remove pixels from lesion regions.

Co-registration inaccuracy of MRI modalities is the reason of FPs in [7]. These FPs creates a third class which is far from both lesion and non-lesion classes. So, the FPs are eliminated by measuring the distance of their attribute vector from lesion and non-lesion classes in Hilbert space. In [8] the lesions are considered as outliers of normal appearing brain tissue model. Other voxel/pixels than white mater hyper-intense signals, are considered as false outliers because of noises, vessels, and partial volumes. To remove the FPs, experts rules are applied; these rules are derived from MRI intensity and voxel/pixel connectivity.

In [9], the lesion regions with the volume smaller than a certain threshold are considered as FPs. This threshold is selected empirically by applying different minimum lesion volumes; the selected threshold is the one with highest overlap with the ground truth. In another effort again the small lesion regions are considered as FPs and they are automatically removed. But, in a further step in this method user/application preferences can be integrated for removing FPs; this leads the method to focus on hyper-intensities in a certain region [10].

In [11] the lesion mask which is the classification result was refined to eliminate FPs caused by bony and flow artifacts by defining the regions of interest. In a further step some voxel/pixels belongs to either white matter mask or pure corticospinal fluid mask are removed from lesion mask. In [12] clustering of normal brain into its three dominant tissue types and background is used as a prior knowledge for elimination of FPs. The elimination is supported by morphological operators. In another effort by [13], to correct the FPs that appears because of FLAIR artifacts two parameters have been considered. The first parameter is the threshold of corticospinal fluid membership image and the second parameter is the size of structuring element used for creating FP mask from the corticospinal fluid segmentation. So, the FPs are corrected by selecting the optimized parameters.

As have been reported by literatures, the better methods for FP reduction are the ones that mimic expert reduction process like the reduction done by [6]. To do it, the experts knowledge must be formalized as a set of rules and also the rules must be applied properly. The next section will described how the proposed conceptual method removes FPs by incorporating atlas information using FIS.

## 3   The Proposed Conceptual Model

In order to automatically and accurately reduce the FPs on segmented MS lesions in MR images, a model has been proposed. In this model, FPs are reduced using FIS aided by atlas prior anatomical knowledge. As illustrated in figure 2, the block diagram of proposed model has four components namely Model Inputs, Pre-Processing, Fuzzy False Positive Reduction (FFPR), and Model Output. The main burden of reduction is carried out by FFPR component which is a complete fuzzy expert system with three processes namely Fuzzification, Inferencing using experts rule base, and Defuzzification. In the following sections the details of each component will be explained.

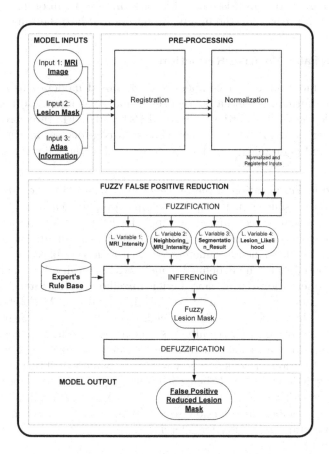

**Fig. 2.** Block diagram of proposed conceptual model for reducing false positives on segmented MS lesions using FIS by incorporating atlas prior anatomical knowledge

## 3.1  Model Inputs

To eliminate the FP voxel/pixels, three inputs are considered in this model. These inputs are the intensity of original MRI, initial lesion mask of MRI, and atlas prior knowledge. The entire required linguistic variables for inferencing process are derived from the inputs.

## 3.2  Pre-processing

Pre-processing component in the proposed model undertakes two important Registration and Normalization tasks. As atlas information shows the probability of occurring lesion in a certain location of brain, it is necessary to align MRI and its corresponding initial lesion mask on the lesion probability map. As mentioned previously, mis-registration is one of the reasons for causing the FPs; so, the accuracy of registration in this method is very important. In order to facilitate the fuzzification process, all the inputs are normalized into the range of [0,1].

## 3.3  Fuzzy False Positive Reduction

FFPR is the most important component of proposed model. This component receives three normalized and atlas registered images as input and its output is FP reduced lesion mask. In order to remove FPs based on FIS three processes like all fuzzy expert systems must be done. These three processes are fuzzification, inferencing, and defuzzification which are described in the following sections.

**Fuzzification.** Fuzzification is the process of deriving linguistic values for linguistic variables using crisp inputs by defining proper membership functions. The parameters that are considered by expert for FP reduction are size of initial lesion, lesion likelihood, intensity of individual voxel/pixel and its neighboring intensity. Based on mentioned parameters four linguistic variables that inferencing will be done based on them are defined namely MRI-Intensity, Neighboring-MRI-Intensity, Segmentation-Result, and Lesion-Likelihood. MRI-Intensity variable considers the intensity of individual voxel/pixel; Neighboring-MRI-Intensity variable considers the average intensity of neighboring voxel/pixels; Segmentation-Results variable considers the results of segmentation methods including lesion size; and finally Lesion-Likelihood variable considers the likelihood of occurring lesion in the location of considering voxel/pixel. Figure 3 illustrate linguistic variables, linguistic values, and their corresponding membership functions. The number of linguistic values that are defined for each linguistic variable and their corresponding membership function seems to be enough to cover the expert opinion on a certain parameter for inferencing.

**Inferencing.** Inferencing is the process of evaluating the rules defined by expert and stored in the experts rule base. The rules are actually if-then structures. The fuzzified inputs are applied on the rule antecedents and consequently the result of

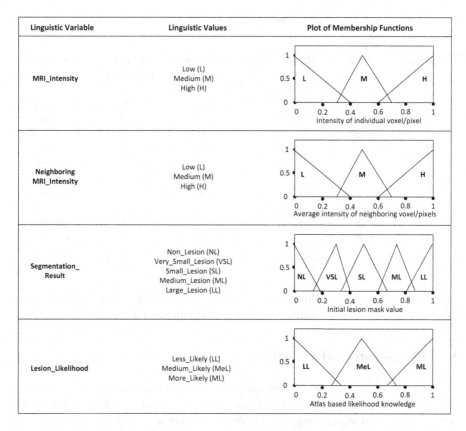

**Fig. 3.** Linguistic variables, Linguistic values and plot of their corresponding membership functions

antecedent is applied to the consequents membership function. After evaluation of all the rules, they are aggregated into one membership function. As table 1 shows eleven fuzzy rules have been defined by experts by observation on the data. Six of these rules keep the lesion voxel/pixels and the rest removes the FPs.

**Deffuzification.** In order to have a crisp output, it is essential to defuzzify fuzzy output of model. The input of defuzzification process is the aggregated results of all evaluated rules. In the proposed model defuzzification component carries out this task.

### 3.4   Model Output

The output of proposed model is a FP reduced lesion mask. This corrected lesion mask increase the performance of segmentation methods. A sample output could

**Table 1.** Expert defined fuzzy rules for Removing or Keeping lesion voxel/pixels from initial lesion mask

| Rule No. | MRI-Intensity | Neighboring-MRI-Intensity | Segmentation-Result | Lesion-Likelihood | Decision |
|---|---|---|---|---|---|
| 1 | × | × | LL | ML | Keep |
| 2 | × | H | LL | × | Keep |
| 3 | H | × | LL | × | Keep |
| 4 | × | × | ML | ML | Keep |
| 5 | × | H | ML | × | Keep |
| 6 | H | × | ML | × | Keep |
| 7 | × | L | VSL | × | Remove |
| 8 | × | M | VSL | × | Remove |
| 9 | × | × | SL | LL | Remove |
| 10 | × | L | SL | × | Remove |
| 11 | × | × | NL | × | Remove |

be a lesion mask in between figure 1 b and figure 1 c. If the output is more close to ground truth, it implies more accuracy of the proposed model.

## 4   Discussion and Conclusion

As FPs are inevitable part of segmentation results, we proposed a FP reduction model based on fuzzy inference system to increase both sensitivity and specificity of MS lesion segmentation methods using fuzzy inference system. The main segmentation method can be set to segment all the potential lesion voxel/pixels; then, the developed version of proposed model can correct the initial segmentation mask by reducing FPs. The proposed model mimics experts inferencing using the rules have been defined by expert. Here a variety of FPs can be detected and removed because of proper definition of linguistic variables and their respective membership functions. For example: Neighboring-MRI-Intensity and Segmentation-Result can consider lesion size and eliminate small lesion regions by considering the surrounding voxel/pixels. The membership functions can be tuned for better performance under experiment. Atlas prior information aid the reduction by defining less probable and more probable lesion regions. In future works we are going to do an experiment and evaluate the proposed model.

**Acknowledgments.** The authors would like to thanks Research Management Centre (RMC), Universiti Teknologi Malaysia (UTM) for the support in R & D for the inspiration in making this study a success.

# References

1. Mortazavi, D., Kouzani, A.Z., Soltanian-Zadeh, H.: Segmentation of multiple sclerosis lesions in MR images: a review. Neuroradiology 54, 299–320 (2012)
2. Llad, X., Oliver, A., Cabezas, M., Freixenet, J., Vilanovab, J.C., Quiles, A., Valls, L., Rami-Torrent, L., Rovirae, A.: Segmentation of multiple sclerosis lesions in brain MRI: A review of automated approaches. Information Sciences 186, 164–185 (2012)
3. Garca-Lorenzo, D., Francis, S., Narayanan, S., Arnold, D.L., Collins, D.L.: Review of automatic segmentation methods of multiple sclerosis white matter lesions on conventional magnetic resonance imaging. Medical Image Analysis 17, 1–18 (2013)
4. Khastavaneh, H., Haron, H.: A Conceptual Model for Segmentation of Multiple Sclerosis Lesions in Magnetic Resonance Images Using Massive Training Artificial Neural Network. In: 5th International IEEE Conference on Intelligent Systems, Modelling and Simulation, pp. 273–278. IEEE Press (2014)
5. Negnevitsky, M.: Artificial Intelligence: A Guide to Intelligent Systems, Addison-Wesley (2005)
6. Abdullah, B.A., Younis, A.A., Pattany, P.M., Saraf-Lavi, E.: Textural Based SVM for MS Lesion Segmentation in FLAIR MRIs. Open Journal of Medical Imaging 1, 26–42 (2011)
7. Lao, Z., Shen, D., Liu, D., Jawad, A.F., Melhem, E.R., Launer, L.J., Bryan, R.N., Davatzikos, C.: Computer-Assisted Segmentation of White Matter Lesions in 3D MR Images Using Support Vector Machine. Academic Radiology 15, 300–313 (2008)
8. Garcia-Lorenzo, D., Prima, S., Morrissey, S.P., Barillot, C.: A robust Expectation-Maximization algorithm for Multiple Sclerosis lesion segmentation. In: The MIDAS Journal - MS Lesion Segmentation, MICCAI 2008 Workshop (2008)
9. Steenwijk, M.D., Pouwels, P.J.W., Daams, M., Dalen, J.W., Caan, M.W.A., Richard, E., Barkhof, F., Vrenken, H.: Accurate white matter lesion segmentation by k nearest neighbor classification with tissue type priors (kNN-TTPs). NeuroImage: Clinical 3, 462–469 (2013)
10. Admiraal-Behloul, F., van den Heuvel, D.M.J., Olofsen, H., van Osch, M.J.P., van der Grond, J., van Buchem, M.A., Reiber, J.H.C.: Fully automatic segmentation of white matter hyperintensities in MR images of the elderly. NeuroImage 28, 607–617 (2005)
11. Souplet, J., Lebrun, C., Ayache, N., Malandain, G.: An Automatic Segmentation of T2-FLAIR Multiple Sclerosis Lesions. In: The MIDAS Journal - MS Lesion Segmentation, MICCAI 2008 Workshop (2008)
12. Kok, H.O., Dhanesh, R., Rajeswari, M., Ibrahim, L.S.: Automatic white matter lesion segmentation using an adaptive outlier detection method. Magnetic Resonance Imaging 30, 807–823 (2012)
13. Simes, R., Mnninghoff, C., Dlugaj, M., Weimar, C., Wanke, I., Walsum, A.V.C.V., Slump, C.: Automatic segmentation of cerebral white matter hyperintensities using only 3D FLAIR images. Magnetic Resonance Imaging 31, 1182–1189 (2013)

# Fuzzy Splicing Systems

Fariba Karimi[1], Sherzod Turaev[2], Nor Haniza Sarmin[3], and Wan Heng Fong[4]

[1] Royal Society Wolfson Biocomputation Research Lab,
Biocomputation School of Computer Science,
University of Hertfordshire,
Hatfield, Hertfordshire AL10 9AB, UK
fk.karimi@gmail.com
[2] Department of Computer Science,
Kulliyyah of Information and Communication Technology,
International Islamic University Malaysia,
53100 Kuala Lumpur, Malaysia
sherzod@iium.edu.my
[3] Department of Mathematical Sciences, Faculty of Science,
Universiti Teknologi Malaysia,
81310 UTM Johor Bahru, Johor, Malaysia
nhs@utm.my
[4] Ibnu Sina Institute for Fundamental Science Studies,
Universiti Teknologi Malaysia,
81310 UTM Johor Bahru, Johor, Malaysia
fwh@ibnusina.utm.my

**Abstract.** In this paper we introduce a new variant of splicing systems, called *fuzzy splicing systems*, and establish some basic properties of language families generated by this type of splicing systems. We study the "fuzzy effect" on splicing operations, and show that the "fuzzification" of splicing systems can increase and decrease the computational power of splicing systems with finite components with respect to fuzzy operations and cut-points chosen for threshold languages.

## 1 Introduction

Though computers have gained such a dominant position in our life, they have many drawbacks: there are numerous intractable problems, which cannot be solved with their help. *DNA computing* appears as a challenge to develop new types of algorithms and to design new types of computers which differ from classical notions of algorithms and computers in fundamental way. DNA computing models use *Watson-Crick complementary* of DNA molecules that are double stranded structures composed of four nucleotides *A* (*adenine*), *C* (*cytosine*), *G* (*guanine*) and *T* (*thymine*) always presenting in pairs *A–T* and *C–G*. Another feature of DNA molecules is the *massive parallelism* of DNA strands, which allows constructing many copies of DNA strands and carrying out operations on the encoded information simultaneously. The use of these two fundamental features of DNA molecules has already illustrated that DNA based computers can

D. Hwang et al. (Eds.): ICCCI 2014, LNAI 8733, pp. 20–29, 2014.

solve many computationally intractable problems: Hamiltonian Path Problem [1], the Satisfiability Problems [2,3], etc.

A concept of *splicing system*, one of the early theoretical proposals for DNA based computation, was introduced by Head [4] using a *splicing operation* – a formal model for DNA recombination under the influence of restriction enzymes. This process works as follows: two DNA molecules are cut at specific subsequences and the first part of one molecule is connected to the second part of the other molecule, and vice versa. This process can be formalized as an operation on *strings*, described by a so-called *splicing rule*, which are the basis of a computational model called a *splicing system*. A system starts from a given set of strings (*axioms*) and produces a *language* by iterated splicing according to a given set of splicing rules. Because of practical reasons, the case when the components of splicing systems are finite is of special interest. But splicing systems with finite sets of axioms and rules generate only regular languages (see [5]). Consequently, several restrictions in the use of rules have been considered (for instance, see [6]), which increase the computational power up to the Turing equivalent languages.

The treatment of splicing systems as language-generating devices allows using concepts, methods and techniques of formal language theory to study the properties of splicing systems. One can easily adapt many extension and restriction mechanisms associated with grammars and automata for splicing systems. In this paper we focus on the study of "fuzzified" splicing systems, whose grammar and automata counterparts have widely been investigated recent years (for details, see the monograph [7]). The concept of fuzzy splicing systems is introduced as follows: we associate the truth values from the closed interval $[0,1]$ with each axiom, and calculate the truth value of a string $w$ resulted from strings $u$ and $v$ applying a fuzzy operation over their truth values. We select a subset of the language generated by a fuzzy splicing system according to some cut-points in $[0,1]$, which is called a *threshold language*. We show that some threshold languages with the selection of appropriate cut-points can generate non-regular languages.

This paper is organized as follows. Section 2 contains some necessary definitions and notations from the theories of formal languages and splicing systems. In Section 3, the concepts of fuzzy splicing systems and threshold languages generated by fuzzy splicing systems are introduced. Section 4 shows the power of fuzzy splicing systems: some fuzzy splicing systems of finite components can generate context-free and context-sensitive languages. Section 5 discusses some open problems and possible topics for future research in this direction.

## 2    Preliminaries

In this section we recall some prerequisites, by giving basic notions and notations of the theories of formal languages and splicing systems which are used in sequel. The reader is referred to [8,6,9] for further information.

Throughout the paper we use the following general notations. The symbol $\in$ denotes the membership of an element to a set while the negation of set membership is denoted by $\notin$. The inclusion is denoted by $\subseteq$ and the strict (proper) inclusion is denoted by $\subset$. The empty set is denoted by $\emptyset$. The cardinality of a set $X$ is denoted by $|X|$. The families of recursively enumerable, context-sensitive, context-free, linear, regular and finite languages are denoted by **RE, CS, CF, LIN, REG** and **FIN**, respectively. For these language families, the next strict inclusions, named *Chomsky hierarchy* (see [9]), hold:

**Theorem 1. FIN $\subset$ REG $\subset$ LIN $\subset$ CF $\subset$ CS $\subset$ RE.**

Further, we briefly cite some basic definitions and results of iterative splicing systems which are needed in the next section.

Let $V$ be an alphabet, and $\#, \$ \notin V$ be two special symbols. A *splicing rule* over $V$ is a string of the form

$$r = u_1 \# u_2 \$ u_3 \# u_4, \text{ where } u_1, u_2, u_3, u_4 \in V^*.$$

For such a rule $r \in R$ and strings $x, y, z \in V^*$, we write

$$(x, y) \vdash_r z \text{ if and only if } x = x_1 u_1 u_2 x_2, \ y = y_1 u_3 u_4 y_2, \text{ and } z = x_1 u_1 u_4 y_2,$$

for some $x_1, x_2, y_1, y_2 \in V^*$.

The string $z$ is said to be obtained by splicing $x, y$, as indicated by the rule $r$; the strings $u_1 u_2$ and $u_3 u_4$ are called the *sites* of the splicing. We call $x$ the *first term* and $y$ the *second term* of the splicing operation.

An *H scheme* (a *splicing scheme*) is a pair $\sigma = (V, R)$, where $V$ is an alphabet and $R \subseteq V^* \# V^* \$ V^* \# V^*$ is a set of splicing rules. For a given H scheme $\sigma = (V, R)$ and a language $L \subseteq V^*$, we write

$$\sigma(L) = \{z \in V^* \mid (x, y) \vdash_r z, \text{ for some } x, y \in L, r \in R\},$$

and we define

$$\sigma^*(L) = \bigcup_{i \geq 0} \sigma^i(L)$$

by

$$\sigma^0(L) = L,$$
$$\sigma^{i+1}(L) = \sigma^i(L) \cup \sigma(\sigma^i(L)), i \geq 0.$$

An *extended H system* is a construct $\gamma = (V, T, A, R)$, where $V$ is an alphabet, $T \subseteq V$ is the *terminal* alphabet, $A \subseteq V^*$ is the set of *axioms*, and $R \subseteq V^* \# V^* \$ V^* \# V^*$ is the set of *splicing rules*. The system is said to be *non-extended* when $T = V$. The language generated by $\gamma$ is defined by

$$L(\gamma) = \sigma^*(A) \cap T^*.$$

**EH**$(F_1, F_2)$ denotes the family of languages generated by extended H systems $\gamma = (V, T, A, R)$ with $A \in F_1$ and $R \in F_2$ where

$$F_1, F_2 \in \{\mathbf{FIN}, \mathbf{REG}, \mathbf{CF}, \mathbf{LIN}, \mathbf{CS}, \mathbf{RE}\}.$$

**Theorem 2 ([6]).** *The relations in the following table hold, where at the intersection of the row marked with $F_1$ with the column marked with $F_2$ there appear either the family* $\mathbf{EH}(F_1, F_2)$ *or two families* $F_3$, $F_4$ *such that* $F_3 \subset \mathbf{EH}(F_1, F_2) \subseteq F_4$.

|      | FIN     | REG | LIN | CF | CS | RE |
|------|---------|-----|-----|----|----|----|
| FIN  | REG     | RE  | RE  | RE | RE | RE |
| REG  | REG     | RE  | RE  | RE | RE | RE |
| LIN  | LIN, CF | RE  | RE  | RE | RE | RE |
| CF   | CF      | RE  | RE  | RE | RE | RE |
| CS   | RE      | RE  | RE  | RE | RE | RE |
| RE   | RE      | RE  | RE  | RE | RE | RE |

# 3   Main Results

In this section, we introduce the concept of *fuzzy splicing system*, initially assigning the truth values (i.e., the fuzzy membership values) from the closed interval $[0, 1]$ to the axioms of splicing systems. Then, we calculate the truth value of every generated string $z$ from strings $x$ and $y$ using a fuzzy operation over their truth values.

**Definition 1.** *A fuzzy extended splicing system (a fuzzy H system) is a 6-tuple* $\gamma = (V, T, A, R, \mu, \odot)$ *where* $V, T, R$ *are defined as for a usual extended H system,* $\mu : V^* \to [0, 1]$ *is a (fuzzy) membership function, $A$ is a subset of $V^* \times [0, 1]$ and* $\odot$ *is a fuzzy operation over* $[0, 1]$.

A fuzzy splicing operation is defined as follows.

**Definition 2.** *For* $(x, \mu(x)), (y, \mu(y)), (z, \mu(z)) \in V^* \times [0, 1]$ *and* $r \in R$,

$$[(x, \mu(x)), (y, \mu(y))] \vdash_r (z, \mu(z))$$

*if and only if* $(x, y) \vdash_r z$ *and* $\mu(z) = \mu(x) \odot \mu(y)$.

Then, for a fuzzy splicing system $\gamma = (V, T, A, R, \mu, \odot)$, we define the fuzzy set of strings obtained by splicing strings in $A$ according to splicing rules in $R$ and the fuzzy operation $\odot$.

**Definition 3.** *Let* $\gamma = (V, T, A, R, \mu, \odot)$ *be a fuzzy splicing system. Then*

$$\sigma_f(A) = \{(z, \mu(z)) : (x, y) \vdash_r z \wedge \mu(z) = \mu(x) \odot \mu(y)$$
$$\text{for some } (x, \mu(x)), (y, \mu(y)) \in A \text{ and } r \in R\}.$$

Further, for a fuzzy splicing system $\gamma = (V, T, A, R, \mu, \odot)$, we define the closure of $A$ under splicing with respect to rules in $R$ and the fuzzy splicing operation $\odot$.

**Definition 4.** *Let $\gamma = (V, T, A, R, \mu, \odot)$ be a fuzzy splicing system. Then*

$$\sigma_f^*(A) = \bigcup_{i \geq 0} \sigma_f^i(A)$$

*where $\sigma_f^0(A) = A$ and $\sigma_f^i(A) = \sigma_f^{i-1}(A) \cup \sigma_f(\sigma_f^{i-1}(A))$ for $i \geq 1$.*

**Definition 5.** *The fuzzy language generated by a fuzzy splicing system $\gamma = (V, T, A, R, \mu, \odot)$ is defined as $L_f(\gamma) = \{(z, \mu(z)) \in \sigma_f^*(A) : z \in T^*\}$.*

We also define the "crisp" languages generated by fuzzy splicing systems.

**Definition 6.** *The crisp language generated by a fuzzy splicing system $\gamma = (V, T, A, R, \mu, \odot)$ is defined as $L_c(\gamma) = \{z : (z, \mu(z)) \in L_f(\gamma)\}$.*

*Remark 1.* It is clear that for every fuzzy splicing system $\gamma = (V, T, A, R, \mu, \odot)$, $L(\gamma') = L_c(\gamma)$ where $\gamma' = (V, T, A', R)$ with $A' = \{x : (x, \mu(x)) \in A\}$.

*Example 1.* We consider the fuzzy splicing system $\gamma$ with multiplication operation as following,

$$\gamma = (\{a, b\}, \{a, b\}, \{(aa, 1/2), (aba, 1/3)\},$$
$$\{r_1 = a\#\lambda\$\lambda\#b, r_2 = a\#\lambda\$\lambda\#a, r_3 = b\#\lambda\$\lambda\#b, r_4 = b\#\lambda\$\lambda\#b\}).$$

One can easily show that $L_c(\gamma) = a\{a, b\}^*a$. Let us analyze the truth values of the strings obtained by splicing the strings in $A$:

$$\sigma^1(A) = \{(aa, 1/2), (aa, 1/4), (aa, 1/6), (aba, 1/3), (aba, 1/6), (aba, 1/9),$$
$$(aaba, 1/6), (aaba, 1/9), (aaa, 1/4), (aaa, 1/6), (a^3ba, 1/6),$$
$$(ababa, 1/9), (abaaba, 1/9), (abba, 1/9)\}.$$

We can see that string $aa$ resulted from different strings has different truth values $1/2, 1/4, 1/6$, and strings $aba$, $aaba$ and $aaba$ have also different truth values. In order to overcome the ambiguity of truth values of strings, we can consider another fuzzy operation.

Another approach for the elimination of ambiguity is to define threshold languages, i.e., the selection of the "successful" subset of the crispy language generated by a fuzzy splicing system with respect to some cut-points. In fact, the fuzzy membership value of each string in the successful subset must satisfy the selected threshold mode. Hereby, we consider two interpretation of the threshold modes: in *strong interpretation* all fuzzy membership values of a string must satisfy the threshold condition and in *weak interpretation* at least one fuzzy membership value of a string must satisfy the threshold condition.

Further, we give formal definitions of threshold languages with respect to cut-points and relations of fuzzy membership values to these cut-points. We consider numbers $\alpha$, subintervals and discrete subsets $\Omega$ (i.e., finite or countable subsets) of $[0, 1]$ as cut-points, and $=, \neq, <, >, \leq, \geq, \in, \notin$ as relations, which are called *threshold modes*.

**Definition 7.** *Let $\gamma = (V, T, A, R, \mu, \odot)$ be a fuzzy extended splicing system. Then, strong threshold languages generated by $\gamma$ are defined as*

$$L_s(\gamma, *\alpha) = \{z : (z, \mu(z)) \in L_f(\gamma) \text{ and for all } \mu(z), \mu(z) * \alpha\},$$
$$L_s(\gamma, \star\Omega) = \{z : (z, \mu(z)) \in L_f(\gamma) \text{ and for all } \mu(z), \mu(z) \star \Omega\}$$

*where $* \in \{=, \neq, >, \geq, <, \leq\}$ and $\star \in \{\in, \notin\}$.*

**Definition 8.** *Let $\gamma = (V, T, A, R, \mu, \odot)$ be a fuzzy extended splicing system. Then, weak threshold languages generated by $\gamma$ are defined as*

$$L_w(\gamma, *\alpha) = \{z : (z, \mu(z)) \in L_f(\gamma) \text{ and for some } \mu(z), \mu(z) * \alpha\},$$
$$L_w(\gamma, \star\Omega) = \{z : (z, \mu(z)) \in L_f(\gamma) \text{ and for some } \mu(z), \mu(z) \star \Omega\}$$

*where $* \in \{=, \neq, >, \geq, <, \leq\}$ and $\star \in \{\in, \notin\}$.*

We denote the family of strong and weak threshold languages generated by fuzzy extended H systems of type $(F_1, F_2)$ by $sf\mathbf{EH}(F_1, F_2)$ and $wf\mathbf{EH}(F_1, F_2)$, respectively, where $F_1, F_2 \in \{\mathbf{FIN}, \mathbf{REG}, \mathbf{CF}, \mathbf{LIN}, \mathbf{CS}, \mathbf{RE}\}$.

**Lemma 1.** *For all families $F_1, F_2 \in \{\mathbf{FIN}, \mathbf{REG}, \mathbf{CF}, \mathbf{LIN}, \mathbf{CS}, \mathbf{RE}\}$,*

$$\mathbf{EH}(F_1, F_2) \subseteq xf\mathbf{EH}(F_1, F_2)$$

*where $x \in \{s, w\}$.*

*Proof.* Let $\gamma = (V, T, A, R)$ be an extended splicing system generating the language $L(\gamma) \in \mathbf{EH}(FIN, F)$ where $F \in \{\mathbf{FIN}, \mathbf{REG}, \mathbf{CF}, \mathbf{LIN}, \mathbf{CS}, \mathbf{RE}\}$. Let $A = \{x_1, x_2, ..., x_n\}$, $n \geq 1$. We associate the fuzzy splicing system $\gamma'$ with $\gamma$ where $\gamma' = (V, T, A', R, \mu, \odot)$, $A' = \{(x_i, 1) : x_i \in A, 1 \leq i \leq n\}$ and $\odot$ is a fuzzy operation (e.g., the multiplication operation, *max* or *min*) with the identity element 1. Then, it is not difficult to see that $L(\gamma) = L(\gamma')$. $\square$

**Lemma 2.** *Let $\gamma = (V, T, A, R, \mu, \times)$ be a fuzzy extended splicing system with multiplication operation $\times$, where $0 < \mu(x) < 1$ for all $x \in A$. Let the sets $A$ and $R$ are finite. Then, for $x \in \{s, w\}$, $\alpha \in [0, 1]$ and $I \subseteq [0, 1]$,*

1. *$L_x(\gamma, > \alpha)$ is a finite language.*
2. *$L_x(\gamma, \leq \alpha)$ is a regular language.*
3. *$L_x(\gamma, \in I)$ is a regular language.*

*Proof.* Case 1. Let $\gamma = (V, T, A, R, \mu, \times)$ be a fuzzy splicing system where

$$A = \{(x_1, \mu_1), (x_2, \mu_2), \ldots, (x_n, \mu_n)\}$$

and $0 < \mu_i < 1$ for all $1 \leq i \leq n$. Then, it is clear that

$$\prod_{j=1}^{k} \mu_{i_j} > \prod_{j=1}^{k+1} \mu_{i_j}, \ \mu_{i_j} \in \{\mu_1, \ldots, \mu_n\}, \ 1 \leq j \leq m.$$

Hence, there exists $m \in \mathbb{N}$ such that

$$\prod_{j=1}^{m} \mu_{i_j} < \alpha, \ \mu_{i_j} \in \{\mu_1, \ldots, \mu_n\}, \ 1 \le j \le m.$$

Thus, a finite number of $\mu(x)$s, $x \in L_f(\gamma)$, can satisfy the inequality $\mu(x) > \alpha$.

Case 2. It is clear that for $x \in \{s, w\}$, $L_c(\gamma) = L_x(\gamma, > \alpha) \cup L_x(\gamma, \le \alpha)$. Since $L_c(\gamma)$ is regular and $L_x(\gamma, > \alpha)$ is finite then $L_x(\gamma, \le \alpha)$ is regular.

Case 3. Let $I = (\alpha_1, \alpha_2)$. Then $L_x(\gamma, \in I) = L_x(\gamma, > \alpha_1) \cap L_x(\gamma, < \alpha_2)$, $x \in \{s, w\}$. From (i) and (ii), it follows that $L_x(\gamma, \in I)$ is regular.      □

**Lemma 3.** *Let $\gamma = (V, T, A, R, \mu, \odot)$ be a fuzzy splicing system and $L_w(\gamma, *\alpha)$ be a threshold language where $\odot \in \{min, max\}$, $* \in \{>, <, =\}$ and $\alpha \in [0, 1]$. Let the sets $A$ and $R$ are finite. Then,*

1. *$L_w(\gamma, *\alpha)$ is a regular language.*
2. *If $\alpha$ is large enough then $L_w(\gamma, > \alpha) = \emptyset$ and $L_w(\gamma, \le \alpha) = L_c(\gamma)$.*
3. *If $\alpha$ is small enough then $L_w(\gamma, > \alpha) = L_c(\gamma)$ and $L_w(\gamma, \le \alpha) = \emptyset$.*
4. *If $I$ is a subsegment of $[0, 1]$ then $L_w(\gamma, \in I)$ is regular.*

*Proof.* Let $\gamma = (V, T, A, R, \mu, \odot)$ be a fuzzy splicing system with

$$A = \{(x_1, \mu_1), (x_2, \mu_2), \ldots, (x_n, \mu_n)\}.$$

We denote by $A'$ the crispy part of the set of axioms, i.e., $A' = \{x : (x, \mu(x)) \in A\}$.

*Case 1.* Consider $max$ as the fuzzy operation and $>$ as the threshold mode. Then, the set $\sigma_f^*(A)$ can be represented as $\sigma_f^*(A) = \sigma_{f,1}^*(A) \cup \sigma_{f,2}^*(A)$ where

$$\sigma_{f,1}^*(A) = \{(x, \mu(x) \in \sigma_f^*(A) : \mu(x) > \alpha\}$$

and

$$\sigma_{f,2}^*(A) = \{(x, \mu(x) \in \sigma_f^*(A) : \mu(x) \le \alpha\}.$$

Let $\sigma_{f,i}^0 = A_i$ and $A_i' = \{x : (x, \mu(x)) \in A_i\}$, $i = 1, 2$. Obviously, $A = A_1 \cup A_2$ and $A' = A_1' \cup A_2'$. Let $\sigma_{c,i}^*(A) = \{x : (x, \mu(x)) \in \sigma_{f,i}^*(A)\}$, $i = 1, 2$.

We construct the splicing system $\gamma' = (V, T, A_2', R)$ where $L(\gamma') = \sigma^*(A_2') \cap T^*$ is regular. Moreover, we show that $\sigma_{c,2}^*(A) = \sigma^*(A_2')$.

First, $\sigma^*(A_2') \subseteq \sigma_{c,2}^*(A)$ since $A_2 \subseteq A$. On the other hand, $\sigma_{c,2}^*(A) \subseteq \sigma^*(A_2')$. Let $x \notin \sigma^*(A_2')$. Then, there is an axiom $(x_1, \mu(x_1)) \in A_1$ such that

$$((x_1, \mu(x_1)), (x_2, \mu(x_2))) \vdash (z_1, \mu(z_1)),$$

$$((z_1, \mu(z_1)), (z_2, \mu(z_2))) \vdash (z_3, \mu(z_3)),$$

$$\vdots$$

$$((z_k, \mu(z_k)), (z_{k+1}, \mu(z_{k+1}))) \vdash (x, \mu(x))$$

where $(x_2, \mu(x_2)) \in A$ and $(z_i, \mu(z_i)) \in \sigma_f^*(A)$. Then,

$$max\{\mu(x_1), \mu(x_2)\} = \mu(z_1) > \alpha,$$

$$\vdots$$

$$max\{\mu(z_k), \mu(z_{k+1})\} = \mu(x) > \alpha.$$

Consequently, $(x, \mu(x)) \notin \sigma_{f,2}^*(A)$, i.e., $x \notin \sigma_{c,2}^*(A)$. Thus, $\sigma_{c,2}^*(A) = \sigma^*(A_2')$. It follows that the language $L_w(\gamma, \le \alpha) = \sigma_{c,2}^*(A) \cap T^*$ is regular.

In its turn, $\sigma_{c,1}^*(A) = \sigma_c^*(A) - \sigma_{c,2}^*(A)$, and the language $L_w(\gamma, > \alpha) = L_c(\gamma) - L_w(\gamma, \le \alpha)$ is also regular.

Similarly, if the fuzzy operation is $min$, it can also be proved that $L_w(\gamma, > \alpha)$ and $L_w(\gamma, \le \alpha)$ are regular.

*Case 2.* We choose $\alpha > max\{\mu_1, \mu_2, \ldots, \mu_n\}$.

*Case 3.* We choose $\alpha < min\{\mu_1, \mu_2, \ldots, \mu_n\}$.

*Case 4.* $L_w(\gamma, \in I) = L_w(\gamma, > \alpha_1) \cap L_w(\gamma, < \alpha_2)$ where $I = (\alpha_1, \alpha_2)$. From (i), $L_w(\gamma, > \alpha_1)$ and $L_w(\gamma, < \alpha_2)$ are regular. Therefore, their intersection is also regular. □

*Remark 2.* It should be noted that the arguments of the proof in Lemma 3.13 cannot be used for the strong case; because $L_s(\gamma, \le \alpha) \subseteq \sigma_{c,2}^*(A) \cap T^*$, and it is not necessary the equality holds.

From the lemmas above we obtain the following theorem.

**Theorem 3.** *Every fuzzy splicing system with the fuzzy operation: multiplication, max or min, and the cut-point: any number in $[0,1]$ or any subinterval of $[0,1]$ generates a regular language.*

Although the threshold languages with numbers and subsegments of $[0,1]$ are regular, the generative power of fuzzy splicing systems can be increased using discrete subsets of $[0,1]$, i.e., functions whose codomains are subintervals of $[0,1]$ as cut-points. The following examples show that, with this restriction, the generative capacity of fuzzy splicing systems can be increased up to context-sensitive languages.

*Example 2.* Let

$$\gamma = (\{a, b, c, d\}, \{a, b\}, \{(cad, 1/3), (dbc, 1/2)\},$$
$$\{r_1 = a\#d\$c\#ad, r_2 = db\#c\$a\#b, r_3 = a\#d\$d\#b\})$$

be a fuzzy splicing system with multiplication operation.

Then, by applying rule $r_1$ to axiom $cad$, we obtain strings $ca^n d$, $n \ge 1$, with $\mu(ca^n d) = 1/3^n$. Similarly, by applying rule $r_2$ to axiom $dbc$, we obtain strings $db^m c$, $m \ge 1$, with $\mu(db^m c) = 1/2^m$. The application of rule $r_3$ to these strings results in $ca^n b^m c$ with $\mu(ca^n b^m c) = 1/3^n \cdot 1/2^m$.

Then $L_c(\gamma) = \{a^n b^m : n, m \geq 1\} \in \mathbf{REG}$ and

$$L_w(\gamma, = 1/5) = \emptyset \in \mathbf{FIN},$$

$$L_w(\gamma, > 1/3) = \{b\} \in \mathbf{FIN},$$

$$L_w(\gamma, \in \{1/6^n : n \geq 1\}) = \{a^n b^n : n \geq 1\} \in \mathbf{CF} - \mathbf{REG}.$$

One can see that the last threshold language generated by the fuzzy splicing system is not regular. However, if we consider *min* or *max* as fuzzy operations with the splicing system above, then the threshold languages are not more than regular. In this case for the generated strings $ca^n b^m c$ we have

$$\mu(ca^n b^m c) = \begin{cases} 1/3, \ n > 0, \\ 1/2, \ n = 0. \end{cases}$$

Therefore,

$$L_w(\gamma, \in \{1/6^n : n \geq 1\}) = \emptyset \in \mathbf{FIN},$$

$$L_w(\gamma, > 1/3) = \{b^n : n \geq 1\} \in \mathbf{REG}.$$

*Example 3.* Consider the following fuzzy splicing system with the multiplication operation

$$\gamma = (\{a, b, c, w, x, y, z\}, \{a, b, c\}, \{(xay, 1/3), (ybz, 1/5), (zcw, 1/7)\},$$
$$\{r_1 = xa\#y\$x\#a, r_2 = yb\#z\$y\#b, r_3 = zc\#w\$z\#c,$$
$$r_4 = a\#y\$y\#b, r_5 = b\#z\$z\#c\}).$$

By rule $r_1$ to the initial string $xay$, we obtain $(xa^k y, 1/3^k), k \geq 1$, by rule $r_2$ to the initial string $ybz$, we get $(yb^m z, 1/5^m), m \geq 1$, by rule $r_3$ to the initial string $zcw$, we have $(zc^n w, 1/7^n), n \geq 1$. The rules $r_4$ and $r_5$ the strings above result in

$$[(xa^k y, 1/3^k), (yb^m z, 1/5^m)] \vdash_{r_4} (xa^k b^m z, 1/3^k 5^m)$$

and

$$[(xa^k b^m z, 1/3^k 5^m), (zc^n w, 1/7^n)] \vdash_{r_5} (xa^k b^m c^n w, 1/3^k 5^m 7^n).$$

Then, the fuzzy language generated by $\gamma$ is

$$L_f(\gamma) = \{(a^k b^m c^n, 1/3^k 5^m 7^n) : k, m, n \geq 1\}.$$

Further, we consider the following threshold languages:

$$L_w(\gamma, > 0) = \{a^k b^m c^n : k, m, n \geq 1\} \in \mathbf{REG},$$

$$L_w(\gamma, > 1/105^5) = \{a^k b^m c^n : 1 \leq k, m, n \leq 5\} \in \mathbf{REG},$$

$$L_w(\gamma, \in \{1/105^n : n \geq 1\}) = \{a^n b^n c^n : n \geq 1\} \in \mathbf{CS} - \mathbf{CF}.$$

# 4   Conclusions

In this paper, we have introduced the concept of fuzzy splicing system and established their preliminary properties. When fuzzy splicing systems are considered with multiplication, *max* or *min* operations and subintervals of $[0, 1]$, they cannot increase the generative power of splicing systems. The regularity of fuzzy splicing systems under strong interpretation remains open. If we choose discrete sets from $[0, 1]$, the power can be increased up to some context-sensitive languages. On the one hand, fuzzy splicing systems allow modeling molecular uncertainty processes appearing in molecular biology, systems biology and medicine. On the other hand, the study of fuzzy splicing systems in particular and the fuzzy variants of other theoretical models of DNA computing makes a significant contributions to formal language and automata theories.

**Acknowledgement.** This work has been supported through the Research University Grant (RUG) **07J41**, Universiti Teknologi Malaysia and Fundamental Research Grant Scheme **FRGS13-066-0307**, International Islamic University Malaysia, Ministry of Education, Malaysia.

# References

1. Adleman, L.: Molecular computation of solutions to combinatorial problems. Science 266, 1021–1024 (1994)
2. Boneh, D., Dunworth, C., Lipton, R., Sgall, J.: On the computational power of DNA. Discrete Applied Mathematics. Special Issue on Computational Molecular Biology 71, 79–94 (1996)
3. Lipton, R.: Using DNA to solve NP–complete problems. Science 268, 542–545 (1995)
4. Head, T.: Formal language theory and DNA: An analysis of the generative capacity of specific recombination behaviors. Bull. Math. Biology 49, 737–759 (1987)
5. Pixton, D.: Regularity of splicing languages. Discrete Applied Mathematics 69, 101–124 (1996)
6. Păun, G., Rozenberg, G., Salomaa, A.: DNA computing. New computing paradigms. Springer-Verlag (1998)
7. Mordeson, J., Malik, D.: Fuzzy Automata and Languages. Theory and Applications. Chapman & Hall/CRC (2002)
8. Dassow, J., Păun, G.: Regulated rewriting in formal language theory. Springer-Verlag, Berlin (1989)
9. Rozenberg, G., Salomaa, A.: Handbook of formal languages, vol. 1-3. Springer, Heidelberg (1997)

# A Preference Weights Model for Prioritizing Software Requirements

Philip Achimugu, Ali Selamat[*], and Roliana Ibrahim

UTM-IRDA Digital Media Centre,
K-Economy Research Alliance & Faculty of Computing,
Universiti Teknologi Malaysia, Johor Bahru, 81310, Johor, Malaysia
check4philo@gmail.com, {aselamat,roliana}@utm.my

**Abstract.** Software requirements prioritization is the act of ranking user's requirements in order to plan for release phases. The essence of prioritizing requirements is to avoid breach of contract, trust or agreement during software development process. This is crucial because, not all the specified requirements could be implemented in a single release due to inadequate skilled programmers, time, budget, and schedule constraints. Major limitations of existing prioritization techniques are rank reversals, scalability, ease of use, computational complexities and accuracy among others. Consequently, an innovative model that is capable of addressing these problems is presented. To achieve our aim, synthesized weights are computed for criteria that make up requirements and functions were defined to display prioritized requirements based on the global weights of attributes across project stakeholders. An empirical case scenario is described to illustrate the adaptability processes of the proposed approach.

**Keywords:** Software, requirements, weights, prioritization, model.

## 1    Introduction

Prioritization is the act of determining an ordered set of requirements based on their perceived importance by project stakeholders so as to plan for software release phases [1-3]. It is considered to be a multi-criteria decision making process. Essentially, the basic components of decision-making problems are: goal/objective goal, criteria/factors or alternatives/actions. During decision making, many unambiguous criteria are used to elect best alternatives from a pool. These criteria could either be quantitative, qualitative or both. The ranking of alternatives are finally achieved through weighting scales which contain values that decision makers use to determine the relative importance of alternatives.

Decision making in software development process is inevitable if the proposed software must satisfy or meet user's requirements and delivered within time and budget. This is a crucial aspect of software development because; clients specify too many

---

[*] Corresponding author.

D. Hwang et al. (Eds.): ICCCI 2014, LNAI 8733, pp. 30–39, 2014.

requirements for implementation without considering the availability of time and resource constraints. Therefore, a meticulously selected set of requirements must be considered for implementation with respect to available resources [4]. The process of selecting preferential requirements for implementation is the most prominent attributes of requirements prioritization techniques. This process aims at determining the most valued or prime requirements from a set or pool of specified requirements [5].

The main aim of this study is to develop a preference based multi-criteria decision making (MCDM) model, capable of enhancing stakeholder's quest for prioritizing requirements with respect to multiple criteria and diverse criteria priorities. A rank is computed by collating all the weights for the requirements in a set with respect to the total number of stakeholders involved in the ranking process. Software products that are developed based on prioritized requirements can be expected to have a lower probability of being rejected.

To prioritize requirements, stakeholders will have to relatively compare them in order to determine their relative value through preference weights [6]. These comparisons grow with increase in the number of requirements [7]. State-of-the-art prioritization techniques such as AHP and CBRanks seem to demonstrate high capabilities [8]. These techniques have performed well in terms of ease of use and accuracy but, still lacking in various areas. In this paper, an enhanced approach for software requirements prioritization is proposed based.

The rest of the article is structured as follows: Section 2 discusses the related work while section 3 enumerates the proposed approach. Section 4 presents an empirical example in order to evaluate the performance of the proposed approach. Section 5 deals with the experimental result which lead to conclusion and future work in section 6.

## 2    Related Work

In literature, many techniques have been proposed for prioritizing software requirements during development processes but most of these techniques are not easily adoptable due to one limitation or the other. Analytic hierarchy process (AHP) seems to be the most widely adopted technique for prioritizing alternatives including software requirements but this prominent technique suffer serious scalability challenges. It cannot prioritize large number of requirements and it is said to be time consuming [9-11]. However, attempts have been made to address the scalability challenges inherent in AHP. For example techniques like binary priority list [12], Case based ranking [13], EVOLVE [14], fuzzy based requirements prioritization approaches [15, 16], Pair wise analysis [17], TOPSIS [18, 19] and fuzzy based MCDM approaches [20, 21] among others. Again, none of these techniques is flawless. Techniques like binary priority list and fuzzy AHP did not cater for dependencies that could exist among requirements before prioritizing them. Case based rank is limited in its inability to support coordination among different stakeholders through negotiations, EVOLVE is reported to be computationally complex along with pair wise comparisons which is also known for producing unreliable results. TOPSIS on the other hand do not possess the ability of updating rank status whenever requirements evolves while the fuzzy based approaches suffer generally from three major setbacks which include: requirements dependency issues as well as lack of implemented tool to support the proposed approaches. Also, validations of these techniques with real-world projects have not been achieved yet. A detailed analysis and descriptions of existing prioritization techniques with their limitations can be found in [22]. Nonetheless, obvious limitations that cut across existing techniques ranges from rank reversals to scalability, inaccurate rank results, increased computational complexities and unavailability of efficient support tools.

## 3    The Proposed Approach

The main objective of this study is to propose an approach that supports stakeholders in ranking software requirements. In this context, the proposed steps are described as follows:

Step 1: Decompose the problem into a hierarchy of requirements and their interrelated attributes;

Step 2: Generate input data consisting of preference weights of attributes across all stakeholders;

Step 3: Synthesize each subjective judgments and compute the global weights;

Step 4: Calculate the final weights to display prioritized requirements.

The proposed prioritization approach consists of defining a common hierarchy of requirements with their respective attributes to aid comparison by all the stakeholders involved in the software development project (Figure 1).

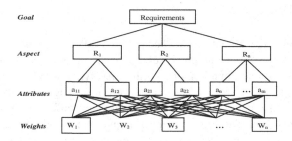

**Fig. 1.** Representation of the 4-level structure of ranking process

Let us assume that, we have $x$ and $y$ requirements and the aim is to rank them based on the weights of attributes $a_{11},...,a_{1n}, a_{21},...,a_{2n}$ provided by the respective stakeholders, each attribute will then have to be ranked based on a weight scale. Therefore, the prioritization process consists of finding the weights that engenders the determination of relative importance of requirements. The structure describing the proposed approach is shown in Figure 2.

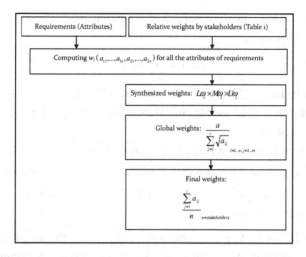

**Fig. 2.** Structure of the prioritization process

The relative value of each requirement is measured on the basis of the accrued weights across project stakeholders. The prioritized output will then be the cumulative sums of the accrued weights.

**Definition 3.3.1:** Let $X$ be a measurable requirement set that is endowed with attributes of $\sigma$-functionalities, where $N$ is all subsets of $X$. A prioritization function $g$ defined on the measurable space $(X, N)$ is a set function $g : N \rightarrow [0,1]$ which satisfies the following properties:

$$g(\phi) = 0, g(X) = 1 \tag{1}$$

But for requirement sets $X, Y;$ the equation for the prioritization process will be:

$$X \subseteq Y \in N \rightarrow [0,1] \tag{2}$$

From the above definition, $X, Y, N, g$ are said to be the parameters used to measure or determine the relative weights of requirements. This process is monotonic. Consequently, the monotonicity condition is obtained as:

$$g(X \cup Y) \geq \max\{g(X), g(Y)\}; g(X \cap Y) \leq \min\{g(X), g(Y)\} \tag{3}$$

In the case where $g(X \cup Y) \geq \max\{g(X), g(Y)\}$, the prioritization function $g$ attempts to determine the total number of requirements being prioritized and if $g(X \cap Y) \leq \min\{g(X), g(Y)\}$, the function attempts to compute the relative weights of requirements provided by the relevant stakeholders.

**Definition 3.3.2:** Let $h = \sum_{i=1}^{n} X_i \cdot 1_{X_i}$ be a simple function.

$1_{X_i}$, is the attribute function of the requirement set $X_i \in N, i = 1, ..., n$; the sets $X_i$ are pairwise disjoints, but if $M(X_i)$ is the measure of the weights between all the attributes contained in $X_i$, then the integral of $h$ which is used to find the local weights of requirements is given as:

$$\int h.dM = \sum_{i=1}^{n} M(X_i) \cdot x_i \qquad (4)$$

**Definition 3.3.3:** Let $X, Y, N, g$ be the measure of weights between two sets of requirements, the integral of weights measure $g : N \rightarrow [0,1]$ with respect to a simple function $h$ is defined by:

$$\int h(r) g(r) = \vee (h(r_i) \wedge g(X_i)) = \max \min \{r'_i, g(Y_i)\} \qquad (5)$$

Where $h(r_i)$ is a linear combination of an attribute function $1r_i$ such that $X_1 \subset Y_1 \subset ... \subset X_n \subset Y_n$ and $X_n = \{r \mid h(r) \geq Y_n\}$. This is used to determine the global weights between requirements.

**Definition 3.3.4:** Let $X, N, g$ be a measure space. The integral of a measure of weights by the prioritization process $g : N \rightarrow [0,1]$ with respect to a simple function $h$ is defined by

$$\int h(r).dg \cong \sum [(h(r_i) - h(r_{i-1})].g(X_i) \qquad (6)$$

Similarly, if $Y, N, g$ is a measure space; the integral of the measure of the weights with respect to a simple function $h$ will be:

$$\int h(r').dg \cong \sum [(h(r_i') - h(r'_{i-1})].g(Y_i) \qquad (7)$$

However, if $g$ measures the relative weights of requirements, defined on a power set $P(x)$ and satisfies the definition 3.3.1 as above; the following attribute is evident:

$$\forall X, Y \in P(x), X \cap Y = \phi \Rightarrow g_2(X \cup Y) = g_2(X) + g_2(Y) + \lambda g_2(X) g_2(Y) \qquad (8)$$

For $0 \leq \lambda \leq \infty$

Therefore, for requirement set $X = \{r_1, r_2, ..., r_n\}$, the density of weights measure $w_i = \{r_i\}$ can be formulated as follows:

$$w(\{r_1, r_2, ..., r_n\}) = \sum_{i=1}^{n} w_i + \lambda \sum_{i_1=1}^{n-1} \sum_{i_2=i_1+1}^{n} w_{i1} . w_{i2} + ... + \lambda^{n-1} . w_{i1} . w_{i2} ... w_n = \frac{1}{\lambda} \langle \prod_{i=1}^{n} (1 + \lambda . w_i) - 1 \rangle \qquad (9)$$

For $0 \leq \lambda \leq \infty$

However, $h$ is a measurable set function defined on the certain measurable space of requirement weights $X, N$ and assuming $h(r_1) \geq h(r_2) \geq ... \geq h(r_n)$, then the integral of the weights measure of requirements $g(\cdot)$ with respect to $h(\cdot)$ can be defined as follows:

$$\int h.dg = h(r_n).g(X) + [h(r_{n-1}) - h(r_n)].g(Y) + ... + [h(r'_{n-1}) - h(r'_n)] \qquad (10)$$

$$= h(r_n).[g(X) - g(r_{n-1})] + h(r_{n-1}).[g(Y) - g(r'_{n-1})] + h(r'_{n-1}) \qquad (11)$$

$$\text{where } X = (r_1, r_2, ..., r_n); Y = (r'_1, r'_2, ..., r'_n).$$

Therefore, the computation of relative weights across all the requirements in the given sets is a dependent relation between attributes and the stakeholders. Multi-attributes multiplicative utility function known as non-additive multi-criteria evaluation technique can be used to refine the situations that do not conform to the assumption of independence between attributes criteria [23].

In practical application of the prioritization process, $X = (r_1, r_2, ..., r_n)$; $Y = (r'_1, r'_2, ..., r'_n)$ probably represents two sets of requirements with their respective attributes that are to be ranked. In these sets, attributes are not necessary mutually independent. In order to drive the synthetic utility values, we first exploit the factor analysis technique to extract the attributes that possess common functionalities using Equation 1. This caters for requirement dependencies challenges during the prioritization process. The attributes with the same functionalities are considered to be mutually dependent. Therefore, before relative weights are assigned to the requirements by relevant stakeholders, attention should be paid to requirement dependencies issues in order to avoid redundant results.

However, when requirements evolve, it becomes necessary to add or delete from a set. The algorithm should also be able to detect this situation and update rank status of ordered requirements instantly. This is known as rank reversals. It is formally expressed as follows: (1) failure of the type $0 \rightarrow 1$ or $1 \rightarrow 0$; (2) failures of the type $0 \rightarrow \phi$ or $1 \rightarrow \phi$ (where $\phi$ = the null string) (called *deletions*); and (3) failures of the type $\phi \rightarrow 0$ or $\phi \rightarrow 1$ (called *insertions*). A weight metric $w$, on two requirement sets $(X, Y)$ is defined as the smallest number of edit operations (deletions, insertions and updates) to enhance the prioritization process. Three types of rank updates operations on $X \rightarrow Y$ are defined as: a *change* operation ($X \neq \phi$ and $Y \neq \phi$), a *delete* operation ($Y = \phi$) and an *insert* operation ($X = \phi$). The weights of all the requirements can be computed by a *weight function* $w$. An arbitrary weight function $w$ is obtained by computing all the assigned non-negative real number $w$ $(X, Y)$ on each requirement sets. This is achieved by Equations 2-7. However, in additive and non-additive measurement (rank updates) cases, Equations 8-11 is utilized to find the synthetic utilities of each attribute in the set within the same factor. On the other hand, there is mutual independence between attributes, and the measurement is an additive case, so we can utilize the additive aggregate method to conduct the synthetic utility values for all the attributes in the entire requirement sets.

Before requirements prioritization is performed, it is expected to ensure that all the attributes and requirements are mutually independent. Thereafter, the relative weights and performance score of each attribute corresponding to each requirement set is computed across all the stakeholders. Then, these scores are aggregated to obtain the final ranks of requirements. The relative weight of the $j$-th attribute is calculated by obtaining the subjective weights of stakeholders using weight scale shown in Table 1.

**Table 1.** Weight scale

| Variables | Rank | Relative numbers |
|---|---|---|
| Extermely important (EI) | 1 | (0.75, 0.90, 1.00) |
| More important (MI) | 2 | (0.25, 0.50, 0.65) |
| Less important (LI) | 3 | (0.15, 0.30, 0.45) |

# 4    Empirical Example

To illustrate the concept of our approach, an electronic health records system is considered in this case. A hospital would like to develop new software to replace the existing one that do not support distributed healthcare delivery services. The new system should allow a medical practitioner to administer quality healthcare from any geographical location across the three tiers of healthcare institutions (primary, secondary and tertiary). The system must be flexible enough to enable physicians gain access into the system and administer appropriate healthcare. It is required that the system be scalable and interoperable. The project consists of nine stakeholders, with four requirements sets denoted as $P$, $F$, $U$, and $M$ representing Performance, Flexibility, Usability and Maintainability respectively containing fourteen attributes all together. Tables 2-4 show the relative variables, synthesized and global weights of the various attributes for the specified requirements.

**Table 2.** Relative variables

|       | $P_{11}$ | $P_{12}$ | $P_{13}$ | $P_{14}$ | $P_{15}$ | $F_{21}$ | $F_{22}$ | $F_{23}$ | $U_{31}$ | $U_{32}$ | $M_{41}$ | $M_{42}$ | $M_{43}$ | $M_{44}$ |
|-------|------|------|------|------|------|------|------|------|------|------|------|------|------|------|
| $S_1$ | EI | EI | EI | EI | EI | MI | MI | MI | MI | MI | EI | EI | EI | EI |
| $S_2$ | EI | EI | EI | EI | EI | EI | EI | EI | EI | EI | EI | EI | EI | EI |
| $S_3$ | MI | MI | MI | MI | MI | EI | EI | EI | EI | EI | LI | LI | LI | LI |
| $S_4$ | MI | MI | MI | MI | MI | LI | LI | LI | EI | EI | EI | EI | EI | EI |
| $S_5$ | EI | EI | EI | EI | EI | MI | MI | MI | LI | LI | MI | MI | MI | MI |
| $S_6$ | MI | MI | MI | MI | MI | EI | EI | EI | LI | LI | MI | MI | MI | MI |
| $S_7$ | EI | EI | EI | EI | EI | EI | EI | EI | LI | LI | MI | MI | MI | MI |
| $S_8$ | MI | MI | MI | MI | MI | LI | LI | LI | MI | MI | MI | MI | MI | MI |
| $S_9$ | EI | EI | EI | EI | EI | MI | MI | MI | LI | LI | EI | EI | EI | EI |

**Table 3.** Synthesized weights

|       | $P_{11}$ | $P_{12}$ | $P_{13}$ | $P_{14}$ | $P_{15}$ | $F_{21}$ | $F_{22}$ | $F_{23}$ | $U_{31}$ | $U_{32}$ | $M_{41}$ | $M_{42}$ | $M_{43}$ | $M_{44}$ |
|-------|-------|-------|-------|-------|-------|-------|-------|-------|-------|-------|-------|-------|-------|-------|
| $S_1$ | 0.675 | 0.675 | 0.675 | 0.675 | 0.675 | 0.081 | 0.081 | 0.081 | 0.081 | 0.081 | 0.675 | 0.675 | 0.675 | 0.675 |
| $S_2$ | 0.675 | 0.675 | 0.675 | 0.675 | 0.675 | 0.675 | 0.675 | 0.675 | 0.675 | 0.675 | 0.675 | 0.675 | 0.675 | 0.675 |
| $S_3$ | 0.081 | 0.081 | 0.081 | 0.081 | 0.081 | 0.675 | 0.675 | 0.675 | 0.675 | 0.675 | 0.020 | 0.020 | 0.020 | 0.020 |
| $S_4$ | 0.081 | 0.081 | 0.081 | 0.081 | 0.081 | 0.020 | 0.020 | 0.020 | 0.675 | 0.675 | 0.675 | 0.675 | 0.675 | 0.675 |
| $S_5$ | 0.675 | 0.675 | 0.675 | 0.675 | 0.675 | 0.081 | 0.081 | 0.081 | 0.020 | 0.020 | 0.081 | 0.081 | 0.081 | 0.081 |
| $S_6$ | 0.081 | 0.081 | 0.081 | 0.081 | 0.081 | 0.675 | 0.675 | 0.675 | 0.020 | 0.020 | 0.081 | 0.081 | 0.081 | 0.081 |
| $S_7$ | 0.675 | 0.675 | 0.675 | 0.675 | 0.675 | 0.675 | 0.675 | 0.675 | 0.020 | 0.020 | 0.081 | 0.081 | 0.081 | 0.081 |
| $S_8$ | 0.081 | 0.081 | 0.081 | 0.081 | 0.081 | 0.020 | 0.020 | 0.020 | 0.081 | 0.081 | 0.081 | 0.081 | 0.081 | 0.081 |
| $S_9$ | 0.675 | 0.675 | 0.675 | 0.675 | 0.675 | 0.081 | 0.081 | 0.081 | 0.020 | 0.020 | 0.675 | 0.675 | 0.675 | 0.675 |

**Table 4.** Global weights

|       | $P_{11}$ | $P_{12}$ | $P_{13}$ | $P_{14}$ | $P_{15}$ | $F_{21}$ | $F_{22}$ | $F_{23}$ | $U_{31}$ | $U_{32}$ | $M_{41}$ | $M_{42}$ | $M_{43}$ | $M_{44}$ |
|-------|------|------|------|------|------|------|------|------|------|------|------|------|------|------|
| $S_1$ | 0.40 | 0.40 | 0.40 | 0.40 | 0.40 | 0.16 | 0.16 | 0.16 | 0.20 | 0.20 | 0.41 | 0.41 | 0.41 | 0.41 |
| $S_2$ | 0.40 | 0.40 | 0.40 | 0.40 | 0.40 | 0.33 | 0.33 | 0.33 | 0.58 | 0.58 | 0.41 | 0.41 | 0.41 | 0.41 |
| $S_3$ | 0.13 | 0.13 | 0.13 | 0.13 | 0.13 | 0.33 | 0.33 | 0.33 | 0.58 | 0.58 | 0.07 | 0.07 | 0.07 | 0.07 |
| $S_4$ | 0.13 | 0.13 | 0.13 | 0.13 | 0.13 | 0.10 | 0.10 | 0.10 | 0.58 | 0.58 | 0.41 | 0.41 | 0.41 | 0.41 |
| $S_5$ | 0.40 | 0.40 | 0.40 | 0.40 | 0.40 | 0.16 | 0.16 | 0.16 | 0.10 | 0.10 | 0.14 | 0.14 | 0.14 | 0.14 |
| $S_6$ | 0.13 | 0.13 | 0.40 | 0.40 | 0.40 | 0.10 | 0.10 | 0.10 | 0.10 | 0.10 | 0.14 | 0.14 | 0.14 | 0.14 |
| $S_7$ | 0.40 | 0.40 | 0.40 | 0.40 | 0.40 | 0.10 | 0.10 | 0.10 | 0.10 | 0.10 | 0.14 | 0.14 | 0.14 | 0.14 |
| $S_8$ | 0.13 | 0.13 | 0.13 | 0.13 | 0.13 | 0.10 | 0.10 | 0.10 | 0.20 | 0.20 | 0.14 | 0.14 | 0.14 | 0.14 |
| $S_9$ | 0.40 | 0.40 | 0.40 | 0.40 | 0.40 | 0.16 | 0.16 | 0.16 | 0.10 | 0.10 | 0.41 | 0.41 | 0.41 | 0.41 |

# 5    Experimental Results

The results displayed in Table 5 shows the summary of output executed for the prioritized requirements, obtained from preference weights of stakeholders. The overall result is shown in Table 6.

**Table 5.** Execution output

| Requirements | Stakeholders | Mean | Std. deviation |
|---|---|---|---|
| QoS ($P_{11}$) | 9 | 0.411 | 0.313 |
| Scalability ($P_{12}$) | 9 | 0.411 | 0.313 |
| Security ($P_{13}$) | 9 | 0.411 | 0.313 |
| Data communication ($P_{14}$) | 9 | 0.411 | 0.313 |
| Data redundancy ($P_{15}$) | 9 | 0.411 | 0.313 |
| Installation ease ($F_{21}$) | 9 | 0.329 | 0.329 |
| User friendly ($F_{22}$) | 9 | 0.329 | 0.329 |
| Compatibility ($F_{23}$) | 9 | 0.331 | 0.327 |
| Code change ($U_{31}$) | 9 | 0.252 | 0.318 |
| File change ($U_{32}$) | 9 | 0.252 | 0.318 |
| Documentation quality ($M_{41}$) | 9 | 0.338 | 0.320 |
| Maintenance plan ($M_{42}$) | 9 | 0.338 | 0.320 |
| Installation manual ($M_{43}$) | 9 | 0.338 | 0.320 |
| User training ($M_{44}$) | 9 | 0.338 | 0.320 |

The proposed approach has the capacity to accurately address rank reversal and dependency issues as against the existing techniques. For example, in Table 6, $R_1$ and $R_4$ emerged as prime requirements even though $R_1$ had more attributes than $R_4$. It is also applicable to large numbers of requirements. Determining the weights of stakeholder's requirements was achieved by synthesizing the priorities over all levels obtained by varying numbers of requirements.

**Table 6.** Prioritized requirements

| Requirements | Final rank |
|---|---|
| $R_1$ | 1.40 |
| $R_2$ | 0.51 |
| $R_3$ | 0.56 |
| $R_4$ | 1.00 |

# 6    Conclusion and Future Work

Many software development projects fail not because there are no skillful programmers but because there are no skillful elicitors who have the capacity of acquiring and ranking requirements in an efficient and precise manner in order to plan for software releases. This research has proposed an approach that will help guide developers, elicitors, architects and other stakeholders in their quest to develop systems that meet the requirements of the users. Surely, when requirements are vaguely elicited, the resulting system will not function as expected even when the codes are free of errors. In conclusion, this research proposed a preference weights model for prioritizing software requirements. Four user's requirements with fourteen respective attributes were described to describe the application of the proposed approach. By using this approach, the subjective judgments can be quantified to make comparison more efficiently and reduce assessment biasness. These efforts will aid developers in designing an architecture and software with preferential requirements of stakeholders. For the future work, the implementation of the proposed approach and its application in real-world project with large number of requirements and stakeholders is underway. Also, there is need to minimize the disagreement rate between final rank weights.

**Acknowledgement.** This work is supported by the Research Management Centre (RMC) at the Universiti Teknologi Malaysia under Research University Grant (Q.J130000.2510.03H02), the Ministry of Science, Technology & Innovations Malaysia under Science Fund (R.J130000.7909.4S062) and the Ministry of Higher Education (MOHE) Under Exploratory Research Grant Scheme (R.J130000.7828.4L051).

# References

1. Svensson, R., Gorschek, T., Regnell, B., Torkar, R., Shahrokni, A., Feldt, R., Aurum, A.: Prioritization of quality requirements: State of practice in eleven companies. In: 19th IEEE International Requirements Engineering Conference (RE), pp. 69–78 (2011)
2. Kassel, N.W., Malloy, B.A.: An approach to automate requirements elicitation and specification. In: Proc. of the 7th IASTED International Conference on Software Engineering and Applications, Marina Del Rey, CA, USA (2003)
3. Ramzan, M., Jaffar, A., Shahid, A.: Value based intelligent requirement prioritization (VIRP): expert driven fuzzy logic based prioritization technique. International Journal of Innovative Computing 7(3), 1017–1038 (2011)
4. Aasem, M., Ramzan, M., Jaffar, A.: Analysis and optimization of software requirements prioritization techniques. In: International Conference on Information and Emerging Technologies (ICIET), pp. 1–6. IEEE (2010)
5. Tonella, P., Susi, A., Palma, F.: Interactive requirements prioritization using a genetic algorithm. Information and Software Technology 55(1), 173–187 (2013)
6. Ahl, V.: An experimental comparison of five prioritization methods. Master's Thesis, School of Engineering, Blekinge Institute of Technology, Ronneby, Sweden (2005)
7. Berander, P., Andrews, A.: Requirements prioritization. In: Engineering and managing software requirements, pp. 69–94. Springer, Heidelberg (2005)
8. Kobayashi, A., Maekawa, M.: Need-based requirements change management. In: Proceedings. Eighth Annual IEEE International Conference and Workshop on Engineering of Computer Based Systems, pp. 171–178. IEEE (2001)

9. Karlsson, J., Wohlin, C., Regnell, B.: An evaluation of methods for prioritizing software requirements. Information and Software Technology 39(14), 939–947 (1998)
10. Duan, C., Laurent, P., Cleland-Huang, J., Kwiatkowski, C.: Towards automated requirements prioritization and triage. Requirements Engineering 14(2), 73–89 (2009)
11. Karlsson, J., Ryan, K.: A cost-value approach for prioritizing requirements. IEEE Software 14, 67–74 (1997)
12. Bebensee, T., van de Weerd, I., Brinkkemper, S.: Binary priority list for prioritizing software requirements. In: Wieringa, R., Persson, A. (eds.) REFSQ 2010. LNCS, vol. 6182, pp. 67–78. Springer, Heidelberg (2010)
13. Perini, A., Susi, A., Avesani, P.: A Machine Learning Approach to Software Requirements Prioritization. IEEE Transactions on Software Engineering 39(4), 445–460 (2013)
14. Thakurta, R.: A framework for prioritization of quality requirements for inclusion in a software project. Software Quality Journal 21, 573–597 (2012)
15. Lima, D.C., Freitas, F., Campos, G., Souza, J.: A fuzzy approach to requirements prioritization. In: Cohen, M.B., Ó Cinnéide, M. (eds.) SSBSE 2011. LNCS, vol. 6956, pp. 64–69. Springer, Heidelberg (2011)
16. Achimugu, P., Selamat, A., Ibrahim, R., Mahrin, M.N.: An adaptive fuzzy decision matrix model for software requirements prioritization. In: Sobecki, J., Boonjing, V., Chittayasothorn, S. (eds.) Advanced Approaches to Intelligent Information and Database Systems. SCI, vol. 551, pp. 129–138. Springer, Heidelberg (2015)
17. Karlsson, J., Ryan, K.: A cost-value approach for prioritizing requirements. IEEE Software 14, 67–74 (1997)
18. Kukreja, N., Payyavula, S., Boehm, B., Padmanabhuni, S.: Value-based requirements prioritization: usage experiences. Procedia Computer Science 16, 806–813 (2012)
19. Kukreja, N.: Decision theoretic requirements prioritization: a two-step approach for sliding towards value realization. In: Proceedings of the 2013 International Conference on Software Engineering, pp. 1465–1467. IEEE Press (2013)
20. Ejnioui, A., Otero, C., Otero, L.: A simulation-based fuzzy multi-attribute decision making for prioritizing software requirements. In: Proceedings of the 1st Annual Conference on Research in Information Technology, pp. 37–42. ACM (2012)
21. Gaur, V., Soni, A.: An integrated approach to prioritize requirements using fuzzy decision making. IACSIT International Journal of Engineering and Technology 2(4), 320–328 (2010)
22. Achimugu, P., Selamat, A., Ibrahim, R., Mahrin, M.N.R.: A systematic literature review of software requirements prioritization research. Information and Software Technology 56(6), 568–585 (2014)
23. Chen, Y., Tzeng, G.: Using fuzzy integral for evaluating subjectively perceived travel costs in a traffic assignment model. European Journal of Operational Research 130(3), 653–664 (2001)

# Fuzzy Logic-Based Adaptive Communication Management on Wireless Network

Taeyoung Kim[1], Youngshin Han[2,*], Jaekwon Kim[1], and Jongsik Lee[1]

[1] Dept. of Computer and Information Engineering, Inha University, Inchon, Republic of Korea
silverwild@gmail.com, jaekwonkorea@naver.com, jslee@inha.ac.kr
[2] Dept. of Computer Engineering, Sungkyul University, Anyang, Republic of Korea
hanys@sungkyul.ac.kr

**Abstract.** This paper presents a fuzzy logic-based adaptive communication management on a wireless network. A combination of both wireless network and handheld device is most widely used in the world today. The wireless network depends on the radio signal to communicate with the device. And the handheld device is the mobile node, which is difficult to determine the certain location. These unstable features have a negative influence on the communication QoS (quality of service). Therefore, we adopt the fuzzy logic to improve the communication efficiency. The access point (AP) may evaluate the communication state with the fuzzy logic. Through this, the relay station utilizes the evaluation result to handle the communication throughput. The simulation demonstrates the efficiency of our proposed model.

**Keywords:** Fuzzy Logic, Rule-based Inference, Adaptive Queue Management, Wireless Network.

## 1 Introduction

A mobile device technology has evolved over the past decade. And a wireless technology is also rapidly advanced with the prevalence of mobile devices. Both mobile device and wireless system become an integral element of the modern world. Moreover, both technologies are still evolving with various researchers around the world [1].

A wireless network transmits a radio signal to communicate with other devices. The wireless signal travels between the atmospheres without regard to the physical interruption. Thus, the wireless device is able to connect to the network, regardless of the environments [2]. Surely, the wireless signal has an obvious boundary to identify the signal. The problem of the coverage arises from the strength of the signal. This kind of boundary may be easily solved with multiple relay stations. However, this feature makes some side effect for communication reliability. The mobile device is frequently moved as its name. In addition, the connection's quality may be different according to the environment. Hence, the wireless connection is more unstable than the wired. And much study has been done to guarantee the wireless QoS. Most of the

---

* Corresponding author.

D. Hwang et al. (Eds.): ICCCI 2014, LNAI 8733, pp. 40–48, 2014.

study focuses on the transport protocol level. The transmission communication proto- col (TCP) is the dominant protocol on the common network. However, the TCP may be a double-edged sword for wireless system. The TCP is the protocol, which is based on the wired environment. Hence, the existing study tries to improve on the TCP scheme for wireless network. For instance, some study focuses on the TCP control message, which ensures the reliable delivery. On the other hand, the other study adopts the supplement method like the packet snooping [3-6].

As above, there is much effort to improve the communication quality. This paper also has an interest in the similar issue. Our study is based on the packet snooping approach. And we adopt the fuzzy logic to manage the communication through the wireless network. The base station (BS) captures and holds the packet, which is going to the mobile host (MH). In this process, we attach the fuzzy logic to estimate the communication reliability between each mobile node. And the estimation result de- termines the communication cycle and buffer size. Through this, the BS controls the communication throughput according to the link's state.

The rest of this paper is organized as follows: In section 2, we briefly review re- lates works. And we describe the main idea of our proposed model in section 3. Sec- tion 4 presents the simulation design and its measured result. And finally we conclude in section 5.

## 2    Related Works

Existing studies try to optimize the TCP over the wireless network. The TCP provides the reliable connection by using its own flow control. This TCP scheme utilizes the control packet to ensure the delivery. However, the wireless communication frequent- ly faces some failure such as the high bit error rate. This problem comes from the environment of the wireless link. Nevertheless, the traditional TCP scheme just fol- lows the wired-based design.

The mobile TCP (M-TCP) is one of the improved versions for the wireless link. The M-TCP is to ensure the countermeasure for unstable connections. If some packet cannot reach to the MH, the M-TCP notifies this information to the fixed host (FH). However, the M-TCP does not establish the snooping buffer. Besides, the notify mes- sage is also one of the TCP packets. It may aggregate the network traffic, and induce the performance degradation. The split connection is the alternative option for this problem. It means; the network may classify the connection into two groups. First connection is to provide the wired communication between the FH and the BS. And second connection is to manage the wireless link between the BS and the MH. How- ever, it also has another problem from the relay agent [4].

The snooping TCP is also one of the indirect strategies for wireless connection. The snooping is just to capture and analyze the packet over the network. However, this scheme may provide the buffer and retransmission process. With the snooping module, The AP may recognize the packet flow, and recover the message on its own responsibility [6].

Our study is based on the snooping TCP, and attaches the fuzzy logic to improve the wireless QoS.

# 3    Fuzzy Logic-Based Adaptive Communication Management

In this paper, we apply the fuzzy logic to estimate the wireless link's status. And the BS controls the buffer size and the communication cycle with the estimation result. The better state ensures the high priority for communication between the BS and the MH. More detailed is as follows:

## 3.1    Fuzzy Logic-Based State Estimation

It is very difficult to judge the communication state clearly. Of course, there are some general factors to measure the network performance. However, the performance does not fully represent the actual states of the MH. We have also few factors with the uncertain characteristics. We adopt the fuzzy logic in order to solve this uncertainty [7].

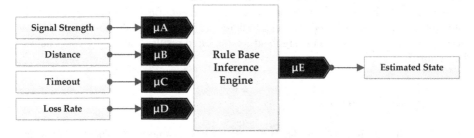

**Fig. 1.** Fuzzy model to estimate the link's state

**Table 1.** Fuzzy Input and Output Parameter

| Parameter Name | Function | Set of Fuzzy |
|---|---|---|
| Signal Strength | $\mu A$ | { Very Good, Good, Weak, Very Weak } |
| Distance | $\mu B$ | { Near, Middle, Far } |
| Timeout | $\mu C$ | { Very Rare, Rare, Normal, Frequent, Very Frequent } |
| Loss Rate | $\mu D$ | { Very Small, Small, Large, Very Large } |
| Estimated State | $\mu E$ | { Excellent, Acceptable, Questionable, Disappointed } |

Fig. 1 shows the designed fuzzy model to estimate the link's state. As shown in Figure 1, there are four input parameters for our fuzzy model. Both the signal strength and the distance are closely connected with the MH's physical factor. On the other hand, both the timeout and the loss rate are to evaluate the communication state. Each input is a crisp value, which directly measures from the environments. Hence, each input may translate to the fuzzy value. The fuzzy logic only handles a fuzzy value, not a crisp value. Each input parameter has different fuzzy value sets as shown in Figure 1.

The inference engine deduces the fuzzy output with the input parameter. To infer the result, the fuzzy logic includes its own rule base system.

**Table 2.** Sample of fuzzy inference rule

| µA | µB | µC | µD | µE |
|---|---|---|---|---|
| Very Good | Near | Very Rare | Very Small | Excellent |
| Good | Middle | Very Rare | Very Small | Excellent |
| Weak | Near | Very Rare | Very Small | Excellent |
| Very Good | Near | Frequent | Large | Acceptable |
| Very Weak | Middle | Rare | Very Small | Acceptable |
| Very Good | Near | Frequent | Very Large | Questionable |
| Very Weak | Far | Normal | Small | Questionable |
| Good | Middle | Very Frequent | Very Large | Disappointed |
| Weak | Near | Very Frequent | Very Large | Disappointed |
| Very Weak | Far | Very Frequent | Very Large | Disappointed |

Table 2 shows the example of the fuzzy inference rule. The inference rule includes the all combination of input parameters. Table 2 is just a sample to present the composition of the inference rule. The actual logic includes total 240 individual rules to deduce the µE. In our study, the inference is based on the Mamdani's min-max method. And we apply the center of gravity (CoG) to defuzzify the output value [8].

## 3.2    Adaptive Communication Management

Fig. 2 shows the brief appearance of our proposed model. Every packet from the FH or the MH may pass the flow analyzer. The flow analyzer not only analyzes incoming packets, but also measures the current state of the links. And the flow analyzer sends both packets and measured information to the flow controller. According to the current link's state, the packet goes to the destination, or the packet buffer. The flow controller determines this process with the state estimation module. And the state estimation module is the fuzzy system, which shown in previous section. Among these modules, the flow controller is the supervisor of our model. Our model is based on the snooping TCP [4, 6], and includes two major functions. The flow controller not only directs the snooping process, but also controls the throughput for flow.

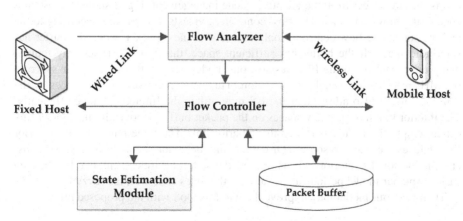

**Fig. 2.** Brief model for adaptive communication management

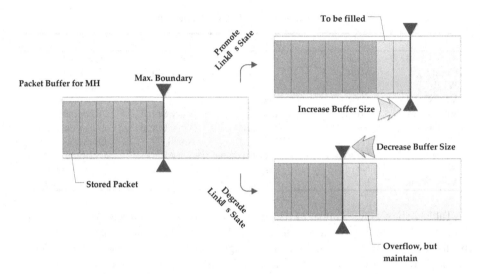

**Fig. 3.** Adaptive queue management according to the link's state

First, the flow controller handles the maximum queue size on the packet buffer. Our model does not ensure the static buffer for every MH. The flow controller allocates the buffer size according to the link's state. Fig. 3 describes this adaptive queue management of our model. The flow controller gives more buffer space to the better connection. For instance, we assume the degradation situation from the former link's state. The flow controller sends the "decrease" order to the packet buffer. In this case, some packet might be overflowed from the buffer. However, the packet buffer does not drop these packets, and ensure the delivery. And also the flow controller waits until the buffer has some idle space. On the other case, the packet buffer increases the idle space for target MH's link. And the flow controller requests more packets to the FH.

And last, the flow controller manages the packet transmission throughput. In fact, this is the side effect from the adaptive queue management. Fig. 4 shows the pseudo-code of the flow controller. The flow controller requests next packets according to the buffer's state. The flow controller only requests the next packet, when the buffer has some idle space. If the buffer has sufficient space, the flow controller may request many more packets to the FH. Besides, our model ensures the large buffer space to the MH, who has the reliable connection. This behavior gives more chances to the superior link. It also allows that the BS may utilize the limited resource efficiently. The inferior link has only few spaces on the packet buffer. Above all, the interior link cannot empty the buffer as fast as the superior link. The large buffer for the interior link induces the waste resource. Of course, this behavior seems to be a severer discrimination for the interior link. However, the state of wireless link can be changed easily. And our model periodically evaluates the link's state with the fuzzy logic.

Therefore, our model may improve the wireless QoS with the proposed process.

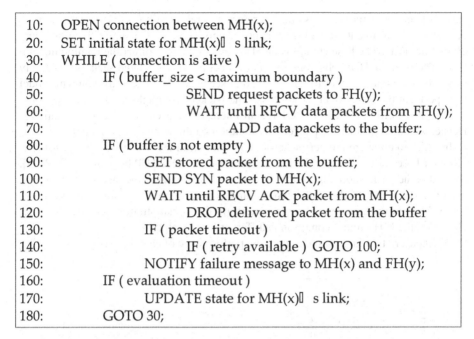

```
10:    OPEN connection between MH(x);
20:    SET initial state for MH(x)'s link;
30:    WHILE ( connection is alive )
40:        IF ( buffer_size < maximum boundary )
50:                SEND request packets to FH(y);
60:                WAIT until RECV data packets from FH(y);
70:                    ADD data packets to the buffer;
80:        IF ( buffer is not empty )
90:            GET stored packet from the buffer;
100:           SEND SYN packet to MH(x);
110:           WAIT until RECV ACK packet from MH(x);
120:               DROP delivered packet from the buffer
130:           IF ( packet timeout )
140:               IF ( retry available ) GOTO 100;
150:               NOTIFY failure message to MH(x) and FH(y);
160:       IF ( evaluation timeout )
170:           UPDATE state for MH(x)'s link;
180:   GOTO 30;
```

**Fig. 4.** Pseudo-code of the flow controller's behavior

## 4    Simulation Design and Results

We design the simulation environment to prove the efficiency of our proposed model. The TCP's behavior may translate to the discrete event. Hence, we apply the DEVS methodology for our simulation design [9].

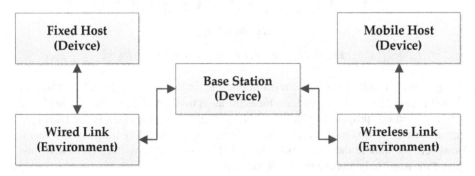

**Fig. 5.** Brief model design for simulation

Fig. 5 shows the brief form of our simulation model. We classify the simulation model into the device model, and the environment model. The device model is to imitate the network device as its name. On the other hand, the environment model is to describe the virtual network environment. Our model and scenario are reduced version from the actual full system. There are several fixed hosts and mobile hosts in the actual environment. However, our model only contains the single fixed host and 20 mobile hosts. The base station is an also single model to direct the communication flow. The most of TCP behaviors does not reflect in our model and scenario. We only apply the essential element to our model and scenario. And each environment model is to perform the intentional packet loss. The packet loss might be occurred with the random probability. However, we have an interest in the wireless network's failures. Hence, we only consider the packet loss situation on the wireless link model.

In order to compare the performance, our simulation includes three different behaviors for base station. First is the no snooping and indirect TCP behavior (I-TCP) [4]. Second is the basic snooping TCP behavior (S-TCP) [6]. And the last is our proposed behavior (FLACM-TCP). Both scenario and condition are equal for each behavior. Every MH shows the same variable pattern on the signal strength and the distance. However, the MH's initial condition is different from each other.

We measure two factors to compare the performance of each behavior.

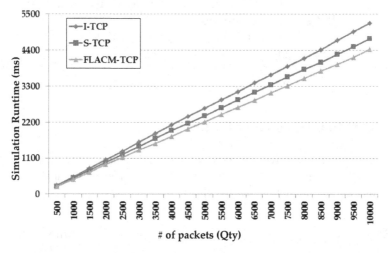

**Fig. 6.** Result graph for simulation runtime

Fig. 6 is the result graph for simulation runtime. In this scenario, the FH sends total 10,000 packets to the MH. And we measure an actual simulation time to finish the delivery. And our proposed model records 4428 milliseconds, which is the least runtime to deliver 10,000 packets. The record seems too late to apply for the real system. However, we handle the simulation time ratio as slow as possible. Otherwise, we cannot measure and compare the exact runtime.

Fig. 7 is the result graph for communication throughput. And we calculate the throughput with the following equation:

$$Thruput(x) = \frac{\sum P_a(x)}{\sum P_s(x)} \qquad (1)$$

Equation (1) is the formula to calculate the x-th MH's throughput. $P_s$ in equation (1) indicates all sent and received packets with the MH. In contrast, $P_a$ indicates the successfully received packets for MH. And we don't count the error or duplicated packet for $P_a$. Hence, the lower error leads the high throughput. We run the simulation for 10,000 milliseconds, and repeat the 256 times to estimate the average for throughput. And our proposed method records relatively high throughput than other model.

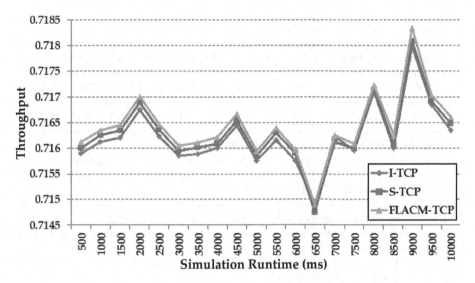

**Fig. 7.** Result graph for throughput

Both results demonstrate that, our proposed method ensures reasonable communication efficiency on the wireless network.

## 5    Conclusion

This paper presents the fuzzy logic-based adaptive communication management on the wireless network. The wireless network is most widely used in the real world. However, the wireless link shows poor stability to communicate with the mobile devices. In order to improve the wireless efficiency, we estimate the mobile host's state with the fuzzy logic. And the estimation result determines the buffer size for each wireless link. The better MH link's state induces the more chance for communication service. We simulate our proposed model to prove the efficiency. And the simulation result shows the reasonable performance of our proposed model. Therefore, the fuzzy logic-based adaptive communication management ensures the appropriate efficiency for wireless link. However, we don't consider the handover situation in this study. The future work will extend the behavior for multiple base stations.

**Acknowledgements.** This research was supported by Basic Science Research Program through the National Research Foundation of Korea(NRF) funded by the Ministry of Education, Science and Technology (2012R1A1A2002751), and funded by the Ministry of Science, ICT and Future Planning (NRF-2013R1A1A3A04007527).

# References

1. Raychaudhuri, D., Mandayam, N.B.: Frontiers of wireless and mobile communications. Proceedings of the IEEE 100(4), 824–840 (2012)
2. Avestimehr, A.S., Diggavi, S.N., Tse, D.N.: Wireless network information flow: A deterministic approach. IEEE Transactions on Information Theory 57(4), 1872–1905 (2011)
3. Shin, K., Kim, J., Choi, S.B.: Loss recovery scheme for TCP using MAC MIB over wireless access networks. IEEE Communications Letters 15(10), 1059–1061 (2011)
4. Maisuria, J.V., Patel, R.M.: Overview of Techniques for Improving QoS of TCP over Wireless Links. In: 2012 International Conference on Communication Systems and Network Technologies (CSNT), pp. 366–370. IEEE (2012)
5. Nguyen, T.H., Park, M., Youn, Y., Jung, S.: An improvement of TCP performance over wireless networks. In: 2013 Fifth International Conference on Ubiquitous and Future Networks (ICUFN), pp. 214–219. IEEE (2013)
6. Tiyyagura, S., Nutangi, R., Reddy, P.C.: An improved snoop for TCP Reno and TCP sack in wired-cum-wireless networks. Ind. J. Comput. Sci. Eng. 2, 455–460 (2011)
7. Rajasekaran, S., Pai, G.V.: Neural networks, Fuzzy logic and Genetic algorithms. PHI Learning Private Limited (2011)
8. Lee, C.C.: Fuzzy Logic in Control Systems: Fuzzy Logic Controller. IEEE Trans. Systems, Man and Cybernetics 20, 404–435 (1990)
9. Zeigler, B., Moon, Y., Kim, D., Ball, G.: The DEVS environment for high performance modeling and simulation, Computational Science and Engineering. IEEE CS&E, 61–71 (1997)

# Application of Self-adapting Genetic Algorithms to Generate Fuzzy Systems for a Regression Problem

Tadeusz Lasota[1], Magdalena Smętek[2], Zbigniew Telec[2],
Bogdan Trawiński[2], and Grzegorz Trawiński[3]

[1] Wrocław University of Technology, Institute of Informatics,
Wybrzeże Wyspiańskiego 27, 50-370 Wrocław, Poland
[2] Wrocław University of Environmental and Life Sciences, Dept. of Spatial Management
ul. Norwida 25/27, 50-375 Wrocław, Poland
[3] Wrocław University of Technology, Faculty of Electronics,
Wybrzeże S. Wyspiańskiego 27, 50-370 Wrocław, Poland
{zbigniew.telec,magdalena.smetek,bogdan.trawinski}@pwr.edu.pl,
tadeusz.lasota@up.wroc.pl, grzegorz.trawinsky@gmail.com

**Abstract.** Six variants of self-adapting genetic algorithms with varying mutation, crossover, and selection were developed. To implement self-adaptation the main part of a chromosome which comprised the solution was extended to include mutation rates, crossover rates, and/or tournament size. The solution part comprised the representation of a fuzzy system and was real-coded whereas to implement the proposed self-adapting mechanisms binary coding was employed. The resulting self-adaptive genetic fuzzy systems were evaluated using real-world datasets derived from a cadastral system and included records referring to residential premises transactions. They were also compared in respect of prediction accuracy with genetic fuzzy systems optimized by a classical genetic algorithm, multilayer perceptron and radial basis function neural network. The analysis of the results was performed using statistical methodology including nonparametric tests followed by post-hoc procedures designed especially for multiple $N \times N$ comparisons.

**Keywords:** self-adaptive GA, mutation, crossover, genetic fuzzy systems.

## 1 Introduction

The execution time of genetic algorithms constitute a big challenge, especially when they are used in hybrid methods to create and optimize different classification and prediction models such as genetic fuzzy systems and genetic neural networks. Many researchers have developed numerous techniques for speeding up the convergence of Genetic Algorithms (*GA*) or Evolutionary Algorithms (*EA*) for above two decades. The methods for adapting the values of various parameters to optimize processes in evolutionary computation has been extensively studied and the issue of adjusting *GA/EA* to the problem while solving it still seems to be a promising area of research. The probability of mutation and crossover, the size of selection tournament, or the

D. Hwang et al. (Eds.): ICCCI 2014, LNAI 8733, pp. 49–61, 2014.

population size belong to the most commonly set parameters of *GA/EA*. Three taxonomies of parameter setting forms in evolutionary computation have been devised by Angeline [1], Smith and Fogarty [2], and Eiben, Hinterding, and Michalewicz [3]. The first determines three different adaptation levels of *GA/EA* parameters: population-level where parameters that are global to the population are adjusted, individual-level where changes affect each member of the population separately, and component-level where each component of each member may be modified individually. The second classification is based on three division criteria: what is being adapted, the scope of the adaptation, and the basis for change which is further split into two categories: evidence upon which the change is carried out and the rule or algorithm that executes the change.

The third taxonomy [3] is a general one distinguishing two major forms of parameter value setting, i.e. parameter tuning and parameter control. The first consists in determining good values for the parameters before running *GA/EA*, and then tuning the algorithms without changing these values during the run. However, this approach stands in contradiction to the dynamic nature of *GA/EA*. The second form is an alternative and relies in dynamic adjusting the parameter values during the execution. The third can be categorized into three classes deterministic, adapting and self-adapting parameter control. Deterministic parameter control is applied when the values of evolutionary computation parameters are modified according to some deterministic rules without using any feedback from the optimization process. In turn, adaptive parameter control is employed when some form of feedback from the process is used to determine the trend or strength of the change to the *GA* parameter. Self-adaptive parameter control takes place when the parameters to be adapted are encoded into the chromosomes and undergo mutation and recombination.

Numerous parameter control methods have been proposed in the literature [4], [5], [6]. Several mechanisms of mutation and crossover adaptation and self-adaptation have been developed and experimentally tested [7], [8], [9], [10].

For several years we have been developing and evaluating techniques for building regression models to aid in property valuation based on various machine learning algorithms. Our study included genetic fuzzy systems and artificial neural networks as both single models [11], [12], [13] and ensembles built using different resampling techniques [14], [15], [16], [17], [18], [19]. An especially good performance revealed evolving fuzzy models applied to cadastral data [20], [21]. Evolving fuzzy systems are suitable for modelling the real estate market dynamics because they can be regularly updated on demand based on new incoming samples and the data of property sales ordered by the transaction date can be treated as a data stream. We have also explored the methods to predict from a data stream of real estate sales transactions based on ensembles of genetic fuzzy systems [22], [23], [24], [25]. Our former investigations on the use of evolutionary algorithms to optimize fuzzy systems, which included the generation of rule base and tuning the parameters of membership functions, showed it is an arduous and computationally expensive process. For this reason we attempted to incorporate self-adapting techniques into genetic fuzzy systems aimed to generate regression models for property valuation.

The research presented in this paper is also a continuation of our former study on self-adaptive genetic algorithms [26], [27]. We developed genetic algorithms with self-adaptive mutation and crossover based on an idea developed by Maruo et al. [28] and tested them using several selected multimodal benchmark functions. The algorithms employing self-adaptive mutation and crossover revealed better performance than a traditional genetic one. We also applied the self-adapting genetic algorithms to compose heterogeneous bagging ensembles [29].

## 2    SAGA Techniques Used to Construct Fuzzy Systems

Six variants of self-adapting genetic algorithms (*SAGA*) with varying mutation (*M*), crossover (*C*), and selection (*T*) were developed. They were named in the paper *SAM, SAC, SAMC, SACT, SAMT,* and *SAMCT*, respectively. In all variants of *SAGA* algorithms constant length chromosomes were used and their structures are illustrated in Figure 1. To implement self-adaptation the main part of a chromosome which comprised the solution was extended to include mutation rates, crossover rates, and/or tournament size. The solution part comprised the representation of a fuzzy system and was real-coded whereas to implement the proposed self-adapting mechanisms binary coding was employed. The mutation rate could be set to values from the bracket 0 to 0.3, and crossover rate from the range 0.5 to 1.0. Therefore, to encode the mutation rate 5 genes and crossover rate 7 genes were used. In turn, the tournament size was encoded using 3 binary genes to represent the range from 1 to 7.

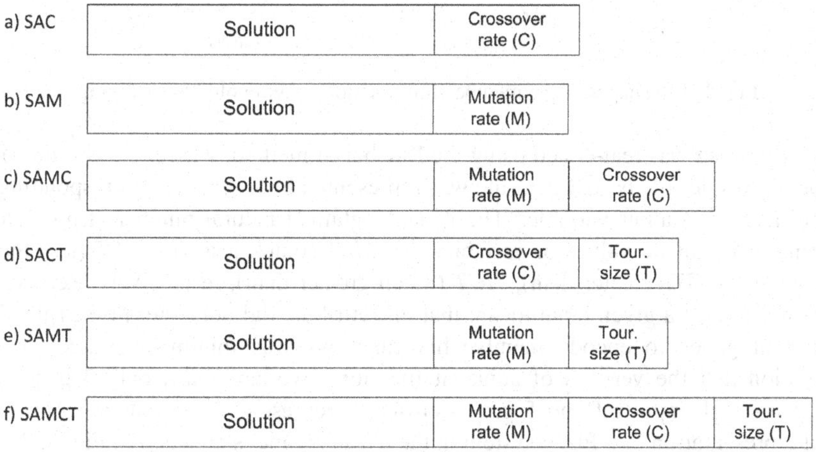

| a) SAC | Solution | Crossover rate (C) | | |
|---|---|---|---|---|
| b) SAM | Solution | Mutation rate (M) | | |
| c) SAMC | Solution | Mutation rate (M) | Crossover rate (C) | |
| d) SACT | Solution | Crossover rate (C) | Tour. size (T) | |
| e) SAMT | Solution | Mutation rate (M) | Tour. size (T) | |
| f) SAMCT | Solution | Mutation rate (M) | Crossover rate (C) | Tour. size (T) |

**Fig. 1.** Chromosome structures of individual self-adaptive genetic algorithms

*Solution.* For each input variable three triangular and trapezoidal membership functions, and for output - five functions, were automatically determined by the symmetric division of the individual attribute domains. The evolutionary optimization process combined both learning the rule base and tuning the membership functions using real-coded chromosomes. Similar designs are described in [30], [31]. The shapes of the triangular and trapezoidal membership functions after evolutionary tuning are presented in Figure 2.

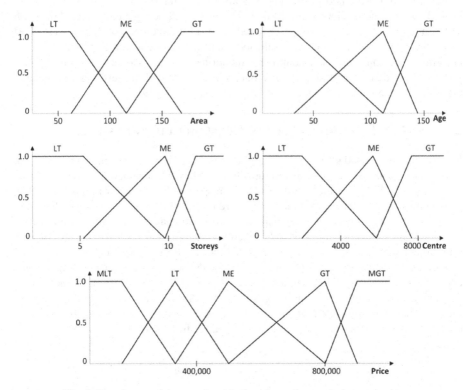

**Fig. 2.** The shapes of the membership functions after evolutionary tuning

The rule base was real-coded using the Pittsburgh method, where one chromosome comprised whole rule base. Each rule was represented by $n+1$ genes corresponding to $n$ input and one output variables The genes contained natural numbers from 1 to 5 referring to linguistic values of variables, i.e. *MLT* (much less than), *LT* (less than), *ME* (medium), *GT* (greater than), *MGT* (much greater than), respectively. Zero value on the position of a given input meant that this attribute did not occur in the rule. The number of genes to encode membership functions was minimized based on the assumption that the vertices of adjacent functions overlap. Thus, only 3 genes are needed for each input mf and 5 genes suffice to represent the output mf. The real-coded representation of a fuzzy system in the chromosome is depicted in Figure 3.

| Rule 1 | | | Rule 2 | | | Rule 3 | | | ... | Rule 15 | | |
|---|---|---|---|---|---|---|---|---|---|---|---|---|
| 1 | 3 | 1 | 2 | 3 | 2 | 0 | 0 | 0 | 2 | 0 | 1 | 1 | 2 | 1 | ... | 2 | 1 | 0 | 3 | 2 |

| Input mf 1 | | | Input mf 2 | | | ... | Output mf | | | | |
|---|---|---|---|---|---|---|---|---|---|---|---|
| 18.1 | 53.9 | 117.4 | 31.7 | 103.3 | 120.5 | ... | 149.4 | 336.9 | 466.8 | 641.3 | 757.6 |

**Fig. 3.** Encoding rules and membership functions in the solution part of the chromosome

*Self-adaptive crossover.* The self-adaptive crossover, which is depicted in Figure 4, is different from a traditional *GA* crossover. A special *K×1* table with real, randomly selected values from the brackets 0.5 to 1.0 is created, where *K* is the number of chromosomes in the population. Each chromosome is connected with one real value in the table. The self-adaption of the crossover goes on in the following way. For each chromosome from population:

- extract the genes representing the crossover rate from the chromosome,
- calculate the value of crossover rate extracted from chromosome,
- if the value from the table is lower than the value of crossover rate from the chromosome, then the chromosome is selected to a classic crossover process,
- the *K×1* table remains unchanged during the execution of the *SAGA* algorithm.

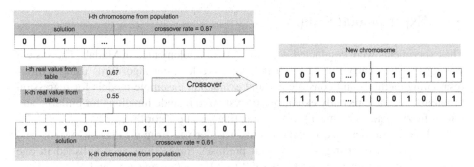

**Fig. 4.** Self-adaptive crossover in *SAGA* algorithms

*Self-adaptive mutation.* The self-adaptive mutation applied in *SAGA* algorithms is illustrated in Figure 5. It differs from a standard *GA* mutation which rate remains constant during the run. Each chromosome from the population can be subject to the mutation. A special *K×L* matrix with real, randomly selected values from the range 0 to 0.3 is created, where *K* is the number of chromosomes in the population, and *L* stands for the number of genes in a chromosome. Each gene in each chromosome is connected with one real value in the matrix. The self-adaptation of the mutation proceeds as follows. For each chromosome from population:

- extract the genes representing the mutation rate from the chromosome,
- calculate the value of mutation rate extracted from chromosome,
- if the value from the matrix is lower than the value of the mutation rate taken from the chromosome, then the chromosome mutates in a traditional way,
- the *K×L* matrix remains unchanged during the execution of the *SAGA* algorithm.

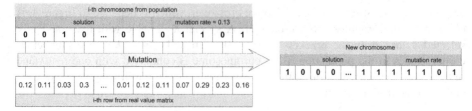

**Fig. 5.** Self-adaptive mutation in *SAGA* algorithms

*Self-adaptive selection.* Three genes were added to the chromosome to encode the tournament size which could be set to integer number from 1 to 7. Therefore, before selection average tournament size in the population was calculated and this value was used as the final tournament size in the selection operation.

## 3    Experimental Setup

The experiments were conducted with our system implemented in Matlab. The system was designed to carry out research into machine learning algorithms using various resampling methods and constructing and evaluating ensemble models for regression problems. We have recently extended our system to include functions for building and tuning fuzzy systems by means of self-adapting genetic algorithms.

Real-world dataset used in experiments was derived from a cadastral system  and included records referring to residential premises transactions accomplished in one Polish big city within 14 years from 1998 to 2011. After selection and cleansing the final dataset counted 9795 samples. Four following attributes were pointed out as main price drivers by professional appraisers: usable area of a flat (*Area*), age of a building construction (*Age*), number of storeys in the building (Storeys), the distance of the building from the city centre (*Centre*), in turn, price of premises (*Price*) was the output variable.

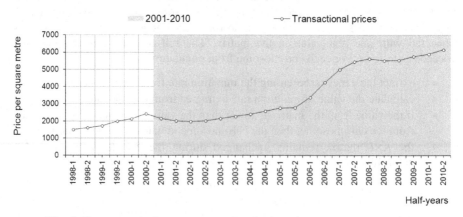

**Fig. 6.** Change trend of average transactional prices per square metre over time

The evaluating experiments were conducted for 36 points of time from 2002-01-01 to 2010-10-01. Single models were built over training data delineated by the time span of 12 months. In turn, the test datasets were determined by the interval of 3 months, current for a given time point. The selected dataset ensured the variability of individual points of observation due to the dramatic rise of the prices of residential premises during the worldwide real estate bubble (see Fig. 6).

For comparative tests we took the results of our previous investigations into genetic fuzzy systems optimized by a classical genetic algorithm (*GA*) [32] (see parameters in Table 1) and artificial neural networks: multilayer perceptron (*MLP*) and radial basis function neural networks (*RBF*) [33] conducted over the same datasets. To examine the better convergence of *SAGA* algorithms fuzzy systems were built within 50 generations whereas with a classical *GA* 100 generations were executed. The number of epochs to learn each neural network was equal to 100. As the performance measure the root mean square error (*RMSE*) was used.

**Table 1.** Parameters of GA to optimize fuzzy sets used in experiments

| Fuzzy system | Genetic Algorithm |
|---|---|
| Type of fuzzy system: Mamdani | Chromosome: rule base and mf, real-coded |
| No. of input variables: 4 | Population size: 100 |
| Type of membership functions (mf): triangular | Fitness function: MSE |
| No. of input mf: 3 | Selection function: tournament |
| No. of output mf: 5 | Tournament size: 4 |
| No. of rules: 15 | Elite count: 2 |
| AND operator: prod | Crossover fraction: 0.8 |
| Implication operator: prod | Crossover function: two point |
| Aggregation operator: probor | Mutation function: custom |
| Defuzzyfication method: centroid | No. of generations: 100 |

The analysis of the results was performed using statistical methodology including nonparametric tests followed by post-hoc procedures designed especially for multiple $N \times N$ comparisons [34], [35], [36], [37]. The routine starts with the nonparametric Friedman test, which detect the presence of differences among all algorithms compared. After the null-hypotheses have been rejected the post-hoc procedures are applied in order to point out the particular pairs of algorithms which produce differences. For $N \times N$ comparisons nonparametric Nemenyi's, Holm's, Shaffer's, and Bergmann-Hommel's procedures are employed.

# 4    Analysis of Experimental Results

## 4.1    Performance of SAGA Fuzzy Systems

The performance of single *SAGA* fuzzy models over 36 points of observation is depicted in Figure 7. The values of *RMSE* are given in thousand PLN. We can see unstable behaviour of *SAMT* models which produce excessive errors for two points.

However, the differences among the models are not visually apparent, therefore one should refer to statistical tests of significance.

Average rank positions of single *SAGA* models determined during Friedman test are shown in Table 2, where the lower rank value the better model. Adjusted p-values for Nemenyi's, Holm's, Shaffer's, and Bergmann-Hommel's post-hoc procedures for *N×N* comparisons for all possible pairs of *SAGA* methods are shown in Table 3. For illustration the p-values of the paired Wilcoxon test are also given. The significance level considered for the null hypothesis rejection was 0.05. Significant differences were observed only for one pair of SAGAs: *SAC* ensembles surpassed the *SAMCT* ones. The paired Wilcoxon test can lead to over-optimistic decisions because it allows for rejection of a greater number of null hypotheses.

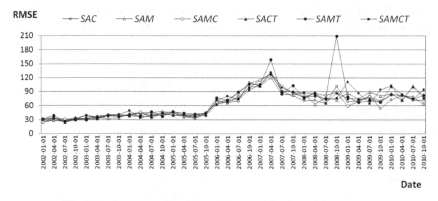

**Fig. 7.** Performance of *SAGA* models over 36 points o observations

**Table 2.** Average rank positions of *SAGA* models determined during Friedman test

| 1st | 2nd | 3rd | 4th | 5th | 6th |
|-----|-----|-----|-----|-----|-----|
| SAC (2.86) | SAM (3.22) | SACT (3.22) | SAMT (3.69) | SAMC (3.78) | SAMCT (4.22) |

**Table 3.** Adjusted p-values for *N×N* comparisons of *SAGA* models for all 15 hypotheses

| Method vs Method | pWilcox | pNeme | pHolm | pShaf | pBerg |
|------------------|---------|-------|-------|-------|-------|
| *SAC vs SAMCT* | *0.0184* | *0.0304* | *0.0304* | *0.0304* | *0.0304* |
| SAM vs SAMCT | *0.0192* | 0.3501 | 0.3268 | 0.2334 | 0.2334 |
| SACT vs SAMCT | *0.0443* | 0.3501 | 0.3268 | 0.2334 | 0.2334 |
| SAC vs SAMC | *0.0169* | 0.5645 | 0.4516 | 0.3764 | 0.3764 |
| SAC vs SAMT | *0.0095* | 0.8817 | 0.6466 | 0.5878 | 0.4115 |
| SAM vs SAMC | 0.5094 | 1.0000 | 1.0000 | 1.0000 | 1.0000 |
| SAMC vs SACT | 0.3300 | 1.0000 | 1.0000 | 1.0000 | 1.0000 |
| SAMT vs SAMCT | 0.5297 | 1.0000 | 1.0000 | 1.0000 | 1.0000 |
| SAM vs SAMT | 0.4321 | 1.0000 | 1.0000 | 1.0000 | 1.0000 |
| SACT vs SAMT | 0.1126 | 1.0000 | 1.0000 | 1.0000 | 1.0000 |
| SAMC vs SAMCT | 0.3300 | 1.0000 | 1.0000 | 1.0000 | 1.0000 |
| SAC vs SACT | 0.2784 | 1.0000 | 1.0000 | 1.0000 | 1.0000 |
| SAC vs SAM | 0.3705 | 1.0000 | 1.0000 | 1.0000 | 1.0000 |
| SAMC vs SAMT | 0.7296 | 1.0000 | 1.0000 | 1.0000 | 1.0000 |
| SAM vs SACT | 0.8628 | 1.0000 | 1.0000 | 1.0000 | 1.0000 |

## 4.2    Comparison of *SAGA* Fuzzy Models with Other Approaches

For comparison we took the results of our previous investigations into genetic fuzzy systems optimized by a classical genetic algorithm (*GA*) [32] and artificial neural networks: multilayer perceptron (*MLP*) and radial basis function (*RBF*) [33] conducted over the same datasets. For statistical tests we selected the *SAGA* fuzzy models providing the best performance, namely *SAC, SAM,* and *SACT* ones. The *RMSE* of examined methods was computed for the same 36 observation time points. The Friedman test values showed that there were significant differences among models. Average ranks of compared methods produced by the test are shown in Table 4, where the lower rank value the better model. Adjusted p-values for the paired Wilcoxon test as well as Nemenyi's, Holm's, Shaffer's, and Bergmann-Hommel's post-hoc procedures for *N*×*N* comparisons for all possible pairs of algorithms are shown in Table 5. The p-values indicating the statistically significant differences between given pairs of algorithms are marked with italics. The significance level considered for the null hypothesis rejection was 0.05.

**Table 4.** Average rank positions of compared models determined during Friedman test

| 1st | 2nd | 3rd | 4th | 5th | 6th |
|-----|-----|-----|-----|-----|-----|
| GA (2.06) | SAC (2.58) | SAM (2.69) | SACT (2.75) | RBF (5.36) | MLP (5.56) |

**Table 5.** Adjusted p-values for N×N comparisons of SAGA models for all 15 hypotheses

| Method vs Method | pWilcox | pNeme | pHolm | pShaf | pBerg |
|------------------|---------|-------|-------|-------|-------|
| *GA vs MLP* | *1.68E-07* | *3.10E-14* | *3.10E-14* | *3.10E-14* | *3.10E-14* |
| *GA vs RBF* | *1.68E-07* | *9.85E-13* | *9.19E-13* | *6.56E-13* | *6.56E-13* |
| *SAC vs MLP* | *1.83E-07* | *2.37E-10* | *2.05E-10* | *1.58E-10* | *1.58E-10* |
| *SAM vs MLP* | *1.99E-07* | *1.30E-09* | *1.04E-09* | *8.68E-10* | *6.07E-10* |
| *SACT vs MLP* | *1.68E-07* | *2.98E-09* | *2.18E-09* | *1.99E-09* | *1.19E-09* |
| *SAC vs RBF* | *1.68E-07* | *4.48E-09* | *2.99E-09* | *2.99E-09* | *1.79E-09* |
| *SAM vs RBF* | *2.17E-07* | *2.21E-08* | *1.32E-08* | *1.03E-08* | *5.89E-09* |
| *SACT vs RBF* | *1.68E-07* | *4.79E-08* | *2.55E-08* | *2.23E-08* | *1.28E-08* |
| SACT vs GA | *0.003476* | 1.000000 | 0.807034 | 0.807034 | 0.807034 |
| SAM vs GA | *0.014252* | 1.000000 | 0.884254 | 0.884254 | 0.807034 |
| SAC vs GA | 0.123650 | 1.000000 | 1.000000 | 1.000000 | 0.807034 |
| MLP vs RBF | 0.292521 | 1.000000 | 1.000000 | 1.000000 | 1.000000 |
| SAC vs SACT | 0.278352 | 1.000000 | 1.000000 | 1.000000 | 1.000000 |
| SAC vs SAM | 0.370519 | 1.000000 | 1.000000 | 1.000000 | 1.000000 |
| SAM vs SACT | 0.862796 | 1.000000 | 1.000000 | 1.000000 | 1.000000 |

Following main observations could be done: despite *GA* took the first position in the Friedman test rank, the post hoc procedures indicated no significant differences among *GA, SAC, SAM,* and *SACT* models. It should be noted that *SAGA* algorithms were run for 50 generations whereas the classical *GA* for 100 generations. All these four methods outperformed significantly *MLP* and *RBF* neural networks.-The paired

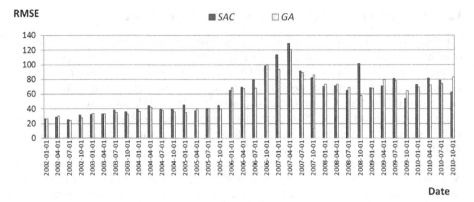

**Fig. 8.** Illustration of *SAC* and *GA* fuzzy system performance over 36 points of observation

Wilcoxon test again allowed for rejection of a greater number of null hypotheses which could lead to over-optimistic decisions. For illustration the performance of *SAC* and *GA* models over 36 points of observation is shown in Fig. 8. The values of *RMSE* are given in thousand PLN.

## 5     Conclusions and Future Work

In the paper we proposed to apply self-adapting genetic algorithms to generate rule base and tune membership functions of fuzzy systems build over cadastral data to predict the values of real estates. We devised and implemented six self-adapting genetic algorithms with varying mutation, crossover, and selection. To implement self-adaptation the main part of a chromosome which comprised the solution, i.e. the definition of a fuzzy system, was extended to include mutation rates, crossover rates, and tournament size. The solution part was real-coded whereas to reflect the parameters of self-adaptive mechanisms binary coding was employed. The resulting self-adaptive genetic fuzzy systems were evaluated using records of sale and purchase transactions of residential premises taken from a cadastral system. They were also compared in respect of prediction accuracy with multilayer perceptron and radial basis function neural networks as well as with genetic fuzzy systems optimized by a classical genetic algorithm. The analysis of the results was performed using nonparametric methodology including the Friedman tests followed by the Nemenyi's, Holm's, Shaffer's, and Bergmann-Hommel's post-hoc procedures designed especially for multiple $N \times N$ comparisons.

The preliminary results showed that there were not statistically significant differences in accuracy among the majority of *SAGA* models. We compared three *SAGA* models with three competing methods, namely genetic fuzzy systems optimized by a classical genetic algorithm (*GA*) and two artificial neural networks: multilayer perceptron (*MLP*) and radial basis function (*RBF*). No significant differences among *GA, SAC, SAM,* and *SACT* models were observed It should be noted that *SAGA* algorithms were run for twice less number of generations than the classical *GA* was. Moreover, all *SAGA* models outperformed significantly the neural networks. It is planned to continue the investigation of *SAGA* algorithms applied to generate and tune fuzzy systems.

We intend continue our study to find optimal parameters of self-adapting mechanisms as well as explore the convergence of *SAGA* algorithms compared to the classical ones. Further experiments will be conducted using greater number of benchmark datasets taken from the UCI repository.

**Acknowledgments.** This work was partially supported by the National Science Centre under grant no. N N516 483840 and the "Młoda Kadra" funds of Wrocław University of Technology.

# References

1. Angeline, P.J.: Adaptive and self-adaptive evolutionary computations. In: Palaniswami, M., Attikiouzel, Y. (eds.) Computational Intelligence: A Dynamic Systems Perspective, pp. 152–163. IEEE Press, New York (1995)
2. Smith, J.E., Fogarty, T.C.: Operator and parameter adaptation in genetic algorithms. Soft Computing 1(2), 81–87 (1997)
3. Eiben, E., Hinterding, R., Michalewicz, Z.: Parameter control in evolutionary algorithms. IEEE Transactions on Evolutionary Computation 3(2), 124–141 (1999)
4. Bäck, T., Schwefel, H.-P.: An Overview of Evolutionary Algorithms for Parameter Optimization. Evolutionary Computation 1(1), 1–23 (1993)
5. Meyer-Nieberg, S., Beyer, H.-G.: Self-Adaptation in Evolutionary Algorithms. In: Lobo, F.G., et al. (eds.) Self-Adaptation in Evolutionary Algorithms. SCI, vol. 54, pp. 47–75. Springer, Heidelberg (2007)
6. Hinterding, R., Michalewicz, Z., Eiben, A.E.: Adaptation in Evolutionary Computation: A Survey. In: Proceedings of the Fourth International Conference on Evolutionary Computation (ICEC 1997), pp. 65–69. IEEE Press, New York (1997)
7. Deb, K., Beyer, H.-G.: Self-adaptive genetic algorithms with simulated binary crossover. Evolutionary Computation 9(2), 197–221 (2001)
8. Hansen, N., Ostermeier, A.: Completely derandomized self-adaptation in evolution strategies. Evolutionary Computation 9(2), 159–195 (2001)
9. De Jong, K.: An analysis of the behavior of a class of genetic adaptive systems. PhD thesis, University of Michigan (1975)
10. Lobo, F.: The parameter-less genetic algorithm: rational and automated parameter selection for simplified genetic algorithm operation. PhD thesis, Nova University of Lisboa (2000)
11. Król, D., Lasota, T., Nalepa, W., Trawiński, B.: Fuzzy system model to assist with real estate appraisals. In: Okuno, H.G., Ali, M. (eds.) IEA/AIE 2007. LNCS (LNAI), vol. 4570, pp. 260–269. Springer, Heidelberg (2007)
12. Król, D., Lasota, T., Trawiński, B., Trawiński, K.: Comparison of Mamdani and TSK Fuzzy Models for Real Estate Appraisal. In: Apolloni, B., Howlett, R.J., Jain, L. (eds.) KES 2007, Part III. LNCS (LNAI), vol. 4694, pp. 1008–1015. Springer, Heidelberg (2007)
13. Graczyk, M., Lasota, T., Trawiński, B.: Comparative Analysis of Premises Valuation Models Using KEEL, RapidMiner, and WEKA. In: Nguyen, N.T., Kowalczyk, R., Chen, S.-M. (eds.) ICCCI 2009. LNCS, vol. 5796, pp. 800–812. Springer, Heidelberg (2009)

14. Lasota, T., Telec, Z., Trawiński, B., Trawiński, K.: Exploration of Bagging Ensembles Comprising Genetic Fuzzy Models to Assist with Real Estate Appraisals. In: Corchado, E., Yin, H. (eds.) IDEAL 2009. LNCS, vol. 5788, pp. 554–561. Springer, Heidelberg (2009)

15. Lasota, T., Telec, Z., Trawiński, B., Trawiński, K.: A Multi-agent System to Assist with Real Estate Appraisals Using Bagging Ensembles. In: Nguyen, N.T., Kowalczyk, R., Chen, S.-M. (eds.) ICCCI 2009. LNCS, vol. 5796, pp. 813–824. Springer, Heidelberg (2009)

16. Graczyk, M., Lasota, T., Trawiński, B., Trawiński, K.: Comparison of Bagging, Boosting and Stacking Ensembles Applied to Real Estate Appraisal. In: Nguyen, N.T., Le, M.T., Świątek, J. (eds.) Intelligent Information and Database Systems. LNCS, vol. 5991, pp. 340–350. Springer, Heidelberg (2010)

17. Krzystanek, M., Lasota, T., Telec, Z., Trawiński, B.: Analysis of Bagging Ensembles of Fuzzy Models for Premises Valuation. In: Nguyen, N.T., Le, M.T., Świątek, J. (eds.) Intelligent Information and Database Systems. LNCS, vol. 5991, pp. 330–339. Springer, Heidelberg (2010)

18. Kempa, O., Lasota, T., Telec, Z., Trawiński, B.: Investigation of bagging ensembles of genetic neural networks and fuzzy systems for real estate appraisal. In: Nguyen, N.T., Kim, C.-G., Janiak, A. (eds.) ACIIDS 2011, Part II. LNCS, vol. 6592, pp. 323–332. Springer, Heidelberg (2011)

19. Lasota, T., Telec, Z., Trawiński, G., Trawiński, B.: Empirical Comparison of Resampling Methods Using Genetic Fuzzy Systems for a Regression Problem. In: Yin, H., Wang, W., Rayward-Smith, V. (eds.) IDEAL 2011. LNCS, vol. 6936, pp. 17–24. Springer, Heidelberg (2011)

20. Lasota, T., Telec, Z., Trawiński, B., Trawiński, K.: Investigation of the eTS Evolving Fuzzy Systems Applied to Real Estate Appraisal. Journal of Multiple-Valued Logic and Soft Computing 17(2-3), 229–253 (2011)

21. Lughofer, E., Trawiński, B., Trawiński, K., Kempa, O., Lasota, T.: On Employing Fuzzy Modeling Algorithms for the Valuation of Residential Premises. Information Sciences 181, 5123–5142 (2011)

22. Trawiński, B., Lasota, T., Smętek, M., Trawiński, G.: An Attempt to Employ Genetic Fuzzy Systems to Predict from a Data Stream of Premises Transactions. In: Hüllermeier, E., Link, S., Fober, T., Seeger, B. (eds.) SUM 2012. LNCS, vol. 7520, pp. 127–140. Springer, Heidelberg (2012)

23. Trawiński, B., Lasota, T., Smętek, M., Trawiński, G.: An Analysis of Change Trends by Predicting from a Data Stream Using Genetic Fuzzy Systems. In: Nguyen, N.-T., Hoang, K., Jędrzejowicz, P. (eds.) ICCCI 2012, Part I. LNCS, vol. 7653, pp. 220–229. Springer, Heidelberg (2012)

24. Trawiński, B., Lasota, T., Smętek, M., Trawiński, G.: Weighting Component Models by Predicting from Data Streams Using Ensembles of Genetic Fuzzy Systems. In: Larsen, H.L., Martin-Bautista, M.J., Vila, M.A., Andreasen, T., Christiansen, H. (eds.) FQAS 2013. LNCS, vol. 8132, pp. 567–578. Springer, Heidelberg (2013)

25. Trawiński, B.: Evolutionary Fuzzy System Ensemble Approach to Model Real Estate Market based on Data Stream Exploration. Journal of Universal Computer Science 19(4), 539–562 (2013)

26. Smętek, M., Trawiński, B.: Investigation of Genetic Algorithms with Self-adaptive Crossover, Mutation, and Selection. In: Corchado, E., Kurzyński, M., Woźniak, M. (eds.) HAIS 2011, Part I. LNCS, vol. 6678, pp. 116–123. Springer, Heidelberg (2011)

27. Smętek, M., Trawiński, B.: Investigation of Self-adapting Genetic Algorithms using Some Multimodal Benchmark Functions. In: Jędrzejowicz, P., Nguyen, N.T., Hoang, K. (eds.) ICCCI 2011, Part I. LNCS, vol. 6922, pp. 213–223. Springer, Heidelberg (2011)
28. Maruo, M.H., Lopes, H.S., Delgado, M.R.: Self-Adapting Evolutionary Parameters: Encoding Aspects for Combinatorial Optimization Problems. In: Raidl, G.R., Gottlieb, J. (eds.) EvoCOP 2005. LNCS, vol. 3448, pp. 154–165. Springer, Heidelberg (2005)
29. Smętek, M., Trawiński, B.: Selection of Heterogeneous Fuzzy Model Ensembles Using Self-adaptive Genetic Algorithms. New Generation Computing 29(3), 309–327 (2011)
30. Cordón, O., Herrera, F.: A Two-Stage Evolutionary Process for Designing TSK Fuzzy Rule-Based Systems. IEEE Tr. on Sys., Man and Cyber., Part B 29(6), 703–715 (1999)
31. Król, D., Lasota, T., Trawiński, B., Trawiński, K.: Investigation of evolutionary optimization methods of TSK fuzzy model for real estate appraisal. International Journal of Hybrid Intelligent Systems 5(3), 111–128 (2008)
32. Trawiński, B., Smętek, M., Lasota, T., Trawiński, G.: Evaluation of Fuzzy System Ensemble Approach to Predict from a Data Stream. In: Nguyen, N.T., Attachoo, B., Trawiński, B., Somboonviwat, K. (eds.) ACIIDS 2014, Part II. LNCS, vol. 8398, pp. 137–146. Springer, Heidelberg (2014)
33. Telec, Z., Trawiński, B., Lasota, T., Trawiński, K.: Comparison of Evolving Fuzzy Systems with an Ensemble Approach to Predict from a Data Stream. In: Bădică, C., Nguyen, N.T., Brezovan, M. (eds.) ICCCI 2013. LNCS, vol. 8083, pp. 377–387. Springer, Heidelberg (2013)
34. Demšar, J.: Statistical comparisons of classifiers over multiple data sets. Journal of Machine Learning Research 7, 1–30 (2006)
35. García, S., Herrera, F.: An Extension on "Statistical Comparisons of Classifiers over Multiple Data Sets" for all Pairwise Comparisons. Journal of Machine Learning Research 9, 2677–2694 (2008)
36. Graczyk, M., Lasota, T., Telec, Z., Trawiński, B.: Nonparametric Statistical Analysis of Machine Learning Algorithms for Regression Problems. In: Setchi, R., Jordanov, I., Howlett, R.J., Jain, L.C. (eds.) KES 2010, Part I. LNCS, vol. 6276, pp. 111–120. Springer, Heidelberg (2010)
37. Trawiński, B., Smętek, M., Telec, Z., Lasota, T.: Nonparametric Statistical Analysis for Multiple Comparison of Machine Learning Regression Algorithms. International Journal of Applied Mathematics and Computer Science 22(4), 867–881 (2012)

# Analysis of Profile Convergence in Personalized Document Retrieval Systems

Bernadetta Maleszka

Institute of Informatics, Wroclaw University of Technology,
Wybrzeze Wyspianskiego 27, 50-370, Wroclaw, Poland
Bernadetta.Maleszka@pwr.edu.pl

**Abstract.** Modeling user interests in personalized document retrieval system is currently a very important task. The system should gather information about the user to recommend him better results. In this paper a mathematical model of user preference and profile is considered. The main assumption is that the system does not know the preference. The main aim of the system is to build a profile close to user preference based on observations of user activities. The method for building and updating user profile is presented and a model of simulation user behaviour in such system is proposed. The analytical properties of this method are considered and two theorems are presented and proved.

**Keywords:** user profile, user preference, profile convergence, evaluating retrieval systems.

## 1 Introduction

In modern information retrieval systems modeling user interests is a very important task. User has a preference that is changing with time. The system should guess the user preference and build his profile. User profile is used to recommend for the user a list of documents that are relevant for him. System should modify user profile based on his current activities. The better profile the system has, the better result the user can obtain.

Evaluation of many personalization systems is performed only on experimental level [2], [3]. It is understandable while the user is only person who can judge the obtained result. In real system it is impossible to collect data about a large group of users that are using a new system for some time. Authors of paper [5] claim that the minimum number of testing users is 30. Usually, a few volunteers are testing if the system is effective, eg. [1], [7], [8]. The additional difficulties is that system should be tested for a long time when user preference are changing. Mathematical properties of personalization system are rarely considered.

In this paper a problem of user profile building and updating is described. A methodology of modeling user interests as the basis for user profile is proposed. In our previous papers [12] and [11] the experimental simulations of personalization system and statistical analysis for effectiveness were presented for 4 adaptation methods. In this paper the method for profile adaptation based on current user

D. Hwang et al. (Eds.): ICCCI 2014, LNAI 8733, pp. 62–71, 2014.

activities is presented and analytical properties are considered. The effectiveness measure is distance between the user preference and the built and adapted profile in subsequent series. The proofs of two convergence theorems are included.

The rest of the paper is organized as follows. In Section 2 we present a short survey of classical measures to evaluate effectiveness of personalization methods. The model of documents set, user preference and profile are presented in Section 3. The algorithm for profile adaptation method is also described. Section 4 contains theorems of profile convergence. In the last Section 5 we gather the main conclusions and future works.

## 2   Related Works

Evaluating the effectiveness of information retrieval system is a significant part of building such system. Zhou et al. [15] present the survey of evaluating methods. The most popular are as follows:

- coverage of documents' collection;
- response time;
- form of presenting results;
- work that user should do to obtain satisfactory results;
- recall;
- precision;
- multilevel evaluating methods that use correlations between two rankings (Spearman or Kendall-Tau correlation, etc).

The authors of paper [15] present methods that compare two documents' ranking. The first ranking is obtained from the system. The second one is a model ranking that should be prepared for the user. The most effective measure to compare those rankings is the normalized distance measure. The disadvantage of this approach is as follows: to compare rankings, both of them should be complete – it is not enough to know which documents are relevant and which are not.

The most popular measures are precision and recall but they do not take user preference into account. Let us consider the situation when two users ask the same query. The same result document can be relevant for one of them, and not relevant for the second one. Jarvelin [6] notes that analysis of the results does not explain user behaviour. It is not clear why user chooses only a few documents from results' ranking.

To evaluate the system Sieg et al. [13] propose the method based on user profile convergence. They consider stability of the profile. Profile is stable when the rate of increase in interest scores stabilizes over incremental updates. The main assumption for convergence is as follows. Initially, the interest scores for the concepts in the profile will continue to change. However, once enough information has been processed for profiling, the amount of change in interest scores should decrease. Finally, the concepts with the highest interest scores should become relatively stable. Similar idea is presented by Trajkova and Gauch [14]. They also assume that although the number of concepts in ontological user profile will monotonically increase, eventually the highest weighted concepts should become relatively stable, reflecting the user's major interests. This approach has the following disadvantages: it is not obvious how stable profile corresponds to real user interests and even if the user interests are stable, his information needs can change over time and then the profile is not stable.

In this paper the author presents a novel approach to evaluating information retrieval system. The user has preference and the system builds user profile. The profile is better when the distance to preference is smaller. When profile is adapted based on current user activities, the profile should converge to preference, despite of the fact that user preference is dynamic.

## 3   User Preference and User Profile

In the proposed model of information retrieval system, user preference is a set of terms that the user is interested in. If the user is asking about some term, it means that he is interested in it. The system finds documents that are connected with the user queries and generates a ranking list of the best results. Observing user feedback, the system can modify the user profile. The more information about user activities system has, the better profile is built.

The system should seek a situation when the user profile is the same as his preference. User profile is built and adapted based on current user activities. In this section we present mathematical formulas for user preference and profile and we describe a method for profile adaptation. The main goal of the adaptation method is to create a profile that gets closer to the preference in subsequent series of profile updates. Distance between user preference and profile is an effectiveness measure. The value of this measure becomes smaller and smaller when profile is updated so the profile converges to the user preference.

### 3.1   User Preference

User preference is a set of weighted terms that the user is interested in. Each term has a weight that represents the degree of user interest in this term. Determining user preference is not a trivial problem as the user can have a problem judging which interest area is more or less important that others. The system observes and saves information about documents that user has chosen as relevant or asks user to fill some questionnaire about his interests [4], [10].

In our system we assume the following document representation:

$$d_i = \{(t_j^i, w_j^i) : t_j^i \in T \wedge w_j^i \in [0.5, 1), j = 1, 2, \ldots, m_i\} \tag{1}$$

where $t_j^i$ is a term from terms set $T$, $w_j^i$ is weight of this term, and $m_i$ is a number of terms is document $d_i$.

User preference is the finite set of weighted terms:

$$Pref = \{(t_i, v_i) : t_i \in T_U \wedge v_i \in [0.5, 1), i = 1, 2, \ldots, k\} \tag{2}$$

where $t_i$ is a term, $v_i$ is appropriate weight (degree of user interests in this term), and $k$ is a number of terms in preference.

The set of terms $T_U$ contains the terms that occur in the user preference. The terms come from the set of all terms $T$ ($T_U \cap T \neq \emptyset$). Number of terms can change between the blocks of sessions.

In our approach we simulate real user activities in an information retrieval system. User preference is determined using the following procedure. A large collection of documents is considered. The user chooses the set of relevant documents $D_r$. The precision and recall for this set is equal to 1. User preference is obtained based on this set of documents: for each term the average weight is calculated when the term occurs in many documents or its weight is close to 1.

User preference is changing with time. In the discussed system, we propose to change a part of relevant documents (e.g. 10% of $D_r$ in each 5 blocks of sessions) and to recalculate the weights. New terms can also be added to the preference.

## 3.2   User Profile

User profile is build and adapted by the system based on information about user queries and relevant documents in subsequent sessions.

$$D(s) = \{(q_i^{(s)}, d_{i_j}^{(s)}) : i = 1, 2, \ldots, I; i_j = 1, 2, \ldots, i_J\} \qquad (3)$$

where $i$ is a subsequent number of the query in the session $s$ and $i_J$ is the quantity of documents that are relevant to query $q_i^{(s)}$.

User profile $UP(s)$ in session $s$ is a set of weighted terms:

$$UP(s) = \{(t_j, w_p^{(s)}(t_j)) : t_j \in T_U^{profile} \wedge w_p^{(s)}(t_j) \in [0,1), j = 1, 2, \ldots, p_s\}, \qquad (4)$$

where $t_j$ is a term from set of terms $T_U^{profile}$, $w_p^{(s)}(t_j)$ is a weight of this term in session $s$ and $p_s$ is the quantity of terms in user profile in session $s$.

User profile is updated based on user activities. After each session the average weight of term from user queries is calculated. If a new term occurs in user queries, the weight for this term is calculated. If the term is in user profile its weight is recalculated using the following equation:

$$w_p^{(s+1)}(t_j) = \begin{cases} w_p^{(s)}(t_j), & \text{if } t_j \text{ is new term} \\ w_p^{(s)}(t_j) + \gamma \cdot \Delta_1(w_d^{(s)}(t_j)), & \text{in other case} \end{cases} \qquad (5)$$

where $w_p^{(s+1)}(t_j)$ is a weight of $t_j$ term in user profile in the session $s+1$, $\gamma$ is a parameter that was tuned in experiments and $\Delta_1(w_d^{(s)}(t_j))$ is absolute change of degree of user interests in term $t_j$.

The absolute change is calculated based on the following equation:

$$\Delta_1(w_d^{(s)}(t_j)) = \begin{cases} w_d^{(s)}(t_j) - w_d^{(s-1)}(t_j), & \text{if } s > 1 \\ w_d^{(1)}(t_j), & \text{if } s = 1 \end{cases} \qquad (6)$$

where $w_d^{(s)}(t_j)$ is the average weight of term $t_j$ after the current session $s$ and $w_d^{(s-1)}(t_j)$ is the average weight of term $t_j$ after the previous session $s-1$.

User profile is updated after a few blocks of sessions (the block contains 5–10 sessions).

## 4   Analytical Properties of Adaptation Method

In this section we have considered analytical properties of the user profile adaptation method. The weight in user profile takes into account the previous weight and the current user interests (5).

**Theorem 1.** *If the user has the same preference and in each session chooses all relevant documents then user profile is identical with user preference.*

$$(Pref = const \wedge \forall_s D_p(s) = D_r(q)) \Rightarrow UP = Pref$$

**Proof of Theorem 1.** Let us use the following symbols (we consider term $t_k$):

- $w_d^{(j)} \in [0.5, 1)$ – average term weight in relevant documents in current session
- $v^{(j)}$ – weight of considered term in preference (average weight of this term in all relevant documents)
- $w_p^{(j)} \in [0, 1)$ – weight of considered term in user profile.

Using recursive formula 5, we can perform the following calculations:

- I session:
  - average weight after this session: $w_d^{(1)}$
  - weight in user profile: $w_p^{(1)} = w_d^{(1)}$
- II session:
  - average weight after this session: $w_d^{(2)}$
  - weight in user profile: $w_p^{(2)} = w_p^{(1)} + \gamma \cdot (w_d^{(2)} - w_p^{(1)}) = w_d^{(1)} \cdot (1-\gamma) + \gamma \cdot w_d^{(2)}$
- III session:
  - average weight after this session: $w_d^{(3)}$
  - weight in user profile: $w_p^{(3)} = w_p^{(2)} + \gamma \cdot (w_d^{(3)} - w_p^{(2)}) =$
    $w_d^{(1)} \cdot (1-\gamma)^2 + w_d^{(2)} \cdot \gamma \cdot (1-\gamma) + w_d^{(3)} \cdot \gamma$

$\vdots$

- $s$-th session:
  - average weight after this session: $w_d^{(s)}$
  - weight in user profile: $w_p^{(s)} = w_p^{(s-1)} + \gamma \cdot (w_d^{(s)} - w_p^{(s-1)}) =$

$$w_d^{(1)} \cdot (1-\gamma)^{s-1} + w_d^{(2)} \cdot \gamma \cdot (1-\gamma)^{s-2} + \ldots + w_d^{(s-1)} \cdot \gamma \cdot (1-\gamma) + w_d^{(s)} \cdot \gamma \quad (7)$$

In considered case (user chooses all relevant documents in each session) we can use the following dependencies:

$$w_d^{(1)} = w_d^{(2)} = \ldots = w_d^{(s)} = w_d.$$

Finally, we obtain the result in equation 8. Using formula for sum of geometric sequence with parameters: $a_1 = 1$ and $q = 1 - \gamma$, we obtain:

$$
\begin{aligned}
w_p^{(s)} &= w_d \cdot (1-\gamma)^{s-1} + w_d \cdot \gamma \cdot (1-\gamma)^{s-2} + \ldots + w_d \cdot \gamma \cdot (1-\gamma) + w_d \cdot \gamma = \\
&= w_d \cdot (1-\gamma)^{s-1} + w_d \cdot \gamma \cdot \left[ \frac{(1-(1-\gamma)^{s-1})}{1-(1-\gamma)} \right] = \\
&= w_d \cdot (1-\gamma)^{s-1} + w_d \cdot (1-(1-\gamma)^{s-1}) = w_d
\end{aligned}
\tag{8}
$$

For continuous user preference, user profile is identical as user preference.

In the general case, user preference can change with time: the weight of terms can increase or decrease or terms can change (user can have new interests or he can be no longer interested in some other terms).

The proof is performed for single term in user preference. We can prove that the weight of considered term is statistically convergent in distribution to normal distribution with parameters that are the same as parameters in preference distribution.

Equation 7 is a linear combination of terms weights in subsequent sessions. Thus it can be written in the following way:

$$Y = \sum_{i=1}^{N} a_i \cdot X_i \tag{9}$$

where $a_i$ is a parameter connected with significance of value $X_i$.

Each random variable $X_i$ has uniform distribution:

$$p_i(x_i) = \begin{cases} \frac{1}{\delta_i}, & \text{if } \alpha_i < x_i < \beta_i \\ 0 & \text{in other case} \end{cases} \tag{10}$$

where $\alpha_i$ and $\beta_i$ are both ends of interval (here we assumed $\alpha_i = 0.5$ oraz $\beta_i = 1$), and $\delta_i$ is calculated with the following formula:

$$\delta_i = \beta_i - \alpha_i \tag{11}$$

Mean and standard deviation for each variable $X_i$ are calculated with the following formulas:

$$\mu_i = \frac{\beta_i + \alpha_i}{2} \tag{12}$$

$$\sigma_i^2 = \frac{\delta_i^2}{12} \tag{13}$$

Let us use $\mu$ as mean and $\sigma^2$ as standard deviation:

$$\mu = \sum_{i=1}^{N} a_i \cdot \mu_i = \sum_{i=1}^{N} a_i \cdot \frac{\beta_i + \alpha_i}{2} \tag{14}$$

$$\sigma^2 = \sum_{i=1}^{N} a_i^2 \cdot \sigma_i^2 = \frac{1}{12} \sum_{i=1}^{N} a_i^2 \cdot \delta_i^2 \tag{15}$$

**Theorem 2.** *If the user has constant preference and in each session he chooses only one relevant document (the system presents him a list of documents), then the weight of this term in user profile is convergent to weight of this term in user preference.*

*If weights $w_d^{(s)}(t_k)$ of term $t_k$ in subsequent sessions $s$ are different, then probability distribution of weight in user profile after standardization is convergent to standard normal distribution $(N(0,1))$.*

$$(Pref = const \wedge \forall_s D_p(s) \cap D_r(q) \neq \emptyset) \Rightarrow lim_{s \to \infty} UP(s) = Pref$$

In the proof of theorem 2 we use Lindeberg lemma, which gives necessary and sufficient condition for distributions convergence [9].

**Lemma 1.** *Let $X_i$, $i = 1, 2, \ldots, M$ be mutually independent random variables with uniform distributions $U(\alpha_i, \beta_i)$ in interval $(\alpha_i, \beta_i)$.*

*The necessary and sufficient condition that random variable distribution $\tilde{Y}$*

$$\tilde{Y} = \left( \frac{\sum_{i=1}^{s} a_i X_i - \mu}{\sigma} \right) \sim N(0,1) \tag{16}$$

*which is normalized weighted sum of random variables $X_i$ ($\mu$ and $\sigma$ are defined by equations (14) and (15), is convergent to standard normal distribution with the following density function:*

$$g(y) \longrightarrow \frac{1}{\sqrt{2\Pi}} e^{-\frac{y^2}{2}} \ for \ s \to \infty \tag{17}$$

*is Lindeberg condition:*

$$\forall_{\epsilon>0} \frac{1}{\sigma^2} \sum_{i=1}^{M} a_i^2 \int_{a_i|x-\mu_i| \leq \epsilon\sigma} dx (x - \mu_i)^2 p_i(x) \to 1 \tag{18}$$

*for $a_i > 0$. $p_i(x)$ is the probability distribution function of variable $X_i$.*

The proof of this lemma was presented in paper [9].

**Proof of Theorem 2.** Let us assume that weights in documents are realizations of random variables with uniform distributions.

We calculate values of mean and standard deviation:

$$\mu = \sum_{i=1}^{N} a_i \cdot \mu_i = \frac{3}{4} \sum_{i=1}^{N} a_i = \frac{3}{4} \tag{19}$$

$$\sigma^2 = \frac{1}{12} \sum_{i=1}^{N} a_i^2 \cdot \delta_i^2 = \frac{1}{48} \sum_{i=1}^{N} a_i^2 = \frac{1}{48} \cdot \left( (1-\gamma)^{s-1} \right)^2 +$$

$$+ \frac{\gamma^2}{48} \cdot \left[ \left( (1-\gamma)^{s-2} \right)^2 + \left( (1-\gamma)^{s-3} \right)^2 + \ldots + (1-\gamma)^2 + 1 \right] =$$

$$+ \frac{1}{48} \cdot (1-\gamma)^{2s-2} + \frac{\gamma^2}{48} \cdot \left[ (1-\gamma)^{2s-4} + (1-\gamma)^{2s-6} + \ldots + (1-\gamma)^2 + 1 \right]$$

Above equation is a sum of geometric sequence with parameters: $b_1 = 1$ and $q = (1-\gamma)^2$. Calculating this sum we obtain:

$$\sigma^2 = \frac{1}{48} \cdot (1-\gamma)^{2s-2} + \frac{\gamma^2}{48} \cdot \left[ \frac{1 - \left( (1-\gamma)^2 \right)^{s-1}}{1 - (1-\gamma)^2} \right] =$$

$$= \frac{1}{48} \cdot \left[ (1-\gamma)^{2s-2} + \gamma \cdot \left[ \frac{(1 - (1-\gamma)^{2s-2})}{2 - \gamma} \right] \right] =$$

$$= \frac{1}{48} \cdot \frac{(1 - (1-\gamma)^{2s-2}) \cdot (2 - \gamma) + \gamma - \gamma \cdot (1-\gamma)^{2s-2}}{2 - \gamma} =$$

$$= \frac{1}{48} \cdot \frac{2 \cdot (1-\gamma)^{2s-2} \cdot (1-\gamma) + \gamma}{2 - \gamma} \tag{20}$$

Going to the infinity $s \to \infty$ and calculating infinite sum of geometric sequence with the following parameters: $a_1 = (1-\gamma)^4$ and $q = (1-\gamma)^2$, we have:

$$\lim_{s \to \infty} \sigma^2 = \frac{1}{48} \cdot \lim_{s \to \infty} \sum_{i=1}^{s} a_i^2 =$$

$$= \lim_{s \to \infty} \frac{1}{48} \cdot \frac{2 \cdot (1-\gamma)^{2s-2} \cdot (1-\gamma) + \gamma}{2 - \gamma} \equiv$$

$$\equiv \lim_{s \to \infty} \frac{1}{48} \cdot (1-\gamma)^{2s-2} + \frac{1}{48} \lim_{s \to \infty} \gamma^2 \cdot \sum_{i=3}^{s} (1-\gamma)^{2(i-1)} =$$

$$= \frac{\gamma^2}{48} \cdot \sum_{j=1}^{s} (1-\gamma)^{2(j+1)} =$$

$$= \frac{\gamma^2}{48} \cdot \frac{(1-\gamma)^4}{1 - (1-\gamma)^2} = \frac{\gamma(1-\gamma)^4}{48 \cdot (2-\gamma)} \tag{21}$$

Assuming that $g$ is the function of $\gamma \in (0,1)$, we can calculate the minimum and maximum values of variance:

$$\lim_{s \to \infty} \sigma^2 = \frac{\gamma(1-\gamma)^4}{48 \cdot (2-\gamma)} = g(\gamma) \tag{22}$$

Calculating the first derivation of $g$ function, we can compare it with 0.

$$\frac{dg}{d\gamma} = \frac{d}{d\gamma}\left\{\frac{\gamma(1-\gamma)^4}{48 \cdot (2-\gamma)}\right\} =$$
$$= \frac{[(1-\gamma)^4 - \gamma \cdot 4 \cdot (1-\gamma)^3] \cdot 48 \cdot (2-\gamma) - \gamma \cdot (1-\gamma)^4 \cdot (-48)}{48^2 \cdot (2-\gamma)^2} = 0 \tag{23}$$

We have:

$$[(1-\gamma)^4 - 4\gamma \cdot (1-\gamma)^3] \cdot (2-\gamma) + \gamma \cdot (1-\gamma)^4 = 0$$
$$(1-\gamma)^4 \cdot (2-\gamma+\gamma) - 4\gamma(1-\gamma)^3(2-\gamma) = 0$$
$$(1-\gamma)^3[2(1-\gamma) - 4\gamma(2-\gamma)] = 0$$
$$2(1-\gamma)^3(2\gamma^2 - 5\gamma + 1) = 0 \tag{24}$$

We calculate the roots for $\gamma \in (0,1)$:
$\gamma_1 = \frac{10-\sqrt{(68)}}{2}$ oraz $\gamma_2 = \frac{10+\sqrt{(68)}}{2} \notin (0,1)$

The maximum value of variance is equal to $f(\gamma) = \frac{85\sqrt{(17)}-349}{1536} \cong 0.00095$ for $\gamma = \frac{5-\sqrt{(17)}}{4} \cong 0.21922$. Maximum standard deviation is $\sigma \cong 0.0308$.

The minimal value of function $g(\gamma) \to 0$ is reached in the ends of interval: for $\gamma \to 0^+$ or $\gamma \to 1^-$. Using these values in our system is not desirable, while for $\gamma \to 0^+$ the current user activity is not taken into account and for $\gamma \to 1^-$ means that history of user activity is omitted.

To sum up, a random variable that is a linear combination of random variables $X_i$ with uniform distribution in the interval of $\alpha_i < x_i < \beta_i$, has the following parameters:

- mean: $\mu = 0.75$;
- standard deviation: $\sigma \in (0; 0.0308)$;
- variance: $\sigma^2 \in (0; 0.00095)$.

## 5   Summary and Future Works

In this paper the authors consider mathematical model of user preference and profile in an information retrieval system. The system does not know the user preference. The system tries to build a profile close to the user preference based on observations of user activities. The method for building and updating the user profile is presented and a model of user behaviour simulation in such system is proposed. The analytical properties of this method are considered and two theorems are presented and proven.

**Acknowledgments.** This research was partially supported by Polish Ministry of Science and Higher Education.

# References

1. Ahmed, E.B., Nabli, A., Gargouri, F.: Group extraction from professional social network using a new semi-supervised hierarchical clustering. Knowledge Information System (2013), doi: 10.1007/s10115-013-0634-x
2. Arapakis, I., Athanasakos, K., Jose, J.: A Comparison of General vs Personalised Affective Models for the Prediction of Topical Relevance. In: ACM SIGIR 2010, pp. 371–378 (2010)
3. Bobadilla, J., Ortega, F., Hernando, A., Bernal, J.: A collaborative filtering approach to mitigate the new user cold start problem. Knowledge Based Systems 26, 225–238 (2012)
4. Clarkea, C.L.A., Cormackb, G., Tudhope, E.A.: Relevance ranking for one to three term queries. Information Processing & Management 36, 291–311 (2000)
5. Ingwersen, P.: The User in Interactive Information Retrieval Evaluation. In: Melucci, M., Baeza-Yates, R. (eds.) Advanced Topics in Information Retrieval. The Information Retrieval Series, vol. 33, Springer, Heidelberg (2011)
6. Järvelin, K.: Explaining user performance in information retrieval: Challenges to IR evaluation. In: Azzopardi, L., Kazai, G., Robertson, S., Rüger, S., Shokouhi, M., Song, D., Yilmaz, E. (eds.) ICTIR 2009. LNCS, vol. 5766, pp. 289–296. Springer, Heidelberg (2009)
7. Law, E.L.-C., Klobučar, T., Pipan, M.: User Effect in Evaluating Personalized Information Retrieval Systems. In: Nejdl, W., Tochtermann, K. (eds.) EC-TEL 2006. LNCS, vol. 4227, pp. 257–271. Springer, Heidelberg (2006)
8. Li, L., Yang, Z., Wang, B., Kitsuregawa, M.: Dynamic Adaptation Strategies for Long-Term and Short-Term User Profile to Personalize Search. In: Dong, G., Lin, X., Wang, W., Yang, Y., Yu, J.X. (eds.) APWeb/WAIM 2007. LNCS, vol. 4505, pp. 228–240. Springer, Heidelberg (2007)
9. Kamgar-Parsi, B., Kamgar-Parsi, B., Brosh, M.: Distribution and moments of the weighted sum of uniforms random variables, with applications in reducing monte carlo simulations. Journal of Statistical Computation and Simulation 52(4), 399–414 (1995)
10. Kiewra, M.: Hybrid method for document recommendation in hypertext environment. PhD dissertation. Wroclaw University of Technology (2006)
11. Maleszka, B., Nguyen, N.T.: Evaluating Profile Convergence in Document Retrieval Systems. In: Nguyen, N.T., Attachoo, B., Trawiński, B., Somboonviwat, K. (eds.) ACIIDS 2014, Part I. LNCS, vol. 8397, pp. 163–172. Springer, Heidelberg (2014)
12. Mianowska, B., Nguyen, N.T.: Tuning User Profiles Based on Analyzing Dynamic Preference in Document Retrieval Systems. Multimedia Tools and Applications 65(1), 93–118 (2013)
13. Sieg, A., Mobasher, B., Burke, R.: Learning Ontology-Based User Profiles: A Semantic Approach to Personalized Web Search. IEEE Intelligent Informatics Bulletin 8(1), 7–18 (2007)
14. Trajkova, J., Gauch, S.: Improving Ontology-Based User Profiles. In: RIAO, pp. 380–390 (2004)
15. Zhou, B., Yao, Y.: Evaluating information retrieval system performance based on user preference. Journal of Intelligent Information System 34, 227–248 (2010)

# SciRecSys: A Recommendation System for Scientific Publication by Discovering Keyword Relationships

Vu Le Anh[1,2], Hai Vo Hoang[3,*], Hung Nghiep Tran[4], and Jason J. Jung[5]

[1] Nguyen Tat Thanh University, Ho Chi Minh city, Vietnam
[2] Big IoT BK Project Team, Yeungnam University, Gyeongsan, Korea
[3] Information Technology College, Ho Chi Minh city, Vietnam
[4] University of Information Technology, Ho Chi Minh city, Vietnam
[5] Chung-Ang University, Seoul, Korea
lavu@ntt.edu.vn, {vohoanghai2,j2jung}@gmail.com,
nghiepth@uit.edu.vn

**Abstract.** In this work, we propose a new approach for discovering various relationships among keywords over the scientific publications based on a Markov Chain model. It is an important problem since keywords are the basic elements for representing abstract objects such as documents, user profiles, topics and many things else. Our model is very effective since it combines four important factors in scientific publications: content, publicity, impact and randomness. Particularly, a recommendation system (called SciRecSys) has been presented to support users to efficiently find out relevant articles.

**Keywords:** Keyword ranking, Keyword similarity, Keyword inference, Scientific Recommendation System, Bibliographical corpus.

## 1 Introduction

Keyword-based search engines (Google, Bing Search, and Yahoo) have emerged and dominated the Internet. The success of these search engines are based on the study of keyword relationships and keyword indexes. The task of measuring keywords' relationships is the basic operation for building the related network of abstract objects which are applied in many problems and applications, such as document clustering, synonym extraction, plagiarism detection problem, taxonomy, search engine optimization, recommendation system, etc. In this work, we focus on two problems. First, we study the rank, inference and similarity of keywords over scientific publications on assuming that the keywords belong to a *virtual ontology of keywords*. The inference relationship will help us find *parents, children*, the similarity relationship will help us find *siblings* of a given keyword and the rank of keywords will determine how important they are. Second, we apply these relationship in our scientific recommendation system, SciRecSys. The first problem is not easy since we have to consider four main factors:

i- *Content factor*. The meaning of the keyword should be considered in the context of its paper. The paper itself is determined by its keywords;

---

* Corresponding author.

D. Hwang et al. (Eds.): ICCCI 2014, LNAI 8733, pp. 72–82, 2014.

ii-   *Publicity factor.* The hotness of a topic (or keyword) depends on how popular it is. People always find the hot topic for reading.

iii-  *Impact factor.* In the world of scientific publications the citation is a very important factor. People often follow the citations of a scientific paper for finding the necessary information;

iv-   *Randomness factor.* Randomness is very important factor in many real complex systems. Readers sometimes find something for reading quite randomly.

SciRecSys recommendation system is designed to be a search engine for scientific publications. We want to apply the rank, inference and similarity of keywords to solve three following problems: (i) Ranking papers matched to the given keyword; (ii) Recommending additional keywords related to a given keyword to help the reader navigating the corpus effectively; (iii) Suggesting papers for *"Reading more"* function in the context the reader is reading a given topic.


# 2  Related Works

The similarity of keywords is often used as a crucial feature to reveal the relation between objects and many methods to measure it have been proposed. One baseline for similarity measures is using distance functions such as squared Euclidean distance, cosine similarity, Jaccard coefficient, Pearsons correlation coefficient, and relative entropy. Anna Huang et al. [6] has compared and analyzed the effectiveness of these measures in text clustering problem. She represent a document as an m-dimensional vector of the frequency of terms and uses a weighting scheme to reflect their importance through frequencies tf/idf. Singthongchai et al. [12] make keywords search more practical by calculating keyword similarity by combining Jaccard's, N-Gram and Vector Space. Probabilistic models for similarity measure has been studied in language speech [4,3]. Ido Dagan et al. have proposed a bigram similarity model and used the relative entropy to compute the similarity of keywords [4]. Sung-Hyuk Cha has conducted a comprehensive survey on probability density functions for similarity measures [3]. Ontology-based methods are also exploited to measure the similarity of keywords [2,11,10]. Bollegala et al. [2] use the Wordnet database - ontology of words to measure keywords' relatedness by extracting lexico-syntactic patterns that indicate various aspects of semantic similarity and modifying four popular co-occurrence measures, including Jaccard, Overlap (Simpson), Dice, and Pointwise mutual information (PMI). Vincent Schickel-Zuber and Boi Falting [11] present a novel similarity measure for hierarchical ontologies called Ontology Structure based Similarity (OSS) that allows similarities to be asymmetric. Snchez et al. [10] presents an ontology-based method relying on the exploitation of taxonomic features available in an ontology.

Markov Chain model which properties are studied in [7,8] is used for computing the similarity and ranking [5,1]. Fouss et al.[5] use a stochastic random-walk model to compute similarities between nodes of a graph for recommendation. Vu et al. [1] introduce an N-star model, and demonstrate it in ranking conference and journal problems. Finally, Lops et al. [9] do a thorough review on state-of-the-art and trends of content-based recommender systems.

## 3  Backgrounds

### 3.1  Basic Definitions

Suppose $\mathcal{K}$, $\mathcal{P}$ are the sets of keywords and papers respectively. $p \in \mathcal{P}$ is a paper and $A \in \mathcal{K}$ is a keyword. $K(p) \subseteq \mathcal{K}$ is the set of keywords belonging to $p$. $P(A) = \{q \in \mathcal{P} | A \in K(q)\}$ is the set of papers containing $A$. We assume that $K(p) \neq \emptyset \wedge P(A) \neq \emptyset$. Finally, $C(p) \subseteq \mathcal{P}$ is the set of papers cited by $p$. In the case $p$ has no citing, we assume that $C(p) = \mathcal{P}$. It guarantees that $C(p) \neq \emptyset$.

A couple $(\mathcal{A}, R)$ is called a *ranking system* if: (i) $\mathcal{A} = \{a_1, \ldots, a_n\}$ is a finite set, and (ii) $R$ is a non-negative function on $\mathcal{A}$ (i.e., $R : \mathcal{A} \to [0, +\infty)$). A *ranking score* on $\mathcal{A}$ can be represented as $R = (R(a_1), \ldots, R(a_n))^T$. Furthermore, $(\mathcal{A}, R)$ is *normalized* if $R^T$ is normalized ($\| R^T \| = 1$).

Suppose $A$ and $B$ are two events. The *inference* and *similarity* of two events $A$ and $B$ are denoted by $I(A, B)$ and $S(A, B)$, respectively. They can be computed as follows:

$$I(A, B) = \frac{Pr(A \text{ and } B)}{Pr(A)} \qquad S(A, B) = \frac{Pr(A \text{ and } B)}{Pr(A \text{ or } B)} \qquad (1)$$

We have $S(A, B) \leq 1$, $I(A, B) \leq 1$. $S(A, B)$ is symmetric. $I(A, B) = Pr(B|A)$.

### 3.2  Relationships of Keywords Based on the Occurrences

Let us introduce two normalized ranking scores $R^c$, $R^p$ on set of keywords, $\mathcal{K}$, based on the occurrences. $R^c$ ($c$ stands for *counting*) is based on counting the documents containing the given keyword.

$$R^c(A) = \frac{|P(A)|}{\Sigma_{B \in \mathcal{K}} |P(B)|} \qquad (A \in \mathcal{K}) \qquad (2)$$

$R^p$ ($p$ stands for *probability*) is based on the probability of the occurrence of the given keyword in the papers.

$$R^p(A) = \frac{1}{|\mathcal{P}|} \Sigma_{p \in P(A)} \frac{1}{|K(p)|} \qquad (A \in \mathcal{K}) \qquad (3)$$

$\frac{1}{|\mathcal{P}|}$ is the probability of choosing a paper from the corpus. $\frac{1}{|K(p)|}$ is the probability of choosing keyword $A$ from the paper $p$ containing $|K(p)|$ keywords.

We propose following formulas for measuring the inference and similarity of two keywords $A$, $B$ based on the occurrences:

$$I^c(A, B) = \frac{|P(A) \cap P(B)|}{|P(A)|} \qquad S^c(A, B) = \frac{|P(A) \cap P(B)|}{|P(A) \cup P(B)|} \qquad (A, B \in \mathcal{K}) \qquad (4)$$

$P(A) \cap P(B)$ is the set of papers containing both keywords $A$ and $B$. $P(A) \cup P(B)$ is the set of papers containing keywords $A$ or $B$.

## 4  Relationships of Keywords Based on Graph of Keywords

### 4.1  Markov Chain Model of the Reading Process

We propose a Markov Chain model to simulate the reading process which can combine four factors: content, publicity, impact and randomness. We assume that the reader reads a topic (keyword) $A$ in some paper $p$ at any time ($A \in K(p)$). $\mathcal{S} = \{(A, p) \in \mathcal{K} \times \mathcal{P} | A \in K(p)\}$ is the set of states. From current state $\theta = (A, p)$, the reader will move to new state $\xi = (B, q) \in \mathcal{S}$ with the conditional probability $Pr(\theta \to \xi)$ by applying 4 following actions:

- $A_1$. *Same paper - some topic.* The reader is interested in current paper, and choose randomly a topic in the same paper with the probability equal to $\alpha_1$. Thus, $p = q$.

$$Pr(\theta \to_{A_1} \xi) = if(p = q, \frac{\alpha_1}{|K(p)|}, 0) \tag{5}$$

- $A_2$. *Same topic - some paper.* The reader is interested in current topic, and choose randomly another paper which have the same topic with the probability equal to $\alpha_2$. Thus, $A = B$.

$$Pr(\theta \to_{A_2} \xi) = if(A = B, \frac{\alpha_2}{|P(A)|}, 0) \tag{6}$$

- $A_3$. *Some topic - cited paper.* The reader is interested in some cited paper with the probability equal to $\alpha_3$. First, he choose randomly a cited paper and then choose randomly a new topic belonging to the chosen paper. Thus, $q \in C(p)$.

$$Pr(\theta \to_{A_3} \xi) = if(q \in C(p), \frac{\alpha_3}{|C(p)||K(q)|}, 0) \tag{7}$$

- $A_4$. *Some paper - some topic.* The reader stop reading the current paper and choose randomly a new paper with the probability equal to $\alpha_4$.

$$Pr(\theta \to_{A_4} \xi) = \frac{\alpha_4}{|\mathcal{P}||K(q)|} \tag{8}$$

$\alpha_i > 0$ are constants and $\Sigma_{i=1}^{4} \alpha_i = 1$. From the assumptions, we have:

$$Pr(\theta \to \xi) = \Sigma_{i=1}^{4} Pr(\theta \to_{A_i} \xi)$$

The necessary condition of the Markov Chain model is that total of the output conditional probability from any state is equal to 1. Here is the formula of the conditions:

$$O(\theta) = \Sigma_\xi Pr(\theta \to \xi) = 1$$

Our readers can check it by apply the formula 5, 6, 7 and 8.

The probability of a reader in state $\theta = (A, p)$ is denoted by $Pr(\theta)$. We have:

$$Pr(\theta) = \Sigma_{\xi \in \mathcal{S}} Pr(\xi) \times Pr(\xi \to \theta) \tag{9}$$

**Proposition 1.** *There exists a unique stationary score $\{Pr(\theta)\}_{\theta \in \mathcal{S}}$ satisfying (9).*

*Proof.* Since a state $\theta$ can jump to any state (and itself too) by apply action $A_4$, the Markov Chain is irreducible and aperiodic (see more [7,8]). The Perron-Frobenius theorem [7] states that there exists a unique stationary score $\{Pr(\theta)\}_{\theta \in \mathcal{S}}$.

$\{Pr(\theta)\}_{\theta \in S}$ is determined by following algorithm:

---

**Algorithm :** Computing Stationary probability $\{Pr(\theta)\}_{\theta \in S}$

---

1. **begin**
2.    $k = 0$
3.    **Foreach**  $\theta \in S$   **do**  $Pr(\theta)^{(0)} = \frac{1}{|S|}$
4.    **repeat**
5.       $k = k + 1$   $stop = true$
6.       **Foreach**  $\theta \in S$   **do**
7.          $Pr(\theta)^{(k)} = 0$
8.          **Foreach**  $\xi \in S$   **do**  $Pr(\theta)^{(k)} += Pr(\xi)^{(k-1)} Pr(\xi \to \theta)$
9.          **if**  $|Pr(\theta)^{(k)} - Pr(\theta)^{(k)}| > \epsilon$   **then**   $stop = false$
10.   **until** $stop$
11.   **Foreach**  $\theta \in S$   **do** $Pr(\theta) = Pr^{(k)}(\theta)$
12. **end**

---

## 4.2   Ranking, Inference and Similarity of Keywords

The rank score of keyword $A$ based on the transition graph, $R^g(A)$ ($g$ stands for *graph*), is equal to the probability of user reads keyword $A$. We have:

$$R^g(A) = \Sigma_{p \in K(A)} Pr(\theta) \quad (\theta = (A, p) \in S) \tag{10}$$

Let $n_0$ be a positive integer. For each sequence $s = \theta_1 \theta_2 \ldots \theta_{n_0} \in S^{n_0}$, let $Pr(s)$ is the probability of $s$ occurs in $S^{n_0}$ generated by the Markov Chain model. $F \subseteq S^{n_0}$, we denote $Pr(F) = \Sigma_{s \in F} Pr(s)$. For each keyword $A$, let

$$P^g(A) = \{s = \theta_1 \theta_2 \ldots \theta_{n_0} \in S^{n_0} | \exists i \in \{1, 2, \ldots, n_0\}, p \in \mathcal{P} : \theta_i = (A, p)\}$$

The formulas for measuring the inference and similarity of two keywords $A$, $B$ based on the Markov Chain model:

$$I^g(A, B) = \frac{Pr(P^g(A) \cap P^g(B))}{Pr(P^g(A))} \quad S^g(A, B) = \frac{Pr(P^g(A) \cap P^g(B))}{Pr(P^g(A) \cup P^g(B))} \tag{11}$$

We will apply Monte Carlo method for Markov Chain to compute the formulas (11). First, we generate a large enough number of $n_0$-length sequences of states by our Markov Chain model. Then we apply counting techniques for approximating the probabilities and computing the formulas.

## 5   SciRecSys – Recommendation System

In this paper, we want to present three scenarios by using SciRecSys.

**Universal ranking vs. keyword based ranking** The reader chooses a keyword $A$ to find some related papers belonging to $P(A)$. Question is "What is the order for sorting $P(A)$?". There are two ways for ranking: (i) All results are sorted by only one *universal ranking*, $\{R_u^g(p)\}_{p \in \mathcal{P}}$ or (ii) The order of the results depend on $A$ with the *keyword based ranking*, $\{R_A^g(p)\}_{p \in P(A)}$. We propose $R_u^g(p)$ is equal to the probability of user reads $p$. Hence,

$$R_u^g(p) = \Sigma_{A \in K(p), \theta = (A,p)} Pr(\theta) \tag{12}$$

We propose $R_A^g(p)$ is equal to the probability of state $(A, p)$. Hence,

$$R_A^g(p) = Pr(\theta) \quad (\theta = (A, p), p \in P(A)) \tag{13}$$

**Recommending keywords** The user is in keyword (topic) $A$. What are the next recommended topics for following situations: (i) similar topics, $Sibling(A)$? (ii) more detail topics, $Child(A)$? (iii) more general topics, $Parent(A)$? Let $m_c$, $m_p$, $m_s$ be the parameters of the system. We denote: $Child(A) = \{B \in \mathcal{K} | I^g(B, A) > m_c\}$; $Parent(A) = \{B \in \mathcal{K} | I^g(A, B) > m_f\}$; $Sibling(A) = \{B \in \mathcal{K} | S^g(A, B) > m_b\}$. We remind our readers that keyword $A$ may have many parents or his parent can be his sibling or his child.

**Recommending papers** The user is in paper $p$. The user can choose the most interesting keyword $A \in K(p)$ to read more. What are the next recommended papers for him? The system will request the user to refine recommended papers by choosing a keyword $A$ and then give the recommendation based on the conditional probability change from state $\theta = (A, p)$ to the papers by applying actions $A_2$ and $A_3$. Suppose:

$$R_\theta^g(q) = \Sigma_{B \in K(q), \xi = (B,q)}(Pr(\theta \to_{A_2} \xi) + Pr(\theta \to_{A_3} \xi)) \tag{14}$$

Finally, the recommended papers are chosen based on $R_\theta^g$ ranking function.

## 6   Experiments

We collect data from DBLP[1] and Microsoft Academic Search[2] (MAS) to conduct experiments. We choose three datasets for experiments from three different domains: (i) $D_1$ is the publications of ICRA - International Conference on Robotics and Automation; (ii) $D_2$ is the publications of ICDE - International Conference on Data Engineering (iii) $D_3$ is the publications of GI-Jahrestagung - Germany Conference in Computer Science. ICRA and ICDE conferences are one of the most famous and biggest ones in their area. They are chosen for rich publications and citations. GI-Jahrestagung is a smaller conference but its topic is quite various.

**Table 1.** Experiment datasets

| Datasets | Paper No. | Keyword No. | Citation No. | State No. |
|---|---|---|---|---|
| ICRA | 9291 | 4676 | 125610 | 28736 |
| ICDE | 3254 | 2755 | 99492 | 12365 |
| GI-Jahrestagung | 1335 | 1640 | 5 | 2932 |

---

[1] http://dblp.uni-trier.de accessed on December 2013.

[2] http://academic.research.microsoft.com/ accessed on December 2013.

## 6.1   Experiments for Rank Scores

We do the experiments for three different ranking scores $R^c$, $R^p$ and $R^g$. The values on $\{\alpha_i\}_{i=1}^4$ are chosen for testing different contexts. The rank scores are scaled with the same rate for the convenience. For each two ranking scores $i$ and $j$, we do examine: (i) the Spearman's rank correlation coefficient (denote $\rho_{ij}$) for monotone checking; (ii) the differences of values: $\Delta_{ij}(A) = R^i(A) - R^j(A)$, $\%\Delta_{ij}(A) = \frac{\Delta_{ij}(A)}{R^j(A)}$. Here are some interesting observations:

- *The rank scores are monotone but quite different to each others.* It comes from that $\rho_{ij}$ are close to 1 and the average values of $\%\Delta_{ij}$ are very high. For instance with dataset $D1$, we have: $\rho_{cg} = 0.97$, $\rho_{pg} = 0.99$, $\rho_{cp} = 0.97$. The average of $|\%\Delta_{cg}|$, $|\%\Delta_{pg}|$, $|\%\Delta_{cp}|$ are $47.13\%, 30.06\%$ and $51.76\%$ respectively.

**Table 2.** Top 5 keywords most different using $R^g$ vs. $R^c$ on $D_1$.

| Keyword | $R^c$ | $R^p$ | $R^g$ | $\Delta_{cg}$ | $\Delta_{cg}\%$ |
|---|---|---|---|---|---|
| Mobile Robot | 924 | 1082 | 1221 | 297 | 32.16 |
| Robot Hand | 140 | 230 | 276 | 136 | 97.40 |
| Path Planning | 313 | 389 | 442 | 129 | 41.19 |
| Motion Planning | 393 | 483 | 511 | 118 | 29.96 |
| Visual servoing | 248 | 316 | 350 | 102 | 41.02 |

(a) Top 5 increasing values

| Keyword | $R^c$ | $R^p$ | $R^g$ | $\Delta_{cg}$ | $\Delta_{cg}\%$ |
|---|---|---|---|---|---|
| Indexing Terms | 419 | 228 | 251 | -168 | 40.19 |
| Satisfiability | 109 | 66 | 83 | -26 | 23.77 |
| Real Time | 422 | 366 | 397 | -25 | 6.02 |
| Extended Kalman Filter | 70 | 37 | 45 | -25 | 35.04 |
| Computer Vision | 58 | 37 | 35 | -23 | 39.11 |

(b) Top 5 decreasing values

- *$R^g$ reflects how hot a keyword is.* Let us see the result shown in Tab. 2. All *increasing* keywords are hot topics now and all *decreasing* keywords seem not being currently hot topics for conferences on robotics and automation. Let us see the example of two keywords *Robot Hand* and *Satisfiability*. They have quite the same $R^c$ values but $R^g$ values are four times difference. The explanation is that the hot topics have rich citations which increase the $R^g$ values.
- *$R^p$ is more acceptable to $R^g$ than $R^c$.* Let us see Tab. 2 again. For top 5 *increasing* keywords, $R^c < R^p < R^g$. For top 5 *decrease* keywords, $R^p < R^g < R^c$. The explanation is that the probability occurrences reflect the world more exactly than the counting but they still neglect the latent information like the citation graphs in the data corpus.

## 6.2   Experiments for Inference and Similarity

We do experiments to compare $(I^c, S^c)$ vs. $(I^g, S^g)$. For both inference and similarity, we find out some amazing results:

- *$I^g$ & $S^g$ exploits the citation network and the results are domain dependent.* Our approach seems to give "hot" results. Let us see Tab. 3(a), which shows similar keywords to *Real Time* over dataset $D_1$. $S^g$ generates "hot" keyword *Humanoid Robot* instead of *Indexing Terms*. This phenomenon also occurs in dataset $D_2$. We assume the cause is $I^g$ & $S^g$ exploit citation network and the paper–keyword relationships to generate more attractive keywords. Additionally, similar keywords to *Real Time* are quite different between two datasets $D_1$ and $D_2$, so we can say that these methods are domain dependent.

- *$I^g$ & $S^g$ could help overcome the missing co-occurrence problem.* The traditional methods based on occurrences of keywords could not work when there is no co-occurrence in the dataset. For instance, Tab. 4(b) shows that $S^c$ generates only one similar keyword to *Sensor Network* in small dataset $D_3$. In contrast, $S^g$ could help us find more acceptable similar keywords.

**Table 3.** Top 5 similar keywords of *"Real Time"* using $S^c$ vs. $S^g$

| $D_1$ | $S^c$ | $S^g$ |
|---|---|---|
| 1 | Mobile Robot | Path Planning |
| 2 | **Indexing Terms** | Mobile Robot |
| 3 | Path Planning | Motion Planning |
| 4 | Obstacle Avoidance | Obstacle Avoidance |
| 5 | Motion Planning | **Humanoid Robot** |
| (a) $D_1$ in robotics domain, $|K| = 4676$ | | |

| $D_2$ | $S^c$ | $S^g$ |
|---|---|---|
| 1 | Stream Processing | Data Stream |
| 2 | Data Stream | Stream Processing |
| 3 | Database System | Database System |
| 4 | Sensor Network | Object Oriented |
| 5 | Query Processing | Data Model |
| (b) $D_2$ in database domain, $|K| = 2755$ | | |

**Table 4.** Top 5 similar keywords of *"Sensor Network"* using $S^c$ vs. $S^g$

| $D_2$ | $S^c$ | $S^g$ |
|---|---|---|
| 1 | Query Processing | Query Processing |
| 2 | Data Management | Database System |
| 3 | Real Time | Data Management |
| 4 | Data Stream | Data Stream |
| 5 | Satisfiability | Query Evaluation |
| (a) Top 5 on $D_2$, $|K| = 2755$. | | |

| $D_3$ | $S^c$ | $S^g$ |
|---|---|---|
| 1 | Distributed System | Mobile Device |
| 2 | - | Web Service |
| 3 | - | Data Warehouse |
| 4 | - | Open Source |
| 5 | - | Middleware |
| (b) Top 5 on $D_3$, $|K| = 1640$. | | |

We also find out some interesting observations about inference particularly:

- *The inference between two keywords is asymmetric and $I^g$ exploits the citation network to generate a clear inference.*
  Unlike similarity between two keywords, inference is asymmetric, i.e., the inference between A and B differs from the one between B and A. Let us see Tab. 5, the inference from *Robot Hand* and *Robot Arm* are almost different. More interestingly, *Robot Hand* could infer *Robot Arm* but *Robot Arm* does not infer *Robot Hand*, so there is a flow of inference here. Further examination shows that, *High Speed* is repeated at top 1 on two inference lists generated by $I^c$. Whereas, *High Speed* only appears on $I^g$'s *Robot Arm* inference list, not on *Robot Hand* inference list. So, we have a clear inference flow from *Robot Hand* to *Robot Arm* then to *High Speed* with $I^g$, not with $I^c$. To explain this difference, we notice that when we infer from *Robot Hand*, *High Speed* shares more common publications than *Robot Arm* , 4 vs. 2. However, *Robot Hand* has more citations from *Robot Arm* than *High Speed*, 9 vs. 5. So we assume that our approach exploits the citation network, which is an asymmetric structure, to generate a clearer inference.

## 6.3 SciRecSys Recommendation System

We compare universal ranking scores $R_u^g(p)$ and local ranking score $R_A^g(p)$ and conduct experimental models for recommending keywords. We get some interesting results:

**Table 5.** Top 5 inference keywords over $D_1$ using $I^c$ vs. $I^g$.

| $D_1$ | $B$ from $I^c(A, B)$ | $B$ from $I^g(A, B)$ |
|---|---|---|
| 1 | **High Speed** | **Robot Arm** |
| 2 | Control System | Force Control |
| 3 | Robot Arm | Three Dimensional |
| 4 | Force Control | Autonomous Robot |
| 5 | Parallel Manipulator | Parallel Manipulator |

(a) $A = Robot\ Hand$.

| $D_1$ | $B$ from $I^c(A, B)$ | $B$ from $I^g(A, B)$ |
|---|---|---|
| 1 | **High Speed** | **High Speed** |
| 2 | Obstacle Avoidance | Parallel Manipulator |
| 3 | Control System | Three Dimensional |
| 4 | Configuration Space | Control System |
| 5 | Parallel Manipulator | Autonomous Robot |

(b) $A = Robot\ Arm$.

**Table 6.** Top 5 recommended papers for the keyword *"Real Time"* using $R_u^g$ vs. $R_A^g$ ($|K|$: keyword no., $|C_i|$: citing no., and $|C_e|$: cited no.).

| $D_1$ | Local Ranking $R_A^g(P)$ | | | | Universal Ranking $R_u^g$ | | | |
|---|---|---|---|---|---|---|---|---|
| No. | PaperID | $|K|$ | $|C_i|$ | $|C_e|$ | PaperID | $|K|$ | $|C_i|$ | $|C_e|$ |
| 1 | 395393 | 10 | 23 | 336 | 395393 | 10 | 23 | 336 |
| 2 | 1749067 | 2 | 11 | 21 | 1653515 | 9 | 19 | 272 |
| 3 | 1867302 | 1 | 0 | 3 | 278613 | 6 | 35 | 141 |
| 4 | 1791141 | 2 | 4 | 14 | 1662450 | 4 | 23 | 171 |
| 5 | 1662450 | 4 | 23 | 171 | 1451499 | 8 | 13 | 105 |

(a) Top 5 on $D_1$, $|P("Real\ Time")| = 422$.

| $D_2$ | Local Ranking $R_A^g(P)$ | | | | Universal Ranking $R_u^g$ | | | |
|---|---|---|---|---|---|---|---|---|
| No. | PaperID | $|K|$ | $|C_i|$ | $|C_e|$ | PaperID | $|K|$ | $|C_i|$ | $|C_e|$ |
| 1 | 2158744 | 1 | 12 | 26 | 2848 | 8 | 19 | 65 |
| 2 | 1209177 | 2 | 34 | 1 | 3045031 | 12 | 22 | 12 |
| 3 | 1142083 | 2 | 10 | 0 | 832819 | 9 | 32 | 69 |
| 4 | 2848 | 8 | 19 | 65 | 3396132 | 17 | 6 | 0 |
| 5 | 1340498 | 2 | 18 | 1 | 308381 | 10 | 20 | 22 |

(b) Top 5 on $D_2$, $|P("Real\ Time")| = 49$.

– *Universal ranking prefers most popular papers, whereas local ranking prefers papers being not only popular but also focused on a small set of topics.*

Let us see Tab. 6 (a). Top recommended papers returned by *Universal ranking* shows a large number of citing/cited papers. Whereas, top recommended papers returned by *Local ranking* have less citing/cited papers but their keywords are more specific. The average number of keywords in each recommended papers of *Local ranking* and *Universal ranking* are 2.25 and 6.75, respectively, except the common top ranked paper with $ID = 395393$. These numbers in $D_2$ are 3.00 and 10.12. We notice that the average number of keywords in one paper is around 2 and 3 for those two datasets.

– *Keywords recommended with the inference relationship depend on both latent information and data domain.*

Let us see Tab. 7(a) for ICDE, conference specialized in database. Three inference methods *Children, Sibling* and *Parent* generate three lists of specialized keywords in database domain. On the other hand, for GI-Jahrestagung, small conference with quite various topics in Tab. 7(b), the recommendation lists are quite diversity. We also notice that $D_2$ have much more citations information than $D_3$.

– *The missing co-occurrence problem can be overcome by using inference relationship.*

**Table 7.** Top 5 recommended keywords for the keyword *"Sensor Network"*

| $D_2$ | Children | Sibling | Parent |
|---|---|---|---|
| 1 | Query Processing | Query Processing | Data Management |
| 2 | Database System | Database System | Query Processing |
| 3 | Indexation | Data Management | Data Stream |
| 4 | Data Stream | Data Stream | Query Evaluation |
| 5 | Data Management | Query Evaluation | Multi Dimensional |

(a) Top 5 on $D_2$, $|K| = 2755$.

| $D_3$ | Children | Sibling | Parent |
|---|---|---|---|
| 1 | Open Source | Mobile Device | Mobile Device |
| 2 | Mobile Device | Web Service | Data Warehouse |
| 3 | Web Service | Data Warehouse | Self Organization |
| 4 | Middleware | Open Source | Distributed System |
| 5 | Data Warehouse | Middleware | P2P |

(b) Top 5 on $D_3$, $|K| = 1640$.

As mention above, our approach using stationary probability graph to generate new states follow Markov Chain Monte Carlo method. This helps to create new states that have the co-occurrence of keywords. So, we agree that the recommended lists computed by inference relationship showed in Table 7 are acceptable.

## 7 Conclusion and Future Works

We have introduced and studied a new approach on discovering various relationships among keywords from scientific publications. The proposed Markov Chain model combined four main factors of the problem: content, publicity, impact and randomness. The stationary probability of the states in the model helps us ranking and measuring the inference and similarity of keywords.

We have proposed SciRecSys recommendation system for navigating the scientific publications efficiently. By applying the relationships among keywords, we have suggested the solutions for the ranking results of a given keyword, related keywords and recommended papers. The experiments have shown that our methods can reflect how hot keyword is and overcome the missing co-occurrence problem. Moreover, our approach can exploit the latent information of citation network effectively.

As future work, we are planning to $i$) do experiment on big dataset to upgrade the quality of our ranking and measuring system, $ii$) study how to combine inference relationship by stationary probability graph with a given ranking systems, $iii$) investigate the time series in the keyword relationship and the trend prediction problem, and $iv$) apply model of recommendation systems in various problems, e.g., product recommendation, event recommendation, and so on.

## References

1. Le Anh, V., Vo Hoang, H., Le Trung, K., Le Trung, H., Jung, J.J.: A general model for mutual ranking systems. In: Nguyen, N.T., Attachoo, B., Trawiński, B., Somboonviwat, K. (eds.) ACIIDS 2014, Part I. LNCS, vol. 8397, pp. 211–220. Springer, Heidelberg (2014)
2. Bollegala, D., Matsuo, Y., Ishizuka, M.: Measuring semantic similarity between words using web search engines. In: Williamson, C.L., Zurko, M.E., Patel-Schneider, P.F., Shenoy, P.J. (eds.) Proceedings of the 16th International Conference on World Wide Web (WWW 2007), Banff, Alberta, Canada, May 8-12, pp. 757–766. ACM (2007)
3. Cha, S.H.: Comprehensive survey on distance/similarity measures between probability density functions. International Journal of Mathematical Models and Methods in Applied Sciences 1(4), 300–307 (2007)
4. Dagan, I., Lee, L., Pereira, F.C.N.: Similarity-based models of word cooccurrence probabilities. Machine Learning 34(1-3), 43–69 (1999)
5. Fouss, F., Pirotte, A., Renders, J.M., Saerens, M.: Random-walk computation of similarities between nodes of a graph with application to collaborative recommendation. IEEE Transactions on Knowledge and Data Engineering 19(3), 355–369 (2007)
6. Huang, A.: Similarity measures for text document clustering. In: Proceedings of 6th In New Zealand Computer Science Research Student Conference, Christchurch, New Zealand, pp. 49–56 (2008)
7. Keener, J.P.: The perron-frobenius theorem and the ranking of football teams. SIAM Review 35(1), 80–93 (1993)

 8. Kien, L.T., Hieu, L.T., Hung, T.L., Vu, L.A.: Mpagerank: The stability of web graph. Vietnam Journal of Mathematics 37(4), 475–489 (2009)
 9. Lops, P., de Gemmis, M., Semeraro, G.: Content-based recommender systems: State of the art and trends. In: Ricci, F., Rokach, L., Shapira, B., Kantor, P. (eds.) Recommender Systems Handbook. ch. 3, pp. 73–105. Springer, Heidelberg (2011)
10. Sánchez, D., Batet, M., Isern, D., Valls, A.: Ontology-based semantic similarity: A new feature-based approach. Expert Systems with Applications 39(9), 7718–7728 (2012)
11. Schickel-Zuber, V., Faltings, B.: Oss: A semantic similarity function based on hierarchical ontologies. In: Veloso, M.M. (ed.) Proceedings of the 20th International Joint Conference on Artificial Intelligence (IJCAI 2007), Hyderabad, India, January 6-12, pp. 551–556 (2007)
12. Singthongchai, J., Niwattanakul, S.: A method for measuring keywords similarity by applying jaccard's, n-gram and vector space. Lecture Notes on Information Theory 1(4), 159–164 (2013)

# Grouping Like-Minded Users
# Based on Text and Sentiment Analysis

Soufiene Jaffali, Salma Jamoussi, and Abdelmajid Ben Hamadou

MIRACL Laboratory Higher Institute of Computer Science and Multimedia,
University of Sfax,
Sfax, BP 1030 - Tunisia

**Abstract.** With the growth of social media usage, the study of online communities and groups has become an appealing research domain. In this context, grouping like-minded users is one of the emerging problems. Indeed, it gives a good idea about group formation and evolution, explains various social phenomena and leads to many applications, such as link prediction and product suggestion. In this dissertation, we propose a novel unsupervised method for grouping like-minded users within social networks. Such a method detects groups of users sharing the same interest centers and having similar opinions. In fact, the proposed method is based on extracting the interest centers and retrieving the polarities from the user's textual posts.

**Keywords:** Social network, like-minded users, interest center, sentiment analysis.

## 1 Introduction

Building relationships is one of the principal activities in social networks as this allows the interaction between users having something in common (ethnicity, locality, interest center, etc.). Since people are selectively connected to others, the interactions between users leads to social groups (communities). Therefore, identifying and understanding groups of users sharing similar interests are emergent tasks of Social Network Analysis (SNA) leading to many applications such as the friend suggestion systems, the collaborative filtering, etc. Most of the works concerned with this issue deals with it as a graph-distribution problem, in which the users are represented by nodes and the relationships between them by edges [22]. These relationships are generally explicit friendship links ("friend" on Facebook, "follower/followee" on Twitter, etc.). According to the big tail distribution of social networks [21], most of the social media users have only few links. Therefore, it is hard to find like-minded people who are several steps away from each other within the same social network. In addition, regarding the huge number of social network users (over 645,750,000 active registered Twitter users according to Statistic Brain[1]), mining only explicit relations within the

---

[1] http://www.statisticbrain.com/twitter-statistics

D. Hwang et al. (Eds.): ICCCI 2014, LNAI 8733, pp. 83–93, 2014.
© Springer International Publishing Switzerland 2014

network do not provide a complete vision. This implies a limitation of link based approach.

In this study, we propose to group the users sharing the same interests by analyzing their textual posts. The main goal is to retrieve the interest centers from the users posts and, then, to group those having the same interests. At this stage, we can find, in a given group, users having opposite opinions about the same subject. So, they cannot be considered as like-minded users. To overcome this problem, we add a sentiment-analysis to know whether the user has a positive or a negative opinion about the interest center. Thus, we obtain two sub-groups by interest center. Grouping like-minded people based on their interest centers and their polarities, is a very interesting task. Indeed, it improves the quality of recommendation and social marketing systems, and leads to many applications like poll systems and familiar stranger recommendation.

## 2   Related Work

Several approaches to community extraction based on link information have been proposed [7]. Moreover, many kinds of information are used to retrieve significant communities, such as the mutual awareness [16], comments and like actions [23]. Adamic and Glance [3] use the link patterns to measure the degree of interaction between two political communities. Also, tags are deeply used to construct the user's profiles [8], and to classify the interest centers [14]. Abrouk et al. [1] create tag communities using the Principal Component Analysis (PCA) and assign the users to the closest communities. Wang et al. [32] connect the like-minded users using the tag network inference. Given the fact that some users do not employ tags in their posts and that the same subject can be described by more than one tag (e.g. "#WMA", "#MusicAward", "#DiamondAward" describing the "World Music Awards" event), the use of tags for community detection may not succeed or yield to unoptimized results. Therefore, we suggest retrieving the latent interest centers from textual posts, and then, using the retrieved centers in order to group the users into communities. In the literature, just a few works deal with extracting social relations between individuals from text [20].

In this context, the Dirichlet Dynamic Allocation (LDA) and the probabilistic Latent Semantic Analysis (pLSA) are largely implemented to generate the subject models being used to regroup the tweets [10]. Similarly, Sachan et al. [28] applies LDA to identify the subjects of discussion based on the interactions between the users. These subjects are used to create the communities in a second stage. Using LDA, Hannachi et al. [10] extract the subjects from the published tweets to build a model directed by them. Tsur et al. [31] propose the scalable multi-stage clustering algorithm (SMSC) in order to categorize the tweets. The SMSC algorithm had been tested on a collection of tweets and presented a high performances.

Our algorithm differs from the classic text based community extraction systems, in the fact that it allows generating signed communities (positive and negative) according to user polarities. The closest works to our algorithm are

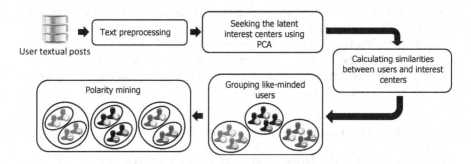

**Fig. 1.** Proposed model of grouping like-minded users

those of Dragomir Radev and his co-authors [2,11], in which, they aim to identify the sub-groups in on-line discussions. The common principle is to use the text mining tools to locate the texts comprising opinions, and to specify their targets. According to the opinion targets, the researchers gather the similar texts. By using a grouping algorithm, they subdivide the users in sub-groups by taking into account the polarity and the target of the opinion.

## 3  Grouping Like-Minded People Algorithm

Given a group of users $E$, we intent to find the optimal distribution of the group $E$ in $K$ clusters. The optimal distribution maximizes the correlation between the intra-group users and minimizes the similarity between the inter-group users. In our approach, the grouping is based on the content of the messages. The main goal here is to find the latent centers of interest around which the input data are concentrated. Then, we calculate the distances between the users and the centers that we found. Ultimately, we gather the users according to the distances which separate them from the interest centers. Thus, the users close to the same center belong to the same group. Each group is then divided into two sub-groups according to the polarity of the users. The five principal steps of our GLIO algorithm (Grouping Like-minded users based on Interest-centers and Opinions) are hereafter detailed. Fig. 1 presents the proposed approach.

### 3.1  Text Preprocessing

This phase consists of filtering the raw data and displaying them in a new representation in order to use them later. In the present work, we deal with the Twitter text messages known as "tweets". Those latter are limited to 140 characters allowing users to share their status. Tweets may contain certain meta-data such as words prefixed with '#' (known as"hashtag") to describe their subjects, with '@' to mention or answer another user, and with 'RT' to 'Re-Tweet' (Republish) another tweet. Tweets are neither structured nor written in a formal language which may make their exploitation very difficult. To remedy this problem, we start by eliminating the stop words(personal pronouns, prepositions, etc.) and converting all upper-case letters to lower-case ones. Next, we eliminate the words appearing with less than a prefixed threshold. Then, we group the tweets of each user and calculate the occurrence-appearance terms in their publications. Thus, we obtain a representative vector of each user. Finally, we combine these vectors to form a matrix $users \times terms$.

## 3.2   Seeking The Latent Centers of Interest

In this step, we seek the latent interest centers within the input data. To do this, we use the Principal Component Analysis (PCA)[24]. This latter proved its effectiveness in the themes detection and textual documents clustering [12]. PCA is generally used to reduce the data space or to determine the axes where data are concentrated as is the case in our study. In our case, the data are the terms used by the users in their tweets. Consequently, the principal components are the axis around which the terms are concentrated. In other words, these components are the latent interest centers in the tweet collection.

Let $T = \{t_1, t_2, ..., t_m\}$ the bag of words used in the users posts, and $U = \{u_1, u_2, ..., u_i, ..., u_n\}$ the occurrence matrix, with $u_i$ is the vector representing the $user_i$, and $n$ is the number of users. Each user's vector is of the form $u_i = \{o_{i1}, o_{i2}, ..., o_{ih}, ..., o_{im}\}$, with $o_{ih}$ is the occurrence number of the term $t_h$ ($t_h \in T$) in the posts of the user $u_i$. We calculate the covariance matrix of $U$ and its eigenvalues and eigenvectors. The obtained eigenvectors present the latent interest centers within the users posts. Each interest center is of the form $C_j = \{c_{j1}, c_{j2}, .., c_{jl}, .., c_{jm}\}$ with $c_{jl}$ is the weight of the term $l$ in the component $j$. The terms having the highest weights in $C_j$ are those reflecting its subject [12]. We use the eigenvalues to determine the number of clusters in 3.4.

## 3.3   Calculating the Similarities between the Users and the Interest Centers

This phase consists in defining a new representation of the users by taking into account the interest centers identified in the previous step. Given that the users who share similar interests are close to the same principal components, we represent each user by his distance from all the centers of interest (principal components). To do this, we calculate the distances between the user's word occurrence vector (calculated in the first step) and each eigenvector (principal component). In this study, we use the cosine measure as a metric for the deviation between the vectors.

## 3.4   Assigning the Users to the Clusters

In this step, we employ the kMeans algorithm to regroup the like-minded users in categories. Indeed, kMeans is a grouping algorithm which classifies the objects in a number $K$ of groups by taking into account the attribute values. It is recognized by its simplicity and effectiveness. As inputs, kMeans takes a set of data $D$ and the number of groups to be identified $K$. Then, $K$ centers of gravity are calculated randomly. In the third step, each object of the data set is allocated to a cluster $C$ having the closest center. After each allocation of an object to a cluster, the center of this latter is recomputed by taking account of the allocated objects. The algorithm reaches its aims when no change is observed and, thus, we obtain $K$ groups as the output.

**Finding the Number of Clusters.** The main problem of the kMeans algorithm lies in the difficulty in finding the number of cluster $K$, which is an NP-complete problem [19]. Most of the works using the kMeans algorithm either set the value of $K$ arbitrary or vary it empirically, which is the case of [30,31]. To overcome this problem, we use the eigenvalues calculated in the second step. Since the principal components correspond to the latent interest centers in the group of users, the knowledge of the component number is related to the knowledge of the number of interest centers and whence the clusters. In the literature, many methods are suggested to determine the component number to be considered [13,6]. In this work, we adopt the Scree Test Acceleration Factor proposed by [25] as a non-graphical solution of the method of Cattel. In fact, this solution consists in considering the principal components which precede the coordinate where the acceleration factor is maximum. These components must also have eigenvalues higher than 1 or than the Location Statistic criterion LS (generally the median, the average or one of the 0.05, 0.50 or 0.95 centile). Thus, we obtain the following system:

$$\begin{cases} n_{af} = Count[(\lambda_i \geq 1 \ \& \ i < k) & with \ k \equiv argmax(af)] \\ \qquad\qquad\qquad or \\ n_{af} = Count[(\lambda_i \geq LS_i \ \& \ i < k) \ with \ k \equiv argmax(af)] \end{cases} \tag{1}$$

With $\lambda_i$ is the $i^{th}$ eigenvalue, and $af$ is the acceleration factor. This latter is indicated by the abrupt change on the curve slope. It can be given for any $i$ (between 2 and $p - 1$ with $p$ is the number of eigenvalues) by the second derivative $f''(i)$. The simplified function of the second derivative can be written as follows:

$$f''(i) = f(i + 1) - 2 * f(i) - f(i - 1) \tag{2}$$

## 3.5 Polarity Mining

To evaluate the polarity of the users toward their interest centers, we use the method of [4]. This algorithm is proposed to determine the sentimental orientation (positive or negative) of a Facebook comment by using the linguistic approach. Ameur and Jamoussi [4] propose to dynamically create the positive and negative dictionaries by exploiting the emotion symbols present in the comments. Thus, they predict the positive and negative polarities of the comment using these prepared sentiment dictionaries. Furthermore, throughout the step of dictionaries construction, the authors assumed that the emotion symbols (emoticons, Acronyms and exclamation words) reflect the sentiment expressed by the words that precede them. In other words, each word has the same polarity as the first encountered emotion symbol. As our corpus tweets often contain emoticons, we focus on this assumption. In this step, we divide each cluster, obtained in the previous stage, into positive and negative sub-groups, according to the user polarities.

## 4    Experimentations

### 4.1    Baseline

To evaluate and position our algorithm compared to the literature, we use three reference grouping algorithms; namely, kMeans, LDA and SMSC. To evaluate the implemented sentiment analysis algorithm, we use the Sentiment140[2] API.

**kMeans :** As a first reference, we use the classical KMeans [18], one of the most used algorithms for the clustering. Liu [17] uses kMeans to detect the communities in the networks. We use kMeans to cluster vectors representing users.

**LDA :** LDA was introduced by [5]. It is conceived to analyze the latent thematic structures in the data with large scales, including large collections of text or web documents. To implement LDA, we use GibbsLDA++, which is a C/C++ implementation of LDA by using the sampling technique of Gibbs to estimate the parameters and the inference.

**SMSC :** In [31], the authors propose the SMSC algorithm (Scalable Multi-Stage Clustering). This algorithm (1) starts by creating a whole of virtual documents $D'$ (a transformation of the messages containing at least a hashtag) based on a set of document $D$. The number of messages in $D'$ is equal to the number of hashtags in $D$; (2) classifies $D'$ messages by applying kMeans; (3) re-transforms each virtual document into its original version by assigning each message containing a hashtag to the cluster of the virtual document with which it is associated. Finally, it assigns the messages which are short of hashtags to the closest clusters.

### 4.2    Datasets

Two reference corpora are used in the experiments to evaluate the performances of the GLIO algorithm vis-a-vis the previously quoted reference algorithms. In this sub-section, we describe the two used reference bases.

**Sander :** The Sander corpus is created by [29], and is composed of 5513 tweets classified by the author into positive, negative, neutral and without importance. In addition, the corpus is labeled by topic: Apple, Google, Microsoft and Twitter. The corpus is available on the Sananalytics site[3].

---

[2] http://help.sentiment140.com/api
[3] http://www.sananalytics.com/lab/twitter-sentiment/

**Citeseer :** The CiteSeer corpus [9] is composed of 3312 scientific publications classified into six classes. Each publication in the dataset is described by a 0/1-valued word vector indicating the absence/presence of the corresponding word from the dictionary. The dictionary consists of 3703 single words. Given its sparseness, the data presented in the Citeseer collection is very close to the social networks data. We consider every scientific publication as a user publication, and we use this corpus to evaluate the performances of our algorithm for grouping like minded users. Since the documents in this collection do not contain hashtags, we do not apply the SMSC algorithm to the Citeseer corpus.

# 5  Results

In this section, we present the results obtained using our algorithm and the three references kMeans, LDA and SMSC. To apply the reference algorithms, we must manually provide the number of the desired clusters in input, While, our GLIO algorithm finds this value automatically (see paragraph 3.4).

For a significant evaluation of the clustering systems, several evaluation metrics are proposed in the literature. We use four evaluation methods namely: Recall, Precision, F-Measure of Rijsbergen [27] and Rand-Index (RI) of Rand [26].

## 5.1  Distribution of the Users by Classes and Clusters

For a complete evaluation and a better interpretation of the used systems, we compare the user distributions by cluster. To evaluate the grouping algorithms, we consider two principal criteria: integrality and homogeneity. Thus, the system must put all the users of a given class $C$ in a same cluster $K$, which contains only the class $C$ users.

Fig. 2 presents the Sander corpus user distributions by cluster respectively obtained with the algorithms kMeans, LDA, SMSC and GLIO. We notice that the clusters obtained with kMeans suffer from a great heterogeneity. Moreover, the users are almost arranged in only three clusters while four subjects are initially considered. Each group obtained with LDA presents the four subjects with close proportions, whereby we can not affirm the dominating subject of each cluster. We note that the SMSC algorithm succeeded in releasing three clusters whose users belong to only one class (Cluster 2= Twitter, Cluster 3 = Google and Cluster 4 = Microsoft). Yet, Cluster 1 contains messages belonging to the four classes. Finally, it is clear that our GLIO algorithm generates four very homogeneous clusters and each one of them contains the users of only one class (Cluster 1= Google, Cluster 2= Microsoft, Cluster 3 = Twitter and Cluster 4 = Apple).

## 5.2  Results of Grouping Like-Minded Users

Table 1 shows the results of the users' grouping obtained for the two corpora (Citeseer and Sander) by applying the algorithms kMeans, LDA, SMSC and

90    S. Jaffali, S. Jamoussi, and A. Ben Hamadou

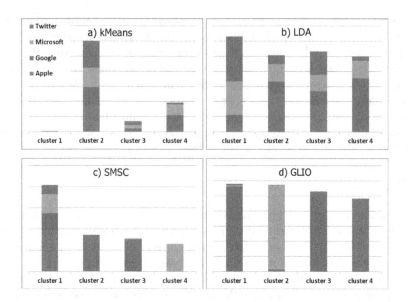

**Fig. 2.** Message distribution by topics and clusters

**Table 1.** Results of grouping Like-Minded users

|  | Citesser | | | | Sander | | | |
|---|---|---|---|---|---|---|---|---|
|  | Recall | Precision | F | RI | Recall | Precision | F | RI |
| kMeans | 0.27 | 0.25 | 0.26 | 0.68 | 0.34 | 0.37 | 0.35 | 0.5 |
| LDA | 0.59 | 0.53 | 0.55 | 0.79 | 0.5 | 0.5 | 0.5 | 0.68 |
| SMSC | - | - | - | - | 0.85 | 0.75 | 0.79 | 0.76 |
| GLIO | **0.65** | **0.66** | **0.65** | **0.81** | **0.98** | **0.97** | **0.98** | **0.97** |

GLIO. We notice the low values obtained with the classical approaches (kMeans and LDA). These low values highlight the limits of the classical approaches vis-a-vis the sparseness which characterizes the data in the social networks. We also notice that the SMSC algorithm provides good performances which exceed those obtained with LDA and kMeans for the grouping of like-minded users. The values of recall, precision, F-Measure and the Rand-Index prove that the results of our GLIO algorithm exceed those of the reference algorithms. These results can be explained by the fact that the PCA in step 2 of our algorithm reduces the data noise. This latter is due to the weak density which characterizes the corpora of the short texts exchanged in the social networks.

As we do not have information about the "Citeseer" documents' sentiment, we use only the "Sanders-Twitter Sentiment" set of data to evaluate the sentiment analysis process. To assess the used algorithm for the sentiment analysis, we compare its results with those obtained with the Sentiment 140 API. The used algorithm assigns the right polarity to 72.36% of the users, when only 64.1%

of them are successfully assigned using the Sentiment 140 API. For a complete vision, we test the ability of grouping the users having the same opinions. Thus, we consider only the positive and negative posts in the Sanders collection. Then, we run our model. We obtain eight groups of users. Each group contains users dealing with the same subjects and having the same polarity (positive/negative). The values of the four evaluation metrics for the obtained clusters are over 0.72. We obtain a Rand-Index and F-Measure equal to 0.86 and 0.75 respectively. Those results imply that our system, for the most part, successfully groups the users having similar opinions and sharing the same interests.

## 6   Conclusion

In this work, we present an algorithm of Grouping Like-minded users based on Interest-centers and Opinions GLIO. Such algorithm aims to seek the latent centers in a group of users based on their publications, and to assign each user to the nearest center. Compared with three reference algorithms, the experimental results using two datasets (Citeseer and Sander) prove the effectiveness and the high quality of our algorithm. Additionally, unlike the reference algorithms, our algorithm allows finding automatically the number of clusters. Moreover, the proposed algorithm is flexible which allows replacing Kmeans or the used sentiment analysis algorithm by other algorithms as required.

In the long run, a classification part of online users may be added in order to study the network evolution. Amongst the prospects which can be considered is to add a semantic layer by integrating an ontology or Folksonomies such as ODP. Information about the links between the users can also be enhanced to improve the grouping results. Finally, it is actually motivating to consider the subjects which can have more than two opinions.

**Acknowledgement.** The first author is grateful to Hanen Ameur, the author of [4], for his support and instructive help.

## References

1. Abrouk, L., Gross-Amblard, D., Leprovost, D.: Découverte de communautés par analyse des usages. In: EGC 2010 Workshops, pp. A5-5–A5-16 (2010)
2. Abu-Jbara, A., King, B., Diab, M.T., Radev, D.R.: Identifying opinion subgroups in arabic online discussions. In: ACL , vol. (2), pp. 829–835 (2013)
3. Adamic, L., Glance, N.: The political blogosphere and the 2004 u.s. election: divided they blog. In: LinkKDD 2005, pp. 36–43. ACM Press, New York (2005)
4. Ameur, H., Jamoussi, S.: Dynamic construction of dictionaries for sentiment classification. In: ICDM Workshops, Dallas, Texas, U.S.A (2013)
5. Blei, D.M., Ng, A.Y., Jordan, M.I.: Latent dirichlet allocation. J. Mach. Learn. Res. 3, 993–1022 (2003)
6. Cattell, R.B.: The scree test for the number of factors. Multivariate Behavioral Research 1(2), 245–276 (1966)

7. Fortunato, S.: Community detection in graphs. Physics Reports 486(3-5), 75–174 (2010)
8. Gemmell, J., Shepitsen, A., Mobasher, B., Burke, R.: Personalizing navigation in folksonomies using hierarchical tag clustering. In: Song, I.-Y., Eder, J., Nguyen, T.M. (eds.) DaWaK 2008. LNCS, vol. 5182, pp. 196–205. Springer, Heidelberg (2008)
9. Giles, C.L., Bollacker, K.D., Lawrence, S.: Citeseer: an automatic citation indexing system. In: ICDL, pp. 89–98. ACM Press (1998)
10. Hannachi, L., Asfari, O., Benblidia, N., Bentayeb, F., Kabachi, N., Boussaid, O.: Community Extraction Based on Topic-Driven-Model for Clustering Users Tweets. In: Zhou, S., Zhang, S., Karypis, G. (eds.) ADMA 2012. LNCS, vol. 7713, pp. 39–51. Springer, Heidelberg (2012)
11. Hassan, A., Abu-Jbara, A., Radev, D.R.: Detecting subgroups in online discussions by modeling positive and negative relations among participants. In: EMNLP-CoNLL, pp. 59–70. ACL (2012)
12. Jaffali, S., Jamoussi, S.: Principal component analysis neural network for textual document categorization and dimension reduction. In: SETIT 2012, pp. 835–839 (2012)
13. Kaiser, H.F.: The Application of Electronic Computers to Factor Analysis. Educational and Psychological Measurement 20(1), 141–151 (1960)
14. Li, X., Guo, L., Zhao, Y.E.: Tag-based social interest discovery. In: Proceedings of WWW 2008, pp. 675–684. ACM Press, New York (2008)
15. Liang, H., Xu, Y., Li, Y.: Mining users' opinions based on item folksonomy and taxonomy for personalized recommender systems. In: ICDM Workshops, pp. 1128–1135. IEEE Computer Society (2010)
16. Lin, Y.R., Sundaram, H., Chi, Y., Tatemura, J., Tseng, B.: Discovery of blog communities based on mutual awareness. In: Proceedings of the 3rd Annual WWE (2006)
17. Liu, J.: Comparative analysis for $k$-means algorithms in network community detection. In: Cai, Z., Hu, C., Kang, Z., Liu, Y. (eds.) ISICA 2010. LNCS, vol. 6382, pp. 158–169. Springer, Heidelberg (2010)
18. MacQueen, J.B.: Some methods for classification and analysis of multivariate observations. In: The Fifth Berkeley Symposium on Mathematical Statistics and Probability, vol. 1, pp. 281–297. University of California Press (1967)
19. Mahajan, M., Nimbhorkar, P., Varadarajan, K.: The planar k-means problem is np-hard. Theor. Comput. Sci. 442, 13–21 (2012)
20. McCallum, A., Wang, X., Corrada-Emmanuel, A.: Topic and role discovery in social networks with experiments on enron and academic email. J. Artif. Int. Res. 30(1), 249–272 (2007)
21. McGlohon, M., Akoglu, L., Faloutsos, C.: Statistical properties of social networks. In: Social Network Data Analytics, pp. 17–42 (2011)
22. Newman, M.E.J., Girvan, M.: Finding and evaluating community structure in networks. Physical Review E 69(026113) (2004)
23. Palsetia, D., Patwary, M.A., Zhang, K., Lee, K., Moran, C., Xie, Y., Honbo, D., Agrawal, A., Liao, W.K., Choudhary, A.: User-interest based community extraction in social networks. In: The 6th SNA-KDD Workshop. ACM (2012)
24. Pearson, K.: On lines and planes of closest fit to points in space. Philosophical Magazine 2 (1901)
25. Raîche, G., Walls, T.A., Magis, D., Riopel, M., Blais, J.G.: Non-graphical solutions for cattell's scree test. Methodology: European Journal of Research Methods for the Behavioral and Social Sciences 9(1), 23–29 (2013)

26. Rand, W.M.: Objective criteria for the evaluation of clustering methods. Journal of the American Statistical Association 66(336), 846–850 (1971)
27. Van Rijsbergen, C.J.: Information Retrieval, 2nd edn. Butterworth-Heinemann, Newton, MA, USA (1979)
28. Sachan, M., Contractor, D., Faruquie, T.A., Subramaniam, L.V.: Using content and interactions for discovering communities in social networks. In: WWW 2012, pp. 331–340. ACM Press, New York (2012)
29. Sanders, N.J.: Sanders-Twitter Sentiment Corpus. Sanders Analytics LLC (2011)
30. Tang, L., Wang, X., Liu, H.: Community detection via heterogeneous interaction analysis. Data Min. Knowl. Discov. 25(1), 1–33 (2012)
31. Tsur, O., Littman, A., Rappoport, A.: Efficient clustering of short messages into general domains. In: ICWSM. AAAI Press (2013)
32. Wang, X., Liu, H., Fan, W.: Connecting users with similar interests via tag network inference. In: CIKM, Glasgow, Scotland, UK (2011)

# A Preferences Based Approach for Better Comprehension of User Information Needs

Sondess Missaoui and Rim Faiz

LARODEC, ISG, University of Tunis,
Bardo, Tunisia
sondes.missaoui@yahoo.fr
LARODEC, IHEC Carthage University
Carthage Presidency, Tunisia
Rim.Faiz@ihec.rnu.tn

**Abstract.** Within Mobile information retrieval research, context information provides an important basis for identifying and understanding user's information needs. Therefore search process can take advantage of contextual information to enhance the query and adapt search results to user's current context. However, the challenge is how to define the best contextual information to be integrated in search process. In this paper, our intention is to build a model that can identify which contextual dimensions strongly influence the outcome of the retrieval process and should therefore be in the user's focus. In order to achieve these objectives, we create a new query language model based on user's preferences. We extend this model in order to define a relevance measure for each contextual dimension, which allow to automatically classify each dimension. This latter is used to compute the degree of change in result lists for the same query enhanced by different dimensions. Our experiments show that our measure can analyze the real user's context of up to 8000 of dimensions. We also show experimentally the quality of the set of contextual dimensions proposed, and the interest of the measure to understand mobile user's needs and to enhance his query.

**Keywords:** Mobile search, User's context, Relevance, User's Preferences.

## 1 Introduction

We live in an information society and expanding technologies provide faster and broadband Internet connections which make users able to access information anywhere at any time in their daily lives. This has encouraged the use of mobile devices as one of the most important web search tools. Since, it is therefore natural to suggest new approaches of Information Retrieval (IR) in order to meet the special information needs of mobile users. Often, with mobile applications, some aspects of the user's context are available, and this context can affect what sort of information is relevant to the user. The context can include a wide range of dimensions that characterize the situation of the user. But the question is:

D. Hwang et al. (Eds.): ICCCI 2014, LNAI 8733, pp. 94–103, 2014.
© Springer International Publishing Switzerland 2014

What contextual dimensions reflect better the mobile user's need and lead to the appropriate search results?

In this paper, we focus our research efforts on this area that has received less attention which is the context filtering. We have brought a new approach that has addressed this issue. How to define the relevant contextual dimension accurately and rapidly?

In fact, our hypothesis is that an accurate and relevant contextual dimension is the one that provides an interesting improvement in both query profiles (Preferences and Content). Those dimensions can improve the quality of search by proposing to the user results tailored to the user's current context.

The remainder of this paper is organized as follows. In section 2, we give an overview of related work which address Context-centered mobile web search. We describe in section 3, the Context adaptation approach to user's preference. In Section 4, we discuss experiments and obtained results. Finally, section 5 concludes this paper and outlines future work.

## 2   Related Works

The mobile users enter limited number of terms in a query. This creates a big challenge to the IR systems which called "query mismatch problem" [9]. So many studies integrate different context fields to enhance the query such as [2], and especially to modelize context, allowing to identify information that can be usefully exploited to improve search results such as [6], [20], [1], [13] and [22]. In fact, the Related work in the domain can be summarized in terms of two categories. Firstly, approaches which are using a set of contextual dimension to personalize all search queries. In this category, several research efforts are proposed in the literature to modelize the current user's context. Some approaches such as [3] and [12] have build models able to categorize queries according to their geographic intent. When [1] operate including Time and Location as main dimensions besides others to automatically infer the user's current context. The previous works propose to use a set of contextual dimensions for all queries and do not offer any context adaptation models to the specific goals of the users. In fact, only a subset of dimensions can be relevant and have the potential to influence the outcome search results Secondly, approaches that are performed to the aim of filtering the user's context and exploit only the relevant information to personalize the mobile search. This category of approaches such as [7], [11], [4] are proposed to identify the appropriate contextual information in order to tailor the search engine results to the specific user's context. In this category, our work has proceeded in terms of adapting the mobile context to user's preferences and identifying relevant contextual dimensions. We propose a new approach allows to define the most relevant and influential user's context dimensions for each search situation. In the next section, we describe our definition of this problem.

# 3    Context Adaptation to Preferences: CAP Approach

In this section, we focus our efforts on evaluating th user's context, in order to leave only a subset of relevant contextual dimensions. These, which go with the user's preferences and are able to enhance the search process.

## 3.1    User's Current Context

The key notion of user's context may have multiple interpretations. In this paper, the context is modeled through a finite set of special purpose attributes, called contextual dimension $p_i$, where $p_i \in C$ and $C$ is the user's context defined by a set of n dimensions $\{p_1, p_2..., p_n\}$. A dimension is a contextual information which is represented by a unique value.

Contextual dimension, will be defined by computing their capacity to enhance the type of retrieved documents. We evaluate their capacity to enhance the query in order to generate results with respect to user's preferences (Preference Query Profile). In this section, we will describe in details this profile.

## 3.2    Preferences Model

Some recent papers have investigated language modeling approach to define the user's intention behind the query. In our work we use the language modeling approach as described in [10] to filter the context. We offer a new query language model.

1. We build a language model based preferences. For each user's preference, we estimate a distribution of terms associated with the user's preference. We can then estimate the probability that a query was issued from a given preference by sampling from the term distribution of that preference.
2. We use the query preference profile to measure the relevance of a contextual dimension.

**Preferences Query Profile.** According to [5]: "One way to analyze a query is to look at the type of documents it retrieves". On basis of this rules, we infer that the best way to analyze a context dimension is to look at its effect on the query. So, its effect on the type of documents the query retrieves. Specifically, it can be accomplished by examining the top N documents of retrieval results. The context dimensions can then be ranked by the probability that they "generated" best results after being integrated in the search process. In language model approach [8] define the document likelihood of having generated the query formally as presented by the following equations:

$$P(Q \backslash D) = \prod_{w \in Q} P(w \backslash D)^{q_w} \tag{1}$$

Given a query Q and a document D, $q_w$ is the number of times the word w occurs in query Q. According to Croft and lafferty [18], document language models $P(w\backslash D)$, are estimated using the words in the document. We use this ranking to build a new query language model, $P(Pre\backslash Q)$, out of the top N documents. It is a new query feature in language model called "Preferences Query Profile" that helps us to define the effectiveness of the query to overcome user's interests. This query language model is named "the Preference profile of the query Q". In fact, a relevant retrieved result is a ranking list which meets, in a better way, the individual user needs according to their preferences. In this same spirit of thinking, we are interested in describing the personalized nature of a query. That's mean the effectiveness of the query to overcome user's interests when it retrieving a precise topic. E.g., Searching for "Music", the mobile search system must take into account the user's preference "Jazz". Therefore, we build a preferences query profile where documents can be ranked by the probability that they have been generated depending on the user's preferences. More concretely, given a set of preferences "Pre", and a query Q, our goal is to rank the preferences by $P(Pre\backslash Q)$ which is initially defined as:

$$\hat{P}(Pre\backslash Q) = \sum_{D \in R} \hat{P}(Pre\backslash D) \frac{P(Q\backslash D)}{\sum_{D \in R} P(Q\backslash D)} \tag{2}$$

Where "Pre" is the name of the user's preference. It's a term that describes a user preferences category from a data base containing all user's interest (his profile). For example if a user is interested by "Sport" a set of terms such as (Football, Tennis, Baseball,...) are defined as "Pre".

$$P(Pre\backslash D) = \{ \begin{matrix} 1 \ if \ Pre \in Pre_D \\ 0 \quad Otherwise \end{matrix} \tag{3}$$

Where $Pre_D$ is the set of categories names of interests contained in document D (e.g. Sport, Music, News, Cinema, Horoscope ...). The profile, that describes the user's interests and preferences could be explicitly set by the user or gathered implicitly from the user search history. In our experiments, a profile is collected explicitly before starting the search session.

A very helpful step is about smoothing maximum likelihood models such as $\hat{P}(Pre\backslash Q_{in})$. We used Jelinek-Mercer process created by [19] for smoothing. We use the distribution of the initial query $Q_{in}$(reference-model) over preferences as a background model. Such background smoothing is often helpful to handle potential irregularities in the collection distribution over preferencs. Also, it replaces zero probability events with a very small probability. Our aim is to assign a very small likelihood of a topic where we have no explicit evidence. This reference-model is defined by:

$$\hat{P}(Pre\backslash Q_{in}) = \frac{1}{|N|} \sum_{D} \hat{P}(Pre\backslash D) \tag{4}$$

Our estimation can then be linearly interpolated with this reference model such that:

$$P^{'}(Pre\backslash Q) = \lambda \hat{P}(Pre\backslash Q) + (1 - \lambda) \hat{P}(Pre\backslash Q_{in}) \tag{5}$$

Given $\lambda$ as a smoothing parameter.

The assumption of the Preference Profile analysis (cf. Fig1) is that irrelevant contextual dimension's can't improve the 'Preference Query Profile'. In fact, When we integrate an irrelevant dimension into search process, the query preference profile show no variance comparing to the initial query profile. Given that, this contextual dimension is not important for a query and shouldn't be selected. In contrast, a relevant dimension provides query preferences profile with at least one peak. Therefore, to define the general effect of a dimension on the search process, we need a measure to specify the relevance of each dimension according to its effect on the search outcomes. In the following, we present this measure.

## 3.3    Preference Score Measure

Our principal objective is to adapt the user's context to his preferences automatically by filtering it. For this purpose, we need to define the most and least influential mobile context dimensions. Indeed, there is no existing measurement method that allows the quantification of the mobile contextual information pertinence especially using a statistical property of retrieved result lists. Hence, the task to be accomplished is to build a relevance metric measure for contextual dimension. Our metric measure is based KL divergence [24] as an essential component to build this metric. We chose Kullback-Leibler divergence as a divergence measure issued from the domain of probability theory. It will be introduced in our approach to define the influence of each dimension on the search results. The KL divergence gives us a test of similarity to the preferences background model $P(Pre\backslash Q_{in})$. Our measure 'Preference Score' is defined basically on the comparison of two result rankings. It allows to identify whether a mobile context dimension enhancing the user query at his preferences profile. We build the "Preference Score" as a new metric to measure the relevance degree of each dimension.

The "Preference Score" is defined as:

$$PreferenceScore\,((P,Q)) = D_{kl}\,(P\,(Pre\backslash Q_p)\,,P\,(Pre\backslash Q_{in})) \qquad (6)$$

Where Q is the mobile query, we denote the appearance of a dimension P in a mobile context C (cf. section 3.1) as $P \in \{C\}$. Let $P(Pre\backslash Q_{in})$ the language model of the initial query used as a background distribution. And $P(Pre\backslash Q_p)$ the language model of the enhanced query using contextual dimension $P$. The proposed context-based measurement model can be expressed in a formal manner with the use of basic elements toward mathematic interpretation that build representative values from 0 to 1, corresponding to the intensity of dimension's relevance. Being null values indicative of non importance for that dimension (it should not be integrated in personalization of mobile information retrieval process). In the experiment, we will try to define the threshold that a dimension should obtain to be classified as relevant or irrelevant information. In the next section, we will evaluate the effectiveness of our metric measure 'Preference Score' to classify the contextual dimensions.

# 4    Experimental Evaluation

Our goal is to evaluate the "Preference Score" metric to predict the type of user's context dimension.

## 4.1    Dataset

We present our training and test collection, our evaluation protocol and then describe and discuss the obtained results. For the experiments reported in this work, we used a sample of real queries submitted to the America Online search engine. We had access to a portion of the 2006 query log of AOL[1]. We randomly selected initial sets of 2000 queries for training, development and testing purposes. After a filtering step to eliminate duplicate and navigational queries, we obtained a set of 1700 queries. While the AOL log is not a mobile search log, so we simulate the user's current mobile situation for each query. Thence, we use three contextual dimensions (Time, Location and Nearby people) to indicate the mobile search context. Then, for each query in the test set we classified manually their related contextual dimensions. Each dimension is associated to a label to indicate whether it is noise, irrelevant or relevant. The criterion to assess whether a given dimension is relevant, is based on whether the mobile user expects to see search results related to this contextual information ordered high on the results list of a search engine. These steps left us, in our sample test queries, with 10% noise dimensions, 24% irrelevant dimensions and 65,6% relevant dimensions.

## 4.2    Classification Performance of Our Metric Measure

Our experimental design allows us to evaluate the effectiveness of our technique to identify user's relevant contextual fields. For this purpose, we propose an evaluation methodology of obtained results using manually labeled contextual dimensions. In fact, a contextual dimension's class is correct only if it matches the labeled results. Using the Rp as a classification feature, we build a context intent classifier.

In order to compute the performance of the classifiers in predicting the dimensions classes, we use standard precision, recall and F-measure measures. We use also classifiers implemented as part of the Weka [2] software. We test the effectiveness of several supervised individual classifiers (Decision trees, Naive Bayes, SVM,and a Rule-Based Classifier) in classifying contextual fields using "Preference Score" as classification feature.

## 4.3    Results and Discussion

**Analysis of Rp Measure.** At this level we analyze the "Preference Score" distribution for each category of contextual dimensions. Fig.1 shows distribution

---

[1] http://www.gregsadetsky.com/aol-data/

[2] http://www.cs.waikato.ac.nz/ml/weka/

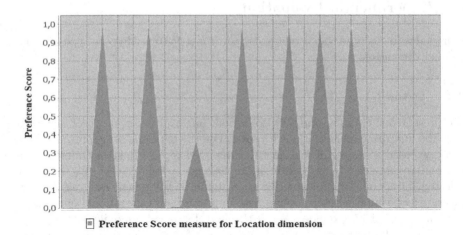

▣ **Preference Score measure for Location dimension**

**Fig. 1.** Distribution of "Preference Score" measure for geographic dimension (Location)

of our measure over different values of Location dimension for different queries. In this figure we notice that there is a remarkable drops and peaks in the value of "Preference Score". Indeed, the relevance of a contextual dimension is independent on his type or value but it depends on the query and the intention of mobile user behind such query. Hence, the measure hasn't a uniform distribution for those contextual dimensions.

**Effectiveness of Contextual Dimension Classification.** Our goal in this evaluation is to assess the effectiveness of our classification attribute "Preference Score" to identify the type of dimension from classes: relevant, irrelevant, and noise.

As discussed above, we tested different types of classifiers and Table 1 presents the values of the evaluation metrics obtained by each classifier. In fact, all the classifiers were able to distinguish between the three contextual dimension classes. Fmeasures, Precision and Recall ranging from 96% to 99%. But "SVM" classifier achieves the highest accuracy with 99% for the F-measure. This first experiment implies the effectiveness of our approach to accurately distinguish the three types of user's current contextual levels. It especially allows to correctly identify irrelevant contextual information with an evaluation measure over 1. When relevant and noise, achieving over 97% classification accuracy.

In a second experiment, we evaluated the classification effectiveness of our approach comparatively to DIR approach developed by Kessler [7]. By using the DIR measure, contextual information is only classified as relevant or irrelevant. It enables distinguishing between irrelevant and relevant context using a threshold value $\delta$. Whence, we compared the two approaches only on this basis. We implemented the DIR approach using the SVM classifier which achieves one of

**Table 1.** Classification performance obtained using a classifier with "Preference Score"

| Classifier | Class | Precision | Recall | F-measure | Accuracy |
|---|---|---|---|---|---|
| SVM | relevant | 0.978 | 0.989 | 0.981 | |
| | irrelevant | 1 | 1 | 1 | |
| | noise | 0.981 | 1 | 0.991 | 99% |
| | average | 0.991 | 0.99 | 0.99 | |
| JRIP rules | relevant | 0.911 | 0.953 | 0.924 | |
| | irrelevant | 1 | 1 | 1 | |
| | noise | 1 | 0.964 | 0.926 | 96.3% |
| | average | 0.965 | 0.962 | 0.962 | |
| Bayes | relevant | 1 | 0.933 | 0.966 | |
| | irrelevant | 1 | 1 | 1 | |
| | noise | 0.946 | 1 | 0.972 | 97% |
| | average | 0.973 | 0.971 | 0.971 | |
| J48 | relevant | 1 | 0.933 | 0.966 | |
| | irrelevant | 1 | 1 | 1 | |
| | noise | 0.946 | 1 | 0.972 | 97% |
| | average | 0.973 | 0.971 | 0.971 | |

the best classification performance using one simple rule: analyzing the individual results in two rankings for the same query expanded by different contextual dimensions. Intended or relevant contextual information must have an impact that goes beyond a threshold value. Hence, we should obtain a high value of DIR measure to classify a context as relevant. Table 2 presents the precision, recall, F-measure and accuracy achieved by the SVM classifier according to the both approaches. The result of comparison show that, our approach gives higher classification performance than DIR approach with an improvement of 1% at accuracy. This improvement is mainly over Relevant context dimensions with 1.3% at Recall.

**Table 2.** Classification performance on Relevant and Irrelevant dimensions: comparison between CAP approach and DIR measure approach

| Approach | DIR approach | | | CAP approach | | | | | |
|---|---|---|---|---|---|---|---|---|---|
| Class | Relevant | Irrelevant | Average | Relevant | Impro | Irrelevant | Impro | Avrege | Impro |
| Precision | 1 | 0.968 | **0.982** | 1 | **0%** | 0.984 | **1.7%** | **0.991** | **1%** |
| Recall | 0.956 | 1 | **0.981** | 0.978 | **2.3%** | 1 | **0%** | **0.99** | **1%** |
| F-measure | 0.977 | 0.984 | **0.981** | 0.989 | **1.3%** | 0.992 | **0.9%** | **0.99** | **1%** |
| Accuracy | 98% | | | 99,5% | | | | | **1,5%** |

## 5   Conclusion

We proposed in this paper a new approach for mobile context adaptation to the user's preferences. It is evaluates the relevance of contextual dimensions using different features. This approach is based a new metric "Preference Score", that allows to classify the contextual dimensions according to their relevance to enhance the search results. Our experimental evaluation show the classification performance of our metric measure comparatively to a cognitively plausible dissimilarity measure namely DIR. For future work, we plan to exploit our proposed approach to personalize mobile Web search. We will customize the search results for queries by considering the determined user's contextual dimension classified as relevant.

# References

1. Aréchiga, D., Vegas, J., Redondo, P.F.: Ontology Supported Personalized Search for Mobile Devices. In: ONTOSE 2010. LNCS, pp. 1–12. Springer, Heidelberg (2009)
2. Bouidghaghen, O., Tamine, L., Boughanem, M.: Context-Aware User's Interests for Personalizing Mobile Search. In: Proc. 12th IEEE International Conference on Mobile Data Management, Sweden, June 6-9, pp. 129–134. IEEE Computer Society (2011)
3. Chirita, P., Firan, C., Nejdl, W.: Summarizing local context to personalize global Web search. In: Proc. of CIKM International Conference on Information and Knowledge Management, Arlington, Virginia, USA, November 6-11, pp. 287–296. ACM (2006)
4. Coppola, P., Della Mea, V., Di Gaspero, L., Menegon, D., Mischis, D., Mizzaro, S., Scagnetto, I., Vassena, L.: The Context-Aware Browser. J. IEEE Intelligent Systems 25(1), 38–47 (2010)
5. Jones, R.: Temporal profiles of queries. J. ACM Transactions on Information Systems (TOIS) 25((3)14) (July 2007)
6. Tsai, F.S., Etoh, M., Xie, X., Lee, W.C., Yang, Q.: Introduction to Mobile Information Retrieval. J. IEEE Intelligent Systems 25(1), 11–15 (2010)
7. Kessler, C.: What is the Difference? A Cognitive Dissimilarity Measure for Information Retrieval Result Sets. J. Knowledge and Information Systems 30(2), 319–340 (2012)
8. Lavrenko, V., Croft, W.B.: Relevance-based language models. In: Proc. of SIGIR 2001 the 24th Annual International ACM SIGIR Conference on Research and Development in Information Retrieval, New Orleans, Louisiana, USA, September 9-13, pp. 120–127. ACM (2001)
9. Arias, M., Cantera, J.M., de la Fuente, P., Llamas, C., Vegas, J.: Knowledge-Based Thesaurus Recommender System in Mobile Web Search. In: Proc. of CERI 1st Spanish Conference on Information Retrieval, Madrid, Spain, June 15-16 (2010)
10. Ponte, J.M., Croft, W.B.: A language modeling approach to Information Retrieval, in. In: Proc. the 21st. International ACM SIGIR Conference on Research and Development in Information Retrieval, Melbourne, Australia, pp. 275–281. ACM (1998)
11. Stefanidis, K., Pitoura, E., Vassiliadis, P.: Adding Context to Preferences. In: Proc. of ICDE IEEE 23rd International Conference on Data Engineering, Istanbul, Turkey, April 15-20, pp. 846–855 (2007)
12. Gravano, L., Hatzivassiloglou, V., Lichtenstein, R.: Categorizing web queries according to geographical locality. In: Proc. of CIKM 2003 the Twelfth International Conference on Information and Knowledge Management, New Orleans, Louisiana, USA, November 2-8, pp. 325–333. ACM (2003)

13. Yau, S., Liu, H., Huang, D., Yao, Y.: Situation-aware personalized Information Retrieval for mobile internet. In: Proc. of COMPSAC 27th Annual International Computer Software and Applications Conference, Dallas, TX, USA, November 3-6, pp. 639–644. IEEE Computer Society (2003)
14. Welch, M., Cho, J.: Automatically identifying localizable queries. In: Proc. of 31st Annual International ACM SIGIR Conference on Research and Development in Information Retrieval, Singapore, July 20-24, pp. 1185–1186. ACM (2008)
15. Cronen-Townsend, S., Zhou, Y., Croft, W.B.: Predicting query performance. In: Proc. the 25th Annual International ACM SIGIR Conference on Research and Development in Information Retrieval, Tampere, Finland, August 11-15, pp. 299–306. ACM (2002)
16. Vadrevu, S., Zhang, Y., Tseng, B., Sun, G., Li, X.: Identifying regional sensitive queries in web search. In: Proc. of WWW 2008 the 17th International Conference on World Wide Web, Beijing, China, April 21-25, pp. 1185–1186 (2008)
17. Poslad, S., Laamanen, H., Malaka, R., Nick, A., Buckle, P.: Zipf.A. Crumpet, Creation of user-friendly mobile services personalised for tourism. In: Proc. of the Second International Conference on 3G Mobile Communication Technologies, Conf. Publ. No. 477, March 26-28, pp. 28–32. IEEE Computer Society, London (2001)
18. Croft, W.B., Lafferty, J.: Language Modeling for Information Retrieval. J. Kluwer Academic Publishers (2003)
19. Jelinek, F., Mercer, R.L.: Interpolated estimation of markov source parameters from sparse data. In: Proc. of the Workshop on Pattern Recognition in Practice, pp. 381–397. North-Holland, Amsterdam (1980)
20. Ahn, J., Brusilovsky, J., He, D., Grady, J., Li, Q.: Personalized Web Exploration with Task Modles. In: Proc. of WWW 2008 the 17th international conference on World Wide Web, Beijing, China, April 21-25, pp. 1–10 (2008)
21. Pitkow, J., Schutze, H., Cass, T., Cooley, R., Turnbull, D., Edmonds, A., Adar, E., Breuel, T.: Personalized search. Communications of the ACM Journal 45(9), 50–55 (2002)
22. Ingwersen, P., Jarvelin, K.: The Turn: Integration of Information Seeking and Retrieval in Context. J. Springer-Verlag Eds, vol. 18, p. 448 (2005)
23. Hollan, J.D., Sohn, T., Li, K.A., Griswold, W.G.: A Diary Study of Mobile Information Needs. In: Proc. of SIGCHI Conference on Human Factors in Computing Systems, Florence, Italy, April 5-10, pp. 433–442. ACM (2008)
24. Eguchi, S., Copas, J.: Interpreting Kullback-Leibler divergence with the Neyman-Pearson lemma. J. Multivariate Anal. 97, 2034–2040 (2006)

# Interlinked Personal Story Information and User Interest in Weblog by RSS, FOAF, and SIOC Technology

Nurul Akhmal binti Mohd Zulkefli and Baharum bin Baharudin

Computer and Information Science,
Universiti Teknologi PETRONAS,
Tronoh, Malaysia
skygur85@gmail.com, baharbh@petronas.com.my

**Abstract.** Interlinked Personal Story and user interest in Weblogs to re-use information in the blog contents is a new way of communication in a Weblog field. With many existing vocabularies such as Rich Site Summary (RSS), Friend of a friend (FOAF) and Semantically Interlinked Online Community (SIOC), interlinked among blogs can successfully help users especially bloggers to find the relationship that occurs inside the contents of the blogs itself. Furthermore, nowadays, personal blog contents are more useful to serve as the answer to the Internet users' search for information rather than existing search engines where personal blog contents are always updated. Our proposed framework system is designed to accomplish the motivation to interlink personal information weblogs and also to improve the uses of the blog among the online community.

**Keywords:** weblog, interlinked, information retrieval, social network.

## 1  Introduction

Personal Story Information can be found in almost any online community such as Weblogs, Facebook, and Twitter. However, the most personal story information sharing in social media can be found from the weblogs. The information sharing is wide and composed of any type of user background such as a traveler, student and even a housewife. In the variety of personal story information, we trust this information to be useful for future search engines. Meanwhile, with regards to the rapidly growing social media, an Internet user may have developed a lot of social media friends. It means that, every internet user may have approved more than one person as a 'friend' in his online community. In Weblog, a blogger will also have 'friends' in his blog site, and they are known as 'followers' or a 'list of friends'. Our motivation is to implement the visualization blog as a new decentralization of information between the owner of the blog and the friends following the blogs where the information extracted from the blog can be interlinked and integrated. Previously, the user would use the existing search engines like Google, Yahoo and Bing to find the information they want, but nowadays with the growing number of blogs, freshly updated information is becoming more and more common. The advantage of the

D. Hwang et al. (Eds.): ICCCI 2014, LNAI 8733, pp. 104–113, 2014.

motivation is that everyone can re-use the information from the updated blog. To accomplish the motivation, Really Simple Syndication (RSS), Friend of a friend (FOAF) and Semantically Interlinked Online Social (SIOC) will be implemented in the proposed system. The combination of these new technologies that covers semantic and ontology will enable the sharing of the contents in the blog.

The RSS document also known as a feed contains a summarized text, metadata such as publishing dates and authorship [1]. It is one of the technologies that underpin the rapid pace of publishing. RSS generates a predictable way for particles of content to be summed, sequenced, and searched. By implementing the RSS in the proposed system, the system can automatically derive newly updated data for content and comment on the site. It is a simple way to identify the current content items and the most important is the way we can connect our content to the larger Web by enabling others to find our contents more easily.

The FOAF allows the representation of private data and relationships. It is a useful building block for developing information systems that support online communities. It is simply an RDF vocabulary, and its typical use is akin to that of the RSS. The FOAF can help propose the system in identifying user's friend in the blog site. Each blogger's FOAF data will be decentralized within blogger's control and by this manner it will give the benefit to blogger to find out a friend's content easily with the help of the RSS. [2] FOAF supports bloggers by allowing provenance tracking and accountability [3]. Nowadays, the source of information is important to ensure that the information is trustworthy. The RDF tool tracker is used to find out the information by implementing FOAF to RDF sources. Besides, FOAF is used to match people with similar interests. People will bring together persons with similar interests [4]. Relationships that are based on common interests can help people with a similar interest to collaborate on that interest. The SIOC is used to facilitate the synthesis of information, and it includes the Semantic Web ontology for describing rich data in the RDF. Meanwhile, SIOC and FOAF vocabulary is integrated to get the user profile and social network information. Implementing the SIOC in proposed system can help the user identify the relationship between a blogger's contents and his or her friend's contents as the SIOC vocabulary that includes SIOC ontology can describe the user-created content by way of moving data from one platform to another [5].

In this paper, we will discuss the proposed system in the next section that includes a brief review of related works. In Section III, we provide the overview of the Interlinked Weblog design framework. A proposed system is presented in Section IV. Finally, Section V establishes the inevitable conclusion and discusses potential future works.

## 2    Related Work

An analysis on blogs [6] studied 25,000 blog sites and 750,000 links to the sites and focused on clusters of blogs connected via hyperlinks named blogspaces and investigated the extraction of blog communities and the evolution of the communities. From the 2013 report [7] stated that 83% people within the age range of 18 to 29 and

77% people at the age from 30 to 49 are involved in online social networks. Gruhl et al. studied the diffusion of information through blogspace [8]. They examined 11,000 blog sites and 400,000 links in the sites, and tried to characterize macro topic diffusion patterns in blogspaces and micro topic-diffusion patterns between blog entries. They also attempted to model topic diffusion by criteria called Chatter and Spikes. Previously, there are several research studies that detect the topic of information and track the information based on a number of categories [9]. First story detection (FSD) has been used to identify the new topic that has yet to be submitted. [10] Proposed a method that used a collection of FOAF and RSS from the web to generate the OnLine Analytical Processing (OLAP). The proposed method is claimed to provide improvement result for millions of users' interests by adapting the automatic FOAF to the social network services. In [7] research, they had figured out the usefulness of the FOAF and Social Network Analysis (SNA) as a platform to improve the result of recommending contents to the users. This research used common tags and characteristics from the content and they verified that when there are more users in the social network, the contents would be more effective and have better quality.

The Content Recommendation Method (CRM) is proposed by [11] based on both the FOAF and RSS. The FOAF is used to analyze the SNS and RSS for evaluating the contents. The researchers have found that their proposed CRM provides more suitable and trustworthy contents compared to the traditional CRSs. In this paper, we use a similar method to recognize a new topic. [12] When the new story is published, the FSD will check and compare the contents with the previous stories. If they detect any difference in the contents, the story will be designed as the first story; and vice versa.

## 3    Scenario

Anyone, from novice internet users to the expert internet users can use the blog to update their personal story information or to share his / her experience to be read by the public. Put simply, it is a storehouse that contains millions of information to be shared with people. The blog has its own term and the details include the site, entry, comments and user profile [13]. A standard blog has a site containing the URL, RSS (really simple syndication), site name, and entries. It is managed by one or more bloggers. A sample scenario (Fig.1) is provided below, where Blogger $A$ uses the system to decentralize his blog with other bloggers. Blogger $A$ starts to log in her blog link address into the system and the system will extract the information from the address link. Then, the system will extract the category which includes the entry itself by using the RSS and all the categories will be collected and computed by a feature vector. Here, the category will be matched with the template ontology to *make it easily defined by the system* and afterwards, it will be saved in the blogger database. A blogger has many choices, whether he or she wants to visualize the data by category, friends or interest topic. In the above example, a blogger wants to display the contents via category so here, the system will decentralize all the updated contents from the blogger and his friends. In this case, he or she can view their friend's blog in the visualization map where their blog is laid out at the center and by categorizing his or her content based on the template ontology, this blogger can also receive other information (entry) from their blogger-friends. This decentralization can make the integration between blogs to be defined by Blogger $A$ (or user of the system). From the above figure, Blogger $A$ can view the categories he or she has defined such as 'holiday', 'INHA', 'MOVIE', 'TRAVEL', 'research', and 'study'. Under the study and research category, for instance, Blogger $A$ can view the new information extracted by her blog where her friends, who are $B$ and $C$ have published new entries similar to entries under his or her category (research and study). In this system, Blogger $A$ can recognize the relationship that exists between Blogger $B$ and $C$ that might be able to establish this Probability:

1. Blogger $B$ and Blogger C are friends.
2. Blogger $B$ and Blogger $C$ are working in the same research area, $Pr[2]$.
3. Blogger $A$, $B$ and $C$ can share interesting information with each other if the $Pr[2]$ is true.

The probability of relationship exists between other categories too, such as 'holiday':

1. Probability of '$eh1$' and '$eh2$' ($Pr[eh1] = Pr[eh2]$) are in the same place or share the same time or they both share the same place and time.
2. Probability of Blogger '$eh1$' and Blogger '$eh2$' are going for a holiday together is true.
3. Probability for new updates of information from $eh1$ and $eh2$ to be true.

Next is a friend's decentralization using the FOAF. In this case, Blogger A will use the same way to start the system. However, Blogger A should determine whether she wants to decentralize the information via category or friend. In this case, Blogger A chooses decentralization by a friend, and she wants to find out some information that she can collect from her friends. Once the information is extracted by the system using the FOAF, it will proceed to extract the feature vector again from blogger entries and categories. This extraction is important to measure blogger's interest and match it with other bloggers. From figure 1, we show an example of group interest in the semantic ontology. After computation by the feature vector, the system can analyze that Blogger E, F and G have the same interest in semantic ontology. A result here will show the highest computation ranking for each group. The advantages include the fact that blogger A can identify her friend's interest from blogs (including their entries and categories) and blogger A can easily decentralize her interest by referring to the visualization results of the blogger's interest topic. Besides, blogger A can also integrate and make an interlink between her friends based on users' decentralized topics of interest.

# 4     Design and System Architecture (Extracting Personal Story Information From Blog Contents for Integration)

Ideally, the entries in the blog contain some hidden personal story information that may be important either for the blogger or internet users. This information is useful for many fields such as the fields of science and technology, travel, documentary, motivation or entertainment. With the grammatical contents increasing every day and bloggers making frequent updates, the information in the content can be extracted based on user's definition.  This section shows how blog contents are extracted from real blog data.

## 4.1     Crawling through Blog Entry

First of all, our system surfs through the blogger's blogs to extract blog entries.

1.  The system finds the RSS feeds in the blog.
2.  The system registers the categories and friends in the blog. Each category is analyzed to make sure that there is no same category defined by bloggers and also there is no similar category in the blog. This process is explained more in section B.
3.  In the first visualization, the system visualizes the data by categories in the blog. In the second visualization, the system visualizes the data by friends where the blogger can identify the contents from all his or her blogger friends.
4.  Return to process 1.

### Visualization Data by Category
From the first step in section 4.1, the RSS is used to collect the categories in the blog. Then, each category is analyzed to prevent similar categories and redundant categories. The step for this process is explained in [14]. In this paper, we explain how the RSS is used to extract the category in the blog.

1.  Firstly, we make an index file for the blog that is collected through the ping server [15]. Each blog entry is assumed to have a unique user ID.
2.  Secondly, all the collected blog categories will be classified into the match template ontology [15]. In this part, the content taxonomy such as genres of travel, hobbies or movies can be learned. For example, a blogger has a category 'Japan' in his or her blog. From the content the readers will know about the blogger's experience traveling to Japan. In this case, by matching the categories into the template ontology, we can filter some categories and they will become more general. Refer to Fig. 1.

Steps to match the category:

1.  Designer chooses the Travel domain to create the category visualizations.'
2.  Choosing metadata for extracting users' interests
3.  Classifying Travel into classes as instances

To match the same categories, firstly, we used the RSS to extract the contents and find the vector features using TF IDF to measure the category that should be matched. In the figure above, there are two different categories published by two bloggers. Let say Blogger *B* post entries about traveling in Japan and define the category 'Japan'. The visualization user who is Blogger *A* defines his traveling category using a more general keyword, which is 'Travel'. In the matching categories of Travel and Japan, the template ontology is used (refer to Fig. 1 and Fig. 2).

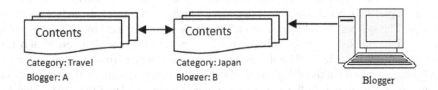

**Fig. 1.** Example of categories in template ontology

After extracting the contents and measuring the feature vector, we can identify that the category Japan is matched with the Category Travel because Travel is defined as the Domain while Japan is defined as one of the properties for the Domain Travel. As a result, Blogger A can automatically find out the relationship between his entry and Blogger B's entry (in the same category). At the same time, Blogger A can get more accurate information after the system classifies the entries into a specific category.

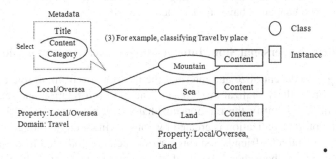

**Fig. 2.** Example of match category problem

## 4.2    Crawling through the Blog Friend

In this paper, we will describe in more detail the visualization blog by a friend and the friend's interest decentralization. The steps are as follows:

1.  The system will analyze the blog to find the friend list using the FOAF ontology.
2.  The system extracts the information from the friend list including the entry and category for each friend. RSS is used to extract the entry and category (section A)

3. For the first time, the system will compute the feature vector for each entry because we want to find the user's topic of interest. More explanation is given in section D.
4. Repeat the process to get more updated information using the RSS concept.

**Features Vector for User Interest Topic**

To calculate a feature vector, we use the TF-IDF. There are two ways to determine blogger interest. The first is by examining the highest interest defined by the blogger. For example, Blogger A has four categories, which are Travel, Research, Music and Family. The system should identify which category is mostly used by Blogger A using the TF-IDF and RSS. The second way is by examining blogger's interest by top five new update entries published by the blogger. It is a new way to find the blogger's interest because we find out that Bloggers tend to have their way to express their interests via blogging. For example, in the first week, blogger always updated the entry about her favorite singer John Elton and the description of a song by John Elton. However, after one week, blogger changed her interest where she moved on to talk about another singer. In this case, it is easy to figure out these changes especially in entertainment and traveling. However, in the research area, although we can find out about different interests, blogger still   demonstrated the difficulty to change their interest gradually; for example, blogger was researching in the field of semantic ontology and the week after, she changed into similar research like the semantic social network or ontology social network. In conclusion, the relationship between users' interests still proves to be useful for bloggers in a wide range of fields because nowadays, bloggers are free to determine their interests in the blog and by centralizing user interest, they should be able to know other bloggers who happen to have similar interests to her.

### 4.3    Visualize the Personal Story Information and Relationship in Blog

The final step is to decentralize and interlink the information from all related blogs in visualization. From the visualization, bloggers can gain information and on the other hand, they can also identify the relationship that exists between the entries, friends and blogs. The blog visualization is created by considering a target blogger who plays a central role. The target's friends and categories which are collected from the target's blog are counted as the second level of the network. Continuously the target's categories are considered as target one if the first targets are friends. The complete visualization is decided by n levels where n>1. The blogger profiles are generated. Relational linkages among the bloggers are produced by matching the corresponding personal profiles with the categories. Notice that there may be many edges tying between bloggers or categories. The number of the edges is equal to the number of the matches between their corresponding profiles and categories. Here, we define the blog visualization as a directed loop graph, as can be referred to Fig. 3 (V1 and V2, where we use Blogger A as user). Blog visualization is a directed loop graph with quadruple [16]:

$$G = (C^*; R^*; N; M) \tag{1}$$

Where, C* is a set of nodes representing bloggers and R* is a set of arcs representing the relations between Blogger A: Friend, Category and Interesting entry. Each arc is associated by a numerical value being weight (w) of a relation as represented by the arc.

$$\bullet \quad w_{ij}^{co} = \frac{n_{ij}^{co}}{N_i^{co}} \qquad (2)$$

where $w_{ij}^{co}$ is the weight of the friend relation from the friend $j$ to $i$'s; $n_{ij}^{co}$ is the number of collaborative time of the friend $j$ with $i$'s who is a blogger; $N_i^{co}$ is number of friends of $i$'s.

$$\bullet \quad w_{ij}^{ref} = \frac{n_i^{ref}}{N_i^{ref}}, \qquad (3)$$

where $w_{ij}^{ref}$ is the weight of the category relation from the blogger $j$ to $i$'s; $n_i^{ref}$ is the number of category time of the author $j$ in $i$'s blogs; $N_i^{ref}$ is the number of category-blogger of $i$'s.

- The weight of each interesting topic relation is equal to the similar degree between two feature vectors representing the corresponding topics.
- $N$ is an adjacency matrix of $G$, written $N(G)$, is the n-by-n matrix in which n is the number of nodes in $G$, entry is the number of arcs in $G$ with endpoints $(v_i, v_j)/$ $v_i \neq v_j$ .
- $M$ is the incidence matrix of $G$, written $M(G)$, is the n-by-m matrix in which m is the number of edges (relations) in $G$.

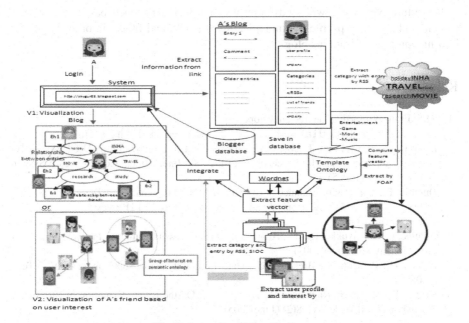

**Fig. 3.** Scenario of Visualization

The visualization naturally represents such a social network. It is useful to deduce an indirect relationship among bloggers. The visualization can be reduced by considering only a specific relation or combining all relations. For example, blogger visualization is generated by taking only blogger and friends' relationships. The network combining all the relations (called combined blogs) is effective in order to find relevant information in blogs. The relationship combines all the relations: Blogger, Blogger's friend, Category and Interesting topic. In particular, we need to estimate the weight of a relation in order to blend all the relations. The weight of the relation can be calculated using Equation 1, where $Er$ is the combined weight of the blogger visualization graph, Cr is the weight of Blogger's friend relation, $Rr$ is the weight of Category, $Ir$ is the weight of interesting topic. To calibrate the coefficients $\alpha, \beta, \delta$, we can generate them by giving consideration to the consensus of suggestions put forth by experts.

$$Er = \alpha * Cr + \beta * Rr + \delta * Ir \qquad (4)$$

## 5    Conclusion and Future Work

In this paper, we have presented a framework for the decentralized personal story information and user interest from blogs. By implementing the RSS we can get the updated information automatically whereas implementing FOAF and SIOC ontology technology enables us to identify the relationship that exists, whether in the user profile or blog's content. Through this implementation, it can improve the information retrieval based on user's interest.

For future works, we will study how user interest is matched based on the blog content and also user profile by implementing the RSS and FOAF in order to get the information needed from the blog.

## References

1. RSS, From Wikipedia, the free encyclopedia,
   http://en.wikipedia.org/wiki/RSS (accessed on March 20, 2014)
2. Ding, L., Zhou, L., Finin, T., Joshi, A.: How the Semantic Web is Being Used: An Analysis of FOAF Documents. In: Proceedings of the 38th International Conference on System Sciences, p. 113 (2005)
3. Dumbill, E.: Tracking provenance of rdf data. IBM's XML Watch (2003),
   http://www-106.ibm.com/developerworks/xml/library/
   x-rdfprov.html
4. Adamic, L.A., Buyukkokten, O., Adar, E.: A social network caught in the web. Journal of First Monday 8(6) (June 2003)
5. Bojars, U., Passant, A., Breslin, J.G.: Data Portability with SIOC and FOAF. XTech (2008)
6. Kumar, R., Novak, J., Raghavan, P., Tomkins, A.: On the Bursty Evolution of Blogspace. Journal of World Wide Web 8(2) (June 2005)

7. Kang, D., Kyunglag, K., Jongsoo, S., Bok-Gyu, J.: Content Recommendation Method Using FOAF and SNA. In: Huang, Y.-M., et al. (eds.) Advanced Technologies, Embedded and Multimedia for Human-centric Computing. LNEE, vol. 260, pp. 93–104. Springer Science+Business Media, Dordrecht (2014)
8. Gruhl, D., Guha, R., Liben-Nowell, D., Tomkins, A.: Information Diffusion Through Blogspace. In: The 13th International World Wide Web Conference (2004)
9. Allan, J.: Topic Detection and Tracking. Kluwer Academic Publishers (2002), http://download.springer.com
10. Jong-Soo, S., In-Jeong, C.: Dynamic FOAF management method for social networks in the social web environment. Journal Supercomputer 66, 633–648 (2013)
11. Jong-Soo, S., Un-Bong, B., In-Jeong, C.: Contents Recommendation Method Using Social Network Analysis. Wireless Personal Communications 73(4), 1529–1546 (2013)
12. Allan, J., Lavrenko, V., Jin, H.: First Story Detection In TDT Is Hard. In: Proceedings of the 9th International Conference on Information and Knowledge Management (CIKM 2000), pp. 374–381 (2000)
13. Nakajima, S., Tatemura, J., Hino, Y., Hana, Y., Tanaka, K.: Discovering Important Bloggers based on Analyzing Blog Threads. In: The 14th International World Wide Web Conference (2005)
14. Zulkefli, N.A.M., Ha, I., Jo, G.-S.: Visualization framework of information map in blog using ontology. In: Nguyen, N.T., Katarzyniak, R.P., Janiak, A. (eds.) New Challenges in Computational Collective Intelligence. SCI, vol. 244, pp. 107–118. Springer, Heidelberg (2009)
15. Nakatsuji, M., Yoshida, M., Hirano, M.: Expanding User Interests by Recommending Innovative Blog Entries. NTT Technical, 2007 Review 5 (8) (August 2007)
16. UCINET: Social Network Analysis Software (accessed on March 2014), http://analytictech.com/

# Sustainable Social Shopping System

Claris Yee Seung Chung*, Roman Proskuryakov, and David Sundaram

Department of Information Systems and Operations Management,
University of Auckland, Auckland, New Zealand
{yee.chung,d.sundaram}@auckland.ac.nz, rpro3000@gmail.com

**Abstract.** Shopping is one of the key activities that humans undertake that has an overwhelming influence on the economic, environmental, and health facets of their life and ultimately their sustainability. More recently social media has been used to connect vendors and consumers together to discover, share, recommend and transact goods and services. However there is a paucity of academic literature on sustainable social shopping as well as systems in industry to support the same. There is no single online shopping system that provides a holistic shopping experience to customers that allows them to balance financial, health, and environmental dimensions. To address this lacuna we propose Sustainable Social Shopping Systems as a means by which we can practically support individuals to become more sustainable and ultimately transform their lives. In this paper we propose and implement concepts, models, processes and a framework that are fundamental for the design of such systems.

**Keywords:** Individual Sustainability, Finance, Health, Environment, Decision making, Habit formations, Online Shopping.

## 1    Introduction

Living a well-balanced and sustainable life is a long-cherished desire of people. In order to transform our lives to be well balanced and sustainable, three issues should be considered and addressed; Firstly we should understand true relationships among life facets; secondly habits for sustainable transformation should be motivated and formed; lastly individuals should be able to keep their sustainable lives even there are life change events. One of the key activities that humans undertake that have an overwhelming influence on their sustainability as an individual as well as a family is shopping. Shopping is closely interconnected with multi-dimensional individual life values such as financial, health, philosophical and environmental values, and is often carried out by individual's habitual behaviors [1, 2]. Nowadays e-Commerce has become one of the normal ways to shop and social shopping features are essential for e-Commerce businesses to be successful [3, 4]. Also personal data, such as financial and health data, are readily available for a personalized service due to the development of wearable and mobile smart technologies. Despite the clear opportunities in

---

* Corresponding author.

D. Hwang et al. (Eds.): ICCCI 2014, LNAI 8733, pp. 114–124, 2014.
© Springer International Publishing Switzerland 2014

the overlapping area of these three concepts, that is supporting multi-dimensional individual sustainability through social shopping features in online shopping and integration of personal data from various sources, no attempts have been made to combine them together in supporting individual sustainability. To address these problems and issues, we propose Sustainable Social Shopping (SSSS) as a pathway to individual sustainability, by synthesizing concepts, models, processes, and frameworks from sustainability, shopping, social shopping, decision-making, and habit formation.

This paper firstly reviews the literature and studies related to sustainability, shopping and online social shopping. Secondly, we propose concepts, and models that could become the foundation for the design of Sustainable Social Shopping Systems (SSSS). Thirdly we propose a framework for the design of SSSS. Finally we describe a prototypical implementation of an SSSS in terms of the process and system views.

## 2     Sustainability, Shopping and Online Social Shopping

"Sustainability" became a challenging pressure to businesses over decades. Nawroth states that from 1997 to 2012, 24% to 46% of British consumers considered organizations' social and environmental contributions and responsibilities in addition to economic factors when they were making purchasing decisions [5]. As these indicate, "sustainability" became not only organizations' tasks to consider but also individuals' concerns. Individuals are showing sustainable behaviors by taking pro-ecological, frugal, altruistic and equitable actions, and they perceive these actions are closely related to their well-being [6]. Therefore individual sustainability should be understood under the holistic point of view rather than narrowed ecological issue.

For supporting individual sustainability, shopping is an important human activity as it is a common decision-making process that people or households make on a daily basis habitually [7]. Also shopping is the activity that can bring fundamental changes in our life, because it is interconnected with various life facets and often reflects life values [1]. United Nations Environment Program explained that sustainable consumption aims at "doing more and better with less, increasing net welfare gains from economic activities by reducing resource use, degradation and pollution along the whole lifecycle, while increasing quality of life" [8]. As such, initially "Sustainability" concepts have been approached and developed by incorporating ecological and environmental issues at organizational level. However it is still stated to be the most "obdurate challenge" for sustainability [9] due to a lack of information about sustainability. Seyfang argues that "a lack of information about environmental and social implications of consumption decisions" left consumers feeling powerless by "the thought that individual action will not make any difference" [7].

This practical issue can be addressed by e-Commerce, which has become a general shopping tendency along with brick and mortar shops. And social shopping is "a subtype of e-Commerce that uses social media to support and enhance social interaction between customers" [3]. In a shopping behavior context, consumers are spending more time and money if family or friends are shopping together [10]. This is because consumers can share and get opinions on products that they are looking for and enjoy interactions with other people who have similar interests [11]. Therefore social shopping is rapidly gaining attention and popularity from the marketplace. Also with the rise of mobile technologies, various personal data can be easily captured and used for a personalized service. For instance, wearable devices track specific health data by capturing physical activities or sleep patterns, and accounting mobile app can log individual's expenditures easily. In summary, individual sustainability can be understood and supported through e-Commerce social shopping which can provide enough sustainability information and recommendations to consumers based on their personal data captured by various sources.

## 3     Research Problems and Practical Issues

In the recent past the online shopping industry has mushroomed from supplementing offline shopping to a mainstream powerhouse [3]. It has expanded rapidly based on its unique characteristics which can provide richness of information and personalized interactions [12]. Many researchers are attempting to develop effective shopping tools while businesses are spending enormous effort in improving their online services in order to increase their sales. Most of the effort so far from academia and industry has focused on online shopping aid tools that incorporate a recommendation agent (RA) and a comparison matrix (CM). RAs are utilized to help consumers in screening for potential products they could purchase based on their personal profile. CMs are algorithms that help consumers to compare multiple products attributes for taking effective decisions. It has clearly been proven that these two tools do significantly impact on consumers decision making process [12] and have become common features in most online shopping sites. With RA and CA features most grocery-shopping sites provide beneficial financial information such as amount that the customer has saved during the shopping session. For example, mySupermarket.co.uk, an online shopping and price comparison website shows different prices from different grocery shopping sites with insightful information like recent price trend. Shopping related health and environmental information are offered from independent non-commercial apps and websites. Packaged food nutrition information app "FoodSwitch" (http://www.foodswitch.co.nz/) shows calories, salt, sugar and saturated fat information in easy traffic light form. By suggesting better nutritional products within similar product range, this app helps individual and households to make better food choices. However the problem is all these services and information are offered separately in silos in monolithic systems. Integrated and balanced information that interweaves financial, environmental, and health dimensions together is virtually non-existent. Due to this information asymmetry, individuals and households are often powerless in transforming themselves to be sustainable [13]. If consumers get informed health, financial and environmental sustainable information all together from a shopping system, they can then choose a better product with a holistic understanding of the relationship among those dimensions. This enables customers to make quality, efficient, and sustainable purchase decisions [12, 14].

Another closely related issue is social shopping. As previously discussed, social shopping and features are currently popular. Generally speaking social shopping connects shops and consumers together to discover, share, recommend and purchase products. This means social features can interact with the entire shopping procedures and affect customers' decision making process [4]. Sommer et al. observed consumer's behaviors and they found that consumers with their family or friends were spending more time and money, when they could share and get support or solicit opinions about products that they were looking for [10]. Hence social shopping features are useful not only in increasing sales but also in enhancing customer satisfaction.

It quite clear that online shopping systems can offer a significant pathway to individual sustainability if they are able to offer: (a) appropriate multi-dimensional information to customer to choose products for improving their sustainability and (b) support their sustainable transformation through social shopping features. However there is a paucity of academic literature on sustainable social shopping as well as systems in industry to support the same. There is no single online shopping system that provides such a holistic shopping experience to customers. This raises three important research and practical issues in designing and implementing Sustainable Social Shopping Systems (SSSS): integration of heterogeneous data from various devices and systems, supporting sustainable shopping habit formation and personalization of SSSS. To address this lacuna we propose SSSS as a means by which we can practically support individuals to become more sustainable and ultimately transform their lives (Fig. 1). In the following section, we propose concepts and models that are fundamental for the design of SSSS. We then suggest a framework and architecture components of SSSS. Finally we illustrate a process and system view of a prototypical SSSS that we are implementing in the real world context of an online supermarket.

## 4    Design of Sustainable Social Shopping Systems

Langley et al. [15] propounded the Interwoven Decision Making model, which is the concept that most decisions are interrelated to other decisions or situations that a decision maker is surrounded by. Interrelated decisions can be sequential (one decision making followed by another), lateral (unrelated decisions but considered simultaneously) and precursive (unrelated decisions at different times which affected each other). Like the interwoven decision model, a "good" decision should be made by considering multiple dimensions and suitability for these boundaries and constraints [16]. Interwoven Decision Making model and consideration of multiple dimensions for decision-making are useful concepts in sustainable social shopping system, because it should help consumers to make an integrated and balanced choice when they purchase a product.

It is important to understand customers and their behaviors with regard to shopping decision-making process to develop the online shopping system. Traditionally, multiple cognitive steps were broadly adapted to understand consumers' behavior in marketing studies [17]. For example, the five-stage buying decision process model explains that when consumers make a purchase of an item, they start the process from need recognition, followed by information search, evaluation, purchase decision and post-purchase behavior. Need recognition can be stimulated both internally and externally; information search step can be divided into light information search stage (general attention to internal and external stimuli) and active information search stage (actively engaged with information search); evaluation process is involved with consumers rules, restrictions and life values; purchase step happens fairly quickly after the evaluation step; post-purchase behavior also can be divided into satisfaction and post-purchase action phases, which determine whether consumers are satisfied with their purchase or not, and their consequent actions like loyalty or word of mouth actions [18].

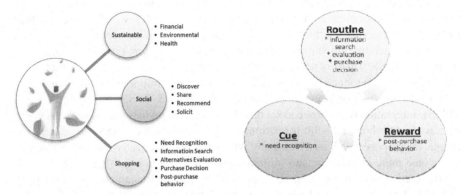

**Fig. 1.** Sustainable Social Shopping          **Fig. 2.** Decision Making and Habit Formation

Transformation process is often achieved by sequential behaviors within an individual's social context [19]: what you eat, and exercise, often determine your health; how you spend money significantly impacts your financial status; what environmental values you have contribute to environmental issues. Duhigg [19] suggests that habit is formed through a three-step loop; cue, routine and reward (Fig. 2). Cue is the intrinsic or extrinsic stimuli that trigger routine. Routine is the habitual activities to achieve reward. Reward is the desirable result of the loop, which makes people to crave for it. Within these steps, we need to change routine to life-affirming transformation. SSSS follows the habit model, but each step is incorporated with the details of the shopping decision processes. Under the cue step, the need recognition stage begins with intrinsic and extrinsic stimuli like craving feeling, time based needs or commercial promotions. In the routine step, information search, evaluation and purchase decision stages are processed. In the reward step, post-purchase behavior stage is carried out and this step will feed the cue step again. Customers will experience these three steps through online shopping, and over time they will form a habit relating to shopping online.

## 5     A Framework for Sustainable Social Shopping Systems

A framework is an abstract form that helps interpret the system concepts for development and visualization of the relationship among components. In order to support an individual's sustainability, SSSS should be developed based on framework which supports sustainable, social, shopping concepts through multi-dimensional personal decision making processes, shopping processes, social engagement modeling, system dynamics, data and information integration modeling. However, there is no current framework that embraces all these concepts, models and processes completely. This research therefore proposes a sustainable social shopping framework (Fig. 3) that integrates concepts, data, models, processes, solvers, and visualizations together in developing a Sustainable Social Shopping System. SSSS constitutes several components: (1) Individual or household profiles and shopping histories, (2) health, financial and environmental information of products, (3) prediction and suggestion models, (4) visualizations, (5) social network data.

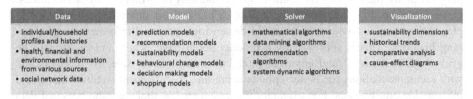

**Fig. 3.** Sustainable Social Shopping System Framework

**Individual or Household Profiles and Shopping Histories**. To initiate shopping activity by notification and generate pre-defined shopping list, the system must allow individuals to store cross-domain information such as their shopping preferences, family demographic, budget information etc.

**Health, Financial and Environmental Information from Various Sources and Social Network Data.** Due to the development of smart, wearable and mobile technologies, these data are easily measured precisely and readily available for personalized services. Some information related to these can be contradicting each other. For example, a purchasing fine health food often negatively affects the financial aspect. However by understanding overall expenses (including medical expenses) or increase of work productivity due to being healthy may influence customers to be win-win situation in both life aspects. Therefore to be able to fully assist customers' choice, the system needs to have the ability to connect, integrate and analyze various databases in order to provide information related to products. Also in order to support individuals' transformation with entertainment, the system should be able to communicate with social network systems and their data.

**Prediction and Suggestion Models.** A variety of models must be provided to users, giving "pre-defined shopping list", and "proper alternatives" must be suggested based on individual's multi-dimensional life information. These models should be able to inter-relate data such as user profiles, shopping histories and multi-dimensional personal information captured by various external systems. The system will provide these suggestions based on combination of models and solvers like Analytical Hierarchy Process (AHP), System Dynamics and prediction and recommendation algorithms.

**Visualizations.** Generating intuitive visualization is essential in this system. Health information will be shown in traffic lights format [20], financial information will be represented through graphical chart and environmental information will be in meter-type visualization.

# 6    Implementation of a Sustainable Social Shopping System

## 6.1    A Process View

SSSS will provide a shopping experience to consumers through several steps. Shopping process can be initiated by various stimuli. Customers may recognize their shopping needs by their desire for something, promotions or prompted by notifications. Once customers log onto the shopping site, a pre-defined shopping list will be displayed based on customer profiles, multi-dimensional personal information and shopping histories. This is followed by a product detail page, where customers will be given health, financial and environmental information about selected products and with recommendations for better options according to customers' multi-dimensional information and profiles. For example, if a customer has diabetes (the personal data from profile page or external wearable smart systems) SSSS will integrate various data and understand their relationships then try to find a product which is the best for the customer. Based on health, financial and environmental dimensions of information, customers can make a quality choice for a sustainable life. When a customer makes a payment, the customer will be informed on the overall health, financial and environmental information via a sustainability dashboard for future references, and it would also serve to give them a sense of accomplishment. For post-purchase step, SSSS will provide customers with a feature to share their purchase with their families and friends through social networking services. Conversely, for SSSS, the shopping data will feed to the customer's shopping history to refine customized suggestions (Fig. 4). On the background of these shopping experiences, SSSS has three supporting features portraying the habit loop.

**Fig. 4.** Sustainable Social Shopping Process

The first feature is a notification. Notification can increase the sense of user control and it motivates and satisfies online buyers [21]. Notification can be based on commercial promotions like email notification of new products and special deals, or time based shopping prompt based on shopping history like "time to buy new shampoo". As notification initiate customers' need recognition, it belongs to the "Cue" step. The second set of features is part of the "Routine" step, prediction and recommendation. Based on customers' profile, data and shopping histories, SSSS will predict customers' needs, then show pre-defined shopping list. These features are crucial in allowing customers to transform themselves. According to Duhigg, in order to change our habits, routine step should be changed [19]. The third set of features is confirmation and social shopping functions that support the last step of shopping experience, namely "Reward". In this step, SSSS needs to provide overall health, financial and environmental shopping result that can be evaluated, to customers; social shopping functions like, sharing and getting support with regard to their purchases should also be provided. Over time, customers can form a life-affirming shopping habit and transform themselves through SSSS.

## 6.2    A System View

Unlike an ordinary online shopping cart system, SSSS will provide at least three life dimensions information (financial, health and environmental aspects) in an integrated manner. The system is a tablet-friendly application, which helps businesses to target not only PC users, but also mobile device users. The system consists of 5 pages: Featured, Search, Stats, Shopping Confirmation, and Profile. A product or a package appearing in the featured page (Fig. 5) is a result of a process run by a sophisticated recommendation engine. Featured page infrastructure is flexible, which means sections (like Top-Charts, Social Choice etc.) are easily altered and updated. Within the Featured page, customers can click a detailed product information page (Fig. 6), and view multi-dimensional sustainability information. Information for each life dimension will be guided by either commonly adapted method or government regulations. For example, information on the health dimension will be shown using traffic lights (green, amber and red) [20]. Search page provides basic search functionality for the user to be able to find products, categories of products or packages. Stats page is responsible for aggregating transactional data, produced by the user, and presenting useful information acquired from this data to the user in three dimensions (Health, Finance, and Environmental Footprint). This page is designed to let the user understand his/her online shopping behavior as well as the level of personal sustainability. All chosen products will be shown in the Shopping Confirmation page (Fig. 7).

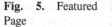

**Fig. 5.** Featured Page    **Fig. 6.** Sustainability of Product    **Fig. 7.** Shopping Cart with Aggregated Health, Finance, and Environmental Scores

Sustainable Social Shopping System will show rich sustainability information which can be personalized for each customers and provide social shopping features. In the shopping confirmation page, health meter will show the number of healthy and unhealthy products purchased in current shopping and overall shopping based on the consumer's particular situation. The financial information dimension will provide information on whether a chosen product is in fact cheaper than other options by understanding a relationship with health and environmental product information. Lastly environmental information will be shown based on ecological footprint standard (http://www.footprintnetwork.org/en/index.php/GFN/page/application_standards/).

Having information of three life dimensions in one page, providing an intuitive graphical presentation while they are making a purchase decision will enable consumers to have a holistic idea on their decisions. The Profile page includes user's basic personal information. The user also is able to log in using existing social networks or wearable device networks, which can provide more useful information about health dimension of the user's sustainability status.

# 7    Conclusion

Sustainability is one of the most often discussed topics in our society. Although no one argues that individuals are the main players in changing society and the environment, individuals have always been treated as just actors and decision makers who transform the organizational, societal, national, and/or global sustainability practices. Shopping can bring fundamental changes in individual lifestyle, and relationships between various life dimensions need to be understood to support individual sustainability. To be sustainable, integrated and balanced information should be offered to consumers. Therefore SSSS can be a very attractive system for both consumers and e-Commerce businesses, as it not only supports individual sustainability but also has the potential of becoming a promising business model. This research proposes concepts, models, and processes that have the potential to be the foundation for sustainable social shopping and for the formation of sustainable and life transforming habits. Furthermore we also propose a framework and architectural components of a sustainable social shopping system to realize the aforementioned concepts, models, and processes. We are also in the process of prototyping a sustainable social shopping system in the context of a purely online supermarket. This system enables customers to understand sustainable purchase choices for their quality of life, and aids them to transform sustainably through shopping experiences. The system has design features based on traditional purchasing decision-making model as well as habit-forming models. Shopping aid features support the conceptual models and system framework proposed. Links with social media and networks enable customers to transform their lives sustainably by measuring various inputs, understanding a relationship among those inputs, and benchmarking outcomes historically, with their friends, and with others in their social groups. At this stage, the system provides information on three life dimensions: health, finance and environment. The data is currently sourced from product suppliers, government regulations and studies from expert organizations.

However in order to support a holistic individual sustainability, the system needs to be flexible to incorporate other aspects of life dimensions and connect to a larger variety of data from outside sources.

# References

1. Young, C.W., Quist, J., Green, K.: Strategies for Sustainable shopping, Cooking and Eating for 2050 – Suitable for Europe. In: International Sustainable Development Research Conference (2000)
2. Gilg, A., Barr, S., Ford, N.: Green Consumption or Sustainable Lifestyles? Identifying the Sustainable Consumer. Futures 37, 481–504 (2005)
3. Kim, H., Suh, K.-S., Lee, U.-K.: Effects of Collaborative Online Shopping on Shopping Experience through Social and Relational Perspectives. Inf. Manag. 50, 169–180 (2013)
4. Olbrich, R., Holsing, C.: Modeling Consumer Purchasing Behavior in Social Shopping Communities with Clickstream Data. Int. J. Electron. Commer. 16, 15–40 (2011)
5. Nawroth, C.: CSR and Sustainability in Times of Crisis: Are Consumers Voting with Their Wallets and Are Companies Putting Their Money Where Their Mouth Is? uwf Umwelt Wirtschafts Forum 21, 75–81 (2013)
6. Tapia-Fonllem, C., Corral-Verdugo, V., Fraijo-Sing, B., Durón-Ramos, M.: Assessing Sustainable Behavior and its Correlates: A Measure of Pro-Ecological, Frugal, Altruistic and Equitable Actions. Sustainability 5, 711–723 (2013)
7. Seyfang, G.: Shopping for Sustainability: Can Sustainable Consumption Promote Ecological Citizenship? Env. Polit. 14, 290–306 (2005)
8. United Nations Environment Program: ABC of SCP: Clarifying Concepts on Sustainable Consumption and Production (2010)
9. Jones, P., Hillier, D., Comfort, D.: Shopping for Tomorrow: Promoting Sustainable Consumption within Food Stores. Br. Food J. 113, 935–948 (2011)
10. Sommer, R., Wynes, M., Brinkley, G.: Social Facilitation Effects in Shopping Behavior. Environ. Behav. 24, 285–297 (1992)
11. Pfeiffer, J., Benbasat, I.: Social Influence in Recommendation Agents: Creating Synergies between Multiple Recommendation Sources for Online Purchase. In: European Conference on Information Systems, Barcelona, Spain (2012)
12. Häubl, G., Trifts, V.: Consumer Decision Making in Online Shopping Environments: The Effects of Interactive Decision Aids. Mark. Sci. 19, 4–21 (1999)
13. Newton, P., Meyer, D.: Exploring the Atitudes-Action Gap in Household Resource Consumption: Does "Environmental Lifestyle" Segmentation Align With Consumer Behaviour? Sustainability 5, 1211–1233 (2013)
14. Wu, J., Rangaswamy, A.: A Fuzzy Set Model of Search and Consideration with an Application to an Online Market. Mark. Sci. 22, 411–434 (2003)
15. Langley, A., Mintzberg, H., Pitcher, P., Posada, E., Saint-Macary, J.: Opening up Decision Making: The View from the Black Stool. Organ. Sci. 6, 260–279 (1995)
16. Marakas, G.M.: Decision Support Systems In the 21st Century. Prentice Hall, New Jersey (2003)
17. Comegys, C., Hannula, M., Va, J.: Longitudinal Comparison of Finnish and US Online Shopping Behaviour Among University Students: The Five-stage Buying. J. Targeting, Mes. Anal. Mark. 14, 336–356 (2006)

18. Amstrong, G., Kotler, P., Harker, M., Brennan, R.: Marketing: An Introduction. Prentice Hall, London (2012)
19. Duhigg, C.: The Power of Habit. Random House Books, London (2012)
20. Department of Health: Guide to Creating a Front of Pack (FoP) Nutrition Label for Pre-packed Products Sold through Retail Outlets (2013), http://www.dh.gsi.gov.uk
21. Wolfinbarger, M., Gilly, M.: Consumer Motivations for Online Shopping. In: Americas Conference on Information Systems, pp. 1362–1366 (2000)

# Grey Social Networks

## A Facebook Case Study

Camelia Delcea[1], Liviu-Adrian Cotfas[1], and Ramona Paun[2]

[1] The Bucharest Academy of Economic Studies, Bucharest, Romania
`camelia.delcea@csie.ase.ro`,
`liviu.cotfas@ase.ro`
[2] Webster University, Bangkok, Thailand
`paunr@webster.ac.th`

**Abstract.** Facebook is one of the largest socializing networks nowadays, gathering among his users a whole array of persons from all over the world, with a diversified background, culture, opinions, age and so on. Here is the meeting point for friends (both real and virtual), acquaintances, colleagues, team-mates, class-mates, co-workers, etc. Also, here is the land where the information is spreading so fast and where you can easily exchange your opinions, feelings, traveling informations, ideas, etc. But what happens when one is reading the news feed or is seeing his Facebook friends' photos? Is he thrilled, excited? Is he feeling that the life is good? Or contrary: he is feeling lonely, isolated? Is he doing a comparison with his friends? These are some of the questions this paper in trying to answer and shaping some of these relationships, the grey system theory will be used.

**Keywords:** grey incidence analysis, social networks, Facebook, correlation analysis.

## 1 Introduction

Facebook is one of the largest socializing networks nowadays, gathering among his users a whole array of persons from all over the world, with a diversified background, culture, opinions, age and so on. Here is the meeting point for friends (both real and virtual), acquaintances, colleagues, team-mates, class-mates, co-workers, etc. Also, here is the land where the information is spreading so fast and where you can easily exchange your opinions, feelings, traveling informations, ideas, etc. That is just one of the reasons why people are becoming more and more attached to this social network.

But, as this network implies first of all the people, the way all this information is transferred from a person to another is very different. Sometimes one can enjoy and be happy for someone else's success, but, in the same time it can be annoyed by another persons' activities, social life, professional life, etc.

This paper tries to see whether the people from a randomly chosen sample are comparing themselves with the ones in their own network by considering the posts their friends are making on Facebook (including here the information posted on news

D. Hwang et al. (Eds.): ICCCI 2014, LNAI 8733, pp. 125–134, 2014.

feed, their photos, etc.) and whether there is an incidence between the social comparison orientation and the appearance of a negative feeling about themselves. Moreover, the connection between the number of Facebook friends and the frequency of using this social network is analyzed. For this, the correlation analysis will be used along with the grey theory incidence analysis.

## 2     Now-a-Days Social Networks Analysis

One of the phenomena encountered in the now-a-days reality in social networks is the spreading speed of any type of information within the social network and can be very good explained through the so-called "go viral" property. [1] Accordingly to this property, not only that the information flow has an enormously highly increased speed, but also, the services broadcasted through the social networks are getting very fast to their end user.

As becoming part of our every-day- life, the social networks have an important impact on our behavior, thoughts, ways of action, state of being, etc. For this reason, the studies on different aspects related to the social networks have increased recently, while these are focusing on different aspects such as:

- Networks' stability [2];
- Social comparison [3];
- Social capital [4];
- Social well-being [5];
- Personality and Facebook use [6, 7];
- Self-presentation in social networks [8], etc.

This study continues the idea of social comparison presence in Facebook relationships presented by Lee in his work [3] and replicate its study in order to see whether this relationship still exists even on another continent on the persons which have similar characteristics with the ones considered, such as age, background, interests, etc.

## 3     Grey Incidence Analysis

Starting from the grey systems general definition as being a mix of information, partly known and partly unknown, it can easily be transferred this grey property to the social networks, especially because in these kind of networks the main component is the human one, greatly characterized by uncertainty in behavior and decisions.

Grey incidence analysis is a central piece of grey system theory and it also can be considered the foundation for grey modeling, decision making and control. [9]

Over time, a permanent interest manifested on this method led researchers from different parts of the world to study and extended it, which has conducted to the development of different other types of grey incidence. [10]

The classical degrees of grey incidence are computed as in the following.

## 3.1    The Absolute Degree of Grey Incidence

Considering two sequences of data with non-zero initial values and with the same length, data X0 and Xj, j=1...n , with t = time period and n = variables: [11]

$$X_0 = (x_{1,0}, x_{2,0}, x_{3,0}, x_{4,0}, \ldots, x_{t,0}),$$  (1)

$$X_j = (x_{1,j}, x_{2,j}, x_{3,j}, x_{4,j}, \ldots, x_{t,j}),$$  (2)

The zero-start points' images are:

$$X_j^0 = (x_{1,j} - x_{1,j}, x_{2,j} - x_{1,j}, \ldots, x_{t,j} - x_{1,j}) = (x_{1,j}^0, x_{2,j}^0, \ldots, x_{t,j}^0)$$  (3)

The absolute degree of grey incidence is:

$$\varepsilon_{0j} = \frac{1 + |s_0| + |s_j|}{1 + |s_0| + |s_j| + |s_0 - s_j|}$$  (4)

with $|s_0|$ and $|s_j|$ computed as follows:

$$|s_0| = \left| \sum_{k=2}^{t-1} x_{k,0}^0 + \frac{1}{2} x_{t,0}^0 \right|$$  (5)

$$|s_j| = \left| \sum_{k=2}^{t-1} x_{k,j}^0 + \frac{1}{2} x_{t,j}^0 \right|$$  (6)

## 3.2    The Relative Degree of Grey Incidence

Having two sequences of data with non-zero initial values and with the same length, $X_0$ and $X_j$, j=1...n, with t = time period and n = variables: [11]

$$X_0 = (x_{1,0}, x_{2,0}, x_{3,0}, x_{4,0}, \ldots, x_{t,0}),$$  (7)

$$X_j = (x_{1,j}, x_{2,j}, x_{3,j}, x_{4,j}, \ldots, x_{t,j}),$$  (8)

The initial values images of $X_0$ and $X_j$ are:

$$X_0' = (x_{1,0}', x_{2,0}', \ldots, x_{t,0}') = (\frac{x_{1,0}}{x_{1,0}}, \frac{x_{2,0}}{x_{1,0}}, \ldots, \frac{x_{t,0}}{x_{1,0}})$$  (9)

$$X_j' = (x_{1,j}', x_{2,j}', \ldots, x_{t,j}') = (\frac{x_{1,j}}{x_{1,j}}, \frac{x_{2,j}}{x_{1,j}}, \ldots, \frac{x_{t,j}}{x_{1,j}})$$  (10)

The zero-start points' images calculated based on (9) and (10) for $X_0$ and $X_j$ are:

$$X_0^{0'} = (x'_{1,0} - x'_{1,0}, x'_{2,0} - x'_{1,0}, \dots, x'_{t,0} - x'_{1,0}) = (x_{1,0}^{'0}, x_{2,0}^{'0}, \dots, x_{t,0}^{'0}) \qquad (11)$$

$$X_j^{0'} = (x'_{1,j} - x'_{1,j}, x'_{2,j} - x'_{1,j}, \dots, x'_{t,j} - x'_{1,j}) = (x_{1,j}^{'0}, x_{2,j}^{'0}, \dots, x_{t,j}^{'0}) \qquad (12)$$

The relative degree of grey incidence is computed as:

$$r_{0j} = \frac{1 + \left|s'_0\right| + \left|s'_j\right|}{1 + \left|s'_0\right| + \left|s'_j\right| + \left|s'_0 - s'_j\right|} \qquad (13)$$

with $\left|s'_0\right|$ and $\left|s'_j\right|$:

$$\left|s'_0\right| = \left|\sum_{k=2}^{t-1} x_{k,0}^{'0} + \frac{1}{2} x_{t,0}^{'0}\right| \qquad (14)$$

$$\left|s'_j\right| = \left|\sum_{k=2}^{t-1} x_{k,j}^{'0} + \frac{1}{2} x_{t,j}^{'0}\right| \qquad (15)$$

### 3.3    The Synthetic Degree of Grey Incidence

The synthetic degree of grey incidence is based on both the absolute and the relative degrees of grey incidence: [11]

$$\rho_{0j} = \theta \varepsilon_{0j} + (1 - \theta) r_{0j}, \qquad (16)$$

with $j = 2, \dots, n$, $\theta \in [0,1]$ and $0 < \rho_{0j} \le 1$.

With these, the grey incidence will be applied in the next section to the data gathered through a questionnaire regarding the Facebook activity.

## 4    Case Study

Starting from a recent case study conducted by Lee [3] on how the people are comparing themselves with others on social network sites, with application on the Facebook network, this paper is redoing the same analysis in similar conditions. The purpose of this study is to see whether the results obtained in [3] can be generally valid within any Facebook community which has almost the same characteristics.

For this, in period $1^{st} - 15^{th}$ March 2014, the students of The Bucharest University of Economic Studies, Faculty of Economic Cybernetics, Statistics and Informatics, have voluntary participated on this survey and they were ask to sincerely answer to a series of questions presented below, most of this questions being similar to the ones used in Lee's study.

Even though the number of questions was quite large, in this paper it is only presented the set of questions that are similar with the one used in the mentioned study. The types of questions was mixed: there have been both open and closed questions, multiple choice questions and yes-no questions. A 5-point Likert scale was used to evaluate the answers received, ranged from 1 (strongly disagree) to 5 (strongly agree). Some of these questions were:

- Personal data:
  - Age;
  - Sex;
  - Study year.
- Split question: Do you use Facebook? – for a "no" answer, the questionnaire was over, while for an "yes" answer it continues with the following questions:
- Number of Facebook friends:
  - How many friends do you actually have on Facebook?
  - Approximately, how many among these were/are your college colleagues?
  - How many of your Facebook friends are also friends with you in the real life?
  - With how many among all your Facebook friends you are usually communicating frequently? (at school, on Facebook, in your spare time, etc.)
  - How many of them, do you consider to be your close friends?
  - How often do you communicate with your close friends?
  - How often do you communicate face-to-face with your close friends?
- Social comparison on Facebook (expressed through frequency) and in real life:
  - I often compare myself with my other Facebook friends while I am reading news feed.
  - I often compare myself with my other Facebook friends while I am viewing their photos and visited places.
  - I usually make comparisons between my dearest ones and the other persons in my group of friends.
  - I observe my behaviour in different situations and I often compare it with others' behaviour in similar situations.
- Self-esteem, uncertainty, anxiety and believes:
  - I think I am worthy person, at least as my friends.
  - I think I have plenty of qualities.
  - I think others are usually appreciating me for my work.
  - I think others are trusting my decisions and ideas.
  - In general, my opinions about myself are different from the others.
  - My opinion about myself is different from one day to another.
  - I usually change the opinion about myself during a day, depending on the encountered events.
  - Unpredictable events are irritating me.
  - I feel frustrated when I have lack of information.
  - I get nervous quite easily.
  - When dealing with unexpected situations, I become angry and irritated.
  - Last week, things that usually are not irritating me, bothered me.
  - Lately, I felt extremely unhappy, even though the dearest ones tried their best to get me out of this situation.

- Depression and negative feelings:
  - Sometimes, while reading the news feed on Facebook, I think the others have a better life than me.
  - Sometimes, while watching my friends' photos on Facebook, I think the others have a better life than me, that they are more happily and are more enjoying their lives.
  - Sometimes, while reading the postings on Facebook, I think the others are doing so much better than I do.
  - Sometimes, while reading the posting on Facebook, I feel lonely and isolated from the world.
- Facebook use and expectations:
  - Facebook is a part of my daily routine.
  - I feel "out of reality" when I stay away from Facebook for a period of time.
  - Usually, I connect on Facebook:
    o I stay connected all day long;
    o A couple of times a day;
    o Once a day;
    o Once a week;
    o Once at every few weeks.
  - When I post something, I expect that the others will positively respond to it.
  - If none of my friends reacts to my posting, I feel sad.

The number of respondents was 144, with an age distribution range between 19 and 28 years old, 33.33% of them being male and 66.67% of them being females. Their distribution on study year is: 57% first year, 16% second year and 27% third year.

By processing the personal data, the following results have been gathered: the medium age of the sample is 20.7 years old, and only 1 person from 144 is not using Facebook, representing less than 1% of the whole sample.

Moreover, by analysing the answers gathered on the questionnaire, it seems that a person has, in medium, almost 671 friends on Facebook, 118 among them being college colleagues. Also, in medium, the respondents have said that approximately 134 of the Facebook friends are persons they have met in real life. The respondents also pointed that only with 52 of these friends they succeed to communicate frequently (at school, on Facebook, etc). Even more, the respondents, are considering, that in medium, only 12 of these friends are close friends, their range being between 1 and 150.

As for the communication with close friends using Facebook, the respondents have said that, in medium, they are communicating quite often with these ones: 3.75 points on a 5 point Likert scale, while the face-to-face communication with the close friends is a more frequent communication way, reaching, in medium, 4.19 points from a 5-point Likert scale.

A number of ten items was constructed based on the questionnaire, eight of these being structured as in Lee's study, while the other two (FNF and FUI) are new and are replacing the PSC (Private Self-Consciousness) and PUSC (Public Self-Consciousness) indicators in the mentioned study. It has been decided to proceed to this modification as these two indicators (namely PSC and PUSC) were not considered so strongly related to the research's purpose and they could easily be integrated in the SCC indicator: SCF - Social Comparison Frequency on Facebook; SCO - Social Comparison Orientation; SE - Self-Esteem; SCC - Self-Concept Clarity; IU - Intolerance of Uncertainty; AXT -Anxiety; DPR - Depression; FNF - Frequency of a Negative Feeling (when seeing others activity on Facebook); FUI - Facebook Use Intensity; EXP - Expectations to others' responses.

The obtained results can be seen in the Table 1 and Table 2 below.

The obtained correlation coefficients are stating that in both case studies, the social comparison frequency on Facebook is positively correlated to the social comparison orientation (0.470 and 0.461) and negatively correlated with the self-esteem level (-0.290 and -0.148).

Moreover, a person's self-uncertainty measured through four indicators, namely SCC, IU, AXT and DPR is also positively correlated with the social comparison frequency in our study, while in Lee's study the SCC indicator seems to be negatively correlated with the SCF indicator.

Also for the expectation to others' responses, the values obtained through both studies are quite similar, both of them being positively correlated to the social comparison frequency.

**Table 1.** Correlation coefficients obtained in Lee's study [3]

|      | SCF    | SCO    | SE     | SCC    | IU     | AXT   | DPR   | PSC   | PUSC  |
|------|--------|--------|--------|--------|--------|-------|-------|-------|-------|
| SCF  | 1.000  |        |        |        |        |       |       |       |       |
| SCO  | 0.470  | 1.000  |        |        |        |       |       |       |       |
| SE   | -0.290 | -0.070 | 1.000  |        |        |       |       |       |       |
| SCC  | -0.540 | -0.360 | 0.560  | 1.000  |        |       |       |       |       |
| IU   | 0.250  | 0.180  | -0.320 | -0.380 | 1.000  |       |       |       |       |
| AXT  | 0.320  | 0.160  | -0.330 | -0.430 | 0.620  | 1.000 |       |       |       |
| DPR  | 0.310  | 0.100  | -0.560 | -0.600 | 0.380  | 0.500 | 1.000 |       |       |
| PSC  | 0.450  | 0.460  | -0.200 | -0.590 | 0.250  | 0.250 | 0.370 | 1.000 |       |
| PUSC | 0.450  | 0.430  | -0.280 | -0.540 | 0.300  | 0.340 | 0.360 | 0.600 | 1.000 |
| EXP  | 0.490  | 0.400  | -0.080 | -0.400 | 0.290  | 0.260 | 0.210 | 0.330 | 0.390 |

**Table 2.** Correlation coefficients

|      | SCF    | SCO    | SE     | SCC    | IU     | AXT   | DPR   | FNF   | FUI   |
|------|--------|--------|--------|--------|--------|-------|-------|-------|-------|
| SCF  | 1.000  |        |        |        |        |       |       |       |       |
| SCO  | 0.461  | 1.000  |        |        |        |       |       |       |       |
| SE   | -0.148 | 0.006  | 1.000  |        |        |       |       |       |       |
| SCC  | 0.473  | 0.394  | -0.142 | 1.000  |        |       |       |       |       |
| IU   | 0.227  | 0.345  | 0.069  | 0.338  | 1.000  |       |       |       |       |
| AXT  | 0.247  | 0.199  | -0.036 | 0.321  | 0.500  | 1.000 |       |       |       |
| DPR  | 0.353  | 0.161  | -0.162 | 0.392  | 0.130  | 0.239 | 1.000 |       |       |
| FNF  | 0.666  | 0.365  | -0.137 | 0.504  | 0.243  | 0.271 | 0.333 | 1.000 |       |
| FUI  | 0.414  | 0.427  | 0.143  | 0.378  | 0.331  | 0.186 | 0.140 | 0.280 | 1.000 |
| EXP  | 0.521  | 0.384  | 0.093  | 0.381  | 0.350  | 0.262 | 0.180 | 0.395 | 0.522 |

As for the PSC and PUSC indicators in Table 1, it can be easily observed that their correlation values with the other indicators are almost the same and they can be gathered, in the future, in a single indicator that can reflect the self-consciousness. In Table 2, these two indicators are mission as they have been incorporated in the SCC indicator, and it can easily be observed that this indicator's value is almost the same with the one obtained in Table 2 (0.473 vs. 0.450).

The two new indicators introduced in our study, FNF and FUI, are positively correlated to the social comparison on frequency on Facebook, which means that a person that compares frequently with others on Facebook is more probably to feel a negative feeling about herself and also that a person that compares frequently with others on Facebook is more likely to use more often the Facebook page.

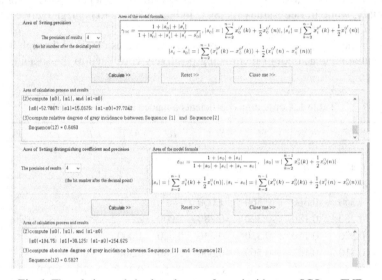

**Fig. 1.** The relative and absolute degree of grey incidence – SCO vs. FNF

As for the SCC concept in comparison with SCO, SE, IU, AXT, DPR the results obtained in the two studies are opposite: in the Table 1 study it has been detected a negative correlation, while in the Table 2 study there is a positive one. The cause of this contradiction can be the way in which the questions were addressed in the questionnaire. For example, in our study, one of the questions regarding self-concept clarity was: "In general, my opinions about myself are different from the others" and the answers were 1 for strongly disagree and 5 for strongly agree. The person who answers 1 is, in fact, declaring that his opinion about herself is the same as the others, which means that he has a good self-concept clarity. So, as the grades are decreasing, the self-concept clarity is better, and the correlation between the grades obtained for this question are negative correlated with the ones obtained for the SCO, SE, IU, AXT or DPR.

Having all these observations, it can be said, that, in general, the social comparison is correlated with the other considered variables.

For better shaping the incidence of the social comparison on someone's every-day life, a grey incidence analysis has been proposed between SCO and FNF.

The results are presented below in Fig. 1.

Therefore, the synthetic degree of grey incidence among SCO and FNF is 0.61435 which denotes a highly incidence of social comparison on Facebook on the frequency of a negative feeling appearance on each person's state.

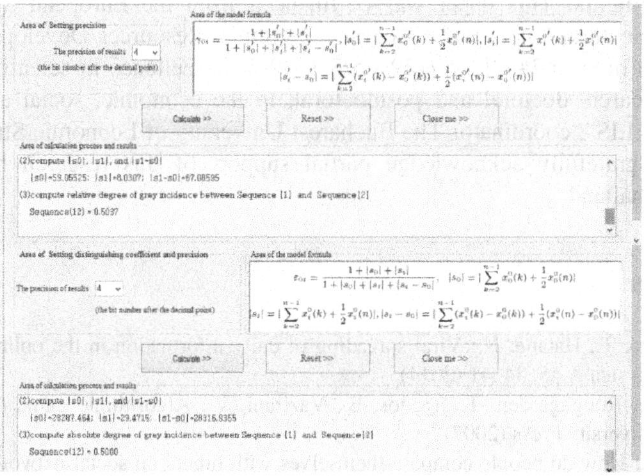

**Fig. 2.** The relative and absolute degree of grey incidence – no. of friends vs. FUI

Even more, by applying an incidence analysis on the number of Facebook friends and the Facebook use intensity, it can be found that the number of friends on Facebook is influencing the usage of this social network by its users as the value determined for the synthetic degree of grey incidence is 0.5018. The values for the absolute and relative degrees of grey are in Fig. 2.

## 5    Concluding Remarks

Based on the study conducted on 144 students, it can be seen that there is a positive correlation between the social comparison orientation and the other analyzed factors: social comparison orientation, self-concept clarity, intolerance of uncertainty, anxiety, depression, frequency of a negative feeling, Facebook use intensity and expectations to others' responses. Even more, the respondents to this study have said, that, in medium, they are comparing more with others on Facebook when they are visualizing their photos than when they are reading their fiends post. As a future study, it can be tried to see whether the group of friends to whom a person is comparing is compounded by all his Facebook friends or only a certain part of them (colleagues, close friends, co-workers, persons that may have the same background, etc).

Also, the respondents have declared, in medium, that Facebook is an integrating part of their daily routine (4.38 of 5 points) and that from a certain point they are feeling out of reality when they are not connected (3.76 of 5 points). In medium, a person is visiting Facebook a couple times a day. Therefore, it has been analysed the incidence on the number of Facebook friends and the Facebook use intensity, and it has been discovered a positively non-neglected value.

Note: the authors' contribution to this paper is equal.

**Acknowledgments.** This paper was co-financed from the European Social Fund, through the Sectoral Operational Programme Human Resources Development 2007-2013, project number POSDRU/159/1.5/S/138907 "Excellence in scientific interdisciplinary research, doctoral and postdoctoral, in the economic, social and medical fields -EXCELIS", coordinator The Bucharest University of Economic Studies. Also, the authors gratefully acknowledge partial support of this research by Webster University Thailand.

# References

1. Kawamoto, T., Hatano, N.: Viral spreading of daily information in the online social networks. Physica A 45, 34–41 (2014)
2. Nisan, N., Roughgarden, T., Tardos, E., Vazirani, V.: Algorithmic game theory. Cambridge University Press (2007)
3. Lee, S.Y.: How do people compare themselves with others on social network sites?: The case of Facebook. Computers in Human Behavior (32), 253–260 (2014)
4. Steinfield, C., Ellison, N.B., Lampe, C.: Social capital, self-esteem and use of online social network sites: A longitudinal analysis. Journal of Applied Developmental Psychology 6(29), 434–445 (2008)
5. Burke, M., Marlow, C., Lento, T.: Social network activity and social well-being. In: Proceeding of the 28th Int. Conf. on Human Factors in Computing Systems, Atlanta, Georgia (2010)
6. Correa, T., Hinsley, A.W., De Zuniga, H.G.: Who interacts on the web?: The intersection of users' personality and social media use. Computers in Human Behavior 26(2), 247–253 (2010)
7. Bachrach, Y., Kosinski, M., Grapel, T., Kohli, P., Stillwell, D.: Personality and patents of Facebook usage. In: Preeedings of the 3rd Conference ACM Web Science Conference (2012)
8. Mehdizadeh, S.: Self-presentation 2.0.: Narcissism and self-esteem on Facebook. Cyberpsychology, Behavior and Social Networking 13(4), 357–364 (2010)
9. Xie, N.-M., Liu, S.-F.: The Parallel and Uniform Properties of Several Relational Models. Systems Engineering 25, 98–103 (2007)
10. Delcea, C.: Not Black. Not even White. Definitively Grey Economic Systems. The Journal of Grey System 26(1), 11–25 (2014)
11. Liu, S.F., Lin, Y.: Grey Systems Theory and Applications, Understanding Complex Systems. Springer, Heidelberg (2010)

# Event Detection from Social Data Stream Based on Time-Frequency Analysis

Duc T. Nguyen, Dosam Hwang, and Jason J. Jung[*]

Department of Computer Engineering,
Yeungnam University,
Gyeongsan, 712-749, Korea
{duc.nguyentrung,dosamhwang,j2jung}@gmail.com

**Abstract.** Social data have been emerged as a special big data resource
of rich information, which is raw materials for diverse research to analyse
a complex relationship network of users and huge amount of daily exchanged data packages on Social Network Services (SNS). The popularity
of current SNS in human life opens a good challenge to discover meaningful knowledge from senseless data patterns. It is an important task
in academic and business fields to understand user's behaviour, hobbies
and viewpoints, but difficult research issue especially on a large volume
of data. In this paper, we propose a method to extract real-world events
from Social Data Stream using an approach in time-frequency domain to
take advantage of digital processing methods. Consequently, this work is
expected to significantly reduce the complexity of the social data and to
improve the performance of event detection on big data resource.

**Keywords:** Social Network Analysis, Event Detection, Big data, Data
Transformation.

## 1 Introduction

Event detection issue, which is close relevant to Topic Trend Detection and
Tracking (TDT) [3], is an interesting research topic recently but it is a hard issue.
Particularly, if the input resource is huge and confused by various topics and also
noise data, where traditional approach to determine events using content-based
analysis reach a limitation in performance and efficiency. So that in this paper
we focus on how to extract real-world events from Social Data of SNS using a
new data representation rather than its original textual data. By representing
input textual data into sequences of numeric values, we convert them to signals
which are materials for digital processing methods in time and frequency domain.
The event detection issue is solved through processes on these pre-constructed
signals. The proposed approach has shown an equivalent efficiency comparing to
methods using content-based analysis.

SNS have been playing an important role in human communication environments, which is a mediate media for sharing information between friends or from

---

[*] Corresponding author.

D. Hwang et al. (Eds.): ICCCI 2014, LNAI 8733, pp. 135–144, 2014.

news agents (News magazines, Manufacturers, Business Companies and so on) to subscribers with variety of purposes. At a certain time, if a real-world event happens, its subject has an opportunity to draw attention from users if its novelty or impression is interesting enough. By using smart personal devices with wireless connection, these involved persons can immediately share the news on SNS to their subscribers. From the original source, the news is diffused quickly on Internet via SNS user's responses such as re-share, comment, like, re-tweet, etc. So that the equivalence indication of an event occurrence in Social Data can be identified by a sudden increment of the number of user's responses toward a certain news. The symptom is a main feature for detecting topics or events from a given corpus. For example, number of tweets and re-tweet action on Twitter is often increased around actual facts such as The visiting of President Barack Obama to Asia 2014, The final football match of Champions League Cup 2014, Earthquakes near Japan and so on. Fig. 1(a) is an illustration of a distribution of messages which contains keyword 'Goal'. The discrete signal is constructed by measuring the meaningful degree of the keyword at certain time ranges continuously. Each strong peak in the signal is possible to be a candidate of an actual event expressed by using the given keyword.

Using the sudden variation feature and similarity of oscillation pattern in signals, we proposed a method to identify real-world events from Social Data in the following sections: Sect. 2 we show several related works of event detection on Social Data and also big data issue; The model of data representation is expressed in Sect. 3, then Sect. 4 shows how to detect events from the given

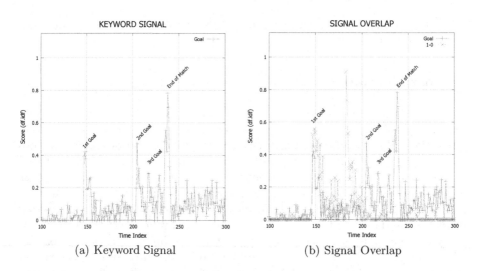

(a) Keyword Signal                    (b) Signal Overlap

**Fig. 1.** Demonstration of keyword signals and the overlap of signals related to the final football match between Chelsea and Liverpool in the FA Cup 2012. (a) Signal of keyword 'Goal', its peaks correspond to the timestamps of the goals and the end of match; (b) Signal overlap of two keywords 'Goal' and '1-0' at the time period of the first goal.

data with signal representation form; Sect. 5 is our experimental result and comparison with other methods, the conclusion is written in the Sect. 6.

## 2    Related Work

In the rest of this paper, we attempt to extract possible real-world events from a dataset containing messages/news of several real-world events which receives attention from SNS users in a finite time period. With a special case study on Twitter, the main materials for analysing is tweets collected from Twitter Data Stream. However, instead of parsing the data using its original textual format, we re-express them in a discrete signal form. Some recent works [1], [9] have been also drawing attention on investigating the issue. They analyse the social data by applying some preprocessing methods which transforms the data to a new dimension space to provide input data for the final analysis tasks.

He, et al. [5] analyse a collected data from SNS by representing each word as a distribution sequence of number of its occurrences by time, then a frequency analysis method is applied for detecting events. They use Discrete Fourier Transformation (DFT) method to generate the basic materials for tracking peaks in the frequency series, power spectrum strength and also periodicity of signals. Based on characteristic of the features, each word is clustered into one of four feature types to identify periodic or aperiodic events. The method is extended in a research of Weng, et al.[8], where authors suggest to use wavelet transformation to keep time information and frequency together as the result of preprocessing tasks. They assumed that several words are used more frequently when an event happens, the number of its occurrences is increased rapidly in a short time period, these words are called 'burst' words. Only signals of burst word are captured, their cross-correlation measurement is used to find subgroup of words which has strong relevant degree, each group is considered as an event.

However two above approaches just attempt to identify events by grouping these words using the similarity of signal patterns, but they do not consider relationship between them in the context of given corpus, so that the result will be distorted if the dataset is mixed from several events. In other word, each word is possible related to multiple events, hence one event can be expressed by a set of meaningful keywords which have totally different signal oscillations. For example in Fig. 1(b), the signal of keyword 'goal' and '1-0' are not similar at the whole time, even they are relevant to the event of the first goal in the final football match between Chelsea and Liverpool in the FA Cup 2012. Two signals are only overlapped at the timestamps when the first goal was scored.

For measuring the efficiency of event detection approaches, Aiello et al. compared several classic event/topic detection methods on three Twitter datasets, which are different in their time scale and topic churn rate [2]. They found that classic topic models such as Latent Dirichlet Allocation (LDA) [4] can well capture the events with narrow topical scope, while methods based on n-gram are much more suitable for broader events. Besides that, an evaluation framework is also released to measure the efficiency of each method based on a list of ground truth topic files. The comparison features are Topic Recall - Percentage of ground truth topic successfully detected by a method; Keyword Precision - Percentage of correctly detected keywords out of the total number of keywords for the matched topic and Keyword Recall - Percentage of correctly detected keywords over the total number of keywords of the ground truth topics that have been matched to some candidate topic. We will use the tool and also the databases from [2] for evaluating our method in the experiments.

## 3    Data Representation

In this section, we discuss how to transform a textual dataset into dataset of discrete signals. Given temporal corpus of news or messages generated by users on SNS, these messages are distributed in a range of time $T = [t1 : t2]$, we count the number of messages with a fixed rate $\Delta T$. $\Delta T$ is the interval gap between two adjacent sampling positions, unit of $\Delta T$ will decide the size of signals in the considered period $T$. It also shows how quickly an event can be detected. The unit can be a number of minutes, hours or days depend on the time scale of events included in the corpus. Smaller $\Delta T$ makes the detected timestamps of an event is more accurately but it requires more memory for a signal and time for processing. Obviously, at the beginning we can use a large value $\Delta T$ for quickly detecting major events, then use other smaller values in a same action on the reduced groups of messages for determining sub events. Value of $\Delta T$ also affects to the total time needed for analysing all the corpus, so that it should be chosen in balance between the sensitivity of event detection result and the performance. The total number of samples in each signal in calculated as $N = \lceil \frac{t2-t1}{\Delta T} \rceil$.

### 3.1    Keyword Signal

Assuming that $KW$ is a set of distinct keywords in a given corpus, we represent an individual keyword signal based on its meaningful degree at each sample $i^{th}$.

**Definition 1 (Keyword Signal).** *Signal of an individual keyword $w$ is expressed as a finite sequence of its meaningful degree*

$$w(i) = d_w(i) \ \forall i \in [0 : N] \tag{1}$$

where $d_w(i)$ is a score of $w$ at the sample $i^{th}$, $d_w(i)$ is defined as an extension of TF.IDF score [6] as following

$$d_w(i) = \frac{DF_w(i)}{M(i)} \times \log \frac{M}{DF_w} \tag{2}$$

where $DF_w(i)$, $M(i)$ respectively are the total number of microposts containing keyword $w$ and the total number of microposts collected at the sample $i^{th}$. $DF_w$ is total number of microposts containing $w$ and $M$ is total number of microposts in the given corpus.

By using this approach, each word in the given corpus has a new representation as a sequence of numeric values along the sampling time-indexes. Particularly, if $N$ is an exponential value of 2, we can apply Fast Fourier Transformation (FFT) method for transforming the signal to a series of complex values. Because the input data is a pure real sequence, so we only need to consider $\frac{N}{2} + 1$ values in the left half of the transformation sequence to take advantage of the symmetry property of FFT. According to the property, $F(x)$ and $F(N-x)$ are the complex conjugates differing only at the sign of their imaginary part [7]. Consequently, the whole transformation series can be fully generated by using the left half. We use this feature to reduce the storage size of the power spectral density of signals as showing in the equation 4.

Besides that, it is an advantage if we design the event detection model using a fixed number $N$, but $\Delta T$ is variable depend on the time period of input data. The system for detecting events can be installed as a chain of integrated processing units as showing in Fig. 2, where output of an unit can be input for another for retrieving a smooth result in term of time.

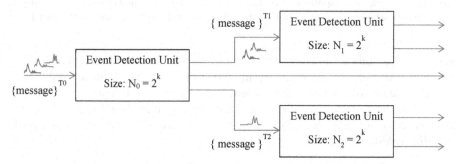

**Fig. 2.** Architecture for event detection system using a chain of predefined processing unit. $\{message\}^T$ is list of messages distributed in the time range $T$.

## 3.2   Signal of a Message

We have two choices for extracting event from signals. The first approach is using peaks in keywords signals, where each peak is an event candidate. However it is not enough, because some time a peak of one event can be blurred by adjacent peaks, e.g. the peak corresponds to the third goal in Fig. 1(a) is an illustration. The second choice is the similarity of signal oscillation patterns, an event can be formed from a group of keywords which shares a similar variation pattern in their signals. The overlap of signals as shown in Fig. 1(b) give us a sign to determine event occurrences at these area. However one keyword can be related to multiple events, hence signal of keywords in one event are quite different, e.g. the signal in Fig. 1(b) obviously is an evidence. So we propose to consider how do users combine these words in a complete sentence to express their opinion. The co-occurrences between keywords in the corpus give us a better image about strong

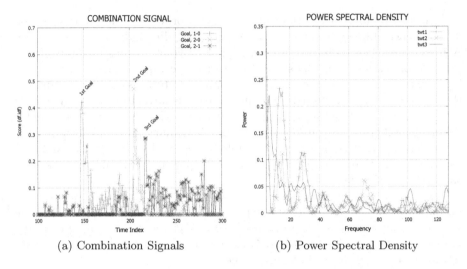

(a) Combination Signals          (b) Power Spectral Density

**Fig. 3.** Combination signal of words and an example of Signal Power Spectral Density. (a) Three signals correspond to events of the goals (b) The power spectral density of three tweets where $twt1, twt2$ refer to the first goal and $twt3$ mention to the third one.

relationships among the keywords. So that, instead of using signal of individual words we attempt to solve the event detection issue based on a combination signal of keywords, where each message in the corpus has its own signal as defined in the Def. 2. The efficiency of this approach is shown in Fig. 3(a), the peak corresponding to the third goal is enhanced while the adjacent noise peaks are eliminated.

**Definition 2 (Signal of a message).** *Signal of a message 'twt' is a combination from signals of each keywords contained in the message's content.*

$$twt(i) = \min_{\forall w_j \in twt} w_j(i) \tag{3}$$

The power spectral density $S_{twt}(x)$ of a message signal is identified via its Fast Fourier Transformation $TWT(x)$.

$$S_{twt}(x) = \frac{|TWT(x)|^2}{N}, \ \forall x \in \left[0 : \frac{N}{2}\right] \tag{4}$$

where $x$ is a component frequency and $N$ is the size of the input signal. This formula is also applied for calculating power spectral density of keyword signals.

Fig. 3(a) is a sample of combination between signal of keyword 'Goal' with signals of keywords '1-0', '2-0' and '2-1' respectively. The result show that the amplitude of the final signal is maintained at the overlapped areas, but it is weakened at other areas. So that if a signal is formed from a keyword list which contains any irrelevant words, its power is attenuated to zero. However in the opposite case, its power still remained at the overlapped areas. So, messages are belonged to one event's message cluster if their signals are oscillated with a similar pattern. It also mean that, the messages' power distribution density in one cluster are quite equivalent. Fig. 3(b) is the power spectral density chart of three messages $tw_1, twt_2, twt_3$ where $twt_1, twt_2$ refer to the first goal between Chelsea and Liverpool in the FACup 2012 final match, $twt_3$ is mentioned to the third goal in that match. Obviously, the oscillation patterns of $tw_1$ and $twt_2$ are quite similar while the message $twt_3$ has a distinguishable oscillation. This characteristic is used to extract list of messages of an event as presenting in the next section.

## 4    Event Detection

In this work we consider an event in the Social Data Stream is a subject which continuously attracts a large number discussion from SNS users within a short time. If we draw the distribution of its messages by time, we can easy find out there are some peaks at certain positions in the chart, each peak is able to correspond to the timestamps of the event candidates.

**Definition 3 (Event in The Social Data Stream).** *Event e in the Social Data Stream is a strong discussed subject on SNS in short time periods.*

$$e = \{bunch_e(t_i, t_j) \; \forall t_i \leq t_j\} \tag{5}$$

*where $bunch_e(t_i, t_j)$ is a list of messages which refers to the major subject of the event e in a time range $[t_i : t_j]$.*

*Property 1 (Event's score). is a degree value for ranking events in a list of event candidates extracted from the given corpus.*

$$score(e) = \frac{1}{|e|} \sum_{\forall bunch_e} \frac{|bunch_e(t_1, t_2)|}{t_2 - t_1 + 1} \tag{6}$$

*Property 2 (Characteristic keywords). $CW(e)$ is a top list of keywords contained in the event's messages, the list is ranked by its meaningful degree using the df.idf method.*

According to the Def. 3, each bunch is a set of messages which contains information related to the major subject at a specified time range. Depending on the number of bunches and the occurrence frequency, the corresponding event can be a periodic or aperiodic kind. It also shows that, the combination signals maintain its power at the areas contained these bunches, hence the power spectral density of signals in the same cluster share a similar distribution pattern. By using the cross-correlation method to compare similarity between two power spectral density sequences, we have a weight measurement to identify which message is a member of a given cluster. However the oscillation pattern in these message signals are not equalled perfectly so that a similarity threshold should be chosen for clustering these message signals. Algorithm 1 is implemented to cluster message signals and generate information of event candidates in the corpus, it uses a membership function named *ccl* which return a value of cross-correlation coefficient between the two sequences.

---

**Algorithm 1:** Event Extraction from Social Data Stream

---

**Data:**
L - is a set of messages;
$K$ - Required minimum number of members;
$\gamma$ - is a similarity threshold
**Result:** List of events ranked by its score

**begin**
    $Events = \emptyset$;
    **for** $\forall m \in L$ **do**
        Finding $Q = \{m_i \mid ccl(m, m_i) \geq \gamma\}$
        Grouping messages in $Q$ based on its timestamps.
        Eliminating groups if its members less than $K$
        Appending a new event into *Events* using the groups as its bunches
    **end**
    Calculating score for the extracted events
    Sorting the list *Events* from largest score to smallest.
**end**

---

## 5    Experimental Result

We evaluated our method on datasets of tweets collected from Twitter, these tweets are often contains noise data with a lot of common hash-tag, special symbols and misspelling word such as: '#FACup', '@Obama', 'Goalllllll', 'lol' and so on. So that the datasets need to be cleaned up before doing any analysis processes. For that goal, we remove all the special symbols and famous slang words from the messages because it is not useful for determining events, then these tweets are converted into lower case form. Signal form is constructed for each individual word with a predefined time interval between two consecutive samples, actually $\Delta T$ is adjusted to a value to make total number of samples $N$ is equal to an exponential value of 2. In our implementation $N = 512$ is an expected value for applying FFT, so that zero value is filled up into extra sample positions at the tail of a signal to make its size to be reached 512 if its size is smaller. The final signal of keywords and messages in corpus are transformed into frequency domain, however only $\frac{N}{2} + 1$ values are stored for each signal based on the symmetric property of FFT sequences. During our implementation process, we recognize that there are a lot of trivial keywords in our dataset which is identified by low power or small value of its *df.idf* score. Even in some cases, keywords have high power or score if they are used in many messages overtime but actually it do not have much meaning, especially in a dataset of focused topic. For filtering all trivial keywords, we use a heuristic function to keep only necessary elements. A keyword $w$ is remained if it is satisfied two following conditions:

1. $df.idf(w)$ is in range from $\frac{(\alpha-1)*min_s+avg_s}{\alpha}$ to $\frac{(\alpha-1)*max_s+avg_s}{\alpha}$
2. Signal power of $w$ is in range from $\frac{(\alpha-1)*min_p+avg_p}{\alpha}$ to $\frac{(\alpha-1)*max_p+avg_p}{\alpha}$

where $min_s$, $max_s$, $avg_s$, $min_p$, $max_p$ and $avg_p$ are aggregated from all keywords' score and their signal power respectively. This trick is also applied to eliminate messages which has power smaller than a threshold determined automatically via all rest message signals with a fixed value $\alpha \geq 5$.

In [2], authors prepared three dataset of tweets focused on there real-world activities. They also released an evaluation framework to measure event detection result from these dataset based on a list of ground truth files. Ground truth files include sets of meaningful keywords, each set is relevant to one actual latent subject/event in the messages' content. Because of limitation in term of time, we just chose to compare our proposed method with three other methods listed in the article [2] and only two dataset 'FACup' and 'SuperTuesday' are used. The similarity threshold $\gamma$ is set to 0.9, and minimum number of messages in a bunch of an event is $K = 3$. We recognize that our approach gives out a good result on both two datasets, number of detected events is quite equivalent with the best method 'BNgram' as suggested in the experimental result of the article [2]. However the keyword precision and recall are a bit better as description in the Table 1.

**Table 1.** Comparison of Topic Detection based on ground truth topics. T-REC, K-PREC, K-REC refers to Topic Recall, Keyword Precision and Keyword Recall respectively.

| Dataset | FACup | | | Super Tuesday | | |
|---|---|---|---|---|---|---|
| Method | T-REC | K-PREC | K-REC | T-REC | K-PREC | K-REC |
| BNgram | **0.769** | 0.355 | **0.587** | **0.500** | 0.628 | 0.647 |
| Our method | **0.769** | **0.453** | 0.533 | 0.455 | **0.652** | **0.714** |
| LDA | 0.692 | 0.230 | 0.511 | 0.182 | 0.325 | 0.52 |
| Doc-P | 0.615 | 0.311 | 0.559 | 0.182 | 0.325 | 0.52 |

# 6   Conclusion

In this work we proposed a method to extract real-world events from Social Data Stream using an approach based on time and frequency analysis. Our event detection model is possible to be implemented as an integrated processing unit for re-usage purpose. The experimental result shows that the method is valuable with a promising result. However the event detection and tracking issue is a hard topic and complex in processing phases, we need a to continue improve the achieved results for a more effective event detection model on Social Data Stream. In the future work we should consider other significant features such as: the structure of SNS network; the speed of information propagation and so on.

**Acknowledgments.** This work was supported under the framework of international cooperation program managed by National Research Foundation of Korea (NRF-2013K2A1 A2055213). Also, this work is supported by BK21+ of National Research Foundation of Korea.

# References

1. Aggarwal, C.C., Subbian, K.: Event detection in social streams. In: Proceedings of the Twelfth SIAM International Conference on Data Mining, pp. 624–635 (2012)
2. Aiello, L., Petkos, G., Martin, C., Corney, D., Papadopoulos, S., Skraba, R., Goker, A., Kompatsiaris, I., Jaimes, A.: Sensing trending topics in twitter. IEEE Transactions on Multimedia 15(6), 1268–1282 (2013)
3. Allan, J.: Introduction to topic detection and tracking. In: Allan, J. (ed.) *Topic Detection and Tracking*. The Information Retrieval Series, vol. 12, pp. 1–16. Springer, US (2002)
4. Blei, D.M., Ng, A.Y., Jordan, M.I.: Latent dirichlet allocation. Journal of Machine Learning Research 3, 993–1022 (2003)
5. He, Q., Chang, K., Lim, E.-P.: Analyzing feature trajectories for event detection. In: Proceedings of the 30th Annual International ACM SIGIR Conference on Research and Development in Information Retrieval, SIGIR 2007, pp. 207–214. ACM (2007)
6. Manning, C.D., Raghavan, P., Schütze, H.: Introduction to Information Retrieval. Cambridge University Press (2008)
7. Proakis, J.G., Manolakis, D.K.: Digital Signal Processing: Principles, Algorithms and Applications, 4th edn. Prentice Hall (2006)
8. Weng, J., Lee, B.-S.: Event detection in twitter. In: Proceedings of the Fifth International Conference on Weblogs and Social Media, ICWSM 2011, Barcelona, Catalonia, Spain. The AAAI Press (2011)
9. Zhou, X., Chen, L.: Event detection over twitter social media streams. The VLDB Journal 1–20 (2013)

# Understanding Online Social Networks' Users – A Twitter Approach

Camelia Delcea[1], Liviu-Adrian Cotfas[1], and Ramona Paun[2]

[1] Bucharest University of Economic Studies, Bucharest, Romania
camelia.delcea@csie.ase.ro,
liviu.cotfas@ase.ro
[2] Webster University, Bangkok, Thailand
paunrm@webster.ac.th

**Abstract.** Twitter messages, also known as tweets, are increasingly used by marketers worldwide to determine consumer sentiments towards brands, products or events. Currently, most existing approaches used for social networks sentiment analysis only extract simple feedbacks in terms of positive and negative perception. In this paper, TweetOntoSense is proposed - a semantic based approach that uses ontologies in order to infer the actual user's emotions. The extracted sentiments are described using a WordNet enriched emotional categories ontology. Thus, feelings such as happiness, affection, surprise, anger, sadness, etc. are put forth. Moreover, compared to existing approaches, TweetOntoSense also takes into consideration the fact that a single tweet message might express several, rather than a single emotion. A case study on Twitter is performed, also showing this approach's practical applicability.

**Keywords:** sentiment analysis, twitter, ontology, text-mining.

## 1 Introduction

While initially designed as a mobile platform on which people could share various status messages, Twitter has gradually become the most commonly used micro-blogging service. Its main feature is that it allows users to broadcast 140 character status messages, also known as tweets. With over 240 million monthly active users, who post more than 500 million tweets every day, as reported in April 2014, Twitter contains news and opinions on virtually everything. Previous studies have shown that micro-blogging posts frequently contain emotional indicators that can be used for sentiment analysis [1].

Sentiment analysis, also known as opinion mining, is a growing area of Natural Language Processing, commonly used to determine whether a text has a positive, negative or neutral meaning. While it has been previously extensively explored for analyzing product and movie reviews [2], forums and blogs [3], it has only recently been applied for extracting opinions out of social media websites.

While many approaches for analyzing Twitter messages have been proposed in the scientific literature, most of them only provide a simple feedbacks, in terms of

D. Hwang et al. (Eds.): ICCCI 2014, LNAI 8733, pp. 145–153, 2014.

positive, negative and neutral perception. The existing literature on Twitter sentiment analysis uses various feature sets and methods, many of them either taken directly or adapted from traditional text classification problems [4]. The approach presented in [5] combines the usage of n-gram analysis with dynamic artificial neural networks in order to analyze twitter brand sentiments. Another approach that focuses on brand sentiments, using an expert-predefined lexicon of seed adjectives is shown in [6]. Besides evaluating whether the perception is positive or negative, many papers have also investigated how the strength of the perception should be evaluated [7].

Taking into account the increased popularity of social media websites, various commercial sentiment analysis tools have also been launched, such as uberVU and AlchemyAPI.

In [8], an interesting approach for using ontologies in twitter sentiment analysis is presented. While still evaluating emotions in terms of negative and positive perception, it takes into consideration the fact that users express opinions about the various characteristics of the analyzed subject and not only about the subject as a whole. For accomplishing this, semantics are used for analyzing the user's sentiments on the different aspects of the studied subject. The authors also present a case study in which they compare user's opinions on several recent smartphones. They conclude that ontologies can be successfully used on studying sentiment analysis on Twitter.

In this paper, it has been considered that while knowing the perception of the user is definitely important, analyzing the categories of emotions contained in twitter messages would bring invaluable information that is not currently taken into account. Such an approach could provide far more information for the marketing teams, which are trying to determine the perception of current and potential customers from the social-media websites. Moreover, compared to many approaches in the field which are analyzing the tweet as a whole, here, it has been considered that a single twitter message can convey multiple feelings [9, 10]. Therefore, in the rest of the paper we present the TweetOntoSense, a sentiment analysis engine that uses an emotion ontology for determining feelings rather than simple positive or negative feedbacks.

The paper is organized as follows. The second section describes the approach used for choosing and enriching the emotion ontology. In the third section of the paper, the steps taken for extracting the user's emotions from raw tweets are presented, while the forth section focuses on the development platform, TweetOntoSense. The last section summarizes the paper and shows some of the future research directions.

## 2    Sentiment Ontology

An ontology represents a predefined vocabulary of concepts. The purpose of such a "vocabulary" is to assure a higher level of knowledge based on semantics. By doing so, the concepts are better defined, while the semantic relations between them are better shaped, this conducting directly to the reduction of ambiguity in a specific domain.

While several sentiment ontologies, currently exist, see [11] for further details, here, it has been chosen the emotional categories Ontology presented in [12], for determining the emotions expressed by the analyzed tweets.

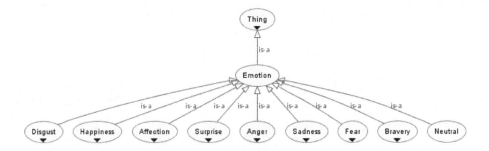

**Fig. 1.** Ontology of emotional categories

As shown in Fig. 1, the selected ontology, written using the Ontology Web Language – OWL, structures the different emotion categories in a taxonomy. It was inspired by recognized psychological models and includes sentiments from several other well-known models. Even though the ontology currently supports only English and Spanish, it can easily be extended with other languages as shown in [13], where the ontology was extended to include concepts in Italian.

The ontology is organized on several levels as shown in Fig. 2. For each class it contains a number of individuals, representing words associated with a particular concept.

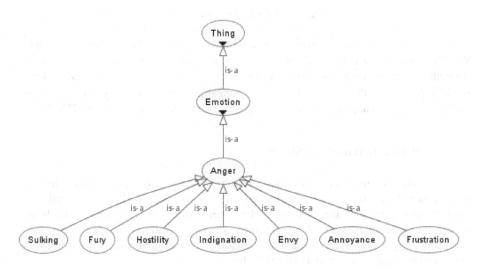

**Fig. 2.** Levels in the ontology of emotion categories

In order to obtain a better coverage of the words used to express emotions, we have chosen to enrich the ontology using some of the values in the corresponding WordNet synsets [14]. Fig. 3. shows the WordNet synset for the word "fear", corresponding to the concept of "Fear" in the emotion categories ontology.

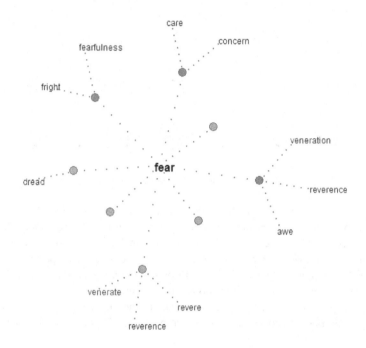

**Fig. 3.** WordNet synset for "fear"

Several extensions of WordNet for sentiment analysis, like SentiWordNet and WordNet-Affect, currently exist and could be used in the future to also add sentiment strength detection to the proposed approach. They provide for each synsets three scores representing the notions of positivity, negativity and objectivity and have been used extensively in approaches that associate a score to the analyzed tweets [13, 14].

## 3    Sentiment Analysis Steps

Sentiment analysis represent a new step in understanding what is happening at micro-level in social networks. As the human component the main part of any social network, focusing on its sentiments and feeling seems to be the right way to do. Even though the individuals' state of being is quite strictly related to a whole array of factors such as: past experiences, culture, education, background, etc., some improvements can be made here for modeling their feeling.

Moreover, [9] argues that the human's characteristics and feelings are the expression of the free-will, imagination, conscience and self-awareness, elements that are inducing even more their appearance's unpredictability.

A lot of discussions can emerge from here: the formation of a feeling, how these feeling are desirable outcomes and how they are related one to another, how the mind can be controlled, guided and directed and also the relation between the mental appearance of a feeling and its impact to a person's behavior. But, all of these will be a further research step.

Meanwhile, the sentiment analysis of these varied feeling should be performed in order to offer the necessary field for new advances.

Among the difficulties there were encountered while performing sentiment analysis, it can be mentioned the huge variety of topics covered, the informality of the language as well as the extensive usage of abbreviations and emoticons. The concise nature of the twitter messages can be considered both an advantage and a drawback. Given the limited length of the messages, users are constrained to be more concise and expressive. On the other hand, the limited number of characters also promotes the use of abbreviations that can make the text harder to correctly analyze.

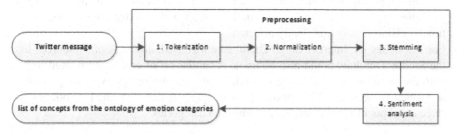

**Fig. 4.** Sentiment analysis steps

For retrieving the Twitter messages, the Twitter Stream API and the Twitter REST API have been used. The steps implied in the sentiment analysis are shown in Fig. 4. and are described in further details in the following subsections.

### 3.1    Preprocessing

In the preprocessing stage tokenization, normalization and stemming are performed. Given the fact that many users write messages using a casual language, the normalization process includes:

- Removing duplicated letters, which frequently occur in twitter messages and emphasize a particular word, in order not to interfere with the stemmer. For example the first tweet in the Sentiment140 corpus, presented in [15] is:

"I looooooooovvvvvveee my Kindle2. Not that the DX is cool, but the 2 is fantastic in its own right."

- All-caps words are converted to lower case. While it can be argued that further information regarding the intensity of a sentiment could be extracted from the use of all-caps [7], in this paper it has been chosen to only focus on extracting the associated emotions. An example of a tweet from the test corpus that uses All-caps is:

*"My Kindle2 came and I LOVE it! :)"*.

- All user tags and URLs are also removed. Hashtags are not taken into consideration in the current version of TweetOntoSense, although they could be included in future versions, as they were shown to contain useful information for sentiment analysis [16].
- Abbreviations are replaced with the corresponding regular words taken from the Internet Lingo Dictionary.

---

AHH YES LOL IMA TELL MY HUBBY TO GO GET ME SUM MCDONALDS

---

- The Google Translate API was used for allowing TweetOntoSense to extract sentiments from other languages than English. While other options such as extending the ontology with new languages and using MultiWordNet hold the promise of providing more accurate results, our initial test have shown that the emotion detection precision is only marginally affected.
- For emoticons the Internet Lingo Dictionary was used, although other sources like the Smiley Ontology could also be used. A similar set of emoticons is used in [9], with the mention that they are only divided into emoticons for expressing positive and negative feelings. The following tables present the associated emoticons for three concepts inside the ontology of emotion categories.

Table 1 includes the emoticons that are mapped to the word happiness during the preprocessing phase.

**Table 1.** Emotions mapped to happiness

| :) | : ) | :-) | :-)) | :-))) | ;) | ;-) | ^_^ | :-D | :D |
|----|-----|-----|------|-------|-----|-----|-----|-----|-----|
| =D | C:  | =)  |      |       |     |     |     |     |     |

Table 2 includes the emoticons that are mapped to the word surprise during the preprocessing phase.

**Table 2.** Emotions mapped to surprise

| :0 |   |   |   |   |   |   |   |   |   |
|----|---|---|---|---|---|---|---|---|---|

Table 3 includes the emoticons that are mapped to the word sadness during the preprocessing phase.

**Table 3.** Emotions mapped to sadness

| :-( | :( | :((  | : ( | D:  | Dx  | 'n' | :\ | /: | ):-/ |
|-----|----|------|-----|-----|-----|-----|----|----|------|
| :'  | ='[ | :_( | /T_T | TOT | ;_; | (:-( |    |    |      |

The last operation of the preprocessing phase consists in applying the Porter stemmer on the resulting sequence of words.

## 3.2    Sentiment Analysis

In this step, the stemmed words are verified in order to determine whether they belong to the enriched Ontology of emotional categories. If a match is found, the concept is added to the list of emotions associated with the analyzed tweet. The user interface of TweetOntoSense is shown in Fig. 5.

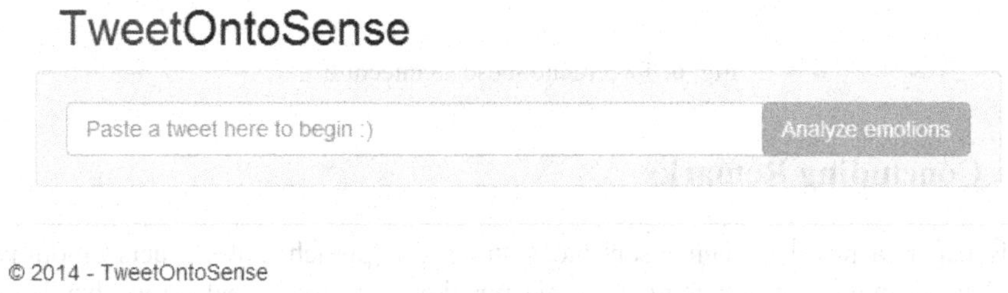

# TweetOntoSense

Paste a tweet here to begin :)                                   Analyze emotions

© 2014 - TweetOntoSense

**Fig. 5.** TweetOntoSense User Interface

The results of the sentiment analysis on a dataset of 100 manually classified tweets, extracted from the corpus presented in [15], have been evaluated. Approximately 78% of the tweets were found to be correctly analyzed.

Three types of detection can be underlined:

- The correct detection category includes the tweets for which all the correct emotions have been detected;
- The partial detection category includes the cases in which not all the correct emotions were detected, but also no incorrect emotion has been detected.
- The incorrect detection category includes the number of tweets for which either no emotion was detected or at least one incorrect emotion has been detected.

Despite the limited size of the evaluation dataset, the proposed approach is expected to provide similar results on larger datasets as well as in real world usage.

# 4    TweetOntoSense

The developed application can be used to constantly analyze all the tweets retrieved from the Twitter Stream API, to filter the stream of tweets by particular words or to query the Twitter REST API.

The architecture of the platform is shown in Fig. 6. The platform was developed using Node.JS which provides both efficient computation and portability across different operating systems and cloud hosting providers.

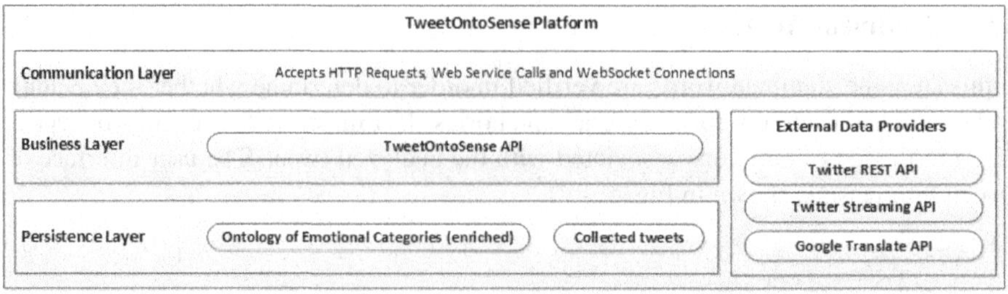

**Fig. 6.** TweetOntoSense architecture

# 5    Concluding Remarks

In this paper, a novel semantic sentiment analysis approach that extracts emotions from Twitter messages, rather than simple positive or negative indicators, has been presented and applied. The starting point in our research was the fact that the human component plays an important role within any social network and, in most of the cases, it usually shows to have unpredictable and sometimes irrational behavior and feelings. This situation is strictly related to the way humans are thinking and acting, combined with a whole array of factors. It has been concluded that in every-day life people tend to have a more complex array of feelings, including: happiness, affection, anger, sadness and their derivatives. For modeling this complex situation, TweetOntoSense proves to be a good step towards a better understanding of the sentiments expressed in the millions of tweets written every day, as it takes into consideration a more wide sphere of elements. By using it, interesting results have been reached and an important step in this direction has been made.

For this, the further research directions include evaluating the strength of the emotions, as well as applying the proposed approach to other social-networking sites such as Facebook and MySpace or to multi-media sharing platforms such as YouTube and Flicker. Another research direction could be the creation of a complex semantic social analysis platform that combines the approach presented in this paper with the one described in [8]. This will provide a more in-depth analysis based on ontologies for both the distinct notions mentioned in tweets and for the expressed sentiments.

Note: the authors' contribution to this paper is equal.

**Acknowledgments.** This paper was co-financed from the European Social Fund, through the Sectorial Operational Programme Human Resources Development 2007-2013, project number POSDRU/159/1.5/S/138907 "Excellence in scientific interdisciplinary research, doctoral and postdoctoral, in the economic, social and medical fields -EXCELIS", coordinator The Bucharest University of Economic Studies. Also, the authors gratefully acknowledge partial support of this research by Webster University Thailand.

# References

1. Pak, A., Paroubek, P.: Twitter as a Corpus for Sentiment Analysis and Opinion Mining. In: Proceedings of the Seventh International Conference on Language Resources and Evaluation, pp. 1320–1326 (2010)
2. He, Y., Zhou, D.: Self-training from labeled features for sentiment analysis. Inf. Process. Manag. 47, 606–616 (2011)
3. Prabowo, R., Thelwall, M.: Sentiment analysis: A combined approach. J. Informetr. 3, 143–157 (2009)
4. Hastings, J., Ceusters, W., Smith, B., Mulligan, K.: Dispositions and processes in the emotion ontology. In: Proceedings of the 2nd International Conference on Biomedical Ontology (2011)
5. Ghiassi, M., Skinner, J., Zimbra, D.: Twitter brand sentiment analysis: A hybrid system using n-gram analysis and dynamic artificial neural network. Expert Syst. Appl. 40, 6266–6282 (2013)
6. Mostafa, M.M.: More than words: Social networks' text mining for consumer brand sentiments. Expert Syst. Appl. 40, 4241–4251 (2013)
7. Thelwall, M., Buckley, K., Paltoglou, G.: Sentiment Strength Detection for the Social Web 63, 163–173 (2012)
8. Kontopoulos, E., Berberidis, C., Dergiades, T., Bassiliades, N.: Ontology-based sentiment analysis of twitter posts. Expert Syst. Appl. 40, 4065–4074 (2013)
9. Delcea, C.: Not even White. Definitively Grey Economic Systems. The Journal of Grey System 26(1), 11–25 (2014)
10. Scarlat, E., Chirită, N., Bradea, I.A.: Grey Knowledge and Intelligent Systems Evolution. In: 11th Conference on Economic Informatics, pp. 47–52 (2012)
11. Kouloumpis, E., Wilson, T., Moore, J.: Twitter sentiment analysis: The good the bad and the omg! In: ICWSM. pp. 538–541 (2011)
12. Francisco, V., Hervás, R., Peinado, F., Gervás, P.: EmoTales: creating a corpus of folk tales with emotional annotations. Language Resources and Evaluation 46(3), 341–381 (2012)
13. Baldoni, M., Baroglio, C., Patti, V., Rena, P.: From tags to emotions: Ontology-driven sentiment analysis in the social semantic web 6, 41–54 (2012)
14. Montejo-Ráez, A., Martínez-Cámara, E., Martín-Valdivia, M.T., Ureña-López, L.A.: Ranked WordNet graph for Sentiment Polarity Classification in Twitter. Comput. Speech Lang. 28, 93–107 (2014)
15. Go, A., Bhayani, R., Huang, L.: Twitter Sentiment Classification using Distant Supervision
16. Petrović, S., Osborne, M., Lavrenko, V.: The Edinburgh Twitter Corpus. In: Proceedings of the NAACL HLT 2010 Workshop on Computational Linguistics in a World of Social Media, pp. 25–26 (2010)

# Intelligent e-Learning/Tutoring – The Flexible Learning Model in LMS Blackboard

Ivana Simonova, Petra Poulova, Pavel Kriz, and Michal Slama

University of Hradec Králové, Rokitanského 62, Hradec Králové, Czech Republic
{ivana.simonova,petra.poulova,pavel.kriz,michal.slama}@uhk.cz

**Abstract.** An insight into a current concept of teaching/learning is introduced in this paper. The article encompasses two main areas which inherently blend together: didactic area representing the theoretical background and the practical level covering the real current educational situation in teaching/learning through online courses. The paper introduces an example of smart solution of e-learning system adjusting to individual learning preferences of each student.

**Keywords:** Cognitive Modeling, eLearning, Intelligent e-learning/tutoring.

## 1 Introduction

Fast technical development and new technologies, globalization of the world, the need of unlimited access to education for everybody – these are some of the reasons which enabled and caused information and communication technologies (ICT) were brought to all spheres of everyday life, including the field of education.

The process of instruction, especially the technology enhanced learning, is highly appreciated by most learners of all age-groups. For adult students, it is often the only way how to study and work, and reach new competences. The younger the learners are, the more easily they accept this approach. Being called ´digital natives´, compared to ´digital immigrants´ [1], today´s students have changed radically. They have not changed their behavior (slang, clothes, body adornments etc.) only, as it happened between generations before, but a really big discontinuity has taken place which could be even called singularity, caused by the arrival and rapid dissemination of digital technology in the last decades of the 20th century. As a result of this ubiquitous environment and the sheer volume of their interaction with it, today´s learners think and process information fundamentally differently from their predecessors. These differences go far further and deeper than most educators suspect or realize. "Different kinds of experiences lead to different brain structures. ... Today´s students' brains have physically changed as a results of the environment they grew up, their thinking patterns have changed", Berry says (in [2]).

## 2 Individual Learning Preferences

This result provides impact not only on a single learner but the whole education systems are affected, curricula reflect this state, particularly in the field of teaching methods where the above mentioned ICT can help substantially.

D. Hwang et al. (Eds.): ICCCI 2014, LNAI 8733, pp. 154–163, 2014.
© Springer International Publishing Switzerland 2014

People vary in the view upon the same situation; they do not do things and see the world in the same way as the others do. They differ in the way of perceiving a situation, evaluating it, judging its consequences, making decisions. In spite of these differences, each person is clever and may be right in his/her own manner. These different strategies, called the cognitive and/or learning styles, are commonly defined as an individual´s characteristic and consistent approach to perceiving, remembering, processing, organizing information and problem solving [3]. Despite some conflicts in the field of learning style stability, reliability and validity of measurements, researching this field is expected to be of great importance for the didactics.

Experience gained in the process of technology enhanced instruction opened discussions on the theory of learning and teaching styles and their application in the technology enhanced process of instruction which provides a wide range of tools to accommodate preferences of all learning style learners. As generally accepted the instructor´s teaching style should match the student´s learning style. Felder and Silverman say mismatching can cause a wide range of further educational problems. It favors certain students and discriminates others, especially if the mismatches are extreme [4]. On the other hand, if the same teaching style is used repeatedly, students become bored. Gregorc claimed that only individuals with very strong preferences for one learning style do not study effectively, the others may be encouraged to develop new learning strategies [5]. Mitchell concluded that making the educational process too specific to one user may restrict the others [6]. Only limited numbers of studies have demonstrated that students learn more effectively if their learning style is accommodated [7]. The question is whether tailoring the process of instruction running within the LMS to student´s individual learning style results in increasing the knowledge.

# 3    Research Project

To discover this was the main objective of the three-year project "A flexible model of the technology enhanced educational process reflecting individual learning styles" was solved at the Faculty of Informatics and Management, University of Hradec Kralove, Czech Republic, in 2010-12. The pedagogical experiment based on the pre-test / instruction / post-test concept was held within the online course intentionally designed for this purpose in three versions

- Reflecting the learner´s style (experimental group 1) where students were offered such study materials, exercises, assignments, ways of communication and other activities which suit their individual learning styles; the selection was made electronically by an e-application which automatically generates the "offer"; this smart solution provides each student with types of materials appropriate to his/her learning style;
- Providing all types of study materials to the learner, the process of selection is the matter of individual decision, the choices are monitored and compared to expected preferences defined by the LCI (experimental group 2);

- Reflecting the teacher´s style (control group) where participants study under traditional conditions, when their course is designed according to the teacher´s style of instruction which they are expected to accept.

The on-line course was designed in the LMS Blackboard. The content focused on library services, which is a topic students have to master before they start studying but they often have hardly any system of knowledge and skills in this field. The e-course was structured into eight parts covering the crucial content, i.e. Basic terminology, Library services, Bibliographic quotations, Electronic sourccs, Bibliographic search services, Writing professional texts, Writing bachelor and diploma theses and Publishing ethics.

The sample group consisted from 530 university students of bachelor Applied Informatics and master Information Management study programs. Three groups were formed by the random choice method. The whole process of instruction (teaching/learning) was tracked by the LMS.

## 4 Application Generating the Course Content: Background

The application (plug-in) supporting the flexible model of instruction within the LMS was designed. Its main objective is to re-organize the introductory page of the e-course where the course content is presented to students. The criterion under which the application works is the student´s individual learning style. Categorization of learner´s preferences (i.e. his/her individual learning style pattern) to a certain type of learning style, which is mentioned below (sequential, precise, technical confluent processors), is not presented in the binary way (yes/no) but it is described by the fuzzy value expressing the relevance rate of each learner to a given group. Single items of the course content, i.e. study materials, exercises, assignments, assessments, communication and other activities applied within the process of instruction, are presented in such order which accommodates student´s preferences, i.e. the plug-in arranges single items of the course content on the introductory page in such order which reflects the student´s individual learning style pattern.

Students´ learning preferences were detected by Johnston´s Learning Combination Inventory (LCI) which consists of 28 statements evaluated on five-level Likert scale and three open-answer questions. The responses are categorized into four groups as follows [8]:

- Sequential Processors, defined as the seekers of clear directions, practiced planners, thoroughly neat workers.
- Precise Processors, identified as the information specialists, info-details researches, answer specialists and report writers.
- Technical Processors, specified as the hands-on builders, independent private thinkers and reality seekers.
- Confluent Processors, described as those who march to a different drummer, creative imaginers and unique presenters.

The LCI differs from other widely used inventories (e.g. by Kolb, Honey and Mumford etc.) in emphasizing not the product of learning, but the process of learning. It focuses on how to unlock and what unlocks the learner's motivation and ability to learn, i.e. on the way how to achieve student´s optimum intellectual development. This was the main reason why the LCI, not any traditional tool was applied for detecting respondents´ individual learning styles.

Johnston [9] designed the concept "Unlocking the will to learn" saying that the traditional learning process is based on belief that all learning occurs as part of learner´s intelligence. The greater the intelligence, the more a child can learn. Johnston attracts attention to the verb can, as no one says will learn. For centuries, the will has been closely aligned with the concept of motivation, being described as the passion, the energy that moves individuals to actions. To work effectively, the will must be supported by the why-question. It can show the learner whether the learning content is relevant, meaningful and applicable to real life. In other words, learners want to discover the wholeness of learning, and it will spark their will to learn. To describe the whole process of learning, Johnston uses the metaphor of a combination lock saying that cognition (processing), conation (performing) and affectation (developing) work as interlocking tumblers; when aligned they unlock an individual´s understanding of his/her learning combination. The will lies in the center of the model, and interaction is the key. She compares human learning behavior to a patterned fabric, where the cognition, conation and affectation are the threads of various colors and quality. It depends on individual weaver (learner) how s/he combines them and what the final pattern is.

## 5    Application Generating the Course Content: Description

To design the online course reflecting learner´s preferences as described above (experimental group 1), not only data on each student´s learning style were required but also single items of the course content and relating activities were classified according to the suitability (appropriateness) for a certain style of learning, i.e. whether the material is preferred, accepted or refused by the student. Finally, single types of study materials and activities were matched to each student´s learning style pattern and the course was tailored to the individual student´s needs. The final phase was carried out by the e-application (plug-in).

As mentioned above, the main objective of the plug-in was to re-organize the entry page of the online course where the course content was presented so that the provided study materials and relating activities were displayed in such order which reflect student´s individual preferences. This objective to be reached (1) learner´s preferences were detected by the LCI and (2) learning objects were classified under the criterion of in/adequacy for single types of processors (technical, sequential, precise, confluent).

## 5.1  Process of Implementation

The plug-in is implemented as the extension of Building Block type for the Black-Board Learn system. The administration rights are required for installing the extension in the system.  The plug-in is distributed in the form of WAR file and script in JavaScript language. The WAR file having been installed by the administrator, the plug-in is available to course designers to be inserted the Course Content page by Add Interactive Tools. The plug-in creates a course item with static HTML code which contains:

— link to jQuery library hosted at ajax.googleapis.com;
— JavaScript code `jQuery.noConflict()` preventing from collision between the Prototype library, internally used by BlackBoard, and the `jQuery` library, used by the plug-in;
— link to `data/script.jsp` file which is part of the `Building Block` (plug-in);
— `HTML DIV` element where the new learning content reflecting learner´s preferences is dynamically generated.

A new item `Table of Contents` is added to the main course menu. This item opens the entry page of the course where the plug-in is inserted, i.e. where study materials and learning activities are structured reflecting learner´s preferences. Then, the original course content (folders with study materials and activities) is available under another item (Course Content) in the main course.

## 5.2  How the Plug-in Works

The plug-in is activated in learner´s browser after accessing the `Table of Contents` page (where the plug-in is inserted). Then, the plug-in runs following activities:

— it downloads the jQuery library from the Internet (the `jQuery` library supports further activities);
— it downloads the JavaScript code generated by the `data/script.jsp` file;
   • this file generates `JSON` data providing information on the classification of files with learning objects, learner´s preferences detected by LCI, evaluation of in/adequacy of learning objects to learner´s preferences and adds the Java-Script code read from the `/uhk-flexible-learning/script.js` file within the given course;
— it activates the JavaScript code which calculates the in/adequacy rate of each learning object for each learner reflecting the LCI results;
— generates the `Table of Contents` and displays it in the place of DIV element, i.e. single learning objects (study materials and related activities) are presented in such order which reflects learner´s   LCI results, i.e. individual preferences.

If in case of error the `Content page` is not generated, the error notice appears.

## 5.3     Requirements for Plug-in Work

Under the `Control Panel - Content Collection`, resp. Files in the latest version, the `uhk-flexible-learning` folder should be created and the script.js and students.csv files uploaded.

Single topics in the `Table of Contents` are structured into folders, one topic per folder, and the link to each learning object (study material, activity) is included.

Each learning object in the folder is described by four figures of the value of -1, 0, 1 which correspond to four types of processors by Johnston´s concept (sequential, precise, technical and confluent) as follows:

— minus one (-1) means this type of study material, activity, assignment, communication etc. is refused, i.e. does not match the given learning style;
— zero (0) is the middle value, i.e. the student neither prefers, nor refuses, but accepts this type;
— one (1) means   this type is preferred, it matches the given learning style.

This three-state model could be extended to a wider scale of fuzzy values reflecting the Johnston´s model in deeper detail. The above mentioned file students.csv contains the classification of students´ individual learning patterns (i.e. fuzzy values reflecting the relevance rate of each learner pattern to a given group of processors – sequential, precise, technical confluent) as displayed in table 1.

**Table 1.** classification of students´ individual learning patterns.

| User name | Classification 1 | Classification 2 | Classification 3 | Classification 4 |
|---|---|---|---|---|
| krizpa1 | 25 | 18 | 14 | 20 |
| webct_demo_69259477001 | 20 | 12 | 18 | 27 |

Data are taken from the spreadsheet (e.g. MS Office Excel) in the CSV format, separated by semicolon, e.g. `krizpa1;25;18;14;20`.

For designer view the designer user name is included in the `students.csv` file. If missing, the error notice is displayed saying the `Table of Contents` does not reflect student´s preferences.

The plug-in requires student access to the Internet so that the `jQuery` library could be downloaded from the ajax.googleapis.com server.

The plug-in was designed for and tested in the BlackBoard Learn system, version 9.1 and uses the Application Programming Interface (API) for detecting the Course Content, metadata for learning object classification, student´s user name and for reading the `script.js` and `students.csv` files. Some functions of the API are not documented (e.g. reading metadata) and changes in BlackBoard version are expected to require modification of Java code using the API.

## 5.4     Implementation Details

The process of plug-in implementation in the course is structured in several steps represented by activities of single files.

**Step 1** is created by `create.jsp`, `create_proc.jsp` and `modify.jsp` files. This step is applied only once, in the moment of plug-in implementation in the course. The `create.jsp` and `create_proc.jsp` files contain codes in Java language which work for creating the above described item `Table of Contents`. Following classes of BlackBoard API are used:

— `Content` – the learning object in the BlackBoard system; the object is always assigned to a given course and it may belong to the folder;
— `FormattedText` – serves for creating the Table of Contents  formatted by HTML code;
— `ContentDbPersister` – serves for saving the learning objects.

The `modify.jsp` file displays information the plug-in generated content (`Table of Contents`) cannot be adjusted but deleted if needed.

**Step 2** is introduced by `script.jsp` file containing the code in Java language generating the necessary code in JavaScript which is subsequently interpreted by user browser. The `script.jsp` file processes the original `Course Content` and information on the user and generates data in the `JSON` format. The `script.js` file is then appended to the generated `JSON` data to be submitted together to the user´s browser.

Compared to the previous version, which was designed for previous generations of the LMS WebCT/Blackboard, the current version is better implemented regarding to API, which was missing in WebCT. Thus the main problem of version 1 was eliminated, i.e. the plug-in dependence on concrete structure of HTML pages, as the Learning Content was detected by parsing HTML pages.

**Step 3** works with `script.js` file which contains the main JavaScript code. The `script.js`  is a common file saved in the course which can be easily adjusted by course designer so that higher flexibility was reached – if small changes in plug-in functionality are required, adjustments in script.js file are made and the plug-in re-installation is not necessary. The plug-in can be also tailored to the course requirements as each course has its own the `script.js` file.

The key part of the `script.js` file is the algorithm calculating the appropriateness of a learning object for the given student which is based on both the learning object and student classification (LCI pattern).  The core of algorithm in JavaScript language is described below:

```
var totalEval = 0;
for (var i = 0; i < topicData.classification.length; i++)
{
    // rejected
    if (userData[i + 1] < refuseValue)
        totalEval += topicData.classification[i] *
                    (userData[i + 1] - refuseValue);
    // appreciated
    if (userData[i + 1] > acceptValue)
        totalEval += topicData.classification[i] *
                    (userData[i + 1] - acceptValue);
}
```

The algoritm in the cycle goes through single values of the given learning object classification in array `topicData.classification` (indexed from 0) and reflecting the `userData` (indexed from 1) it detects for each value whether the student refuses, accepts or prefers material of this type. The threshold values for decision-making process of accepting/refusing the type are saved in constants `refuseValue` and `acceptValue`. The appropriateness value of the type for the learner´s LCI pattern is added to `totalEval` value (being 0 at the beginning). The final appropriateness rate is expressed by the `totalEval` variable. Then, the script.js file ranks learning objects in each folder according to the calculated rate and displays them to the student – the preferred types of learning materials and activities are on the top of the list, underlined, written in bold font of large size and black colour.

The `script.js` file uses the `jQuery` library version 1.4.2 mainly for manipulating with the page content. In `jQuery` library the "$" function cannot be used as it colligates with the same one in the Prototype library used by BlackBoard. That is why the `jQuery` function instead of $ function is used in the `script.js` code.

### 5.5    Currently Known Limits and Future Work

While designing the plug-in, several limits have been discovered for the time being. We have not succeed in hiding the item with original course content (which does not reflect individual learning preferences) in the main menu therefore we have renamed it `Course Content` (as mentioned in chapter Process of implementation) and shifted it on the bottom position so that students did not primarily use it.

Currently the plug-in supports two-level hierarchy of learning objects, i.e. single learning materials and activities are presented in the form of files which are clustered into folders, one folder per topic. The more-level hierarchy requires changes in data structure in `JSON` format (in the `Table of Contents`) generated by the `data/script.jsp` file and in algorithms creating the new learning content reflecting learner´s preferences (`Table of Contents`).

# 6     Conclusion

Current orientation of university education, which is changing under the influence of latest technology development and new key competences, can be researched from various, different points of view. The technology enhanced learning has been spreading because of growing popularity of digital technologies in general. Another reason is it enables easier and more complex realization of the process of instruction, offers the choice of place, time and pace for studying, allows an individual approach to students preferring a certain learning style. These are the key values important for the efficiency of the educational process. Material and technical requirements having been satisfied, strong attention must be paid to didactic aspects of instruction. To contribute to this process was the main objective of the above described project and this paper.

The project having been finished, the-application is still tested in subjects Database Systems I and II. From the pedagogical experiment focusing on the increase in learners´ knowledge in online courses reflecting learner´s preferences it can be seen there is no definite solution and students´ sensitivity to "facilitating" the process of learning widely differs [8]. Unlikely Prensky [1] and Berry (in [2]), whose results formed the background of our project, our results proved most students of IT study programs (Applied Informatics, Information Management) were flexible in learning to such extent they reached the same results either the process of instruction reflected their learning preferences, or not. Gulbahar and Alper [10] developed the e-learning style scale, collecting feedback from 2,722 student of distance study programmes. Starting with 56 items categorized in eight groups, they finally defined 38 criteria structured in seven groups. As our research did not prove statistically significant differences in learners´ knowledge in the experimental 1, 2 and control groups, we are going to use the Gulbahar and Alper´s scale in our future research of the technology enhanced instruction. Two decades ago Mehrlinger (in [11]) emphasized that a variety of learning styles influenced the teacher-designer´s teaching methods and choice of media in a given course/lesson and predicted that technology of the future would be more integrated, interactive, and intelligent. Integration continued to escalate through the development of advanced multimedia systems and interactivity occurred with increased distance learning and Internet interaction, followed by individualized knowledge addressing the learning styles of each student. That has been imperative for teachers to keep abreast of technological changes to empower their students.

As stated in the EC-TEL 2013 conference there is no doubt that technology enhanced learning has created enormous changes in educational institutions of all levels and at the workplaces. However, these innovations have tended to be unsustainable – they need a high degree of effort to be sustained, i.e. mainly funded. At the same time, the technology (mobile and social information and communication technologies) makes impact on everything and everybody around. And, above all, most of educational institutions have taken these technologies up in a systematic way to include them into their learning strategy, sustain them and develop by reflecting feedback provided by research activities.

**Acknowledgment.** This paper is supported by the SPEV Project N. 2110.

# References

1. Prensky, M.: Digital natives, digital immigrants (2001),
   http://www.marcprensky.com/writing/Prensky%20-
   %20Digital%20Natives%20Digital%20Immigrants%20-%20Part1.pdf
2. Prensky, M.: Sapiens Digital: From Digital Immigrants and Digital Natives to Digital Wisdom Innovate (2009),
   http://www.marcprensky.com/writing/Prensky%20-
   %20Digital%20Natives,%20Digital%20Immigrants%20-%20Part2.pdf
3. Stash, N.: Incorporating cognitive/learning styles in a general-purpose adaptive hypermedia system, http://alexandria.tue.nl/extra2/200710975.pdf

4. Felder, R.M., Silverman, L.K.: Learning/Teaching styles in engineering education. Journal of Engineering Education 78(8), 674–681 (1998)
5. Gregorc, A.F.: Learning/teaching styles: potent forces behind them. Educational Leadership 36, 234–2387 (1979)
6. Mitchell, D.P.: Learning style: a critical analysis of the concept and its assessment. Kogan Page, London (1994)
7. Coffield, F., et al.: Learning styles and pedagogy in post-16 learning. A systematic and critical review. Newcatle University report on learning styles (2004)
8. Šimonová, I., Poulová, P.: Learning style reflection within tertiary e-education. WAMAK, Hradec Kralove (2012)
9. Johnston, C.: A Unlocking the will to learn. Corwin Press, Inc., Thousand Oaks (1996)
10. Gulbahar, Y., Alper, A.: Development of e-learning styles scale for electronic environments. Education and Science 39(171), 421–435 (2014)
11. Rogers, P.L.: Designing instruction for technology-enhanced learning. Idea Group Publishing, London (2002)

# Building Educational and Marketing Models of Diffusion in Knowledge and Opinion Transmission

Marcin Maleszka[1], Ngoc Thanh Nguyen[1],
Arkadiusz Urbanek[2], and Miroslawa Wawrzak-Chodaczek[2]

[1] Institute of Informatics, Wroclaw University of Technology, St. Wyspianskiego 27,
50-370 Wroclaw, Poland
[2] Faculty of Historical and Pedagogical Sciences, University of Wroclaw, St.
Uniwersytecki 1, 50-137 Wroclaw, Poland
{marcin.maleszka,ngoc-thanh.nguyen}@pwr.edu.pl,
urbanek.arkadiusz@wp.pl, mwa@pedagogika.uni.wroc.pl

**Abstract.** Group communication and diffusion of information and opinion are important but unresearched aspect of collective intelligence. In this paper a number of hypotheses are proposed in discussed. Each hypothesis proven would be a considerable step towards creating a complete and coherent model of group communication, that could be used both in computer and human sciences. This paper also discusses some methodology that may be used by researchers to determine the hypotheses.

## 1 Introduction

Communication processes are an important aspect of social groups dynamic. They also have a large impact on collective intelligence in human groups and thus influence the collective intelligence models in computer science. Information diffusion in social groups is an especially interesting area, as it may be used in global education, e-learning, business flow of information and more. Group communication is becoming a forecasting field of knowledge which is especially important in the case of unlimited abilities of information flow. Recognizing theoretical mechanisms of diffusion allows one to foresee the directions of spreading opinions, information about the products and purchasers. It also allows one to foresee some force structures after a political group disintegration.

In this paper we point out and discuss a series of open questions and hypotheses based on them. These question and hypotheses differ from basic ones (with time groups adopt similar language) to very practical ones (better communications reduces physical stress). Some of them were considered in other areas of research, but in this paper we point towards the need to observe the real communications in order to determine the real world characteristics of the process. Much of the paper may be understood as an outline of a large experiment, where computer scientists and sociologists would cooperate to determine the real world aspects of collective intelligence. Gathered data could then be used to create new

D. Hwang et al. (Eds.): ICCCI 2014, LNAI 8733, pp. 164–174, 2014.

models of information diffusion in social groups – we outline possible approaches to creating such models. Finally, the models may be again tested in real world situations, by using them to improve the communication skills of educators and others.

This paper is organized as follows: Section 2 provides a short overview of relevant literature in human and computer science; Section 3 provides a group of open research questions and discusses them shortly; the last Section 4 provides a discussion of possible approaches to solving the problems with computer science tools.

## 2  Related Works

This paper is based in part on theoretical assumption of network communication made in [3,4,12], which also suggest this problem niche.

One of the basic social capital definition has been suggested in [10]. It states that social capital is formed by those social organization features such as networks (arrangements) of individuals or households, as well as related with them norms and values which create outside effects for the whole community [9]. The main advantage resulting from high social capital is reduction of transactional costs e.g. the costs connected with concluding contracts, court proceedings, and other formal actions. It concerns an economic sphere of life. High level of social capital is associated with sound civil state functioning, as well as with creating groups and associations being a fulfillment between the state and family. Lack of social capital causes social dysfunctions (corruption, terrorism etc.) and may lead to an economic shortage or decrease. The crucial element of social capital is trust.

Voluntary cooperation is also dependent on the social capital. Norms of generalized reciprocation and social involvement networks are favorable to social trust and cooperation because they decrease the advantages connected with breaking off. They also reduce uncertainty and provide future cooperation patterns. The trust itself is a newly formed social systems property and in the same degree a personal quality. Individuals are able to trust (its not only naivety) thanks to social norms and interdependence networks covered by their actions. The social capitals components are: communication and participation network, trust, divisible norms and values [11].

Through social phenomena examination, both in a real world and in virtual space, social groups features can by observed, as well as the relations between them. The group should be consider in categories of relations between individuals features and between the individuals (attributes themselves are not enough). Social research requires taking into account the two of possible attitudes- individuals attributes, and the context they exist [1].

Relations in the Internet space may be examined by social networks analysis, using the method set created to test formed in the Net structures, including persons or objects which are related [5,12].

The theory of consensus which some of proposed research is based on was developed based on sociological sciences and is based in Consensus Theory [8].

The basic problem solved with it may be defined as follows: There is a given set X (of alternatives or objects) that is a subset of some universe U. The selection is to determine some subset of X based on some criteria. If the selection is deterministic, then repeating it for the same X should always give the same result. In that case it may be called some selection function. In the consensus theory the assumption that the result is a subset of X is eliminated [8]. Furthermore, consensus for X may even have a different structure than X, which is a given in the selection theory. At the beginning of the consensus theory research, the authors were concerned with simple structures of the universe U, like the linear order, or partial order. With more advanced techniques, more complex structures of U were researched (i.e. n-trees, divisions, coverage, complex relations, etc.). Most commonly a homogeneity of elements in U is assumed.

## 3    Research Hypotheses

Communication as the process has different levels and components that had been already examined within subject literature. Here the basic issue concerned is the information diffusion in the process of collective communication. We assume the information transmission and relation quality between interlocutors has its own dynamics and evolves under the influence of the group. On the way of mutual communication it generates a linguistic, mental and relational agreement level, and meaning of those issues results from the current review of theories giving grounds for the studies direction. In addition, that process of communication has cultural connotations, resulting from superior level of announcement understanding in a broader, collective context in which a cultural cohesion of social realitys picture gives them sense. This area of communication dynamics is where we propose the main research hypotheses on the groups influence. The hypotheses were based on following questions:

- What is the dynamics of information diffusion process within a weak ties group? In what way does the sequence of recurrent linguistic categories within collective communication change?

**Hypothesis 1.** With increasing time of group communication the frequency of repeatable sequences is increasing.

Theoretical bases for such hypothesis come from concept assumptions of logical relation between the agreement and common canon of used linguistic categories. Additionally, theoretical assumptions show that the group agreement results from a common conviction about obligations, activities and sense shared by the group members. Therefore, the hypothesis is a logical consequence of theoretical assumptions, because the length of time of group discussion should enable information diffusion focusing around common areas of meanings that is a process of terminological communication cohesion consultation.

- Does common implication of social order in the course of communication model a collective agreement process?

**Hypothesis 2.** Mental similarity in the picture of social order causes similar people to use common linguistic sequences.

The theoretical basis for this hypothesis is the homophile mechanism and the objective concept of hermeneutics, which assume that in the process of communication people express not only their own intentions, but implicate a certain idea about social reality. Logical consequence of such unintended message occurrence is a new agreement level i.e. superior mental agreement. Announcers in the group find common ground not only thus they have common argumentation, but first of all through interpreting common level of reality representation. Therefore on a theoretical level the term of hidden sense structures has been introduced because they show mental level of agreement. Interlocutors accept argumentation and language style of those ones who in their communication have a similar vision of surroundings. So far such types of qualitative studies havent been carried out in groups, alike the mental agreement as a cohesion factor of collective communication forms an essence of cultural diversity of communication processes, thus it is a factor indication which forms meta-communication within the groups that is verified by the basic project studies.

– What is the dynamics of relations in community where people communicate on the Internet in a certain forum? Knowing the mechanisms of social influence including people connected through the Web acquaintance relation, one might ask a question: Which of the social influence mechanisms is the factor responsible for forming consumer attitudes?

**Hypothesis 3.** It is assumed that even at established level of interaction simplifications it is possible to achieve stable communication systems, pointing out many interesting properties because the net of interpersonal relations is a significant information within social phenomena analysis. The researchers attention focused on the relations linking individuals and not on the actors features allows to record complex phenomena an different analysis levels including microscale of individuals relations with the surroundings and macroscale of complex social system.

Constructed topology of social network may be an important source of information on emergent features of complex social system, often impossible to catch in other form. Through referring the information to marketing studies we can collect data concerning social relation among researched group. The results of manifested attitude are connected with the net of social relations. Feature similarity between the researched people connected through relations may be the result of three mechanisms: homophile, external factor effect and social contagiousness. An important role in consumer attitudes formation may be also played by social capital, therefore examining individuals behaviors in the network and relations taking place among the structures forming themselves in the Internet can be used for example for modeling public opinion or forecasting changes of customers behaviors and requirements in a long-term prospect.

– What is the correlation between frequency of used linguistic categories in collective communication and verbal accent?

**Hypothesis 4.** In the course of collective communication linguistic categories corresponding with the social order implication are going to be more stressed verbally.

This hypothesis is coherent with hypothesis 2 because it assumes that interlocutors not only pass on information or arguments, but also communicate their personal attitudes to them. Therefore, if the above hypothesis assumes that social order implication is the key factor of group understanding and agreement. In that way the accents expressed through a louder saying of those categories should entail it. This is the way the interlocutors being in a direct relation want to pay attention not to argumentation, but to information of mental similarity. On that level certain subgroups secondary accepting the language style, arguments, phrases etc. are formed. Following this issue we will carry out an analysis of sound spectrum recorded in the course of direct communication.

– Does exist a correlation between a group leaders communication activity and psychophysical reactions of his body?

**Hypothesis 5.** Stimulation of muscle activity increases proportionally to collective communication difficulty.

A specific role considered will be a teacher as group leader. The difficulty here is twofold: the quality of passed on information and the persuasion task when the teacher is to persuade the group to his arguments. In these levels the strength of psychophysical relation will be measured with the use of thermal cameras. On the basis of taken pictures there will be made a statistical correlation of three variables: duration of communication process, intensity of graphic stimulation with higher temperature, stimulation area of certain part of muscles. The hypothesis is strengthen by Polish nationwide studies concerning teachers occupational diseases. Its been known for a long time they are voice dysfunction diseases. Nationwide data by Nofer Institute of Occupational Medicine (NIOM) showed that the risk group is enormous. In the USA nearly 3 million people per year are on sick leaves connected with occupational diseases of vocal organs. Among British the amount of people with such problems reaches 5 mln, and in Poland it is nearly 800 thousand people annually. The research on this phenomenon points a strong dependence between occupational diseases of vocal organs and muscle overactivity. If the scope of transmitted content and the way of their organizing is chaotic then the teachers muscle tension is additionally increased, because hes got more difficult communication job to do.

– What are mathematical models of diffusion process in the group?

An important part in the process of knowledge processing in societies is an effective process of transmitting knowledge and beliefs from one entity to the other. The second entity may further transmit the knowledge only if it is fully assimilated by it. On this level this process is different than the simple information transmission. An important goal of the research is thus creating such a mechanism of transmission from one group member to the others that allows maximization of the entities convinced to the new knowledge. These methods are important for example in the process of introducing new products to consumer markets.

Collective intelligence methods are tools to unify, integrate and solve conflicts in distributed environments of autonomic entities. In general, they allow creating the intelligence of the whole group based on the intelligence of its members. In previous research on the knowledge diffusion process in environments of autonomic and connected entities (like social networks) no collective intelligence methods were applied. The use of multi-agent technology for simulation and research into knowledge processing is also infrequent. One of the proposed models is for diffusing and integrating knowledge in different representations (hierarchical, logical, relational) from autonomous objects in a collective. The model for diffusing knowledge should contain the criteria and procedures for this process. For this purpose a member of the collective should be modeled with such elements as: knowledge and belief base; trust distribution; communication possibility. The general criterion for the diffusion process is based on achieving the knowledge integrated as similar as possible to an assumed state.

We also consider the analysis of the dependence between the knowledge state of a collective and the knowledge states of its members. We deal with the problem of determining the knowledge of a collective where the states of knowledge of its members are given. We assume that collective members have their own knowledge bases, which can generate states of knowledge referring to the same real world, but not necessarily in a complete and certain way. The reason for this phenomenon can follow from the restrictions the members have, or from the nondeterministic and complex character of the real world. Thus we assume that the collective members knowledge states reflect the proper state of knowledge (the real knowledge state) to some degree because of their incompleteness and uncertainty.Using methods for knowledge integration we can determine a state of knowledge of the collective. However, this state may reflect the real state only to a certain degree. We set the following problem: How to evaluate the quality of the knowledge of the collective? In other word, what is the distance from the collective knowledge to the real state? We will build a mathematical model for this aim, in which we will investigate the relationships between the distance from the collective knowledge to the real state and the distances from members knowledge states to the real state. The influence of the knowledge inconsistency in the collective on the distance from the collective knowledge to the real state will also be investigated.

As one of possible approaches we propose a framework for belief merging (with integrity constraint) by negotiation in which a set of rational postulates for belief merging result are proposed, a model to construct merging operators is presented, and a representation theorem to characterize the connection between the set of postulates and the operators is stated. We aim to provide the rational and effective approaches for belief merging working on pure knowledge bases, which are including all beliefs or all goals but not both, based on the negotiation mechanisms. This approach can overcome the disadvantages of traditional approaches such as agents do not need to expose all their beliefs, the merging work do not need the arbitrator, and the merging result is a strong consensus from participants. Following this part we introduce another framework for negotiation including axiomatic and strategic models for bargaining and investigate the link between them. In this framework, the knowledge base of each agent is considered to include both beliefs and goals. The agents join in the negotiation process to reach the consensus about goals while referring to their private beliefs. The purpose of this phase is to work out a new approach for negotiation with the presence of beliefs, which is more rational and human-like than the pre-existing approaches, which work on only pure goals.

Finally, we intend to develop a model for ontology alignment in communicating tribes. In this part we will build and analyze an approach to aligning ontologies, which enables flexible representation of knowledge about the real world. They can be treated as a decomposition of the considered domain into objects (concepts or classes) along with defining their inner structures and correspondences between them. The alignment itself is a task of providing a procedure for reliable transformation of one ontology into another, which assures their interoperability. Therefore, the problem of aligning ontologies can be described as follows: For given two ontologies $O1$ and $O2$, one should determine a set containing tuples $< c, c', M_c(c, c') >$, where $c$ is a concept in $O1$ and $c'$ is a concept in $O2$, and real value $M_c(c, c')$ representing the degree to which concept $c$ can be aligned to concept $c'$. Methods for designating such mappings between ontologies become useful when two or more systems need to migrate contents of their knowledge bases (expressed as ontologies) maintaining their independence and separation.

– In what range do probabilistic models of diffusion correspond with the collective mental similarity of group members?

**Hypothesis 6.** The mathematical models of diffusion are subjected to typology determined by the cultural similarity of communication group members.

Mathematical models are the basis for objectification of collective communication processes and form the scope for scientific theory. However, using transformation sequence of triangulation, mathematical models need to be confronted with culturally created similarities of group members. These similarities are on the level of meta-analysis and concern a certain cohesion in scope of social reality vision. On the theoretical basis those assumptions are strengthened in the objective concept of hermeneutics.

Three concepts are going to be used for interpreting the research results in virtual space. First of them is Weak Ties Theory, as the instrument facilitating establishing and keeping up weak ties is the Internet. The second is the theory of intergroup relations [2], and the third one concerns the social capital. All of the theories will be fundamental in formulating aims and instruments in the case of present research. They form complementary explanations of social integration/ segregation processes.

– What is the correlation between an ordered model of collective communication leading and information diffusion effectiveness?

**Hypothesis 7.** The arrangement of information transmission process in the group, with following the structure of ordered diffusive model, increases communication quality.

Theoretical bases of the hypothesis are fully strengthened on the level of intergroup relations theory in [2], because on the level of different groups contact, a specific negotiating of similarities ordering the communication process appears. The hypothesis assumes the diffusive mode which takes into account direct and general factors of group communication favors a systematized communication. Acknowledgment of such thesis would be of key importance to optimize information management systems, and become economically implemented potential for the whole research.

- What is the correlation between the directed model of collective communication and the level of psychophysical stimulation of human body?

**Hypothesis 8.** In the case of person implementing communication tasks, the process of communication basing on diffusive models reduces the level of muscle tension.

Referring to the research on a teachers muscle management it is reasonable to test the ways of reduction of communication difficulties in a group of students, thus directed diffusive model implementation becomes the way of more effective management in the process of information diffusion and it reduces muscle tension. We consider transformation of quantitative and qualitative communication parameters in groups with weak ties into theoretical mathematical models. Acquired data is subjected to quantitative parameters: communication structure, frequency of used linguistic categories, intensity of linguistic category use, verbal accent intensity for expressing category, signs of nonverbal language or the level of psychophysical relations, measured by strength of muscle tension termography. With the use of these quantitative parameters it is possible to create diffusive models that are based on mathematical logical algorithms and statistical correlations. The other group of qualitative data isnt subjected to quantitative parametrization. The assumption for the research on communication in groups with weak ties is a cult (centroid) character of this communication. Current studies show that communication is a product of a certain culture and its superior level of nonverbal message or hidden meanings is being understood in this cultures space, and proposed investigation of its culturally consolidated patterns of agreement is much broader than words or gestures. Another important aspect is a common idea of social reality, strengthened in a certain social groups tradition and culture, which is a meta-level of human mental agreement. Since the researched groups with weak ties are being formed spontaneously thus the analysis of communication process exceeds traditional attribute understanding such as: education, social position, experience etc. An important question appears: as the person passing on information within the group with weak ties is unknown so why is that some people are more convincing? Lack of group member acquaintance excludes the use of described in social psychology attributes of position in such group, also spontaneity of group conversation decreases chances for the use of manipulative techniques. Therefore we should assume the existence of other factors in collective communication which cause that recipients establish agreement with a certain person and her arguments seem to be accepted or reliable.

## 4  Research Methodology

Social networks analysis is generally focused on the structure of occurred relations between social subjects (people, organizations, regions), and realization network is illustrated in form of the graphs. This method enables for network structures assessment, making analyzes of different level relations, identify the structural gaps, estimate particular persons importance in researched social processes, define network parameters (density, coherence). Large part of research required for proving the proposed hypotheses comes from the Human Sciences area and will not be discussed in detail in this paper. We instead focus on the mathematical part.

The need for knowledge integration at the collective level arises i.e. as a consequence of having to use different autonomic sources of knowledge. In such situations the integration allows: eliminating knowledge inconsistency, creating a single complex knowledge base. Integration may be done on two levels: syntactic level and semantic level. On the syntactic level the integration process only operates on a set of some non-interpreted logical formulas. The purpose of this type of integration is selecting such a formula that for the given syntactic criteria is the best representation of the integrated set. On the semantic level, the knowledge structures have some given interpretation (in the considered real world) and their integration may also lead to determining the median interpretation, or to determining a consistent representation of all the integrated knowledge structures.

Useful tools for knowledge integration in distributed environments are consensus selection methods. In general, the consensus theory (like data exploration methods) is mainly focused on data analysis problems for extracting useful information. The difference between the methods lays in the goals. Were the data exploration methods look for patterns in the data and determining cause-effect relationships in it, the consensus theory methods look for a best representative of the set of data or a compromise between different data, best accepted by all the sided of the conflict.

The problems that are solved by the consensus theory may be divided into following groups: problems related to determining the hidden structure of the object; and problems related to solving data inconsistency problems about the same object.

The first group contains problems related with determining the structure of some complex model [6]. The object may be for example a set of elements, and the structure to discover is the similarity function between the elements. The data that are the base for the discovery are usually the product of some experiment, observation or an imprecise model of the structure. The second group contains problems where for the same domain experts (or agent, or autonomic programs) give different models (or vote differently). In such case the task is to find a method to determine a single version of the model (single result of the vote) for further processing.

For problems in the second group, three basic approaches may be used: axiomatic, constructive and optimization. In the axiomatic approach some axioms are used to determine the requirements for the consensus or for consensus selection function. Axiomatic approach is a natural and easy method to solve consensus problems. Usually a set of postulates for consensus selection function is defined and the relationships between those postulates are analyzed. The results of the analysis are used to obtain the most representative selection criteria. Due to the postulates, the most basic properties of selection criterion are available, which has a high practical importance. The approach allows to determine a broader group of different criteria for selecting the consensus, which also makes it easier to use in certain practical applications. In constructive approach, the consensus problems are solved on two levels: microstructure level and macrostructure level of the universe U. The microstructure of the universe U is the structure of its elements. Such structure is for example the linear order in some set or a division or coverage of some set. The macrostructure of the universe U is the structure of the universe itself, which may be i.e. a preference relation or a distance (similarity) function between elements in the universe. As mentioned before, in consensus theory all the elements in universe have the same structure, but the consensus may have a different one. In the researched consensus problems the most common macrostructure of U is some binary relation in U or a distance functioned defined based on the microstructure [7]. In constructive approach the criteria defined in axiomatic approach are also used. The selection is based on the microstructure and macrostructure of the universe U. This method is used to solve the problems of determining a representative of a set of documents in the area of information retrieval and classification. The research in this area is aimed at developing new algorithms for determining the consensus (or the representative). In optimization approach, the consensus function is defined based on some optimization theory rules. A broad survey of consensus methods and some algorithms for consensus selection may be found in [8]. Here it is necessary to use proper methods of consensus selection based on the microstructure of the values of specific attributes, as well as the whole profile.

**Acknowledgement.** This research was co-financed by a Ministry of Higher Education and Science grant.

# References

1. Batorski, D.: Analiza Sieci Spoecznych Pajek, Instytut Socjologii UW (2005)
2. Blau, P.: Structural Contexts of Opportunities. The University of Chicago Press, Chicago (1994)

3. de Nooy, W., Mrvar, A., Batagelj, V.: Exploratory Social Network Analysis with Pajek. Cambridge University Press, Cambridge (2005)
4. Hanneman, R., Riddle, M.: Introduction to Social Network Methods. Online textbook. Department of Sociology, University of California (2005)
5. Knoke, D., Kuklinski, J.: Network analysis. Quantitative Applications in the Social Sciences, Sage University Paper (1982)
6. McMorris, F.R., Mulder, H.M., Powers, R.C.: The median function on median graphs and semilattices. Discrete Applied Mathematics 101, 221–230 (2000)
7. Nguyen, N.T.: Consensus systems for conflict solving in distributed systems. Journal of Information Sciences 147(1-4), 91–122 (2002)
8. Nguyen, N.T.: Metody wyboru consensusu i ich zastosowania w rozwizywaniu konfliktw w systemach rozproszonych. Monograph, Oficyna Wydawnicza PWr (2002)
9. Pogonowska, B.: Kapita spoeczny prba rekonstrukcji kategorii pojciowej. In: Kapita spoeczny aspekty teoretyczne i praktyczne, Januszek H. (ed.), Wyd. Akademii Ekonomicznej w Poznaniu, Pozna (2004)
10. Putnam, R.D.: Bowling Alone: Americas Declining Social Capital (1995)
11. Sierociska, K.: Kapita spoeczny. Definiowanie, pomiary, typy. Economic Studies no. 1 (LXVIII), 70–71 (2011)
12. Wasserman, S., Faust, K.: Social Network Analysis: Methods and Applications. Cambridge University Press (1994)

# Semantic Model of Syllabus and Learning Ontology for Intelligent Learning System

Hyun-Sook Chung[1] and Jung-Min Kim[2]

[1] Dept of Computer Engineering, Chosun University, Gwangju, Korea
hsch@chosun.ac.kr
[2] Dept of Computer Engineering, Daejin University, Pocheon, Korea
jmkim@daejin.ac.kr

**Abstract.** The syllabus is a blueprint of course for teaching and learning because it contains the important meaning of promise between instructor and students in higher education and university. However, the current most of all syllabus management systems provide simple functionalities including creation, modification, and retrieval of the unstructured syllabus. In this paper, our approach consists of a definition of the ontological structure of the syllabus and semantic relationships of syllabuses, classification and integration of the syllabus based on ACM/IEEE computing curriculum, and formalization of learning goals, learning activity, and learning evaluation in syllabus using Bloom's taxonomy for improving the usability of the syllabus. Also, we propose an effective method for enhancing the learning effect of students through the construction of subject ontology, which is used in discussion, visual presentation, and knowledge sharing between instructor and students. We prove the retrieval and classification correctness of our proposed methods according to experiments and performance evaluations.

**Keywords:** curriculum, e-learning, learning path, syllabus, ontology.

## 1 Introduction

The evolution of World Wide Web technology has been influencing on the development of e-Learning technology continuously, for instance, Web 1.0 and 2.0 led e-Learning 1.0 and 2.0 respectively. Recently, the semantic web and ontology engineering technology have been applied in order to conceptualize knowledge of many different domains including education to lead e-Learning 3.0. In e-Learning 3.0, learning will be socially achieved and will shift from what to learn to how to learn[1],[2].

The syllabus is created by an instructor for students in order to introduce the teaching course and provide useful information and learning materials. Most students may use the course syllabus to recognize course purpose, policies, assignments, tests, outcomes, and so on. The course syllabus, however, doesn't contain the essential concepts and relations, which must be learned while a semester. The most syllabus only organizes general information about the course like title, description, instructor, grading policy, textbook, schedule, etc.

D. Hwang et al. (Eds.): ICCCI 2014, LNAI 8733, pp. 175–183, 2014.

In this paper, we propose the semantic model of the course syllabus, which is called syllabus ontology, in order to represent learning concepts and semantic relations between them. In addition, we introduce our learning ontology model and propose an effective method for enhancing the learning effect of students through constructing learner-based ontologies in which knowledge discovered by students is conceptualized and organized. Learner-based ontologies can be merged into teacher-based ontologies which conceptualize teaching contents in classes. Thus, our subject ontology is composed of teacher-based ontologies and learner-based ontologies. Teachers and students share and understand knowledge of learning materials based on learning ontologies.

This paper is structured as follows. Section 2 introduces some related work. Section 3 represents the revised syllabus structure for supporting adaptive learning of students. Section 4 provides an overview of the layered structure of our learning ontologies and describes the hierarchical structure of the subject ontology. Section 5 shows the experimental result and in the end the paper presents our conclusion in Section 6.

## 2     Related Work

The researches applying ontology technology to education field are classified into curriculum or syllabus ontology creation[3],[4], ontology-based learning object organization, and ontology-based learning content retrieval. The studies for the creation of education-related ontology include curriculum ontology creation[5] and personal subject ontology creation[6]. Mizoguchi[7],[8] proposed an ontology-based solution to solve several problems caused by intelligent instructional systems. Other works define the metadata of learning objects and learning path based on ontology engineering technology[9],[10].

These works concentrated on the management of learning objects and materials and performance enhancement of instructional systems. Ontology technology, however, can be used to make the knowledge structure, which improves the interaction among teachers and students and enables spontaneous learning of students, of teaching contents and learning materials for students based on semantic information[11]. Yu *et al*[12] propose a method to construct a syllabus repository storing the structured syllabus. They collect freely available unstructured syllabus from Internet, extract topics and convert to the structured format. In order to do, they define an entity mapping table and hierarchy structure of the syllabus.

## 3     Semantic Modeling of the Syllabus

In this paper, we design the semantic model of the syllabus to represent the semantic relationships of entities. Using this semantic model, the e-learning system can manage a standard integrated format of the syllabus and provide the intelligent services including adaptive learning path creating. As shown in Fig. 1, a class named *Syllabus* is a main class, which have many different semantic relationships for connecting to other component classes. Adaptive learning path generation refers to the organization of learning objects in a proper order so that students can effectively study a subject area. In a learning graph a node denotes a learning object or learning element. However, effective assessment for learning activities of students is required in order to support adaptive learning of students. In other words, a node in a learning graph should be composed of lectures, learning goals, learning activities and assessment. Thus, we have considered a syllabus as a node in a learning graph because it includes course description, learning goals, lectures, activities, and learning materials also rather than a course.

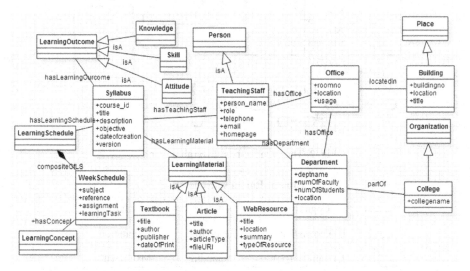

**Fig. 1.** A portion of the syllabus model represented by the UML class diagram

Fig. 2 shows the syllabus-based learning graph in which learning graphs can be generated in two levels, i.e. course-level and concept-level. A *Syllabus* class has a link to a *LearningSchedule* class, which represents learning subjects to be learned weekly for a semester. We define a leaning map for a particular subject and create a relationship with *LearningSchedule* class in order to enable students can identify their learning sequences. In addition, we define systematic models of learning goal, learning activity and assessment should be represented in the syllabus based on Bloom's taxonomy, which classifies behaviors of students to six cognitive levels of complexity. Table 1 shows cognitive, attitude, and skill domains of Bloom's taxonomy.

**Definition 1.** Learning goal can be defined as a set of tuples in which each tuple is consisting of four items, learning goal, cognitive level, attitude level, and skill level.

$$< sentence_g, C(i), A(j), S(k) > \qquad (1)$$

In expression (1), $goal_p$, $C_i$, $A_j$, and $S_k$ denotes $p$-th learning goal, $i$-th cognitive complexity level, $j$-th attitude complexity level, and $k$-the skill complexity level respectively. For example, a teacher defines a learning goal like as <"Understanding class inheritance in JAVA", $C_3$, $A_3$, $S_2$>.

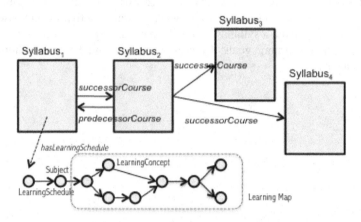

**Fig. 2.** The syllabus-based learning graph

**Table 1.** Bloom's taxonomy

| Levels | Cognitive | Attitude | Skill |
|--------|-----------|----------|-------|
| 1 | Knowledge | Receiving | Imitation |
| 2 | Comprehension | Responding | Manipulation |
| 3 | Application | Valuing | Precision |
| 4 | Analysis | Organizing | Articulation |
| 5 | Synthesis | Characterizing | Naturalization |
| 6 | Evaluation | Cooperating | Representation |

**Definition 2.** A learning activity can be defined as a set of tuples in which each tuple is consisting of four items, learning activity, cognitive level, attitude level, and skill level. The types of learning activity performed by students defined as a set of (R)eading, (E)ssay, (P)resentation, (D)iscussion, pr(A)ctice, e(X)ercise, (H)omework, and (T)eamwork.

$$< LA_p, C(i), A(j), S(k) > \qquad (2)$$

In expression (2), $LA_p$ denotes one of the elements in a learning activity set. One or more learning activities should be mentioned in every week on lecture schedule in a syllabus.

**Definition 3**. Learning assessment can be defined by making connection to one or more learning goals.

$$< QE_p, sentence_g >$$    (3)

In expression (3), $QE_p$ denotes one of activities for learning assessment, such as exercise, assignment, quiz, and exam. The connection between assessment and learning goals enables teachers estimate outcomes of students more precise.

# 4    Learning Ontology Design

As a core component of the intelligent learning system, we design the learning ontology which covers several kinds of knowledge of e-learning environment from curriculum to learning materials. Commonly, a curriculum can be represented as a set of description of courses and syllabuses. Therefore, the curriculum ontology conceptualizes the knowledge of curriculum-related concepts, i.e. *ProgramOfStudy*, *Course*, *KeyConcept*, *AttainmentGoal*, *AttainmentLevel*, and includes the direct semantic connections between courses and their syllabus ontologies. The syllabus ontology conceptualizes the internal and external structures of syllabuses. A *syllabus* class, which is the core concept of syllabus ontology, has 9 data type properties, i.e. *titleOfCourse*, *description*, *gradingPolicy*, *goalOfCourse*, and 12 object type properties, i.e. *oldVersionOf*, *hasInstructor*, *hasMaterial*, *hasSchedule*, *hasLectureRoom*, to describe the content and relationships extracted from traditional textual syllabus templates. Fig. 3 shows the relationships between syllabus ontology and each of other ontologies, top-level ontology, curriculum ontology, and subject ontology. Syllabus ontology has one or more subject ontologies because a conventional syllabus represents multiple concepts to be taught during a school semester. The *LearningConcept* class is a top level concept in the subject ontology. The *LearningConcept* class has responsibilities to collect lower level topics and link to syllabus ontology.

Subject ontology is composed of one or more of teacher-based ontology, several learner-based ontologies and learning materials. Teacher-based ontology contains learning concepts and knowledge structure to be studied in a class. Learner-based ontology contains concepts and knowledge structure created by students. When a teacher presents learning subjects, students investigate the subjects and extract mea-ningful concepts and knowledge structure to create a new learner-based ontology or extend existing learner-based ontology during their learning process. Entities of teacher-based ontology are classified into following 3 categories:

- Learning Concept – Main topics will be described in a class for a semester. This category includes fundamental concepts, advanced concepts, related concepts, examples and exercises.

- Learning Structure – Learning concepts organized as a semantic network to describe knowledge structures of topics. In addition, learning path and schedule represented in syllabus added to the learning structure.
- Learning Material – Teacher collects useful resources like web pages, images, audios, and videos and creates lecture notes using the resources. These lecture notes have connections to relevant concepts.

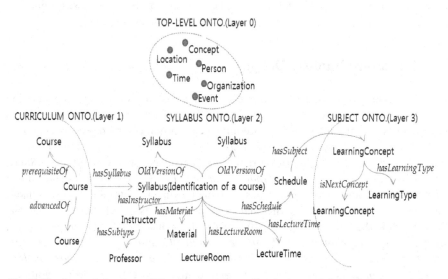

**Fig. 3.** The layered structure of Top-level ontology, Curriculum ontology, Syllabus ontology, and Subject Ontology

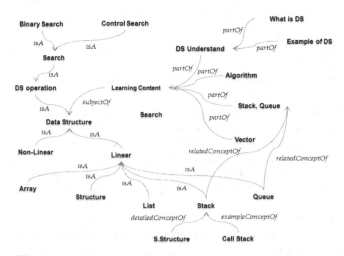

**Fig. 4.** A partial example of the 'Data Structure' subject ontology

Subject ontology is described as a set of tuples, $<C, P, I, R_H, R_C>$. The symbol C, P, I, $R_H$ and $R_C$ represent class, property, instance, the hierarchical relation between classes and association between classes individually. Table 2 represents classes, properties, and relations defined in subject ontology.

**Table 2.** Classes, properties and relations defined in the subject ontology

| Type | Name | Description |
|---|---|---|
| CLASS | LearningConcept | Root class |
| | FundamentalConcept | Conceptualization of fundamental topics of learning subjects |
| | AdvancedConcept | Conceptualization of advanced topics of learning subjects |
| | RelatedConcept | Conceptualization of additional topics of learning subjects |
| | Example | Conceptualization of example topics of learning subjects |
| PROPE RTY | Name | Concept name |
| | AuxiliaryName | Auxiliary name of concept name |
| | Definition | Definition of concept |
| | Description | Description of concept |
| RELATI ON | Fundamental-Concept-Of | A is fundamental class of B |
| | | Reversed relation is Has-Fundamental-Concept |
| | Advanced-Concept-Of | A is advanced class of B |
| | Related-Concept-Of | A is related concept with B |
| | | Reversed relation is Has-Related-Concept |
| | Example-Of | A is example class of B |
| | | Reversed relation is Has-Example |
| | Exercise-Of | A is exercise class of B |
| | | Reversed relation is Has-Exercise |
| | Same-Concept | Both concepts have same semantic |

# 5    Experiments

We applied our method to classes, Understanding Data Structure and Java Programming, to evaluate the effectiveness of learning ontology-based education. We collect and analyze two kinds of experimental data like feedbacks from students and test data, such as midterm exam, final exam, quiz, homework, and so on. Feedbacks of students are acquired by the interview with the students. From the analysis of the feedbacks of the students we know that students understand the fundamental concept of ontologies and the way of applying ontologies to learning. However, creating of subject ontology is somewhat a difficult work, but it is useful to present, discuss, and share of studying subjects of students.

The graph depicted in Fig. 5 shows the values of learning outcomes, which are understanding concepts(LO01), organizing relations(LO02), implementing concepts(LO3), finding the related concepts(LO4), application to computer programs(LO5), adding new related concepts(LO6) and sharing knowledge(LO7), before and after applying learning ontologies to class. We compute the values of learning outcomes of students through evaluating of quiz, exams, homework, and so on.

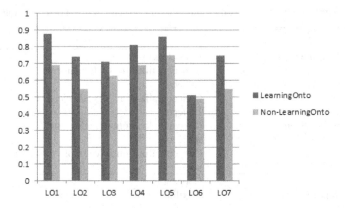

**Fig. 5.** Learning outcomes before and after applying subject ontology to class

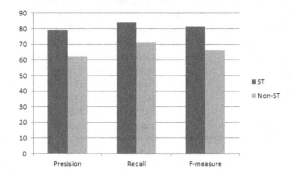

**Fig. 6.** Comparison of retrieval performance between ST and Non-ST

Another experiment evaluates the retrieval performance of elements from syllabuses before and after transformation to our proposed syllabus template in section 3. Syllabus transformation and retrieval have been performed on 45 syllabuses in the computer engineering field collected from the Web. As the result of our retrieval experiments, we know that precision, recall and f-measure averaged for 10 test sets is 0.78, 0.87 and 0.82 respectively. In addition, we know that our syllabus model is well structured and conceptualized than current syllabus formats from the result depicted in Fig. 6.

# 6 Conclusion

The objective of our designated learning ontology model is the provision of the integrated semantic model covering multiple domains from curriculum to learning materials. Our ontology model's main entity is a syllabus because it is used to identify and describe the content of a course in detail. We designed curriculum ontology and subject ontology to be connected into syllabus ontology for supporting adaptive learning and knowledge sharing of students. In adaptive learning path creation, the definition of a knowledge unit or learning unit is important. We consider a syllabus as a knowledge unit because a syllabus includes learning outcome, assessment, learning concepts of a course. Therefore, our structured semantic model of the syllabus has acceptable values in integrating the unstructured available syllabuses and generating the effective learning path in the course-level and subject-level. Our future work will be adaptive learning path generation and recommendations based on the proposed learning ontology.

**Acknowledgement.** This work was supported by the Korea Research Foundation Grant funded by the Korean Government (No. 2010-0006521).

# References

1. Rubens, N., Kaplan, D., Okamoto, T.: E-Learning 3.0: anyone, anywhere, anytime, and AI. In: International Workshop on Social and Personal Computing for Web-Supported Learning Communities (2011)
2. Yu, D., Zhang, W., Chen, X.: New Generation of e-learning Technologies. In: Proceedings of First International Multi-Symposiums on Computer and Computational Sciences, pp. 455–459 (2006)
3. Chi, Y.: Developing curriculum sequencing for managing multiple texts in e-learning system. In: Proceedings of International Conference on Engineering Education (2010)
4. Libbrecht, P.: Cross curriculum search through the Geoskills' Ontology. In: Proceedings of SEAM 2008, pp. 38–50 (2008)
5. Shackelford, R., McGettrick, A., Sloan, R., Topi, H., Davies, G., Kamali, R., Cross, J., Impagliazzo, J., LeBlanc, R., Lunt, B.: Computing Curricula 2005: The Overview Report. ACM SIGCSE Bulletin 38(1) (2006)
6. Apple, W.P.F., Horace, H.S.I.: Educational Ontologies Construction for Personalized Learning on the Web. Studies in Computation Intelligence 62, 47–82 (2007)
7. Mizoguchi, R.: Tutorial on ontological engineering-Part2: Ontology development, tools and languages. New Generation Computing, OhmSha & Springer 22(1) (2004)
8. Mizoguguchi, R., Bourdeau, J.: Using Ontological Engineering to Overcome AI-ED Problems. International Journal of Artificial Intelligence in Education 11(2), 107–121 (2000)
9. Marco, R., Joseph, S.: Curriculum Management and Review - an ontology-based solution. Technical Report # DIT-07-021, University of Trento (2007)
10. Nilsson, M., Palmer, M., Brase, J., The, L.R.: binding - principles and implementation. In: Proceedings of The 3rd Annual ARIDNE Conference, Lueven, Belgium (2003)
11. Sampson, D.G., Lytra, M.D., Wagner, G., Diaz, P.: Guest Editorial: Ontologies and the Semantic Web for E-learning. Educational Technology and Society 7(4), 26–28 (2004)
12. Yu, X.-Y., Tungare, M., Fan, W., Pérez-Quiñones, M.A., Fox, E.A., Cameron, W., Cassel, L.N.: Using Automatic Metadata Extraction to Build a Structured Syllabus Repository. In: Goh, D.H.-L., Cao, T.H., Sølvberg, I.T., Rasmussen, E. (eds.) ICADL 2007. LNCS, vol. 4822, pp. 337–346. Springer, Heidelberg (2007)

# Creating Collaborative Learning Groups in Intelligent Tutoring Systems*

Jarosław Bernacki and Adrianna Kozierkiewicz-Hetmańska

Institute of Informatics, Wroclaw University of Technology, Poland
172811@student.pwr.edu.pl, adrianna.kozierkiewicz@pwr.wroc.pl

**Abstract.** Intelligent Tutoring Systems offer an attractive learning environment where learning process is adapted to students' needs and preferences. More than 20 years of academic research demonstrates that learning in groups is more effective than learning individually. Therefore, it is motivating to work out procedure allowing a collaborative learning in Intelligent Tutoring Systems. In this paper original algorithm for creating collaborative learning groups is proposed. The research showed that students working in groups (generated by the proposed algorithm) achieved 18% better results than students working in randomly generated groups. It proves the effectiveness of the proposed algorithm and demonstrates that creating suitable learning groups is very important.

## 1    Introduction

The Intelligent Tutoring System (ITS) is a computer software that provides management of teaching process without the intervention of a teacher [16]. ITS can be also called a "computer teacher" [2]. ITS systems are a combination of the latest achievements of artificial intelligence (i.e. in the management of the teaching process) and multimedia learning environments. The fundamental assumption in the design of ITS systems is to most faithfully reflect teacher's work [17]. Furthermore, ITS should be able to fit the teaching process to student's individual requirements, needs and abilities.

ITS systems can consist of multiple modules; their number is dependent on a particular design and implementation. The typical ITS architecture consists of four modules: subject's knowledge representation, student module, teacher module and communication interface [17]. Subject's knowledge representation module is responsible for teaching resources presentation. It contains different types of materials such as: statements, charts, illustrations, examples, tasks [18]. The student module contains information about users. The teacher module plays a very important role in the system. It allows to define and manage teaching materials. Students interaction with the system is possible by means of communication interface module. Figure 1 presents a general ITS architecture.

This work focuses on the teacher module which is a very important part of an intelligent tutoring systems. This module is responsible for providing adequate learning material that fits students' needs, preferences, learning styles and abilities. Moreover,

---

* This research was financially supported by the Polish Ministry of Science and Higher Education.

D. Hwang et al. (Eds.): ICCCI 2014, LNAI 8733, pp. 184–193, 2014.

it evaluates student's knowledge and allows student to work in groups. More than 20 years of academic research demonstrates the fact that studying in groups is more effective than learning alone [19]. Therefore, intelligent tutoring systems should allow users to learn in groups.

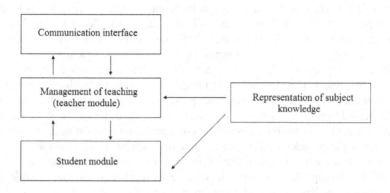

**Fig. 1.** Architecture of ITS system [17]

This paper focuses on a method which creates an effective collaborative learning group. If we take into consideration a diversity of users' needs and preferences it seems a difficult task. The proposed procedure will be conducted in two stages. Each student in intelligent tutoring system is represented by a set of user's data which contains information about student's demographic data, abilities, personal character traits, interests and learning style (according to Felder and Silverman model). Students characterized by a particular learning style (for example active students), demonstrate stronger needs and preferences for working in groups than others. That is why only those attributes were taken into consideration while creating effective collaborative learning groups. In the first step we choose attributes which determine that student prefers working in groups or working alone. Next, the $k$-means algorithm is used for creating clusters of students with similar learning styles that support group studying.
The rest of the paper is organized as follows: Section 2 contains a short overview of methods creating learning groups together with systems where those were applied. In Section 3 there is described proposed method based on learning style and *k-means* algorithm. Section 4 presents the results of experiment and their statistical analyses. Section 5 contains conclusions and further work.

## 2  Related Work

### 2.1  Educational Agents

"Educational agent" is an intelligent program which aim is to help students to gain knowledge and skills [14]. Such agents should be reactive, which means that they should be able to respond to signals received from the environment. They should also be able to communicate with a user in his natural language. Agents are also characterized by adaptability, which is the ability to adapt to changing environmental conditions, such as the emotions of the learner. Software agents are usually introduced as a static or dynamic visualization of the form, similar to a human. The important feature of educational agents is the role of "colleague" which assist the student during learning process and could imitate the learning in pairs (peer learning).

Interesting example of an educational agent is *AutoTutor* [3]. Learning process focus on individual conversation with a student in his natural language. Teaching is based on the initial presentation of the considered problem and learner's "conversation" with the system. *AutoTutor* uses predefined patterns of conversations which are developed by experts. However, this agent does not support typical collaborative learning.

Another learning agent is *Duffy*, which can be described as a learning partner [4]. This agent plays two roles - a "colleague", which helps studying, and a troublemaker. To force learner to reflection (during answering the questions), *Duffy* (knowing the correct answer) deliberately gives the wrong answer. This causes the enforcement of the learner to verify the correctness of his own position. Using such a strategy, the agent is able to assess effectively the confidence of the learner and to recognize his/her emotions.

*Coach Steve* is a part of system that allows to move and navigate through a virtual engine room of a ship [10]. It can fulfill two roles: a teacher (tutor) and a "colleague". In both cases *Steve* has the expert knowledge. In teacher mode, the agent is able to demonstrate different activities, for example how to repair equipment. It can also provide guidance for learners by detecting and correcting made by them errors. In the colleague mode, *Steve* can execute tasks or track actions performed by learners.

## 2.2    Collaborative Learning and Groups Creation

Collaborative learning is an effective method of knowledge acquisition. It is even more effective with the use of computer software [1], especially in fields like mathematics [15]. Unfortunately, the number of tools enabling collaborative learning, is small. Many ITS systems are dedicated to the individual learning and do not have mechanisms for effective collaborative learning. Recently, peer learning [1, 11] became very popular and researches showed the effectiveness of such solution.

The problem of the optimal selection of work group is investigated by many researchers. The two factors play important role in effective collaborative learning: optimal number of a group and personal features of team members. It is often assumed, that the optimal number of group members is 4-5 people [5]. Recent experiments have been carried out for peer learning (learning groups consist of two persons) [10, 14]. A very important aspect of effective collaboration is personal features of its team members. According to many researchers, a group is able to work successfully, when people are characterized by similar personality [7, 11], and learning styles [6].

## 3    Method for Creating Collaborative Learning Groups

The general idea of the proposed algorithm is to check whether students are suitable for working in groups by analysing their learning styles. If they do not, they should study alone. Otherwise, suitable students (with similar learning styles) should be clustered by using *k-means* algorithm. Next, students belonging to the same cluster should be paired/grouped in a random way (depending on availability in the intelligent tutoring system). The process of collaborative learning will be conducted with the use of an implemented intelligent tutoring system. Grouped/paired students will have possibility to use a tool such as "chat", that will make it possible to contact each other and allow to discuss about presented learning materials and tasks.

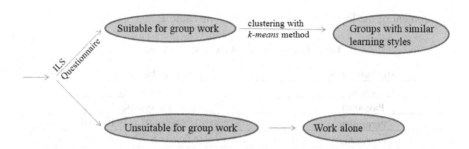

**Fig. 2.** The general idea of creating collaborative learning group method

### 3.1    The Felder / Silverman Model and Learner Profile

A first step in creating learning groups is to verify whether students are suitable for working in groups. For this task the model proposed by Richard Felder and Linda Silverman [6] describing learning styles is used. The learner's behaviour is considered in four dimensions: perception, reception, processing and understanding. Every dimension is bipolar:

- processing: *active* or *reflective*
- perception: *sensitive* or *intuitive*
- receiving: *visual* or *verbal*
- understanding: *sequential* or *global*

In order to determine learning styles, the *Index of Learning Styles* (ILS) [9] can be used. ILS is a questionnaire that contains 44 questions (11 questions for each of four dimensions). The results obtained in the ILS test are presented as a pair, where first element refers to learner's preferred direction, and the second is a score on a scale 1-11 that points the intensity of student's behaviour in a given direction. If one dimension equals more than zero, it implies zero intensity degree of the second value. In [8] it is showed, that ILS questionnaire can be brought to 20 questions, because of each of four dimensions, there are exactly 5 most representative questions.

The analysis of learning style showed that, if based on the processing dimension, the intelligent tutoring system could decide whether a student is suitable (or not) for learning in groups. An active student should cooperate with other students, whereas a reflective student prefers working alone.

If a student register to the intelligent tutoring system and fill in the ILS questionnaire, then system creates a learner profile. In this paper we assumed that the learner's profile is represented as a tuple of values defined as follows:

$$t : A \rightarrow V ,$$

where:

$$A \text{ - finite set of profile attributes, } V \text{ - attribute values, } V = \bigcup_{a \in A} V_a \text{ ,}$$

$$\underset{a \in A}{\forall} (t(a) \in V_a) .$$

The content of the learner profile is presented in Table 1.

**Table 1.** The content of the learner profile

| Attribute name | Attribute domain |
|---|---|
| Login | sequence of symbols |
| Password | sequence of symbols |
| Perception | $\{(sensitive,i),(intuitive,j)\}$ $i, j \in \{0,...,11\}$ |
| Processing | $\{(active,i),(reflective,j)\}$, $i, j \in \{0,...,11\}$ |
| Understanding | $\{(sequential,i),(global,j)\}$, $i, j \in \{0,...,11\}$ |
| Receiving | $\{(viusal,i),(verbal,j)\}$, $i, j \in \{0,...,11\}$ |
| Result of the initial test | $[0\%,100\%]$ |
| Result of the final test | $[0\%,100\%]$ |

## 3.2     Creating Groups by Using *k-means* Algorithm

According to [7, 18, 19] it is recommended that people working with each other in groups should be characterized by the same (or if it is not possible) similar personality features. In [13] authors claim that similar students should learn in the same or very similar way. Based on those assumptions, intelligent tutoring systems should ensure effective co-operation between such people.

Groups are created as clusters which consist of students with similar learning styles. Such solution allows to personalize learning process. For example students, which are characterized as sequential and sensitive, should be supported by intelligent tutoring system by set of tasks, experiments, practical exercises.

For creating groups *k-means* algorithm is used. This algorithm works by creating $K$ clusters and distributes them randomly, then every value of the input dataset is assigned to the closest cluster by using a distance function. In our case the input dataset are the learner's profiles stored in the intelligent tutoring system. The distance between learner's profiles could be calculated in the following way:

$$d(t_1,t_2) = \sum_{i=1}^{4} \delta_a$$

where: $\delta_a = \begin{cases} 0 & if \ t_1(a) = t_2(a) \\ 1 & otherwise \end{cases}$ ;

$a \in \{perception, receiving, processing, understanding\}$

When all values have been assigned to one given cluster, the cluster position is re-calculated. The new centroid is calculated as the mean location of the values that are assigned to this cluster. Next, values move from one cluster to another until there are no significant changes - then algorithm is finished. The output clusters contain students with the same learning styles. Next, students belong to the same cluster, are paired/grouped in the random way. They start the learning process by discussing about learning materials and tasks proposed by intelligent tutoring system.

The learning recommendation procedure material suitable for a student's learning styles was described in [12].

## 4    Experimental Results

The main goal of our experiment is to examine the efficiency of the proposed algorithm in creating learning groups. We try to assess how the collaborative learning influences on the learning process. The experiment was conducted on a group of 84 people (students) who used the specially implemented for this task a prototype of intelligent tutoring system. First, students learned about basics of computational complexity and next they tried to solve a list of tasks. In the second part of the experiment students were paired by proposed algorithm or in random way. They learned in groups and solved the proposed list of tasks. The list of tasks contained 5 questions with 4 possible answers, where only 1 answer was correct. For each correct answer student got 1 point.

The verification of the efficiency of the algorithm was based on the statistical comparison that:

- Students working in pairs generated by the algorithm achieved better learning results than those who worked individually;
- The results of work in randomly generated pairs differed from the results of the individual work;
- Results of the work in pairs generated by the algorithm are better than the results of pairs generated randomly.

In our experiments students were divided into following groups:

- 32 students working in pairs generated by the proposed algorithm (16 pairs);
- 32 students working in pairs generated randomly (16 pairs);
- 20 students working individually.

The statistical analysis used the following data (samples):

(1) The results of individual work;
(2) The results obtained by groups generated by the algorithm;
(3) The results obtained by randomly generated groups ;
(4) The results of individual work of students who were paired by the algorithm;
(5) The results of individual work of students who were paired randomly.

Some of the results are presented in Figures below.

Only 2 students correctly answered all questions. Both 4 and 3 points were achieved by 3 students. One student gave 2 correct answers. The most common result was 1 point, obtained by 8 students. Three students did not give a single correct answer.

Students working in pairs generated by the algorithm, obtained the following results:

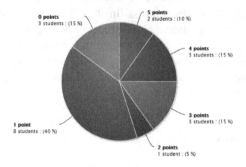

Fig. 3.   The results of individual work

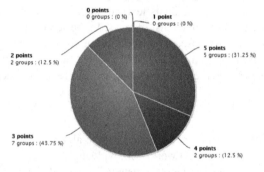

Fig. 4. The results of work in pairs generated by algorithm

Five pairs achieved the maximum number of points, 2 pairs gave only 1 wrong answer. The most common results was 3 points. The worst result was 2 points, obtained by 2 pairs. None of the pairs received both 1 point and 0 points.

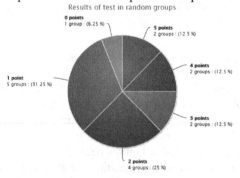

Fig. 5. The results of work in random groups

Diagram 4 presents the results of work in random groups. Only 2 groups gave correct answers for all questions in the test. Both 4 and 3 point were achieved by 2 groups. Four groups obtained 2 points. The most common result, which was obtained by 5 groups, was 1 point. One pair gave wrong answers to all questions, so their result was 0 points.

The whole analysis was made at significance level $\alpha = 0.05$. Before selecting a proper test, the distribution of each of mentioned samples was analyzed by Lilliefors test. The obtained results are presented in the Table 2:

**Table 2.** The results of the Lilliefors test

| Sample | Statistical test value | p-value |
|---|---|---|
| (1) | 0.285317 | 0.000159 |
| (2) | 0.279702 | 0.001588 |
| (3) | 0.205555 | 0.069299 |
| (4) | 0.179788 | 0.010007 |
| (5) | 0.21013 | 0.000995 |

The analysis showed that only one out of five considered samples (sample containing results by randomly generated groups) was from a normal distribution. In other cases the null hypothesis was rejected, therefore we assumed that examined data do not come from a normal distribution. Therefore, to perform further analysis, a nonparametric tests were used.

In order to compare medians, all of the above samples were tested with the Kruskal-Wallis one-way analysis of variance. The statistical test value and p-value were equal 17.877712 and 0.001304, respectively. This means that medians of considered sequences differed significantly.

Next, it was analysed, which samples differ significantly. For this purpose the U Mann-Whitney test was used. Results are presented in Table 3.

**Table 3.** The results of U Mann-Whitney test

| Tested samples | Statistical test value | p-value |
|---|---|---|
| (1) and (2) | 74.5 | 0.004662 |
| (1) and (3) | 139.5 | 0.516567 |
| (2) and (3) | 62.0 | 0.010122 |
| (2) and (4) | 233.5 | 0.623410 |
| (3) and (5) | 251.0 | 0.921121 |

U Mann-Whitney test showed that two sample pairs were statistically different. Table 4 presents the sum of ranks and the mean of the ranks for three samples.

The best values come from the second sample, which contains the results of students working in groups generated by the proposed algorithm. It is clearly visible that creating suitable learning groups is very important. Students working in pairs generated by the algorithm achieved better results than users working in random groups or students working alone. This proves that proposed algorithm works and confirms the hypothesis that establishing an effective learning group increases the learning results.

**Table 4.** Sum of ranks and the mean of the ranks for three sample

| Sample | Sum of rank | Average rank group |
|--------|-------------|--------------------|
| (1) | 284.5 | 14.225 |
| (2) | 381.5 | 23.84375 |
| (3) | 316.5 | 19.78125 |

# 5    Conclusions and Future Works

In this paper we proposed a method for creating learning groups and analyzed its effectiveness. The experiments showed that assigning a learner to a suitable group can increase learning effectiveness. The result of experiments demonstrated that students working in groups generated by the proposed algorithm, achieved significantly better results than students working in randomly created pairs or students working alone. Students, who were working in groups created by the algorithm, achieved 18% better results than students working in groups generated randomly. This confirms, that suitable classification to groups is very important and influence on the effectiveness of the learning process.

In future work it is planned to prepare prototype of ITS system allows a collaborative learning in groups bigger than two students, for example in groups of 4-5 students. Additionally, we would like to consider other aspects of student's characteristics, like personal character traits or interests in creating learning groups. Moreover, the strategy of creating groups could be changed, i.e. instead of creating groups of people with similar learning preferences, match people with a diverse learning styles.

# References

1. Aleven, V., Belenky, D.M., Olsen, J.K., Ringenberg, M., Rummel, N., Sewall, J.: Authoring Collaborative Intelligent Tutoring Systems. In: Looi, C.-K., Walker, E. (eds.) Proceedings of the Workshops at the 16th International Conference on Artificial Intelligence in Education AIED 2013, Memphis, USA (2013)
2. Biedrycki, A.: The modular organization of learning content in e-learning training system Koszalin University of Technology, pp. 81–88 (2011) (in Polish)
3. Chipman, P., Graesser, A.C., Haynes, B.C., Olney, A.: AutoTutor: An intelligent tutoring system with mixed-initiative dialogue. IEEE Transactions in Education 48, 612–618 (2005)
4. Chou, C.Y., Chan, T.W., Lin, C.J.: Redefining the learning companion: the past, present, and future of educational agents. Computers & Education 40, 255–269 (2003)
5. Clifford, M.: Facilitating Collaborative Learning: 20 Things You Need to Know From the Pros (2012),
   http://www.opencolleges.edu.au/informed/features/
   facilitating-collaborative-learning-20-things-you-need-
   to-know-from-the-pros/ (last access: April 18, 2014)
6. Felder, R.M., Silverman, L.K.: Learning and teaching styles in engineering education. Engr. Education 78(7), 674–681 (2002)

7. Garcia, E., Romero, C., Ventura, S.: Data mining in course managements systems: Moodle case study and tutorial. Elsevier Science 51(1), 368–384 (2008)
8. Graf, S., Leo, T., Viola, S.R.: In-Depth Analysis of the Felder-Silverman Learning Style Dimensions. Journal of Research on Technology in Education 40(1), 79–93 (2007)
9. ILS Questionnaire,
   http://www.engr.ncsu.edu/learningstyles/ilsweb.html
   (last access: March 31, 2014)
10. Johnson, L., Rickel, J.: STEVE: An animated pedagogical agent for procedural training in virtual environments. Sigart Bulletin 8(1-4), 12–16 (1997)
11. Koedinger, K.R., Rummel, N., Walker, E.: Integrating collaboration and intelligent tutoring data in evaluation of a reciprocal peer tutoring environment. Res. Practice Tech. Enhanced Learning 4, 221 (2009)
12. Kozierkiewicz-Hetmańska, A.: A method for scenario recommendation in intelligent e-learning systems. Cybernetics and Systems 42(2), 82–99 (2011)
13. Kukla, E., Nguyen, N.T., Daniłowicz, C., Sobecki, J., Lenar, M.: A model concep-tion for optimal scenario determination in an intelligent learning system. W: ITSE - International Journal of Interactive Technology and Smart Education 1(3), 171–184 (2004)
14. Landowska, A.: The role of education agents in distance learning environments, Scientific Papers of the Faculty of Electrical and Control Engineering at Gdańsk University of Technology, No. 25 (2008) (in Polish)
15. McLaren, B.M., Rummel, N., Tchounikine, P.: Computer Supported Collabo-rative Learning and Intelligent Tutoring Systems, pp. 447–463. Springer, Berlin (2010)
16. Mutter, S.A., Psotka, J.: Intelligent Tutoring Systems: Lessons Learned, vol. 14(6), pp. 544–545. Lawrence Erlbaum Associates Inc. (1988)
17. Rybak, A.: Intelligent Multimedia Learning Systems [in Polish], Work realized under the project "Graduate in computer science or mathematics specialist at labor market", University of Białystok (2009) (in Polish)
18. Rybak, A.: Examples of remote education systems realizing selection of teaching strategies to learning style of user. In: Dąbrowski, M., Zając, M. (eds.) Foundation for the Promotion and Accreditation of Economic Fields, pp. 127–133 (2010)
19. Schoenherr, N.: Discovering why study groups are more effective,
    http://news.wustl.edu/news/Pages/5642.aspx
    (last access: April 18, 2014)

# Method of Driver State Detection for Safety Vehicle by Means of Using Pattern Recognition

Masahiro Miyaji

Aichi Prefectural University, Nagakute, Aichi, Japan
masahiro@toyota.ne.jp

**Abstract.** Evolution of preventive safety devices for vehicles is highly expected to reduce the number of traffic accidents. Driver's state adaptive driving support safety function may be one of solutions of the challenges to lower the risk of being involved in the traffic accident. In the previous study, distraction was identified as one of anormal states of a driver by introducing the Internet survey. This study reproduced driver's cognitive distraction on a driving simulator by imposing cognitive loads, which were arithmetic and conversation. For classification of a driver's distraction state, visual features such as gaze direction and head orientation, pupil diameter and heart rate from ECG were employed as recognition features. This study focused to acquire the best classification performance of driver's distraction by using the AdaBoost, the SVM and Loss-based Error-Correcting Output Coding (LD-ECOC) as classification algorithm. LD-ECOC has potential to further enhance the classification capability of the driver's psychosomatic states. Finally this study proposed next generation driver's state adaptive driving support safety function to be extendable to Vehicle-Infrastructure cooperative safety function.

**Keywords:** ITS, preventive safety, Internet survey, driver distraction, ECOC, AdaBoost, pattern recognition.

## 1    Introduction

Traffic fatalities in Japan as of 2013 have declined for fourteen years by the comprehensive countermeasure [1]. However the number of traffic injuries still exceeds some 0.7 mwwwsillion as shown in Fig. 1. Reducing the number of traffic accident remains a key issue for creation of the sustainable mobility society. In the area of passive safety to mitigate an injury in an event of vehicle crashes, seatbelt system and airbag system have developed and introduced into ordinary vehicle for over twenty years. According to a statistical study on the effect of passive safety devices in the traffic accident, the reduction rate of traffic fatalities which is brought by the airbag system was figured as around 20% in 1990s [2,3]. In the area of preventive safety system, electronic stability control system (ESC), lane departure warning, pre-crash safety system with functions that detects the direction of a driver's face or movement of eyes have been developed and put into ordinary vehicle [4,5]. Next generation driving support safety function is highly expected to evolve into an enhancement in safety performance of traffic accident's prevention or avoidance. Around 90% of the traffic accident is

D. Hwang et al. (Eds.): ICCCI 2014, LNAI 8733, pp. 194–203, 2014.

thought to be caused by human errors [6]. Therefore establishing technologies which detect driver's psychosomatic states is highly expected. By executing Internet survey, driver's distraction was identified as one of anormal states while driving [7]. This study reproduced driver's cognitive distraction on a driving simulator by means of imposing cognitive loads such as arithmetic and conversation while driving. To identify driver's psychosomatic state, capturing physiological feature is indispensable. According to previous study visual features such as gaze direction and head orientation, pupil diameter and heart rate from ECG were employed as recognition features to detect cognitive distraction [8]. This study proposes a method of rapidly detecting cognitive distraction using the AdaBoost [9], which is widely used in pattern recognition. The SVM was adopted to compare with this proposed method. Furthermore ECOC [10] in combination with the AdaBoost as a binary classifier was introduced to obtain the best classification performance in accuracy of detecting driver's cognitive distraction.   Finally the concept idea was proposed for a driving support safety function [11].

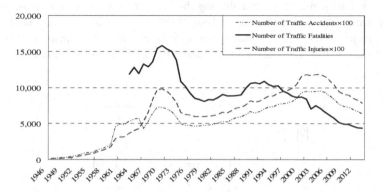

**Fig. 1.** Transition of road traffic accidents, fatalities, injuries as of 2013 in Japan

## 2    Internet Survey of the Traffic Incident

Accident investigation may be effective means to clarify root cause of the traffic accident which would establish countermeasures to reduce the number of traffic accident. This study reviewed the Internet-based questionnaire survey to collect information concerning traffic incidents in normal driving [7] [11]. Screening test was carried out to select proper respondents and to avoid bias of age, gender, driving experiences, driving usage and area. The system controlled not to appear missing value by manually checking answer. The questionnaire contained the picture of seven traffic incidents scene including near-miss accidents, which are right turn, left turn, crossing path, person to vehicle, head-on, rear end, and lane change [12]. These accident models were defined as potential accident risks in the ASV Promotion Project conducted by the MLIT in Japan. The respondents were asked whether they encountered any such experiences within the past two to three years. From the results of the questionnaire survey, the number of respondents was 2,000 (1,117 men and 883 women) and their average age was 41.1 years; their average driving experience was 19.9 years. The analysis clarified that traffic incidents occurred on an average of 2.39 times during the last three years. The twenty-eight questionnaire items were set based on driver behavior just before the traffic incident being classified by the Road Traffic Law violations (e.g., no safety confirmation, inappropriate assumption, and desultory driving) ,which was defined by National Police Agency in Japan (NAPJ). In the same manner, the ten items were set concerning psychosomatic states (e.g., hasty, lowered concentration, and drowsiness). Major answers for the driver behavior were "No safety confirmation" (30.9%), "Inappropriate assumption" (23.2%), "Distracted driving" (12.5%), "Not looking ahead carefully" (3.7%), "Not looking movement carefully" (2.1%), and "Inappropriate operation" (1.2%). From the analysis of the Internet survey, the human error related violence of the Road Traffic Law has occurred for at least 74%, which agreed with the analytical report of NAPJ. Therefore the Internet survey of this study is judged as suitable in use to analyze the root cause of the traffic accident. The result means that detecting the driver's psychosomatic state just before encountering a traffic incident is indispensable for establishing the counter-measure to reduce the number of the traffic accident. From the analysis of the collected answer, the author analyzed that major psychosomatic states immediately before traffic incident obtained were *haste* (22%), *distraction* (21.9%), *and drowsiness* (5.3%) as shown in Fig.2.

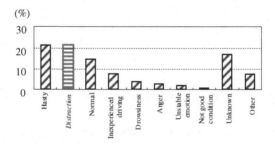

**Fig. 2.** Driver's psychosomatic states

# 3    Acquisition of Physiological Information

## 3.1    Characteristics of a State of Distraction

When a driver is in a state of cognitive distraction by imposing mental loads of conversation and thinking while driving, its affection may appear in eye movement, resulting in pupil dilation which causes reduction of the area of focal point of gaze direction by acceleration of the autonomic nerve. This also influences heart rate activity (hereinafter; HR), which decreases heart rate RRI (HR-RRI) [13]. Therefore capturing the change of movement of eyes and head, and heart rate may be effective means to detect driver's cognitive distraction when a driver is engaged in conversation and thinking [8] [14, 15].

## 3.2    Reproduction of a State of Distraction

By using a mock-up driving simulator, a state of driver's distraction was reproduced. Subjects were instructed to operate on a course that was projected on the frontal screen. The driving course was a rural road without traffic signals, which may allow the reproduction of cognitive distraction imposed by mental load. They were two types of cognitive loads, one was arithmetic and the other was conversation. Arithmetic loads involved verbally subtracting prime number (for example 7) from 1,000 successively. The number of subjects was 10 (7 males and 3 females) who ordinarily drove as part of their daily commute and consented to participate in the experiment.

## 3.3    Acquisition of Physiological Signal

This study used the tracking unit composed of stereo camera with data processor (The faceLAB, Seeing Machines, Australia) shown in Fig. 3, which measures physiological signals from images by tracking movement of eyes and head, and, pupil diameter. The tracking unit stores the signal on real time basis. Gaze angle and head rotation angle (hereinafter; visual information) were both output as a vertical rotation "pitch angle" and a lateral rotation "yaw angle" as shown in Fig. 4. A standard deviation of gaze angle and head rotation angle were adopted as features to detect cognitive distraction.

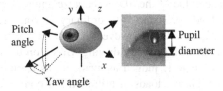

**Fig. 3.** Tracking unit (Stereo camera)          **Fig.4.** Pitch angle and Yaw angle

The standard deviation (SD) of gaze direction and head direction was derived as recognition features by using the data from the preceding five seconds. $x(i)$, $\sigma(i)$ were calculated by the equation (1), (2) respectively.

$$x(i) = \sqrt{x_{pitch}(i)^2 + x_{yaw}(i)^2} \tag{1}$$

$$\sigma(i) = \sqrt{(1/4)\sum\nolimits_{j=i-4}^{i}(x(j)-\bar{x})^2} \tag{2}$$

Here, $x(i)$ is combined gaze (head rotation) angle, $x_{pitch}(i)$ is pitch angle, and $x_{yaw}(i)$ is yaw angle. $\sigma(i)$ indicates the standard deviation of the gaze (head rotation) angle.

## 3.4    Acquiring ECG Waveform

Heart rate (HR) and heart rate RRI (HR-RRI) were calculated by measuring an interval between R waves (RRI) in an ECG waveform as shown in Fig. 5. A monitor lead method involving standard limb lead (II) and measurement with 3 chest electrodes was used. The data was acquired every 5 seconds, and sampling of data set was done at 60 Hz. A 1 – 30 Hz band-pass filter was used to remove the noise. HR was calculated by the equation (3), where $t$ is time duration of HR-RRI in Fig. 5.

$$HR = 60 / t \qquad (3)$$

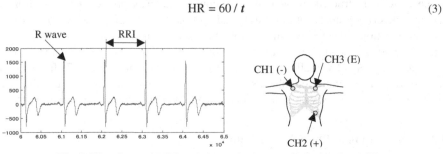

**Fig. 5.** Waveform of ECG and electrodes layout of Monitor Lead (II)

### 3.5    Verification of Physiological Signals

Verification was executed by confirming the differences in features with and without the cognitive loads to physiological signals. Although the frontal focal points were scattered widely to the peripheral area during ordinary driving, the frontal focal points were concentrated within a narrower range when the cognitive load was imposed. Average value of the SD of the gaze angle decreased by 9% while driving with arithmetic load compared with ordinary driving. This agreed with the trend of previous study [8] [14, 15]. However the SD of the head rotation angle decreased by 54% compared with ordinary driving. Based on the results, the SD of gaze angle and head rotation angle is judged as available as features to classify the cognitive distraction. When cognitive loads of arithmetic or conversation were imposed to the subject, the pupil dilated by acceleration of the autonomic nerve. An average value of the pupil diameter by cognitive load of arithmetic increased by 28.1% compared with ordinary driving. Trend of the above results agreed with the previous study [13]. From the results, the SD of the combined gaze angle and head rotation angle and pupil diameter were concluded available as features for classification of cognitive distraction. The average heart rate increased approximately by seven beats per minute when cognitive loads were imposed. The order of this result agreed with the previous studies [13]. Heart rate RRI of arithmetic decreased by 10.8% compared with ordinary driving. This change is believed to be a result of the higher heart rate caused by the cognitive loads. Based on the results, the average value of the heart rate RRI was available as a feature for classification of cognitive distraction.

## 4    Classification of Psychosomatic State

### 4.1    Machine Learning    Algorithms to Detect Distraction

The AdaBoost is one of the widely used machine learning algorithms for pattern recognition called Boosting [9]. Because it has many advantages such as high classification performance and rapid recognition process of time as well extendibility of recognition features, this study adopted the AdaBoost for detection of cognitive distraction. Machine learning of the AdaBoost algorithm involves creating different classifiers while successively changing weighting of each training data. A weighted majority decision is then made of the multiple classifiers in order to obtain the final classifier function as shown in Fig. 6. Each individual classifier is defined as a "weak classifier (e.g. $h_1(x)$, $h_2(x)$, $h_3(x)$)," while the combination of classifiers is defined as a "strong classifier (e.g. $H_t(x)$ in Fig. 6.)". This study introduced the GML Matlab Toolbox for the AdaBoost and introduced the stump which is usually adopted for a Boosting framework as a weak classifier. This stump is the simplest with only one classification node on each stump. CPU speed was 2.8GHz (Intel Core i7), memory capacity was 4.8GB, OS was Windows Vista SP1, and software was ver. 7.4 of the Matlab. The SVM was referred to compare the detection performance. The software of the SVM was SVM $^{light}$. This study adopted the Gauss kernel for the kernel function.

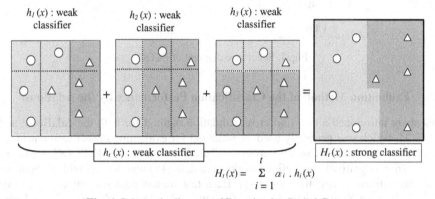

$$H_t(x) = \sum_{i=1}^{t} \alpha_i \cdot h_i(x)$$

**Fig. 6.** Schematic diagram of Boosting by the AdaBoost

## 4.2    Enhancement of Capability of    Distraction Detection by ECOC

Many multi-class identification methods has been developed. One method uses loss function which treats more than three labels at the same time, and minimizes by some methods. Typical approach is said as Neural Network and $k$-Nearest Neighbor ($k$-NN) algorithm. Although the approaches are easier to analyze on theorem bases, the computational calculation is not easy for a large number of data. The other is to combine a multi-class identification method with a binary classifiers because its generalization is easy. Error-correcting output codes (ECOC) extends binary classifiers to multiclass identification. It divides multi-class classification problem into some binary classification problems by an encoding rule and decodes binary classification results to multiple classes by a decoding rule. In this study, ECOC was used to obtain the best identification capability in combination with the AdaBoost. Loss-Based ECOC (hereinafter; LD-ECOC) has potential to obtain the best detection accuracy of driver's cognitive distraction. The schematic diagram of calculation procedure for LD-ECOC (or HD-ECOC) is shown in Fig. 7.

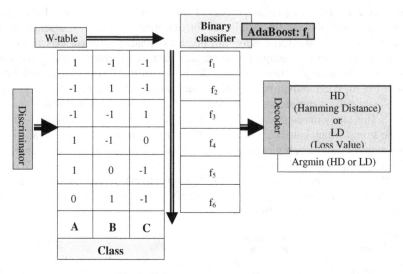

**Fig. 7.** Schematic diagram of ECOC

## 4.3    Evaluation Method of the Classification Performance by the AdaBoost

This study introduced a two-fold cross validation method for both the AdaBoost and the SVM, which was generally used as a method for evaluating classification accuracy of unknown data. A five-fold cross validation was used for LD-ECOC. Ordinary driving (non-cognitive) was defined as positive data (+1) and driving with a cognitive load was defined as negative data (−1). Each test subject data was divided into two sets, with test set $X_1$ used to evaluate performance when test set $X_2$ was used for learning. In the same way, test set $X_2$ was used for evaluation when test set $X_1$ was used for learning. To obtain classification performance, $O_p$ is defined as the positive output data, and $O_n$ is defined as the negative output data. $T_p$ is defined as true data of the positive, and $T_n$ is defined as true data of the negative. Then *Accuracy*, *Precision*, *Recall*, and the overall classification performance index *F* were defined as the classification indexes following (4) - (7) in order to calculate those classification indexes.

$$Accuracy = \frac{\left(O_p \cap T_p\right) \cup \left(O_n \cap T_n\right)}{T_p \cup T_n} \tag{4}$$

$$Precision = \frac{O_p \cap T_p}{O_p} \tag{5}$$

$$Recall = \frac{O_p \cap T_p}{T_p} \tag{6}$$

$$F = \frac{2 \times Precision \times Recall}{Precision + Recall} \tag{7}$$

## 5    Detection Performance of Driver's Distraction

Detection performance of pattern recognition by using the SD of combined gaze angle and head rotation angle (Visual information), pupil diameter, and heart rate RRI as recognition features are shown in Table 1, where the arithmetic load was imposed. The top common result in the average *accuracy* was 95.5 percent by LD-ECOC, which were combination of all the feature of Visual information plus Pupil diameter plus HR-RRI. From the results improvement was achieved by using LD-ECO. Average *accuracy* of Visual information of the AdaBoost was 81.6 percent, while the *accuracy* of the SVM in Visual information was 77.1 percent. From the comparison between the AdaBoost and the SVM, the *accuracy* and *F value* of the AdaBoost was higher than that of the SVM.

**Table 1.** Detection Performance in arithmetic load (Units: %)

| Algorithm | Recognition features | Average  Accuracy | F |
|---|---|---|---|
| SVM | Visual information | 77.1 | 67.4 |
| AdaBoost | Visual information | 81.6 | 83.1 |
| | Pupil diameter | 84.5 | 85.0 |
| | HR- RRI | 85.0 | 85.1 |
| | Visual information + Pupil diameter + HR-RRI | 91.4 | 92.3 |
| HD-ECOC | Visual information + Pupil diameter + HR-RRI | 89.6 | - |
| LD-ECOC | Visual information + Pupil diameter + HR-RRI | 95.5 | - |

## 6    Concept of Driver's State Adaptive Driving Support Function

Next generation driving support safety system may be composed by means of using driver's psychosomatic states monitoring feature as shown in Fig.8.

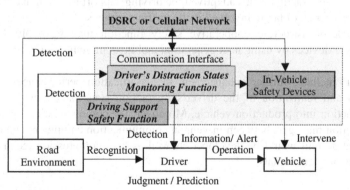

**Fig. 8.** Schematic diagram of next generation driving support safety function

For example in a front-end traffic incident, to explain its potential ability, the driver's psychosomatic state monitoring function detects the driver's distraction as well as improper recognition of the driving condition on real time. After the moment, the system may deliver sufficient information to the driver, warning alerts, or intervenes into the driver's activity to help lower the risk of incidents. When the driver's psychosomatic state is normal but the driver makes improper assumptions, traffic safety information from road infrastructures such as represented by ITS services (e.g.; VICS in Japan) [16] and the driving safety support system (e.g.; DSSS in Japan) may be employed [17]. To realize this ITS service, DSRC (Dedicated Short Range Communication) and Cellular Network (or Mobile communication network) could be employed.

# 7     Summary

The Internet survey was conducted to the traffic incident. The survey identified one of major root cause of the traffic incident was driver's distraction. The driver's psychosomatic state was reproduced by using the driving simulator. The physiological signals of the movement of eyes and head, pupil diameter as well as HR-RRI in ECG was obtained as the alternative of driver's distraction state. The AdaBoost, the SVM and LD-ECOC were introduced for the classification of driver's distraction state. Finally this study proposed the concept of driver's distraction adaptive type driving support function, which is expandable to a next generation vehicle-infrastructure cooperative safety function in preventive safety area.
The conclusion is as follows;

A) Distraction may be one of major root causes of human error which may be involved    in the traffic accident. Internet based survey is effective means to collect real world experiences of the traffic incidents as well as traffic accidents.
B) Physiological information of driver's psychosomatic states may be collectable
C) The AdaBoost is effective means to classify driver's distraction states. Furthermore LD-ECOC showed best performance of classification of driver's psychosomatic states in combination with the AdaBoost
D) Driver's distraction states adaptive type driving support function may help lower the risk of being involved in the traffic accidents
E) Vehicle-infrastructure cooperative safety function may be evolved for in-vehicle devices by using next generation mobile communication technology

Future issue includes further enhancement of classification performance of pattern recognition and realization of the driver's distraction states adaptive type driving support function into production vehicle. Moreover, the communication interface (or communication manager) between in-vehicle communication equipment and roadside communication unit would be developed as well as data compression technology of related physiological signals.

**Acknowledgements.** Hearty appreciation for the dedicated support of Professor K. OGURI, Dr, and Associate Professor H. KAWANAKA, Dr, the Graduate School of Information Science and Technology, Aichi Prefectural University in Japan.

# References

1. National Police Agency of Japan.: Road traffic accidents as of 2013 in Japanese (January 2014),
   http://www.e-stat.go.jp/SG1/estat/
   Pdfdl.do?sinfid=000023614425
2. Viano, D.C.: Effectiveness of Safety-belts and Airbags in Preventing Fatal Injuries. SAE 910901 (1991)
3. Edwards, W.R.: An Effectiveness Analysis of Chrysler Driver Airbags after Five Years Exposure. In: 14th ESV Conference, 94-S4-0-09 (1994)
4. Hattori, A., Tokoro, S., Miyashita, M.: Development of forward collision warning system using the driver behavioral information. In: SAE Technical Paper Series, 2006 SAE World Congress, vol. 115(7), pp. 818–827 (2006)
5. Nishina, T., Moriizumi, K.: Development of new pre-crash safety system using driver monitoring sensor. In: 15th World Congress on ITS, TS135-10315, pp.1–12 (2008)
6. Klauer, S.G., Dingus, T.A., et al.: The Impact of Driver Inattention on Near-Crash/Crash Risk. An Analysis Using the 100-Car Naturalistic Study Data, US-DOT HS-810-594 (2004)
7. Miyaji, M., Kawanaka, H., Oguri, K.: Analysis of Driver Behavior based on Experiences of Road Traffic Incidents investigated by means of Questionnaires for the Reducing of Traffic Accidents. International Journal of ITS Research 6(1), 47–56 (2008)
8. Kutila, M., Jokela, M., Markkula, G.: Driver Distraction Detection with a Camera Vision System. Proceeding of IEEE International Conference, Image Processing, pp. 201–204 (2007)
9. Freund, Y., Schapire, R. E.: A decision-Theoretic Generalization of On-line Learning and an Application to Boosting. In: Computational Learning Theory, Eurocolt, pp.23–37(1995)
10. Allwein, E.L., Schapire, R.E., Singer, Y.: Reducing multiclass to binary: a unifying approach for margin classifiers. The Journal of Machine Learning Research 1, 113–141 (2001)
11. Miyaji, M., Kawanaka, H., Oguri, K.: Internet-Based Survey on Driver's Psychosomatic State for the Creation of Driver Monitor Function. In: Proceeding of 2011 3rd International Congress on Ultra Modern Telecommunication, ICUMT 2011 (2011)
12. Toji, R.: Advanced Safety Vehicle Promotion Project: Phase3 and 4. Journal of Automotive Engineers of Japan 60(12), 10–13 (2006)
13. Kahneman, D., Tursky, B.: Pupillary, heart rate, and skin resistance changes during a mental task. Journal of Experimental Psychology 79(1) (1969)
14. Engstrom, J., Johansson, E., Ostlund, J.: Effects of visual and cognitive load in real and simulated motorway driving, Transportation Research, Technical Report, 8, part F (2005)
15. Victor, T.: Sensitivity of Eye-movement Measures to in-vehicle Task Difficulty. Part F Transportation Research 8, 167–190 (2005)
16. VICS in JAPAN.: Vehicle Information Communication Service,
    http://www.vics.or.jp/index1.html
17. MTS Society of Japan.: Driving Safety Support System,
    http://www.utms.or.jp/english/pre/dsss.html

# Motion Segmentation Using Optical Flow for Pedestrian Detection from Moving Vehicle

Joko Hariyono, Van-Dung Hoang, and Kang-Hyun Jo

Graduate School of Electrical Engineering, University of Ulsan, Ulsan 680–749 Korea
{joko,hvzung}@islab.ulsan.ac.kr, acejo@ulsan.ac.kr

**Abstract.** This paper proposes a pedestrian detection method using optical flows analysis and Histogram of Oriented Gradients (HOG). Due to the time consuming problem in sliding window based, motion segmentation proposed based on optical flow analysis to localize the region of moving object. A moving object is extracted from the relative motion by segmenting the region representing the same optical flows after compensating the ego-motion of the camera. Two consecutive images are divided into grid cells 14x14 pixels, then tracking each cell in current frame to find corresponding cells in the next frame. At least using three corresponding cells, affine transformation is performed according to each corresponding cells in the consecutive images, so that conformed optical flows are extracted. The regions of moving object are detected as transformed objects are different from the previously registered background. Morphological process is applied to get the candidate human region. The HOG features are extracted on the candidate region and classified using linear Support Vector Machine (SVM). The HOG feature vectors are used as input of linear SVM to classify the given input into pedestrian/non-pedestrian. The proposed method was tested in a moving vehicle and shown significant improvement compare with the original HOG.

**Keywords:** Pedestrian detection, Optical flow, Motion Segmentation, Histogram of oriented gradients.

## 1    Introduction

Detecting pedestrian as moving object is one of the essential tasks for understanding environment. In the past few years, moving object and pedestrian detection methods for mobile robots/vehicles have been actively developed. For real-time pedestrian detection system, Gavrila et al. [1] were employed hierarchical shape matching to find pedestrian candidates from moving vehicle. Their method uses a multi-cues vision system for the real-time detection and tracking of pedestrians. Nishida et al. [2] applied SVM with automated selection process of the components by using Ada-Boost. These researches show that the selection of the components and the combination of them are important to get a good pedestrian detector.

Many local descriptors are proposed for object recognition and image retrieval. Mikolajczyk et al. [3], [14] compared the performance of the several local descriptors and showed that the best matching results were obtained by the Scale Invariant

D. Hwang et al. (Eds.): ICCCI 2014, LNAI 8733, pp. 204–213, 2014.
© Springer International Publishing Switzerland 2014

Feature Transform (SIFT) descriptor [4]. Dalal et al. [5] proposed a human detection algorithm using histograms of oriented gradients (HOG) which are similar with the features used in the SIFT descriptor. HOG features are calculated by taking orientation histograms of edge intensity in a local region. They extracted the HOG features from all locations of a dense grid on an image region and the combined features are classified by using linear SVM. They showed that the grids of HOG descriptors significantly out-performed existing feature sets for human detection. Kobayashi et al. [6] proposed selected feature of HOG using PCA to decrease the number of feature. It could reduce the number of features less than half without lowering the performance

Moving object detection and motion estimation methods using the optical flow for a mobile robot also have been actively developed. Talukder et al. [7] proposed a qualitative obstacle detection method was proposed using the directional divergence of the motion field. The optical flow pattern was investigated in perspective camera and this pattern was used for moving object detection. Also real-time moving object detection method was presented during translational robot motion.

Several researchers also developed methods for ego-motion estimation and navigation from a mobile robot using an omnidirectional camera [8], [9]. They used Lucas Kanade optical flow tracker and obtained corresponding features of background in the consecutive two omnidirectional images. Use analyzing the motion of feature points, camera ego-motion was calculated. They obtained camera ego-motion compensated based on an affine transformation of two consecutive frames where corner features were tracked by Kanade-Lucas-Tomasi (KLT) optical flow tracker [10]. However using corner feature for tracking, the detecting moving objects resulted in a problem that only one affine transformation model could not represent the whole background changes. For this problem, our previous work [11] proposed each affine transformation of local pixel groups should be tracked by KLT tracker. The local pixel groups are not a type of image features such as corner or edge. We use grid windows-based KLT tracker by tracking each local sector of panoramic image while other methods use sparse features-based KLT tracker. Therefore we can segment moving objects in panoramic image by overcoming the nonlinear background transformation of panoramic image.

Proposed method is inspired by the works on pedestrian detection from moving vehicle [1], [7], using optical flow [10] and ego-motion estimation [8], which is ego-motion compensated [11]. Pedestrian as a moving object is extracted from the relative motion by segmenting the region representing the same optical flows after compensating the ego-motion of the camera. To obtain the optical flow, feature extracted from an image by divided into grid cells 14x14 pixels. Then, track corresponding cells in the next frame. At least using three corresponding feature cells, affine transformation is performed according to each corresponding cells in the consecutive frame, so that conformed optical flows are extracted. The regions of moving object are detected as transformed objects are different from the previously registered background. Morphological process is applied to get the candidate human region. In order to recognize the object, the HOG features are extracted on the candidate region and classified using linear Support Vector Machine (SVM) [13]. The HOG feature vectors are used as input of linear SVM to classify the given input into pedestrian/non-pedestrian. For the performance evaluation comparative study was presented in this paper.

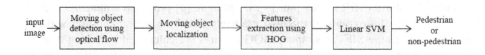

**Fig. 1.** The overview of the pedestrian detection algorithm

# 2  Motion Segmentation

This section presents the method to detect object motion from the camera, which is mounted on the moving vehicle. In order to obtain moving object regions from video or sequent images, it is not easy to segment out only moving object region, because it faced two kinds of motion, the first is independence motion caused of object movement and the second is motion caused by camera ego-motion. So, we proposed a method to deal with this situation [11], [15]. We used optical flow analysis to segmenting independent motion of object movement from ego-motion caused by camera, which is ego-motion compensated.

Human eyes will be easy to see and analyze which one is object of interest, such as moving object, and otherwise are static environment. However, robot will understand the environment base on the mathematical model. The optical flow analysis needed to proof that the motion caused of the independent motion of object movement will have difference pattern compare with flow caused by ego-motion from camera. These cues then will give us the region of moving objects and should be localized. Then, this region will be a candidate of detected human/pedestrian after we apply features extraction using HOG and linear SVM as a classifier.   The overview of the pedestrian detection algorithm is shown in Fig. 1.

## 2.1  Ego-motion Compensated

In our previous work [11], we apply KLT optical flow feature tracker [10] in order to deal with several conditions. Brightness constancy which is projection of the same point looks the same in every frame, small motion that points do not move very far and spatial coherence that points move like their neighbors.   However, using frame difference will not solve the problem, because it represents all motions caused by the camera ego-motion and moving object in scenes together. It needs to compensate this effect from frame difference to segment out only the independent motion of the object movement, so how much the image background has been transformed in two sequent images. The affine transformation represents the pixel movement between two sequent images as in (1),

$$P' = AP + t \tag{1}$$

where P and P' are the pixel location in the first and second image respectively. A is transformation matrix and t is translation vector. Affine parameters can be calculated by the least square method using at least three corresponding features in two images.

In this work, the original input images converted to grayscale images, and obtain one channel intensity pixel value from the input images. Then, using two consecutive images are divided into grid cells size 14x14 pixels, then compare and track each cell in current image to find corresponding cell in the next image. The cell has most similar intensity value in a group will selected as corresponding value. Using method from [10], then find the motion distance of each pixel in a group of cell, the motion $d$ in x and $y-$ axis of each cell $g_{t-1}(i,j)$ by finding most similar cell $g_t(i,j)$ in the next image.

$$g_{t-1}(i,j) = g_t(i + d_x, j + d_y) \qquad (2)$$

where $d_x$ and $d_y$ are motion distances in x and $y-$ axis respectively. Using at least three corresponding features in two images, affine parameters can be calculated by the least square method. So, equation (2) can represented as affine transformation of each pixel in the same cell as (3)

$$I_t(x,y) = A\,I_{t-1}(x,y) + d \qquad (3)$$

where $I_t(x,y)$ and $I_{t-1}(x,y)$ are vector 2x1 represent pixel location in the current and previous frame respectively, $A$ is 2x2 projection matrix and $d$ is 2x1 translation vector.

To obtain the camera ego-motion compensated, frame difference is applied in two consecutive input images by calculated based on the tracked corresponding pixel cells using (4)

$$I_d(x,y) = |I_{t-1}(x,y) - I_t(x,y)| \qquad (4)$$

where $I_d(x,y)$ is a pixel cell located at $(x,y)$ in the grid cell.

Suppose two consecutive images shown in Fig. 2 (a) and (b) can not segment out moving object using frame difference (c), however when we apply frame difference with ego-motion compensate could obtain moving objects area shown in Fig. 2 (d).

## 2.2    Motion Segmentation

Each pixel output from frame difference with ego-motion compensated cannot show clearly as silhouette. It just gives information of motion area of object movement. Those moving area are applied morphological process to obtain region of moving object and noise removal. Ideally, we would seek to devise a region segmentation algorithm that accurately locates the bounding boxes of the motion regions in the difference image. Given the sparseness of the data, however, accurate segmentation would involve the enforcement of multiple constraints, making fast implementation difficult. To achieve faster segmentation, we assumed the fact that humans usually appear in upright positions, and conclude that segmenting the scene into vertical strips is sufficient most of the time. In this work we define detected moving objects are represented by the position in width in x axis. Using projection histogram $h_x$ by pixel voting vertically project image intensities into $x$ − coordinate.

Adopting the region segmentation technique by [12], we define the region using boundary saliency. It measures the horizontal difference of data density in the local neighborhood. The local maxima correspond to where maximal change in data density occur, are candidates for region boundaries of pedestrian in moving object detection.

**Fig. 2.** From two consecutive images (a) and (b), then we applied frame difference (c) and comparing when we applied frame difference with ego-motion compensated (d)

## 3    Feature Extraction

In this section present how we extract feature from candidate region obtained from previous section. In this work we use Histogram of Oriented Gradients (HOG) to extract features from moving object area localization. Local object appearance and shape usually can be characterized well by the distribution of local intensity gradients or edge direction. HOG features are calculated by taking orientation histograms of edge intensity in local region.

### 3.1    HOG Features

In this work, we extract HOG features from 16×16 local regions as shown in Fig.3. The first, we use Sobel filter to obtain the edge gradients and orientations were calculated from each pixel in this local region. The gradient magnitude $m(x, y)$ and orientation $\theta(x,y)$ are calculated using directional gradients $d_x(x,y)$ and $d_y(x,y)$ which are computed by Sobel filter as follow (5),

$$m(x,y) = \sqrt{dx(x,y)^2 - dy(x,y)^2} \tag{5}$$

$$\theta(x,y) = \begin{cases} \tan^{-1}\left(\frac{dy(x,y)}{dx(x,y)}\right) - \pi, & \text{if } dx(x,y) < 0 \text{ and } dy(x,y) < 0 \\ \tan^{-1}\left(\frac{dy(x,y)}{dx(x,y)}\right) + \pi, & \text{if } dx(x,y) < 0 \text{ and } dy(x,y) > 0 \\ \tan^{-1}\left(\frac{dy(x,y)}{dx(x,y)}\right), & \text{otherwise} \end{cases} \tag{6}$$

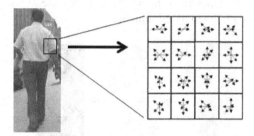

**Fig. 3.** Extraction Process of HOG features. The HOG features are extracted from local regions with 16 ×16 pixels. Histograms of edge gradients with 8 orientations are calculated from each of 4×4 local cells.

The local region is divided into small spatial cell, each cell size is 4×4 pixels. Histograms of edge gradients with 8 orientations are calculated from each of the local cells. The total number of HOG features is $128 = 8 \times (4 \times 4)$ and they constitute a HOG feature vector. To avoid sudden changes in the descriptor with small changes in the position of the window, and to give less emphasis to gradients that are far from the center of the descriptor, a Gaussian weighting function with $\sigma$ equal to one half the width of the descriptor window is used to assign a weight to the magnitude of each pixel.

A vector of the HOG features represent local shape of an object, it have edge information at plural cells. In flatter regions like a ground or a wall of a building, the histogram of the oriented gradients has flatter distribution. On the other hand, in the border between an object and background, one of the elements in the histogram has a large value and it indicates the direction of the edge. Even though the images are normalized to position and scale, the positions of important features will not be registered with same grid positions. It is known that HOG features are robust to the local geometric and photometric transformations. If the translations or rotations of the object are much smaller than the local spatial bin size, their effect is small.

Dalal *et al.* [5] extracted a set of the HOG feature vectors from all locations in an image grid and are used for classification. In this work, we extract the HOG features from all locations on the candidate region from an input image as shown in Fig. 4.

## 3.2    Linear SVM Classifier

In the human detection algorithm proposed by [5], the HOG features are extracted from all locations of a dense grid and the combined features are classified using the linear SVM. The HOG shows significantly outperformed existing feature sets for human detection. This work also used the linear SVM to perform work in various data classification tasks. Let $\{fi, ti\}_{i=1}^{N}$ $(f_i \in R^D, t_i \in \{-1, 1\})$ be the given training sample in D-dimensional feature space. The classification function is given as

$$z = sign(\omega^T f_i - h) \tag{7}$$

where $w$ and $h$ are the parameters of the model. For the case of soft-margin SVM, the optimal parameters are obtained by minimizing

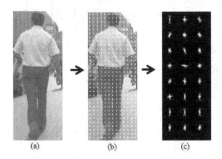

(a)              (b)              (c)

**Fig. 4.** From a candidate input image size 150x382 (a), HOG features are extracted from all locations on the candidate region of an input image with 16x16 pixels region (b), and the result shown in (c)

$$L(\omega, \xi) = \frac{1}{2}\|\omega\|^2 + C \sum_{i=1}^{N} \xi_i \tag{8}$$

under the constraints

$$\xi_i \geq 0, t_i(\omega^T f_i - h) \geq 1 - \xi_i (i = 1, ..., N) \tag{9}$$

where $\xi i$ ($\geq 0$) is the error of the $i$-th sample measured from the separating hyperplane and $C$ is the hyper-parameter which controls the weight between the errors and the margin. The dual problem of (8) is obtained by introducing Lagrange multipliers $\alpha = (\alpha 1, ...., \alpha N)$, $\alpha k \geq 0$ as

$$L_D(\alpha) = \sum_{i=1}^{N} \alpha_i - \frac{1}{2}\sum_{i,j=1}^{N} \alpha_i \alpha_j t_i t_j f_i^T f_i \tag{10}$$

under the constraints

$$\sum_{i=1}^{N} \alpha_i t_i = 0, \quad 0 \leq \alpha_i \quad (i = 1, ... N) \tag{11}$$

By solving (10), the optimum function is obtained as

$$z = sign(\sum_{i \in S} \alpha_i^* t_i f_i^T f_i - h^*) \tag{12}$$

where $S$ is the set of support vectors. To get a good classifier, we have to search the best hyper-parameter $C$. The cross-validation is used to measure the goodness of the linear SVM classifier.

## 4    Experimental Results

In this work, our vehicle system is run in outdoor environment with varies speed and detected object moving surround its path. Proposed algorithm was programmed in MATLAB and executed on an Intel Pentium 3.40 GHz, 32-bit operating system with 8 GB Random Access Memory. The proposed algorithm was evaluated by using five images sequences from ETHZ pedestrian datasets which contains around 5,000 images of pedestrians in city scenes [12]. It contains only front or back views with relatively limited range of poses and the position and the height of human in the image are almost adjusted. The size of the image is 640 × 480 pixels. For the training process, we used person INRIA datasets in [5]. These images were used for positive samples in the following experiments. The negative samples were originally collected from images of sky, mountain, airplane, building, etc. The number of negative images is 3,000. From these images, 1,000 person images and 2,000 negative samples were used as training samples to determine the parameters of the linear SVM. The remaining 100 pedestrian images and 200 negative samples were used as test samples to evaluate the recognition performance of the constructed classifier. When we implemented the original HOG, which proposed by Dalal *et. al* using those dataset, the recognition rate for test dataset is 98.3%. We used ego-motion compensated and HOG feature to evaluate performance improvement.

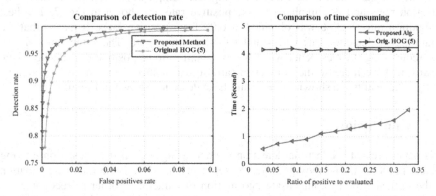

**Fig. 5.** Comparison result when we tested our proposed method and Original HOG by Dalal *et. al.* (a) comparison of detection rate and (b) comparison of time consuming

  HOG feature vectors were extracted from all locations of the grid for each training sample. Then, the selected feature vectors were used as input of the linear SVM. The selected subsets were evaluated by cross validation. Also we evaluated the recognition rates of the constructed classifier using test samples. The relation between the detection rates and the number of false positive rate are shown in Fig. 5. The best recognition rate 99.3 % was obtained at 0.09 false positive rates. It means that we obtain higher detection rate with smaller false positives rate. The computational cost also reduces eight times better when we use small ratio of positive to evaluated data. However, if we increase the number of ratio it also reduces time consuming significantly. The results are shown in Fig 6 and false detection shown in Fig. 7

**Fig. 6.** Successful moving objects detection results

(a)                    (b)

**Fig. 7.** (a) False positives detection and (b) False negative detection

HOG feature vectors were extracted from all locations of the grid for each training sample. Then, the selected feature vectors were used as input of the linear SVM. The selected subsets were evaluated by cross validation. Also we evaluated the recognition rates of the constructed classifier using test samples. The relation between the detection rates and the number of false positive rate are shown in Fig. 7. The best recognition rate 99.3 % was obtained at 0.09 false positive rates. It means that we obtain higher detection rate with smaller false positives rate. The computational cost also reduces eight times better when we use small ratio of positive to evaluated data. However, if we increase the number of ratio it also reduces time consuming significantly. The results are shown in Fig 8 and false detection shown in Fig. 9.

## 5    Conclusion

This paper presents pedestrian detection method using optical flow based on moving vehicle properties. The moving object is segment out through the relative evaluation of the optical flow to compensate ego-motion of camera. In order to recognize the object, the HOG features were extracted on a candidate region and classified using the SVM. The HOG feature vectors are used as an input of linear SVM to classify the given input into pedestrian/non-pedestrian. The proposed algorithm achieved comparable results comparing with the original HOG, and also reduces computational cost significantly using moving object localization.

**Acknowledgement.** This research was supported by the MOTIE (The Ministry of Trade, Industry and Energy), Korea, under the Human Resources Development Program for Convergence Robot Specialists support program supervised by the NIPA (National IT Industry Promotion Agency) (H1502-13-1001)

# References

1. Gavrila, D.M., Munder, S.: Multi-cue Pedestrian Detection and Tracking from a Moving Vehicle. International Journal of Computer Vision 73(1), 41–59 (2007)
2. Nishida, K., Kurita, T.: Boosting soft-margin SVM with feature selection for pedestrian detection. In: Oza, N.C., Polikar, R., Kittler, J., Roli, F. (eds.) MCS 2005. LNCS, vol. 3541, pp. 22–31. Springer, Heidelberg (2005)
3. Mikolajczyk, K., Schmid, C.: A performance evaluation of local descriptors. IEEE Transactions on Pattern Analysis and Machine Intelligence (PAMI) 27, 1615–1630 (2005)
4. Lowe, D.G.: Distinctive Image Features from Scale-Invariant Keypoints. International Journal of Computer Vision 60(2), 91–110 (2004)
5. Dalal, N., Triggs, B.: Histograms of Oriented Gradients for Human Detection. In: IEEE Conference on Computer Vision and Pattern Recognition, San Diego, pp. 886–893 (2005)
6. Kobayashi, T., Hidaka, A., Kurita, T.: Selection of Histograms of Oriented Gradients Features for Pedestrian Detection. In: Ishikawa, M., Doya, K., Miyamoto, H., Yamakawa, T. (eds.) ICONIP 2007, Part II. LNCS, vol. 4985, pp. 598–607. Springer, Heidelberg (2008)
7. Talukder, S.G., Matthies, L., Ansar, A.: Real-time detection of moving objects in a dynamic scene from moving robotic vehicles. In: Proc. of Int. Conf. Intelligent Robotics and Systems, pp. 1308–1313 (2003)
8. Vassallo, R.F.: Santos-Victor and H. Schneebeli, A General Approach for Egomotion Estimation with Omnidirectional Images. In: Proceedings of the Third Workshop on Omnidirectional Vision, Copenhagen, pp. 97–103 (2002)
9. Liu, H., Dong, N., Zha, H.: Omni-directional Vision based Human Motion Detection for Autonomous Mobile Robots. Systems Man and Cybernetics 3, 2236–2241 (2005)
10. Tomasi, C., Kanade, T.: Detection and Tracking of Point Features. International Journal of Computer Vision 9, 137–154 (1991)
11. Hariyono, J., Hoang, V.-D., Jo, K.-H.: Human detection from mobile omnidirectional camera using ego-motion compensated. In: Nguyen, N.T., Attachoo, B., Trawiński, B., Somboonviwat, K. (eds.) ACIIDS 2014, Part I. LNCS, vol. 8397, pp. 553–560. Springer, Heidelberg (2014)
12. Ess, A., Leibel, B., Gool, L.V.: Depth and Appearance for Mobile Scene Analysis. In: IEEE International Conference on Computer Vision, ICCV 2007 (2007)
13. Hoang, V.-D., Le, M.-H., Jo, K.-H.: Hybrid Cascade Boosting Machine using Variant Scale Blocks based HOG Features for Pedestrian Detection. Neurocomputing 135, 357–366 (2014)
14. Lu, Y.-Y., Huang, H.-C.: Adaptive reversible data hiding with pyramidal structure. Vietnam Journal of Computer Science, 1–13 (2014)
15. Hariyono, J., Kurnianggoro, L., Wahyono, Hernandez, D.C., Jo, K.-H.: Ego-motion compensated for moving object detection in a mobile robot. In: Ali, M., Pan, J.-S., Chen, S.-M., Horng, M.-F. (eds.) IEA/AIE 2014, Part II. LNCS, vol. 8482, pp. 289–297. Springer, Heidelberg (2014)

# Articular Cartilage Defect Detection Based on Image Segmentation with Colour Mapping

Jan Kubicek[1], Marek Penhaker[1], Iveta Bryjova[1], and Michal Kodaj[2]

[1] VSB–Technical University of Ostrava, FEI, K450
17. listopadu 15, 708 33, Ostrava–Poruba, Czech Republic
{jan.kubicek,marek.penhaker,iveta.bryjova}@vsb.cz
[2] Nemocnice Podlesí, a.s., Konská 453, 739 61 Třinec, Czech Republic
michal.kodaj@gmail.com

**Abstract.** This article addresses a possible approach for a higher quality diagnosis and detection of the pathological defects of articular cartilage. The defects of articular cartilage are one of the most common pathologies of articular cartilage that a physician encounters. In clinical practice, doctors can only estimate visually whether or not there is a pathological defect with the use of magnetic resonance images. Our proposed methodology is able to accurately and precisely localize ruptures of cartilaginous tissue and thus greatly contribute to improving a final diagnosis. When analysing MRI data, we work only with grey-levels, which is rather complicated for producing a quality diagnosis. Our proposed algorithm, based on fuzzy logic, brings together various shades of grey. Each set is assigned a colour that corresponds to the density of the tissue. With this procedure, it is possible to create a contrast map of individual tissue structures and very clearly identify where cartilaginous tissues have been interrupted. The suggested methodology has been tested using real data from magnetic resonance images of 60 patients from Podlesí Hospital in Třinec and currently this method is being put into clinical practice.

**Keywords:** Fuzzy modelling, Image segmentation, Soft thresholding, Membership function, MRI, MATLAB.

## 1 Introduction

Cartilage is a specialized type of fibrous tissue. It is composed of different substances, each of them responsible for its overall integrity, deformability, hardness and the ability to repair itself. Cartilage is created from mesenchymal cells at the ends of the epiphyses of bones during embryonic development in human foetuses.

Histologically, it is classified into three basic types: elastic, fibrous and hyaline. The contact surfaces of joints are covered with hyaline cartilage. Cartilage is organized into a layered structure, which is functionally and structurally divided into four layers. The surface layer is responsible for its smoothness and is resistant to friction. It makes up about 10-20% of the total depth of the cartilage. It can compress about 25 times more than the middle layer.

D. Hwang et al. (Eds.): ICCCI 2014, LNAI 8733, pp. 214–222, 2014.

Adult cartilage is composed of 75% water and 25% solid compounds. The metabolism of cartilage is mainly anaerobic. Due to cartilage's shortage of its own vascular supply and nerve fibres, it has a very low ability to repair itself or remove metabolites. Despite its low metabolic turnover, the replacement and the continuous exchange of cells occurs in cartilage. [13] [14] [15]

## 2    Chondromalacia

Chondromalacia or chondropathy is a pathological condition in which there is softening of the cartilage and damage to the lattice, often with its erosion and fissuration caused by damage to adjacent bone. Chondrocytes have only limited ability to self-repair. When there is a defect, they do not travel into the affected site and are able to only synthesize new cartilage in their immediate vicinity. According to the extent of macroscopic disability, the degrees of disability of chondromalacia are most often judged according to the Outerbridge classification:

Degree 0 – physiological cartilage;
Degree I – cartilage with swelling and softening;
Degree II – partial rupture with a crack on the surface that does not interfere with
        the subchondral bone
Degree III – a crack extending to the subchondral bone with a diameter of up to 1.5 cm;
Degree IV - exposed subchondral bone.
    [13], [14], [15]

## 3    Displaying Cartilage with Magnetic Resonance Imaging

Magnetic resonance imaging (MRI) is the most common investigative method. Thanks to its high distinctive ability and spatial resolution, it is the optimal non-invasive method for viewing the soft tissues of joints and cartilage. Chondral separations manifest as vertical defects in cartilage that extend deeply to the subchondral bone and are sharply outlined against the surrounding cartilage. The best results in chondral pathology imaging are achieved via a proton density weighted sequence with fat suppression and a gradient spin-echo sequence. During readings and evaluations by radiologists or orthopaedists, small chondral lesions may remain undiagnosed. Post-processing methods based on colour coding can significantly contribute to more accurate diagnostic conclusions. [13], [14], [15], [16], [17], [18]

## 4    The Proposed Algorithm of Image Segmentation for
        Articular Cartilage

The main contribution of this work is to create an appropriate segmentation algorithm that can detect changes in the density of cartilage tissue in order to accurately capture ruptures in the surface of cartilage. The main objective is to allow various tissue structures that are represented by shades of grey to be clearly separated into isolated output sets. Each set is assigned a different colour. The algorithm produces a colour map that represents individual analysed structures. Based on these colour maps, it is possible to identify not only areas where there is a change in the density of the analysed cartilage, but also with suitable thresholding, it can bring out the subject of our focus and mark the rest as background. The key factor of the proposed algorithm is sensitivity. Sensitivity can control a minimum slope of detected intensity [1], [2], [4], [6], [19].

### 4.1    A Fast Thresholding Algorithm

The proposed fast thresholding algorithm can be used, in particular, for segmentation of tissues that contain rather different structures. The core of this approach is to determine a suitable membership function for each pixel in the input image and then match pixels with the same properties into output classification classes. This approach represents the main difference from standard hard thresholding methods where the input decision-making criterion is a fixed thresholding value.

The algorithm can be divided into two main parts:

pre-processing of the input image data
outputting for creating colour mapping of tissue structures.
It is assumed that the image signal, whitch is defined by function f(x,y) that is represented by the histogram H(i) is the input for the algorithm,
where:
i is the image intensity,
x, y - coordinates of individual pixels
The first step in segmentation is dividing the input image into N corresponding different regions.

The second step in segmentation is defining partial membership functions, typically for the $n^{th}$ output region:

$$n = 1, 2, ... N \tag{1}$$

The next step is normalising the input image's histogram. The approach is practically identical to all other thresholding methods because the histogram is the basis for identifying the minimum and maximum values for the decisive threshold. Sometimes, problems can appear in the separation of tissues in medical images because individual levels of tissue structures are quite often not known, this being, for instance, the case with articular cartilage mapping. In that case, it is recommended to unify the input image as much as possible – of course, all remote values should be, ideally, removed. Thus, it is possible to restrict the interval in a closed interval [0,1].

An important segment of the algorithm is the detection and analysis of maximum values. The basic assumption for this step is that the number of histogram peaks corresponds to the number of output classes. When mapping complex tissue structures, it frequently occurs that the tissue being mapped is comprised of many different intensities that should be distinguished reliably and efficiently. This is a big advantage of the proposed solution. The number of histogram peaks can be determined as follows:

$$N = NUM_{MAX} \tag{2}$$

Where:

NUM$_{MAX}$ is number of histogram peaks.

The final part in detecting the maximum values is filtration of the output image. A low-pass filter was used for filtering. The output histogram downstream from the filter can be described as follows:

$$H_{LP}(i) = H(i) * K \tag{3}$$

After the maximum values are detected, it is necessary to adjust N known distributions of image intensities. This adjustment is made using the formula that is made up of contributions from all partial probability distributions:

$$H_{LP}(i) \approx \sum_{n=1}^{N} \Omega_n p_n(x) \tag{4}$$

where:

$\Omega_n$ is a specific weight function

$p_n(x)$ is the distribution probability

An optimising procedure should be performed to adjust each image intensity. In order to minimise errors in image intensity, a gradual iteration algorithm has been used. The optimising issue can be described as follows:

$$\min_{l \in N} |H_{LP}(I) - \sum_{n=1}^{N} \Omega_n p_n(x)|^2 \tag{5}$$

The next step derives the membership function for each output region. This key phase is divided into several parts. First, it is necessary to estimate the probability that the x pixel in the input image is a part of the REG$_n$ region. Using the standard definition of probability, the estimate can be formulated as follows:

$$p(REG_n) = \frac{p(x|REG_n)p(REG_n)}{\sum_{n=1}^{N} p(x|REG_n)p(REG_n)} \tag{6}$$

The key objective is to use the membership function. For this, regularities of the histogram are used to model the resulting Fuzzy sets. The sets provide us information about the membership. The standard Gauss function of relevance appears to be a very efficient approach.

Now it is necessary to define and assign the membership function for the respective region. The membership function of the input image (x) in the respective output $REG_n$ region can be defined as $\mu_n(f(x))$. Let us assume that the following restrictions apply with respect to that function:

$$\sum_{n=1}^{N} \mu_n(f(x)) = 1 \qquad (7)$$

If the input image is segmented in iteration steps, the first iteration step will be:

$$TI(y) = \max_n \mu_n(f(x)) \qquad (8)$$

TI(y) is the output image with thresholding. Of course, this is the example with the simplest thresholding. More detailed segmentation needs the neighbouring membership values, so it is advisable to carry out another step in the process.

A common issue that should be kept in mind is the maximum possible noise invariance. Colour mapping close to object edges where the intensity of pixels changes rather dramatically appears which, of course, impairs the segmentation results. For this reason, it is recommended to optimise the process in order to achieve greater noise resistance, especially close to the object edges. These artefacts appear typically in bone - muscle tissue borders. This approach is linear. For a non-linear process, a median can be used for each channel. The final mathematical model is:

$$\mu_n(f(x)) = med_n(\mu_n(f(x))) \qquad (9)$$

A median is normally used as a robust indicator of position which is robust against remote observations. The median is calculated for the environment of each pixel. In some cases, a rather good solution has been to replace the median with another averaged operation. But here, only the medial with the best noise resistance is used. Using the non-linear approach, the final thresholding image can be formulated using the formula below:

$$TI(y) = \max_n(med_n(\mu_n(f(x)))) \qquad (10)$$

[10], [11], [12],[16], [17, [18]

## 5    Testing of Algorithm

The following outputs show the use of the proposed algorithm for real patient data. A very important aspect for doctors is the identifying regions of interest (ROI) of the analysed image of the suspected pathology. Consequently, it is necessary to interpolate the selected area in order to improve the quality of the analysed record. It is important to note that the quality of the input image largely affects the quality of segmentation and hence the relevance of the results obtained.

- Pathology 1:

An obvious cartilage defect of the medial femoral condyle affecting the subchondral bone - Outerbridge grade IV. The resulting segmented image (Error: Reference source not found) clearly indicates the defect in the cartilage and compared to the native MR image (Error: Reference source not found), it significantly displays anatomical structures.

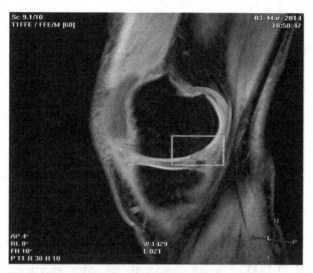

**Fig. 1.** A Sagittal T1 WATS-c sequence selectively displaying articular cartilage with the maximum suppression of signals of the surrounding tissues

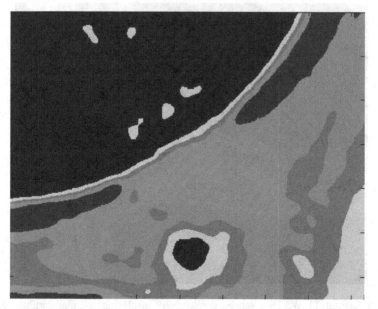

**Fig. 2.** A Sagittal T1 WATS-c sequence selectively displaying articular cartilage with the maximum suppression of signals of the surrounding tissues - colour mapping with interpolation

• Pathology 2:

In the native MR image, the cartilage defect of the medial femoral condyle is apparent (Fig. 3). The image with interpolated colour coding (Fig. 4) also shows chondropathy in the dorsal section of the medial femoral condyle - Outerbridge grade II, which was not recognized in the native image.

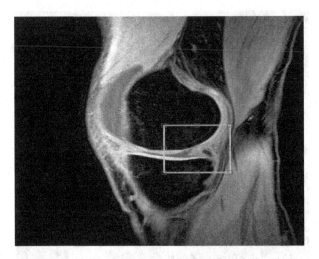

**Fig. 3.** A Sagittal T1 WATS-c sequence selectively displaying articular cartilage with the maximum suppression of the signals of surrounding tissues

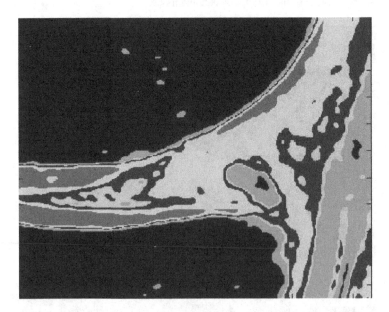

**Fig. 4.** A Sagittal T1 WATS-c sequence selectively displaying articular cartilage with the maximum suppression of signals of the surrounding tissues - colour mapping with interpolation

Pathological changes in the cartilage in the marked area (ROI) are shown on data obtained from an MRI. These changes are often indistinctly represented by visible changes of luminance values. The output of the proposed algorithm is a selective colour map that duplicates individual tissue structures according to their density. For pathology no 1, the cartilage tissue is marked with a deep red colour. In the upper part of the image, the interruption of cartilage is clearly visible, which was not very clear in the original image.

## 6   Conclusion

Image segmentation has a very wide application for medical image analysis. Images from magnetic resonance form the core basis for the analysis and detection of pathological changes in knee cartilage. The main disadvantage is that the pathologies of cartilage are often poorly identifiable. It is often a pressing issue for the physician to detect locations subject to pathological changes as those locations are typically presented by a minor change in the brightness scale which is almost impossible to recognise for a human eye. The only chance to improve visibility of pathological changes is supply of a contrast substance which, however, loads a human organisms with a radiation dose and effects of such examination are not too satisfactory. The proposed segmentation method can efficiently separate each tissue structure and identify locations subject to pathological changes. This is of a major benefit for the physicians who may perform a better diagnosis even with pathological changes being in an early stage. The proposed methodology is able to successfully identify individual structures that appear in the analysed tissue under varying densities. A colour mapping algorithm has been developed in collaboration with Podlesí Hospital in Třinec. MRI images have been used as test data. After selecting the region of interest, an interpolation of image data has been carried out in order to soften and smooth the image map. With this segmentation methodology, we are able to very effectively identify the interruption of cartilage and to suggest further treatment plans.

**Acknowledgment.** The work and the contributions were supported by the SP2014/194 'Biomedicínské inženýrské systémy X' project, and the paper was written within in the Framework of the IT4Innovations Centre of Excellence project, reg. no. CZ.1.05/1.1.00/02.0070 supported by Operational Programme 'Research and Development for Innovations' funded by the Structural Funds of the European Union and the state budget of the Czech Republic. This paper was written within in the Framework of the "Support for Research and Development in the Moravian-Silesian Region 2013 DT 1 - International Research Teams" (RRC/05/2013) project financed from the budget of the Moravian-Silesian Region. The paper was written within in the framework of BIOM (reg. č. CZ.1.07/2.3.00/20.0073).

222    J. Kubicek et al.

# References

1. Hlaváč, V., Sedláček, M.: Zpracování signálů a obrazů, skripta ČVUT Praha, Vydavatelství ČVUT Praha (2005) ISBN 80-01-03110-1
2. Horová, I., Zelinka, J.: Numerické metody (2. vyd.) Masarykova univerzita v Brně, 294 (2004) ISBN 8021033177
3. Klíma, M., Bernas, M., Hozman, J., Dvořák, P.: Zpracování obrazové informace, skripta ČVUT Praha, Vydavatelství ČVUT Praha (1999) ISBN 80-01-01436-3
4. Marr, D., Hilderth, E.: Theory of edgedetection. Proc. Royal Soc. Lond. B 207, 187–217 (1999)
5. Smith, S.M., Brady, J.M.: SUSAN - A newapproach to lowlevel image processing. International Journal of Computer Vision (1997)
6. Šonka, M., Hlaváč, V., Boyle, R.: Image ProcessingAnalysis, and Machine Vision. PWS Publishing, Pacific Grove (1999) ISBN 0-534-95393-X
7. Otsu, N.: A threshold selection method from gray-scale histogram. IEEE Trans. on Sys., Man and Cyb. 9(1), 62–66 (1979)
8. Szczepaniak, P., Lisboa, P.J.G., Kacprzyk, J.: Fuzzy systems in medicine. Physica-Verlag, New York (2000)
9. Ville, D.V.D., Nachtegael, M., der Weken, D.V., Kerre, E.E., Philips, W., Lemahieu, I.: Noise reduction by fuzzy image filtering. IEEE Trans. Fuzzy Sys. 11(4) (2003)
10. Vegas-Sanchez-Ferrero, G., et al.: On the influence of interpolation on probabilistic models for ultrasonic images. In: Proc, of the ISBI, Rotterdam, Netherlands (2010)
11. Collins, D., et al.: Design and construction of a realistic digital brain phantom. IEEE Trans. Med. Imaging 17(3), 463–468 (1998)
12. Fernández, S., et al.: Soft tresholding for medical image segmentation. IEEE EMBS (2010)
13. Štouračová, A., et al.: Možnosti zobrazení artikulární chrupavky včetně volumetrických měření. Česká radiologie 65(1), 61–69 (2011) ISSN 1210-7883
14. Junqueira, L.C.U., Carneiro a Robert O Kelley, J.: Základy histologie. 1. vyd. v ČR. Jinočany: H, vi, 502 s (1997) ISBN 80-857-8737-7
15. Višňa, Petr a Radek Hart. Chrupavka kolena. 1. vyd. Praha: Maxdorf, 205 (2006) ISBN 80-734-5084-4
16. Kasturi, R., Jain, R.C.: Computer Vision: Advances & Applications. IEEE Comput. Society Press, Los Alamitos (1991)
17. Bezdek, J.C., Pal, S.K.: Fuzzy Models for Pattern Recognition. IEEE Press, New York (1992)
18. McAuliffe, M.J., Eberly, D., Fritsch, D.S., Chaney, E.L., Pizer, S.M.: Scale-space boundary evolution initialized by cores, In: Höhne, K.H., Kikinis, R. (eds.) VBC 1996. LNCS, vol. 1131, pp. 173–182. Springer, Heidelberg (1996)
19. Falcão, A.X., Udupa, J.K., Samarasekera, S., Sharma, S.: User-steered image segmentation paradigms: live wire and live lane. Graphical Models Image Process 60, 233–260 (1998)
20. Udupa, J.K., Saha, P.K., Lotufo, R.A.: Fuzzy-connected object definition in images with respect to co-objects. In: Proc. of SPIE: Medical Imaging, vol. 3661, pp. 236–245 (1999)

# Enhanced Face Preprocessing and Feature Extraction Methods Robust to Illumination Variation

Dong-Ju Kim, Myoung-Kyu Sohn, Hyunduk Kim, and Nuri Ryu

Dept. of Convergence, Daegu Gyeongbuk Institute of Science & Technology (DGIST)
50-1 Sang-Ri, Hyeongpung-Myeon, Dalseong-Gun, Daegu, 711-873, Korea

**Abstract.** This paper presents an enhanced facial preprocessing and feature extraction technique for an illumination-roust face recognition system. Overall, the proposed face recognition system consists of a novel preprocessing descriptor, a differential two-dimensional principal component analysis technique, and a fusion module as sequential steps. In particular, the proposed system additionally introduces an enhanced center-symmetric local binary pattern as preprocessing descriptor to achieve performance improvement. To verify the proposed system, performance evaluation was carried out using various binary pattern descriptors and recognition algorithms on the extended Yale B database. As a result, the proposed system showed the best recognition accuracy of 99.03% compared to other approaches, and we confirmed that the proposed approach is effective for consumer applications.

**Keywords:** Face recognition, Preprocessing, illumination variation.

## 1    Introduction

Numerous face recognition methods have been developed for face recognition in the last few decades [1]. However, an illumination-robust face recognition system is still a challenging problem due to difficulty in controlling the lighting conditions in practical applications [2], [3]. Recently, numerous approaches have been proposed to deal with this problem. Basically, these approaches can be classified into three main categories: preprocessing, illumination invariant feature extraction, and face modeling [4]-[6]. Among them, local binary pattern (LBP) has recently received increasing interest to overcome the problem caused by illumination variation on the face [7], [8]. More recently, a centralized binary pattern (CBP) [9] and a center-symmetric local binary pattern (CS-LBP) [10] were introduced for face representation.

In this paper, we propose a novel face recognition method using a preprocessing descriptor and facial feature robust to illumination variation. We first devise an enhanced center-symmetric local binary pattern (ECS-LBP) descriptor emphasizing the diagonal component of previous CS-LBP to make a more illumination-robust binary pattern image. Here, the diagonal components are emphasized because facial textures along the diagonal direction contain much more information than those of other directions. Next, we introduce a fusion method based on a facial feature, i.e., differential two-dimensional principal component analysis (D2D-PCA). The proposed D2D-PCA can be simply derived from 2D-PCA, in which 2D-PCA is line-based local features. Since

D. Hwang et al. (Eds.): ICCCI 2014, LNAI 8733, pp. 223–232, 2014.

differential components between lines rarely vary in relation to illumination direction, we expect the proposed feature to be able to cope with illumination variation.

## 2     Illumination-Robust Face Recognition

### 2.1     System Architecture

This paper proposes a novel face recognition system that uses an enhanced facial pre-processing technique, i.e., ECS-LBP, and an illumination-robust facial feature, i.e., D2D-PCA. The ultimate aim of the proposed approach is to improve the overall recog-nition performance under harsh illumination conditions. The whole architecture of the proposed system is depicted in Fig. 1. The face image first undergoes the enhanced preprocessing procedure using the ECS-LBP descriptor, producing an illumination-robust image. The binary pattern image is then partitioned based on the vertical center line which is the eye center of the face region. Next, D2D-PCA is performed on the left and right images, and each distance score is computed using Euclidian distance mea-surement. Finally, the score normalization and fusion procedures are applied, and the nearest neighbor classifier is utilized to recognize an unknown user.

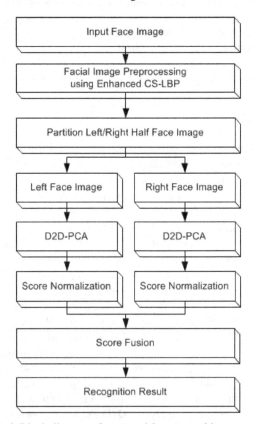

**Fig. 1.** Block diagram of proposed face recognition approach

## 2.2    Preprocessing

To make a more significant pattern, we modified the CS-LBP operator by reordering the bit priorities as pre-defined directions in this work. Generally, the decimal value of most binary pattern operators is created by combining each binary code toward continuous direction. In this view, we can suppose that each binary unit has a different characteristic in terms of facial texture. Thus, we rearrange the bit priorities in time of pattern generation as follows:

$$ECS - LBP(P,R) = \sum_{p=0}^{(P/2)-1} s(g_p - g_{p+(P/2)}) \times 2^{w_{P,R}(p)}, \qquad s(x) = \begin{cases} 1, & x \geq 0 \\ 0, & x < 0, \end{cases} \tag{1}$$

where $w_{P,R}(p)$ means a weighting function to decide the bit priority. Here, suppose that the 3x3 neighborhood pixel positions are set as shown in ref [10]. When $P$ and $R$ are set 8 and 1, respectively, $w(p)$ is defined by

$$w(p) = (3,1,2,0), \quad p = 0,1,2,3. \tag{2}$$

In the proposed ECS-LBP descriptor, we assign the high weight to components of diagonal directions, and we then assign weight to components of the vertical and horizontal directions as sequential steps. Fig. 2 shows facial texture images transformed by various binary pattern operators, such as LBP, CBP, CS-LBP, and ECS-LBP. As seen Fig. 2, we can confirm that the ECS-LBP operator achieves a more significant facial texture than other operators, since we set the smallest bit priority to component of the horizontal direction.

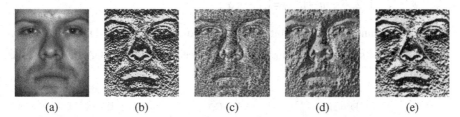

| (a) | (b) | (c) | (d) | (e) |

**Fig. 2.** Example of various binary pattern images; (a) original image, (b) binary pattern image obtained by LBP operator, (c) binary pattern image obtained by CBP operator, (d) binary pattern image obtained by CS-LBP operator, (e)   binary pattern image obtained by proposed ECS-LBP operator.

## 2.3    Differential 2D-PCA

The basic idea of D2D-PCA is that 2D-PCA [11] is a line-based local feature set; therefore, the differential components between line features will be more robust against illumination variation than the original feature. In the face recognition using principal component analysis (PCA) [12], 2D face image matrices were previously transformed into 1D image vectors column by column or row by row fashions. However, concatenating 2D matrices into 1D vector often leads to a high-dimensional vector space, where it is difficult to evaluate the covariance matrix accurately due to its large size. To overcome these problems, a new technique called 2D-PCA was proposed. 2D-PCA, which directly computes eigenvectors of the so-called image covariance matrix without matrix-to-vector conversion was proposed to decrease the computational cost of the standard PCA. Because the size of the image covariance matrix is equal to the width of images, 2D-PCA evaluates the image covariance matrix more accurately and computes the corresponding eigenvectors more efficiently than PCA. It was reported that the recognition accuracy of 2D-PCA on several face databases was higher than that of PCA, and the feature extraction method of 2D-PCA is computationally more efficient than PCA. Consider an $m$ by $n$ image matrix $A$. Let $X \in R^{n \times d}$ be a matrix with orthonormal columns, $n \geq d$. Projecting $A$ onto $X$ yields a $m$ by $d$ matrix $Y = AX$. Then, the optimal projection matrix $X$ is obtained by computing the corresponding eigenvectors of the image covariance matrix. As a result, the feature vector $Y$ of 2D-PCA, in which $Y$ has a dimension of $m$ by $d$, is obtained by projecting the images, $A$ into the eigenvectors as follows:

$$Y_k = (A - \overline{A}) X_k, \quad k = 1, 2, \cdots d. \tag{3}$$

Next, the D2D-PCA feature can be simply obtained from the corresponding 2D-PCA feature vector. Since 2D-PCA is local feature matrices against each $m$-line, the differential components are easily calculated by subtracting each 2D-PCA feature between horizontally neighboring lines. Therefore, D2D-PCA feature is computed by

$$\begin{aligned} dy_{i,j} &= y_{i+1,j} - y_{i,j}, \\ i &= 1, 2, \cdots, m-1, \quad j = 1, 2, \cdots, d, \end{aligned} \tag{4}$$

where $y_{i,j}$ means 2D-PCA feature matrices, and $dy_{i,j}$ denotes D2D-PCA feature matrices. Fig. 3 shows the feature extraction procedure of D2D-PCA from the given 2D image.

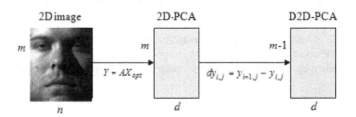

**Fig. 3.** Feature extraction of D2D-PCA

We also divide a face image into two sub-images based on the vertical center line to slightly decrease illumination effects. Since the illumination effects in the two partitioned regions of the face are different from each other, we applied the D2D-PCA to each half-face image and integrated the results generated by each half-face image at the score level. Furthermore, some performance degradation occurred in the approach using a whole face image, because line features contain horizontal information of the face, and the horizontal pixel-level intensities of the left and right regions differ according to the illumination directions. Thus, we separately considered partial face images as shown in Fig. 4, and integrated corresponding score results from sub-images. Next, we applied a sigmoid function to normalize these raw-scores from 0 to 1, since the distance scores from the left and right half-face images have different numerical ranges and statistical distributions. Then, the score fusion phase is performed using two normalized-scores, and the nearest neighbor classifier is utilized to recognize an unknown user.

(a)                                               (b)

**Fig. 4.** Region partitioning of sample face images; (a) original images, (b) partitioned images

## 3    Experimental Results

Performance evaluation was carried out using the extended Yale face database B which consists of 2,414 face images for 38 subjects representing 64 illumination conditions under the frontal pose [13]. An example images from the extended Yale face database B are shown in Fig. 5. In this work, we partitioned the extended Yale face database B into training and testing sets. Each training set comprised five images per subject, and the remaining images were used to test the proposed system. Note that illumination-invariant images were used for training, and the illumination-variant images were employed for testing.

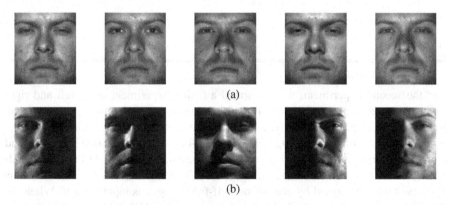

(a)

(b)

**Fig. 5.** Some face images from the extended Yale face database B; (a) training images, (b) test images

In the first experiment, we investigated the recognition performance of the proposed ECS-LBP descriptor using various recognition algorithms, such as PCA, linear discriminant analysis (LDA) [14], 2D-PCA, and D2D-PCA. The experimental results were also evaluated using several binary pattern descriptors, such as LBP, CBP and CS-LBP, for performance comparison. The recognition results obtained using the PCA, LDA, 2D-PCA, and D2D-PCA recognition algorithms with whole-face images are shown in Table 1. From the experimental results, the recognition rates were found to be 85.86%, 73.06%, 96.37% and 98.26% for PCA, LDA, 2D-PCA and D2D-PCA, when an ECS-LBP image was employed. For overall recognition algorithms, the approach using the ECS-LBP descriptor outperformed methods using the other binary pattern descriptors in terms of recognition accuracy. Also, the D2D-PCA approach with an ECS-LBP operator showed performance improvements of 13.83%, 2.38%, 19.98%, and 8.33% compared to raw, LBP, CBP and CS-LBP images, respectively. Consequently, the proposed method using the ECS-LBP descriptor and D2D-PCA feature showed better recognition accuracy than other approaches. These results confirm that the proposed ECS-LBP descriptor and D2D-PCA feature is robust to illumination variations.

**Table 1.** Summary of Recognition Accuracies using Whole-Face Images

| Input Image | Recognition Algorithms | | | |
|---|---|---|---|---|
| | PCA | LDA | 2D-PCA | D2D-PCA |
| Raw | 49.11% | 55.26% | 64.58% | 84.43% |
| LBP | 73.24% | 51.87% | 91.46% | 95.88% |
| CBP | 50.67% | 46.57% | 66.55% | 78.28% |
| CS-LBP | 61.33% | 49.06% | 75.81% | 89.93% |
| ECS-LBP | 85.86% | 73.06% | 96.37% | **98.26%** |

In the second experiment, we performed a fusion experiment using left and right sub-images to minimize the illumination effect, leading to performance improvement. Here, we only employed 2D-PCA and D2D-PCA in the fusion experiment, and the fusion process utilizes the sigmoid function-based normalization method and weighted-summation rule. Fig. 6 shows each recognition result of 2D-PCA and D2D-PCA obtained when a raw image and an ECS-LBP image were used. Note that this experiment was performed by employing half-face images as input images. When the ECS-LBP operator was applied, the recognition rates of D2D-PCA were 92.62% and 97.70% for left and right images, respectively. In addition, the recognition rates of 2D-PCA were 91.41% and 94.10% for left and right images, respectively. In addition, Note that the recognition rates of D2D-PCA were better than those of 2D-PCA. In particular, the 2D-PCA approach with an ECS-LBP image achieved performance improvements of 25.72% and 31.65% compared to the results of left raw images and right raw images, respectively. Also, the D2D-PCA approach with an ECS-LBP image showed performance improvements of 5.26% and 8.45% compared to left raw images and right raw images, respectively. These results also confirm that the proposed ECS-LBP operator is an effective preprocessing method against illumination variation.

Also, we performed the fusion experiments with different weights of the left face score against 2D-PCA and D2D-PCA using raw images and ECS-LBP images. The fusion results along with different weights are shown in Fig. 7, and the maximum recognition results are summarized in Table 2. From these results, we can notice that the proposed approach using the ECS-LBP images and D2D-PCA feature achieved better accuracy than the other approaches over the entire range of weights. Also, the corresponding maximum recognition rates of methods using D2D-PCA were 95.59% and 99.03%, for raw images and ECS-LBP image, respectively. In other words, the proposed fusion approach with ECS-LBP images showed performance improvement of 3.44% in comparison to the method with raw images when D2D-PCA were employed. Consequently, we confirmed the effectiveness of the proposed face recognition system under illumination-variant conditions from the experimental results.

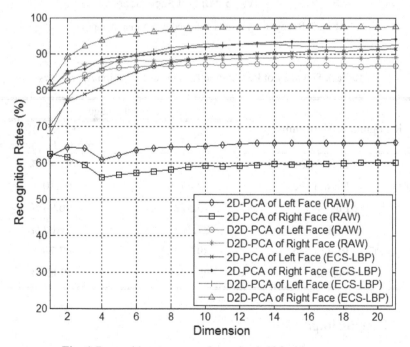

**Fig. 6.** Recognition accuracy when using half-face images

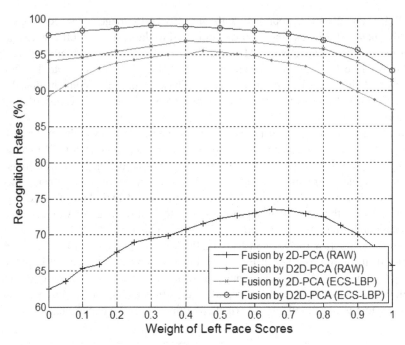

**Fig. 7.** Recognition results obtained using fusion approach

**Table 2.** Summary of Recognition Accuracies for Fusion Approaches

| Recognition Algorithms | | Input Image | |
|---|---|---|---|
| | | Raw | ECS-LBP (Proposed Approach) |
| 2D-PCA | Left Face | 65.69 % | 91.41 % |
| | Right Face | 62.45 % | 94.10 % |
| | Fusion | 73.57 % | 96.94 % |
| D2D-PCA | Left Face | 87.36 % | 92.62 % |
| | Right Face | 89.25 % | 97.70 % |
| | Fusion | 95.59 % | **99.03 %** |

## 4    Conclusions

This paper presented an enhanced facial preprocessing and feature extraction technique for an illumination-roust face recognition system. To minimize illumination effects and maximize performance improvements, the proposed system employed a novel ECS-LBP operator, D2D-PCA feature, and a fusion technique integrating two half-face images. Performance evaluation of the proposed approach was carried out with the extended Yale B database, and the corresponding recognition results confirmed that the proposed approach achieves the best recognition rate of 99.03%. Through the experimental results, we were able to confirm the effectiveness and performance improvement of the proposed system under illumination-variant conditions.

**Acknowledgement.** This work was supported by the DGIST R&D Program of the Ministry of Education, Science and Technology of Korea (14-IT-03), and Ministry of Culture, Sports and Tourism (MCST) and Korea Creative Content Agency (KOCCA) in the Culture Technology (CT) Research & Development Program (Immersive Game Contents CT Co-Research Center).

# References

1. Zuo, F., de With, P.H.N.: Real-time embedded face recognition for smart home. IEEE Trans. Consum. Electron. 51(1), 183–190 (2005)
2. Zhao, W., Chellappa, R., Phillips, R.J., Rosenfeld, A.: Face recognition: A literature survey. ACM Comput. Surv. 35(4), 399–458 (2003)
3. Abate, A.F., Nappi, M., Riccio, D., Riccio, G.: 2D and 3D face recognition: A survey. Pattern Recognit. Lett. 28(14), 1885–1906 (2007)
4. Chen, W., Er, M.J., Wu, S.: Illumination compensation and normalization for robust face recognition using discrete cosine transform in logarithm domain. IEEE Trans. Syst. Man Cybern. Part B-Cybern. 36(2), 458–466 (2006)
5. Ruiz-del-Solar, J., Quinteros, J.: Illumination compensation and normalization in eigenspace-based face recognition: A comparative study of different pre-processing approaches. Pattern Recognit. Lett. 29(14), 1966–1979 (2008)
6. Hsieh, P.C., Tung, P.C.: Illumination-robust face recognition using an efficient mirror technique. In: International Congress on Image and Signal Processing, pp. 1–5 (2009)
7. Ahonen, T., Hadid, A., Pietikainen, M.: Face description with local binary patterns: Application to face recognition. IEEE Trans. Pattern Anal. Mach. Intell. 28(12), 2037–2041 (2006)
8. Zhang, W., Shan, S., Chen, X., Gao, W.: Local Gabor binary patterns based on mutual information for face recognition. Int. J. Image Graph. 7(4), 777–793 (2007)
9. Fu, X., Wei, W.: Centralized binary patterns embedded with image Euclidean distance for facial expression recognition. In: International Conference of Neural Computation, vol. 4, pp. 115–119 (2008)
10. Heikkilä, M., Pietikäinen, M., Schmid, C.: Description of interest regions with center-symmetric local binary patterns. In: Kalra, P.K., Peleg, S. (eds.) ICVGIP 2006. LNCS, vol. 4338, pp. 58–69. Springer, Heidelberg (2006)

11. Jian, Y., David, Z., Alejandro, F., Yang, J.Y.: Two-dimensional PCA: A new approach to appearance-based face representation and recognition. IEEE Trans. Pattern Anal. Mach. Intell. 26(1), 131–137 (2004)
12. Turk, M., Pentland, A.: Eigenfaces for recognition. J. Cogn. Neurosci. 3(1), 71–86 (1991)
13. Georghiades, A., Belhumeur, P., Kriegman, D.: From few to many: Illumination cone models for face recognition under variable lighting and pose. IEEE Trans. Pattern Anal. Mach. Intell. 23(6), 643–660 (2001)
14. Belhumeur, P.N., Hespanha, J.P., Kriegman, D.J.: Eigenfaces vs. Fisherfaces: Recognition using class-specific linear projection. IEEE Trans. Pattern Anal. Mach. Intell. 19(7), 711–720 (1997)

# Facial Expression Recognition Using Binary Pattern and Embedded Hidden Markov Model

Dong-Ju Kim, Myoung-Kyu Sohn, Hyunduk Kim, and Nuri Ryu

Dept. of Convergence, Daegu Gyeongbuk Institute of Science & Technology (DGIST)
50-1 Sang-Ri, Hyeongpung-Myeon, Dalseong-Gun, Daegu, 711-873, Korea

**Abstract.** This paper proposes a robust facial expression recognition approach using an enhanced center-symmetric local binary pattern (ECS-LBP) and embedded hidden Markov model (EHMM). The ECS-LBP operator encodes the texture information of a local face region by emphasizing diagonal components of a previous center-symmetric local binary pattern (CS-LBP). Here, the diagonal components are emphasized because facial textures along the diagonal direction contain much more information than those of other directions. Generally, feature extraction and categorization for facial expression recognition are the most key issue. To address this issue, we propose a method to combine ECS-LBP and EHMM, which is the key contribution of this paper. The performance evaluation of proposed method was performed with the CK facial expression database and the JAFFE database, and the proposed method showed performance improvements of 2.65% and 2.19% compared to conventional method using two-dimensional discrete cosine transform (2D-DCT) and EHMM for CK database and JAFFE database, respectively. Through the experimental results, we confirmed that the proposed approach is effective for facial expression recognition.

**Keywords:** Facial Expression Recognition, Binary Pattern, EHMM.

## 1    Introduction

A challenging research issue and one that has been of growing importance to those working on human-computer interactions are to endow a machine with an emotional intelligence. Such a system must be able to create an affective interaction with users: it must have the ability to perceive, interpret, express and regulate emotions [1]. In this case, recognizing the user's emotional state is one of the main requirements for computers to successfully interact with humans [2]. Facial expression recognition is one of the most powerful, natural and immediate means for human beings to communicate their emotions. Automatic facial expression analysis is an interesting and challenging problem, and impacts important applications in many areas such as human–computer interaction and data-driven animation. There are two common approaches to extract facial features: geometric feature-based methods and appearance-based methods. Geometric features present the shape and locations of facial components, which are extracted to form a feature vector that represents the face geometry. Facial action cod-

D. Hwang et al. (Eds.): ICCCI 2014, LNAI 8733, pp. 233–242, 2014.
© Springer International Publishing Switzerland 2014

ing system (FACS) introduced by Ekman and Friesen [3] is one of the most popular geometric feature-based methods that represents facial expression using a set of action units (AU), where each action unit corresponds to the physical behavior of a specific facial muscle. On the other hand, appearance-based methods employ image filter or filter bank on the whole face or some specific regions of the facial image in order to extract changes in facial appearance. Recently, facial expression analyses based on local binary pattern (LBP) [4] and its variants such as centralized binary pattern (CBP) [5] and CS-LBP [6] have gained much popularity for their superior performances.

In this paper, we propose a robust facial expression recognition approach using ECS-LBP and EHMM. In fact, the methodology using 2D-DCT and EHMM was previously employed in the face recognition fields. However, this paper applied the EHMM in a different manner for successful facial expression recognition. In particular, we devise a novel feature descriptor, i.e., enhanced center-symmetric local binary pattern (ECS-LBP), to achieve better performance compare to conventional features such as 2D-DCT, LBP, CBP and CS-LBP. The ECS-LBP descriptor is the modified binary pattern of emphasizing the diagonal component of previous CS-LBP to make a more illumination-robust binary pattern. Here, the diagonal components are emphasized because facial textures along the diagonal direction contain much more information than those of other directions. Consequently, we implemented a novel facial expression recognition system with ECS-LBP feature descriptor and EHMM. Performance evaluation of the proposed system was carried out using an extended Yale B database which consists of 2,414 face images for 38 subjects representing 64 illumination conditions under the frontal pose. In the experiments, we will demonstrate the effectiveness of the proposed approach by comparing it with various other approaches.

## 2    Feature Descriptor for Facial Expression Recognition

### 2.1    Conventional 2D-DCT

Two-dimensional discrete cosine transform has been employed in face recognition to reduce dimensionality. The advantage of 2D-DCT is that it is data independent. That is, the basis images are only dependent on one image instead of on the entire set of training images. It can be also implemented using a fast algorithm. Feature extraction of a face image using 2D-DCT consists of two steps [7].

In the first step, the face image is divided in small block images. Let $P \times L$ be the window size of 2D-DCT, and $Q \times M$ be the overlap size in the horizontal and vertical directions of the image. Then, the number of blocks is calculated by the following equation for an image with $W$ rows and $H$ columns.

$$T = (\frac{W-Q}{P-Q}) \times (\frac{H-M}{L-M}) \tag{1}$$

In the next step, the 2D-DCT coefficients of the image block $f(x, y)$ are calculated. If we assume that $P$ and $L$ are equal to $N$ ($P = L = N$), then 2D-DCT coefficients, $C(u, v)$ is computed defined by

$$C(u,v) = \alpha(u)\alpha(v)\sum_{x=0}^{N-1}\sum_{y=0}^{N-1} f(x,y)\beta(x,y,u,v) \qquad (2)$$

for $u,v = 0, 1, 2, ...., N-1$,

where $\alpha(u), \alpha(v) = \begin{cases} 1/\sqrt{N} & \text{for } u,v = 0 \\ 2/\sqrt{N} & \text{for } u,v = 1,2,...,N-1 \end{cases}$

and $\beta(x,y,u,v) = \cos\left[\dfrac{(2x+1)u\pi}{2N}\right] \times \cos\left[\dfrac{(2y+1)v\pi}{2N}\right]$ .

## 2.2  Conventional Binary Patterns

Recently, the LBP has received increasing interest for face representation to overcome the problem of performance degradation caused by illumination variation. The LBP operator labels the pixels of an image by thresholding a 3x3 neighborhood of each pixel with the center value, and considering the results as a binary number, of which the corresponding decimal number is used for labeling. The LBP code is derived by

$$LBP = \sum_{i=0}^{7} s(g_i - g_c) \times 2^i, \quad s(x) = \begin{cases} 1, & x \geq 0 \\ 0, & x < 0, \end{cases} \qquad (3)$$

where $g_c$ and $g_i$ denote the center pixel value and neighborhood pixel values, respectively. Also, the CBP operator compares pairs of neighbors which are in the same diameter of the circle, and compares the central pixel with the mean of all the pixels as shown in Fig. 1. Furthermore, the CS-LBP operator can be computed by only considering the corresponding patterns of symmetric pixels as shown in Fig. 1.

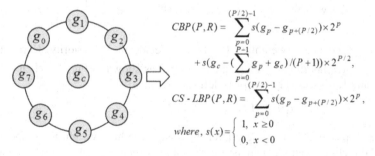

$$CBP(P,R) = \sum_{p=0}^{(P/2)-1} s(g_p - g_{p+(P/2)}) \times 2^P$$
$$+ s(g_c - (\sum_{p=0}^{P-1} g_p + g_c)/(P+1)) \times 2^{P/2},$$
$$CS\text{-}LBP(P,R) = \sum_{p=0}^{(P/2)-1} s(g_p - g_{p+(P/2)}) \times 2^P,$$
$$where, s(x) = \begin{cases} 1, & x \geq 0 \\ 0, & x < 0 \end{cases}$$

**Fig. 1.** Symmetric based-binary patterns

## 2.3  Proposed Binary Pattern

To make a more significant pattern, we modified the CS-LBP operator by reordering the bit priorities as pre-defined directions in this work. Generally, the decimal value of most binary pattern operators is created by combining each binary code toward continuous direction. In this view, we can suppose that each binary unit has a different

characteristic in terms of facial texture. In other words, previous binary pattern operators have not considered the relation between each bit position and facial texture. Thus,
we first investigated their corresponding relation by composing a texture image, in
which each texture image is created using only one pair of symmetric pixels. The resultant texture images are shown in Fig. 2, while $P$ and $R$ are 8 and 1, respectively.
As seen in Fig. 2, the composed texture image (see Fig. 2 (e)) using only horizontal
pixels contains more indistinguishable texture than other composed images (see Fig. 2
(b), (c), and (d)). This component has negative effects on the complete binary pattern
image of the CS-LBP. Thus, we assign low priority to component of the horizontal
directions. In addition, notice that other composed images have similar textures as seen
in Fig. 2 (b), (c), and (d).

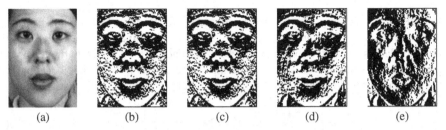

    (a)            (b)            (c)            (d)            (e)

**Fig. 2.** Relation of bit position and facial texture; (a) original image, (b) composed image using
only $g_0$ and $g_4$ terms, (c) composed image using only $g_1$ and $g_5$ terms, (d) composed image using
only $g_2$ and $g_6$ terms, (e) composed image using only $g_3$ and $g_7$ terms.

Due to the lack of significant facial textures in the composed image using only horizontal pixels, we rearrange the bit priorities in time of pattern generation as follows:

$$ECS-LBP(P,R) = \sum_{p=0}^{(P/2)-1} s(g_p - g_{p+(P/2)}) \times 2^{w_{P,R}(p)}, \qquad s(x) = \begin{cases} 1, & x \geq 0 \\ 0, & x < 0, \end{cases} \qquad (4)$$

where $w_{P,R}(p)$ means a weighting function to decide the bit priority. Here, suppose
that the 3x3 neighborhood pixel positions are set as shown in Fig. 1. When $P$ and $R$
are set 8 and 1, respectively, $w(p)$ is defined by

$$w(p) = (3,1,2,0), \quad p = 0,1,2,3. \qquad (5)$$

In the proposed ECS-LBP descriptor, we assign the high weight to components of
diagonal directions, and we then assign weight to components of the vertical and horizontal directions as sequential steps. Fig. 3 shows facial texture images transformed by
various binary pattern operators, such as LBP, CBP, CS-LBP, and ECS-LBP. As seen
Fig. 3, we can confirm that the ECS-LBP operator achieves a more significant facial
texture than other operators, since we set the smallest bit priority to component of the
horizontal direction. Consequently, the proposed ECS-LBP operator seems more stable
than other binary pattern images, as shown in Fig. 3, since it has fewer noise components than other images. Similar to 2D-DCT, each histogram feature of all binary pattern descriptors is used in observation vector of EHMM by dividing small image
blocks.

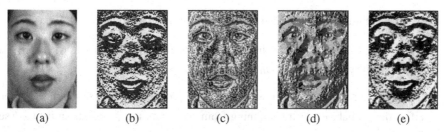

(a)                 (b)                 (c)                 (d)                 (e)

**Fig. 3.** Example of various binary pattern images; (a) original image, (b) binary pattern image obtained by LBP operator, (c) binary pattern image obtained by CBP operator, (d) binary pattern image obtained by CS-LBP operator, (e) binary pattern image obtained by proposed ECS-LBP operator

## 3    Embedded Hidden Markov Model

An HMM is a Markov chain with a finite number of unobservable states. Although the Markov states are not directly observable, each state has a probability distribution associated with the set of possible observations. EHMM is an extension of the one-dimensional HMM to deal with 2-D data such as images and videos. EHMM was first introduced for character recognition by Kuo and Agazz [8]. It was applied as a new approach to face recognition by Nefian et al. [7]. Since this method showed the best performance, we adopt Nefian's EHMM for modeling face images in this paper. EHMM comprises a set of super-states and each super-state is associated with a set of embedded-states. The super-states represent the primary image regions along the vertical direction while the embedded-states within each super-state describe the image regions along the horizontal direction in detail. From this structure, we know that the sequence of super-states is used to model a horizontal slice of the image along the vertical direction and the sequence of embedded-states in a super-state is used to model a block image along the horizontal direction. The elements of EHMM are defined as follows [7].

1) $N_0$: The number of super-states in the vertical direction.

2) $\Pi_0$: The initial super-state probability distribution, i.e., $\Pi_0 = \{\pi_{0,i} : 1 \leq i \leq N_0\}$, where $\pi_{0,i}$ is the initial probability of being in $\Lambda_0$ super-state.

3) $A_0$: The super-state transition probability matrix, i.e., $A_0 = \{a_{0,ij} : 1 \leq i, j \leq N_0\}$ where $a_{0,ij}$ is the probability of transition from $i$-$th$ super-state to $j$-$th$ super-state.

4) $\Lambda_0$: The set of one-dimensional HMM in each super-state, i.e., $\Lambda_0 = \{\lambda^i : 1 \leq i \leq N_0\}$, where $\lambda^i$ indicate the model parameters of embedded-states in $i$-$th$ super-state. Each $\lambda^i$ is represented by the one-dimensional HMM parameters as follows.

· $N_1^k$ is the number of embedded-states in the $k$-$th$ super-state.

· $\Pi_1^k = \{\pi_{1,i}^k : 1 \le i \le N_1^k\}$ is the initial state probability distribution, where $\pi_{1,i}^k$ is the probability of being in $i$-$th$ state of $k$-$th$ super-state.

· $A_1^k = \{a_{1,ij}^k : 1 \le i, j \le N_1^k\}$ is the state transition probability matrix, where $a_{1,ij}^k$ specifies the probability of transitioning from $i$-$th$ state to $j$-$th$ state in the $k$-$th$ super-state.

· $B_1^k = \{b_i^k(O_{t_0,t_1}) : 1 \le i \le N_1^k\}$ is the observation probability matrix, where $O_{t_0,t_1}$ represents the observation vector at row $t_0$ and column $t_1$, and $b_i^k(O_{t_0,t_1})$ denotes the probability of being observed the vector, $O_{t_0,t_1}$ in the $i$-$th$ state of $k$-$th$ super-state. In a continuous density HMM, the states are characterized by a continuous observation density function. The probability density function that is typically represented in terms of a mixture of Gaussian functions, i.e.,

$$b_i^k(O_{t_0,t_1}) = \sum_{m=1}^{M} c_{i,m}^k \ N(O_{t_0,t_1}, \mu_{i,m}^k, U_{i,m}^k), \qquad (6)$$

where $1 \le i \le N_1^k$, $c_{i,m}^k$ denotes the mixture coefficient for the $m$-$th$ mixture in the $i$-$th$ embedded-state of the $k$-$th$ super-state, and $N(O_{t_0,t_1}, \mu_{i,m}^k, U_{i,m}^k)$ is a Gaussian probability density function with mean vector $\mu_{i,m}^k$ and covariance matrix $U_{i,m}^k$.

Since $\Lambda_0$ denotes the set of one-dimensional HMM in each super-state, the EHMM can be completely specified by the following parameter set,

$$\lambda = \{\Pi_0, A_0, \Lambda_0\}, \qquad (7)$$

where $\Lambda_0 = \{\lambda^i : 1 \le i \le N_0\}$, and $k$-$th$ super-state is defined by the set of parameters as $\lambda^k = \{\Pi_1^k, A_1^k, B_1^k\}$. Although EHMM is more complex than a one-dimensional HMM, EHMM is more suited to 2-D images.

## 4     Experiments

The experiments were performed with two well-known data sets which are collected from the CK facial expression database [9] and the JAFFE database [10]. A set of prototypic emotional expressions includes anger, disgust, fear, happiness, sadness, and surprise. This six-class expression set is further extended as a seven-class expression set by adding a neutral expression. The CK database consists of sequences of 100 university students aged from 18 to 30 years, of which 65% are female, 15% are African-American, and 3% are Asian or Latino. Each sequence begins with a neutral expression and proceeds to a peak expression. In our setup, we selected 320 sequences from 96 subjects, each of which was labeled as one of the six basic emotions. For each sequence, the neutral emotion and three peak frames are used for expression recognition, resulting 1280 images (99 Anger, 144 Disgust, 138 Fear, 273 Happiness, 96 Sadness, 210 Surprise and 320 Neutral). The Jaffe database contains 213 images of 7 facial expressions (6 basic + 1 neutral) posted by 10 Japanese female. Facial images are cropped from original images and normalized to 150×110 pixels using two eyes position. Fig. 4 show an example of sample face images for the CK and the JAFFE facial expression database, respectively.

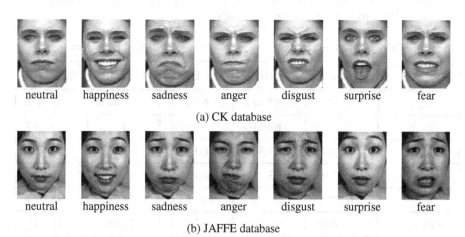

| neutral | happiness | sadness | anger | disgust | surprise | fear |

(a) CK database

| neutral | happiness | sadness | anger | disgust | surprise | fear |

(b) JAFFE database

**Fig. 4.** Sample facial expression images

In the experiment, each database is sequentially divided into five and ten groups. And 5-fold and 10-fold cross-validation was carried out to measure the average recognition rate for each database. Based on these databases, the performance evaluation of proposed approach was carried out using conventional feature descriptors such as 2D-DCT, LBP, CBP, and CS-LBP together with EHMM. The general recognition process of EHMM-based approach was presented as follows. Each emotion was previously modeled with their observation vectors by a doubly embedded Viterbi segmentation algorithm of EHMM, and these models were then saved in the database. After extracting the observation vectors corresponding to the testing face images, the probability of the observation sequence given an EHMM model was computed via a doubly embedded Viterbi recognizer. The model with the highest likelihood was selected and this model revealed the class of unknown emotional states. To extract the facial feature, we use following parameters: window size is 16×16 and the moving step is 4×4. In 2D-DCT, we get the 5×5 low frequency coefficients as the observation vector for each window block.

For the CK facial expression database, facial expression recognition results are depicted in Table 1 and 2 for 5-fold and 10-fold cross-validation, respectively. As a result, the maximum recognition rates showed 71.24% and 70.38% for 5-fold and 10-fold cross-validation, respectively. Here, we can observe that maximum rates were revealed in the proposed approach that uses ECS-LBP feature descriptor. Compare to 2D-DCT, the proposed method showed performance improvement of 2.65%.

In addition, the proposed method showed a better recognition rates compared to LBP, CBP, and CS-LBP feature descriptors. Also, we performed the experiments for JAFFE facial expression database. As a result, the recognition results are shown in Table 3 and 4 for 5-fold and 10-fold cross-validation, respectively. The maximum recognition rates showed 57.65% and 61.77% for 5-fold and 10-fold cross-validation, respectively. Similar to CK facial expression database, the maximum rates were revealed in the proposed approach that uses ECS-LBP feature descriptor. Also, the proposed method showed performance improvement of 2.19% compare to 2D-DCT. Consequently, we confirmed the effectiveness of the proposed approach using ECS-LBP and EHMM from the experimental results.

**Table 1.** Five-fold facial expression recognition results for CK database

|         | 2D-DCT | LBP    | CBP    | CS-LBP | ECS-LBP |
|---------|--------|--------|--------|--------|---------|
| Set 1   | 71.83% | 57.94% | 65.48% | 69.44% | 67.86%  |
| Set 2   | 70.47% | 68.11% | 63.39% | 70.47% | 73.23%  |
| Set 3   | 64.98% | 64.59% | 60.31% | 67.32% | 69.65%  |
| Set 4   | 68.34% | 63.71% | 66.02% | 69.88% | 69.50%  |
| Set 5   | 67.83% | 75.97% | 80.62% | 78.29% | 75.97%  |
| Average | 68.69% | 66.06% | 67.14% | 71.08% | 71.24%  |

**Table 2.** Ten-fold facial expression recognition results for CK database

|         | 2D-DCT | LBP    | CBP    | CS-LBP | ECS-LBP |
|---------|--------|--------|--------|--------|---------|
| Set 1   | 65.87% | 55.56% | 48.41% | 57.94% | 76.19%  |
| Set 2   | 58.73% | 63.49% | 68.25% | 70.63% | 68.25%  |
| Set 3   | 68.50% | 64.57% | 62.20% | 69.29% | 65.35%  |
| Set 4   | 72.44% | 61.42% | 61.42% | 64.57% | 68.50%  |
| Set 5   | 64.34% | 59.69% | 51.16% | 67.44% | 72.09%  |
| Set 6   | 63.28% | 62.50% | 62.50% | 64.84% | 64.06%  |
| Set 7   | 72.09% | 66.67% | 61.24% | 71.32% | 68.99%  |
| Set 8   | 74.62% | 72.31% | 75.38% | 74.62% | 71.54%  |
| Set 9   | 68.22% | 70.54% | 79.07% | 77.52% | 70.54%  |
| Set 10  | 66.67% | 82.17% | 77.52% | 83.72% | 78.29%  |
| Average | 67.47% | 65.89% | 64.71% | 70.18% | 70.38%  |

**Table 3.** Five-fold facial expression recognition results for JAFFE database

|          | 2D-DCT  | LBP     | CBP     | CS-LBP  | ECS-LBP |
|----------|---------|---------|---------|---------|---------|
| Set 1    | 52.38%  | 61.90%  | 52.38%  | 57.14%  | 59.52%  |
| Set 2    | 59.52%  | 57.14%  | 45.24%  | 54.76%  | 57.14%  |
| Set 3    | 59.52%  | 47.62%  | 45.24%  | 45.24%  | 52.38%  |
| Set 4    | 57.14%  | 33.33%  | 50.00%  | 54.76%  | 54.76%  |
| Set 5    | 55.56%  | 71.11%  | 57.78%  | 66.67%  | 64.44%  |
| Average  | 56.82%  | 54.22%  | 50.12%  | 55.71%  | 57.65%  |

**Table 4.** Ten-fold facial expression recognition results for JAFFE database

|          | 2D-DCT  | LBP     | CBP     | CS-LBP  | ECS-LBP |
|----------|---------|---------|---------|---------|---------|
| Set 1    | 52.38%  | 61.90%  | 71.43%  | 57.14%  | 61.90%  |
| Set 2    | 61.90%  | 57.14%  | 57.14%  | 80.95%  | 76.19%  |
| Set 3    | 66.67%  | 57.14%  | 66.67%  | 61.90%  | 57.14%  |
| Set 4    | 57.14%  | 52.38%  | 33.33%  | 47.62%  | 47.62%  |
| Set 5    | 71.43%  | 47.62%  | 57.14%  | 52.38%  | 57.14%  |
| Set 6    | 52.38%  | 42.86%  | 52.38%  | 42.86%  | 52.38%  |
| Set 7    | 47.62%  | 42.86%  | 52.38%  | 42.86%  | 52.38%  |
| Set 8    | 61.90%  | 38.10%  | 52.38%  | 76.19%  | 80.95%  |
| Set 9    | 45.45%  | 36.36%  | 40.91%  | 36.36%  | 36.36%  |
| Set 10   | 65.22%  | 86.96%  | 82.61%  | 69.57%  | 95.65%  |
| Average  | 58.21%  | 52.33%  | 56.63%  | 56.78%  | 61.77%  |

## 5 Conclusions

In this paper, we proposed a facial expression recognition approach using ECS-LBP and EHMM. By devising the ECS-LBP descriptor, we designed the facial expression recognition system. To evaluate the performance of the proposed approach, experiments were performed with the CK facial expression database and the JAFFE database, and the results confirmed that the proposed approach is effective compared to conventional approaches.

**Acknowledgement.** This work was supported by the DGIST R&D Program of the Ministry of Education, Science and Technology of Korea (14-IT-03), and Ministry of Culture, Sports and Tourism (MCST) and Korea Creative Content Agency (KOCCA) in the Culture Technology (CT) Research & Development Program (Immersive Game Contents CT Co-Research Center).

# References

1. Picard, R.: Affective computing. MIT Press, Boston (1997)
2. Cowie, R., Douglas-Cowie, E., Tsapatsoulis, N., Votsis, G., Kollias, S., Fellenz, W., Taylor, J.G.: Emotion recognition in human-computer interaction. IEEE Signal Processing Magazine (2001)
3. Tian, Y., Kanade, T., Cohn, J.: Facial Expression Analysis. In: Handbook of Face Recognition. Springer, Heidelberg (2005)
4. Shan, C., Gong, S., McOwan, P.W.: Facial Expression Recognition based on Local Binary Patterns: A Comprehensive Study. Image and Vision Computing 27(6), 803–816 (2009)
5. Fu, X., Wei, W.: Centralized binary patterns embedded with image Euclidean distance for facial expression recognition. In: Int. Conf. Neural Computation (2008)
6. Heikkilä, M., Pietikäinen, M., Schmid, C.: Description of interest regions with center-symmetric local binary patterns. In: Kalra, P.K., Peleg, S. (eds.) ICVGIP 2006. LNCS, vol. 4338, pp. 58–69. Springer, Heidelberg (2006)
7. Nefian, A., Hayes, M.: An Embedded HMM-based Approach for Face Detection and Recognition. In: Proc. IEEE Int. Conf. on Acoustics, Speech and Signal Processing, vol. 6 (1999)
8. Kuo, S., Agazzi, O.: Keyword spotting in poorly printed documents using pseudo 2-D Hidden Markov Models. IEEE Transactions on Pattern Analysis and Machine Intelligence 16, 842–848 (1994)
9. Kanade, T., Cohn, J., Tian, Y.: Comprehensive Database for Facial Expression Analysis. In: IEEE Int. Conf. Autom. Face Gesture Recog., pp. 46–53 (2000)
10. Lyons, M.J., Budynek, J., Akamatsu, S.: Automatic Classification of Single Facial images. IEEE Trans. Pattern Anal. Mach. Intell. 21(12), 357–1362 (1999)

# Creating a Knowledge Base to Support the Concept of Lean Manufacturing Using Expert System NEST

Radim Dolák, Jan Górecki, Lukáš Slechan, and Michael Kubát

Silesian University in Opava,
The School of Business Administration in Karviná, Karviná, Czech Republic
{dolak,gorecki,O130079,O130076}@opf.slu.cz

**Abstract.** This article deals with lean manufacturing principles and its model. We describe basic principles, metrics and rules for creating lean manufacturing knowledge base. The case study included in this paper deals with creating of a knowledge base that supports an implementation of the concept of lean manufacturing. The knowledge base could be used for identification of waste in each level of production areas. The knowledge base also can be used for a recommendation of appropriate methods and tools of industrial engineering to reduce the waste. The knowledge base is build using the expert system NEST.

**Keywords:** lean company, lean manufacturing, expert systems, knowledge base.

## 1    Introduction

We live in the global market environment, which is characterized by high levels of competition. It is very important to use information systems and BI solutions outputs. Output data provide the managers information support of decision-making processes and help the managers to develop or change corporate strategy [10]. The management teams of business companies have to increase the flexibility and tempo of decision-making in order to maintain pace with market developments [11].

New methods of management principles which achieve a competitive advantage are gaining importance. Getting a competitive advantage is very important for the survival of firms in the global market environment nowadays. Using the principles of lean company concept is one form of gaining a competitive advantage. These principles seek to eliminate all unnecessary processes and activities that do not bring value to the customer and profit for the company. The aim is to streamline the contrary and support processes with business value which deliver profit.

This article deals with possibility of using knowledge base of expert system for support implementing of lean manufacturing concept. We suppose that with expert system should be significantly accelerate the process of implementing lean manufacturing concept. Attention will be focused on basic information about the NEST expert system, which includes general information about the system, its structure, knowledge representation, knowledge base syntax and the rules of inference (inference

D. Hwang et al. (Eds.): ICCCI 2014, LNAI 8733, pp. 243–251, 2014.

mechanism). The case study deals with creating of knowledge base for supporting implementing of lean manufacturing concept. The knowledge base will be used for identification wasting (losses in production efficiency) in each level of production areas and then there will be recommended appropriate methods and tools of industrial engineering to reduce this wasting. Knowledge base will be edited in expert system NEST, which is an empty expert system for diagnostic applications based on rules.

## 2     Lean Company

The management of a company may use several types of efficient management methods and approaches. One of the most effective approaches is effort to implement the lean company concept. The concept of lean manufacturing (lean production) was introduced at Toyota company in 50-60 of the 20th century.

We can consider this concept as a value chain oriented approach. Concentration on a framework based on value flow is very important. The lean philosophy is based on a single principle: all forms of wasting should be identified and eliminated. This seems simplistic, but it is not because recognizing true areas of waste is difficult [7].

The lean manufacturing paradigm is simple. Take a process. Focus on the intent of the process. Eliminate all the parts of the process which do not contribute to meeting the intent, all those that do not contribute to value. Then look at each remaining part and work continually to lower its cost, make it timelier, and improve the quality of results. This focus on eliminating all wasteful effort, the fat that did not contribute to achieving the desired outcome, resulted in Toyota´s lean production system [5].

Benefits from the introduction of the principles of the lean company can be divided into a number of the following groups: operational, administrative and strategic. The most important are operational benefits as follows: reducing the use of space, improving of quality, reduction of unfinished inventory, increasing productivity and decreasing product cycle time. Strategic benefits are for example reducing the time required for implementation, reducing costs and improving of quality.

## 3     Expert Systems

Expert systems are computer programs, designed to make some of the skills of the expert available to non-experts. Since such programs attempt to emulate the thinking patterns of the expert, it is natural that the first work was done in Artificial Intelligence (AI) circles [9]. Expert systems are characterized by separation of knowledge and inference mechanisms for their use. This is significantly different from traditional programs. There are some modern methods for knowledge base development. The most popular ones are for example: fuzzy logic, neural networks and Bayesian networks. Fuzzy logic is a multivalued logic that allows intermediate values to be defined between the two aforementioned conventional evaluations. With crisp logic, it is difficult to represent notions like rather warm or pretty cold mathematically and have them processed by machines. Such linguistic terms help in applying a more human-like way of thinking to the programming of computers. Using fuzzy logic makes the system more flexible, transferable, and user-friendly [1].

# 4     Case Study: Creating a Knowledge Base to Support the Concept of Lean Manufacturing

Manufacturing companies have been shifted from financial strategies to manufacturing strategies to derive competitive strategy and profitability. Although financial strategies are still important to a manufacturing company, these are manufacturing strategies that are being used to increase profitability [8].

Expert systems can be applied in many areas. This case study deals with using knowledge base of expert system to support implementation of lean manufacturing concept. The analysis of lean manufacturing principles and measurement of these principles is usually consulted with the experts. It is also possible to use expert system for this analysis. Using expert system should be faster and less expensive way how to provide managers necessary information for making decision in process of lean manufacturing implementation.

The main objectives of the case study is creation of a knowledge base for assessing the state of the introduction of the concept of lean manufacturing using expert system and evaluation using the knowledge base on data from selected companies in each level of waste production areas and then finally recommend appropriate methods and tools of industrial engineering to reduce this waste. The aim is therefore to help provide recommendations for the implementation of lean manufacturing in an enterprise.

Sub-objective of the case study is to define a model for area of the lean manufacturing and convert model of lean manufacturing in the form of rules of the knowledge base for expert system NEST.

## 4.1     Lean Manufacturing Research in the Czech Republic

What is the situation about using concept of lean manufacturing in the Czech Republic? We performed a questionnaire survey focused on manufacturing companies in the Czech Republic to answer this question. We were trying to get basic information such as knowledge of lean manufacturing concept in companies, using the lean manufacturing concept in company and rate of experts for lean manufacturing concept working in companies.

It was sent about 3500 questionnaires during this research to manufacturing companies. We received responses from 112 companies in the Czech Republic and data were mostly from mechanical, electrical and chemical companies.

We received the following information about using the lean manufacturing concept: 62.5% of companies know the concept of the lean manufacturing and there was implemented training of this concept in 49% companies. The lean manufacturing concept is fully implemented in only 7% of companies, partially implemented is in 36%. There is amount of 20% companies which are planning to implement this concept. 37% of companies claim that there will be no implementation of the concept in the future. Building a knowledge base for identification of the main areas of wasting in production and recommendation of appropriate methods and tools of industrial engineering to reduce this waste can be important for the following reasons: 78.6% companies are sure that there are some types of waste in process of production and there are only 30.4% of companies where is working expert for lean manufacturing concept.

## 4.2    Expert System NEST

NEST is an empty expert system which includes inference mechanism. NEST was developed at the University of Economics in Prague, Czech Republic. The program provides a graphical user interface (GUI) for: creating, editing and loading knowledge bases, setting the access processing of uncertainty, consultations, the target evaluation and recommendation statement with an explanation of the findings. NEST is the program designed primarily for the academic purposes, which puts emphasis not only on the appearance, but also on the functionality of the program aimed at creating a knowledge base, comparing the results of consultation in the selection of various types of work with uncertainty [4].

NEST consists of the following components:

- stand-alone version - a program for consultation,
- editor - creating and editing knowledge bases,
- client-server version.

Knowledge base of NEST is represented by attributes and propositions, rules, contexts and integrity constraints.

## 4.3    Acquisition of Knowledge for Building Knowledge Base about Lean Manufacturing Concept

The most important factor for the quality of each expert system is a good knowledge base including knowledge expressed by the different types of rules. Very important is good cooperation between an expert and knowledge engineer or to study relevant issues to acquire the knowledge. There are many sources and literature for example [3], [5] or [12] which provide a detailed overview about lean manufacturing principles and about methods and tools of industrial engineering, specifying the characteristics and benefits of the various methods and instruments. There will be mention two important groups of necessary information and knowledge for building knowledge base in the expert system: criteria for lean manufacturing and methods and tools of industrial engineering to support lean manufacturing implementation.

## 4.4    Criteria for Lean Manufacturing

There are many criteria for lean manufacturing, but the basic idea is to reduce the basic types of waste in production. The types of waste in different areas of manufacturing can be divided into these 8 different categories:

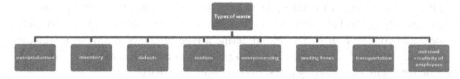

**Fig. 1.** The basic model of types of waste in manufacturing (Source: adapted from Košturiak)

We can define appropriate criteria for lean manufacturing according to the specific type of waste. To identify the rate of the level of waste in lean manufacturing concept, we need at first to define some key elements for lean manufacturing which are describe for example in [6].

We can use some numerical criteria relating to lean manufacturing processes as for example follows: productivity of manufacturing area, percentage of planned performance standards, values of the productive use of the facilities etc. We can also use various questions that can be answered by selecting predetermined scales.

## 4.5    Methods and Tools of Industrial Engineering

Industrial engineering is an important tool to achieve higher business productivity. Implementation of industrial engineering methods is the responsibility of the industrial engineers who are trying to implement appropriate methods to achieve higher productivity and to avoid excessive wastage in the enterprise.

Methods and tools of industrial engineering are final statements of created knowledge base. Inference mechanism will recommend appropriate methods and tools of industrial engineering according to rules in knowledge base and input information about situation in manufacturing processes. Inference mechanism can recommend these  methods and tools of industrial engineering such as: Kanban, MOST (Maynard Operation Sequence Technique), OPF (One Piece Flow), Poka–yoke, 5S, DMAIC, SMED (Single Minute Exchange of Dies), TOC (Theory Of Constraints), TPM (Total Productive Maintenance), VSM (Value Stream Mapping), Pull system, Process layout, team working or workshops.

## 4.6    Building Knowledge Base in the Expert System NEST

There were set out basic criteria that will be used for creating rules in knowledge base in the previous section. Knowledge base is created by the NEST editor and saved in XML file. Knowledge base of expert system NEST is using XML version 1.0 and coding windows - 1250 and has the following basic structure of the elements:

- global properties,
- attributes and propositions,
- contexts,
- rules: apriori rules, logical rules, compositional rules,
- integrity constraints.

First step in process of building knowledge base is to enter the global parameters. Important settings include specifying the range of weights, the threshold of a global context and condition. It is possible to define a type inference mechanism (standard logic, neural network or hybrid). It is also possible to add the name of an expert and knowledge engineer, including a description of the knowledge base.

Knowledge base has the following hierarchical structure:

- queries,
- intermediate statements,
- final statements.

The structure of the knowledge base "lean manufacturing" is shown in the following figure as a viewport of knowledge base for lean manufacturing and specifically in the area of waste of overproduction.

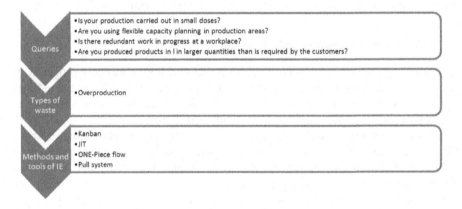

**Fig. 2.** Part of the designed knowledge base focused on overproduction in lean manufacturing (Source: Own source)

The types of rules which we used to create the knowledge base can be systematically classified into the following groups:

- evaluation (basic) rules,
- recommending rules
- specific rules,
- direct rules.

Evaluation (basic) rules represent the part of the rules that can be used to derive the types of waste directly from the basic questions (input data obtained from the questionnaire survey). Recommendation rules are provided in its intermediate questions (types of waste) and at the conclusion of the rule are industrial engineering methods that can be applied to eliminate the waste. Specific rules combine in its assumption intermediate inquiries and basic questions and recommend the choice of specific methods of industrial engineering. Direct application of the rules is used to significant questions that lead directly to the specific recommendation methods of industrial engineering.

It was necessary to make some corrections after testing knowledge base on real data from 20 companies when was the knowledge base finished. There was important to make consultation with industrial engineer and make some corrections. There were subsequently modified some rules for deriving the final recommendations in the form of appropriate methods and tools of industrial engineering to reduce waste in manufacturing according to consultation with industrial engineer.

Statistics of the total number of attributes, propositions and rules used to establish the knowledge base "lean manufacturing" shows the following figure.

| Attributes | | Propositions | |
|---|---|---|---|
| Total | 61 | Total | 67 |
| Binary | 58 | Binary | 58 |
| Single | 0 | Single | 0 |
| Set | 0 | Set | 0 |
| Numeric | 3 | Numeric | 9 |
| question | 33 | question | 39 |
| intermediate | 8 | intermediate | 8 |
| goal | 20 | goal | 20 |
| alone | 0 | alone | 0 |
| Sources | 0 | Actions | 0 |
| Actions | 0 | | |
| **Rules** | | **Other** | |
| Total | 117 | Contexts | 0 |
| Apriori | 0 | Integrity constraints | 0 |
| Logical | 0 | | |
| Compositional | 117 | | |
| Actions | 0 | | |

**Fig. 3.** Knowledge base "lean manufacturing" (Source: Output from NEST editor)

## 4.7    Consultation Process and Testing

We can start consultation process in the NEST expert system after the knowledge base is created. The consultation process is based on acquiring data from the user. There are 33 queries about situation in manufacturing processes. There are derived final results (goal statements) by inference mechanism that works with the knowledge base that contains knowledge in the form of rules and with input data which are based on the responses to questions during consultation process.

We have about 30 data sets from Czech manufacturing companies for testing our knowledge base. Data sets include information for consultation process using created knowledge base. There is an example of results of consultation process using data about manufacturing processes from one real Czech manufacturing company: figure 4 shows the most recommended methods and tools of industrial engineering to reduce waste and following figure 5 shows identified rate of each type of waste in manufacturing. We can see minimal and maximal weight values for propositions from the interval [-3; 3].

Propositions

| Name | Min weight < | Max weight | Status | Type |
|---|---|---|---|---|
| ONE Pice flow | 1,893 | 1,893 | final | goal |
| 5S | 1,476 | 1,476 | final | goal |
| spaghetti diagram | 1,476 | 1,476 | final | goal |
| process layout | 1,476 | 1,476 | final | goal |

**Fig. 4.** Recommended methods and tools of industrial engineering to reduce waste (Source: NEST, knowledge base: "lean manufacturing")

Propositions

| type of waste: overproduction | 1,418 | 1,418 | final | intermediate |
|---|---|---|---|---|
| type of waste: inventory | -1,334 | -1,334 | final | intermediate |
| type of waste: defects | -2,000 | -2,000 | final | intermediate |
| type of waste: motion | 2,952 | 2,952 | final | intermediate |
| type of waste: overprocessing | -2,890 | -2,890 | final | intermediate |
| type of waste: waiting times | -2,991 | 2,357 | final | intermediate |
| type of waste: transportation | 2,784 | 2,784 | final | intermediate |
| type of waste: not used creativi | 0,430 | 0,430 | final | intermediate |

**Fig. 5.** Identified rate of each type of waste in manufacturing (Source: NEST, knowledge base: "lean manufacturing")

# 5    Conclusion

The case study describes the building process of a knowledge base in the expert system NEST that can be used for identification of the losses in production efficiency and lean manufacturing concept implementation. Knowledge base can be used for identification wasting in each level of production areas and for recommendation appropriate methods and tools of industrial engineering to reduce this wasting.

We have tested the knowledge base using real data from Czech manufacturing companies. We want to improve the knowledge base by adding more rules and we want also test knowledge base in a real industrial setting in the future. We suppose that users of this knowledge base will be production managers. We hope that our knowledge base of expert system will be useful for supporting decision making about lean manufacturing concept implementation.

**Acknowledgements.** This paper was supported by the project SGS/21/2014 - Advanced methods for knowledge discovery from data and their application in expert systems.

# References

1. Akerkar, R., Sajja, P.: Knowledge-Based Systems. Jones and Bartlett Publishers, Sudbury (2010)
2. Coppin, B.: Artificial Intelligence Illuminated. Jones and Bartlett Publishers, Sudbury (2004)
3. Dlabač, J.: Cesta ke štíhlému podniku. Úspěch: produktivita a inovace v souvislostech 1, 11–12 (2009)
4. Ivánek, J., Kempný, R., Laš, V.: Znalostní inženýrství. OPF Karviná, Karviná (2007)
5. Jordan, J.A., Michel, F.J.: The lean company: making the right choices. Society of Manufacturing Engineers, Dearborn (2001)
6. Košturiak, J., Frolík, Z.: Štíhlý a inovativní podnik. Alpha Publishing, Praha (2006)
7. Morgan, J.: Creating lean corporations: reengineering from the bottom up to eliminate waste. Productivity Press, New York (2005)
8. Roethlein, C., Mangiameli, P., Beauvais, L.: Components of manufacturing strategy within levels of U.S. manufacturing supply chains. In: E+M Ekonomie a Management, vol. 11, pp. 33–52. Technická univerzita v Liberci, Liberec (2008)
9. Siler, W., Buckley, J.J.: Fuzzy expert systems and fuzzy reasoning. Wiley & Sons, New Jersey (2005)
10. Suchánek, P.: Business Intelligence - The Standard Tool of a Modern Company. In: Proceedings of the 6th International Scientific Symposium on Business Administration: Global Economic Crisis and Changes: Restructuring Business System: Strategic Perspectives for Local, National and Global Actors, pp. 123–132. Silesian University-School of Business Administration, Karviná (2011)
11. Šperka, R.: Application of a Simulation Framework for Decision Support Systems. In: Mitteilungen Klosterneuburg, Hoehere Bundeslehranstalt und Bundesamt fuer Wein und Obstbau, Klosterneuburg, Austria, vol. 64, pp. 134–145 (2014)
12. Wang, J.X.: Lean Manufacturing: Business Bottom-Line Based. CRC Press, Boca Raton (2010)

# A Cognitive Integrated Management Support System for Enterprises

Marcin Hernes

Wrocław University of Economics, Wrocław, Poland
marcin.hernes@ue.wroc.pl

**Abstract.** This paper presents the design and implementation of the scalable and open multi-agent Cognitive Integrated Management Information System (CIMIS) as an application of computational collective intelligence. The system allows for supporting the management processes related with all the domain of enterprise's functioning. The system is based on LIDA cognitive agent architecture, described shortly in the first part of the paper. The main part of article presents the logical architecture of CIMIS. The examples of selected agent's functionality are discussed at the last part of article.

**Keywords:** Integrated Management Information Systems, Enterprise Resource Planning, cognitive agents, decision making, computational collective intelligence applications.

## 1 Introduction

Integrated Management Information Systems (IMIS), equated also with ERP (Enterprise Resource Planning) systems, play an essential role nowadays in the operation of companies, being one of the most important solutions that allow to gain competitive advantage. Note that in the age of information, the entire economy is based on information and knowledge, therefore companies must employ systems which allow to collect, process and send large volumes of information as well as draw conclusions from the information, i.e. create knowledge of an organization. Contemporary IMIS exemplify such features, they are already commonly used by the companies and are characterized by full integration both at the system/application level and the business process level. Note, however, that the properties of contemporary IMIS are becoming more and more inadequate. Apart from collecting and analyzing data and generating knowledge, the system should also be able to understand the significance of phenomena occurring around the organization. It is becoming more and more necessary to make decisions based not only on knowledge but also on experience, thus far regarded as purely human domain [4]. In order to accomplish tasks set by IMIS, a multi-agent system can be used consist of several cognitive agents. They not only enable quick access to information and quick search for the required information, its analysis and conclusions, but also, besides being responsive to environment stimuli, they have cognitive abilities that allow them to learn from empiric experience gained through

D. Hwang et al. (Eds.): ICCCI 2014, LNAI 8733, pp. 252–261, 2014.

immediate interaction with their environments [13], which consequently allows a number of decision versions to be automatically generated and to make and execute decisions. The cognitive information processing is detailed presented at [22].

The purpose of this paper is to present the design and implementation of the scalable and open IMIS based on Learning Intelligent Distribution Agent (LIDA) architecture. The system is named CIMIS (Cognitive Integrated Management Information System) and it is an application of computational collective intelligence[17, 18] methods in form of multi-agent system. The first part of article describes the structure and functioning of the LIDA agent. Next, the logical architecture of CIMIS is presented. The examples of selected agent's functionality are discussed at the last part of article.

## 2    The LIDA Cognitive Architecture

In the study [4] considering the taxonomy of cognitive agent architectures with respect to memory organization and learning mechanism, three main groups of the architectures were distinguished:

1. Symbolic architectures which use declarative knowledge included in relations recorded at the symbolic level, focusing on the use of this knowledge to solve problems. This group of architectures includes, among others: State, Operator And Result [14], CopyCat [12], Non-Axiomatic Reasoning System [23].
2. Emergent architectures using signal flows through the network of numerous, mutually interacting elements, in which emergent conditions occur, possible to be interpreted in a symbolic way. This group of architectures includes, among others: Cortronics [8], Brain-Emulating Cognition and Control Architecture [20].
3. Hybrid architectures which are the combinations of the symbolic and emergent approach, combined in various ways. This group of architectures includes, among others: CogPrime [7], Cognitive Agents Architecture [11], The Learning Intelligent Distribution Agent (LIDA) [6].

The realization of the CIMIS is based on the  LIDA cognitive agent architecture [6], which is of emergent-symbolic nature, owing to which the processing of both structured and unstructured  knowledge is possible. In addition, the Cognitive Computing Research Group established by S. Franklin, elaborated in 2011 the framework (in Java language) significantly facilitating the implementation of the cognitive agent. It should also be emphasized that the whole framework code is open, i.e. the developer has access to the definitions of all methods, as opposed to, for instance, Cougaar architecture framework software, in which the agent's software code constitutes the so-called "blackbox". The LIDA cognitive agent's architecture consist of the following modules [3,6]:

- sensory memory,
- perceptual memory,
- workspace,

- episodic memory,
- declarative memory,
- attentional codelets,
- global workspace,
- action selection,
- sensory-motor memory.

In the LIDA architecture it was adopted that the majority of basic operations are performed by the so-called codelets, namely specialized, mobile programs processing information in the model of global workspace[6]. The functioning of the cognitive agent is performed within the framework of the cognitive cycle and it is divided into three phases: the understanding phase, the consciousness phase and the selection of actions and learning phase. At the beginning of the understanding phase the stimuli received from the environment activate the codelets of the low level features in the sensory memory [3]. The outlets of these codelets activate the perceptual memory, where high level feature codelets supply more abstract things such as objects, categories, actions or events. The perception results are transferred to workspace and on the basis of episodic and declarative memory local links are created and then, with the use of the occurrences of perceptual memory, a current situational model is generated; it other words the agent understands what phenomena are occurring in the environment of the organization. The consciousness phase starts with forming of the coalition of the most significant elements of the situational model, which then compete for attention so the place in the workspace, by using attentional codelets. The contents of the workspace module is then transferred to the global workspace, simultaneously initializing the phase of action selection. At this phase possible action schemes are taken from procedural memory and sent to the action selection module, where there compete for the selection in a given cycle. The selected actions activate sensory-motor memory for the purpose of creating an appropriate algorithm of their performance, which is the final stage of the cognitive cycle [6]. The cognitive cycle is repeated with the frequency of 5 - 10 times per second.

Parallely with the previous actions the agent's learning is performed, which is divided into perceptual learning concerning the recognition of new objects, categories, relations; episodic learning which means remembering specific events: what, where, when, occurring in the workspace and thus available in the awareness; procedural learning, namely learning new actions and action sequences needed for solving the problems set; conscious learning relates to learning new, conscious behaviours or strengthening the existing conscious behaviours, which occurs when a given element of the situational model is often in the module of current awareness. The agent's learning may be performed as learning with or without a teacher.

It is worth emphasizing that LIDA agent have the ability of grounding the symbols, namely assign relevant real world objects to specific symbols of the natural language. This is necessary to correctly process unstructured knowledge saved mainly by means of the natural language and thus, for instance, the clients' opinions on products.

The next part of article describes the architecture of CIMIS based on the LIDA cognitive agents.

## 3    The CIMIS System Architecture

The CIMIS system is dedicated mainly for the middle and large manufacturing enterprises operating on the Polish market (because the user language, at the moment, is a Polish language). There is no unequivocal definition that specifies what sub-systems form an IMIS and, in addition, according the surveys presented at work [16], there are only 326 publications, during the period 1997-2010, relates with enterprise management systems. However, the analysis of subject literature and practical solutions [2], [5], [19], [15], [1], [24] allows to systematize the architecture of the system, concluding that it is composed of the following sub-systems:

- fixed assets,
- logistics,
- manufacturing management,
- human resources management,
- financial and accounting,
- controlling,
- CRM,
- business intelligence.

The fixed assets sub-system  includes support for the realization of processes related to fixed asset and involved their depreciation.

The logistics sub-system has all the main features supporting the employees of logistics department in their effective work [10]. The logistics sub-system enables maintaining optimal stock to meet the needs of production department.

The manufacturing management sub-system support a processes related to a manufacturing execution . It include functions from the scope of the technical preparation of production capacity, production planning, material consumption planning, planning and execution of a manufacturing tasks, manufacturing control, visualization, monitoring and archiving.

The human resources management sub-system supports realization of  such processes, as the employees of the company data and contract registering, recording of working time, wage calculation, creating the tax and social security declaration.

The financial-accounting sub-system supports registering, to the full extent, economic events, also provides important, from the point of view of business management, information, concerning, inter alia, payment capacity, revenues, costs, financial result.

Controlling sub-system is automatically processing data related to  profit and loss account in cooperation with accounting sub-system. The controlling sub-system consist of both a strategic and operational controlling.

The CRM sub-system is engaged in matters connected with ensuring the best company-customer relations and collecting information in the customers' preferences in terms of product purchase in order to increase sales. The enterprise's environment monitoring is also realized by this sub-system.

The purpose of business intelligence sub-system is to enable easy and safe access to information in a company, operation of its analysis and distribution of reports within the company and among its business partners, which in turn enables quick and flexible decision making. In the context of, most of all, the business intelligence sub-system, but other sub-systems as well, the CIMIS makes cognitive visualization features available, meaning it enables a visualization of multi-dimensional data in one picture that allows to find the source of a problem in a short time and contributes to creating new knowledge about an object or problem [25].

Considering the fact, that the structure of CIMIS uses cognitive agents, figure 1 presents the logical architecture of the system.

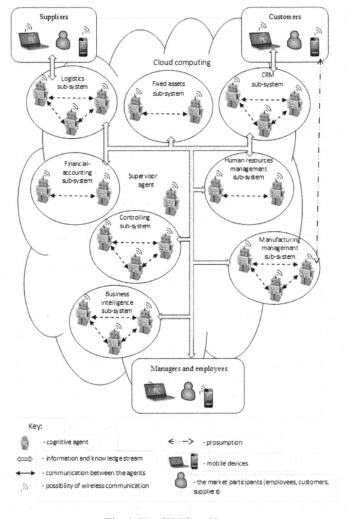

**Fig. 1.** The CIMIS architecture

The system assumes that all agents are at 'not-taught' status in the initial phase. They can be initially grouped according company's needs for sub-systems. For instance, one group of agents is assigned to the logistics sub-system, another one is assigned to the manufacturing management sub-system and yet another one to financial and accounting sub-system. Within the groups, the agents can be initially 'taught' by the company that implements the system. Next stages of learning for both grouped and ungrouped agents are done by the company staff. Agents can also learn without teacher through analyzing the results of their decisions.

The agents of all sub-systems cooperate themselves in order to better business processes realization. For example, the enterprise's environment monitoring results performed by CRM sub-system agent are using by the other agents.

The main operating purpose of the Supervisor agent is to monitor the proper operation of other agents, mainly in the field of detection and solving conflicts of knowledge and experience. The agent analyzes, in close-to-real time, the structures of knowledge and experience of all agents. Whenever a conflict occurs, it employs a solution algorithm based on a method that uses consensus theory [9, 21], and the result of the agent's actions is accepted by the system as current state of knowledge and experience.

Note that all CIMIS sub-systems are connected by a single, coherent stream of information and knowledge available online to the management, because nowadays attention is paid to functional complexity, managing all fields of operation in a company, proper flow of information and knowledge among sub-systems as well as the ability to perform a variety of analyses and to create reports for management. The implementation of this solution is realized as follow:

1. Communication between modules of agents architecture was ensured by using LIDA framework's codelets,
2. Communication between agents is based on Java Message Service (JMS) technology. The representation of information and knowledge (generated in result of agents' operating) in form of XML format document, was adopted (the JMS messaging  is at the text type). The communication is realized in publish/subscribe messaging domains – it guarantee, that information or knowledge generated by one of agents is immediately available for the other agents. The asynchronous message consumption is used.

All of the sub-systems functions are available as a local services or e-services (e.g. e-business, e-procurement, e-payment) by using Web Services technology.

At the physical level, the IMIS is built on the basis of the main  two technologies – the LIDA framework (due to framework is developed at Java language and it is open the implementation of the other Java technologies – mentioned JMS, Java Database Connectivity or Java API for XML Web Services - is possible) and Microsoft SQL Server 2008 database management system.

The examples of functionality of selected CIMIS agents is presented in the next part of article.

# 4    Functionality of the Selected Agents

The CIMIS system is now in the prototyping phase. The two agents was implemented for now: the CRM sub-system agent and the manufacturing management sub-system agent. The examples of functionality of these agents will be presented in the further part of the article.

## 4.1    CRM Agent

The cognitive agent of CRM sub-system perform, for instance, the following tasks (assuming the agent's environment is the company and its environment):

1. The agent receives, in a continuous manner, stimuli from the environment on sales characteristics, such as sales dynamics indexes categorized into each customer, product opinion from customers (e.g. from the data base of the company's online shop), the characteristics of products offered by competitors, the characteristics of actions taken by competitors (e.g. from competitor monitoring agent). Sales characteristics are stored on an ongoing basis in the sensory memory of the agent.
2. Next, sales characteristics are sent to perceptual memory, where they are interpreted. For example, interpretation includes determination whether customer opinions are positive or negative, determination of difference between characteristics of products offered by the analyzed company and offered by the competitors.
3. The results of perception in the form of objects or events are  sent to the workspace. Then, using the events stored in episodic memory (for instance 'last year witnessed a drop in sales', 'two years before competitor introduced a product with better characteristics'), and rules stored in declarative memory (for instance 'if user opinions are negative then sales will drop'), a current situational model is generated in the form of objects (e.g. sales characteristics), events (e.g. competitor's actions) and connections between them (e.g.: 'competitor has offered a product with better characteristics and our company is witnessing a drop in sales').
4. At the next stage, important elements of the situational model are formed (the agent 'rejects' unimportant elements of the situational model, e.g. 'a drop in sales to customer X was noted, because he wound up their business activity' – the element is unimportant because no marketing action can be taken towards customer X).
5. Next, important elements of the situational model are transferred to a global workspace and, with the elements as basis, specific schemes of actions are taken from the procedural memory – for instance 'improve product characteristics' or 'lower product price' or 'introduce a new product that will meet customer expectations'.
6. Next step is to perform a selection of the actions. For instance, the following action will be selected: 'lower product price', 'introduce a new product that will meet customer expectations'. The actions are transferred to the sensory-motor memory, where procedure algorithms are initiated – for example a set of steps to be taken in order to introduce a new product to the market.

## 4.2    Manufacturing Management Agent

The manufacturing management agent perform, for instance, the following tasks:

1. On the basis of orders collected by the agent of CRM sub-system and specific manufacturing capacity (it was assumed, that the enterprise produces in three shift system) the agent creates manufacturing plan.
2. Then agent automatically creates a detailed schedule of manufacturing orders (and thus also the demand for materials), and the automatic transfer of orders to carry out the production line is realized.
3. At 11 pm the CRM sub-system agent, which monitoring the enterprise environment acquired the information that the sale of the manufacturing at this point products to one of the customer is impossible (for example the customer is established in a country which has just been suspended trade).
4. Because at 11 pm in the enterprise there is no person liable to decide on production the orders related to that customer, the manufacturing management subsystem agent take a decision itself. It may, suspend the realization of these orders and to send information (e.g. SMS) to the person managing the company or automatically change the manufacturing plan and execution schedule. This type of action can protect the company against large losses. For example, consider a company that produces animal feed for farm animals. The manufacturing capacity of the production line is 1200 tons per day. Assume further that the foreign company has ordered feed, which failed to sell in our market (for example, to feed antibiotic is added in a proportion of non-compliant by the laws of our country, but is permitted by the law of the customer's country). If deciding on production plan changes will wait until the arrival of the decision-making person, for example to 7 am next day, the losses incurred by the enterprise can reach hundreds of thousands of euros (from 11 pm to 7 am next day will be manufactured 400 tons of feed - the cost of a tone of it around 1000 euro- the loss could reach the amount of 400000 euro).

On the basis of the presented example, it can be noted that the functioning of the cognitive agent allows enterprise to not only make decisions in close to real time, but also to reduce the cost of functioning of an enterprise.

With regard to the described two agents, the sensory memory is implement by using "java.net" packed which classes allow to read information from the internet web pages (for example online shops, competitors web pages, social networks) - currently, the sensory memory of the CIMIS prototype can sense only strings, in the future it is planned to adapt the sensory memory to sense the audio - speech recognition, and graphics – image recognition). The low level codelets perform tokenization process of text documents shallow analysis. The perceptual memory is implemented as a semantic net in a topic map standard. The LIDA represents a topics as a node and associations as a links, with activation level (slipnet). This allows to perform by high level codelets such processes of shallow analysis as morphological analysis, removing the ambiguity, recognize their own names, replacing pronouns, cutting sentences. The results of perception are transferred to workspace and global workspace also as slipnet. The episodic memory rules condition and conclusion are stored in form nodes and links. In episodic memory nodes and links have a timestamp. The procedural memory consist of names of specific schemes of actions and names of java classes, which consist of a procedure algorithms (they are stored in sensory-motor memory).

## 5 Conclusions

The CIMIS as an application of computational collective intelligence allows for group decision supporting related with enterprise's processes management. Cognitive agents that operate in the system replace humans in making decisions on the operational, tactical and strategic level. They can also perform many routine activities instead of human (e.g. receiving an e-mails, actions related to production line). Of course, it is also necessary to apply appropriate actuators.

It is important to emphasize that such approach does not assume making employees redundant in the company, because, while the agent is performing a certain task, they should perform supervision over the agents, improve their knowledge and seek solutions that will expedite the operation of the company in a specific field. Therefore, the benefits from implementing the discussed system in a company will not be found in lower costs of employment, but rather in the following aspects: increasing work efficiency (the agent program can work non-stop while human work is connected with such events as unworked hours), having proper amount and most up-to-date information, drawing conclusions based on the information, accelerating the decision-making process (the agent makes decision in close-to-real time), lack of influence from non-substantive factors (such as fatigue, pressure from third parties) on the decisions made, increase in the automation of business processes, providing information and suggesting solutions to managers and employees, lower risk of work-related accidents (humans do not have to be present in the manufacturing hall).

Due to this article volume, the details algorithms and methods of agents implementation as well as the experimentation and case studies will be presented in subsequent publications. The work on the implementation of the other sub-system agents and multilingualism options is in progress. Using a Microsoft SQL Server 2014 database management system, with In-Memory OLTP technology, is also planned to facilitate the IMIS performance.

## References

1. Better execute your business strategies – with our enterprise resource planning (ERP) solution, http://www.sap.com/pc/bp/erp/software/overview.html (April 28, 2014)
2. Bytniewski, A. (ed.): Architektura zintegrowanego systemu informatycznego zarządzania. Wydawnictwo AE we Wrocławiu. Wrocław (2005) (in Polish)
3. Cognitive Computing Research Group, http://ccrg.cs.memphis.edu/ (April 28, 2014)
4. Duch, W., Oentaryo, R.J., Pasquier, M.: Cognitive architectures: where do we go from here? In: Wang, P., Goertzel, P., Franklin, S. (eds.) Frontiers in Artificial Intelligence and Applications, vol. 171, pp. 122–136. IOS Press (2008)
5. Davenport, T.: Putting the enterprise into the enterprise system. Harvard Business Review, 121–131 (1998)

6. Franklin, S., Patterson, F.G.: The LIDA architecture: Adding new modes of learning to an intelligent, autonomous, software agent. In: Proc. of the Int. Conf. on Integrated Design and Process Technology. Society for Design and Process Science, San Diego (2006)
7. Goertzel, B.: OpenCogPrime: A Cognitive Synergy Based Architecture for Embodied General Intelligence. In: Proceedings of ICCI 2009 (2009)
8. Hecht-Nielsen, R.: Confabulation Theory: The Mechanism of Thought. Springer (2007)
9. Hernes, M., Nguyen, N.T.: Deriving Consensus for Hierarchical Incomplete Ordered Partitions and Coverings. Journal of Universal Computer Science 13(2), 317–328 (2007)
10. Hernes, M., Matouk, K.: Knowledge conflicts in Business Intelligence systems. In: Proceedings of Federated Conference Computer Science and Information Systems, Kraków, pp. 1253–1258 (2013)
11. Hensinger, A., Thome, M., Wright, T.: Cougaar: A Scalable, Distributed Multi-Agent Architecture. In: IEEE International Conference on Systems, Man and Cybernetics (2004)
12. Hofstadter, D.R., Mitchell, M.: The copycat project: A model of mental fluidity and analogy-making. In: Hofstadter, D. (ed.) Fluid Concepts and Creative Analogies. ch. 5. Basic Books (1995)
13. Katarzyniak, R.: Grounding modalities and logic connectives in communicative cognitive agents. In: Nguyen, N.T. (ed.) Intelligent Technologies for Inconsistent Knowledge Processing, Advanced Knowledge International, Australia, Adelaide, pp. 21–37 (2004)
14. Laird, J.E.: Extending the SOAR Cognitive Architecture. In: Wang, P., Goertzel, P., Franklin, S. (eds.) Frontiers in Artificial Intelligence and Applications, vol. 171, pp. 224–235. IOS Press (2008)
15. Mleczko J., Banaszak Z., Kłos S.: Integrated management systems. Management and engeeniering of manufacturing, Polskie Wydawnictwo Ekonomiczne, Warszawa (2011)
16. Nazemi, E., Tarokh, M.J., Djavanshir, G.R.: ERP: a literature survey. International Journal of Advanced Manufacturing Technology 61, 999–1018 (2012)
17. Nguyen, N.T.: Inconsistency of Knowledge and Collective Intelligence. Cybernetics and Systems 39(6), 542–562 (2008)
18. Nguyen, N.T.: Metody wyboru consensusu i ich zastosowanie w rozwiązywaniu konfliktów w systemach rozproszonych. Wroclaw University of Technology Press (2002)
19. Plikynas, D.: Multiagent Based Global Enterprise Resource Planning: Conceptual View. Wseas Transactions on Business And Economics 5(6) (2008)
20. Rohrer, B.: An implemented architecture for feature creation and general reinforcement learning. Workshop on Self-Programming in AGI Systems. In: Fourth International Conference on Artificial General Intelligence, Mountain View, CA (April 11, 2014), http://www.sandia.gov/rohrer/doc/Rohrer11Implemented ArchitectureFeature.pdf
21. Sobieska-Karpińska, J., Hernes, M.: Consensus determining algorithm in multiagent decision support system with taking into consideration improving agent's knowledge. In: Proceedings of Federated Conference Computer Science and Information Systems, pp. 1035–1040 (2012)
22. Tran, C.: Cognitive information processing. Vietnam Journal of Computer Science. Springer, Heidelberg (2014), http://link.springer.com/article/10.1007/s40595-014-0019-4
23. Wang, P.: Rigid flexibility. The Logic of Intelligence. Springer (2006)
24. Xpertis – intelligents systems of enterprise manufacturing, Macrologic (April 15, 2014), http://www.macrologic.pl/rozwiazania/erp
25. Zenkin, A.: Intelligent Control and Cognitive Computer Graphics. In: IEEE International Symposium on Intelligent Control, Montreal, California, pp. 366–371 (1995)

# Combining Time Series and Clustering to Extract Gamer Profile Evolution

Héctor D. Menéndez*, Rafael Vindel, and David Camacho

Departamento de Ingeniería Informática, Escuela Politécnica Superior,
Universidad Autónoma de Madrid,
C/Francisco Tomás y Valiente 11, 28049 Madrid, Spain
{hector.menendez,david.camacho}@uam.es, rafael.vindel@estudiante.uam.es
http://aida.ii.uam.es

**Abstract.** Video-games industry is specially focused on user entertainment. It is really important for these companies to develop interactive and usable games in order to satisfy their client preferences. The main problem for the game developers is to get information about the user behaviour during the game-play. This information is important, specially nowadays, because gamers can buy new extra levels, or new games, interactively using their own consoles. Developers can use the gamer profile extracted from the game-play to create new levels, adapt the game to different user, recommend new video games and also match up users. This work tries to deal with this problem. Here, we present a new game, called "Dream", whose philosophy is based on the information extraction process focused on the player game-play profile and its evolution. We also present a methodology based on time series clustering to group users according to their profile evolution. This methodology has been tested with real users which have played Dream during several rounds.

**Keywords:** Video-games, Gamer profile, User evolution, Time Series, Clustering.

## 1 Introduction

Game extensions are an emergent business over the last few years [4]. Usually, different companies sell optional phases, characters or costumes to the gamers more interested in the games. Gamers which have been immersed in a deep game experience feel attraction for this extra content, however, a high number of options is usually provided by the different business making difficult to the user to select their preference.

From a different perspective, it is also important to understand how the user adapts himself to the video game during the game-play experience. This information provides a general profile of the gamer which allows the developers to

---

* This work has been partly supported by: Spanish Ministry of Science and Education under project TIN2010-19872 and Savier an Airbus Defense & Space project (FUAM-076914 and FUAM-076915).

D. Hwang et al. (Eds.): ICCCI 2014, LNAI 8733, pp. 262–271, 2014.
© Springer International Publishing Switzerland 2014

adjust the game to a concrete user profile, for example, according to their age, gender, play time, etc. [10]. Moreover, the video game can also be used to define the player behaviour, extracting data from different movements and decissions chosen during the game-play [5].

There has been used different approaches, based on Machine Learning and Data Mining, to model the player behaviour. These approaches, usually named human or robot behaviour modelling, have been applied in different domains like Robossocer simulations [1]. Due to this extraction process is initially blind, it is interesting to consider unsupervised Data Mining techniques, such as, clustering, in order to face this problem.

Clustering [2] is an extensive unsupervised Data Mining field. These techniques are based on a blind pattern identification process usually carried out through statistical models [9]. The most classical clustering algorithms are [7]: K-means and Expectation Maximization (EM). Over the last decade, clustering techniques have exploited forecasting fields, and have been specially focused on time series analysis [8]. Measuring the evolution of several time series, which are group by the clustering algorithms, these techniques can be used to predict a new series trend.

This work is focused on the application of time series clustering techniques to the user profile extraction, based not only on the general behaviour, but also on its evolution (because gamer evolution shall also be considered in the profile definition process). In order to consider this information we extract players profile and generate an evolutionary model based on player game-play during different rounds. A whole video game, named "Dream", has been designed using extra data extraction and analysis modules, in order to improve the game-play from several perspectives, and extract the user profiles and its evolution. This game has been tested with some players during different rounds in order to achieve these goals.

The rest of the paper runs as follows. Next Section introduces the Dream game and the data analysis architecture. Section 3 describes the experimental setup for the analysis, while Section 4 presents the experimental results. Finally, in Section 5, conclusions and future work are discussed.

## 2   The Video-Game Analysis Architecture

The game architecture is divided into four modules. This section describes each module and their goals.

### 2.1   Game Module

Dream is a computer game in 3D Action-RPG genre. This game is very similar to the dynamics of games like Skyrim, Oblivion or Dark Souls. The player controls the main character and his goal is to eliminate certain key enemies in order to finish the game. To do this, the user has a set of skills that, together with its basic attack, will help him to defeat enemies. The user will manage its own equipment, inventory and attributes. The enemies have different types of artificial intelligence, forcing user to change his tactic during each phase. The enemy tactics are: pursue the player, pursue the player and call other enemies when his life takes down below a certain threshold, or pursue the player forming groups of enemies. The map is divided in four main phases with enemies (see Fig. 1 right, "Ph. 1" to "Ph.4"), a tutorial (see Fig. 1 right, "Tutorial") and a resting place (see Fig. 1 right, "Rest"). After each phase the player is teleported to the resting place.

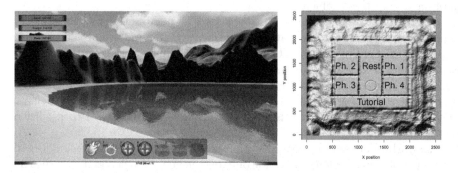

**Fig. 1.** Game-play example of Dream game (left) and the whole map divided by regions (right)

Dream has been developed using Unity3D and C# as the programming language. Unity3D [3] is a game engine platform which incorporates a development environment to create game. Video game creation programs and 3D animation were used for 3D elements. Fig. 1 (left) shows an example of the 3D environment during the game-play.

## 2.2   Data Extraction and Representation Module

This module extracts the statistical data of each player and stores these statistics in the database of the game. This process is divided into two parts: the data compilation within the game and the sending process.

The **Data Compilation** process takes place within the game and it is transparent to the player. Because there are different types of statistics that are extracted from the game, not everything is collected and sent in the same way. There are events that are collected asynchronously: use of objects, acquisition of objects, attacks (from players and enemies), use of skills, level up, increment abilities, ect. Also there are some data which is taken synchronously: enemies and player movements (every 2 seconds). Statistics collected are sent directly to the server, and not stored in the local machine.

The **Sending** process takes place once the communication between the game and the web server is established. The game communicates with the server using the HTTP protocol. To send stats, the game sends these statistics as a POST parameter. The server is able to process these statistics and provides different visualization options such as the player or enemies movement, enemies position, etc. Fig. 2 shows an example of two visualizations: player trajectory and enemies position.

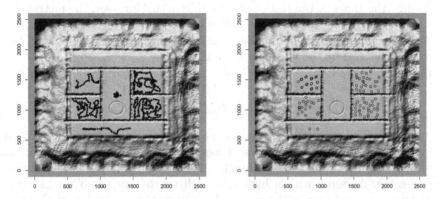

**Fig. 2.** Example of the maps representation of Dream. The left image shows the player movement around the map. The right image shows the enemies position in each region.

## 2.3   Basic User Profile Module

This module will be responsible for calculating and representing profiles of each player. In order to define a clear user profile, different metrics have been defined combining the information extracted in the previous module. Each of these metrics quantify a characteristic of the gamer in order to group the players with similar profiles. All metrics are normalized in range [0,1] (0 is the lowest possible value and 1 the highest). The metrics are defined as follows:

- "**Strength**": This metric indicates whether a player (**pl**) has an aggressive profile. A player with a high value will be the one that prefers physical attacks (**PA**) instead of magical attacks. This kind of user usually increases attributes like strength (**SA**) and agility (**AA**) instead of the other attributes (**tot(A)**). Also, the metric considers the number of enemies that the player has killed (**EK**) over the total of enemies present in the game (**ET**). Finally, the level reached by the player (**LP**) also influences the final value achieved in that metric. The value of the metric is:

$$st_{pl} = \frac{1}{4} \left( \frac{EK_{pl}}{ET} + \frac{SA_{pl} + AA_{pl}}{tot(A)_{pl}} + \frac{max(PA_{pl})}{max(PA_{all(pl)})} + \frac{max(LP_{pl})}{max(LP_{all(pl)})} \right). \tag{1}$$

- **"Agility"**: This metric indicates whether the player **(pl)** solves the puzzles easily and if he has chosen to kill only the minimum number of enemies to overcome each of the phases. A player with a high value will be the one who solves each different phase in the shortest possible time **(PT(i))**, compared against the times of the rest of players **(all(pl))**. Moreover, the metric also considers the number of enemies killed **(EK)** to complete each phase. The lower the number of enemies killed, the higher the value of this metric. The value is defined by the following formula:

$$ag_{pl} = \frac{1}{2} \left( \sum_{i=1}^{4} \frac{PT(i)_{pl}}{max(PT(i)_{all(pl)})} + \frac{1}{EK_{pl}} \right). \tag{2}$$

- **"Items"**: This metric indicates whether the player **(pl)** behavior is based on the acquisition **(ac())** and use **(use())** of objects that can be found in the video game or, on the contrary, he dispenses with the items to complete the game. A gamer with a high value will be the one to collect as many objects scattered around the map and use those objects for his own benefit. The use and acquisitions of potions **(PO)** and equipment **(EQ)** will increase the value of this metric. The value of the metric is:

$$it_{pl} = \frac{1}{4} \left( \frac{ac(PO)_{pl}}{tot(PO)} + \frac{ac(EQ)_{pl}}{tot(EQ)} + \frac{use(PO)_{pl}}{ac(PO)_{pl}} + \frac{use(EQ)_{pl}}{ac(EQ)_{pl}} \right). \tag{3}$$

- **"Defense"**: This metric indicates whether the player **(pl)** chooses to avoid damage and confrontations. A player with a high value increments attributes that increase resistance **(RA)** over other attributes **(tot(A))**. The minimum hit **(min(H))** by an enemy is also considered. Finally, the total number of deaths that the player suffer **(DE)** and the enemies killed **(EK)** are part of the metric. The final value is:

$$df_{pl} = \frac{1}{4} \left( \frac{1}{EK_{pl}} + \frac{RA_{pl}}{tot(A)_{pl}} + \frac{1}{min(H)_{pl}} + \frac{1}{1 + DE_{pl}} \right). \tag{4}$$

- **"Intelligence"**: This metric indicates whether the player **(pl)** bases its offensive strategy in the use of magic skills **(MS)**, and whether he increases his level **(PL)** killing the minimun possible number of enemies **(EK/PL)**. A player with a high value uses offensive skills compared to other players **(all(pl))**. He also increases magical abilities **MA** over other attributes **(tot(A))**. The value of this metrics is:

$$in_{pl} = \frac{1}{3} \left( \frac{MS_{pl}}{max(MS_{all(pl)})} + \frac{MA_{pl}}{tot(A)_{pl}} + \frac{(EK/LP)_{pl}}{max(EK/LP)_{all(pl)}} \right). \tag{5}$$

## 2.4   Profile Evolution Module

This module tries to group the users according their similarity. In this case, the profile metrics and their evolution during the different rounds are considered. This allows the analyser to consider not only the global information of each player, but also his learning abilities. The analysis steps are as follows:

- The time series statistics of the players are generated. The temporal statistics are generated by round, this allows to compare the player evolution during different rounds.
- We clusterized the time series of the players per metric, using this information, we generate a matrix with the metric cluster and the player associated as follows:

$$
\begin{pmatrix}
C_1^{(st)} & C_3^{(st)} & C_4^{(st)} & \dots & C_1^{(st)} \\
C_2^{(in)} & C_3^{(in)} & C_2^{(in)} & \dots & C_4^{(in)} \\
C_1^{(it)} & C_2^{(it)} & C_1^{(it)} & \dots & C_3^{(it)} \\
C_3^{(df)} & C_2^{(df)} & C_3^{(df)} & \dots & C_3^{(df)} \\
C_4^{(ag)} & C_3^{(ag)} & C_1^{(ag)} & \dots & C_4^{(ag)}
\end{pmatrix}
\tag{6}
$$

In this matrix each row represents a metric and each column represents a player. The elements of the matrix $C_i^{(metric)}$ represents the assignation of each user to a determined cluster during the time-series clustering process.
- Using the previous matrix, we generate a dissimilarity matrix around the players, and we will use this dissimilarity matrix to clusterized the players using a medoid based clustering algorithm. The dissimilarity measure applied for players is the following:

$$
diss(p_i, p_j) = 1 - \frac{\sum_{C_q} \delta_{C_q}^i \cdot \delta_{C_q}^j}{M}
\tag{7}
$$

Where $M$ is the number of metrics considered, $p_i, p_j$ are the players to be compared, $C_q$ represents the possible clusters per metric, and $\delta_{C_q}^i$ defines the Dirichlet delta defined by:

$$
\delta_{C_q}^i =
\begin{cases}
1 & \text{if } p_i \in C_q \\
0 & \text{otherwise}
\end{cases}
\tag{8}
$$

- The medoid based clustering algorithm is applied again to determined the most relevant medoids of the datasets which corresponds with the most representative players (due to we have generated a players clustering process).

## 3   Experimental Setup

The experiments have been carried out with around 30 users. These users have been playing to Dream during several rounds (at most 17, depending on the user). Each round has been a complete game which has finished when the user has completed all the phases or he has been defeated. All the data per round has been colected in order to extract the user profile. Also the evolution of the metrics during each round has been used to measure the user profile evolution.

The algorithms which have been used for the analysis are a combination of time-series clustering [8] and Partition Around Medoids (PAM) clustering [6]. The time-series clustering process has been carried out in the following steps:

1. The times series have been set in the search space.
2. The time series dissimilarity is calculated to generate a dissimilarity matrix.
3. PAM is applied to group the time series by their similarities.

Once the time series are clusterized, the similarity matrix among the users is generated for the user profile evolution phase (see Section 3). The final clustering process is also carried out using PAM. This algorithm uses a dissimilarity matrix as a search space and chooses the most relevant instances in this dissimilarity matrix. All the data instances are candidate solutions for the algorithm, and the final solutions are composed by the most representative data instances, called medoids.

The metric used by the time series clustering process is the dissimilarity metric called Autocorrelation-based Dissimilarity [8]. This measure performs the weighted Euclidean distance between the simple autocorrelation coefficients. It is defined as:

$$d(x,y) = \{(\rho_x - \rho_y)^t \Omega (\rho_x - \rho_y)\}^{\frac{1}{2}} \qquad (9)$$

where $\rho_x, \rho_y$ represent the autocorrelation vectors, and $\Omega$ is a inner product which depends on a geometric weights decaying factor $p$, as follows:

$$\Omega = \begin{pmatrix} p(1-p)^1 & 0 & \cdots & 0 \\ 0 & p(1-p)^2 & \cdots & 0 \\ \vdots & \vdots & \ddots & \vdots \\ 0 & 0 & \cdots & p(1-p)^n \end{pmatrix} \qquad (10)$$

The $p$ value which provides more stable solutions in this analysis is 0.075. Also the number of clusters which provides the most stable solutions, according to the Within-Sum quality metric [7], is 3.

## 4    User Profiles Analysis

The application of the time series clustering techniques have shown three relevant profiles as the chosen medoids for the final analysis. Fig. 3 and 4 show the global profiles and the evolutionary profiles, respectively, of the chosen users. These results discriminate three different user behaviours and evolutions:

- **Inexperience Gamer Profile:** This profile is associated with User 15 and covers the 60% of the total users. Analysing the general profile of this kind of player (see Fig. 3) we can discover that the values related to strength,

**Fig. 3.** User global profile extraction based on the five metrics. These results represent the most relevant users of the profile extraction process.

items and intelligent are low, the agility is the lowest and his strategy is only focused on the defense. The evolution of the player (see Fig. 4) shows that this kind of gamer has usually been defeated several times and he only starts to learn after several rounds. In this case, the learning improvement is remarkable for all values but the agility. It means that this kind of gamer does not deeply adapt to the game.

– **Average Gamer Profile**: This profile is associated with User 29 and covers the 20% of the total users. The average profile, in general terms (see Fig. 3), has balanced and low results according to all the metrics except for the intelligence. These users tries to profound in the game-play and interact more with the environment. According to their evolution (see Fig. 4), it is clear that their learning process is fuzzy but there are some trends which indicates that they are trying to improve a metric per match, specially the defense, agility and items used.

– **Hardcore Gamer Profile:** This profile is associated with User 26 and covers the 20% of the total users. The hardcore gamer shows good general statistics (see Fig. 3) according to all metrics (the best for the three representative users). This means that these users quickly adapt to the game enviroment and have a deep game-play experience. Their evolution is a little fuzzy (see Fig. 4), which is usual with this kind of user because the real adaptation becomes during the first or second round. These users specially focused their evolution on the strength and agility trying to optimize their game-play decisions. They modify their behaviour according to satisfy this goal, using a deep knowledge of the game, as it is shown in the high values of each metric.

With the information of the users profile, we are able to choose those users which could be more interested on the different extensions and propose different options to each profile, for example, helpful objects and equipments for average users, new abilities which makes the game easier for inexperienced gamers and more complex phases for hardcore users (which can provide an interesting challenge for them). Also, given a new user, we are able to identify the most accurated profile according to his features and evolution.

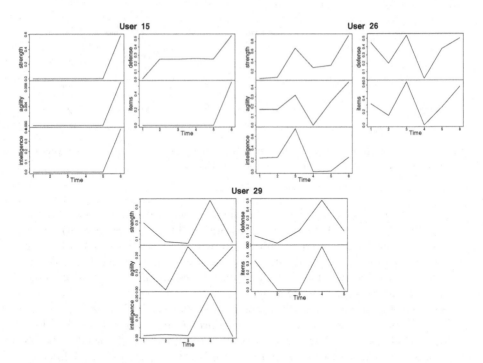

**Fig. 4.** Evolution of each user in the different games that he has played. These results are used for the time-series clustering process.

## 5   Conclusion

This paper has presented a new game, named "Dream", which is an Action-RPG game for individual gamers. During the game-play, the player has to defeat different enemies using abilities and items in order to help him to improve his performance. The game extracts data of the user experience giving information about the enviroment, user position, user decisions, enemies decisions, etc.

Using the information provided by the data extraction process, developers are able to redesign different parts of the game visualizing the different aspects, such as user path, enemies path, user most frequent abilities, hit statistics, etc. Also this work proposes some metrics to extract the user profile using the previous information.

With the information about the profiles, we also propose a methodology (based on time series clustering) to analyse user profile evolution during the game-play experience, in order to discriminate players according to their learning skills. With this information we identify three main profiles: Inexperienced, Average and Hardcore players. Using these profile, future game extensions may be offered to the users, in order to improve their game experience in a personal way.

Future work will be focused on online games and multi-player analysis. Using the profile evolution, we might be able to propose gamers for different teams in order to improve the multi-player experience and also analyse the team as a whole, in order to propose extensions to different teams.

# References

1. Aler, R., Valls, J.M., Camacho, D., Lopez, A.: Programming robosoccer agents by modeling human behavior. Expert Systems with Applications 36(2), 1850–1859 (2009)
2. Bello-Orgaz, G., Menéndez, H.D., Camacho, D.: Adaptive k-means algorithm for overlapped graph clustering. International Journal of Neural Systems 22(05), 1250018 (2012), PMID: 22916718
3. Creighton, R.H.: Unity 3D Game Development by Example: A Seat-of-Your-Pants Manual for Building Fun, Groovy Little Games Quickly. Packt Publishing Ltd. (2010)
4. Ip, B.: Technological, content, and market convergence in the games industry. Games and Culture 3(2), 199–224 (2008)
5. Jiménez-Díaz, G., Menéndez, H.D., Camacho, D., González-Calero, P.A.: Predicting performance in team games. In: II for Systems, C. Technologies of Information, and Communication, ICAART, pp. 401–406 (2011)
6. Kaufman, L., Rousseeuw, P.: Clustering by Means of Medoids. Reports of the Faculty of Mathematics and Informatics. Faculty of Mathematics and Informatics (1987)
7. Larose, D.T.: Discovering Knowledge in Data. John Wiley & Sons (2005)
8. Warren Liao, T.: Clustering of time series data – a survey. Pattern Recognition 38(11), 1857–1874 (2005)
9. Menéndez, H.D., Barrero, D.F., Camacho, D.: A genetic graph-based approach for partitional clustering. International Journal of Neural Systems 24(03),1430008 (2014) PMID: 24552507
10. Williams, D., Yee, N., Caplan, S.E.: Who plays, how much, and why? debunking the stereotypical gamer profile. Journal of Computer-Mediated Communication 13(4), 993–1018 (2008)

# Rehandling Problem of Pickup Containers under Truck Appointment System

Dusan Ku

Department of Information Systems and Operations Management,
Business School, University of Auckland,
12 Grafton Road, Auckland, New Zealand
d.ku@auckland.ac.nz

**Abstract.** This paper studies rehandling strategies for pickup containers in marine container terminals where a truck appointment system (TAS) is in place. The main purpose of the TAS is to address the imbalance of peaks and troughs of truck arrival times, thereby reducing the number of external trucks during peak hours and improving their turnaround time. This study suggests that the TAS can also be used to improve the efficiency of yard handlings for pickup containers, thus improving the productivity of yard handling equipment. To this end, a stochastic dynamic programming (SDP) model was proposed considering the truck appointment information. A branch-and-bound (B&B) approach was shown to be able to provide the exact solution to calculate the expected number of rehandlings in the decision tree. To overcome the computational restriction of the exact solution, a heuristic was proposed and its performance was compared with that of the B&B approach.

**Keywords:** container relocation problem, truck appointment system, stochastic dynamic programming.

## 1 Introduction

Once containers are stacked in a container yard, they can be accessed only from above. Therefore, a trade-off between the efficient use of yard surface and the minimisation of rehandling problem arises. According to [6], the main objectives of a stacking strategy are *i)* efficient use of storage space, *ii)* efficient transportation from quay to stack and vice versa, and *iii)* avoidance of unproductive moves. The stack with only one height would be optimal for the third objective, but would lead to an inefficient use of storage yard, which conflicts with the first objective. Therefore, the container relocation problem arises as a result of stacking containers on top of each other, and reducing the number of rehandles, i.e. unproductive moves, is of practical interest to terminal operators.

### 1.1 The Container Relocation Problem (CRP)

The CRP or the block relocation problem (BRP) ([12]) is a classic optimisation problem in container terminals, which may be formally defined as follows: given

D. Hwang et al. (Eds.): ICCCI 2014, LNAI 8733, pp. 272–281, 2014.
© Springer International Publishing Switzerland 2014

a retrieval order for containers in a given stack configuration, the objective is to retrieve all containers with the minimum number of moves. The basic setting of this problem is *static* in the sense that it assumes all containers are sequenced with the predefined departure order and no containers arrive during the retrieval process. In practice, however, container arrivals and departures overlap; incoming containers are stored in a stack while some containers in the stack are claimed for departure. This setting is referred to as *dynamic* and the associated problem may be referred to as the dynamic container relocation problem (DCRP). While most studies on the relocation problem deal with the static problem, a few studies allow container arrivals during the retrieval process ([18,3]). There are also some variants to this problem; e.g. some studies assume that only the containers above the target container are allowed to be relocated while a few studies remove this restriction ([22,3]). In this paper, we assume no container arrivals as in the static CRP, but the departure order is determined by truck appointment, which gives uncertainty to the containers booked in the same time slot.

Export containers arrive at the terminal with high uncertainty, but leave the terminal with a predetermined sequence by a ship loading plan. This pattern is the opposite for import containers: they arrive with predetermined unloading sequence; but they leave unpredictably over days. For studies of export container stack, [13] derived an optimal stacking strategy for export containers given weight group information whereas [10] studied the same problem under uncertain weight information. [15] developed a system for determining the storage position of an arriving container to minimise the number of reshuffling moves by the reshuffle index. While the previous studies assumed no container arrivals during the retrieval process, [18] allowed container arrivals during the retrieval process. [14] studied the CRP in the range of multiple bays. For studies of import container stack, [19] proposed an accessibility index as an indication of rehandling occurrences and applied it to estimate the expected number of rehandles. [5] studied two storing strategies for import containers: one based on the expected number of moves per container and the other based on segregating containers according to their arrival times. [11] proposed analytic evaluations to estimate the expected number of rehandles to clear all containers in a bay. [12] considered the CRP for import containers in a bay and proposed a heuristic based on the expected number of additional relocations. [20] considered the problem of finding the stacking policy for incoming containers to minimise the total number of rehandles by exploiting the property called *storage demand unit (SDU)*, which can be stored together and retrieved in any order. To evaluate the stacking strategies by the number of rehandling moves generated, [16] developed a mathematical model based on probabilistic distribution functions for container dwell times. [17] extended the CRP by considering the distance travelled by the crane.

## 1.2   The Truck Appointment System (TAS)

The main purpose of the TAS is to address the imbalance of peaks and troughs of truck arrival times, thereby reducing the number of trucks during peak hours and improving their turnaround time. This study suggests that the information available from the TAS can also lend itself to improving the handling for import containers. In this paper, we consider the CRP under the TAS. Then, the revised objective is, given the appointment information, to provide a sequence of moves that minimises the expected number of rehandles to clear the stack.

There is a limited amount of research on the TAS, but only a little addresses the rehandling policy in a container terminal. [8] formulated a problem of finding the appropriate level of capping for the allowable number of trucks given at each zone and in each time window. [9] developed an event-based simulation model that captures interactions among various subsystems, assuming the port must have some method of gathering information about containers' departure time, e.g. the expected dwell time. [21] assumed truck arrival times are obtained after import containers are stored on the yard. [1] applied a discrete-event simulation model to evaluate the impact of a TAS on the performance of online container stacking rules discussed in their prior research ([2]).

The paper is organised as follows: in Section 2, we formulate the problem into a SDP model; in Section 3, a heuristic rule to overcome the restriction of the exact solution is described; in Section 4, the computational experiments are demonstrated and compared; in the last section, we conclude our findings and suggest future research topics.

## 2    The CRP with Time Windows

We develop our model under the following assumptions: *i)* containers are reshuffled *iff* any container below them is to be retrieved; *ii)* each reshuffled container is moved once only for a retrieval of any container; *iii)* no container is arriving in the stack during the retrieval process; *iv)* all containers in the stack are booked by the time slots, which are mapped to retrieval sequences, ascending by the hour. Unlike the cases for groups of blocks introduced in [12], the same sequence does not mean that containers with the same sequence can be retrieved in any order. It means that the actual departure order among them is unknown until the corresponding truck arrives in the stack, and we assume there is equal probability for any departure among the containers with the same sequence.

### 2.1    Stochastic Dynamic Programming (SDP) Model

A decision tree in a SDP model consists of chance and decision nodes. A chance node is a tree structure expressing stochasticity whereas a decision node expresses possible decision. Each node in the decision tree corresponds to a state of the stack. The states in children are produced from the parent by an action, i.e. one or a series of moves. In our B&B approach, each decision node produces a combination of compound moves and a bounding procedure is applied. Following the notations from [7], lower bound (LB) consists of two parts, $Z_{LB}^n = Z_R^n + Z_M^n$: $Z_R^n$ considers the reshuffles that occurred until the node $n$; $Z_M^n$ considers the minimum reshuffles that will occur until emptying the stack. The first part is trivial. The second part is not, but is at least greater than or equal to the number of misplaced containers across columns, i.e. those placed above the earliest departing container in the column. Upper bound (UB) also consists of two parts, $Z_{UB}^n = Z_R^n + Z_F^n$: $Z_R^n$ is the same as above; $Z_F^n$ considers the reshuffle events that will occur until emptying the stack. $Z_F^n$ is not necessarily the minimum number of future rehandles. Different heuristics may yield different number of future rehandles differently. With $Z_{LB}^n$ and $Z_{UB}^n$ in each node, a set of nodes can be pruned *en masse* when the LB of that node is greater than or equal to the best known UB, $Z_{UB}^{best}$, such that $Z_{UB}^{best} = \min(Z_{UB}^{best}, Z_{UB}^n)$. The following notations are used for the formulation of our SDP model:

- $N$: the total number of containers in the initial stack
- $a^k$: the action taken for the removal of the $k^{th}$ container
- $S^k$: the state of the stack after $k$ containers are retrieved from the stack
- $C^k$: the set of the earliest sequenced containers at the state $S^k$.
- $c_k$: the container to be retrieved during action $a^k$. It is a stochastic variable.
- $p_k(c_k)$: the probability of retrieving $c_k$ among $C^k$ given the state $S^k$ such that $\sum_{c_k \in C^k} p_k(c_k) = 1$
- $\pi(S^k, c_k)$: the state transformation function that observes container $c_k$ to retrieve from the state $S^k$.
- $S^{(k')}$: the state transformed by $\pi(S^k, c_k)$. If $|C^k| = 1$, $S^k = S^{k'} = \pi(S^k, c_k)$.
- $r(a^k | S^{k-1'})$: the number of rehandles that occur during action $a^k$ given the state $S^{k-1'}$.
- $f(S^k)$: the expected minimum total number of rehandles to retrieve the remaining containers from the state $S^k$.

The problem can be formulated as the recursive function shown in Equation (1). $c_k$ is a stochastic variable because we do not know with certainty how the state will be transformed before $c_k$ is observed by the external factor, i.e. the truck arrivals. A general recursive function may be derived as in Equation (2).

$$f(S^0) = E[\min_{a^1}[r(a^1) + f(S^1)] | \pi, S^0] = \sum_{c_0 \in C^0} p_0(c_0) \min_{a^1}[r(a^1 | S^{0'}) + f(S^1)]$$

$$\text{(1)}$$

$$where \ \ S^{0'} = \pi(S^0, c_0) \ \ and \ \ S^0 \xrightarrow{S^{0'}, a^1} S^1$$

$$f(S^{k-1}) = \sum_{c_{k-1} \in C^{k-1}} p_{k-1}(c_{k-1}) \min_{a^k}[r(a^k | S^{k-1'}) + f(S^k)]$$

$$f(S^N) = 0 \qquad \text{(2)}$$

$$where \ \ S^{k-1'} = \pi(S^{k-1}, c_{k-1}) \ \ for \ k = 1, ..., N$$

$$and \ \ S^{k-1} \xrightarrow{S^{k-1'}, a^k} S^k \ \ for \ k = 1, ..., N$$

## 2.2 Numerical Case

Fig. 1 illustrates a simple case of 3x3 stack with five containers by our branching method. At the state $S^0$, the number of the earliest sequenced containers is only one. Therefore, it is trivial that $r(a^1|S^{0'})$ is 2, which is the number of containers placed above. Thus, $f(S^0) = 2 + f(S^1)$. Since there are three possible decisions at the $S^0$, some children may be pruned by the bounding procedure while branching out each decision. Table 1 summarises the optimal solution at each possible state $S^1$. As the solutions for $f(S^0)$ in each case indicate, moving to the state [1.1], is the optimal policy for the action $a^1$ by yielding the minimum $f(S^0)$.

Even though the SDP model above may solve the problem exactly, it is computationally prohibitive to be applied to a practical situation. The CRP was already shown to be NP-hard in previous studies ([4]). Allowing containers to be booked in the same time window makes the problem even harder due to the stochastic branches, which cannot be pruned until their expected values are fully computed down the tree. Suppose that we have 10 containers to be retrieved at the initial stack and all containers are booked in the same time window. Such a situation would yield 10! stochastic branches, which may be multiplied by the number of rehandling options during each retrieval stage. Such a task is formidable, thus being called *the curse of dimensionality*. Therefore, we introduce a simple heuristic to overcome this computational restriction.

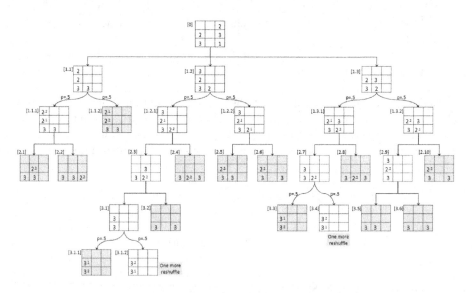

**Fig. 1.** *A decision tree example considering the same time slots*

**Table 1.** *Summary of expected rehandles calculated by each node at level 1 of Fig. 1*

| $S^1$ | $\min r(a^1)$ | $p_2(c_2)$ | $\min r(a^2)$ | $\min r(a^3)$ | $\min r(a^4)$ | $\min r(a^5)$ |
|-------|---------------|------------|---------------|---------------|---------------|---------------|
| [1.1] | [2] | 0.5 via [1.1.1] | 1 | 0 | 0 | 0 |
|       |     | 0.5 via [1.1.2] | 0 | 0 | 0 | 0 |
| $f(S^0) = 2 + 0.5(1 + 0) = 2.5$ | | | | | | |
| [1.2] | [2] | 0.5 via [1.2.1] | 1 | 0 | 0 | 0 |
|       |     | 0.5 via [1.2.2] | 1 | 0 | 0 | 0 |
| $f(S^0) = 2 + 0.5(1 + 1) = 3$ | | | | | | |
| [1.3] | [2] | 0.5 via [1.3.1] | 1 | 0 | 0 | 0 |
|       |     | 0.5 via [1.3.2] | 1 | 0 | 0 | 0 |
| $f(S^0) = 2 + 0.5(1 + 1) = 3$ | | | | | | |

# 3    The Expected Reshuffle Index (ERI) Heuristic

The Reshuffle Index (RI) heuristic, defined by [15], is the index indicating the number of containers that depart earlier than the incoming container in the column. This heuristic chooses the column to be reshuffled by computing the lowest RI, and the tie is broken towards the taller column and arbitrarily afterwards. The RI heuristic cannot calculate the index of a column where there exist containers with unknown precedence relationship. We hereby propose the ERI heuristic, a variant of the RI heuristic, to calculate the expected number of containers that depart earlier than the incoming one in the column where there exist containers with unknown precedence relationship. We introduce the following notations.

- $n$: number of the containers in which precedence relationships are unknown
- $k$: number of earlier departing containers among containers in the column with unknown precedence relationship
- $p_n(k)$: the probability of having $k$ number of earlier departing containers below the incoming container
- $f_n(k)$: the marginal contribution of having $k$ number of earlier departing containers to the ERI
- $f_n$: the marginal expected reshuffling index (ERI) of the column

$$f_n(k) = k p_n(k) \tag{3}$$

$$f_n = \sum_{k=0}^{n} f_n(k) \tag{4}$$

$f_n(k)$ and $f_n$ can be computed by Equation (3) and Equation (4), respectively. Then, the total reshuffling index of the column is the summation of $f_n$ and the already observed number of containers booked in earlier departure time. Initially, there will be $(n+1)$ containers with unknown precedence relationships including the incoming container. Fig. 2(a) illustrates the case of two existing containers, all being booked in the same time slot. At the root node of level 0, three containers including the incoming one contain uncertainty with respect to the retrieval sequence, thus generating three branches with equal probability of $1/3$. At level 1, the first branch yields no earlier container below the first departing container, therefore there is no need to branch out further. Now since there are two containers with unknown precedence relationships, the second and third branch produce two children with equal chance of $1/2$, respectively. This establishes all precedence relationships of the initial column in Fig. 2(a). The ERI for the case $n = 2$ is calculated to be 1 by $f_2 = \sum_{k=0}^{2} f_2(k) = 1/3 + 2/3 = 1$ using Equation (4), the general case of which is summarised in Table 2. The table induces Equation (5), which provides us with a simple heuristic to our problem.

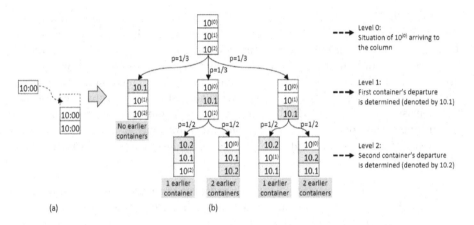

**Fig. 2.** *A case where there exists two unknown precedence relationships (n=2)*

**Table 2.** *The ERI calculation table for general case*

| $k$ | prob. of reaching each leaf with $k$ earlier containers | number of such $k$ node | $p_n(k)$ | $f_n(k)$ |
|---|---|---|---|---|
| 0 | $1/(n+1)$ | 1 | $1/(n+1)$ | 0 |
| 1 | $1/(n+1)n$ | $n$ | $1/(n+1)$ | $1/(n+1)$ |
| 2 | $1/(n+1)n(n-1)$ | $n(n-1)$ | $1/(n+1)$ | $2/(n+1)$ |
| | | ... | | |
| $n$ | $1/(n+1)!$ | $n!$ | $1/(n+1)$ | $n/(n+1)$ |

$$f_n = \sum_{k=0}^{n} \frac{k}{n+1} = \frac{1}{n+1} \sum_{k=0}^{n} k = \frac{1}{n+1} \frac{n(n+1)}{2} = \frac{n}{2} \qquad (5)$$

## 4    Computational Experiment

A computer program was developed to compare the performance between B&B and the ERI heuristic in this problem. It was programmed in Java language and the program was run on Intel i5-2500 3.3 GHz CPU and 8GB RAM using windows 7 enterprise 64-bit.

Since this study is the first attempt to model the CRP with time windows, no benchmarking instance is available. Therefore, we randomly generate thirty instances per stack configuration, the column size of which ranges from 3 to 12. Considering a straddle carrier(SC)-based terminal, we have restricted the height of the stack to 3. The stacked ratio for each instance is about 2/3 and does not vary for the sake of comparison convenience. For each instance generation, we impose that the ratio of the different retrieval sequence groups divided by the number of containers be over 50% since a too low ratio may not provide a meaningful interpretation to our model setting.

The B&B approach was run once per instance since the expected outcome will always be the same in every run. The ERI heuristic was simulated by 500 times per instance, and the overall average rehandles as well as the average of the best rehandles are summarised in Table 3. The best rehandles in the heuristic run mean the minimum number of rehandles observed among the repeated simulation. It turns out that the B&B approach is no longer viable when the number of columns exceeds 6. It is also notable that the standard deviation for the computational time in the B&B approach increases exponentially as the number of columns increases. Among the instances of seven columns, some instances could be solved within 10 seconds whereas others could not be solved even with 10 hours' computation. The last column, (d)/(a), indicates the best-case scenario among the simulated cases. In overall, our conjecture from available comparison is that the heuristic performance is reasonably good by deviating marginally from the exact solution and is applicable to stacks of any practical size.

**Table 3.** *Comparison between the expected B&B and the ERI heuristic simulation*

| col. x tier | average B&B | | | | average of 500 heuristic runs | | | | | performance | |
|---|---|---|---|---|---|---|---|---|---|---|---|
| | (a) rehand. | (b) time(s) | sd. (a) | sd. (b) | (c) rehand. | (d) best rehand. | (e) time(s) | sd. (c) | sd. (e) | (c)/(a) | (d)/(a) |
| 3x3 | 2.16 | 0.03 | 1.03 | 0.08 | 2.26 | 1.65 | 0.15 | 1.18 | 0.14 | 1.03 | 0.71 |
| 4x3 | 2.22 | 0.07 | 1.25 | 0.10 | 2.27 | 1.6 | 0.19 | 1.25 | 0.13 | 1.02 | 0.62 |
| 5x3 | 3.41 | 2.22 | 1.04 | 3.84 | 3.50 | 2.89 | 0.22 | 1.11 | 1.10 | 1.02 | 0.81 |
| 6x3 | 3.65 | 47.6 | 1.11 | 90.45 | 3.7 | 3.2 | 0.30 | 1.08 | 0.01 | 1.02 | 0.86 |
| 7x3 | | | | | 4.35 | 3.8 | 0.40 | 1.47 | 0.02 | | |
| 8x3 | | | | | 4.77 | 4.1 | 0.48 | 1.35 | 0.02 | | |
| 9x3 | | | | | 5.21 | 4.6 | 0.6 | 1.25 | 0.02 | | |
| 10x3 | | | | | 6.17 | 5.5 | 0.73 | 1.42 | 0.02 | | |
| 11x3 | | | | | 6.25 | 5.55 | 0.90 | 1.55 | 0.19 | | |
| 12x3 | | | | | 7.38 | 6.9 | 1.09 | 1.51 | 0.17 | | |

# 5    Conclusion

Since truck appointment allocates groups of containers to predetermined time window, precedence relationships among those in the same time window are not known. This is a typical situation in import container stack, but the appointment system reduces the uncertainty to a certain degree. The difference of this study in comparison to the static CRP is that the retrieval sequences are mapped by the appointed time slots and the same sequence indicates the uncertainty which will likely increase the expected number of rehandles.

We have proposed the ERI heuristic to compute the RI index of a column containing containers booked in the same time window. In the computational experiment, we have run the simulation to calculate the minimum expected number of rehandles in the given stack configuration. The result shows the fast computation as well as its reasonable performance in terms of a marginal gap (of about 2-3% increase on average) from the exact solution. One limitation in this interpretation is that there were relatively a very few instances that could be solved exactly. Therefore, our efforts on the future research will be focused on the exact solution approaches that can expand the boundary of the problem size so that we can improve the validity of the benchmarking result.

We suggest this heuristic is directly applicable to terminals using the TAS. This heuristic can be powerful when choosing a stack for import containers. When the terminal system chooses a grounding stack, there are generally several criteria to evaluate. So far, the expected number of rehandles was hardly one of them for import stack due to the unpredictable retrieval order. That is why mainstream literature on the CRP addresses the relocation problem from the perspective of export stack where retrieval sequences are predetermined by a loading plan. Now that we may have some, though incomplete, information available for truck arrivals by using the appointment system, exploiting such information further will provide us with interesting topics for future studies.

**Acknowledgments.** Support for this work was provided by the University of Auckland Doctoral Scholarship (Code No 43). The author is also grateful for the valuable comments and suggestions made by the three anonymous reviewers.

# References

1. van Asperen, E., Borgman, B., Dekker, R.: Evaluating impact of truck announcements on container stacking efficiency. Flexible Services and Manufacturing Journal 25(4), 543–556 (2013)
2. Borgman, B., van Asperen, E., Dekker, R.: Online rules for container stacking. OR Spectrum 32(3), 687–716 (2010)
3. Borjian, S., Manshadi, V.H., Barnhart, C., Jaillet, P.: Dynamic stochastic optimization of relocations in container terminals. Working paper. MIT (2013)
4. Caserta, M., Schwarze, S., Voß, S.: A mathematical formulation and complexity considerations for the blocks relocation problem. European Journal of Operational Research 219(1), 96–104 (2012)

5. de Castillo, B., Daganzo, C.F.: Handling strategies for import containers at marine terminals. Transportation Research Part B: Methodological 27(2), 151–166 (1993)
6. Dekker, R., Voogd, P., van Asperen, E.: Advanced methods for container stacking. In: Container Terminals and Cargo Systems, pp. 131–154. Springer (2007)
7. Hakan Akyüz, M., Lee, C.Y.: A mathematical formulation and efficient heuristics for the dynamic container relocation problem. Naval Research Logistics, NRL (2014)
8. Huynh, N., Walton, C.M.: Robust scheduling of truck arrivals at marine container terminals. Journal of Transportation Engineering 134(8), 347–353 (2008)
9. Jones, E.G., Michael Walton, C.: Managing containers in a marine terminal: Assessing information needs. Transportation Research Record: Journal of the Transportation Research Board 1782(1), 92–99 (2002)
10. Kang, J., Ryu, K.R., Kim, K.H.: Deriving stacking strategies for export containers with uncertain weight information. Journal of Intelligent Manufacturing 17(4), 399–410 (2006)
11. Kim, K.H.: Evaluation of the number of rehandles in container yards. Computers & Industrial Engineering 32(4), 701–711 (1997)
12. Kim, K.H., Hong, G.P.: A heuristic rule for relocating blocks. Computers & Operations Research 33(4), 940–954 (2006)
13. Kim, K.H., Park, Y.M., Ryu, K.R.: Deriving decision rules to locate export containers in container yards. European Journal of Operational Research 124(1), 89–101 (2000)
14. Lee, Y., Lee, Y.J.: A heuristic for retrieving containers from a yard. Computers & Operations Research 37(6), 1139–1147 (2010)
15. Murty, K.G., Liu, J., Wan, Y.W., Linn, R.: A decision support system for operations in a container terminal. Decision Support Systems 39(3), 309–332 (2005)
16. Sauri, S., Martin, E.: Space allocating strategies for improving import yard performance at marine terminals. Transportation Research Part E: Logistics and Transportation Review 47(6), 1038–1057 (2011)
17. Ünlüyurt, T., Aydın, C.: Improved rehandling strategies for the container retrieval process. Journal of Advanced Transportation 46(4), 378–393 (2012)
18. Wan, Y.W., Liu, J., Tsai, P.C.: The assignment of storage locations to containers for a container stack. Naval Research Logistics (NRL) 56(8), 699–713 (2009)
19. Watanabe, I.: Characteristics and analysis method of efficiencies of container terminal: an approach to the optimal loading/unloading method. Container Age 3, 36–47 (1991)
20. Yang, J.H., Kim, K.H.: A grouped storage method for minimizing relocations in block stacking systems. Journal of Intelligent Manufacturing 17(4), 453–463 (2006)
21. Zhao, W., Goodchild, A.V.: The impact of truck arrival information on container terminal rehandling. Transportation Research Part E: Logistics and Transportation Review 46(3), 327–343 (2010)
22. Zhu, W., Qin, H., Lim, A., Zhang, H.: Iterative deepening a* algorithms for the container relocation problem. IEEE Transactions on Automation Science and Engineering 9(4), 710–722 (2012)

# Emergent Concepts on Knowledge Intensive Processes*

Gonzalo A. Aranda-Corral[1], Joaquín Borrego-Díaz[2], Juan Galán-Páez[2],
and Antonio Jiménez-Mavillard[3]

[1] Universidad de Huelva, Department of Information Technology,
Crta. Palos de La Frontera s/n. 21819 Palos de La Frontera, Spain
[2] Universidad de Sevilla, Department of Computer Science and Artificial Intelligence,
Avda. Reina Mercedes s/n. 41012 Sevilla, Spain
[3] Western University, Department of Modern Languages and Literatures,
1151 Richmond Street, London, Ontario, N6A 3K7, Canada

**Abstract.** An approach to refine and revise the general framework of KiP (Knowledge Intensive Process) is presented. The specific case of collaborative KiP is studied and the prominent role of collaborative KiPs in the general context of Business Processes is revealed. The approach is based on Formal Concept Analysis.

## 1 Introduction

Nowadays there exists a growing interest in deeply understanding business processes based on the use of Knowledge in an intensive way (Knowledge Intensive Process, KiP). This kind of processes are governed by Business Processes (BP) at several levels within knowledge companies. The task of obtaining an adequate integration of KiP into classic BP informational ecosystems represents a challenge that Knowledge Economy needs to solve [9], [12].

The design and extraction of patterns [18] for KiP is inherent to this challenge. BP patterns provide a number of benefits for BP models (BPM). As is it well known, obtaining BP patterns, allows to (see [11]): simplify work, encourage best practices, assist in BP analysis, show inefficiencies, remove redundancies and greatly aid to consolidate interfaces for a proper design of BPs as well as to facilitate their re-use. KiP are particularly complex BP (even more for the collaborative ones) and the study of their patterns is particularly challenging. An analysis of the requirements, characteristics and frameworks for KiPs is mandatory to state a successful formal basis for their modeling and extraction of patterns [8]. The formal basis is the first step in an attempt to answer questions as Which is the adequate pattern for a concrete KiP? Are we facing a new KiP? Which kind of KiPs are the best in a concrete BPM?

In order to formalize the detailed analysis of KiP and to study real-world applications and experiences, the aim of this paper is to show how to use Formal Concept Analysis (FCA) in the refinement of characterizations and requirements for managing and executing KiPs. Specifically we propose a formal refinement of the semantic relationships

---

* Supported by TIC-6064 Excellence project (*Junta de Andalucía*) cofinanced with FEDER funds.

D. Hwang et al. (Eds.): ICCCI 2014, LNAI 8733, pp. 282–291, 2014.

**Table 1.** Characteristics of KiPs [8]

| Characteristic | Explanation |
|---|---|
| C1: Knowledge driven | The status and availability of data and knowledge objects drive human decision making and directly influence the flow of process actions and events. |
| C2: Collaboration-oriented | Process creation, management and execution occurs in a collaborative multi-user environment, where human-centered and process-related knowledge is co-created, shared and transferred by and among process participants with different roles. |
| C3: Unpredictable | The exact activity, event and knowledge flow depends on situation- and context-specific elements that may not be known a priori, may change during process execution, and may vary over different process cases. |
| C4: Emergent | The actual course of actions gradually emerges during process execution and is determined step by step, when more information is available. |
| C5: Goal-oriented | The process evolves through a series of intermediate goals or milestones to be achieved. |
| C6: Event-driven | Process progression is affected by the occurrence of different kinds of events that influence knowledge of workers' decision making. |
| C7: Constraint- and rule-driven | Process participants may be influenced by or may have to comply with constraints and rules that drive actions performance and decision making. |
| C8: Non-repeatable | The process instance undertaken to deal with a specific case or situation which is hardly repeatable, i.e., different executions of the process vary from one to another. |

among characteristics and requirements of KiPs. An analysis of this kind can aid to clarify in which status KIPs are in BPM field. FCA has been successfully proved as an useful tool to analyze phenomenological reconstructions of Complex Systems [5], and BP in Knowledge-based companies are inherent complex systems where the understanding and classification of their elements strongly depends on the features used.

The motivation of this work is based on the fact that this approach clarifies a number of features associated with the evaluation and assessment of KiP, as for example: to refine evaluation and assessment frameworks by means of semantic methods, to compare requirements and characteristics of different KiP models and concrete cases, to decide which sets of requirements represent an innovation niche, etc.

Lastly, the application of FCA allows to revise KiPs associated with specific tools for Knowledge Management (KM). The resuls are applied to the analysis of OntoxicWiki's KiP for collaborative enrichment, extension and refinement of Documentation on ontologies [4] (see Sect. 5).

*Structure of the paper.* The next section introduces the main characteristics and requirements of the KiPs considered in this paper. Sect. 3 succinctly reviews FCA. In Sect. 4 FCA based analysis of KiPs is described, showing a number of interesting features of KiPs from this semantic perspective. In Sect. 5 a first application of this analysis to some processes, assisted by a semantic knowledge externalization tool (OntoxicWiki) is presented. Lastly, related work and some insights on future work are given.

## 2    Characteristics and Requirements for KiPs

In collaborative KiPs three dimensions converge: the Knowledge Dimension, its collaborative nature, and its consideration as Business Process. All these three dimensions have to be modeled. A consensus on KiP definition is a key step in order to understand hidden mechanisms and patterns that operate in this kind of processes. A definition catching the complex nature of KiP could be the following (see [17] and also [8]):

*A KiP is a process whose conduct and execution are heavily dependent on knowledge workers performing various interconnected knowledge intensive decision making tasks.*

**Table 2.** Requirements for KiPs [8]

| Requirements on data (RD) | Requirements on Processes (RG) |
|---|---|
| | R11 Support for different modeling styles |
| | R12 Visibility of the process knowledge |
| R1 Data modeling | R13 Flexible process execution |
| R2 Late data modeling | R14 Deal with unanticipated exceptions |
| R3 Access to appropriate data | R15 Migration of process instances |
| R4 Synchronized access to shared data | R16 Learning from event logs |
| **Requirements on Knowledge Actions** (RK) | R17 Learning from data sources |
| R5 Represent data-driven actions | **Requirements on Knowledge Workers** (RW) |
| R6 Late actions modeling | R18 Knowledge workers' modeling |
| **Rules and constraints** (RR) | R19 Formalize interaction between knowl. workers |
| R7 Formalize rules and constraints | R20 Define knowledge workers' privileges |
| R8 Late constraints formalization | R21 Late knowledge workers'modeling |
| **Requirements on Goals** (RG) | R22 Late privileges modeling |
| R9 Goals modeling | R23 Capture knowledge workers' decisions |
| R10 Late goal modeling | **Requirements on Environment** (RE) |
| | R24 Capture and model external events |
| | R25 External events late modeling |

*KiPs are genuinely knowledge, information and data centric and require substantial flexibility at design- and run-time*

A fine analysis of the main elements to consider in KiP analysis is given in [8]. The authors enumerate two sets of ingredients to describe and study KiP in order to provide a precise characterization. On the one side, it is mandatory to highlight the characteristics that could make KiP different of other BP (see Table 1). On the other side, a list of requirements retrieved for KiPs was considered (see Table 2). Both sets of ingredients were extracted from real-world application scenarios.

Requirements for KiPs are driven to achieve a sound representation and performance of the KiP instance models. In Table 2 a complete requirement list, due to Ciccio et al., [8], is shown. The aim of our work is to refine the analysis given in that article, by means of a systematic treatment of requirements and characteristics and other features which are essential in KiP analysis. In order to devise a robust (semantic based) refinement, a formal analysis is applied to concepts associated to Ciccio et al.'s framework. The analysis is carried out by means of formal concept reasoning.

## 3   Formal Concept Analysis

FCA mathematizes the philosophical understanding of a concept as a unit of thoughts composed of two parts: the extent and the intent. The extent covers all objects belonging to the concept, while the intent comprises all common attributes valid for all the objects under consideration [10].

A *formal context* $M = (O, A, I)$ consists of two sets, $O$ (objects) and $A$ (attributes), and a relation $I \subseteq O \times A$. Finite contexts can be represented by a 1-0-table (identifying $I$ with a boolean function on $O \times A$). Given $X \subseteq O$ and $Y \subseteq A$, it defines

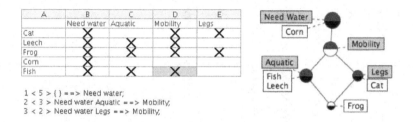

| | A | B | C | D | E |
|---|---|---|---|---|---|
| | | Need water | Aquatic | Mobility | Legs |
| Cat | | X | | X | X |
| Leech | | X | X | X | |
| Frog | | X | X | X | |
| Corn | | X | | | |
| Fish | | X | X | X | |

1 < 5 > { } ==> Need water;
2 < 3 > Need water Aquatic ==> Mobility;
3 < 2 > Need water Legs ==> Mobility;

**Fig. 1.** Formal context of fishes, and its associated concept lattice

$$X' = \{a \in A \mid oIa \text{ for all } o \in X\} \text{ and } Y' = \{o \in O \mid oIa \text{ for all } a \in Y\}$$

The main goal of FCA is the computation of the concept lattice associated with the context. A (formal) concept is a pair $(X, Y)$ such that $X' = Y$ and $Y' = X$. For example, the concept lattice from the formal context of fishes of Fig. 1, left (attributes are understood as *live in*) is depicted in Fig. 1, right. Each node is a concept, and its intension (or extension) can be formed by the set of attributes (or objects) included along the path to the top (or bottom). For example, the bottom concept $(\{eel\}, \{Coast, Sea, River\})$ is the concept *euryhaline fish*. CL contains every concept that can be extracted from the context. As well, concepts are defined but it is possible that no specific term (word) exists to denote it.

Knowledge Bases (KB) in FCA are formed by *implications between attributes*. An implication is a pair of sets of attributes, written as $Y_1 \rightarrow Y_2$. It is true with respect to $M = (O, A, I)$ according to the following definition. A subset $T \subseteq A$ *respects* $Y_1 \rightarrow Y_2$ if $Y_1 \not\subseteq T$ or $Y_2 \subseteq T$. $Y_1 \rightarrow Y_2$ is said to hold in $M$ ($M \models Y_1 \rightarrow Y_2$ or $Y_1 \rightarrow Y_2$ is an implication of $M$) if for all $o \in O$, the set $\{o\}'$ respects $Y_1 \rightarrow Y_2$.

**Definition 31.** *Let $\mathcal{L}$ be a set of implications and $L$ be an implication.*

1. *$L$ follows from $\mathcal{L}$ ($\mathcal{L} \models L$) if each subset of $A$ respecting $\mathcal{L}$ also respects $L$.*
2. *$\mathcal{L}$ is complete if every implication of the context follows from $\mathcal{L}$.*
3. *$\mathcal{L}$ is non-redundant if for each $L \in \mathcal{L}, \mathcal{L} \setminus \{L\} \not\models L$.*
4. *$\mathcal{L}$ is a (implication) basis for $M$ if $\mathcal{L}$ is complete and non-redundant.*

A particular basis is the *Duquenne-Guigues* or so called *Stem* Basis (SB) [13]. The SB for the context of Fig. 1 is shown (down). In this paper no specific property of the SB is used, so it can be replaced by any other basis. In order to reason with implications, a production system can be used (see e.g. [2]).

**Theorem 1.** *Let $S$ be a basis for $M$ and $\{A_1, \ldots, A_n\} \cup Y \subseteq A$. The following statements are equivalent:*

1. *$S \cup \{A_1, \ldots A_n\} \vdash_p Y$ ($\vdash_p$ is the entailment by means of a production system).*
2. *$S \models \{A_1, \ldots A_n\} \rightarrow Y$*
3. *$M \models \{A_1, \ldots A_n\} \rightarrow Y$.*

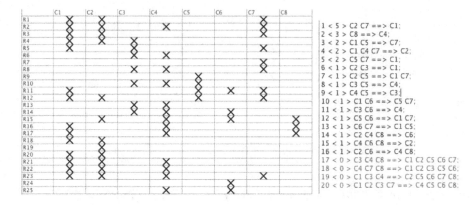

**Fig. 2.** Formal Context extracted from [8] and its associated STEM basis

In FCA, association rules are also implications between sets of attributes. Confidence and support are defined as usual in data mining. The *Stem Kernel Basis* (SKB) is the subset of the SB formed by the implications with nonzero support. To simplify, we assume that $\mathcal{L}_M$ is a concrete basis for $M$ (it is not necessarily the Stem basis). Likewise, the kernel of $\mathcal{L}_M$ is denoted by $\mathcal{L}_M^s$. As it was above-mentioned, in the specific framework of CS observability, the set of implications with nonzero support gives some insights in a number of applications on both, micro and macro levels (cf. [1,3]).

## 4  FCA-Based Analysis of KiPs

In this section a number of results on the nature of (collaborative) KiP, obtained by means of FCA tools, are presented. The following subsections show a number of consequences, obtained from the semantic analysis performed on the concepts involved in Ciccio et al.'s framework, about the analysis performed about the refinement of the set of the characteristics, requirements, characterization of kind of KiPs and formal relationships. A brief analysis of the basis gives some insights about the KiP framework. Due to the lack of space, different applications have to be selected in each subsection. The analysis starts from an excellent and deep review [8]. A natural assumption from this paper is that the deep analysis made on it, is consequence of the study of a great number of KiPs, tools an BPM.

The relationship among KiP requirements (as objects) and characteristics (as features) is described in the formal context $M_1$ depicted in Fig. 2 (left). The Stem basis $\mathcal{L}_C$ associated to the context, is also shown in Fig. 2 (right). Likewise, the dual context $M_2$ is considered (that its, that one built by using characteristics as objects and requirements as attributes). $M_1$ is very useful to understand how characteristics are related in the basis on requirements, while $M_2$ is very useful to relate requirements, showing if they are independent, if there exists subsumption among them, etc.

The concept lattice associated to $M_1$ has 31 concepts. In this case, a concept can be viewed as a set of characteristics with common attributes, which are the common requirements for a set of characteristics. The analysis of $\mathcal{L}_C$ gives a number of interesting insights of the framework. Without being exhaustive, the main consequences are:

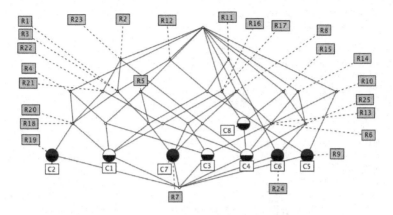

**Fig. 3.** Concept lattice considering requirements as attributes

## 4.1  Any Non-repeatable KiP Has Emergent Nature

This is the interpretation of the fact

$$\mathcal{L}_C \models C8 \to C4$$

Thus, from the point of view of [8] framework, the overall process of non repeatable KiP cannot be univocally determined by step-by-step elements.

## 4.2  Essential Requirements for Characteristics of KiPs

By analyzing the concept lattice associated to $M_2$ (see Fig. 3), it is possible to associate to each characteristic an essential requirement characterizing it. For example, from the point of view of the collaborative dimension, the following facts hold:

- C1 is characterized by R2 and R4: late data modeling and data access to shared data is essential for knowledge-driven KiP.
- C2 is characterized by R19: That is, every KiP accomplishing R19 (the process formalizes the interaction between knowledge workers) is collaboration-oriented.
- C3 is characterized by R6 and R4: KiP which accomplishes synchronized access to shared data and late actions modeling are essentially unpredictable.
- C4 is characterized by R2 and R6: late data modeling and late action modeling are essential requirements of emergent Kips.
- C5 is characterized by R10 and R12: KiP satisfying the visibility of process knowledge and late goal modeling are goal-oriented KiPs.
- C6 is characterized by R24: The capture and modeling of external events is the key feature of event-driven KiPs.
- C7 is characterized by R7: Constraint- & rule-driven KiPs formalize rules and constraints. Another more interesting characterization is "$R2 \land R8$": late data modeling and late constraints formalization.

```
12 < 2 > R5 R11 ==> R1 R2 R3 R12 R23;
13 < 2 > R4 R12 ==> R1 R2 R3 R18 R20 R21 R22 R23;
14 < 2 > R5 R12 ==> R1 R2 R3 R11 R23;
15 < 2 > R13 ==> R6 R8 R10 R14;
16 < 2 > R8 R14 ==> R6 R10 R13;
17 < 2 > R10 R14 ==> R6 R8 R13;
18 < 2 > R18 ==> R1 R2 R3 R4 R12 R20 R21 R22 R23;
19 < 2 > R20 ==> R1 R2 R3 R4 R12 R18 R21 R22 R23;
20 < 2 > R2 R16 R17 R23 ==> R21 R22;
21 < 2 > R2 R11 R23 ==> R1 R3 R5 R12;
22 < 2 > R2 R5 R23 ==> R1 R3 R11 R12;
23 < 2 > R2 R4 R23 ==> R1 R3 R12 R18 R20 R21 R22;
24 < 2 > R1 R2 R3 R12 R21 R22 R23 ==> R4 R18 R20;
25 < 2 > R25 ==> R14 R15;
26 < 2 > R14 R15 ==> R25;
27 < 1 > R7 ==> R1 R2 R3 R5 R8 R11 R12 R23;
28 < 1 > R4 R8 ==> R5 R6 R10 R13 R14;
29 < 1 > R9 ==> R10 R11 R12;
30 < 1 > R4 R10 ==> R5 R6 R8 R13 R14;
31 < 1 > R5 R10 ==> R4 R6 R8 R13 R14;
32 < 1 > R4 R11 ==> R1 R2 R3 R5 R12 R16 R17 R18 R20 R21 R22 R23;
33 < 1 > R8 R11 ==> R1 R2 R3 R5 R7 R12 R23;
34 < 1 > R10 R11 ==> R9 R12;
35 < 1 > R8 R12 ==> R1 R2 R3 R5 R7 R11 R23;
36 < 1 > R10 R12 ==> R9 R11;
37 < 1 > R4 R14 ==> R5 R6 R8 R10 R13;
38 < 1 > R5 R14 ==> R4 R6 R8 R10 R13;
39 < 1 > R12 R16 R17 ==> R1 R2 R3 R4 R5 R11 R18 R20 R21 R22 R23;
40 < 1 > R11 R16 R17 ==> R1 R2 R3 R4 R5 R12 R18 R20 R21 R22 R23;
41 < 1 > R5 R16 R17 ==> R1 R2 R3 R4 R11 R12 R18 R20 R21 R22 R23;
42 < 1 > R4 R16 R17 ==> R1 R2 R3 R5 R11 R12 R18 R20 R21 R22 R23;
43 < 1 > R19 ==> R1 R2 R3 R4 R12 R18 R20 R21 R22 R23;
44 < 1 > R11 R14 ==> R15 R24 R25;
45 < 1 > R8 R15 ==> R2 R6 R10 R13 R14 R16 R17 R21 R22 R23 R25;
46 < 1 > R10 R15 ==> R2 R6 R8 R13 R14 R16 R17 R21 R22 R23 R25;
47 < 1 > R11 R15 ==> R14 R24 R25;
48 < 1 > R14 R16 R17 ==> R2 R6 R8 R10 R13 R15 R21 R22 R23 R25;
49 < 1 > R10 R16 R17 ==> R2 R6 R8 R13 R14 R15 R21 R22 R23 R25;
50 < 1 > R8 R16 R17 ==> R2 R6 R10 R13 R14 R15 R21 R22 R23 R25;
51 < 1 > R2 R15 R23 ==> R6 R8 R10 R13 R14 R16 R17 R21 R22 R25;
52 < 1 > R2 R14 R23 ==> R6 R8 R10 R13 R15 R16 R17 R21 R22 R25;
53 < 1 > R2 R10 R23 ==> R6 R8 R13 R14 R15 R16 R17 R21 R22 R25;
54 < 1 > R2 R8 R21 R22 R23 ==> R6 R10 R13 R14 R15 R16 R17 R25;
55 < 1 > R24 ==> R11 R14 R15 R25;
56 < 0 > R12 R14 ==> R1 R2 R3 R4 R5 R6 R7 R8 R9 R10 R11 R13 R15 R16 R17 R18 R19 R20 R21 R22 R23 R24 R25;
57 < 0 > R4 R15 ==> R1 R2 R3 R5 R6 R7 R8 R9 R10 R11 R12 R13 R14 R16 R17 R18 R19 R20 R21 R22 R23 R24 R25;
58 < 0 > R5 R15 ==> R1 R2 R3 R4 R6 R7 R8 R9 R10 R11 R12 R13 R14 R16 R17 R18 R19 R20 R21 R22 R23 R24 R25;
59 < 0 > R12 R15 ==> R1 R2 R3 R4 R5 R6 R7 R8 R9 R10 R11 R13 R14 R16 R17 R18 R19 R20 R21 R22 R23 R24 R25;
60 < 0 > R1 R2 R3 R4 R5 R11 R12 R16 R17 R18 R19 R20 R21 R22 R23 ==> R6 R7 R8 R9 R10 R13 R14 R15 R24 R25;
```

**Fig. 4.** STEM basis for $M_2$

- C8 is characterized by R15 and R16 (or R15 and R17). That its, non-repeatable KiPs usually provide the migration of process instances -R15- (that would that facilitates the instantiation and number of agents working in the process) and also helps to learn from data (event logs, as R16 or data sources, as R17).

### 4.3   Requirements Refinement

The analysis of the stem basis for $M_2$, $\mathcal{L}_R$ (see Fig. 4), shows a number of redundant requirements. A set of requirements $S \subseteq \{R_1, \ldots R_{2}5\}$ is a **minimal descriptional requirement system (mrds)**. It is possible to describe any requirement by conjunction of requirements of $S$, and any proper subset of $S$ does not. In logic terms,

$$\mathcal{L}_R \cup S \models R_i \text{ for any } 1 \leq i \leq n$$

and $\mathcal{L}_R \cup (S \setminus \{R\}) \not\models R$ for any $R \in S$. The following set of requirements is a mrds:

$$\{R_2, R_4, R_6, R_7, R_{10}, R_{12}, R_{15}, R_{16}, R_{19}, R_{24}\}$$

A fine analysis of the mrds is interesting. On the one side, it contains requirements of each kind, supporting the completeness and non redundancy of the classification of the requirements from [8]. On the other side, any requirement is subsumed by the conjunction of at most two requirements of the mrds. Thus the mrds is a nice set of requirements to accomplish in order to design complete KiPs models and software.

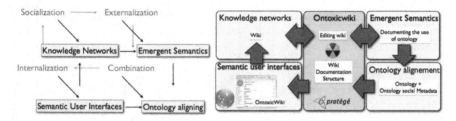

**Fig. 5.** Projection of Nonaka & Takeuchi's cycle (left) and its interpretation in Ontoxicwiki[4]

### 4.4   The Conceptual Nature of Collaborative KiPs

Collaboration oriented (C2) KiPs are important KiPs to consider in current BPMs. The recognizing of the collaborative nature of actions/process is important in order to design proper models as well as to detect patterns. In the concept lattice, C2 represents the intent of a concept that comprises the 48% of requirements, which gives us a sense of the relative complexity of collaborative KiPs within KiP. Moreover, the extent provides a fine definition in terms of requirements by means of FCA of collaborative KiPs:

$$R_1, R_2, R_3, R_4 \text{ (that is, every requirement on data), } R_{12}, R_{15},$$
$$R_{18}, R_{19} R_{20}, R_{21}, R_{22}, R_{23} \text{ (that is, every requirement on knowledge workers)}$$

Therefore, data and knowledge workers are essential elements in collaborative KiPs, allowing to (*a priori*) recognize collaboration in KiPs.

## 5   Case of Study: OntoxicWiki as a Collaborative Tool for KiP

OntoxicWiki [4] was designed to provide a semantic bridge between the knowledge activities of the projection of Nonaka and Takeuchi's cycle (NTC) for Knowledge externalization [15], enhancing both Web2.0 and SW solutions in this context (fig. 5). The tool is designed to satisfy several needs which arise when NTC is adapted for a semantic framework. Specifically, OntoxicWiki aims to bridge the gap between user and ontology. The main objective of this application is to represent ontologies in an intuitive and easy understandable way for any user by providing them with an environment from which the can repair and document ontologies socially, concretely with wiki technologies and associated patterns collaborative methods. KiP activities associated to OntoxicWiki are intimately related with the semantic specialization of NTC, which shows four needs for creating truly SW2.0 communities: *emergent semantics, semantic user interfaces, knowledge networks and ontology alignment* (see fig. 5):

Therefore, KiPs associated to Ontoxicwiki use fit on CKW life cycle from [14]. As KiP tool, OntoxicWiki exploits the social nature of Wiki technologies with the formatting, by means meta-data, of knowledge on use of concepts by knowledge workers. The knowledge processing cycle associated to OntoxicWiki ecosystem is depicted in Fig. 6 (specialized to a Pharmaceutical lab). To study its collaborative nature, authors have analyzed the fourteen requirements associated to collaborative KiP according to FCA interpretation of Ciccio et al.'s framework.

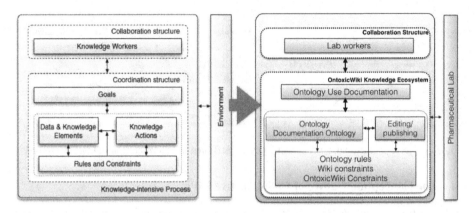

**Fig. 6.** Ontoxicwiki assisted process as a KiP

The requirements on data are accomplished by the Reparation of Ontology Process (ROP) and the Use Documentation Process (UDP). Both process also accomplish the full set of requirements on knowledge workers. However $R_{12}$ and $R_{15}$ are not satisfied by ROP and UDP. Therefore, an interesting consequence of the FCA-based analysis applied to OntoxicWiki's KiP suggest the inclusion of tools for visualizing the knowledge process (which Wiki technologies hide on edition logs and it is affected by user permissions), allowing the fair migration of process instances (mainly, the basic operations of ontology class edition and documentation).

## 6   Related Work

An approach to refine and revise the general framework of KiP has been presented. The specific case of collaborative KiP is has been studied, and the prominent role of collaborative KiPs in the general context was revealed. Other approach to recognize collaboration by means of FCA applications can be found in the scientific literature. In [6] FCA is applied for detecting and recognizing recurring collaborations among software artifacts. Our approach is at general level and it covers general KiPs in BP. An ambitious program would be the application of this approach to the BPs archived in the MIT process handbook (http://ccs.mit.edu/ph). In this case, it is convenient the use of the association rules basis (called *Luxenburger basis*) instead of Stem Basis.

Our approach is bottom-up, but an top-down approach to the semantic analysis of KiP can be achieved by means of ontologies. In [7] an ontology for KiP, KIPO, is presented. This approach is useful due to the fact that ontologies provide reuse of KiPs, and current research in the literature points to the lack of approaches of this kind. Requirements enumerated in that paper highlight the role of tacit knowledge in KiPs. Authors propose an orthogonal approach to the one devised in this paper: KIPO declares internal elements of KiPs whilst our approach only declares *exogenous* requirements for KiPs.

## 7   Future Work

With respect to the preliminary analysis of KiPs associated to the use of OntoxicWiki, it is interesting to remark that OntoxicWiki outputs documented evidence of KiPs (ontology versions). Therefore it is possible to enhance OntoxicWiki in order to accomplish requirements on learning from log and data ($R_{16}, R_{17}$). With respect to the documentation of KiPs, a similar analysis to the one made in this paper can facilitate the reuse as well as to detect patterns in documentation of KiPs [16].

Lastly, a further work is the analysis and characterization of KiPs which autonomous agents execute in Knowledge based tasks. In the case presented in [3], dialogue and argument-based process follow KiP patterns which can be reused in other BPM.

# References

1. Aranda-Corral, G.A., Borrego-Díaz, J., Galán-Páez, J.: Qualitative reasoning on complex systems from observations. In: Pan, J.-S., Polycarpou, M.M., Woźniak, M., de Carvalho, A.C.P.L.F., Quintián, H., Corchado, E. (eds.) HAIS 2013. LNCS, vol. 8073, pp. 202–211. Springer, Heidelberg (2013)
2. Aranda-Corral, G.A., Borrego-Díaz, J., Galán-Páez, J.: Complex Concept Lattices for Simulating Human Prediction in Sport. J. Syst. Sci. and Complexity 26(1), 117–136 (2013)
3. Aranda-Corral, G.A., Borrego-Díaz, J., Giráldez-Cru, J.: Agent-mediated shared conceptualizations in tagging services. J. Multimedia Tools and Applications 65(1), 5–28 (2013)
4. Aranda-Corral, G.A., Borrego-Díaz, J., Jiménez-Mavillard, A.: Social Ontology Documentation for Knowledge Externalization. Comm. in Comp. and Inf. Sci. 108, 137–148 (2010)
5. Aranda-Corral, G.A., Borrego-Díaz, J., Galán-Páez, J.: On the Phenomenological Reconstruction of Complex Systems–The Scale-Free Conceptualization Hypothesis. Systems Research and Behavioral Science 30(6), 716–734 (2013)
6. Arevalo, G., Buchli, F., Nierstrasz, O.: Detecting Implicit Collaboration Patterns. In: Proceedings of WCRE (2004)
7. Carvalho, J.E.S., Santoro, F.M., Baião, F.A., Pimentel, M.A.: KiPO: the knowledge-intensive process ontology. Software & Systems Modeling, 1–31 (2014)
8. Di Ciccio, C., Marella, A., Russo, A.: Knowledge-Intensive Processes–Characteristics, Requirements and Analysis of Contemporary Approaches. Journal on Data Semantics (2014)
9. Fiechter, C.A., Marjanovic, O., Boppert, J.F., Kern, E.-M.: Knowledge management can be lean: Improving knowledge intensive business processes. In: Howlett, R.J. (ed.) Innovation through Knowledge Transfer 2010. SIST, vol. 9, pp. 31–40. Springer, Heidelberg (2011)
10. Ganter, B., Wille, R.: Formal Concept Analysis. Mathematical Foundations. Springer (1999)
11. Glushko, R.J., McGrath, T.: Document engineering. MIT (2008)
12. Gronau, N., Weber, E.: Management of knowledge intensive business processes. In: Desel, J., Pernici, B., Weske, M. (eds.) BPM 2004. LNCS, vol. 3080, pp. 163–178. Springer, Heidelberg (2004)
13. Guigues, J.-L., Duquenne, V.: Familles minimales d'implications informatives resultant d'un tableau de donnees binaires. Math. Sci. Humaines 95, 5–18 (1986).
14. Mundbrod, N., Kolb, J., Reichert, M.: Towards a system support of collaborative knowledge work. In: La Rosa, M., Soffer, P. (eds.) BPM Workshops 2012. LNBIP, vol. 132, pp. 31–42. Springer, Heidelberg (2013)
15. Nonaka, I., Takeuchi, H.: The Knowledge-Creating Company: How Japanese Companies Create the Dynamics of Innovation. Oxford Univ. Press (1995)
16. Scheithauer, G., Hellmann, S.: Analysis and documentation of knowledge-intensive processes. In: La Rosa, M., Soffer, P. (eds.) BPM Workshops 2012. LNBIP, vol. 132, pp. 3–11. Springer, Heidelberg (2013)
17. Vaculin, R., Hull, R., Heath, T., Cochran, C., Nigam, A., Sukaviriya, P.: Declarative business artifact centric modeling of decision and knowledge intensive business processes. In: 15th IEEE International Conference on Enterprise Distributed Object Computing, EDOC 2011 (2011)
18. Verginadis, Y., Papageorgiou, N., Apostolou, D., Mentzas, G.: A review of patterns in collaborative work. In: Proc. 16th ACM Int. Conf. Supporting Group Work (GROUP 2010), pp. 283–292. ACM (2010)

# Optimal Partial Rotation Error for Vehicle Motion Estimation Based on Omnidirectional Camera

Van-Dung Hoang and Kang-Hyun Jo

Graduated School of Electrical Engineering, University of Ulsan, Ulsan, Korea
hvzung@islab.ulsan.ac.kr, acejo@ulsan.ac.kr

**Abstract.** This paper presents a method for robust motion estimation using an optimal partial rotation error based on spirits of the rotation averaging and the minimum spanning tree approaches. The advantage of an omnidirectional camera is that allows tracking landmarks over long-distance travel and large rotation of vehicle motions. The method does not process the optimal rotation at every frame due to the computational time, instead that, the optimal rotation error is applied for each interval of motion called partial motion so that the set of landmarks are tracked in all sequent images. This approach takes advantage of partial optimal error for reducing the divergences of estimated trajectory results in long-distance travel. The global motion of the vehicle is estimated in high accuracy based on utility of the optimal partial rotation error based on the rotation averaging method, which contrasts with traditional bundle adjustment using the minimum Euclid distance of back-projection errors. The experimental results demonstrate the effectiveness of this method under the large view scene in the outdoor environments.

**Keywords:** Fusion sensors, motion estimation, visual odometry, structure from motion, omnidirectional camera, optimal rotation error.

## 1    Introduction

In recent years, many methods have been developed for localization, navigation, visual odometry, which have been applied in modern intelligent systems, autonomous robots, especially intelligent transportation in outdoor environments, surveillance systems, such as [1-9]. The localization estimation, mapping, scene understanding are the key step towards the autonomy. For localization, the onboard GPS devices receive signals from satellites and then plot the absolute positions of vehicle. In partial error, the signals from satellites using cheap GPS receivers are often drifted as comparing with the ground truth. Moreover, under high building regions in urban scenes and eclipse regions, the positional signals may be lost or jump in certain period, vehicle localization information may be lost. Therefore, the position GPS signal is acceptable in global shape of ego-motion but low accuracy in the local position. The improvement of that method is supplemented by other sensors such as wheel odometer, IMU. The wheel odometer can improve the translation but it may work inaccurately if the wheels slip or move on the rough surface of roads. The laser rangefinder (LRF) is also

D. Hwang et al. (Eds.): ICCCI 2014, LNAI 8733, pp. 292–301, 2014.
© Springer International Publishing Switzerland 2014

a good choice in these cases. The signal is weak in the case of objects in far distance or non-reflection. The authors in [1] describes method for localization and reconstruction using multiple sensors of LRF and IMU in outdoor environments. The position and attitude of a robot is estimated by particle filters using the likelihood of positions. Other approaches focus on vision odometry methods using vision devices. Some authors proposed methods using a single camera or binocular cameras [4, 7]. Because of FOV limitation, some authors used the Omni-camera for odometry systems [10-13]. The basic principle of approaches is the corresponding feature points and the epipolar geometry constraints. The main disadvantage of this approach is accumulative error over time. The trajectory is diverged when the vehicle moves under long-distance travel without any prior information. This is also a challenge of the incremental methods. Moreover, monocular vision is process on the scale trajectory. Some researcher groups proposed methods based on the fusion of vision systems and other electromagnetic devices [1, 14, 15]. The results were significantly improved. Due to limited of precision of sensors and additional ambient noise, there are location and mapping estimation errors. Over time, the errors will also be accumulated when a vehicle moves in the large scale scenes. Therefore the final global trajectory will be diverged and distorted. The system can yield the accurate results in a short distance of movements or integrating with information from the GPS receiver.

Therefore, to overcome the drawback mentioned above, this paper presents an optimal solution using the omnidirectional vision could deal with some disadvantages above. The advantages of omnidirectional vision are the wide-angle of view. It is $360^{\circ}$ of view which rich information for long range tracking the landmarks. It is suitable for error correction under long- distance travel. The problem of motion estimation is how to detect and remove outliers as well as optimize the results with the highest accuracy. The main reason of the outlier problem is caused by motion blur and distortion. There is different the formulation optimizations between the omnidirectional camera and the perspective camera. By using the spherical model, the error is optimized in the angular value error instead of the Euclidean distance of back-projection 3D points and image points in the case of the perspective camera.

This paper is organized into six sections. The next section describes the basic of visual odometry based on geometry constraint of omnidirectional images. The motion constraint in the omnidirectional camera based adjustment is presented in section 3. Section 4 describers the optimal motion estimation to minimize partial rotation error. The experimental results are showed in section 5. Finally, conclusions and future works are presented in section 6.

## 2   Omnidirectional Camera Based Visual Odometry

The visual odometry system is composed of consecutive images to determine the position and orientation of a vehicle by analyzing the associated camera images. In this work, the geometry constraints are analyzed directly by the epipolar constraint based on the essential matrix to discovery the motion of a vehicle using a monocular omnidirectional camera. In order to apply epipolar geometry constraint to the omnidirectional image, it should be converted to the spherical model. The method in [16] is used to calibrate the camera system and construct the projection rays from the focal point of the hyperboloid mirror to points in the world space. The geometry constraints are analyzed based on the epipolar constraint using the essential matrix. There is much kind of features that have been considered in recent researches in a feature extraction and matching problem, e.g., SIFT, SURF. In this paper, the corresponding image points in sequent images are extracted by the SIFT method [17]. The SIFT is demonstrated that this feature descriptor is very invariant and robust for feature matching with scaling, rotation, or affine transformation. The corresponding points are also projected to the spherical model and the matching process using RANSAC algorithm[18] for outlier removal. A 3D point $P$ with respect to two corresponding projection rays of $r$ and $r'$ from the center projection of the hyperboloid mirror at two camera poses to the same world point $P$ and that of in the spherical model. Notice that, the omnidirectional camera using the hyperboloid mirror is a single center projection. The rays of $r$ and $r'$ are observed from two camera poses, whose relative geometry satisfies the constraint as follows:

$$r'^T Er = 0 \qquad (1)$$

where the essential matrix $E$ is defined as $E=[T]_xR$. The matrix $[T]_x$ is a skewed symmetric matrix of the translation vector $T=[T_X,T_Y,T_Z]^T$. The rotation matrix $R=R_ZR_YR_X$, where $R_Z$, $R_Y$, $R_X$ are yaw $\alpha$, pitch $\beta$, and roll $\gamma$ rotation matrices, respectively. Notice that, the multiplication of the rotation matrix according to X, Y, Z is not algebraic commutation.

$$R=R_ZR_YR_X = \begin{bmatrix} \cos\alpha\cos\beta & \cos\alpha\sin\beta\sin\varphi - \sin\alpha\cos\varphi & \sin\alpha\sin\varphi + \cos\alpha\sin\beta\cos\varphi \\ \sin\alpha\cos\beta & \cos\alpha\cos\varphi + \sin\alpha\sin\beta\sin\varphi & \sin\alpha\sin\beta\cos\varphi - \cos\alpha\sin\varphi \\ -\sin\beta & \cos\beta\sin\varphi & \cos\beta\cos\varphi \end{bmatrix} \qquad (2)$$

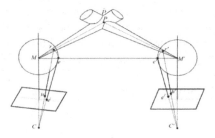

**Fig. 1.** Epipolar geometry of spherical model with triangular back-projection error

The set of corresponding rays of $r$ and $r'$ from two sequent images are computed based on calibration omnidirectional camera parameters by (1). The points in the spherical model of the different types of mirrors, for example parbola or ellipsol, are also computed similar way. The point $\hat{P}$ is the intersection point of two lines of rays $r$ and $r'$. In practical, due to the limited precision, rays $r$ and $r'$ do not intersect each other, point $\hat{P}$ can be estimated by the point which nearest both lines of two rays $r$ and $r'$. The solution problem is that solves (2) to estimate the translation and the rotation parameters. There are several methods to deal with this problem. The canonical solutions to the problem are the eight-point method in [19]. Recently, the one-point RANSAC[12] and the one-point combining with edge feature [10] are considered as the typical algorithm with the vehicle motion cause of the car-like structured model and planar motion assumption. In this paper, the eight-point RANSAC, full constraint method, is used to estimate the homogenous camera transformation.

## 3    Omnidirectional Vision Based Adjustment

The structure from motion is estimated based on sequent images. However, there is error in result due to noises on measured data and computational precision. The estimated result is refined through a local minimization error. As mentioned above, the advantage of an omnidirectional camera is the sufficient tracking landmarks under long-distance motion and large rotation. Taking into account advantage reduces the estimated error of a vehicle motion. In this experiment, one adjustment is applied for each local motion where at least $n_{th}$ corresponding points are tracked in all frames of this interval. The threshold $n_{th}=21$, which is defined by experimental try and test.

Let $C=\{C_i\}$ and $M=\{M_i\}$, with $i=1\dots n$, be $n$ poses of the camera and the center of the spherical model with known homogeneous rigid body transformations. Let $P=\{P_j\}$ and $\hat{P}=\{\hat{P}_j\}$, with $j=1..m$, be the world points and the estimated corresponding 3D points based on the camera motion estimation, respectively. The problem is finding the set of the 3D points in sequent images so that minimizes the error of the transformation estimation. Ideally, the perfections in measurements and motion estimations are that the angles between the rays from the spherical center $M$ to $P$ and the back-projection of the estimated 3D points $\hat{P}$ to $M$ will be zero, e.g., they are coincided each other. Due to noises of the measurement, matching and computational precision, the angle error $\alpha$ is non-zero in reality, see also Fig. 1. Thus, we can estimate the 3D points $\hat{P}$ so that minimize the cost function $\varepsilon$ of angular back-projection error $\alpha$, as follows.

$$\varepsilon = \left|\tan(\alpha)\right| = \frac{\left|r\times(\hat{P}-\hat{M})\right|}{\left|r^T(\hat{P}-\hat{M})\right|} \tag{3}$$

where $[r]_\times$ and $\hat{M}$ are skewed symmetric matrix of $r$ and the center of spherical model.

The optimal problem is that minimizes the error cost function of angular back-projection based on a set of estimated point $\hat{P}$ with respect to ray $r$. The convenience of tangent function is that it monotones with angle $\alpha$, when $\alpha$ is belong in each square of unit circle. In this paper, angular error is considered in $[0, \pi/2]$. That means the quotient of (4) is a positive value, $r^T(\hat{P}-M) \geq 0$. According to the advantages of second order cone programs (SOCP) in multiple view optimazation in [20], the coin with the vertex is the center of sphere intersect when the radius is more sufficiency. The final optimization problem is

$$\min_{P} \frac{\left\| [r]_\times \hat{P} - [r]_\times M ) \right\|}{r^T \hat{P} - r^T M} \qquad (4)$$
$$\text{subject to} \quad r^T \hat{P} - r^T M \geq 0$$

Equation (5) can be written in general form of SOCP as follows:

$$\min_{x} \frac{\left\| A_i x + b_i \right\|}{c_i^T x + d_i} \qquad (5)$$
$$\text{subject to} \quad c_i^T x + d_i \geq 0$$

Kahl *et al.* in [21] point out that the $L_2$-norm error of the cost function in multiple views triangulation results three local minima whereas the $L_\infty$-norm result a single minimum. Taking the advantage of $L_\infty$-norm, it is easily used to minimize the error function. Notice that the problem has some convexity properties. Thus, this problem can be solved by quasiconvex optimization method. Considering of local adjustment to $m$ views and $n$ corresponding points, the criterion is minimalized.

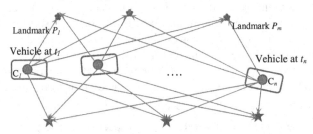

**Fig. 2.** Optimal motion estimation based adjustment in partial travel. The landmarks are tracked in long- distance travel by an omnidirectional camera.

Difference to the method in [22], it considers the corresponding image points in each pairwise images, our method considers to process the corresponding points, which are tracked in all partial sequent frames, see also Fig. 2.

## 4      Optimal Rotation Error

This section presents the optimal partial motion error based on the rotation averaging method. In this work, the corresponding points are considered as the invariant features. The property of the omnidirectional camera is that allows capturing scene in long-distance of travel, which is suitable for optimal estimated error of the partial motion. Since the estimated rotation using pairwise images is archived in each interval of vehicle motion, the goal now is finding the relative rotation of omnidirectional images for minimal error. In the case of a perspective camera, some common approaches for camera registration have been proposed. The cycles in the camera graph and a Bayesian framework was used for incorrect pair-wise detection in [21]. Another linear solution based on the least squares method was presented in [23]. Whereas a branch-and-bound search over the rotation space was used to determine the camera orientation in [19]. In this work, we apply a robust rotation averaging method for omnidirectional images based on the method in [22, 24]. The results demonstrate that the graph-based sampling scheme efficiently removes outliers in the individual relative rotation based on the RANSAC scheme.

The partial motion of the vehicle is represented by a weighted graph. Each positional frame is presented by a vertex. The relative rotation between pairwise images is represented by an edge, which can estimate directly. The weighted edge is estimated based on the rotation error. For initial the weighted graph, all edges are assigned by the identical value. In difference with previous authors, landmarks are tracked in all frame of partial motion, which are considered to minimize the estimated rotation error. The relative rotation $R_{ij}$, which represents the rotation from frame $i$ ($R_i$) to frame $j$ ($R_j$) in the global coordinates, as follows

$$R_j = R_{ij}R_i \qquad (6)$$

The objective of this optimal is that minimizes the error of the global rotation $R_i$ and $R_j$ based on the relative partial rotation $R_{ij}$. The error is represented by the difference between $R_j$ and $R_{ij}R_i$, which is denoted by $d(R_j, R_{ij}R_i)$.

## 5    Experiments

The sequent images were collected by an omnidirectional camera system, which constitutes from the classical perspective camera and the hyperboloid mirror. The experiments were carried out the electric vehicle with the omnidirectional camera mounted on the roof, the GPS receiver, IMU and the laser device mounted on the bumper as shown in Fig. 3. The omnidirectional image direction was defined at the first frame, collinear with the directional head of the vehicle. The parameters of omnidirectional image are 1280×960pixesl resolution, center point (646, 460), inner circular radius 150 pixels, and outer circular radius 475 pixels. The corresponding image points are extracted from sequent images by the SIFT method [17]. The camera system was calibrated by using the toolbox in [16]. The calibration result contents $\gamma=276.26$. The SICK LMS-291 laser worked under the conditional parameters of angular range 180 degree, angular resolution 0.5 degree, and maximum distance measurement 80$m$. The transformation of LRF is estimated by using the Iterative Closest Point (ICP) method [25]. The camera-LRF system was calibrated by using [26]. In this experiment, LRF information is only used for extracting translation magnitude, which supports for absolute motion estimation. The ground truth is measured by using the IMU MTi-G-700. The optimization toolbox of Matlab is used to solve the equation systems.

Geometry information at the starting and the ending positions were computed based on an average of GPS location measurements, which is used as the ground truth for comparison. The GPS position is computed based on an average of ten times of measurements. The error of GPS measurement is ±5$m$ under normal condition. The travel is about 1,220$m$, with 281 key images and laser scans, which were collected under maximum speed 10km/h. GPS locations, images, laser scans were simultaneous collected. An average distance between key sequent frames is about 4.34$m$.

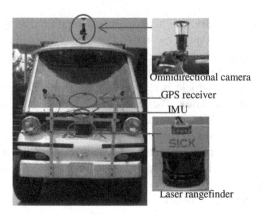

**Fig. 3.** Vehicle equipped with multiple sensors: camera systems, LRF, GPS, and IMU

Fig. 6 shows estimated localization results of vehicle about 1,220*m* traveled trajectory, which is superimposed on the aerial image of the Google Maps. Table 2 shows the compared error results given by visual odometry based on the proposed, method 8-poins non-optimization, and GPS location data. The criterion for comparison is location information at the starting and the ending of vehicle travel. It is computed with an average of GPS location measurements. The starting position is located at (35.54216°, 129.25596°), and the ending travel is located at (35.54276°, 129.25778°). The experimental results show that the error of visual odometry is accumulative over time. The GPS positions indicate that GPS has not accumulated global error. However, the partial error is large, especially at the eclipsed positions, e.g., under high buildings or lack of the number of satellites. The final trajectory is been diverged. The maximal error our method comparing with the ground truth is 9.45*m* (7.7%), while the maximal error of GPS position is 22.7*m* (18.8%).

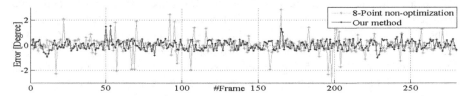

**Fig. 4.** The vehicle rotation error using omnidirectional camera compares with the ground truth

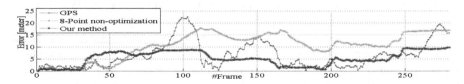

**Fig. 5.** The vehicle location error based on our method using vision+ LRF and GPS results compare with the ground truth

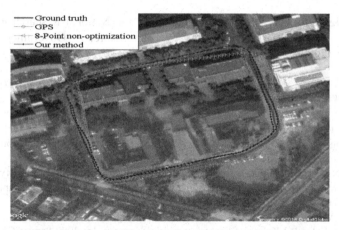

**Fig. 6.** The trajectories of the vehicle motion results based on the proposed method, 8-poins non-optimization, GPS receiver, and the ground truth

**Table 1.** Location errors use GPS and methods using vision comparing with the ground truth

| | Localization error (meter) | | | Rotation error (degree) | |
|---|---|---|---|---|---|
| | *Max* | *Ending local* | *%* | *Max* | *Average* |
| GPS | 22.52 | 15.73 | 1.28 | | |
| 8-Point non-optimization | 17.80 | 16.87 | 1.38 | 3.117 | 0.416 |
| Our method | 9.74 | 9.74 | 0.79 | 1.518 | 0.292 |

# 6    Conclusions

This paper presented the method to estimate vehicle motion using a monocular omni-directional camera based on the optimal error of partial rotation. The system consists of the omnidirectional camera and the LRF mount on the electric vehicle. The LRF is used only for estimating translation magnitude. The rotation averaging is used to optimize the partial rotation error based on the basic idea of the partial adjustment. The advantage of the omnidirectional camera is that allows capturing scene in long-distance travel, especially in large rotation. Based on this characteristic, the high accuracy of estimated rotation is obtained. The method avoids only using incremental approach, which is often diverging in long-distance motion and bundle adjustment approach, which spend high computational time. The experimental results also show that GPS information is drifted or jumped under eclipsed obstacle terrain such as high building, tunnel. On the contrary, the trajectory of the vehicle motion based on vision encounters with cumulative error. The experimental results show that the effectiveness of the method and indicate that this method is suitable for real application. The future works focus on the combination of this method based on cameras with GPS to deal with both of error types for application in long-distance motion estimation and auto-navigation. It is expected to improve this method for the real-time and high accuracy for odometry.

**Acknowledgment.** This work was supported by the National Research Foundation of Korea (NRF) Grant funded by the Korean Government (MOE) (2013R1A1A2009984).

# References

1. Suzuki, T., Kitamura, M., Amano, Y., Hashizume, T.: 6-DOF localization for a mobile robot using outdoor 3D voxel maps. In: IEEE/RSJ International Conference on Intelligent Robots and Systems (IROS), pp. 5737–5743 (2010)
2. Do, T.-N.: Parallel multiclass stochastic gradient descent algorithms for classifying million images with very-high-dimensional signatures into thousands classes. Vietnam Journal of Computer Science 1, 107–115 (2014) (10.1007/s40595-013-0013-2)
3. Hoang, V.-D., Le, M.-H., DaniloCáceres, H., Kang-Hyun, J.: Localization estimation based on Extended Kalman filter using multiple sensors. In: 39th Annual Conference of the IEEE Industrial Electronics Society (IECON), pp. 5498–5503 (2013)
4. García, D.V., Rojo, L.F., Aparicio, A.G., Castelló, L.P., García, O.R.: Visual Odometry through Appearance- and Feature-Based Method with Omnidirectional Images. Journal of Robotics 2012, 13 (2012)
5. Lee, M., Oh, S.: Alternating decision tree algorithm for assessing protein interaction reliability. Vietnam Journal of Computer Science, 1–10 (2014) (10.1007/s40595-014-0018-5)
6. Hoang, V.-D., Le, M.-H., Jo, K.-H.: Hybrid cascade boosting machine using variant scale blocks based HOG features for pedestrian detection. Neurocomputing 135, 357–366 (2014)
7. Konolige, K., Agrawal, M., Solà, J.: Large-Scale Visual Odometry for Rough Terrain. In: Kaneko, M., Nakamura, Y. (eds.) Robotics Research. STAR, vol. 66, pp. 201–212. Springer, Heidelberg (2010)
8. Chohra, A., Kanaoui, N., Amarger, V., Madani, K.: Hybrid intelligent diagnosis approach based on soft computing from signal and image knowledge representations for a biomedical application. Vietnam Journal of Computer Science, 1–13 (2014) (10.1007/s40595-014-0017-6)
9. Hoang, V.-D., Le, M.-H., Jo, K.-H.: Robust Human Detection Using Multiple Scale of Cell Based Histogram of Oriented Gradients and AdaBoost Learning. In: Nguyen, N.-T., Hoang, K., Jędrzejowicz, P. (eds.) ICCCI 2012, Part I. LNCS, vol. 7653, pp. 61–71. Springer, Heidelberg (2012)
10. Hoang, V.-D., Hernández, D.C., Jo, K.-H.: Combining Edge and One-Point RANSAC Algorithm to Estimate Visual Odometry. In: Huang, D.-S., Bevilacqua, V., Figueroa, J.C., Premaratne, P. (eds.) ICIC 2013. LNCS, vol. 7995, pp. 556–565. Springer, Heidelberg (2013)
11. Le, M.-H., Hoang, V.-D., Vavilin, A., Jo, K.-H.: One-point-plus for 5-DOF localization of vehicle-mounted omnidirectional camera in long-range motion. International Journal of Control, Automation and Systems 11, 1018–1027 (2013)
12. Scaramuzza, D.: 1-Point-RANSAC Structure from Motion for Vehicle-Mounted Cameras by Exploiting Non-holonomic Constraints. Int. J. Comput. Vis. 95, 74–85 (2011)
13. Hoang, V.-D., Hernández, D.C., Le, M.-H., Jo, K.-H.: 3D Motion Estimation Based on Pitch and Azimuth from Respective Camera and Laser Rangefinder Sensing. In: IEEE/RSJ International Conference on Intelligent Robots and Systems (IROS), pp. 735–740 (2013)

14. Le, M.-H., Hoang, V.-D., Vavilin, A., Jo, K.-H.: Vehicle Localization Using Omnidirectional Camera with GPS Supporting in Wide Urban Area. In: Park, J.-I., Kim, J. (eds.) ACCV Workshops 2012, Part I. LNCS, vol. 7728, pp. 230–241. Springer, Heidelberg (2013)

15. Hoang, V.-D., Le, M.-H., Jo, K.-H.: Planar Motion Estimation using Omnidirectional Camera and Laser Rangefinder. In: International Conference on Human System Interactions (HSI), pp. 632–636 (2013)

16. Mei, C., Rives, P.: Single View Point Omnidirectional Camera Calibration from Planar Grids. In: IEEE International Conference on Robotics and Automation (ICRA), pp. 3945–3950 (2007)

17. Lowe, D.: Distinctive Image Features from Scale-Invariant Keypoints. Int. J. Comput. Vis. 60, 91–110 (2004)

18. Fischler, M.A., Bolles, R.C.: Random sample consensus: a paradigm for model fitting with applications to image analysis and automated cartography. Communications of the ACM 24, 381–395 (1981)

19. Hartley, R.I., Kahl, F.: Global optimization through rotation space search. Int. J. Comput. Vis. 82, 64–79 (2009)

20. Qifa, K., Kanade, T.: Quasiconvex Optimization for Robust Geometric Reconstruction. IEEE Transactions on Pattern Analysis and Machine Intelligence 29, 1834–1847 (2007)

21. Kahl, F., Hartley, R.: Multiple-View Geometry Under the L_inf-Norm. IEEE Transactions on Pattern Analysis and Machine Intelligence 30, 1603–1617 (2008)

22. Chatterjee, A., Govindu, V.: Efficient and Robust Large-Scale Rotation Averaging. In: International Conference on Computer Vision, pp. 521–528 (2013)

23. Martinec, D., Pajdla, T.: Robust rotation and translation estimation in multiview reconstruction. In: IEEE Conference on Computer Vision and Pattern Recognition, CVPR 2007, pp. 1–8. IEEE (2007)

24. Govindu, V.M.: Robustness in motion averaging. In: Narayanan, P.J., Nayar, S.K., Shum, H.-Y. (eds.) ACCV 2006. LNCS, vol. 3852, pp. 457–466. Springer, Heidelberg (2006)

25. Besl, P.J., McKay, H.D.: A method for registration of 3-D shapes. IEEE Transactions on Pattern Analysis and Machine Intelligence 14, 239–256 (1992)

26. Hoang, V.-D., Cáceres Hernández, D., Jo, K.-H.: Simple and efficient method for calibration of a camera and 2D laser rangefinder. In: Nguyen, N.T., Attachoo, B., Trawiński, B., Somboonviwat, K. (eds.) ACIIDS 2014, Part I. LNCS, vol. 8397, pp. 561–570. Springer, Heidelberg (2014)

# A Smart Mobility System Implemented in a Geosocial Network

Cristopher David Caamana Gómez and Julio Brito Santana

University of La Laguna, 33271 San Cristobal of La Laguna, Spain
{ccaamana,jbrito}@ull.es
http://www.ull.es

**Abstract.** The continuous evolution of internet and web 2.0 technologies facilitates the creation of dynamic content. Social networks with georreference can be helpful to handle information from different sources and provide user-oriented services. Among these applications we can consider the intelligent systems for mobility.In this paper we introduce our geosocial network platform called Vidali. The open source social platform Vidali is developed provides a set of tools for the benefit of interactivity and collaboration between people and the provision of location-based services, which creates an environment that enhances collective intelligence. Starting with this platform as base, we developed a solution to improving mobility in local environments, which includes among other features the management of shared vehicles. We discuss the design and implementation of Vidali and of a smart mobility system.

**Keywords:** geosocial network, smart mobility system, intelligent collective.

## 1 Introduction

The application of information technology and communications to enhance mobility, traffic and logistics of cities is one aspect that contributes to smart cities. Efficient management and sustainable use of infrastructure, transport resources and the mobility needs of citizens requires suitable technologies to handle information from different sources and provide user-oriented services. Current availability of all types of connected mobile devices enables consistent data capture, ongoing communication and exchange of data in real time. This same availability demand from users of online information services allows them to make decisions, resolve incidents and manage appropriately. Among these applications we can consider the intelligent systems for mobility in urban spaces where people, vehicles and infrastructure exchange information in real time, allowing activities such as: efficiently managing public roads by improving accessibility, traffic and parking; reducing the use of private vehicles, strengthening public transport and car sharing; tracking vehicles and people mobility, analyzing and predicting behavior and guiding users. Significant papers addressing the development of systems to improve mobility are found in the literature, and of these many

D. Hwang et al. (Eds.): ICCCI 2014, LNAI 8733, pp. 302–311, 2014.

analyze Intelligent Transportation Systems (ITS). Recent reviews of technologies associated with ITS can be found in [4] and [10]. In this area, planning for travel and transportation are called Traveler Information Systems (TIS), which provide information and knowledge about the means and modes and also offers appropriate path alternatives [13], [1], with multimodal characteristic as the trip planning system proposed by Su J-m. and Chang C-h. [20] or the transport network model developed by Zhan et al. [23]. In [18], the authors proposed an advanced traffic system with collaborative information in real-time. Several ITS are proposed, with use or absence of georeferenced information, which provide information and perform analysis, modeling and decision making for strategic planning [2], [21], [6].

The development and evolution of internet technologies, applications and services oriented towards more interactive and collaborative environments, is currently an indisputable reality. Applications and services in web 2.0 facilitate the collaborative creation of dynamic contents. One of its greatest exponents of Web 2.0 are social networks.Social networks are a set of tools to create virtual spaces to promote communities and social exchange, which can generate new knowledge, learning and collective intelligence [19]. With these platforms, a social, participatory and real time web is a social, economic and business phenomenon which guides the activity to the end user or customer as protagonist.

Moreover, for several decades systems have been developed and technologies, which have allowed geographically localized data to be captured, stored, analyzed, shared, and visualized are found in Geographic Information Systems. These systems handle spatial information and their capabilities are based on geolocation or georeference, location or positioning systems data or mapping coordinates. Developments of these technologies have emerged in the context of the web, the GeoWeb or in the context of Web 2.0, the Web Mapping 2.0 [12]. In recent years location-based service (LBS) have been developed with the popularization of mobile phones, smartphones and tablets. These services make use of the capabilities of mobile devices to facilitate location through GSM phone triangulation, GPS or information sent by the user [22]. Thus through LBS, knowing the geographic position of the device can identify people or objects, and has driven applications in multiple contexts [5].

Geosocial networks are social networks which include capabilities and services based on georeferencing, and geotagging. It uses mapping interfaces systems, capable of geolocation, participatory and collaborative systems, which facilitate collective intelligence. These features provide users with additional social dynamics by integrating the interaction according their location. There are several commercial geosocial platforms available, such as Wikitude, which applies augmented reality to the geolocated content of the users, GPSMess which allows users to leave messages in a location, Foursquare, which allows "check -ins " to be made on visited places and allows the user to receive discounts, Yelp which suggests business places for users, and can evaluate them and respond to their concerns. None of these platforms is open, neither in data nor in development nor are they aimed at the provision of services in the context of intelligent mobility and literature on the engineering and development of these systems is non-existent. Some references in this field on geosocial networks are available. A general survey on social networking in georeferencing can be found in [17] and a preliminary description of the services that geography provides information to social networks is seen in [15].

In this paper we contribute a system for intelligent mobility, dedicated to transportation information, support media selection and routes, allowing inter alia exchange information on user routes which also meets transport demands as well as sharing vehicles. It is a georeferenced system interface that allows real-time communication between devices that facilitate integration and data processing, able to aggregate data from multiple heterogeneous sources and information provided by users about mobility. This system is implemented on an open source social platform based on geosocial networks and their design and development is presented. In addition to open source other features are that it can be installed on any server, it is flexible and can be extended through new applications. From the point of view of applications it is directed to the development of services in environments of local development, urban development, smart cities and smart mobility.

To our knowledge in the literature there are no mobility systems that are oriented towards an integrated and global vision of information processing and services for the stakeholder, end user, transport operator and responsible stewardship and considers public and private transport, as we propose. Various shared services vehicles settings are available in the market, such as Blablacar, Carpooling or Uber. These environments are closed systems that do not have information about existing transport services nor do they allow the information generated to be used.Some references in the literature on systems that share vehicles are [8] and [3], and models of planning support [16] and [14]. In addition, there are proposals to develop platforms which offer services within social communities, such as Carpal [7], offering carpooling systems that share travel information in a decentralized travel mode, allowing users to interact with each other without the need for a home server.

The remainder of this paper is organized as follows. Section 2, it details the platform developed, Vidali, explaining its design and implementation. In Section 3, we describe the intelligent mobility system, a service developed using the capabilities of Vidali. Finally, in Section 4 we present the conclusions and describe future research and developments.

## 2   Vidali

An innovative, modular and scalable geosocial platform with additional features is developed which coincides with the present day evolution and innovation of location-based social networks. Vidali is a service platform supported on an open source geosocial network (AGPLv3 license). The georeference allows knowledge of users with their geographic location and other information from the environment to be published and shared. The integrated social media tools permit interaction with other related and/or nearby users, and offer features such as creating a user profile, creating and joining groups, managing a group, being added as contacts for other users, chatting, uploading files, updating events and status, and positioning it in a georeferenced interface. Each user can add up to 150 contacts, which is the number of individuals that a human being can process within their social skills [9].

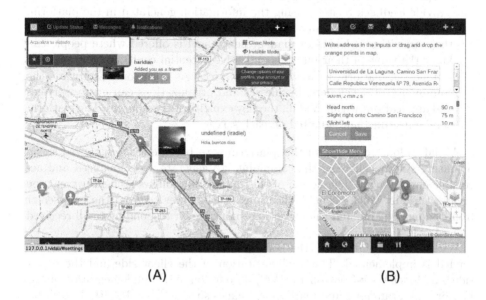

(A)                                    (B)

**Fig. 1.** (A) Platform interface. (B) Service interface.

The main actors of this geosocial service platform include: end users, businesses and organizations of any type, and application developers. Each of them share and exchange information in order to cover a range of needs. Users are now asking and sharing more information from their environment, activities or a specific location. Companies and organizations need to meet their demands and user ratings in relation to its activities or services. Developers need to have a number of available resources to develop new applications. The benefits to these actors are mainly based on feedback, as they bring knowledge to other actors, companies, organizations and developers to meet the demand of users and improve user experience. Moreover, developers can create applications for companies and organizations, which generate a portfolio of beneficial services on the platform for the user.

The technological characteristics of the design and development of the platform lead to the following activities:

- Enhancing the interaction between people with similar likes and characteristics (through groups and messages). This application intuitively represents groups and users nearby, where this is part of their interests. The purpose

306 C.D.C. Gómez and J.B. Santana

of this interaction is to allow members of each group to publish information of interest to its members.

- Harnessing the activity and environmental events. User location and all elements nearby are obtained and represented on the map, which facilitates orientation and search activities.
- Report real-time active service. The location of the user can choose a source of open data to provide additional information about its environment.
- Provide feedback to users using the information generated in the platform. With the purpose that all possible actors of this platform may have data on their activities in Vidali, a management panel is offered where people can know the value of their business and solve potential conflicts.
- Extend the platform with additional services. As open source software, the community of developers, businesses and organizations can create services and applications within Vidali, thereby exploiting its resources. The platform can be installed and adapted on each server.

A mid to long term aim of this platform is to become a decentralized social operating system, allowing the community to generate a personalized and accessible environment. The design of this application follows a client-server type architecture, in which client applications can be accessed from the browser or from the mobile app, and the server is responsible for responding to all received requests.

Within this architecture, the software that is sustainable and can be extended is implemented. The implementation of the client side and the server side is a Model-View-Controller (MVC). In relation to the implementation of this platform programming and markup languages such as PHP, BASH, Javascript, HTML5 are used, and CSS3 is generated by LESS for the UI design. We use also frameworks such as Slim PHP, Underscore, Backbone.JS, Require.JS, JQuery, Bootstrap and Leaflet for facilitate the development. OpenStreetMap is used for get collaborative and editable maps and MySQL is the database engine selected for data storage. These tools are used to develop an adaptable platform to any device and scalable to any need for end users, businesses and government. The platform is also intended to facilitate the modularity of the project and promote its growth, following the principles of reuse and flexibility, and allowing for better control on the evidence of each party and its maintenance.

The use of RequireJS and Backbone in the case of the client application leads to an assumed library structure, so we extend these classes of controllers, views and models. Since this application connects to the server via REST API using BackboneJS, the models must contain the URI which must perform the query to get the data. For the application server, we implement the system APIs through the Slim framework, capturing the URI's that the client sends, and if they are correct, we proceed to load the corresponding server items, previously checking system security. Once everything is verified, we return to the results returned to the client in JSON format.

The scalability and modularity of the developed platform allows applications or fully integrated services to be deployed. These applications can use the

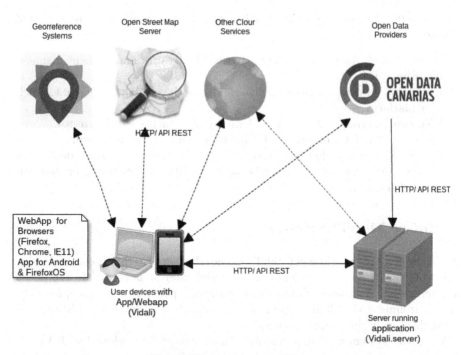

**Fig. 2.** Client-server arquitecture

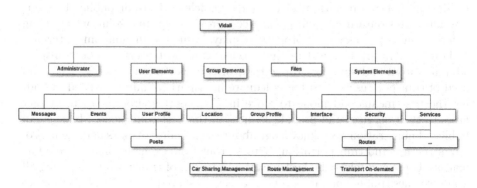

**Fig. 3.** Structure diagram

resources and capabilities of the platform to obtain and manage data, interface, user profiles and groups, system events, notifications and system messages via calling the general API.

As seen in the figure, the main structure of project is made up of the following components:

- Admin Panel: All management activities for network and element management for third parties.
- Elements and user activities: Define the hierarchy of items available for each user platform.
- Elements of groups: Defines the hierarchy of elements within groups.
- Files: Sets the level of activity for file processing on the platform.
- System elements: Defines the hierarchy of system actions, which deals with the control of the user interface, system security and service layer that can be implemented.

## 3  Smart Mobility System

One of the goals of our platform Vidali is to develop services that tackle real needs and problems of local environments. A special interest is to satisfy the needs of different policy areas of Smart cities, economy, people, governance, mobility and environment [11]. This paper, and specifically this section, addresses the demands of emerging smart mobility.

A support system and smart mobility management is developed with Vidali. This service provides a social and collective intelligence layer in order to improve mobility, and focuses on: sharing information on transportation and user trips, identifying the demands of transport and offers that are close to each user and facilitating optimal management of shared vehicles private or public. This service integrates data from multiple sources, provides optimal solutions applying different smart processes and properly displays results in the mapping interface.

Three actors are involved in this service: transport users, public transport administrators and suppliers (operators) of transportation services. The main need of transport users is to find adequate information and be assisted in finding the best means and mode to move in. Transport operators try to attract users to their services and provide services in the best conditions. Managers of public administration need knowledge about the mobility of users and the uses of infrastructure and transportation. Actors profit from exchange and share information. Users can select the best means and modes of transportation among all available ones (public, private, proper, collective, shared, on-demand ...) to move in. Operators (including public or private companies, rental, ...) obtain detailed information about users and demands, but in real time, offer services on demand, complement current offers and new services. Public administrations analyze and evaluate transport services and promote measures to improve mobility.

A smart mobility system has three important features:

- Transport information: multiple data of all actors, which can display ratings, results of routes and stops, reception data of other actors and give feedback.
- Shared Transport: This functionality includes operations of the interaction between actors in the management of shared vehicles, either from a public or

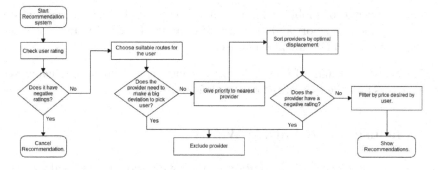

**Fig. 4.** Recommendation system. Flow diagram

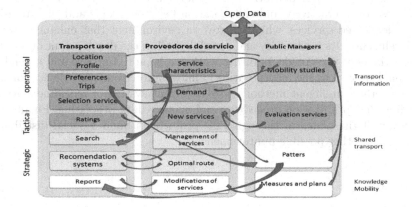

**Fig. 5.** Knowledge map

private service (taxi, private car or other on-demand medium). This system is based on two intelligent search procedures. A variation of the A * heuristic to solve the problem of shortest paths between source and destination and other, GRASP metaheuristics to solve the Travelling Salesman Problem (TSP) and the best route to go through a set of nodes. In both cases, the approach is multicriteria and multimodal. This approach generates the best solutions for user trips and also allows the same for the service supplier. Taking into account the rating of users and providers, the recommendation system uses the following strategy:

The result is a set of optimal solutions that shows the user the right information to select which service he is interested in. Once the user makes the trip, he evaluates the service provider and the quality of the recommendation, a result for future recommendations.

– Knowledge on urban mobility: This feature provides a set of analytical tools and data mining. Its purpose is to extract patterns and new knowledge with the aim of improving mobility decisions. It provides periodic reports of the uses of infrastructure, transport facilities and frequent itineraries. The data mining tools use historical data to uncover different patterns of user behavior and knowledge of mobility for planning, for example, new infrastructure, new bus lines transport and measures to improve efficiency, reduce costs and negative environmental externalities.

In Figure 5 show a map of knowledge with a view of the knowledge generated by the actors and the flow of tacit to explicit knowledge.

# 4   Conclusion and Future Works

The proposed application is a solution to improving mobility in local environments, which includes among other features the management of shared vehicles. The open source social platform that is developed provides a set of tools for the benefit of interactivity and collaboration between people and the provision of location-based services, which creates an environment that enhances collective intelligence. Its modularity and scalability allow the development of services such as the ones presented. As future work, we hope to improve and extend its functionality, such as the incorporation of information from other devices or sensors and other intelligent methods to the current system of recommendations, including the optimization procedures and massive data processing. Finally, the expansion of the platform through P2P protocols t build a social distributed platform is a other possible improvement.

# References

1. Adler, J.L., Blue, V.J.: Toward the design of intelligent traveler information systems. Transportation Research Part C: Emerging Technologies 6(3), 157–172 (1998)
2. Arampatzis, G., Kiranoudis, C.T., Scaloubacas, P., Assimacopoulos, D.: A gis-based decision support system for planning urban transportation policies. European Journal of Operational Research 152(2), 465–475 (2004)
3. Barth, M., Shaheen, S.A.: Shared-use vehicle systems: Framework for classifying carsharing, station cars, and combined approaches. Transportation Research Record: Journal of the Transportation Research Board 1791(1), 105–112 (2002)
4. Bazzan, A.L.: Introduction to intelligent systems in traffic and transportation. Synthesis Lectures on Artificial Intelligence and Machine Learning 7(3), 1–137 (2013)
5. Bellavista, P., Kupper, A., Helal, S.: Location-based services: Back to the future. IEEE Pervasive Computing 7(2), 85–89 (2008)
6. Brown, A.L., Affum, J.K.: A GIS-based environmental modelling system for transportation planners. Computers, Environment and Urban Systems 26(6), 577–590 (2002)
7. Ciancaglini, V., Liquori, L., Vanni, L.: Carpal: Interconnecting overlay networks for a community-driven shared mobility. In: Wirsing, M., Hofmann, M., Rauschmayer, A. (eds.) TGC 2010, LNCS, vol. 6084, pp. 301–317. Springer, Heidelberg (2010)
8. Dobrosielski, J., Gray, T., Nhan, A., Stolen, M.: Carpool. umd: community carpooling. In: CHI 2007 Extended Abstracts on Human Factors in Computing Systems, pp. 2055–2060. ACM (2007)

9. Dunbar, R.I.: Coevolution of neocortical size, group size and language in humans. Behavioral and Brain Sciences 16(04), 681–694 (1993)
10. Elkosantini, S., Darmoul, S.: Intelligent public transportation systems: A review of architectures and enabling technologies. In: 2013 International Conference on Advanced Logistics and Transport (ICALT), pp. 233–238 (May 2013)
11. Giffinger, R., Fertner, C., Kramar, H., Kalasek, R., Pichler-Milanovic, N., Meijers, E.: Smart cities. ranking of european medium-sized cities. Final report of a research project, Centre of Regional Science (SRF), Vienna University of Technology (2007)
12. Haklay, M., Singleton, A., Parker, C.: Web mapping 2.0: The neogeography of the geoweb. Geography Compass 2(6), 2011–2039 (2008)
13. Hall, R.W.: Route choice and advanced traveler information systems on a capacitated and dynamic network. Transportation Research Part C: Emerging Technologies 4(5), 289–306 (1996)
14. de Almeida Correia, G.H., Antunes, A.P.: Optimization approach to depot location and trip selection in one-way carsharing systems. Transportation Research Part E: Logistics and Transportation Review 48(1), 233–247 (2012); select Papers from the 19th International Symposium on Transportation and Traffic Theory
15. Huang, Q., Liu, Y.: On geo-social network services. In: 2009 17th International Conference on Geoinformatics, pp. 1–6 (August 2009)
16. Kek, A.G.H., Cheu, R.L., Meng, Q., Fung, C.H.: A decision support system for vehicle relocation operations in carsharing systems. Transportation Research Part E: Logistics and Transportation Review 45(1), 149–158 (2009)
17. Kelm, P., Murdock, V., Schmiedeke, S., Schockaert, S., Serdyukov, P., Laere, O.: Georeferencing in social networks. In: Ramzan, N., Zwol, R., Lee, J.S., Clüver, K., Hua, X.S. (eds.) Social Media Retrieval. Computer Communications and Networks, pp. 115–141. Springer, London (2013)
18. Lee, W.-H., Tseng, S.-S., Shieh, W.-Y.: Collaborative real-time traffic information generation and sharing framework for the intelligent transportation system. Information Sciences 180(1), 62–70 (2010); special Issue on Collective Intelligence
19. Schoder, D., Gloor, P., Metaxas, P.: Social media and collective intelligence–ongoing and future research streams. KI - Künstliche Intelligenz 27(1), 9–15 (2013)
20. Su, J.-M., Chang, C.-H.: The multimodal trip planning system of intercity transportation in Taiwan. Expert Systems with Applications 37(10), 6850–6861 (2010)
21. Toledo, T., Beinhaker, R.: Evaluation of the potential benefits of advanced traveler information systems. Journal of Intelligent Transportation Systems 10(4), 173–183 (2006)
22. Wang, S., Min, J., Yi, B.K.: Location based services for mobiles: Technologies and standards. In: IEEE international Conference on Communication (ICC), pp. 35–38 (2008)
23. Zhang, J., Liao, F., Arentze, T., Timmermansa, H.: A multimodal transport network model for advanced traveler information systems. Procedia Computer Science 5(0), 912–919 (2011); the 2nd International Conference on Ambient Systems, Networks and Technologies (ANT-2011) / The 8th International Conference on Mobile Web Information Systems (MobiWIS 2011)

# A Prototype of Mobile Speed Limits Alert Application Using Enhanced HTML5 Geolocation

Worapot Jakkhupan

Information and Communication Technology Programme, Faculty of Science,
Prince of Songkla University, Hat Yai Campus, Songkla 90112, Thailand
worapot.j@psu.ac.th

**Abstract.** This study proposes the HTML5 geolocation-based vehicle speed alert application aims to facilitate the passengers who could not see the information on the dashboard of vehicle. The traditional vehicle speed determined from HTML5 geolocation API is improved using haversine distance calculation. The speed limit value is automatically set according to the specify type of vehicle and the current type of road. The prototype was developed and tested under the transportation regulations in Thailand. The result reveals that the enhanced HTML5 geolocation speed determination using haversine distance significantly improves the accuracy of vehicle speed detection compared with the traditional HTML5 geolocation API.

**Keywords:** HTML5 geolocation, vehicle speed alert, mobile devices, GPS, haversine distance.

## 1    Introduction

Road speed limits have been enforced in many countries to set the minimum or maximum speed of which vehicles may legally travel on the road. Speed limits may be vary mostly depends on locations, types of road and types of vehicle. Speed limits are normally indicated on a traffic sign beside the road in which driver can easily acknowledge. There are several reasons to do this, especially to improve the traffic flow and for safety reason. In Thailand, the road speed limits are set by the department of land transport, ministry of transport.

In ages, the development of the vehicle garget that facilitates the driver in conveniently travelling on the road is becoming more important. There are many technologies support the driver built in the vehicle. The vehicle dashboard installed behind the steering wheel gives the information to the driver such as speed of vehicle or the fuel remain. To track the speed of the vehicle, normally, the driver can acknowledge the vehicle speed from dashboard display calculated by speedometer built in the vehicle. However, passengers who could not see the dashboard display might also want to know how fast the vehicle is moving.

Nowadays, mobile device, especially smart devices, is one of the gargets that increasingly being used and change people's lifestyle. One of the great features of mobile device is the location services [1] such as cell site based, WiFi based and GPS satellite based positioning system. The location services determine the current

D. Hwang et al. (Eds.): ICCCI 2014, LNAI 8733, pp. 312–321, 2014.
© Springer International Publishing Switzerland 2014

position of the smart device. Therefore, people be able to have their GPS navigation with more features in hand. There are many location-based applications have been launched, for example, mobile navigation, location tracking system [2], location aware applications, as well as using the location-based services in the transportation system.

GPS is one of the technologies can be used to measure the speed of the moving vehicle. Thus, passengers can use their mobile devices with built in GPS to determine the speed of vehicle they are travelling with. There are many techniques have been proposed to measure the speed of vehicle using GPS [3-6]. One of the possible techniques is HTML5 geolocation API [7]. Anyway, the preliminary experimentation has revealed that the speed taken from the API has low sensitive to the speed change, especially the determined speed usually lower than the actual speed. This might cause from many reasons such as the efficiency of the mobile device, the GPS mechanism or satellite errors [8] or the API itself [9]. To improve the accuracy of the speed, this study propose the mobile vehicle speed alert system using enhanced HTML5 geolocation API. The proposed prototype improves the accuracy of the GPS speed detection in HTML5 geolocation API using the haversine distance calculation. The proposed prototype also introduces the automatic speed limits value setting using Google geocoding API [10]. The accuracy of the speed was tested by comparing the speed determined from traditional API and the enhanced API. The result reveals that the enhanced API significantly improves the accuracy of speed compared with the traditional HTML5 geolocation API.

The rest of this paper is organized as follows. Sect. 2 gives the background literature that related to this study. Sect. 3 describes the purposed system and the experimentation. Sect. 4 reveals the results. Finally, Sect. 5 draws the conclusions and future works.

## 2    Literature Review

### 2.1    Thailand's Vehicle Speed Limits

The speed limits are different in each country. The unit of speed limits is generally measured in kilometers per hour (KPH) or miles per hour (MPH). In Thailand, the road speed limits are set by the department of land transport, ministry of transport. The speed limits in Thailand are defined by types of vehicle, areas, and types of road. Generally, the speed limits between 80 – 90 KPH. In Bangkok and Pattaya, the speed limits for motorcycle and car are 80 KPH, 60 KPH for truck, and 45 KPH for trailer. Outside Bangkok and Pattaya area, the speed limits 90 KPH for motorcycle and car, 80 KPH for truck and 60 KPH for trailer. There are two special types of road; intercity highway and motorway. The speed limits are 100 KPH for truck weight below 1,200 kilograms, 80 KPH for trailer, and 120 KPH for the other types of vehicle. The speed limits are also restricted in some special environments, for example, inside educational institutes, the street in front of the schools, for instance. The drivers have to observe the traffic sign beside the road to control the speed of the vehicles by themselves.

## 2.2    Speed Calculation

Speed is a scalar quantity states how fast an object is moving from one point to another point in an exact duration. The speed can be calculated as follow.

$$s = \frac{d}{t} \tag{1}$$

where $s$ denotes speed, $d$ denotes distance, kilometers or miles, between origin and destination, $t$ denotes total time, seconds, minutes, or hours, spent for travel.

## 2.3    Distance Estimation on the Maps

To measure the speed of vehicle using GPS in Eq. 1, the distance between two points defined by latitudes and longitudes on the maps must be calculated. The haversine formula [11-13] is a well-known equation used to calculate the distance between two points from longitudes and latitudes. The haversine formula (haversine distance) is calculated as follow.

$$d = R \times c \tag{2}$$

where $R$ is earth's radius (mean radius = 6,371 km) and $c$ is calculated from

$$c = 2a \tan 2[\sqrt{a}, \sqrt{(1-a)}] \tag{3}$$

where $a$ is calculated from

$$a = \sin^2(\frac{\Delta lat}{2}) + \cos(lat_1) \times \cos(lat_2) \times \sin^2(\frac{\Delta long}{2}) \tag{4}$$

where $\Delta lat$ is a minus of two latitudes, $\Delta long$ is a minus of two longitudes.

Another possible solution to estimate the distance between two points on the maps is to use Google API named Distance Matrix [14]. To retrieve the result, application sends the latitudes and longitudes of origin and destination using REST web service via URL interface. The current latitude and longitude of GPS device can be retrieve from HTML5 geolocation API using *getCurrentPosition()* or *watchPosition()* method. However, the smart device has to connect with internet, and the result from the API might delay.

## 2.4    HTML5 Geolocation API

HTML5 is the latest version of HyperText Markup Language (HTML). It has been designed to fulfill the gaps between web application and software development. HTML5 enables a cross-platform application development comes with many mobile device plugin connections, for example, camera, graphic, multimedia, as well as geolocation. The geolocation API allows user to provide the current position of the smart device to web applications so applications can provide users the smarter location-based services. The accuracy of HTML5 geolocation API has been continuously improved [9]. To determine user's current location using HTML5 geolocation API, the *getCurrentPosition()* method is called. This method requests the positioning hardware such as GPS, WiFi, GSM, to get the position of the device. When the position is determined, the defined callback function is executed. To monitor the position of the device, the *watchPosition()* method is used. The method is called every exact time to update the location, as well as to calculate speed of device movement. The important properties of HTML5 geolocation API are listed in Table 1.

**Table 1.** The important property of *getCurrentPosition()* and *watchPosition()*

| Property | Description |
| --- | --- |
| coords.latitude | The latitude as a decimal number |
| coords.longitude | The longitude as a decimal number |
| coords.accuracy | The accuracy of position in meters |
| coords.speed | The speed in meters per second |
| timestamp | The date/time of the response |

## 3     Experimental Methodology

### 3.1     System Architecture

The proposed prototype was developed using HTML5, JavaScript, jQuery and jQuery mobile in a cross-platform application development concept. The application was built using Cordova. To automatically set the speed alert value, mobile device must be connected to the internet to get the information from Google geocoding API. The architecture of the proposed prototype is shown in Fig. 1.

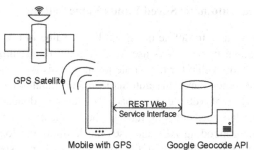

**Fig. 1.** Architecture of the proposed prototype

### 3.2     Procedure of the Proposed Prototype

When the vehicle stops, the application shows a zero speed. When the vehicle is move, the *watchPosition()* is called to update the speed of vehicle. The updated speed is compared with the defined speed alert value. The speed alert value may be set as

Manual or automatic. If the speed is higher than the defined value, application will alert with red screen, danger message and sound, otherwise, application will show safe message with green screen. The procedure of the prototype is drawn in Fig. 2.

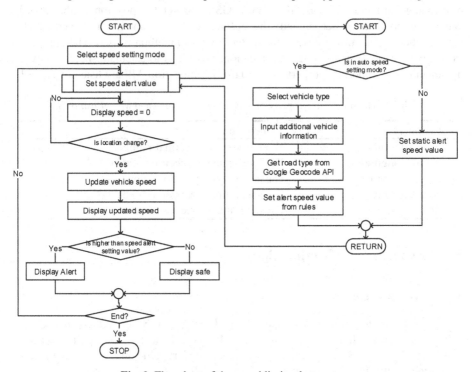

**Fig. 2.** Flowchart of the speed limits alert system

### 3.3    Manual and Automatic Speed Limits Value Setting

The prototype allows user to define the speed limits value in two modes, manual and automatic. The manual mode allows user to define the fixed numeric value in which the application uses this value for the whole journey. This mode does not require data connection. On the other hand, the automatic mode updates the speed limits value according to the type of vehicle and road. To get this done, the *watchPosition()* method returns the latitude and longitude to the application. After that, application connects to Google geocoding API and uses the latitude and longitude to query the name of street and city, which will be compared with the speed limits value as described in Sect. 2.1. The example of Google geocoding request is as follow.

```
https://maps.googleapis.com/maps/api/geocode/json?
latlng=latitute,longitute&sensor=true&key={App Key}
```

Google geocoding looks for the nearest street and address, then return the result in the defined format, in this case JSON, as the following example.

```
"results" : [{ "address_components" : [ { "long_name" :
"Motorway 7", "short_name" : "Motorway 7", "types" : [
"route" ] }, { "long_name" : "Bangkok", "short_name" :
"Bangkok", "types" : [ "locality", "political" ] } ],
"status" : "OK" }
```

Finally, application extracts the result from Google geocoding and compares with the defined rules to get the appropriate speed limits value. There are two important information in the address components; route and political. Attribute "route" returns the name and number of the nearest street if exists. Attribute "political" returns name of city or country. As an example, the street named "Motorway 7" locates the vehicle on the road type motorway in Bangkok. Thus, if the vehicle was set as motorcycle or car, the speed limits will be automatically updated to 120 KPH, for instance.

### 3.4    Vehicle Speed Detection Using HTML5 Geolocation API

As described in Sect. 2.4, the *watchPosition()* method is called to get the current latitude and longitude of the vehicle and to monitor the movement of vehicle. The code is written in JavaScript as the following example.

```
navigator.geolocation.watchPosition(
    hasPosition, null,{
        "enableHighAccuracy":true,
        "timeout":27000, "maximumAge":500
});
Function hasPosition(position) {
  var speed = position.coords.speed;
}
```

As shown in the example source code, when *watchPosition()* is called, the *hasPosition()* function is operated. The *enableHighAccuracy* option is set to true in order to force hardware to get the location from GPS. The *timeout* and *maximumAge* options are manually set, which must be calibrated on the actual devices to get the best accuracy. The *maximumAge* is important to define the time interval to refresh the location. In the prototype, the default *timeout* is set to 27000 and the *maximumAge* is set to 500.

### 3.5    Vehicle Speed Detection Using Enhanced HTML5 Geolocation

From the preliminary experiment, the speed taken from HTML5 geolocation as described in *hasPosition()* method in Sect. 3.4 is lower than actual speed and, moreover, is low sensitive with the speed change. This might cause from many factors such as the delay or the deviated location of GPS, or from the API itself. Thus, the proposed enhanced HTML5 geolocation calculates the new distance using haversine distance calculation mentioned in Sect. 3.3 as the following equation.

$$S_{haversine} = \frac{HaversineDistance}{t_2 - t_1} \tag{5}$$

where $S_{haversine}$ is the speed calculated from haversine distance, $t_2$ is the timestamp at the destination, and $t_1$ is the timestamp at the origin acquired from *position.timestamp*. The latitude and longitude of the origin and destination are located using *position.coords.latitude* and *position.coords.longitude*. The example of the enhanced speed calculation using haversine distance from Eq.2 – Eq.5 and HTML5 geolocation API written in JavaScript is shown as follow.

```
var lat1=0, lat2=0, time_start=0
Function hasPosition (pos) {
  var lat2=pos.coords.latitude;
  var lon2=pos.coords.longitude;
  var diff_time=time_start-pos.timestamp;
  var R=6371; //Earth's radius in Kilometers
  var x1=lat2-lat1; var dLat=x1.toRad();
  var x2=lon2-lon1; var dLon=x2.toRad();
    var a=Math.sin(dLat/2)*Math.sin(dLat/2)+
    Math.cos(lat1.toRad())*Math.cos(lat2.toRad())*
    Math.sin(dLon/2)*Math.sin(dLon/2); //a from Eq.4
  //c from Eq.3
  var c=2*Math.atan2(Math.sqrt(a),Math.sqrt(1-a));
  var d=R*c; //d from Eq.2
  var speed=d/diff_time //Haversine Enhanced Speed
  lat1=pos.coords.latitude;
  lat2=pos.coords.longitude;
  time_start=pos.timestamp;
}
```

Finally, the speed is compared with the speed limits value. If the speed is higher than the limit value, application will alert with danger message, red screen and sound, otherwise, application will show the safe message and green screen.

## 4    Results

The prototype was tested in multi operating systems of smart devices. Firstly, it was tested on the web browsers using Mozilla Firefox, Safari and Google Chrome. After that, the prototype was built in HTML5 hybrid application platform using Cordova PhoneGap. It was installed and tested in Android, iOS operating system. The screenshot of the prototype is shown in Fig. 3.

The accuracy of the prototype was tested by comparing the result on mobile device with the actual vehicle speed shown on the dashboard display. The test speed has been controlled by increasing from 0 to 50 KPH within 5 seconds and then decreasing from 50 to 5 within 5 seconds. The result is shown in Fig. 4.

**Fig. 3.** The example of the speed limits prototype shows the screen shot of the application when the vehicle stops, vehicle moves in safe speed and vehicle moves in danger speed

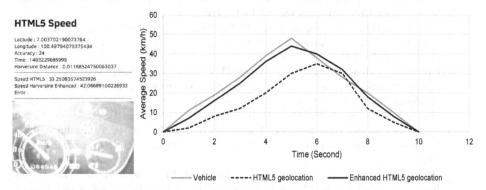

**Fig. 4.** The comparison between prototype and vehicle dashboard display

The result shows that the enhanced algorithm significantly improves the accuracy of HTML5 geolocation API, especially the sensitivity to the speed change. As shown in Fig. 4, the actual speed of vehicle is about 42 – 44 KPH, the HTML5 detects the speed as 32 KPH, but the enhanced algorithm detects 42 KPH, which is more accurate.

## 5    Conclusions and Future Works

This study proposes the vehicle speed alert application aims to facilitate the passenger in order to monitor the speed of vehicle. The traditional speed of vehicle movement calculated from HTML5 geolocation API is lower than actual speed and low sensitivity to the speed change. Therefore, this study improves the accuracy of HTML5 geolocation API by using the haversine distance calculation. The result shows that the speed calculated from haversine distance is more accurate compare with the traditional speed from HTML5 geolocation API. The proposed application allows user to manually or automatically set the speed alert value. The automatic speed alert value is set follow the transportation regulation of Thailand. The prototype was developed based on HTML5 cross-platform development concept, which can be built in multiple mobile operating platform such as Android, iOS or run through the web browser. For the future works, the accuracy of speed detection when the speed is decreased can be improved. The prototype will allow the passenger to track the movement of the vehicle on Google Maps. Finally, the location of the vehicle movement will be stored in the mobile device and the user will be able to share the directions on the social network such as Facebook.

# References

1. Lane, N.D., Miluzzo, E., Lu, H., Peebles, C.D., Campbell, A.T.: A survey of mobile phone sensing. IEEE Commun. Mag. 48(9), 140–150 (2010)
2. Menard, T., Miller, J.: Comparing the GPS Capabilities of the iPhone 4 and iPhone 3G for Vehicle Tracking using FreeSim_Mobile. In: 4th IEEE Intelligent Vehicles Symposium, pp. 278–283. IEEE Press, New York (2011)
3. Liou, R.H., Lin, Y.B., Chang, Y.L., Hung, H.N., Peng, N.F., Chang, M.F.: Deriving the vehicle speeds from a mobile telecommunications network. IEEE Transaction on Intelligent Transportation Systems 14(3), 1208–1217 (2013)
4. Chang, J.-Y., Wang, T.-W., Chen, S.-H.: Multiple Vehicle Speed Detection and Fusion Technology on the Road Network: A Case Study from Taiwan. In: Future Information Technology, Application, and Service. LNEE, vol. 164, pp. 223–230. Springer, Heidelberg (2012)
5. Cortés, C.E., Gibson, J., Gschwender, A., Munizaga, M., Zúñiga, M.: Commercial bus speed diagnosis based on GPS-monitored data. Transportation Research Part C: Emerging Technologies 19, 695–707 (2011)
6. Şimşek, B., Pakdil, F., Dengiz, B., Testik, M.C.: Driver performance appraisal using GPS terminal measurements: A conceptual framework. Transportation Research Part C: Emerging Technologies 26, 49–60 (2013)
7. W3C, http://dev.w3.org/geo/api/spec-source.html
8. Miura, S., Kamijo, S.: GPS Error Correction by Multipath Adaptation. International Journal of Intelligent Transportation Systems Research, 1–8 (2014)
9. Gup, A,
   http://www.andygup.net/how-accurate-is-html5-geolocation-really-part-2-mobile-web/
10. Google Developers,
    https://developers.google.com/maps/documenttation/geocoding/
11. Chris, V.: http://www.movable-type.co.uk/scripts/latlong.html
12. Palmer, M.C.: Calculation of distance traveled by fishing vessels using GPS positional data: A theoretical evaluation of the sources of error. Fisheries Research 89(1), 57–64 (2008)

13. Feng, T., Timmermans, H.J.: Transportation mode recognition using GPS and accelerometer data. Transportation Research Part C: Emerging Technologies 37, 118–130 (2013)
14. Google Developers ,
    https://developers.google.com/maps/documentation/distancematrix/
15. Apache, https://cordova.apache.org/docs/en/3.0.0/cordova_geolocation_geolocation.md.html

# Data-Driven Pedestrian Model:
# From OpenCV to NetLogo

Jan Procházka and Kamila Olševičová

University of Hradec Králové
Rokitanského 62, Hradec Králové 500 03, Czech Republic
{jan.prochazka,kamila.olsevicova}@uhk.cz

**Abstract.** Our objective was to replicate the movement of real pedestrians in NetLogo agent-based model using the video recording of pedestrians as the source of reliable data. To achieve this, it was necessary to develop the video-processing extension for NetLogo. The paper presents the principles of video data transformation, the implementation of the extension and the experiment with a sample video stream that demonstrates the self-organization of bi-directional flows of walkers. The extension builds on the computer vision library OpenCV.

## 1    Introduction

Pedestrians and crowd motion simulations can help us to identify and analyze spatial walking patterns under normal or competitive situations, to design pedestrian facilities or to define evacuation scenarios. Thanks to the computing performance and availability of empirical data, current pedestrian simulations shift from quantitative reproduction of various aspects of crowds towards qualitative reproduction [4]. Technically, pedestrian simulations build on cellular automata principles, social force models, velocity-based models or network models [3]. Various implementation frameworks and platforms are available, for applications see e.g. [1, 2, 7].

Our primary objective was to create a video-to-agent-based pedestrian simulation using the video records from traffic cameras as a source of reliable data about walking patterns of individuals. The idea behind is to transform the behavior of entities from video recording into the behavior of artificial agents, independently on the agent-based simulation platform. To achieve this, we developed the video-processing extension which enables detection and tracking of objects from the video recording. The extension was tested using NetLogo [8], but can be connected with other Java agent-based frameworks too. The extension is presented in the rest of the paper. The process of video data transformation is described in section 2, the implementation of extension using algorithms from open source computer vision library OpenCV [5] is explained in section 3 and the experiment with a sample video stream of bi-directional flows of pedestrians is provided in section 4.

D. Hwang et al. (Eds.): ICCCI 2014, LNAI 8733, pp. 322–331, 2014.

# 2    Video to Agent-Based Model Transformation

## 2.1    Video Processing

Video-to-agent model is a data-driven model where an input video recording supplies the agent-based model with data about agents, their positions and movements. The transformation $T$ takes the input $V$ (sequence composed of $f$ video frames) and transforms it to a sequence $A$ (sequence of sets containing data about agents' positions within the model environment, one set for one time point $t$, t is integer).

$$T: V \mapsto A \tag{1}$$

$$V = \{F_1, F_2, \dots, F_f\}; \quad A = \{A_1, A_2, \dots, A_f\}; \quad |V| = |A| = f \tag{2}$$

A single video frame is formalized as a matrix $F_t$. Elements of the matrix correspond to pixels of the $t$-th frame (3). Matrix size is given by the video frame dimensions (height $h$, width $w$). In case of 8-bit color depth and RGB color model, the matrix elements are represented by ordered triples of integers in range [0, 255].

$$F_t = \begin{pmatrix} (R,G,B)_{1,1} & \cdots & (R,G,B)_{1,w} \\ \vdots & \ddots & \vdots \\ (R,G,B)_{h,1} & \cdots & (R,G,B)_{h,w} \end{pmatrix}; \quad R, G, B \in \langle 0, 255 \rangle \tag{3}$$

Set $A$ defines $x$, $y$ and $z$ coordinates of all agents at the given time point (4).

$$A_t = \{(x, y, z)_{1,t}, (x, y, z)_{2,t}, \dots, (x, y, z)_{a,t}\} \tag{4}$$

The transformation (1) is composed of $T_1$ and $T_2$.

$$T = T_2 \circ T_1; \qquad A = T_2(T_1(V)) \tag{5}$$

In the first transformation $T_1$, a single video frame $F_t$ is transformed to the set $O_t$ of coordinates $x$, $y$, $z$ of objects of our interest (5).

$$T_1: F_t \mapsto O_t; \qquad O_t = \{(x, y, z)_{1,t}, \dots, (x, y, z)_{o,t}\}; \quad |O_t| = o_t \tag{6}$$

The second transformation $T_2$ operates with the set of objects' coordinates $O_t$ at time $t$ and the set $A_{t-1}$ of agents' positions within the model at time $t$-$1$ (6).

$$T_2: (O_t, A_{t-1}) \mapsto A_t; \quad A_t = \{(x, y, z)_{1,t}, \dots, (x, y, z)_{a,t}\}; \quad |A_t| = a_t \tag{7}$$

The splitting of $T$ into $T_1$ and $T_2$ reflects different problems treated on each level. While the objective of $T_1$ is to identify position of specific objects shown in a video frame, $T_2$ solves a problem of agents' lifecycles: recognizing new agents entering the scene, keeping track of agents and recognizing disappearance of agents leaving the scene.

The fig. 1 illustrates both transformation levels and the third step which is optional: storing objects' positions into the data file. The proposed extension is suitable for real time applications, i.e. agent-based models running simultaneously with online camera stream processing. Alternatively, it is possible to run only the first level transformation (to determine positions of objects and to store them) and to use the data file later. This approach allows user to run the same model several times without a need to process the video data repeatedly.

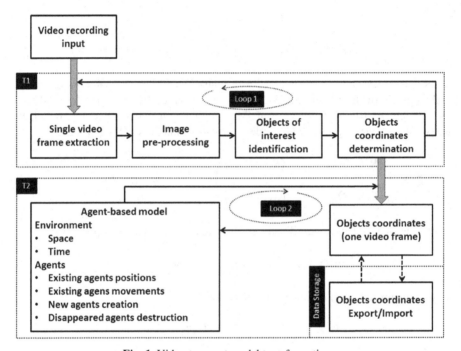

**Fig. 1.** Video-to-agent model transformations

## 2.2    From Video to Objects' Positions

First level transformation (6) is an image processing task. To recognize objects means to identify more or less well defined shapes using appropriate algorithm. We focused on circle finding using Hough transformation function and spot finding, i.e. searching continuous areas of pixels that match certain color characteristics.

Image processing algorithms require pre-processing (simplification) of the input using e.g. Gaussian blur to decrease the noise, or conversion to black and white. The turn of pixel $P$ of the original video frame into white or black is defined by RGB conditions C (8) that specifies color shades of objects to be identified. One RGB subcondition is a triple of intervals for R, G, and B (9). The disjunction (10) or other logical expressions can be used for specification of color characteristics of objects.

$$C = \{C_i : C_i = (R_i, G_i, B_i)\}; \quad P = (R_p, G_p, B_p) \tag{8}$$

$$R_i = \langle R_{i,min}, R_{i,max} \rangle; \; G_i = \langle G_{i,min}, G_{i,max} \rangle; \; B_i = \langle B_{i,min}, B_{i,max} \rangle \tag{9}$$

$$Turn\ pixel\ P\ to\ white? = \bigvee_{i=0}^{|C|}(R_p \in R_i \wedge G_p \in G_i \wedge B_p \in B_i) \tag{10}$$

Hough circles transformation returns a set of circles that were found in the input frame. The circle center and radius represent a position of the object and its size. Hough transform works well even when many circles are presented in the scene, but the rate of false detections is high and the transformation function is sensitive to its parameters. The spot-finder does not operate with any specific object's shape.

## 2.3    From Static Objects to Moving Agents

The second transformation (7) provides data about agents' movements based on objects' changing positions in subsequent video frames. In each time step $t$ new positions of all agents are determined. Objects in a radius $k(A_1, r_k)$, are candidates to be associated with the next agent's position (fig. 2). The radius corresponds to the maximal possible length of the agent's step. The agent's speed is another important parameter. The movement of agent from position $A_1$ to position $A_2$ is:

$$O = \{O_i : |A_1 O_i| < r_k\}; \; A_2 \equiv O_i : \min |A_1 O_i| \tag{11}$$

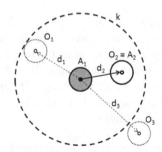

**Fig. 2.** Agent's next position determination

The agent's lifecycle is managed by the following rules:

— An object is associated with an agent already existing in the model → the agent moves to new coordinates corresponding to the object's position in the scene.
— An object is not associated with any agent already existing in the model → a new agent is created in the model with coordinates corresponding to the object's position in the scene.
— An object moves out of the scene → the agent goes out of the model borders and is destroyed.
— An object is not moving for certain time → the agent is destroyed (agent's detection or tracking was wrong).

## 2.4     Time and Space Scales

Both video recording and agent-based model have their own time and space scales which must be taken into account.

The video recording has got the following time parameters: frames per second (*fps*), video footage duration ($t_v$) and the real duration of the process ($t_r$) (the video footage can be slower or faster).

Moreover there is a period $T_f$ in which a single video frame is processed for objects identification in the first level transformation (6). $T_f = 1$ means that every frame is used to identify objects, $T_f = 5$ means that only every 5$^{\text{th}}$ frame is used. The NetLogo model has an internal clock measuring in ticks. The model receives new data about objects' positions at every tick and transforms them according agents' lifecycle rules. Let us use $t_m$ for the number of ticks elapsed since the beginning of the model run. The relationship between the video recording and agent-based model time-related measures is expressed:

$$1 \; video \; frame \; replay \; duration = \frac{1}{fps} \; seconds$$

$$1 \; video \; replay \; second = \frac{t_r}{t_v} \; real \; process \; time \; seconds$$

$$1 \; model \; tick = T_f \cdot \frac{1}{fps} \; seconds \; of \; video \; recording$$

$$1 \; model \; tick = T_f \cdot \frac{t_r}{t_v} \cdot \frac{1}{fps} \; real \; process \; time \; seconds \tag{12}$$

In relation to space scale, the video frame height $H_v$ and width $W_v$ in pixels are given. The real scene is the 3D space. If the video recording is taken from the bird's view and the scene is small enough, the distortion of the image can be neglected and the scene coordinates can be simplified to only height $H_r$ and width $W_r$ (in meters). The NetLogo model environment is defined by its height $H_m$ and width $W_m$ (in patches). The size of patch $e_m$ in pixels corresponds to a real size $e_r$ in meters. The transformation between real and model coordinates is:

$$1 \; pixel = \frac{H_r}{H_v} = \frac{W_r}{W_v} \; [meters]$$

$$e_r = e_m \cdot \frac{H_r}{H_v} = e_m \cdot \frac{W_r}{W_v} \; [meters] \tag{13}$$

When camera view is not orthogonal and the scene cannot by understood as 2D surface, then the transformation of coordinates requires more information (e.g. a referential size of the object when shot from a given distance) and more complex processing (e.g. object shadows identification).

## 3     Implementation

We implemented a video to agent capture system (V2ACapture) as an end-to-end system processing the input video file into the agent-based model in NetLogo. The system composition is shown in the UML component diagram (fig.3).

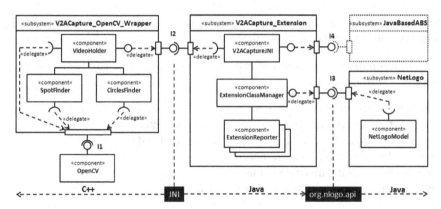

**Fig. 3.** UML component diagram of V2ACapture

The system consists of the following three subsystems:

— *V2ACapture_OpenCV_Wrapper* is a dynamic link library in C++. It provides the functionality of *VideoHolder* for maintaining video replay and manipulation with frames and two objects finders' functionalities: *SpotFinder* and *CirclesFinder*.
— *V2ACapture_Extension* is a NetLogo extension written in Java and deployed into the NetLogo installation as Java archive repository (jar). It provides functions to be used in NetLogo models for manipulating input video files and finding objects.
— *NetLogo* is an instance of NetLogo system in which the NetLogo models using our V2ACapture extension are hosted.

The following interfaces interconnect subsystems into one seamless system:

— *OpenCV-API* (I1) is standard C++ API interface between our wrapper dll and OpenCV libraries.
— *Java-Native-Interface* (I2) interconnects C++ code of VideoHolder, SpotFinder and CircleFinder with Java code of NetLogo extension. V2ACapture.dll provides this interface to enable access to OpenCV functionalities from any Java code.
— *NetLogo-API* (I3) is the interface provided by NetLogo extension to any NetLogo model to call extension functions for manipulating videos and finding objects directly from NetLogo.
— *Interface-to-another-Agent-Based-Systems* (I4) demonstrates the fact that the extension is developed in Java, therefore its functionalities can be simply used by any Java agent-based system. This increases the usability of our extension.

**The SpotFinder component** searches for continuous areas of the same color. We used a simple four-way seed fill algorithm which is sufficient for our purpose and can process video frames in real time (more than 25 fps). Numerous better performing algorithms could be used in the same way. Parameters are:

— *Blur diameter* for the Gaussian blur algorithm,
— *spot_min_size* and *spot_max_size* specifying the spot in pixels.

The SpotFinder returns the spot size in pixels, its minimum and maximum $x$ and $y$ coordinates and the spot centre coordinates $[(x\_max - x\_min) /2),(y\_max - y\_min) / 2]$.

**The CirclesFinder component** encapsulates Hough circles transformation function of OpenCV. Parameters are:

— $dp$ = the inverse ratio of resolution,
— $min\_dist$ = minimum distance between two circles centers,
— $param\_1$ = upper threshold for the internal Canny edge detector,
— $param\_2$ = threshold for center detection,
— $min\_radius$ = minimum circle radius,
— $max\_radius$ = maximum radius to be detected.

The CircleFinder returns a vector of coordinates and radiuses of circles that were found in the input video frame.

# 4    Experiments

The functionality of the system was tested using the sample pedestrian lane formation video that demonstrates the self-organization of bi-directional flows of walkers [6]. The video was shot from the bird's perspective and the pedestrians wear white caps, therefore the 3D scene could be understood as 2D and the identification of individuals was manageable. Parameters of the input MP4 video file were:

— video frame size: 640 x 360 pixels,
— real scene size: 10 x 6 m,
— video duration: 77 seconds, 1937frames, i.e. 25 fps.

NetLogo model parameters for our video were:

— environment size: 64 x 36 patches,
— patch size: 10 x 10 pixels,
— patch real world size: 0.15   x 0.16 m,
— 1 real world second = 25 ticks, i.e. 1 tick = 0.04 seconds.

The Netlogo model interface consists of several sections (fig. 4). The figure 5 shows original video frames of two situations (crowded and not crowded corridor), their pre-processing and the visualization in the agent-based model. The model parameters were:

— $ticks\_to\_move$ = 5 ticks = 0.2 seconds – the period for which the agent can stay still not receiving position updates, before the agent may dismiss from the model.
— $existing\_turtle\_range$ = 2 [patches] = 0.30 m – the maximum distance the agent can move during 1 model tick.
— $RGB$ conditions: R_min = 215, R_max = 256, G_min = 216, G_max = 256, B_min = 215, B_max = 256. These intervals represent color shades close to white.

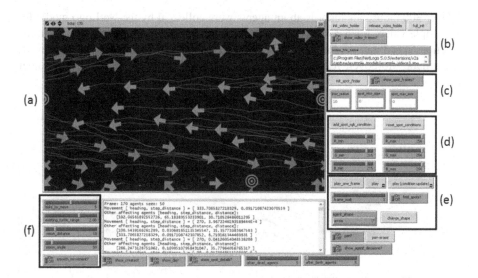

**Fig. 4.** Model interface: (a) visualization of agents, (b) video holder setup, (c) Gaussian blur pre-processing parameter and spot min/max size specification, (d) initial SpotFinder's RGB conditions, (e) video replay control, (f) parameters for association of objects with agents.

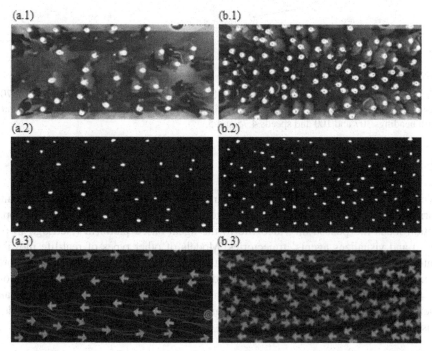

**Fig. 5.** Crowded (a) and not crowded (b) pedestrian situation: original video frame (1), pre-processed frame (2), agent-based model (3)

Over 300 thousands records about agents were collected by the NetLogo model for the sample video. Each record contains data about ticks, current position of agent, its direction, speed and neighbors. The agent's trajectory can be reconstructed from the output data. Moreover, it is possible to observe how agents are affecting each other (fig. 6).

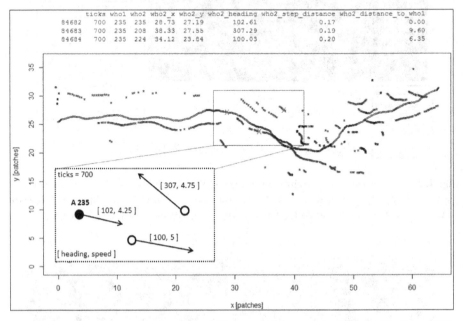

**Fig. 6.** NetLogo model output for agent A 235: whole trajectory and details for tick no.700: heading = 102, speed = 4.25 patch/sec, two neighbors in the agent's vision angle and distance with headings 307 and 100 and speeds 4.75 and 5.

# 5    Conclusion

We presented the proof-of-concept of OpenCV-based application which allows transformation of video data into the reliable agent-based pedestrian simulation. The application loads the pedestrian video file to the agent-based platform (NetLogo in our case) and visualizes agents' trajectories. Models of other types of real-life movable entities can be developed analogically with our extension, if relevant video records were available. Our further work is focused on more complex data-driven simulations based on OpenCV. Two main research challenges are (1) identification of agents and their trajectories from non-orthogonal camera view and (2) combination of video-based agents with artificial ones (3).

**Acknowledgement.** This work was supported by the project No. CZ.1.07/2.2.00/28.0327 Innovation and support of doctoral study program (INDOP), financed from EU and Czech Republic funds.

# References

1. Ballinas-Hernández, A.L., Muñoz-Meléndez, A., Rangel-Huerta, A.: Multiagent System Applied to the Modeling and Simulation of Pedestrian Traffic in Counterflow. Journal of Artificial Societies and Social Simulation 14(3), 2 (2011)
2. Banos, A., Godara, A., Lassarre, S.: Simulating pedestrians and cars behaviours in a virtual city: an agent-based approach. Paper presented at the European Conference on Complex Systems, Paris (2005)
3. Duives, D.C., Daamen, W., Hoogendoorn, S.P.: State-of-the-art crowd motion simulation models. Transportation Research Part C: Emerging Technologies 37, 193–209 (2013)
4. Johansson, A., Kretz, T.: Applied Pedestrian Modelling. In: Heppenstall, A.J., et al. (eds.) Agent-Based Models of Geographical Systems, Springer, Heidelberg (2012) ISBN 978-90-481-8926-7.
5. OpenCV homepage (2014), http://opencv.org/
6. Pedestrian-dynamics experiment: lane formation in counter flow (2010), HERMES project, http://www.youtube.com/watch?v=J4J__lOOV2E
7. Pluchino, A., Garofalo, C., Inturri, G., Rapisarda, A., Ignaccolo, M.: 'Agent-Based Simulation of Pedestrian Behaviour in Closed Spaces: A Museum Case Study'. Journal of Artificial Societies and Social Simulation 17(1), 16 (2014)
8. Wilensky, U.: NetLogo 5.0.5, Center for Connected Learning and Computer-Based Modeling, Northwestern University, Evanston, IL (2013), http://ccl.northwestern.edu/netlogo/

# Extending HITS Algorithm for Ranking Locations by Using Geotagged Resources

Xuan Hau Pham[1], Tuong Tri Nguyen[2,*], Jason J. Jung[3], and Dosam Hwang[2]

[1] Faculty of Technology, QuangBinh University, Vietnam
[2] Department of Computer Engineering, Yeungnam University, Korea
[3] Department of Computer Engineering, Chung-Ang University, Korea
{pxhauqbu,tuongtringuyen,j2jung,dosamhwang}@gmail.com

**Abstract.** The paper focuses on using geotagged resources from the social network service (SNS) for searching the famous places from keyword. We extend the HITS[9] algorithm in order to rank locations which are collected from geotagged resources on SNS. Our approach not only uses the similarity measurement between locations'tags for computing the value of locations but also calculate the term frequency of tags which occur in each location to modify the value of tags for ranking. We implement and show the experimental results with the set of locations from the geotagged resources.

## 1 Introduction

Social network services (e.g, such as Facebook[1], Photobucket[2] and Flickr[3]) own billions of images and videos, which have been annotated and shared among friends, or a community that cover certain interesting topics [3,5,7,8,13]. In fact, users comment on the content in a form of tags, ratings, preferences etc and that these are applied on a daily basis, gives this data source an extremely dynamic nature that reflects events and the evolution of community focus [5]. There are so many studies using tags as well as its attributes for predicting [3], recommendation system [1,10,14], data clustering [6], data classifying, etc.

There are some studies focusing on traveling problem by using geotagged resources [3,11,10,12]. While traveling, the people usually wonders 'How to find the best restaurant to enjoy a delicious meal?' 'Where is the best beach to travel?', etc. In this paper, we indicate that distributed geotagged resources on SNS can solve these issues. Thus, when focusing on a place and provide a solution to find out the ranked locations with each keyword in order to answer the question 'where is the best location for to do something?'. Here, we present the experimental results with the set of tags from the geotagged photos on SNS.

The Fig. 1 show the work flow for searching the famous place related to the resources content based on its tags. Its components are described as

---

* Corresponding author.
[1] www.facebook.com
[2] www.photobucket.com
[3] www.flickr.com

D. Hwang et al. (Eds.): ICCCI 2014, LNAI 8733, pp. 332–341, 2014.

- *Collecting data*: Dataset is collected by using Open's API of SNS;
- *Tag analysis*: The list of tags is refined by removing the stop words and classified based on the geotagged photos (all tags will be determined the number of locations it's belongs to) and determined the set of common tags
- *GeoHITS algorithm*: According to [2,4,9], we apply the HITS algorithm by using a set of nodes which are tags or locations for an undirected graph, called *GeoHITS*;
- *Extend GeoHITS*: We use similarity measurement between tags of each location and a set of common tags (called $GeoHITS_S$ algorithm). Another way, we use the tag frequency (TF) in order to add value for tag nodes (called $GeoHITS_{TF}$ algorithm).

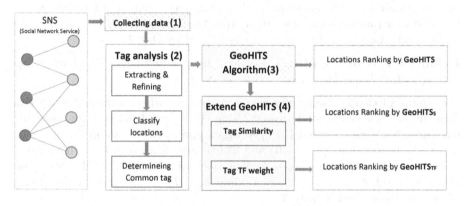

**Fig. 1.** An overview of work flow of locations ranking system

The paper is organized as follows. Sect. 2 introduces Backgrounds. Sect. 3 shows the algorithms for ranking locations. In Sect. 4, experimentation was conducted to evaluate the results. Sec. 5 draws a conclusion and future work of this study.

## 2    Backgrounds

### 2.1    Definitions

When mentioned about a place, there are a list typical things coming from it (e.g., Japan relates to 'sumo', 'sushi', 'flower of cherry'). On the other hand, talking about a keywork, there will have a series of locations that related to it (e.g., 'tower' relates to France, Malaysia, United States, etc). Our approach discovers the famous places which come from geotagged photos on SNS by using ranked method.

We denote $T = \{t_i\}$ is a set of tags, $L = \{l_j\}$ is a set of locations, $L^{t_i}$ is a set of locations which contain tag $t_i$, $T^{l_j}$ is a set of tags of location $l_j$. We use a function $f$ in order to determine tag $t_i$ is contained in location $l_j$ or not, as follows:

$$f(t_i, l_j) = \begin{cases} 1, & \text{if tag } t_i \text{ is contained in location } l_j. \\ 0, & \text{otherwise.} \end{cases} \tag{1}$$

**Definition 1 (Location).** *A location $l_j$ is presented by a name of a country, a region, a city, or a place where is identified by a set of tags $T^{l_j} = \{t_i : t_i \in T, |L^{t_i}| \geqslant 2, f(t_i, l_j) = 1\}$.*

In this issue, we use a set of geotagged photos in order to find out the best location by ranked method. These tags which occur on many locations should be selected for using in this work. And if a tag only occurs in a location that is not worth for ranking locations. So, in order to rank locations, we need to use a set of special tags (called common tag as defined by Def. 2).

**Definition 2 (Common tags).** *Common tags (denoted $T_C$), are a set of tags $T_C = \{t_i : t_i \in T \text{ and } |L^{t_i}| \geqslant \alpha, \quad with \quad \alpha \geqslant 2\}$.*

For example, to help someone who would like to know where is the best place with 'Kimchi' can find out a good answer. Using the keyword 'Kimchi' for collecting data from SNS, we collected 1416 geotagged photos (as shown in 1). In which, there was 10 locations ($|L| = 10$) and 23395 tags. With $\alpha = 5$, we have $T_C = \{$kimchi, spicy, soup, food, lunch, korean, noodles, tofu, rice, salad, restaurant, hot, egg, travel, dinner, pork, market, kimchee$\}$ ($|T_C| = 18$), and the list of tags of locations is showed in Tab. 3.

## 2.2   Using Tag Frequency

Using term frequency to compute the term weight is popular such as classifying document [18]. They can determine the valued class of a document by using term frequency from a set of words in that document.

For this work, we compute the weight of tags with locations (called tag frequency - TF). We calculate the occurrence of tag $t_i$ in location $l_j$ (is denoted $w_{ij}$). The value of $w_{ij}$ is computed as follows:

$$w_{ij} = \frac{tf(t_i, l_j)}{max\{tf(t_k, l_j)|t_k \in T^{l_j}\}} \tag{2}$$

with $tf(t_i, l_j)$ is the number of occurrence of tag $t_i$ in location $l_j$.

## 3   Ranking Locations

### 3.1   The Formalization of Locations Ranking by HITS

According to [9], the HITS algorithm is used for webpages ranking. The hyperlinks from these webpages form a directed web graph $G = \langle V, E \rangle$, where V is the set of nodes representing webpages, and E is the set of hyperlinks. The hyperlink topology of the web graph is contained in the asymmetric adjacency matrix $L = \{l_{ij}\}$, where $l_{ij} = 1$ if $page_i \rightarrow page_j$ and $l_{ij} = 0$ otherwise. And each webpage $p_i$ has both a hub score $p_i^{Hub}$ and an authority score $p_i^{Aut}$.

In the paper, we use the relationship between tags and locations same as hubs and authorities in [9] and [4]. However, we use an undirected graph $G = \langle V, E \rangle$, where V is the set of nodes representing tags or locations, and E is the set of edges. $V = T \cup L$, and $E = \{(t_i, l_j) : f(t_i, l_j) = 1\}$.

At the beginning, all nodes have weight equal to 1. For each iterations, they are computed by the formulas as follows

$$l_j = \sum_{i=1}^{m} \frac{1}{|L^{t_i}|} t_i; \qquad t_i = \sum_{j=1}^{n} \frac{1}{|T^{l_j}|} l_j; \qquad (3)$$

The formula to compute the value of nodes as follows:

$$Tag = ALoc \quad and \quad Loc = A^T Tag \qquad (4)$$

where $A$ is an adjacency matrix $(m \times n)$ and $a_{ij}$ is determined by Equ. 1, $Tag = \{t_1, t_2, \ldots, t_m\}^T$, $Loc = \{l_1, l_2, \ldots, l_n\}^T$.

According to [2], we can compute Equ. 4 by recursive computing (similar to Equ. 3.2 in [2]) as follows

$$Tag = AA^T Tag \quad and \quad Loc = A^T ALoc \qquad (5)$$

For each iteration step, the value of nodes are recomputed and normalized.

## 3.2   Locations Ranking Algorithm

From studying results of [2,9], we have proposed the *GeoHITS* Algorithm (as shown in Alg. 1). In which each node is represented by a tag or a location.

To extend the *GeoHITS* Algorithm, we focus on two aspects:

*(1) Tag similarity:* Using statistics methods [6,17] with the dataset, we compute the set of common tags. To calculate the ranking coefficient for each location based on its tags, we have used the Jaccard method [15] for determining the similarity between the set of common tags and the set of tags of each location. We extend *GeoHITS* Algorithm by using the similarity coefficient for each location node (called *GeoHITS_S* Algorithm).

*(2) Tag term frequency:* Both *GeoHITS* and *GeoHITS_S*, the weight of each tag is not considered. In actually, the value of each tag in each location is distinguished. Consequently, the value of each tag for each location will be different. Thus, we compute the TF weight [16] for each tag and propose the *GeoHITS_{TF}* Algorithm.

We implement three algorithms (*GeoHITS, GeoHITS_S, GeoHITS_{TF}*) with the dataset that is described in Tab. 1 on the next section.

**Data:** $Tag, Loc, \varepsilon$
**Result:** $R$
Initialization;
Determine a set of common tags;
Determine matrix $A(m \times n)$;
Let $Tag = \{t_1, \ldots, t_m\}$ denote the vector $\{1, \ldots, 1\}$;
Let $Loc = \{l_1, \ldots, l_n\}$ denote the vector $\{1, \ldots, 1\}$;
Iterations=0;
**while** *max of* $|Loc_k - Loc_{k-1}| \geq \varepsilon$ *or Iterations=0* **do**

    Computing $t_i = \sum\limits_{j=1}^{n} (a_{ij}.l_j)$ for $\forall i \in [1:m]$;

    Computing $l_j = \sum\limits_{i=1}^{m} (a_{ij}.t_i)$ for $\forall j \in [1:n]$;

    Normalize(Tag);
    Normalize(Loc);
    Iteration+=1;

**end**
$R < -arsort(Loc)$;
Return$(R)$

**Algorithm 1:** GeoHITS Algorithm

## 4     Experimentation

### 4.1     Dataset

We collect data and perform basic processes to get the data in Tab. 1 as the basis dataset to experiment. In Tab. 1, there are 5 keywords (*kimchi, pho, pizza, poutine and sushi*) of 5 traditional dishes which are from 5 different countries: *Sounth Korea, Vietnam, Italy, Canada and Japan.*

**Table 1.** The dataset

| Keyword | #Photos | #Tags | #Locations |
|---------|---------|-------|------------|
| kimchi | 1416 | 23395 | 10 |
| pho | 5141 | 72612 | 46 |
| pizza | 1544 | 23805 | 40 |
| poutine | 1027 | 10352 | 24 |
| sushi | 2615 | 28063 | 57 |

With each keyword, we calculate the number of locations for ranking based on geotagged photos. We classify the dataset for each keyword based on geotagged photos. For example, the value of Tab. 2 shows 10 locations (countries) to rank. We also can rank with regions of a country as an option to implement.

**Table 2.** The list of locations for ranking (by country) with 'kimchi'

| # Location | The list of locations |
|---|---|
| 10 | Canada, China, France, Japan, North Korea, South Korea, Taiwan, Thailand, United Kingdom, United States |

## 4.2   Results on Ranking Locations

Here, we used the *GeoHITS* algorithm for ranking locations based on tags. Giving a set of tags for ranking locations, these tags and locations are described as Def. 1 and Def. 2.

We conduct the experimentation with two methods for ranking which are called *online* and *offline*. With offline ranking, we use the dataset in order to calculate all variables before using iterations of *GeoHITS*. Using the Alg. 1 with $\varepsilon = 10^{-8}$ (with 'kimchi'),we found that the convergence of iterations is very rapid ($k = 9$). The results is showed in Fig. 2.

With online ranking, we use the dataset as the same collecting time. We add more tags at each iteration step as well using $\varepsilon$ value, and the convergence value is obtained at $k = 395$ (with 'kimchi') as shown in Fig. 3.

For the purpose of comparing and solving the feasibility of this approach for ranking locations based on tags, we use TFIDF and determine tags similarity to expand the *GeoHITS* Algorithm. We has tried with 'kimchi' and showed results as shown in Tab. 4. In order to make more clearly for comparing, the values in Tab. 4 are normalized at $[0 \ldots 1]$.

For comparing the results, we focus on Tab. 3 and Tab. 4. Based on the results of Tab. 4, we can comment that the *GeoHITS* got early convergence so that is the best algorithm. However, in this work we are considering using tags for ranking locations. Thus, we consider the analyzed data in Tab. 3 and conclude that it isn't an exact answer. Indeed, the ranking with $GeoHITS_TF$ converge slower than GeoHITS and $GeoHITS_S$ but the result is more suitable than two algorithms above.

The experimental results which are presented in Tab. 5 show that there are four returned values match with the expected results (*'kimchi'-South Korea; 'pho'-Vietnam; 'poutine'-Canada and 'sushi'-Japan*). Additionally, we implemented the dataset and got five top positions that are constant. With these results, we believe that this approach is useful to rank locations based on geo-tagged resources from SNS.

**Table 3.** The dataset for ranking locations with 'kimchi'

| Location | #Photo | #Tag | Top-10 popular tags |
|---|---|---|---|
| Canada | 26 | 244 | kimchi, food, korean, toronto, spicy, banchan, market, scallions, finch, foodie |
| China | 26 | 432 | kimchi, china, beijing, duck, daejeon, travel, noodles, airport, friends, cold |
| France | 10 | 337 | food, kimchi, soup, noodles, pickled, rice, cuisine, ginger, balls, pepper |
| Japan | 32 | 435 | kimchi, japan, food, korean, tokyo, dinner restaurant, pork, hot, lunch, shrimp |
| North Korea | 212 | 2208 | kimchi, pyongyang, korea, dprk, juche, arirang, koryo, travel, cold, fun |
| South Korea | 504 | 10983 | kimchi, korea, food, korean, seoul, restaurant, culture, red, spicy, dinner |
| Taiwan | 16 | 225 | kimchi, taiwan, taipei, geotagged, food, , egg, tomato, soup, pot, watermelon, hot, tainan, shrimp |
| Thailand | 13 | 93 | kimchi, teachingsagittarian, kimchee, korean, thailand, bangkok, delicacies, chilli, food, vegetables |
| United Kingdom | 82 | 607 | kimchi, food, korean, kimchee, london, gimchi, season, extenstion, fermentation, restaurant |
| United States | 495 | 7831 | kimchi, korean, food, bulgogi, dinner, ssam, banchan, shrimp, pork, restaurant |

**Table 4.** The results on locations ranking with 'kimchi' (offline)

| $GeoHITS$ | $(k = 9)$ | $GeoHITS_S$ | $(k = 11)$ | $GeoHITS_{TF}$ | $(k = 30)$ |
|---|---|---|---|---|---|
| SouthKorea | 0.16494845 | SouthKorea | 0.16439629 | SouthKorea | 0.16573591 |
| UnitedStates | 0.15979381 | UnitedStates | 0.15937351 | UnitedStates | 0.16394372 |
| Japan | 0.10824742 | Japan | 0.10835741 | Japan | 0.11062289 |
| UK | 0.10309278 | UK | 0.10056532 | UK | 0.10655475 |
| France | 0.09793814 | Taiwan | 0.09865156 | France | 0.1018892 |
| Taiwan | 0.08247423 | China | 0.0869082 | NorthKorea | 0.07972374 |
| Canada | 0.07731959 | Thailand | 0.08109632 | China | 0.0770678 |
| NorthKorea | 0.07216495 | France | 0.07255086 | Taiwan | 0.06713576 |
| China | 0.06701031 | Canada | 0.06695163 | Canada | 0.06549773 |
| Thailand | 0.06701031 | NorthKorea | 0.06114889 | Thailand | 0.06182849 |

**Table 5.** Top-10 locations ranking with 5 keywords (*GeoHITS$_{TF}$-online*)

| Pos. | kimchi (KR) | pho(VN) | pizza(IT) | poutine(CA) | sushi(JP) |
|------|-------------|---------|-----------|-------------|-----------|
| 1 | **SouthKorea** | **Vietnam** | USA | **Canada** | **Japan** |
| 2 | USA | USA | **Italy** | USA | USA |
| 3 | Japan | Canada | UK | SouthKorea | Canada |
| 4 | UK | Thailand | Canada | France | UK |
| 5 | France | Australia | Australia | UK | Australia |
| 6 | NorthKorea | UK | Germany | Ireland | China |
| 7 | China | SouthKorea | Spain | HongKong | Taiwan |
| 8 | Taiwan | Philippines | China | China | Spain |
| 9 | Canada | HongKong | France | Netherlands | Germany |
| 10 | Thailand | China | Netherlands | Taiwan | Singapore |

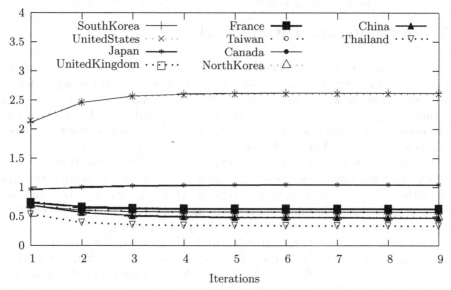

**Fig. 2.** Ranking location with 'kimchi' (*GeoHITS-offline*)

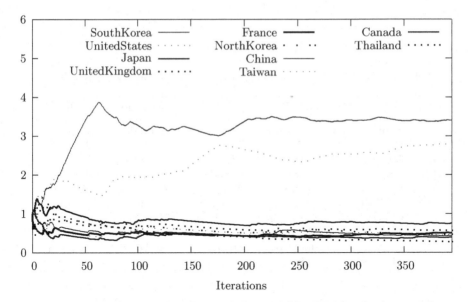

**Fig. 3.** Ranking location with 'kimchi' (*GeoHITS-online*)

## 5   Conclusion and Futrure Work

For the purpose of ranking locations based on geotagged resources, we propose using *GeoHITS* algorithm and modifying value of nodes with tags and relevant locations. Besides, we extend *GeoHITS* by using similarity between the set of common tags and a set of tags of each location, called $GeoHITS_S$ algorithm. Moreover, we use term frequency of tags in each location in order to compute the weight of tags and apply them into the $GeoHITS_{TF}$ algorithm.

In our experiments, more importantly, we empirically showed that the Geo-HITS algorithm (offline case) converge quickly. The obtained results with three algorithms are quite interesting and suitable. Although we could not determine the precision of these results, but based on the reality with keywords belong to countries, our results have obtained the high ranking values.

In spite of the imbalance of dataset with many locations for ranking (a large part of dataset belongs to United States due to users of Flickr), our method found out locations which hold traditional dishes for each keyword as shown in collected dataset. We appreciate using term frequency of tags in each location (called $GeoHITS_{TF}$ algorithm).

As future work, we plan *i*) to propose a location recommendation system for traveler based on tags from SNS; *ii*) to detect events based on geotagged photos from SNS as a our new approach.

**Acknowledgement.** This study is funded by Vietnam National Foundation for Science and Technology Development (NAFOSTED) under grant number 102.01-2013.12. Also, this work is supported by BK21+ of National Research Foundation of Korea.

# References

1. Belém, F.M., Martins, E.F., Almeida, J.M., Gonçalves, M.A.: Personalized and object-centered tag recommendation methods for web 2.0 applications. Information Processing and Management 50(4), 524–553 (2014)
2. Benzi, M., Estrada, E., Klymko, C.: Ranking hubs and authorities using matrix functions. Linear Algebra and its Applications 438(5), 2447–2474 (2013)
3. Clements, et al.: Using flickr geotags to predict user travel behaviour. In: Proceedings of the 33rd International ACM SIGIR Conference on Research and Development in Information Retrieval, pp. 851–852. ACM (2010)
4. Ding, C., He, X., Husbands, P., Zha, H., Simon, H.D.: Pagerank, hits and a unified framework for link analysis. In: Proceedings of the 25th Annual International ACM SIGIR Conference on Research and Development in Information Retrieval, pp. 353–354. ACM (2002)
5. Diplaris, S., Sonnenbichler, A., Kaczanowski, T., Mylonas, P., Scherp, A., Janik, M., Papadopoulos, S., Ovelgoenne, M., Kompatsiaris, Y.: Emerging, collective intelligence for personal, organisational and social use. In: Bessis, N., Xhafa, F. (eds.) Next Generation Data Technologies for Collective Computational Intelligence. SCI, vol. 352, pp. 527–573. Springer, Heidelberg (2011)
6. Giannakidou, E., Koutsonikola, V., Vakali, A., Kompatsiaris, I.: Co-clustering tags and social data sources. In: The Ninth International Conference on Web-Age Information Management, WAIM 2008, pp. 317–324. IEEE (2008)
7. Jung, J.J.: Discovering community of lingual practice for matching multilingual tags from folksonomies. The Computer Journal 55(3), 337–346 (2012)
8. Jung, J.J.: Cross-lingual query expansion in multilingual folksonomies: A case study on flickr. Knowledge-Based Systems 42(0), 60–67 (2013)
9. Kleinberg, J.M.: Authoritative sources in a hyperlinked environment. Journal of the ACM (JACM) 46(5), 604–632 (1999)
10. Kurashima, T., Iwata, T., Irie, G., Fujimura, K.: Travel route recommendation using geotags in photo sharing sites. In: Proceedings of the 19th ACM International Conference on Information and Knowledge Management, pp. 579–588. ACM (2010)
11. Lee, H.C., Liu, H., Miller, R.J.: Geographically-sensitive link analysis. In: Proceedings of the IEEE/WIC/ACM International Conference on Web Intelligence, pp. 628–634. IEEE Computer Society (2007)
12. Lee, I., Cai, G., Lee, K.: Exploration of geo-tagged photos through data mining approaches. Expert Systems with Applications 41(2), 397–405 (2014)
13. Morrison, P.: Tagging and searching: Search retrieval effectiveness of folksonomies on the world wide web. Information Processing and Management 44(4), 1562–1579 (2008)
14. Pham, X.H., Jung, J.J., Le Anh Vu, S.B.P.: Exploiting social contexts for movie recommendation. Malaysian Journal of Computer Science 27(1) (2014)
15. Rokach, L., Maimon, O.: Clustering methods. In: Data Mining and Knowledge Discovery Handbook, pp. 321–352. Springer, Heidelberg (2005)
16. Sebastiani, F.: Machine learning in automated text categorization. ACM Computing Surveys (CSUR) 34(1), 1–47 (2002)
17. Zhang, H., Korayem, M., You, E., Crandall, D.J.: Beyond co-occurrence: discovering and visualizing tag relationships from geo-spatial and temporal similarities. In: Proceedings of the Fifth ACM International Conference on Web Search and Data Mining, pp. 33–42. ACM (2012)
18. Zhang, W., Yoshida, T., Tang, X.: Tfidf, lsi and multi-word in information retrieval and text categorization. In: SMC 2008, IEEE International Conference on Systems, Man and Cybernetics, pp. 108–113. IEEE (2008)

# Solving the Permutation Problem Efficiently
# for Tabu Search on CUDA GPUs

Liang-Tsung Huang[1], Syun-Sheng Jhan[2], Yun-Ju Li[3], and Chao-Chin Wu[3,*]

[1] Department of Biotechnology, Mingdao University, Changhua 523, Taiwan
larry@mdu.edu.tw
[2] Department of Information Technology, Lin Tung University, Taichung 408, Taiwan
janss@teamail.ltu.edu.tw
[3] Department of Computer Science and Information Engineering
National Changhua University of Education, Changhua 500, Taiwan
ccwu@cc.ncue.edu.tw, icecloud6666@gmail.com

**Abstract.** NVIDIA's Tesla Graphics Processing Units (GPUs) have been used to solve various kinds of long running-time applications because of their high performance compute power. A GPU consists of hundreds or even thousands processor cores and adopts (Single Instruction Multiple Threading) SIMT architecture. This paper proposes an approach that optimizes the Tabu Search algorithm for solving the Permutation Flowshop Scheduling Problem (PFSP) on a GPU. We use a math function to generate all different permutations, avoiding the need of placing all the permutations in the global memory. Experimental results show that the GPU implementation of our proposed Tabu Search for PFSP runs up to 90 times faster than its CPU counterpart.

**Keywords:** GPU, CUDA, Parallel algorithm, Tabu Search, Permutation Flowshop Scheduling Problem.

## 1    Introduction

A GPU (Graphics Processing Unit) has hundreds, even more than one thousand, of processing elements, making it very suitable for executing applications with big data and data-level parallelism [1, 2]. Compute Unified Device Architecture (CUDA) [3-5] is proposed by nVIDIA for easier programming on nVIDIA GPUs. Due to the low cost and the popular GPU-inside desktops and laptops, more and more researchers focus on how to parallelize various algorithms on GPU architecture. On the other hand, computational intelligence has been successfully applied to solve many kinds of applications [6-9]. Researchers have investigated how to use GPU computing to accelerate computational intelligence. For example, *Janiak et al.* [10] proposed the GPU implementations of the Tabu Search algorithm for the Travelling Salesman Problem and the Permutation Flowshop Scheduling Problem. Lots of research has reported that the optimized GPU implementations can run tens of times, or even more than one hundred times, faster than their sequential CPU counterparts.

---

* Correpsonding author.

D. Hwang et al. (Eds.): ICCCI 2014, LNAI 8733, pp. 342–352, 2014.

The Tabu Search algorithm is a neighbourhood-based and deterministic metaheuristic, which is proposed to solve many discrete optimisation problems by *Glover* [11, 12]. This algorithm is similar to the function of human's memory. If the solution has been chosen by the previous generation, then it cannot be chosen again until a specified time interval has passed. This way can avoid choosing the local optimal solution to the problems. While computing the flowtime of the permutations, we use the Tabu list to record which permutations have been chosen to produce local optimal solutions during the previous several generations. In addition, users can set an initial value for the so called Tabu value, which determines how many generations the corresponding permutation cannot be used again since the permutation is selected. Whenever a permutation is selected, it is added into the Tabu list and its corresponding Tabu value is set to the user specified input value. Each Tabu value in the Tabu list will be decreased by one whenever proceeding to the next generation. The permutations in the Tabu list cannot be used until its corresponding Tabu value becomes zero. How to optimizing Tabu search on GPUs has been discussed on several projects [13-15].

The Permutation Flowshop Scheduling Problem (PFSP) has been first proposed by Johnson [16] in 1954. The PFSP is to find the best way to schedule many jobs to be processed on several ordered machines, which minimizes the flowtime that is equal to the total processing time of a permutation of the jobs. PFSP can be applied to the manufacturing and resources management in factories and companies. Due to the large number of jobs, the sequential program for PFSP is too slow to be adopted. Therefore, this paper proposes a high performance parallel approach to implement the Tabu Search algorithm for PFSP on CUDA GPU architecture. Compared with the sequential CPU version, our approach can run up to 90 times faster.

This paper is organized as follows. Section 2 introduces the CUDA architecture, the Permutation Flowshop Scheduling Problem, and related parallel methods. In Section 3, our proposed approach for implementing the PFSP on a CUDA GPU is described in detail. Section 4 demonstrates the experimental results and analyse the performance. Finally, conclusions are given in Section 5.

## 2    Related Work

### 2.1    Compute Unified Device Architecture

The CUDA (Compute Unified Device Architecture) development environment is mainly based on a sequential programming language, such as C/C++, and extended with some special functions that hide most issues of GPUs [3-5]. A GPU consists of several streaming multiprocessors (SMs) and each SM has multiple streaming processor cores [1-2]. From the software perspective, a CUDA's device program is organized as a hierarchy of grids, blocks and threads. To design a CUDA device program, programmers must define a C/C++ function, called kernel. While a CPU invokes a kernel to execute the kernel on GPU, the programmer must specify the number of blocks and the number of threads to be created. A block will be allocated to a SM and the threads within a block are able to communicate each other through the shared memory in the SM. Each thread is executed on a streaming processor. One or more blocks can be executed concurrently on a streaming multiprocessor at a time. There are hundreds or even thousands of threads within a block on CUDA. These threads can be organized as a 1-, 2- or 3-dimensional array. However, blocks can be organized as only a 1-, or 2-dimensional array.

There are many types of memory on GPU. They have different size, access time, and whether they can be written or read by blocks and threads. The description of each memory type is as below. Global memory is the main memory on a GPU, it can be allocated and deallocated explicitly through invoking the CUDA APIs in the kernel to communicate the CPU with the GPU. It has the largest memory space on the GPU, but it requires 400-600 clock cycles to complete a read or write operation. Blocks can communicate with each other via the global memory.

Constant memory is accessible as global memory except it is cached. A read operation takes the same time as that for the global memory in the case of a cache miss, otherwise it is much faster. The CPU can write and read the constant memory. It is read-only for GPU threads. Shared memory is a very fast memory on the GPU, it is used to communicate between threads in the same block. Data in the shared memory of a block cannot be directly accessed by other blocks. Accessing shared memory requires only 2-4 clock cycles. Unfortunately, the memory space of shared memory is limited. The maximum space is 16384 bytes per block for Tesla C1060. When a thread needs more space than the shared memory, the thread has to swap out and in the data in shared memory explicitly. Registers are the fastest memory that can only be used in the thread scope. They are for automatic variables. The number of 32-bit register is limited up to 16384 on each streaming multiprocessor on Tesla C1060. Local memory is used for large automatic variables per-thread, such as arrays. Both read and write operations take the same time as that for the global memory.

## 2.2    Permutation Flowshop Scheduling Problem

In the PFSP, a set of $N$ jobs is to be processed on a set of $R$ machines. Each job will be divided into $R$ parts and go through the $R$ machines in a predefined order. Assume $M_i$ is Machine $i$, and $J_k$ is Job $k$. Let $P_{i,j}$ denote the processing time of Job $k$ on Machine $i$. Compute the flowtime, denoted as $C_{i,k}$, for processing $J_k$ on machine $M_i$, which is defined as the following formula. Each permutation has its own flowtime $C_{m,n}$.

$$C_{0,0} = p_{0,0},$$
$$C_{i,0} = p_{i,0} + C_{i-1,0},$$
$$C_{0,k} = p_{0,k} + C_{0,k-1},$$
$$C_{i,k} = p_{i,k} + \max\{C_{i,k-1}, C_{i-1,k}\},$$

*where* $i \in \{1,2,...,m\}$, *and* $k \in \{1,2,...,n\}$

To solve the PFSP is to find the minimum of all flowtimes from all permutations. Let $\omega_i$ is a permutation, then $C_{m,n}(\omega_i)$ denotes the flowtime of the permutation $\omega_i$. $\Omega_x$ denotes the set of all permutations of length $x$.

$$\forall \omega \in \Omega_x, C_{max} = \max\{ C_{m,n}(\omega_1), C_{m,n}(\omega_2),..., C_{m,n}(\omega_x)\}$$

Because the PFSP is a NP problem, it has been parallelized to shorten its execution time. For instance, *Chakroun et al.* [10] used the branch-and-bound algorithm and the inter-task parallel method to improve the performance of the flowshop problem on GPUs. In the inter-task method, each thread calculates the flowtime for a permutation. Each thread is responsible for sequentially computing the flowtime for a permutation. The advantage is that the threads have no data dependency between each other in the block, so they do not need to synchronize with each other or wait for another. The disadvantage is that each thread needs a large amount of the shared memory space for processing a permutation. It has low performance when more jobs and machines have to be processed because threads in the same block contend for the use of the shared memory. Due to the limitation of available shared memory space, the maximum number of threads per block cannot be very large.

On the other hand, the intra-task method let all the threads in a block process a permutation together. *Michael et al.* [11] used the intra-task method by well utilizing the characteristic of the GPU memory, such as memory coalescing for accessing the global memory, and avoiding bank conflict on the shared memory. They let each block be responsible for computing the flowtime of a permutation, where multiple threads in block work together to compute the flowtime for a permutation. The advantage of the method is that a larger number of threads can execute the PFSP concurrently because of using less shared memory when the flowtime of a permutation is processed by a block. In other words, it means the elements in an anti-diagonal have no data dependency between each other. Unfortunately, this method has two drawbacks. First, the number of threads in each phase is not equivalent. It causes the waste of thread resources, due to the idle threads in some phases. Second, the elements in each anti-diagonal have to wait for the results produced by the elements in the previous anti-diagonal. It needs synchronization between threads and blocks, making it necessary to invoke one kernel for each phase.

# 3    Our Tabu Search for PFSP on CUDA

For the Tabu search for PFSP, in each generation, the permutations to be processed are generated based on the best processing order of jobs produced in the previous generation. If there are $N$ jobs, there will be $C^N_2$ permutations at most to be processed in each generation, where any permutation leading to a job processing order the same as one in the Tabu list will be prohibited in the generation.

To compute the flowtime of all permutations, the previous work proposed placing all the permutations in the global memory initially [10]. These permutations are produced by CPU sequentially. In each generation, each thread will read a permutation from the global memory. For efficient global memory access, the authors of Reference [10] proposed a data placement method that enables coalesced global memory accesses. They arrange all the permutations in an interleaving way. In other words, all the $i$-th elements of $C^N_2$ permutations are stored in the global memory contiguously. Following the i-th elements are the contiguous $C^N_2$ $(i+1)$-th elements.

Nevertheless, it takes time to read the permutations from the global memory in each generation. The latency of global memory access is about 300 to 400 cycles. To address the problem, we propose the following approach to generate permutations on the fly, avoiding the need of accessing global memory in each generation.

## 3.1    Distribute Tasks to Threads and Blocks

Our method is similar to the approach proposed in [10], i.e. we adopt the inter-task parallel method. Consequently, the total number of threads is equal to the number of all permutations. If there are $N$ jobs, then we have $N(N-1)/2$ permutations. Generally speaking, there are 512 or 1024 threads in a block on CUDA. However, we aim at generating the permutations on the fly, instead of placing all the permutations on the global memory statically as in [10]. If the number of the permutations is too large, then we divide the permutations into sets of permutations. Each set of permutation will be processed in a phase. That is, we need multiple phases to complete the computations of the flowtimes for all the permutations. For example, there are 4 jobs in Figure 1. If we process the all the permutations at the same time, we need 6 threads, as sown in Figure 1(a). On the other hand, if we use only three threads totally, two phases are required, as shown in Figure 1 (b). Three permutations are processed in each phase.

Although multiple phases are required to complete all the tasks, we can avoid performance degradation because of shared memory contention by excessive number of threads within a block.

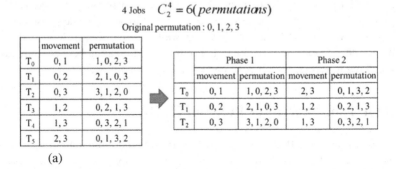

4 Jobs    $C_2^4 = 6(permutations)$

Original permutation : 0, 1, 2, 3

| | movement | permutation |
|---|---|---|
| $T_0$ | 0, 1 | 1, 0, 2, 3 |
| $T_1$ | 0, 2 | 2, 1, 0, 3 |
| $T_2$ | 0, 3 | 3, 1, 2, 0 |
| $T_3$ | 1, 2 | 0, 2, 1, 3 |
| $T_4$ | 1, 3 | 0, 3, 2, 1 |
| $T_5$ | 2, 3 | 0, 1, 3, 2 |

(a)

| | Phase 1 | | Phase 2 | |
|---|---|---|---|---|
| | movement | permutation | movement | permutation |
| $T_0$ | 0, 1 | 1, 0, 2, 3 | 2, 3 | 0, 1, 3, 2 |
| $T_1$ | 0, 2 | 2, 1, 0, 3 | 1, 2 | 0, 2, 1, 3 |
| $T_2$ | 0, 3 | 3, 1, 2, 0 | 1, 3 | 0, 3, 2, 1 |

**Fig. 1.** Subtasks of threads and blocks. (a) The 6 permutations are processed concurrently. Six threads are required. (b) Three permutations are processed at each phase. Three threads are required but two phases are needed.

Because the odd and even numbers of jobs have different characteristics, we propose two different generation methods for these two cases. Figure 2 shows the generation function for the case having odd number of jobs. In the table on Figure 2(a), there are 10 permutations totally if the number of jobs is 5. In the table, the cell in row $i$ and column $j$ means we need to swap the $i$-th job and the $j$-th job in the original job processing order to generate a new job processing order, i.e., a new permutation. If we assign one row to one block, four blocks are required. Furthermore, the workload of each block varies. For instance, the first row needs four threads for the four assigned permutations by different swapping of two jobs. Consequently, we have the problem of load imbalance among blocks. To address the problem, we move the last two rows to the heads of the first rows and merge them one by one, as shown in Figure 2(b). The third row is merged with the second row and the fourth row is merged with the first row. Moreover, the order of each of the last rows is reversed before it is merged with the preceding rows. In this way, the number of permutations in each row is constant. After moving and merging, the number of threads required in each block is increased by one while the number of blocks is half of the original. In the case shown in Figure 2, we need two blocks and each block consists of five threads.

If there are N jobs and N is odd, we need (N-1)/2 blocks and each block is comprised of N threads. For a thread, if its *threadIdx* is larger than its *blockIdx*, it has to generate a new permutation by swapping Job *blockIdx* and Job *threadIdx*. Otherwise, the thread needs to swap Job $(N - 2 - blockIdx)$ and Job $(N - 1 - threadIdx)$.

Similarly, Figure 3 shows how to generate the permutations when the number of jobs is even. Assume there are 4 jobs. Originally, we require three rows and each row consists of three columns, as shown in Figure 3(a). We move the last row to the head of the second row and merge these two rows, resulting two rows in the new table as shown in Figure 3(b). Before merging, we reverse the order of the cells in the third row. Assume there are N jobs and N is even. We need N/2 blocks and each block is comprised of $(N - 1)$ threads. For a thread, if its *threadIdx* is larger than or equal to its *blockIdx*, it has to generate a new permutation by swapping Job *blockIdx* and Job $(threadIdx + 1)$. Otherwise, the thread needs to swap Job $(N - 1 - blockIdx)$ and Job $(N - 1 - threadIdx)$.

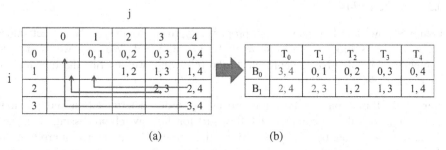

(a)                    (b)

**Fig. 2.** The mapping of odd number of jobs, N=5

**Fig. 3.** The mapping of even number of jobs, N=4

## 3.2   Shared Memory Utilisation

Shared memory is fast memory for the scope of a CUDA block. The number of threads is limited by the available space of shared memory if each thread requires shared memory space. In other words, if each thread uses less shared memory space to process and compute the flowtime of a permutation, the block can have more threads. For PFSP, the number of machines is less than that of jobs in general. To keep the required shared memory as much as possible, the number of shared memory words per thread is equal to the number of machines. As shown in Figure 4, if there are 3 machines and 4 jobs, we allocate 3 shared memory words for a thread. Then, it computes sequentially according to the order of the permutation.

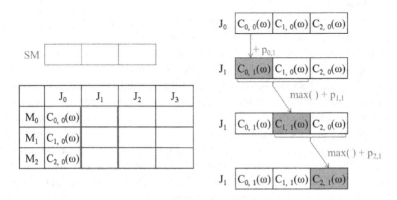

**Fig. 4.** Shared memory usage

## 3.3   System Flow

Figure 5 shows the flow chart of our proposed approach. To ensure the correct values are read across blocks, blocks have to be synchronized. We use two kernels to implement this GPU-based approach. In the following, we describe the functions of the two kernels.

**Kernel 1.** Based on the best job processing order determined in the previous generation, each thread computes the flowtime for the new job processing order, i.e., the new permutation, by swapping two locations in the best permutation produced in the previous generation. However, we need to check if the new permutation matches with any one in the Tabu list. Only those permutations not in the tabu list do we need to compute their flowtimes. Finally, we update the Tabu list by decreasing each Tabu values by one. The tabu list is saved in the global memory. Because CUDA does not implement any primitives for synchronizing blocks, we cannot update the Tabu list right after the block finds out which is the local best permutation. Otherwise, a block may get the wrong data value from the Tabu list. Instead, we compare the flowtimes of all permutations in each block and write the best one to the global memory. Finally, kernel 1 is returned to the host to ensure the solution of each block has been completely written to the global memory.

**Kernel 2.** In this kernel, at first, we use a block to find the best solution from all the local best solutions produced by blocks Kernel 1. The best solution selected in the kernel will become the parent permutation in the next generation. Secondly, the Tabu list is updated again because the best solution found in the generation should be added into the Tabu list. Finally, we get the global best solution by comparing the best solution found in this generation and the original global best solution.

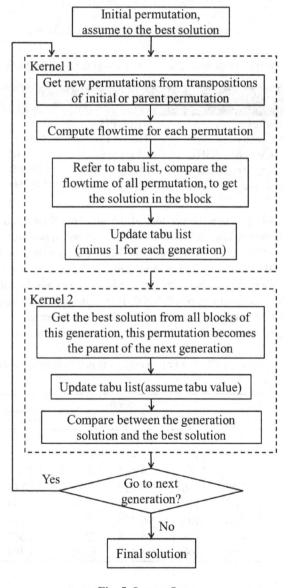

**Fig. 5.** System flow

# 4    Experiment Results

The sequential version of the Tabu Search for PFSP is written in C and evaluated on an AMD Phenom X4 2.5 GHz CPU with 1 GB memory. Detailed configurations are shown in Table 1. The parallel version is developed using CUDA and evaluated on the same workstation equipped with an NVIDIA GeForce GTX460, , as shown in Table 1, which has only one chip, 336 CUDA cores, and 768MB memory.

We use CUDA version 3.2 to implement the parallel approach for the Permutation Flowshop Scheduling Problem using the Tabu Search algorithm. The operating system installed is Linux and its version is openSUSE 11.2, 32-bit.

**Table 1.** The specifications of the AMD CPU and the NVIDIA GeForce 460

| AMD Phenom 9850 Quard-Core | | GeForce GTX 460 | |
|---|---|---|---|
| # of cores | 4 | # of GPU | 1 |
| # of threads | 4 | Thread Processors | 336 |
| Clock speed | 2GHz | Clock speed | 1.35GHz |
| Memory size | 4GB | Memory size | 768MB |
| Cache size | 512KB | Memory clock | 1.8GHz |

As shown in Table 2, when the product of the number of jobs and the number of machines is small, the performance of the parallel evaluation is lower than that of the sequential version. The reason is the required huge transfer time between the CPU memory and the GPU memory, comparing with the small amount of computation workload. The performance is also convergence in the large number of machines and jobs because of the specifications of GTX460. If the number of permutations is larger than the number of thread processors on GTX 460, the performance will be reduced. We are also trying to find other factors of the performance degradation.

**Table 2.** Speedups of Tabu Search for PFSP between the parallel approach and the sequential program

| machines | jobs | m*j | iteration | | |
|---|---|---|---|---|---|
| | | | 10 | 100 | 1000 |
| 4 | 3 | 12 | 0.03 | 0.03 | 0.03 |
| 6 | 13 | 78 | 1.58 | 0.865 | 0.265 |
| 9 | 35 | 315 | 14.551 | 14.386 | 5.68 |
| 12 | 223 | 2676 | 90.413 | 87.577 | 84.118 |
| 15 | 301 | 4515 | 70.398 | 69.926 | 69.099 |
| 18 | 356 | 6408 | 90.258 | 88.825 | 85.913 |
| 22 | 417 | 9174 | 75.468 | 72.766 | 70.667 |
| 30 | 500 | 15000 | 54.11 | 53.18 | 50.795 |

# 5    Conclusions

In this paper, a parallel approach was presented for the Tabu Search algorithm to solve PFSP on a CUDA GPU. We discussed how to create the appropriate numbers of threads and blocks and efficiently manage the shared memory. Moreover, we propose using a math function to generate all the permutations on the fly, without the need of generating all the permutations by CPU and placing them on the global memory. According to our experimental results, the performance can be as high as 90 times faster than a sequential CPU version. Although CUDA programs require data transfer between a CPU and a GPU, they still almost have higher performance than the sequential versions. The reason is that the parallel programs let many subtasks be processed at the same time.

In further work, we will apply more optimization techniques of CUDA and utilize the features of a GPU workstation to optimize the Tabu search algorithm, such as how to efficiently manage device memories, synchronize blocks, and reduce the number of computing subtasks.

**Acknowledgment.** The authors would like to thank the National Science Council, Taiwan, for financially supporting this research under Contract No. NSC102-2221-E-018-014.

# References

1. Owens, J.D., Luebke, D., Govindaraju, N., Harris, M., Kruger, J., Lefohn, A.E., Purcell, T.J.: A survey of general-purpose computation on graphics hardware. Computer Graphics Forum 26, 80–113 (2007)
2. NVIDIA GPU, http://www.nvidia.com/object/cuda_home_new.html
3. NVIDIA GPU Programming Guide, http://docs.nvidia.com/cuda/cuda-c-programming-guide/index.html
4. Kirk, D.B., Hwu, W.W.: Programming Massively Parallel Processors. NVIDIA
5. Oster, B.: Programming the CUDA Architecture: A Look at GPU Computing. Electronic Design 57(7) (2009)
6. Ge, M., Wang, Q.-G., Chiu, M.-S., Lee, T.-H., Hang, C.-C., Teo, K.-H.: An effective technique for batch process optimization with application to crystallization. Chemical Engineering Research and Design 78(1), 99–106 (2000)
7. Precup, R.-E., David, R.-C., Petriu, E.M., Preitl, S., Radac, M.-B.: Novel adaptive gravitational search algorithm for fuzzy controlled servo systems. IEEE Transactions on Industrial Informatics 8(4), 791–800 (2012)
8. Saha, S.K., Ghoshal, S.P., Kar, R., Mandal, D.: Cat swarm optimization algorithm for optimal linear phase FIR filter design. ISA Transactions 52(6), 781–794 (2013)
9. Yazdani, D., Nasiri, B., Azizi, R., Sepas-Moghaddam, A., Meybodi, M.R.: Optimization in dynamic environments utilizing a novel method based on particle swarm optimization. International Journal of Artificial Intelligence, Vol 11(A13), 170–192 (2013)
10. Bożejko, W., Wodecki, M.: Parallel genetic algorithm for the flow shop scheduling problem. In: Wyrzykowski, R., Dongarra, J., Paprzycki, M., Waśniewski, J. (eds.) PPAM 2004. LNCS, vol. 3019, pp. 566–571. Springer, Heidelberg (2004)
11. Glover, F.: Tabu search—part I. ORSA Journal on Computing 1(3), 190–206 (1989)

12. Glover, F.: Tabu search—part II. ORSA Journal on Computing 1(2), 4–32 (1990)
13. Janiak, A., Janiak, W., Lichtenstein, M.: Tabu search on GPU. Journal of Universal Computer Science 14, 2416–2427 (2008)
14. Czapiński, M., Barnes, S.: Tabu Search with two approaches to parallel flowshop evaluation on CUDA platform. J. Parallel Distrib. Comput. 71, 802–811 (2011)
15. Chakroun, I., Bendjoudi, A., Melab, N.: Reducing thread divergence in GPU-based b&B applied to the flow-shop problem. In: Wyrzykowski, R., Dongarra, J., Karczewski, K., Waśniewski, J. (eds.) PPAM 2011, Part I. LNCS, vol. 7203, pp. 559–568. Springer, Heidelberg (2012)
16. Johnson, S.M.: Optimal two- and three-stage production schedules with setup times included. Naval Research Logistics Quarterly 1(1), 61–68 (1954)

# A Genetic Programming Based Framework for Churn Prediction in Telecommunication Industry

Hossam Faris\*, Bashar Al-Shboul, and Nazeeh Ghatasheh

The University of Jordan,
Amman, Jordan
{hossam.faris,b.shboul,n.ghatasheh}@ju.edu.jo

**Abstract.** Customer defection is critically important since it leads to serious business loss. Therefore, investigating methods to identify defecting customers (i.e. churners) has become a priority for telecommunication operators. In this paper, a churn prediction framework is proposed aiming at enhancing the ability to forecast customer churn. The framework combine two heuristic approaches: Self Organizing Maps (SOM) and Genetic Programming (GP). At first, SOM is used to cluster the customers in the dataset, and then remove outliers representing abnormal customer behaviors. After that, GP is used to build an enhanced classification tree. The dataset used for this study contains anonymized real customer information provided by a major local telecom operator in Jordan. Our work shows that using the proposed method surpasses various state-of-the-art classification methods for this particular dataset.

**Keywords:** Churn prediction, Genetic Programming, Self Organizing Maps, Telecommunication.

## 1 Introduction

Nowadays, presence of multiple service providers in the mobile telecommunication industry creates an intensive competition environment. Therefore, it is possible for any customer to have different subscriptions with the same service provider or switch completely to another service provider for cost or quality reasons. The case when a costumer cancels his subscription is called "customer churn". Since the customer is considered as the most important asset for the company, many researchers and practitioners tackled the customer churn problem in the telecommunication sector [1–5].

Service providers are concerned about predicting when a churn might happen. The Customer Relationship Management (CRM) activities intend to maximize the lifetime of a customer, they include acquisition, followup, and retention. For that it is important to target more those customers expected to churn. An

---

\* King Abdullah II School for Information Technology, The University of Jordan, Amman, Jordan.

D. Hwang et al. (Eds.): ICCCI 2014, LNAI 8733, pp. 353–362, 2014.

illustrative example in [2] shows how a CRM contact might extend the lifetime of a customer intending to churn. Due to the huge number of subscriptions for a service provider, CRM personnel need to focus on those expected to churn soon.

In literature, researchers and practitioners have developed wide range of data mining and heuristic based models for industrial and business applications [6–8]. For churn prediction in particular, applied data mining approaches include traditional classification methods like Decision trees algorithms, Naive Bayes and Logistic Regression [2,9,10]. It also includes artificial intelligence based approaches like Artificial Neural Networks, Genetic Programming and Support Vector Machines [11–13]. Among the numerous number of data mining approaches applied in the literature for predicting customer churn, we noticed that Genetic Programming (GP), which is an evolutionary heuristic technique, is much less investigated. We believe that GP has some powerful features which can contribute to the problem domain.

In this paper we propose a framework for predicting customer churn in telecommunication companies. The framework is based on combining Self Organizing Maps (SOM) and GP in a hybrid churn prediction technique. This work is different from previous works in that no one to the best of our knowledge have proposed the idea of hybridizing two different methods similarly. In a nutshell, SOM is used to perform data reduction and eliminate outliers, and then GP is applied to develop the final prediction model to classify a set of testing data. The main goal of the proposed hybrid technique is to enhance the ability of identifying which customer are expected to churn. The hybrid technique is tested using real data obtained from a major Jordanian telecommunication operator. The approach is then evaluated using different criteria and compared to other classical classification techniques.

This paper is structured as follows. In section 2 we present the overall framework proposed in this work. An overview of the data set used in this work is given in section 3. The valuation criteria used in order to evaluate the proposed hybrid approach are listed in section 4. Experiments and results are discussed in section 5. Finally, the conclusion of this work is provided in section 6.

## 2    Proposed Churn Management Framework

The model development method proposed in this research is based on using SOM clustering with GP in two stages. In the first stage, data reduction is performed by applying SOM clustering algorithm on the selected training data set. This process will split the training data set into a number of smaller sets (clusters). Clusters which contain only churners or non-churners are selected to form a new training data set. While other clusters that could not separate churners and non churners are excluded. In the second stage, GP is applied on the joined clustered resulted from stage one. Consequently, GP starts its evolutionary cycle and develops the final classification model. Finally, the developed GP model is assisted using left aside data set. We will denote for this approach as SOM+GP. The proposed framework is illustrated in Figure 1 while SOM and GP are described in the following two subsections.

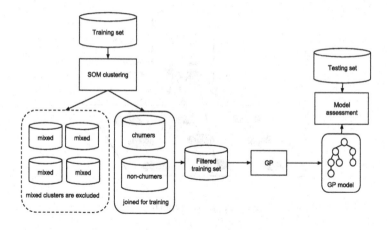

**Fig. 1.** Customer churn prediction framework

## 2.1   SOM Clustering

SOM represent a special type of unsupervised Neural Networks [14–16]. SOM is based on a basic idea of mapping input layer patterns to n-dimensional set of output layer nodes, while preserving the inputs topological organization in the output layer [15, 16]. Therefore one among many applications of the SOM is clustering. Vector Quantization [17] is a data compression technique used in SOM to simplify output space. This makes SOM powerful in representing huge and complex input data in a relatively simple output space [18].

A simple way to represent SOM is the analogy of an input layer nodes mapped to an output matrix, map of nodes, where each node in the output map is linked to every input node [15,18]. Figure 2 shows how "Customer" nodes are connected to all the nodes in the output layer. If $W$ and $H$ denote the width and the height of the output map respectively for $N$ inputs, the number of the links in the SOM is the product of $W$, $H$ and $N$. There are no interconnections between the map nodes, therefore each node in the map is identified with (i, j) width and height coordinates respectively, while having the center node as a reference anchor point [16, 18].

SOM algorithm seeks reducing the neighborhood distance in the output clusters. Normally a hexagonal form represents the neighborhood area. It starts with a large radius and shrinks it over the runs. Having the nodes falling within the radius as neighbors with various weights. Where as closer the nodes to the reference node as more the weight [16,18]. Over time the output layer, output space, becomes smoother with representative clusters of nodes.

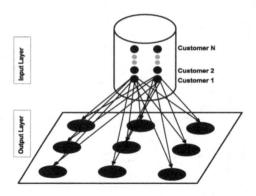

**Fig. 2.** SOM structure

## 2.2   Genetic Programming

Genetic Programming (GP) is an independent domain evolutionary algorithm which automatically creates computer programs. It was first inspired by the biological evolution theories [19, 20]. GP has some advantages when used to model complex problems such as flexibility and interpretability, [21, 22].

GP algorithm performs as an evolutionary cycle which can be summarized in the following five basic steps in order:

1. **Initialization:** GP algorithm starts by creating a predetermined number of individuals which form a population. This number is set by the user. Each individual represents a computer program which is in our case a classification model for customer churn prediction. The output of the model will be 1 for churn or 0 for active customer. GP individuals can be viewed as symbolic tree structures which are graphical representations of their equivalent S-expressions in LISP programming language [23]. Figure 3 gives an example of a simple GP symbolic tree where $X_1$ and $X_2$ are some input variables multiplied by random coefficients.
   The following next three processes are performed while the termination conditions in step 5 are not met.
2. **Fitness Evaluation:** In this process, each individual is evaluated using a specific measurement. In this work, we use mean squared error (MSE) for evaluating all individuals. MSE can be represented by the following equation:

$$MSE = \frac{1}{n} \sum_{i=1}^{n} (y_i - \hat{y_i})^2 \qquad (1)$$

where $y$ is real actual value, $\hat{y}$ it the estimated target value. $n$ is the total number of measurements [23]. Therefore, the fittest individual is the one which has the minimal MSE value. The goal of GP is to minimize the MSE of the evolved individuals.

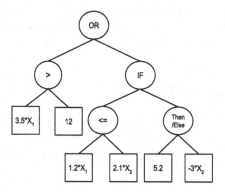

**Fig. 3.** Simple genetic program represented as a structure tree

3. **Selection:** In this process a number of individuals are selected to reproduce new individuals and form new generation. There are different techniques to perform this selection [24]. In this work, Tournament selection is applied which is one of the common and effective selection techniques.
4. **Reproduction:** This evolutionary process creates new individuals using reproduction operators and which replace others. In general, this process makes small random changes to the construction of the individuals. Reproduction operators include the following:
   - *Crossover:* Refers to producing two new individuals (children) by selecting a random subtree in each of the two parents and swapping the resultant subtrees. The new individuals form a new generation or offspring.
   - *Mutation:* This operator is applied on a single GP individual. A random node is selected in the tree of the individual and then the subtree under this node is replaced by a new randomley generated subtree. Usually, the mutation rate is set to be much smaller than the crossover rate.
   - *Elitism:* This operator selects one or more individuals with high fitness values and copies them to the next generation without any modification.
5. **Termination:** GP evolutionary cycle stops iterating when it finds an individual with the required fitness value or when the maximum number of iterations set by the user is reached.

By this process the individual programs evolve and have better fitness values by time. In this work, GP will develop a classification model in order to fit the given data set and minimize the error with respect to the actual value .

## 3    Dataset Description

Data used in this research was provided by a major cellular telecommunication company in Jordan. The data set contains 11 attributes of randomly selected 5000 customers subscribed to a prepaid service for a time interval of three months. The attributes cover outgoing/incoming calls statistics. The data were provided with an indicator for each customer whether the customer churned (left the company) or still active. The total number of churners is 381 (7.6% of total customers).

## 4    Model Evaluation Criteria

In order to evaluate the developed churn prediction model, we refer to the confusion matrix shown in Table 1 which is the primary source for accuracy estimation in classification problems. Based on this confusion matrix, the following four different criteria are used for evaluation:

**Table 1.** Confusion matrix

|  | Actual | |
| --- | --- | --- |
|  | non-churners | churners |
| Predicted non-churners | A | B |
| Predicted churners | C | D |

1. Accuracy: Identifies the percentage of the total number of predictions that were correctly classified.

$$Accuracy = \frac{A + D}{A + B + C + D} \tag{2}$$

2. Actual churners rate (Coverage rate): The percentage of predicted churn in actual churn. It can be given by the following equation:

$$Actual\ churners\ rate = \frac{D}{B + D} \tag{3}$$

3. Hit rate (HR): Shows the percentage of predicted churn in actual churn and actual non-churn:

$$Hit\ rate = \frac{D}{C + D} \tag{4}$$

4. Lift coefficient (LC): shows the precision of model. It can be given by the following equation:

$$Lift\ coefficient = \frac{D}{(C + D).CP} \tag{5}$$

where parameter $CP$ represents the real churn percentage in the data set. The higher the lift is, the more accurate the model is [13].

Since the churn dataset is highly imbalanced, using only accuracy rate is insufficient. Therefore, we refer to more appropriate evaluation criteria; they are the churn rate, hit rate and lift [25]. These criteria give more attention to the rare class which is in our case is the churn class. Our goal is to obtain a prediction model with high churn and hit rates.

# 5   Experiments and Results

Before applying SOM to cluster the training subsets, the number of clusters has to be determined. To find the optimal number, different empirical SOM sizes were applied on all the datset (i.e; 2×2, 3×3, 4×4 and 5×5). Then we found that SOM with the size of 3×3 is the best in identifying two clusters with the highest rates of churners and non-churners. This approach was adopted in [26]. Thus, 3×3 SOM is used in the first stage of our experiments to reduce the training sets as described earlier in section 2.

**Table 2.** Largest two clusters identified churners and non-churners using different SOM sizes

| SOM size | Churners | non-Churners |
|---|---|---|
| 2×2 | 10 (1.2%) | 1232 9 (89.73%) |
| 3×3 | 793 (95.08%) | 6774 (49.03%) |
| 4×4 | 821 (98.44%) | 5099 (37.11%) |
| 5×5 | 784 (94.01%) | 4626 (33.67%) |

In order to give a better indication of how well the developed classification model will perform when it is asked to classify new data, a cross validation with five folds is applied. The data set described in the previous section is split into 5 random subsets of equal sizes. The first four subsets are used for training and the last one is used for testing. Therefore, all criteria described earlier are computed for the testing subset. Then the same process is repeated for different four subsets and another subset is used for testing. Consequently, this process is repeated five times. Finally, the average values for the five different tests is calculated.

Training and testing subsets were loaded into Heuristiclab framework[1] then a symbolic regression via GP was applied with parameters set as shown in Table 3. The best generated GP model tree was evaluated using all the criteria mentioned in section 4 and compared with those for basic GP without SOM, k-Nearest Neighbour (IBK), Naive Bayes (NB) and Random Forest (RF). For IBK, linear search was used to find the best number of neighbours which is found to be 1. For RF, number of trees was set empirically to 10. Comparison results are shown in Figure 4.

Figure 4-a shows that SOM+GP approach is better than the basic GP and NB algorithms. Using SOM to eliminate the outliers enhanced the accuracy of GP by approximately 8%. Although it can be noticed that IBK and RF outperformed SOM+GP in accuracy, they failed in predicting churners with very poor results; 1% and 0.5% respectively as shown in Figure 4-c. On the other side, SOM+GP

---

[1] HeuristicLab is a framework for heuristic and evolutionary algorithms that is developed by members of the Heuristic and Evolutionary Algorithms Laboratory (HEAL), http://dev.heuristiclab.com

**Table 3.** GP parameters

| Parameter | Value |
|---|---|
| Mutation probability | 15% |
| Population size | 1000 |
| Maximum generations | 100 |
| Selection mechanism | Tournament selector |
| Maximum Tree Depth | 10 |
| Maximum Tree Length | 50 |
| Elites | 1 |
| Operators | {+,-,*,/,AND,OR,NOT, IF THEN ELSE} |

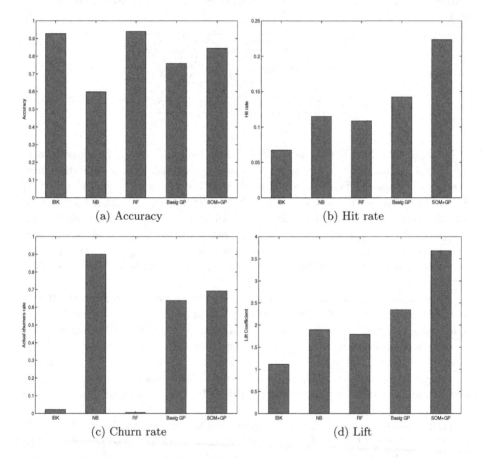

(a) Accuracy

(b) Hit rate

(c) Churn rate

(d) Lift

**Fig. 4.** Evaluation results

achieved the best performance in means of hit rate and lift coefficient as shown in Figures 4-b and 4-d respectively. This shows that taking only the accuracy as a performance measurement does not show the whole picture. According to these results, we can conclude that the SOM + GP model significantly performs better than the basic GP model and the other classical approaches selected from the literature.

## 6   Conclusions

In this paper, we investigated the application of a churn prediction approach based on Self Organizing Maps (SOM) and Genetic programming (GP) for predicting possible churners in a Jordanian cellular telecommunication network. SOM was used to eliminate outliers which correspond to unrepresentative customers' data. While GP was used to develop a final churn prediction model. Performance of the proposed approach was evaluated using different criteria and compared with other common classification approaches. It was found that the SOM with GP approach has a promising capability in predicting possible churns and it outperformed the common classifiers.

## References

1. Blattberg, R.C., Do, K.B., Scott, N.A.: Database Marketing: Analyzing and Managing Customers. In: International Series in Quantitative Marketing, vol. 18, pp. 607–633. Springer, New York (2008)
2. Owczarczuk, M.: Churn models for prepaid customers in the cellular telecommunication industry using large data marts. Expert Systems with Applications 37, 4710–4712 (2010)
3. Kim, N., Lee, J., Jung, K.H., Kim, Y.S.: A new ensemble model for efficient churn prediction in mobile telecommunication. In: 2013 46th Hawaii International Conference on System Sciences, pp. 1023–1029 (2012)
4. Miguéis, V.L., Van den Poel, D., Camanho, A.S., Cunha, J.F.: Modeling partial customer churn: On the value of first product-category purchase sequences. Expert Syst. Appl. 39, 11250–11256 (2012)
5. Miguéis, V.L., Van den Poel, D., Camanho, A.S., Cunha, J.F.: Predicting partial customer churn using markov for discrimination for modeling first purchase sequences. Advances in Data Analysis and Classification 6, 337–353 (2012)
6. Saha, S.K., Ghoshal, S.P., Kar, R., Mandal, D.: Cat swarm optimization algorithm for optimal linear phase fir filter design. ISA Transactions 52, 781–794 (2013)
7. Yazdani, D., Nasiri, B., Azizi, R., Sepas-Moghaddam, A., Meybodi, M.R.: Optimization in dynamic environments utilizing a novel method based on particle swarm optimization. Int'l Journal of Artificial Intelligence 11, 170–192 (2013)
8. Sheta, A., Faris, H., Alkasassbeh, M.: A genetic programming model for s&p 500 stock market prediction. International Journal of Control and Automation 6, 303–314 (2013)
9. Huang, B., Kechadi, M.T., Buckley, B.: Customer churn prediction in telecommunications. Expert Syst. Appl. 39, 1414–1425 (2012)
10. Li, G., Deng, X.: Customer churn prediction of china telecom based on cluster analysis and decision tree algorithm. In: Lei, J., Wang, F.L., Deng, H., Miao, D. (eds.) AICI 2012. CCIS, vol. 315, pp. 319–327. Springer, Heidelberg (2012)
11. Adwan, O., Faris, H., Jaradat, K., Harfoushi, O., Ghatasheh, N.: Predicting customer churn in telecom industry using multilayer preceptron neural networks: Modeling and analysis. Life Science Journal 11 (2014)

12. Obiedat, R., Alkasassbeh, M., Faris, H., Harfoushi, O.: Customer churn prediction using a hybrid genetic programming approach. Scientific Research and Essays 8, 1289–1295 (2013)
13. Yu, X., Guo, S., Guo, J., Huang, X.: An extended support vector machine forecasting framework for customer churn in e-commerce. Expert Systems with Applications 38, 1425–1430 (2011)
14. Kohonen, T.: Clustering, taxonomy, and topological maps of patterns. In: Proceedings of the Sixth International Conference on Pattern Recognition, Silver Spring, MD, pp. 114–128. IEEE Computer Society Press (1982)
15. Kohonen, T.: The self-organizing map. Proceedings of the IEEE 78, 1464–1480 (1990)
16. Bação, F., Lobo, V., Painho, M.: Self-organizing maps as substitutes for K-means clustering. In: Sunderam, V.S., van Albada, G.D., Sloot, P.M.A., Dongarra, J. (eds.) ICCS 2005. LNCS, vol. 3516, pp. 476–483. Springer, Heidelberg (2005)
17. Gray, R.: Vector quantization. IEEE ASSP Magazine 1, 4–29 (1984)
18. Kohonen, T.: Data management by self-organizing maps. In: Zurada, J.M., Yen, G.G., Wang, J. (eds.) Computational Intelligence: Research Frontiers. LNCS, vol. 5050, pp. 309–332. Springer, Heidelberg (2008)
19. Koza, J.: Evolving a computer program to generate random numbers using the genetic programming paradigm. In: Proceedings of the Fourth International Conference on Genetic Algorithms. Morgan Kaufmann, La Jolla (1991)
20. Koza, J.R.: Genetic Programming: On the programming of computers by means of natural selection, vol. 1. MIT press (1992)
21. Espejo, P.G., Ventura, S., Herrera, F.: A survey on the application of genetic programming to classification. IEEE Transactions on Systems, Man, and Cybernetics, Part C: Applications and Reviews 40, 121–144 (2010)
22. Kotanchek, M., Smits, G., Kordon, A.: Industrial strength genetic programming. In: Riolo, R.L., Worzel, B. (eds.) Genetic Programming Theory and Practice, pp. 239–256. Kluwer (2003)
23. Affenzeller, M., Winkler, S., Wagner, S., Beham, A.: Genetic Algorithms and Genetic programming- Modern Concepts and Practical Applications. CRC Press (2009)
24. Miller, W., Sutton, R., Werbos, P.: Neural Networks for Control. MIT Press, Cambridge (1995)
25. Burez, J., Van den Poel, D.: Handling class imbalance in customer churn prediction. Expert Syst. Appl. 36, 4626–4636 (2009)
26. Tsai, C.F., Lu, Y.H.: Customer churn prediction by hybrid neural networks. Expert Systems with Applications 36, 12547–12553 (2009)

# Genetic Programming with Dynamically Regulated Parameters for Generating Program Code

Tomasz Łysek and Mariusz Boryczka

Institute of Computer Science, University of Silesia,
ul.Będzińska 39, Sosnowiec, Poland
{tomasz.lysek,mariusz.boryczka}@us.edu.pl

**Abstract.** Genetic Programming (GP) is one of the Evolutionary Algorithms. There are many theories concerning automatic code generation. In this article we present the latest research of using our dynamic scaling parameter in Genetic Programming to create a code. We have created practically functioning program code with the dynamic instruction set for $L$ language. For testing we have chosen the best known problems. Our investigations of the best range of each parameter were based on our preliminary experiments.

**Keywords:** Genetic Programming, Linear Genetic Programming, Dynamic Parameters, Code generation

## 1 Introduction

Genetic Programming (GP), one of Evolutionary Algorithms has been developed mainly by John Koza and Wolfgang Banzhaf between 1992 and 2007 [4]. The greatest improvement of the algorithm has been suggested by Banzhaf and Bremaier approach to Linear Genetic Programming. They have proposed an proprietary solution that depends on tournament selection and strict machine code for individuals structure. Genetic Programming is an extension of Genetic Algorithm, and one of the population algorithms based on the genetic operations. The main difference between those two is the representation of the structure they manipulate and the meanings of the representation. Genetic Algorithms usually operate on a population of fixed-length binary strings, GP typically operates on a population of parse trees that usually represent computer programs [5]. There are various models of an individual structure in GP population as well as various methods to modify an individual - by crossing-over or mutation. It is common that higher value of the crossover probability will result in better exploitation and high mutation will improve the exploration. Based on our previous research we have created a possibility of dynamic regulation of the parameters, responsible for the probability of the certain manner of modify an individual, in a way that it will increase the speed of finding better results and earlier detect stagnation of population at the same time [2]. We created a dynamic control parameter

D. Hwang et al. (Eds.): ICCCI 2014, LNAI 8733, pp. 363–372, 2014.

- responsible for the course of the algorithm that in fewer iterations receives better value of the fitness function, comparing to the results known from the literature. We have managed to reduce the length of a code of the individual as well. Our aim is to adapt the dynamic control parameter so that it will generate a program code which would be capable of solving programming issues depending on collection of input-output vectors. This article is organized as follows: first we analyse related works and ideas of creating most effective GP in the literature. Afterwards, we present the GP theory, that is the base to our research and experiments. The fourth section is dedicated to our idea of the dynamic control parameter and experiments proving its effectiveness. In the last section we describe the course of the experiment associated with a program code generating and we present its applicability. We compare our proposal with the results achieved by classical GP algorithms. We summarize with short conclusions.

## 2    Related Work

Genetic Programming was applied to various domains by Koza. Further development of this method involves modify of the population structure, of the type of an individual structure and introducing new methods of genetic operations [1]. Classical approach assumes representation of an individual by tree structure (Tree-based GP). There are also some modifications in which an individual is represented through Rule-based GP (Rule-based GP), which is gene expression for Genotype-Phenotype Mappings (GPM) by Ferreira [8], which assumes that individual's structure is a string with a head and a tail. The head is a list of expressions (functions and symbols) and the tail is a list of the arguments. Linear Genetic Programming (LGP) algorithm proposed by Koza and improved by Banzhaf and Bremaier turned out to be the breakthrough. LGP is based on presenting an individual in a graph structure, which vertices are the program instructions. Brameier introduced population divided into two groups and the leaders of those groups are crossed [7]. This is a huge leap from the classical approaches with tree-based code. For the purposes of the experiments classical approach TBGP and linear approach LGP has been tested. An individual modifications (mainly through different mutation types) has been taken from the literature and implemented according to given patterns. Experiments regarding effectiveness of modification and setting probability intervals of choosing control parameters has been conducted in 2012 and 2013.

## 3    Genetic Programming and Linear Genetic Programming

Genetic Programming described by John Koza is an algorithm processing test input vectors into their corresponding output vectors. Koza has defined as well five steps to be performed in order to solve a problem using GP:

1. Define the terminal set,
2. The function set,
3. Define fitness measure,
4. Select control parameters,
5. Define termination and result designation.

Individuals in GP are build with instructions made in defined $L$ language. Collection of programs in $L$ language is called genotype $G$. Phenotype $P$ is defined as a set allowing reflection of input vector into output vector.

$$f_{gp} : I^n \rightarrow O^m : f_{gp} \in P$$
$$gp \in G$$

The defined approximation target is to find the best $T$ from the given collection using the evolution process:

$$T = \{(i, o) | i \in I' \subseteq I^n, o \in O' \subseteq O^m, f(i) = o\}$$

Evaluating of the fitness function is determined by detection of the error size made by each individual. In order to exacerbate selection requirements of the best individual, to the fitness function we add modifications as a penalty e.g. late iterations or a tree depth/length of the generated code. A popular way of determining errors in approximation tasks is the sum of squared errors (SSE) [3]. The mean square error for SSE is evaluated from equation:

$$MSE(gp) = \frac{1}{n} \sum_{k=1}^{n} (gp(i_k) - o_k)^2$$

LGP algorithm differs from the classical approach mainly in an individual construction, a population structure and the number of parameters improving its effectiveness. An individual in LGP algorithm is build in the way resembling a program code in the machine language. LGP structure is a combination of the idea of creating a genotype with the binary code and the idea of programming using genes. Genotype in the form of the binary code (RBGP - Rule-based genetic programming) presents coding the set of symbols (terminals) and operations by means of appropriate binary values. Each individual of the population consists of many classifiers processing $I^n$ into $O^m$. In this algorithm a crossover and a mutation are performed in a classical way, whereas the fitness function is based exclusively on the high value of comparing an individual's genotype with encoded version $o \in O' \subseteq O^m$. Individual's structure presented by genes connections (GPM Genotype-phenotype programming) has been presented by Ferreira. He based it on dividing genotype into two groups: a head and a tail. The head is a list of functions and operators used in an individual. The tail is a list of arguments provided program as a component of the $L$ language. Additional symbols introduced in functional part resemble registers used in LGP algorithm. The algorithm of the linear Genetic Programming combines features of RBGP and GPM, because the structure of the individual based on the programming code contains encoded values in form of registrations. In some modifications, at the beginning of each individual appears a headline containing functions and symbols used in the certain genotype. Thanks to graph-like structure of an individual even at the very first tests conducted by Banzhaf it has been demonstrated that LGP is a better method for solving more elaborated problems. An ultimate advantage of LGP over the classical approach has been revealed in Bremaiers scientific work. In the TBGP there is only a limit of the tree depth. There are two basic parameters restrictive creating genotypes in LGP, these are maximal length of the code (number of code's lines) and maximal length of the single code line (determined on the base of the number of operators and symbols). Main features of LGP are:

- the structure of command list (that can be presented in the form of directed graph), instead of classical approach based on a tree,
- the linear structure of the program performed as a processor machine code,
- acquired values are saved in the dynamic registers which behaves like inner processor registers,
- subgraphs formed inside of an individual, used as functions, registers, are treated as variables,
- inner algorithm searching and deleting inefficient code through individual's evaluation, based on their fitness function,

Assuming that fulfilment of the four out of five point out of five Koza's steps is being represented by the collection of the parameters:

- probability of crossover   differs two individuals by crossing-over chooses parts,
- probability of mutation  changes part of an individual with by specified way,
- maximum numer of individuals in population  restricts the number of individuals,
- maximum tree depth/maximum code lines  restrict individual size,
- probability of changing function/terminal  chenges function or terminal to other from L in tree node/line of code,
- probability of permutation  swap over pieces of the tree (TBGP),
- probability of inserting / deleting  adding randomly generated part from L to individual/deleting random part of individual,
- probability of encapsulation  protection part of individual against further changes,
- probability of automated defined function (ADF) - Recognition of useful fragments of genotype and transfer the parts to the set of available features.

In the classical TBGP, as well as in each modification there is a number of parameters influencing the quality of obtained results. Various parameters and their values can be used depending on the studying problem. Most of the parameters are flexible and can be used in the TBGP algorithm as well as in the LGP. However there are prepared special parameters, that can only be used in the particular versions of genetic programming. For the GP based on tree structure there is a possibility of the mutation through lifting a fragment of the tree for selected number of levels, and putting it in other node's place, while in LGP there is a possibility of limiting the number of the registers. Those solutions cause the loss of some parts of the genotype, shortening its length. Combining that with encapsulation or automatic defined functions leads to improving the results.

## 4    Dynamic Parameters in Genetic Programming

In Luke's and Spector's experiments it has been demonstrated that simple mutation and standard crossover in case of genetic programming algorithms affects results to the same degree [5]. However the crossover works well for large populations, while the mutation allows to obtain better outcomes for smaller populations with a larger number of iterations of the algorithm. When the crossover and the mutation is used the most important problem is selecting a node. A mutation needs to have specified a node to which it is applied, and a crossover needs to have chosen two nodes, from which it starts the operation. In the literature, the most frequently discussed is an example of a random selection. The research was used Weise method, where the base is factor defining the weight of a subtree of the test fragment [8]. Weise weighting factor is based on the assumption that the best selection will be selecting all nodes $c$ and $n$ tree $t$, with the same probability distribution as in the case of random select, eg. $P(nodeSelection(t) = c) = P(nodeSelection(t) = n)\forall s, n \in t$. Weight node $n$ is obtained by the number of nodes in the subtree of $n$:

$$W(n) = 1 + \sum_{i=0}^{l(succ(n))-1} W(succ(n)_i),$$

where $W$ is function that determines weight of node $n$, $succ(n)$ is function that determines set of child nodes of $n$, and $l$ is function determines the lenght of the $n$ subtree.

---

**Algorithm 1:** Setting node weight

---

**begin**

1     $flag = true; c = t;$

2     **while** $flag$ **do**

3        $r = \lfloor random(0, W(c)) \rfloor;$

4        **if** $r \geq W(c) - 1$ **then**

5           $b = false;$

       **else**

6           $i = l(succ(c)) - 1;$

7           **while** $i >= 0$ **do**

8              $r = r - W(succ(c)_i);$

9              **if** $r < 0$ **then**

10                 $c = succ(c)_i; i = -1;$

             **else**

11                 $i = i - 1;$

12     **return** c;

---

The evaluation function has a major impact on the structure of the population in subsequent iterations. We propose the construction of the evaluation function based on the MSE and the additional penalty rates for a large number of iterations of the algorithm and the length of the code of individual:

$$\text{FO}(gp) = \text{MSE}(gp) + \text{KI} + \text{KW}(gp)$$

where:

- KI  designated punishment for a long iteration
- KW($gp$)  designated penalty for a large depth of the tree (TBGP) / large number of lines of code (LGP)

Dynamic parameters involves adding to the algorithm scaling factor based on the results of studies on the extent to which the crossover and mutation parameters allows to get the best possible result. In addition, studies have been carried out concerning the designation of a minimum number of iterations that achieve the high value of fitness function [2]. In table 1 and 2 we present test results ($F$ function value tending to 0; $G$ depth of generated tree for TBGP; $D$ length of generated program for LGP).

**Table 1.** Setting the parameters for the best intervals TBGP and LGP (part 1)

| | function change | | | | terminal change | | | | permutation | | | | inserting | | | |
|---|---|---|---|---|---|---|---|---|---|---|---|---|---|---|---|---|
| | TBGP | | LGP | | TBGP | | LGP | | TBGP | | LGP | | TBGP | | LGP | |
| Value | F | G | F | D | F | G | F | D | F | G | F | D | F | G | F | D |
| 0.1 | 13.9 | 19 | 4.7 | 192 | 14.9 | 20 | 4.3 | 200 | 15.2 | 20 | 6.2 | 200 | **14.5** | **17** | **4.7** | **189** |
| 0.3 | **13.6** | **17** | **3.7** | **185** | 13.8 | 17 | **3.8** | **186** | **14.5** | **18** | **4.9** | **186** | 14.9 | 19 | 5.1 | 194 |
| 0.5 | **13.5** | **17** | **3.7** | **191** | **13.7** | **18** | **3.7** | **194** | 14.3 | 19 | **5.1** | **194** | 15.2 | 20 | 5.3 | 198 |
| 0.7 | 14.1 | 20 | 5.4 | 200 | 14.1 | 20 | 4.9 | 200 | 14.9 | 20 | 5.7 | 200 | 15.7 | 20 | 6.4 | 200 |

**Table 2.** Setting the parameters for the best intervals TBGP and LGP (part 2)

| | cutting | | | | encapsulation | | | | ADF | | | | lifting | |
|---|---|---|---|---|---|---|---|---|---|---|---|---|---|---|
| | TBGP | | LGP | | TBGP | | LGP | | TBGP | | LGP | | TBGP | |
| Value | F | G | F | D | F | G | F | D | F | G | F | D | F | G |
| 0.1 | 14.9 | 16 | 4.2 | 186 | 14.1 | 16 | 4.2 | 188 | 14.2 | 18 | **4.7** | **193** | 13.8 | 18 |
| 0.3 | 15.1 | 18 | 4.9 | 189 | 14.4 | 17 | 4.6 | 193 | **13.8** | **14** | **4.2** | **186** | 13.6 | 16 |
| 0.5 | 15.6 | 18 | **5.3** | **194** | 15.3 | 19 | 4.9 | 197 | 14 | 14 | 5.4 | 182 | 13.9 | 17 |
| 0.7 | 15.9 | 19 | 5.4 | 197 | 15.8 | 20 | 5.1 | 199 | 14.5 | 14.5 | 9.6 | 179 | 14.3 | 20 |

On the basis of these studies was created scaling parameter $\gamma$, that value at the beginning of solving the problem is 1, and in subsequent iterations oscillates between the values that $(0, 1\rangle$ [2]. If the parameter is set to a value of 1 or close to, then occurs the exploitation of searched space solution and the value of the fitness function of the population will converge. If the value of the parameter $\gamma$ would be closer 0, then the population will be a subject of a greater number of mutations and there will be a greater exploration of the solution space. Fitness function value of individuals in successive iterations of the algorithm may get closer to the results from the initial sampling algorithm or possess a better result.

In such a case, the parameter $\gamma$ is changed to the exploratory character not to allow an excessive convergence, and acquired results can be add to the collection of output vectors. In case of a deterioration the $\gamma$ parameter is transformed into the exploitation form.

$$PM(M_i) = \begin{cases} \dfrac{SW_{(M_i)}}{\gamma} & \text{if encapsulation or inserting} \\[2ex] SW(M_i) \cdot \gamma & \text{, for other cases} \end{cases}$$

where:

- $M$ is collection of the possible mutations,
- $PM$ is the function giving a new value of probability for drawing the $M$ mutation,
- $SW$ is a weighted average based on values of the fitness function $FO$, upper and lower edges of the best range of the values of probability for mutation $M_i$.

---

**Algorithm 2:** The algorithm for determining the value of a new mutation

---

1 Determine the degree of the population stagnation
2 Establish a new $\gamma$ value in $(0, 1\rangle$ (progressing stagnation of the population involves $\gamma$ parameter is getting closer to 0),
3 Designate a weighted average of the lower and upper edge of the best range of values for a set of mutation probabilities
4 Fixing a new probability values for each mutation

---

Example:

- The $m$ stagnation appears,
- A new $\gamma$ value is calculated $\gamma = \begin{cases} \gamma - 0.1 \text{ , } \gamma > 0.1 \\ 0.1 \text{ , } \gamma = 0.1 \end{cases}$,
- Cutting mutation value according to table 2 $SW = 0.2$ must be replaced by $PM = SW \cdot \gamma$ in order to achieve new probability.
- For each $M_i$ element it is necessary to determine a new $PM$.

## 5   Experiments and Results

The purpose of experiments was to compare effectiveness of GP and LGP in C++ to the same algorithms written and compiled in authors' platform [6]. To make our algorithm more efficient we added smart code completion that checks if used function in generated individual needs to add functions library from programming language. This mechanism will provide smaller start collection of terminals in language $L$ and in further point of iteration will decrease consumption of adding new libraries. Classical GP-like algorithm with proposed modification and smart code mechanism is as follow:

---

**Algorithm 3:** Modified Genetic Programming algorithm

---

1 Generate population P with random composition of defined functions;
2 **while** *stop criterion is not met* **do**
3      Parse generated individuals (programs) to set value of fitness function;
4      Copy the best individual;
5      Calculate the weight of the obtained results to determine the degree of convergence of the population;
6      Change the value of $\gamma$;
7      Calculate the Weise weight for selected fragments of individuals;
8      Create new programs using mutation and crossover;
9      Check the length of the algorithm;
10      Check the length function;
11      Check the depth of nesting;
12      Append missing libraries;
13 An individual whose genotype achieved the best result of the adaptation function can be exact or approximate solution.

---

Test problems:

- Loop Input vector: value that determine loop stop, value that determine loop step. Output: collection of values generated by loop,
- Factorial - Input vector: value for factorial. Output: factorial for value,
- Fibonacci - Input vector: value of Fibonacci number. Output: value of Fibonacci number,
- GCD - Input vector: number 1, number 2. Output: GCD for two numbers,
- Bubble sort - Input vector: collection on values. Output: sorted collection of numbers,
- Distinct roots Input vector: values of delta, b and a. Output: distinct roots values.

Parameters for experiments were:

- the size of population $N = 500$,
- the crossover parameter $CR = 0.9$,
- the mutation parameter $F = 0.1$,
- the maximum number of iterations is equal to 500,
- the maximum tree depth (TBGP) is equal to 20,
- the maximum operator nodes is equal to 200,
- the maximum program length (LGP) is equal to 200,
- for every testable function the algorithm was run 10 times,
- the maximum algorithm length: $algLen = 100$,
- the maximum function length: $funLen = 10$,
- the maximum depth of nesting: $maxDepth = 2$,
- percent of the population of test subjects: $testBoids = 5$.

**Table 3.** Medium percentage errors of best individuals in the population (part 1)

|  | Loop | | | | Factorial | | | | Fibonacci | | | |
|---|---|---|---|---|---|---|---|---|---|---|---|---|
|  | TGP | TGP ($\gamma$) | LGP | LGP ($\gamma$) | TGP | TGP ($\gamma$) | LGP | LGP ($\gamma$) | TGP | TGP ($\gamma$) | LGP | LGP ($\gamma$) |
| 50 | 28.3 | 29.9 | 21 | 22.1 | 57.9 | 59.1 | 55.9 | 56.8 | 70.3 | 76.1 | 66.1 | 66.4 |
| 100 | 28 | 28.1 | 20.3 | 19.2 | 56.4 | 58.4 | 54.1 | 54.2 | 65.1 | 67.9 | 65 | 64.7 |
| 150 | 27.8 | 27.9 | 19.7 | 17.6 | 53.1 | 55.9 | 49.8 | 50.7 | 61.8 | 63.8 | 59.4 | 59.3 |
| 200 | 27.4 | 25.3 | 19.4 | **16.1** | 49.1 | 49.3 | 46.2 | 46.6 | 56.2 | 55.9 | 53.1 | 53.7 |
| 250 | 26.5 | 24.1 | 18.9 | **14.9** | 43.9 | 43.4 | 41.6 | 39.3 | 49.6 | 49.5 | 47.6 | 46.2 |
| 300 | 25.2 | **22.3** | 18.2 | **14.1** | 41 | 39.2 | 38.4 | 33.9 | 47.3 | 45.1 | 44.4 | 40.9 |
| 350 | 24.3 | **21.6** | 17.8 | **13.5** | 38.4 | 32.8 | 33.9 | 29.1 | 45 | **40** | 39.6 | 37.1 |
| 400 | 23.6 | **21.4** | 17.1 | **13.2** | 32.6 | **28.4** | 29.1 | **25** | 42.9 | **37.4** | 37.5 | **33.8** |
| 450 | 23.1 | **21.4** | 16.8 | **12.9** | 30.2 | **26.1** | 27.8 | **23.4** | 41.2 | **36.1** | 35.1 | **31.4** |
| 500 | 22.8 | **21.4** | 16.7 | **12.9** | 29.4 | **25.9** | 26.1 | **22.5** | 40.3 | **35.7** | 34.9 | **30.2** |

**Table 4.** Medium percentage errors of the best individuals in the population (part 2)

|  | GDC | | | | Bubble sort | | | | Distinct roots | | | |
|---|---|---|---|---|---|---|---|---|---|---|---|---|
|  | TGP | TGP ($\gamma$) | LGP | LGP ($\gamma$) | TGP | TGP ($\gamma$) | LGP | LGP ($\gamma$) | TGP | TGP ($\gamma$) | LGP | LGP ($\gamma$) |
| 50 | 91.6 | 91.4 | 84.7 | 88.2 | 96.4 | 98.6 | 83.7 | 88.9 | 90.2 | 93.9 | 89.4 | 91.6 |
| 100 | 90.9 | 89.6 | 79.1 | 81 | 84.5 | 85.7 | 79.9 | 81.3 | 85.2 | 89.4 | 83.1 | 85 |
| 150 | 89.1 | 81.2 | 73.5 | 75.9 | 82.6 | 82.1 | 76.6 | 75.4 | 80.1 | 80.6 | 73.9 | 74.2 |
| 200 | 84.9 | 74.7 | 67.9 | 68.1 | 80.1 | 76.4 | 72.9 | 70.2 | 76.7 | 78.1 | 64.8 | 61.9 |
| 250 | 70.6 | 61.9 | 55.6 | 55.3 | 76.5 | 72.2 | 68.5 | 62.3 | 72.8 | 72.5 | 59 | 56.4 |
| 300 | 62.7 | 51.8 | 49.4 | 46.8 | 69.8 | 68.5 | 62 | 58.6 | 66.2 | 64.9 | 56.7 | 52.8 |
| 350 | 53.8 | 45.5 | 45.3 | 42.3 | 65.2 | 63.8 | 58.3 | 53.8 | 61.9 | 58.7 | 50.3 | 48.9 |
| 400 | 49.2 | **39.4** | 42.5 | **38.5** | 61.4 | 60.2 | 56.7 | 52.1 | 54.3 | 50.1 | 48.7 | 46.7 |
| 450 | 46.3 | **37.9** | 40 | **36.2** | 59.7 | 58.9 | 52.1 | **51.3** | 49.8 | **44.6** | 45.1 | **42.5** |
| 500 | 44 | **36.1** | 39.1 | **34.9** | 58.4 | **58.1** | 51.6 | **50.9** | 48.3 | **42.1** | 44.5 | **39.8** |

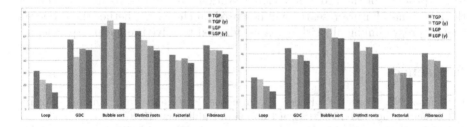

**Fig. 1.** Median error (left) and minimum error (right) in individual

Tables 3 and 4 show the results of the best individuals in a given iteration of the algorithm. Adding a $\gamma$ parameter allowed improving the results obtained in each test. In addition, our algorithm acquires a better result long before the classical approach. As expected from previous studies, when more than 500 iterations stagnation of the results appears[2]. In fig. 1 we presented that the median results of the algorithm with use of $\gamma$ parameter does better than the classical approach (with the exception of bubble sort algorithm). In fig. 1 you can also notice that the minimal error of the best individual is significantly reduced comparing to the classical GP algorithms. In the box plot it an be noticed that in most cases the population is concentrated around better solutions, and individuals with large error are marginalized. Furthermore we have noticed a strong

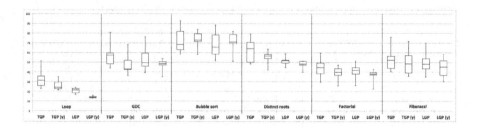

**Fig. 2.** Box plot of iteration 500

tendency of the population with good solutions to concentrate while maintaining the margin to the possibility of exploration.

## 6    Conclusions and Future Work

We have provided a better algorithm for generating a program code. By applying the proposed modification created program code is usable as a template code or the initial solution generator. Proposed $\gamma$ parameter allows significant acceleration for better performance and reduces the size of individuals. The proposed algorithm in accordance with earlier experiments confirms the validity of the application of the populations of not less than 300, but not more than 500 because of constant stagnation results. LGP algorithm is definitely better adapted to the problem of the program code generation, mostly due to the structure of the individual. Future research will be based on checking whether the proposed genetic programming algorithm can be adapted to solve the hashing problem.

## References

1. Banzhaf, W., Nordin, P., Keller, R., Francone, F.: Genetic Programming - An Introduction, pp. 133–134. Morgan Kaufmann Publishers (1998)
2. Łysek, T., Boryczka, M.: Dynamic parameters in GP and LGP. In: Nguyen, N.T., Trawiński, B., Katarzyniak, R., Jo, G.-S. (eds.) Adv. Methods for Comput. Collective Intelligence. SCI, vol. 457, pp. 219–228. Springer, Heidelberg (2013)
3. Brameier, M., Banzhaf, W.: Linear Genetic Programming, pp. 130, 183–185, 186. Springer (2007)
4. Koza, J.: Genetic Programming: On the Programming of Computers by Means of Natural Selection. MIT Press (1992)
5. Luke, S., Spector, L.: A Comparison of Crossover and Mutation in Genetic Programming (1997)
6. Łysek T.: Dedicated language and MVC platform for Genetic Programming algorithms. Journal of Information, Control and Managament Systems (2012)
7. Nedjah, N., Abraham, A., de Macedo Mourelle, L.: Genetic Systems Programming: Theory and Experiences, pp. 16–17. Springer-Verlag (2006)
8. Weise, T.: Global Optimization Algorithms: Theory and Application, pp. 169–174, 191–195, 207–208 (2009)

# A Guidable Bat Algorithm Based on Doppler Effect to Improve Solving Efficiency for Optimization Problems

Yi-Ting Chen, Chin-Shiuh Shieh, Mong-Fong Horng, Bin-Yih Liao,
Jeng-Shyang Pan, and Ming-Te Tsai

Department of Electronics Engineering,
National Kaohsiung University of Applied Sciences, Kaohsiung, Taiwan
mfhorng@kuas.edu.tw

**Abstract.** A new guidable bat algorithm (GBA) based on Doppler Effect is proposed to improve problem-solving efficiency of optimization problems. Three searching polices and three exploration strategies are designed in the proposed GBA. The bats governed by GBA are enabled the ability of guidance by frequency shift based on Doppler Effect so that the bats are able to rapidly fly toward the current best bat in guidable search. Both refined search and divers search is employed to explore the better position near the current best bat and develop new searching area. These searching polices benefit discover the eligible position to upgrade the quality of position with the current best bat in a short time. In addition, next-generation evolutionary computing (EC 2.0) is created to breaks the bottleneck of traditional ECs to create the new paradigm in ECs. In EC 2.0, conflict theory is introduced to help the efficiency of solution discovery. Conflict between individuals is healthful behavior for population evolution. Constructive conflict promotes the overall quality of population. Conflict, competition and cooperation are the three pillars of collective effects investigated in this study. The context-awareness property is another feature of EC 2.0. The context-awareness indicates that the individuals are able to perceive the environmental information by physic laws.

**Keywords:** Guidable bat algorithm, Doppler Effect, EC 2.0, Collective effect, Context-awareness, Conflict behavior.

## 1 Introduction

A powerful swarm-based evolutionary optimization, bat algorithm is developed based on the echolocation behavior of bats [1, 2]. In bat algorithm, the characteristics of bats including frequency, velocity, position, emission pulse rate and loudness are utilized to drive the bats to continuously explore possible solutions in a solution space in order to find the global optimal solution. The ultrasound frequency of bat is determined by a random value with uniform distribution in the original bat algorithm. This randomly generated frequency without the guidance effortlessly causes aimless search of bats. In order to guide the bats toward the direction of global optimal solution, the property of Doppler Effect is adopted to determine the ultrasound frequency of each bat in this

D. Hwang et al. (Eds.): ICCCI 2014, LNAI 8733, pp. 373–383, 2014.
© Springer International Publishing Switzerland 2014

study. Doppler Effect produces a frequency shift ($\Delta f$) caused by the velocity between sound source and observer. The two objects (sound source and observer) continuously moving in various velocities will cause the frequency shift of the sound source received by the observer. The changeable velocity forms the frequency variation in a sound wave emitted by the source. Hence, the observer is in different positions, the received ultrasound frequency is also dissimilar for an identical ultrasound.

Next-generation ECs (EC 2.0), implemented as bio-inspired evolutionary computing with collective-effect and context-awareness is first proposed in this study. And, the conception of EC 2.0 is introduced in BA to designed guidable bat algorithm (GBA). The context-awareness is that the bats can sense the environmental change by physical laws, Doppler Effect. The collective effect is from the individual behavior. The individual behavior is able to affect the development of population. The individual behavior comprises cooperation, competition and conflict. The individual can adopt different search strategies according to their self-behavior to promote the activity of population.

## 2    Related Works

### 2.1    Bat Algorithm

Bat Algorithm (BA) is a nature-inspired algorithm and proposed by Xin-She Yang to solve optimization problems of single objective and multi-objectives [1, 2]. And, BA is investigated in depth and is applied [3-5]. All bats have ability to sense the distance between prey. This ability is called echolocation. Bats randomly fly in the velocity ($v_i$) with a fixed frequency ($f_{min}$), varying wavelength ($\lambda$), adjustable pulse emission rate ($r_i$) and changeable loudness ($A_0$) at position ($x_i$) to search the prey. The frequency $f_i$ in bat $i$ is assumed to range from $f_{min}$ to $f_{max}$. The loudness is assumed between 1 and 2 as well as decreases from a large positive $A_0$ to a minimum constant value $A_{min}$. The pulse emission rate is set between 0 and 1.

In BA, the updates of the frequency ($f_i$), velocity ($v_i$) and position ($x_i$) in a $d$-dimension search space are defined in Eq. (1-3). The new velocity ($v_i^t$) and position ($x_i^t$) of bat $i$ at iteration $t$ are given by

$$f_i = f_{min} + (f_{max} - f_{min})\beta \tag{1}$$
$$v_i^t = v_i^{t-1} + (x_i^{t-1} - x_*)f_i \tag{2}$$
$$x_i^t = x_i^{t-1} + v_i^t \tag{3}$$

where $\beta \in [0,1]$ is a random vector with uniform distribution. The variable, $x_*$, is the location of the current global best bat (solution) derived from a comparison of all solutions discovered by all bats. The frequency $f_i$ of the $i^{th}$ bat is used to adjust the velocity ($v_i^t$) to move bats to the position ($x_i^t$). In an implementation of BA, it is usually assumes $f_{min}=0$ and $f_{max}=100$. Hence, each bat is randomly assigned by a frequency with uniform distribution in [$f_{min}, f_{max}$] initially.

In the procedure of local search, a position is selected from the current best bat. And a new position found by a bat is generated locally by using random walk as follows,

$$x_i^t = x_* + \epsilon A^t \tag{4}$$

$$A^t = \frac{\sum_{i=1}^{PS} A_i^t}{PS} \tag{5}$$

where $\epsilon \in [-1,1]$ is random number. $A^t$ is the average loudness of all the bats in iteration $t$. The population size (PS) is the number of bats in a population. Additionally, the loudness and the emission pulse rate have to be updated accordingly as the iteration. In general, the loudness ($A_i$) and the rate of pulse emission ($r_i$) will be decreased and increased respectively when the bat $i$ found the prey. In [1, 2], $A_0$ and $A_{min}$ are set to 1 and 0, respectively to simplify the case. $A_{min}=0$ means that the bat has found the prey and temporarily stop emitting sound. The updated loudness and emission pulse rate are given by

$$A_i^t = \alpha A_i^{t-1}, \quad r_i^t = r_i^0[1 - exp\,(-\gamma t)] \tag{6}$$

$$A_i^t \to 0, \; r_i^t \to r_i^0, \; as \; t \to \tag{7}$$

where $\alpha$ and $\gamma$ are constants and assigned to be 0.9 in the original papers [1, 2] in order to simplify the implemented simulations. Initially, each bat should have different loudness and emission pulse rates. When the iterations increase gradually, according to Eq. (7), these two parameters will slowly approach to zero and final emission rate ($r_i^0$), respectively. The loudness and emission pulse rate will be updated only if the new position of the current new bat is improved, which means that the bats are moving towards the optimal solution.

## 2.2 Next-Generation Evolutionary Computing (EC 2.0)

The proposed GAB represented an excellent performance in discovery the minimum solution of continuous functions [6]. Based on the previous research work, the conception of EC is investigated in depth to create next-generation ECs (EC 2.0) in this study. In traditional ECs, the individuals are advanced by the cooperation and competition by communicating with each other and tracking the best individual. In this study, the conflict behavior is first studied in ECs to create a novel conception of collective-effect. In the tradition ECs, the individuals cooperate and compete for each other so that the individuals are very similar in approaching to the optimal solution. The higher similarity causes the lower diversity of population to deteriorate the performance efficiency. The individuals in a population cooperate, compete and conflict to advance the population diversity in the proposed next-generation ECs (EC 2.0). The conflict conception increases the population diversity to enhance the performance efficiency. This study is the first to consider conflict behaviors of individuals as a kind of collective-effect in swarms and to realize EC 2.0 close to real world. Additionally, how to adjust the individual behavior with cooperation, competition and conflict to maintain the population advancement is valuable and difficult topic for the technology of evolutionary computing in future.

# 3    A Guidable Bat Algorithm Based on Doppler Effect

In this study, the bats governed by GBA are able to adjust their velocity according to frequency shift caused by Doppler Effect. This frequency shift depends on Doppler Effect between the bats and the current best bat. When the bats are close to the current best bat, the bats should receive the ultrasound with higher frequency. The bats accelerate to fly toward the direction of the current best bat. The bats attempt to find a better position than the current best bat according to this direction. On the contrary, when the bats run away from the current best bat, the bats will receive an ultrasound with lower frequency. The bats should slowly move to explore the better position than their own position along the path of the current best bat. This manner is different from tradition ECs. Therefore, the velocity of bats is adjusted by frequency shift of the received ultrasound to discover a better position than the current best bat or their own position in guidable search. In additional, this proposed GBA is employed to discover the minimum solution of continuous function during evolution. Hence, the smaller the fitness value is, the better the solution quality is. The flowchart of the proposed GBA is as shown in Fig. 1. The detailed operations of GBA are described as follow:

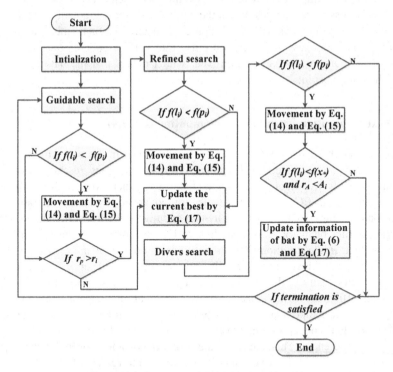

**Fig. 1.** The flowchart of Guidable Bat Algorithm

Step1. Initialization
In this step, there are many parameters to be specified, including iteration number, population size and dimension of search space for algorithm. Then, for the properties of

bat such as the frequency, velocity, previous position, current position, location, loudness and emission pulse rate are initialed. These parameters will affect the performance of the proposed algorithm. In this study, the proposed algorithm is used to discover the minimum solution of continuous functions. Therefore, the smaller the fitness value is, the better the bat position is.

Step2. Guidable Search

In order to improve the movement of bats in the original bat algorithm, there are many novel conceptions to be applied to design a guidable search. For this guidable search, Doppler Effect is employed to establish a regular rule of bat movement overseen in GBA. The bats use their own previous positions and current positions to ascertain close to or away from the current best bat according to Eq. (8) and Eq. (9).

$$d_i^t = \sqrt{(\tilde{x}_i^{t-1} - x_*)^2} \tag{8}$$

$$d_i^t = \sqrt{(x_i^{t-1} - x_*)^2} \tag{9}$$

where $d'^t_i$ is the distance of previous position of bat $i$ and the current best bat. $d_i^t$ is the distance of current position of bat $i$ and the current best bat. The bats obtain a frequency of received ultrasound as shown in Eq. (10). Then, a lowpass filter is utilized to filter out the background noise as shown in Eq. (11). Hence, the bats can more accurate to adjust the velocity given by Eq. (12). If $d_i^t \leq d'^t_i$, the bats fly toward the current best bat. The bats should receive the ultrasound with higher frequency. The bats will quickly fly toward the direction of the current best bat to find a better position than the current best bat according to this direction. On the contrary, if $d_i^t > d'^t_i$, then the bats fly away from the current best bat. The bats receive the ultrasound with lower frequency. The bats should decelerate to explore a better position than their own positions along the path of the current best bat. The bats modify their velocity by the frequency their received sound given by

$$\hat{f}_i^t = \begin{cases} (V + v_i^{t-1})/(V - v_*)]\hat{f}_i, & if\ d_i^t \leq d_i^t \\ [(V - v_i^{t-1})/(V + v_*)]\hat{f}_i, & if\ d_i^t > d_i^t \end{cases} \tag{10}$$

$$f_i^t = (f_i^{t-1} + \hat{f}_i^t)/2 \tag{11}$$

$$v_i^t = v_i^{t-1} + (x_i^{t-1} - x_*)f_i^t \tag{12}$$

$\hat{f}_i^t$ is the frequency caused by Doppler Effect for bat $i$ in iteration $t$. $V$ is the propagation speed of sound in 25°C. In this study, $V$ is set 340 m/s. $v_i^{t-1}$ is the velocity of bat in iteration $(t-1)$. $\hat{f}_i$ is the frequency of emitted ultrasound for the bats. $v_*$ is the velocity of the current best bat. $f_i^t$ is the new frequency obtained by past frequency $(f_i^{t-1})$ and  the frequency $(\hat{f}_i^t)$ caused by Doppler Effect of bat $i$ in iteration $t$. The velocity $(v_i^t)$ is modified by the distance between bat $i$ and the current best bat and the new frequency $(f_i^t)$.

The bats managed by GBA utility the velocity and their own current position to explore the next location. If the quality of an explored location is better than their current positions, the bats will update their own current positions with the new found locations and the original current positions will be the previous positions. Otherwise, the bats stay at their own position as shown in Eq. (14) and Eq. (15). On the whole, this searching policy provides the proper frequency and suitable velocity to guide the bats toward correct direction to follow the current best bat.

$$l_i^t = x_i^{t-1} + v_i^t \tag{13}$$

$$x_i^t = \begin{cases} l_i^t, & if \ f(l_i^t) < f(x_i^{t-1}) \\ x_i^{t-1}, & otherwise \end{cases} \tag{14}$$

$$\tilde{x}_i^t = \begin{cases} x_i^{t-1}, & if \ f(l_i^t) < f(x_i^{t-1}) \\ \tilde{x}_i^{t-1}, & otherwise \end{cases} \tag{15}$$

The bats use the modified velocity $(v_i^t)$ to explore new location $(l_i^t)$ based on the position $(x_i^{t-1})$ at the iteration $(t\text{-}1)$. If the new location is better than their own position $(x_i^{t-1})$, the bat will update its own position $(x_i^t)$ with the new found location $(l_i^t)$. Then, the original position $(x_i^{t-1})$ will be the previous position $(\tilde{x}_i^t)$ according to Eq. (15).

Step3. Refined Search
In this step, the bats attempt to find a better location by slightly explore near the current best bat. In the original bat algorithm, the bats slightly move steps based on the average loudness $(A^t)$ of all the bats at iteration $t$ to search a better location near the current best bat. However, the average loudness decreases as the iterations. The smaller the average loudness is, the narrower the search region is. The search region of bats is centralized near the current best bat. This centralized search easily makes bats falling into the local optimal solution. In order to overcome this weakness, a new approach is proposed to improve the weakness of limited search region. The bats search a better location near the current best bat according to their frequency. The movement steps are not seriously affected by iteration in proposed approach. In this manner, the search region is elastic to enhance the possibility with position updating.

$$l_i^t = x_* + \varepsilon f_i^t, \quad \varepsilon = [-1,1] \tag{16}$$

where $x_*$ is the current best bat. $\varepsilon$ is the random number with uniform distribution. $f_i^t$ is the frequency of bat $i$ in iteration $t$. $l_i^t$ is the found new location by slight movement. If the new location is better than their own position $(x_i^{t-1})$, the bat will update their own position $(x_i^t)$ with the new found location $(l_i^t)$ by Eq. (14). Then, the original position $(x_i^{t-1})$ will be the previous position $(\tilde{x}_i^t)$ by Eq. (15).

Step4. Update the current best bat
All positions found by bats will be ranked according to the quality evaluated by a fitness function. And the current best bat with the best quality is selected after comparing with all positions of bats. If the quality of this selected best bat is better than the current best bat, it will be the current best bat as shown in Eq. (17). This current best

bat will lead other bats toward the search direction of global optimal solution. The derived global optimal solution is the position of the current best bat when the evolution is finished.

$$x_* = x_i^t, \ if \ f(x_i^t) < f(x_*) \tag{17}$$

Step5. Divers search

In order to strengthen the global searching ability of bats, the conflict conception is applied in a divers search. The velocity of bats will be randomly modified to search new location by Eq. (18). There are three exploration strategies including excavate near position, excavate near location and disorder, to be adopted according to a random value $(R)$ for each dimension as shown in Eq. (19). If $R=1$, the bat explores new location based on its current position. If $R=2$, the bat explores new location near its current location. The bats attempt to carefully discover the better position near their own position and location in this two exploration strategies. Otherwise, when $R=3$, the bat randomly selects a location in solution space to exploit new search region.

$$v_{i,j}^t = \sigma v_{i,j}^t, \ \sigma \in [-2,2] \tag{18}$$

$$\tilde{l}_{i,j}^t = f(x) = \begin{cases} x_{i,j}^{t-1} + v_{i,j}^t, & if \ R = 1 \\ l_{i,j}^{t-1} + v_{i,j}^t, & if \ R = 2 \\ random(), & if \ R = 3 \end{cases} \tag{19}$$

where $j$ is the dimension number of bat $i$. $v_{i,j}^t$ is randomly modified by a random value, $\sigma$, from a uniform distribution. $\tilde{l}_i^t$ is the new location of bat by the designed three strategies. Therefore, a new produced location consists of different elements from position, location and random for the bat in order to deviate the search direction of bats. The divers search benefits the bats to exploit new search region to avoid falling into the local optimal solution. Then, the cosine similarity is utilized to analyze the similarity of the new location and original location of a bat as Eq. (20).

$$Cos\_sim\left(l_i^t, \tilde{l}_i^t\right) = \frac{\sum_{j=1}^d l_j^t \times \tilde{l}_j^t}{\sqrt{\sum_{j=1}^d l_j^{t^2}} \times \sqrt{\sum_{j=1}^d \tilde{l}_j^{t^2}}} \tag{20}$$

$Cos\_sim$ is the similarity of new location and original location. If the similarity of bat is greater than the similarity threshold $(T_{sim})$, the bat still search the location around their own location after the search direction of bat suffers violent disturbance by the designed strategies. This bat will discover a new location through tracking for the current best bat to fast approach to the current best bat as shown in Eq. (21).

$$l_i^t = \begin{cases} \tilde{l}_i^t, & If \ c_{sim} \leq T_{sim} \\ x_* c_{sim} + x_i^t(1 - c_{sim}), & If \ c_{sim} > T_{sim} \end{cases} \tag{21}$$

$T_{sim}$ is the similarity threshold. If the new location is better than their own position $(x_i^{t-1})$, the bat will update their position $(x_i^t)$ with the new found location $(l_i^t)$ by Eq. (14). Then, the original position $(x_i^{t-1})$ will be the previous position $(\tilde{x}_i^t)$ according to Eq. (15).

Step 6. Update bat behavior

If the quality of a new location is better than the current best bat in divers search, this bat will update its position according to Eq. (14) and the current best bat will be replaced by Eq. (15). In addition, this bat will update its properties to afresh adjust status, including loudness $(A_i)$ and emission pulse rate $(r_i)$ according to Eq. (6).

This new current best bat with better status will lead all bats toward new search direction to explore new region. The loudness and emission pulse rate of bats will be updated when (1) the current best bat is updated in divers search and (2) the random number ($r_A$) between 0 and 1 is smaller than the loudness ($A_i$) of bat $i$.

## 4     Simulation Results

Two benchmark functions are used to validate the performance of the proposed algorithm [7]. These selected benchmark functions are described in details as follows.

● Rastrigin Function
This function is a variation of De Jong function with cosine modulation in order to produce frequent local minima. The global minimum $f_1(x) = 0$ is obtainable when $x_i = 0$ $for$ $i=1,...,n$. Hence, this function is a multimodal. However, the locations of minima are regularly distributed and an $n$-dimension Rastrigin function is formulated as follows,

$$f_1(x) = 10n + \sum_{i=1}^{n}[x_i^2 - 10\cos(2\pi x_i)] \ where \ -5.12 \leq x_i \leq 5.12, for \ i = 1,...,n$$

(22)

● Griewangk Function
    This function is similar to the Rastrigin function. There are many widespread local minima distributed regularly. The global minimum $f_2(x) = 0$ is obtainable when $x_i = 0$, $i=1,...,n$. There are 191 local minimal as $d=1$ in this function. It is defined as follows,

$$f_2(x) = \frac{1}{4000}\sum_{i=1}^{n}x_i^2 - \prod_{i=1}^{n}\cos\left(\frac{x_i}{\sqrt{i}}\right) + 1 \ where \ -600 \leq x_i \leq 600, for \ i = 1,...,n.$$

(23)

In order to provide an impartial comparison criterion, the patterns are randomly generated. A round number is repeated for the same scenarios to derive the reliable statistics results. Hence, each pattern is evaluated in a round number. A round progresses an iteration limitation to obtain a convergent solution and convergent iteration. These statistics include average error ($Avg$), standard deviation ($Std$) maximum ($Max$) and minimum ($Min$) to appraise the solving efficiency of the proposed algorithm. The average error presents the quality of convergent solution and required iteration of the proposed algorithm in convergence. The higher the $ACU$ is, the smaller the average error is. In other words, the solution with higher $ACU$ has better quality and more approaches to the global optimal solution.

In addition, the standard deviation, maximum and minimum of error are applied to validate the algorithm reliability. The maximum and minimum are the best result and the worst result in a convergent solution and convergent iteration respectively. In this simulation, the main purpose is to validate the accuracy ($ACU$), success rate ($SR$) and problem-solving speed ($PSS$) of solution derived by the proposed algorithm in an iteration limitation. A tolerable threshold ($T_{tor}$) is utilized to determine the acceptance of a solution. If the average error of solution is less than the tolerable threshold, this solution will be as a global optimal solution. Furthermore, if the global optimal solution is found in an iteration limitation, the evolution is successful in this round. The SR is defined as following:

$$SR = \frac{no.\,of\ success}{no.\,of\ round} \tag{24}$$

Hence, the trial patterns with randomly deployed initial position of bats are produced for each scenario. The settings of parameters, such as population size, frequency range, emission pulse rate range and loudness range are set as in [1]. The properties of bats are randomly generated including previous position, velocity, location, frequency, emission pulse rate and loudness in the proposed algorithm. The other parameters, including the numbers of iteration limitation and the number of rounds, are set to 1000 and 100. A round executes 1000 iterations to derive the convergent solution and convergent iteration. The statistics of convergent solution and convergent iteration will be derived from the simulation results of 100 rounds in each scenario. The statistics of convergent solution with the single pattern for all scenarios is shown in Table 1. In Table 1, the *Avg* is the average error of convergent solutions. The *Max* and *Min* are the errors of the worst and the best convergent solution. The *Std* is variance of convergent solutions found in 100 rounds. The smaller the *Std* is, the higher the reliability is. Besides, for another issue, convergent iteration, the statistics of convergent iteration with single pattern for all scenarios are shown in Table 2. The *Avg* is the average number of convergent iterations in rounds. The Max and Min are the maximum convergent iterations and the minimum convergent iterations. These obtained statistics for each pattern are adopted to estimate the indicators proposed in this study to present the solving efficiency of GBA. The solving efficiency of GBA will be expressed by these indicators as follows:

➢ Accuracy (*ACU*)

This indicator mainly appraises the quality of solution. The higher the *ACU* is, the smaller the average errors of converged solution is and the better the performance of algorithm is. In all scenarios, the error of best convergent solution is 0. For the cases of Griewangk function and Rastrigin function as shown in Table 1, the convergent solutions are almost with no error. The errors of these cases slightly rise in the case of high-dimension functions.

➢ Success Rate (*SR*)

The tolerable threshold ($T_{tor}$) is set to 1.00E-6 in this study, rather than 1.00E-05. If the average error of convergent solution in an evolution is less than the tolerable threshold, the bats successfully finds the global optimal solution in a round. The higher the SR is, the better the ability of discovering global optimal solution for bats are. In Table 1, the proposed algorithm well works in the 100% of scenarios. The SR is 100% to indicate that the bats find the global optimal solution in all rounds for a pattern. However, the SR is 91% for the 128-dimension Griewangk function, because the errors of the worst convergent solution are 3.36E-05.

➢ Problem-Solving Speed (*PSS*)

This indicator is used to evaluate the algorithm whether the bats fall into the local optimal solution or not by convergent iteration. If the convergent iteration is less than the iteration limitation and the error of the convergent solution is less than the tolerable threshold, the bats have a better and faster ability of search. On the contrary, when the convergent iteration is equal to the iteration limitation and the error of convergent solution is greater the tolerable threshold, the bats have fallen into the local optimal solution as well as the iteration limitation is insufficient so that the bats are unable to escape the local optimal solution. The bats are able to fast discover the global optimal solution in an iteration limitation in simulation except for the cases in d=128 for Griewangk function as shown in Table 2.

**Table 1.** The error statistics and success rate of the proposed GBA with a single pattern in 2 benchmark functions with various dimensions

| | Griewangk | | | | | |
|---|---|---|---|---|---|---|
| D | 2 | 10 | 20 | 30 | 64 | 128 |
| Avg | 0.00E+00 | 0.00E+00 | 0.00E+00 | 0.00E+00 | 2.90E-16 | 1.31E-06 |
| Std | 0.00E+00 | 0.00E+00 | 0.00E+00 | 0.00E+00 | 0.00E+00 | 4.37E-06 |
| Max | 0.00E+00 | 0.00E+00 | 0.00E+00 | 0.00E+00 | 1.50E-14 | 3.36E-05 |
| Min | 0.00E+00 | 0.00E+00 | 0.00E+00 | 0.00E+00 | 0.00E+00 | 3.23E-18 |
| SR | 100% | 100% | 100% | 100% | 100% | 91% |
| | Rastrigin | | | | | |
| D | 2 | 8 | 10 | 16 | 20 | 30 |
| Avg | 0.00E+00 | 0.00E+00 | 0.00E+00 | 0.00E+00 | 0.00E+00 | 7.96E-15 |
| Std | 0.00E+00 | 0.00E+00 | 0.00E+00 | 0.00E+00 | 0.00E+00 | 0.00E+00 |
| Max | 0.00E+00 | 0.00E+00 | 0.00E+00 | 0.00E+00 | 0.00E+00 | 1.71E-13 |
| Min | 0.00E+00 | 0.00E+00 | 0.00E+00 | 0.00E+00 | 0.00E+00 | 0.00E+00 |
| SR | 100% | 100% | 100% | 100% | 100% | 100% |

**Table 2.** The statistics with a single pattern of the convergent iteration in 2 benchmark functions with various dimensions for GBA

| | Griewangk | | | | | |
|---|---|---|---|---|---|---|
| D | 2 | 10 | 20 | 30 | 64 | 128 |
| Avg | 21.93 | 37.63 | 73.06 | 112.06 | 355.93 | 888.88 |
| Max | 53 | 118 | 174 | 354 | 762 | 1000 |
| Min | 4 | 42 | 12 | 13 | 15 | 335 |
| | Rastrigin | | | | | |
| D | 2 | 8 | 10 | 16 | 20 | 30 |
| Avg | 5.35 | 16.53 | 18.93 | 25.65 | 29.32 | 37.17 |
| Max | 24 | 58 | 47 | 70 | 88 | 111 |
| Min | 2 | 42 | 8 | 10 | 12 | 17 |

# 5    Conclusions

In this study, a guidable bat algorithm based on Doppler Effect is proposed to meliorate the solving efficiency of the original bat algorithm. In order to strengthen the ability of discovering global optimal solution, there are three search polices and three exploration strategies in GBA. The bats reigned by GBA are enabled with the ability of guidance by frequency shift based on Doppler Effect so that the current best bat is able to lead other bats toward the correct direction in guidable search. Furthermore, both refined search and divers search are employed to reinforce the ability of local search and global search. The bats are able to discover the eligible position to upgrade the position with the current best bat in a short time. Therefore, the bats are able to rapidly and precisely to discover the global optimal solution to augment the solving efficiency of the proposed GBA. On the other hand, this study is the first to consider conflict behaviors of individuals as a kind of collective-effect in swarms and to realize EC 2.0 close to real world. Additionally, how to adjust the individual behavior with cooperation, competition and conflict to maintain the population advancement is valuable and difficult topic for the technology of evolutionary computing in future.

**Acknowledgement.** The authors would like to express their sincere thanks to the National Science Council, Taiwan (ROC), for financial support under the grants NSC 102-2221-E-151 -039 - , NSC 102-2221-E-151-004- and NSC 102-2218-E-151 -005 -.

# References

1. Yang, X.-S.: A New Metaheuristic Bat-Inspired Algorithm. In: González, J.R., Pelta, D.A., Cruz, C., Terrazas, G., Krasnogor, N. (eds.) NICSO 2010. SCI, vol. 284, pp. 65–74. Springer, Heidelberg (2010)
2. Yang, X.S.: Bat Algorithm for multi-objective optimization. International Journal of Bio-Inspired Computation 3(5), 267–274 (2011)
3. Wang, G., Guo, L.H., Duan, H., Liu, L., Wang, H.: A Bat Algirthm with Mutation for UCAV Path Planning. The Scientific World Journal 2012, 1–15 (2012)
4. Wang, G., Guo, L.H.: A Novel Hybrid Bat Algorithm with Harmony Search for Global Numberical Optimization. Journal of Applied Mathematics 2013, 1–21 (2013)
5. Chen, Y.T., Lee, T.F., Horng, M.F., Pan, J.S.: An Echo-Aided Bat Algorithm to Support Measurable Movement for Optimization Efficiency. In: Proceeding of IEEE International Conference on Systems, Man, and Cybernetics (SMC 2013), pp. 806–811 (2013)
6. Horng, M.F., Chen, Y.T., Wu, P.L., Liao, B.Y., Pan, J.S., Lee, T.F.: A Guidable Bat Algorithm based on Doppler Effect to Meliorate Solving Efficiency for Optimization Problems. Submitted to Journal of Applied Soft Computing (2014)
7. Molga, M., Smutnicki. C.: Test functions for optimization needs. (2005),
   http://www.zsd.ict.pwr.wroc.pl/files/docs/functions.pdf

# Collective Detection of Potentially Harmful Requests Directed at Web Sites

Marek Zachara

AGH University of Science and Technology,
30 Mickiewicza Av., 30-059, Krakow, Poland
mzachara@agh.edu.pl

**Abstract.** The number of web-based activities and websites is growing every day. Unfortunately, so is cyber-crime. Every day, new vulnerabilities are reported and the number of automated attacks is constantly rising. Typical signature-based methods rely on expert knowledge and the distribution of updated information to the clients (e.g. anti-virus software) and require more effort to keep the systems up to date. At the same time, they do not protect against the newest (e.g. zero-day) threats. In this article, a new method is proposed, whereas cooperating systems analyze incoming requests, identify potential threats and present them to other peers. Each host can then utilize the findings of the other peers to identify harmful requests, making the whole system of cooperating servers "remember" and share information about the threats.

## 1 Introduction

There is little doubt that an increasing part of peoples' activities has moved over to the Internet. From online shopping and digital banking to personal blogs and social networking sites the number of websites is constantly growing, with the recent number of 800 million websites (of which almost 200 million were considered active) reported by the end of 2013 [14],

While the underlying operating systems, web servers, firewalls, and databases are usually well known and tested products that are subject to continuous scrutiny by thousands of users, the applications themselves are often much less tested (if at all). However, even if all the underlying software is secure, a successful attack against the web application may compromise the data in its database. At the same time, the attacks directed at the web application are performed via normal http requests, which pass undetected by typical security means (e.g. a firewall). More about applications vurnelabilities can be found in [9], and the most common ones are periodically published be CWE [5].

## 2 The Unprotected Web Applications

The reason why firewalls do not protect websites from harmful requests is that their ability is only to filter traffic at the lower layers of the OSI model, while the identification of harmful requests is only possible at layer 7. There are specialized

D. Hwang et al. (Eds.): ICCCI 2014, LNAI 8733, pp. 384–393, 2014.

firewalls, known as Web Application Firewalls (WAF), but their use is limited mainly because of the effort required to configure and maintain them. An answer to this issue could be a learning WAF proposed in [15], but it is still in the early stages. The mentioned paper also describes the reasons for the low adoption of the application firewalls in more details.

As a result, each web application has to deal with security issues on its own. This means that each web developer should be familiar with security issues and be able to address them in their application. Experience and data shows however, that this is not the case.

## 2.1  The State of Websites Security

According to the report [8], 99% of the web applications they had tested had at least one vulnerability (35 on average), with 82% of the websites having at least one high/critical vulnerability. This report was based on their customer base of over 300 companies. In another report [18], WhiteHat Security states that 86% of the web applications they had tested had at least one serious vulnerability, with an average of 56 vulnerabilities per web application.

On the other hand, Symantec claims that while running 1400 vulnerability scans per day, they have found approximately 53% of the scanned websites to have at least one unpatched vulnerability [17]. Also, in a single month of May of 2012 the LizaMoon toolkit was responsible for at least a million successful SQL Injections attacks. In the same report, they also state that approximately 63% of websites used to distribute malware were actually legitimate websites compromised by attackers.

Since these companies do not publish details on their methodology it's hard to precisely assess the number of vulnerable websites, but an exact number is not that important. Whether it is 50% or 90% of all, this means that there are hundreds of millions of websites with unpatched vulnerabilities on the Internet.

# 3  Proposed Method for Prevention of Attacks

It does not seem feasible to propose a single, monolithic security system to defend the websites against various attacks. Since the attack vectors often change and evolve, the security systems need to do the same. Actually, an accepted approach to building secure systems is named "defense in depth" [1] and requires each part (layer) of the system to be secured with the best tools and knowledge. In this paper, a method to deal with specific (but popular) attack vectors is proposed, which employs methods similar to those used by the human immune system.

First, let's look at example patterns presented in Figure 1. This figure is made with actual data from a running web server. The requests arriving at the server are grouped into sessions (consecutive requests from the same source), and then presented as a graph, starting at the "Start" node in the center. For reasons of clarity, only a few sessions are presented. The sessions in the top left corner are "legitimate" sessions, namely requests from regular users who visit the web page and do not attempt any harmful actions. The other three sessions, can be noted as actual attack attempts. As can be seen in this figure, these requests try to access a popular database management module [2] via various URLs at which it may be available if installed.

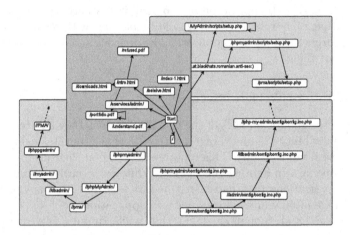

**Fig. 1.** Sample users' sessions presented as a part of a graph, with attack attempts marked by red background

It is hard to identify whether there was a real user behind these presumed attack attempts, or if they were carried out by an automated script or a malware. The latter is actually more likely, since the majority of attack attempts are automated, either directly coming from malware or a large number of "script kiddies" [11] who download and use ready-made attack tool-kits. These two groups have the ability to perform attacks attempts on a large scale. However, these attacks are usually identical, or very similar to each other. Since they are performed by programs and usually rely on one or just a few attack vectors, they generate similar requests to a large number of web servers.

### 3.1 Key Concepts of the Proposed Methodology

For clarity reasons, let's reiterate the features of the attacks performed by malware and "script kiddies":

- Distributed. The attacks are usually targeted towards a large number of web hosts, not at the same time, but within a limited time-frame.
- Similar. Malware and "script kiddies" both use single pieces of software with incorporated attack vectors. This software produces identical or very similar requests to various hosts.
- Unusual. The requests generated by the automated software are not tailored for the specific host, so they vary from its usual usage patterns.

Fortunately, these features can be utilized to detect and neutralize the majority of the threats coming from these sources. In order to do so, a method similar to one of the immunology processes referred to above is used.

First, potential unusual requests are identified whether they seem to be "suspicious" or not. This is done based on the server response and a behavioral evaluation of the request sequence. Requests identified as suspicious are stored locally for reference and also "presented" to other hosts, somewhat mimicking the way antigens are presented by various cells to the T-cells of the immune system. Periodically, each host retrieves the lists of such suspicious requests identified by its peers. The information received from other hosts, together with the local evaluation is then used to assess the request and take an appropriate action.

## 3.2   Identification of Suspicious Requests

The simplest way to identify suspicious requests is to look for HTTP 404 responses from the web server, which are results of requests for nonexistent resources or URLs. For better results, a heuristic behavior-based method is also proposed. Anomaly-based detection is a common approach for security purposes [12], [13], [4]. The proposed method is based on a dynamically build usage graph, which contains information about typical usage of the website.

For the purpose of this task, it is proposed to use an oriented, weighted graph where nodes represent requested URLs and the edges denote transitions between URLs by the users. The weight of each edge is equivalent to the frequency of the particular transition (number of transitions during the recent time window). Obviously, users' session tracking (e.g. using cookies) is necessary for reliable results. In Figure 2 a sample graph is presented.

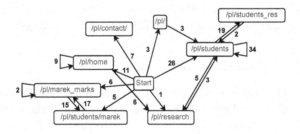

**Fig. 2.** Sample graph representing a time-framed usage of a website

Furthermore, Figure 3, shows the typical situation where there are a number of edges with high weights and a number of edges with very low (even equal to 1) weights. These low-weighted edges usually point to nodes (i.e. URL requests) that can accurately be considered as "suspicious".

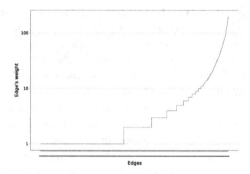

**Fig. 3.** Distribution of edge weights in a usage graph created for a real website

### 3.3   Collective Assessment of the Requests

The result of the process described in the previous section, means that a server hosting a website can label certain incoming requests as potentially suspicious. However, the results of experiments show that such a list contains a large portion of false positives (i.e. legitimate requests that are incorrectly labeled as potentially dangerous. This list must be narrowed down, which can be done by utilizing the distributed knowledge from several websites. If a particular request is identified as suspicious locally and has also been received at other servers (and labeled by them as such), the likeliness of it being a real attack attempt is very high.

To facilitate the exchange of information between web servers, various means can be utilized. It seems that the simplest way would be for the server to maintain a dedicated web page accessible from outside at a certain URL that contains a list of its findings (requests locally identified as "suspicious"). This page obviously needs to be regularly updated. Other servers can then periodically retrieve these lists from their peers and use them to verify their local findings.

### 3.4   Protection of Sensitive Information

Certainly, publishing information about received requests could be a security issue in its own right (including a potential security and/or confidentiality breach). Fortunately, it is not necessary to disclose full requests' URLs to other peers. Since each host just needs to check if the URL it marked as suspicious has also been reported by other hosts, it is enough if the hosts compare the hashes (digests) of the requests' URLs. Hashes (like SHA-256 and MD6) are generated using one-way functions and are generally deemed irreversible. If a host only publishes hashes of the requests that it deems suspicious, it allows other hosts to verify their findings, but at the same time protects its potentially confidential information.

For the proposed method to work, it is enough if hosts just publish a list of hashes of the received requests that they consider suspicious, but other data (e.g. type of anomaly, the time it was detected, etc.) may improve the interoperability and possible future heuristics. Therefore it is suggested that hosts use a structured document format, e.g. based on [10]. A sample published document may look like the one in Listing 1.1. This will be referred to as a *Server List of Suspicious Requests*, abbreviated *SLSR*. The list presented in 1.1 is a part of a list taken from an actual running instance of the reference implementation.

**Listing 1.1.** Sample SLSR with additional debug information

```
{ C:0, T:M, A:57, MD5:2cf1d3c7fe2eadb66fb2ba6ad5864326 }
{ C:0, T:M, A:53, MD5:2370f28edae0afcd8d3b8ce1d671a8ac }
{ C:F, T:M, A:32, MD5:2f42d9e09e724f40cdf28094d7beae0a }
{ C:F, T:M, A:31, MD5:8f86175acde590bf811541173125de71 }
{ C:F, T:M, A:24, MD5:eee5cd6e33d7d3deaf52cadeb590e642 }
{ C:0, T:B, A:17, MD5:bd9cdbfedca98427c80a41766f5a3783 }
```

### 3.5   Maintenance of the Lists

For the process to work as intended, each server must not only identify suspicious requests, but also generate and publish the list of them and retrieve similar lists from other servers. However, exchanging of the SLRS leads to an issue of data retention, and two questions need to be answered:

– How long should an SLSR contain an entry about a suspicious request after such a request was received.
– Should the hashes received from other peers be preserved locally if the originating server does not list them any more, and if so, for how long?

Both issues are related to the load (i.e. number of requests per second) received either by the local or the remote server. Servers with very high loads and a high number of suspicious requests will likely prefer shorter retention times, while niche servers may only have a few suspicious requests per week and would prefer a longer retention period.

It is difficult to currently propose a justified approach to these issues, until the proposed method is more widely accepted, which would provide necessary statistical data. Therefore, as a "rule of thumb" it may be assumed that a record (i.e. a suspicious request) should not be listed for more than a couple of weeks and it may be removed earlier if the SLSR grows too long (e.g. more than a couple of hundred records). Experience shows that some attack attempts last for months or years (e.g. the attempts at phpMyAdmin) but some attacks, particularly if they are malware-based, tend to die off quickly as the vulnerability they exploit is patched.

## 4   Implementation and Deployment

A reference implementation has been prepared to verify the feasibility of the proposed method and to prove this concept. The application has been programmed in Java and executed on a few real web servers. The architecture of the application is presented in Figure 4.

**Fig. 4.** The architecture of the reference implementation

The application monitors the requested access log files of Apache Web Server and retrieves incoming requests. This is one of a well-established security monitoring methods [7], [16], [3]. The requests for unavailable resources are directly forwarded to the aggregation of suspicious requests. Other requests are first analyzed by session extractor, which groups them into user sessions. After filtering out requests for the web page elements (e.g. images), the primary requests for web pages are used to build the behavior graph explained earlier in this article. Requests that do not seem to match typical behavior (i.e. they do not match the built graph) are considered suspicious and forwarded to the "suspicious activity detector" module.

This module aggregates all suspicious requests and matches them against pattern-based rules, primarily to eliminate requests that are well known and harmless, yet for some reason are still reported to this module. Such requests are mainly automated requests from browsers for the website icon *'favicon.ico'* or requests from automated crawlers *"robots.txt"* which are often missing in websites. Finally, the remaining suspicious requests are formatted according to the sample presented in Listing 1.1, and are stored in a document inside the web server directory to be accessible to other hosts.

## 4.1   Reasoning

The last module (named 'Reasoning') receives the current request, which is the list of requests locally considered as suspicious, and, at the same time periodically retrieves similar lists from other hosts. With each request, the module has to decide whether it should report it for human (e.g. administrator's) attention or not. There are various strategies that can be implemented here, depending on the type of application being protected and the number and type of peer servers it receives the data from. Since the reference test environment consisted of only three hosts, a simple algorithm was used. This algorithm reported requests that were both locally considered an anomaly and listed by at least one other peer server. Even this simple strategy provided considerable results, as will be demonstrated later in the paper.

# 5   Test Method and Achieved Results

The prototype application was deployed on three web servers. These were:

- A small commercial B2B application used to coordinate work between one company and its customers.
- A website used by a group of people to inform others about local events.
- A personal (author's) website.

All the web servers were located in the same city. The relatively large number of requests served by the third (personal website) server was due to the fact that it contained teaching materials and was therefore used by the author's students. The number of all received requests is relatively high because it includes all requests for all the resources (not only web pages, but all images, scripts, etc.)

The results received from the application were compared against a semi-manual analysis of log files (negative selection against a set of patterns representing legitimate requests). These results are presented in Table 1.

**Table 1.** Server A (small B2B dynamic application): requests statistics

|  | Server A | Server B | Server C |
| --- | --- | --- | --- |
|  | small B2B | community site | author's site |
| All received requests | 33 706 | 143 211 | 193 241 |
| Reported attempts | 306 | 278 | 377 |
| of which most common |  |  |  |
| /w00tw00t.* | 75 | 75 | 45 |
| /wp-login.php | 48 | 31 | ? |
| phpMyAdmin variations | 47 | 37 | 29 |
| Manually identified attempts | 979 | 612 | 1 134 |
| of which most common |  |  |  |
| /w00tw00t.* | 124 | 115 | 118 |
| /wp-login.php | 50 | 34 | ? |
| phpMyAdmin variations | 359 | 96 | 313 |

In terms of the efficiency of the described method, the reference single-thread implementation (in Java) was able to process over 4,000 requests per second on a typical PC (Intel, 3GHz). This was achieved without any specific optimizations and was a couple orders of magnitude higher than required - considering the real load that the servers experienced.

# 6   Conclusions and Future Work

The method proposed in this paper aims at providing automated detection of potentially harmful requests with the minimum level of involvement from the system administrators. Inspired by some mechanisms that evolved within the human immune system, it works by presenting potentially harmful requests to other peers for verification. A final decision is then reached by considering the votes of other peers along with local judgment. The method may significantly hinder the *modus operandi* of most malware, due to the fact that after initial the attempts to attack a number of servers, it will become increasingly difficult for a malware to successfully attack new servers. These servers will recognize the attacks because of the knowledge acquired from its peers. This way, the whole system will develop an immune response similar to the one observed in living organisms.

The reference implementation shows the benefits for deploying the proposed method, even in a very small test scenario of only three server nodes. The limited detection rate of attacks directed at *phpMyAdmin* results from various permutations of the URL, including package version numbers. This is the area that certainly requires further attention, because introducing some heuristics generalizing the request URLs should greatly boost the detection rate. Also, deploying these application on a larger number of servers will likely greatly increase its effectiveness.

It is important to note, that the proposed method provides unique features, that make it difficult to compare to other anomaly-based methods published so far. While most of these methods require some kind of training with labeled data sets (e.g. [6], [12]) to distinguish the attacks, the method described in this paper does not have such prerequisite. Also, compared to the mentioned methods, which tend to report a number of false positives, this method provides virtually no such false alarms - a feature of great value to potential end users.

Further work that could improve the detection ratio should mainly be focused on the evaluation of decision criteria that result in reporting of a particular request (i.e. the "reasoning" module). These should include: The impact of the number of peers on reasoning accuracy, the retention period for both local and remote SLSRs and the form of the decision algorithm (e.g. input weights, rules, etc.). The generalization of request URLs has already been mentioned.

# References

1. Defense in Depth: A practical strategy for achieving Information Assurance in today's highly networked environments,
   http://www.nsa.gov/ia/_files/support/defenseindepth.pdf,
   http://www.nsa.gov/ia/_files/support/defenseindepth.pdf
2. PhpMyAdmin, http://sourceforge.net/projects/phpmyadmin/
3. Agosti, M., Crivellari, F., Di Nunzio, G.: Web log analysis: a review of a decade of studies about information acquisition, inspection and interpretation of user interaction. Data Mining and Knowledge Discovery 24(3), 663–696 (2012),
   http://dx.doi.org/10.1007/s10618-011-0228-8,
   http://www.bibsonomy.org/bibtex/29b85b7d3c5587c5f0920f0d602ba93b1/sdo
4. Auxilia, M., Tamilselvan, D.: Anomaly detection using negative security model in web application. In: 2010 International Conference on Computer Information Systems and Industrial Management Applications (CISIM), pp. 481–486 (2010)

5. Christey, S.: 2011 CWE/SANS Top 25 Most Dangerous Software Errors. Tech. rep. (2011), `http://cwe.mitre.org/top25/archive/2011/2011_cwe_sans_top25.pdf`

6. Cova, M., Balzarotti, D., Felmetsger, V., Vigna, G.: Swaddler: An Approach for the Anomaly-Based Detection of State Violations in Web Applications. In: Kruegel, C., Lippmann, R., Clark, A. (eds.) RAID 2007. LNCS, vol. 4637, pp. 63–86. Springer, Heidelberg (2007), `http://dx.doi.org/10.1007/978-3-540-74320-0_4`

7. Iváncsy, R., Vajk, I.: Frequent Pattern Mining in Web Log Data. Acta Polytechnica Hungarica 3(1) (2006),
`http://citeseerx.ist.psu.edu/viewdoc/summary?doi=10.1.1.101.4559`;
`http://www.bibsonomy.org/bibtex/2f29f4627c9ae99370fc7ba005982e2e6/sdo`

8. iVIZ: Web Application Vulnerability Statistics Report (2013),
`http://www.securitybistro.com/?p=4966`

9. Johari, R., Sharma, P.: A Survey on Web Application Vulnerabilities (SQLIA, XSS) Exploitation and Security Engine for SQL Injection. In: 2012 International Conference on Communication Systems and Network Technologies (CSNT), pp. 453–458 (2012)

10. JSON: A lightweight data-interchange format, `http://www.json.org`

11. Kayne, R.: What Are Script Kiddies,
`http://www.wisegeek.com/what-are-script-kiddies.htm`

12. Kruegel, C., Vigna, G.: Anomaly Detection of Web-Based Attacks. pp. 251–261. ACM Press (2003),
`http://www.bibsonomy.org/bibtex/`
`2099e1b9a6e57960e4b3e02410e83cb64/liangzk`

13. Kruegel, C., Vigna, G., Robertson, W.: A multi-model approach to the detection of web-based attacks. Computer Networks 48(5), 717–738 (2005),
`http://dx.doi.org/10.1016/j.comnet.2005.01.009`

14. Netcraft: Web Server Survey (2013),
`http://news.netcraft.com/archives/`
`2013/11/01/november-2013-web-server-survey.html`

15. Pałka, D., Zachara, M.: Learning web application firewall - benefits and caveats. In: Tjoa, A.M., Quirchmayr, G., You, I., Xu, L. (eds.) ARES 2011. LNCS, vol. 6908, pp. 295–308. Springer, Heidelberg (2011),
`http://dl.acm.org/citation.cfm?id=2033973.2033999`

16. Sun, Z., Sheng, H., Wei, M., Yang, J., Zhang, H., Wang, L.: Application of web log mining in local area network security. In: EMEIT. pp. 3897–3900. IEEE (2011),
`http://dblp.uni-trier.de/db/conf/emeit/emeit2011.html#SunSWYZW11`;
`http://dx.doi.org/10.1109/EMEIT.2011.6023097`;
`http://www.bibsonomy.org/bibtex/23badf5326d9486b9b17c48ab47576eaa/dblp`

17. Symantec: Internet Security Threat Report (2013),
`http://www.symantec.com/security_response/publications/threatreport.jsp`

18. WhiteHat: Website Security Statistics Report (2013),
`http://info.whitehatsec.com/2013-website-security-report.html`

# Increasing the Efficiency of Ontology Alignment by Tracking Changes in Ontology Evolution

Marcin Pietranik[1], Ngoc Thanh Nguyen[1], and Cezary Orłowski[2]

[1] Institute of Informatics, Wroclaw University of Technology,
Wybrzeze Wyspianskiego 27, 50-370, Wroclaw, Poland
{marcin.pietranik,ngoc-thanh.nguyen}@pwr.wroc.pl
[2] Faculty of Management and Economics, Gdansk University of Technology,
Narutowicza 11/12 80-233 Gdansk
cezary.orlowski@zie.pg.gda.pl

**Abstract.** In this paper we present a development of our ontology alignment framework based on varying semantics of attributes. Emphasising the analysis of explicitly given descriptions of how attributes change meanings they entail while being included within different concepts have been proved useful. Moreover, we claim that it is consistent with the intuitive way how people see the real world and how they find similarities and correspondences between its elements. In this paper we concentrate on the issue of tracking changes that may occur within aligned ontologies and how these potential changes can influence the process of finding new mappings or validating ones that have already been found.

## 1 Introduction

In recent years ontologies have became a frequently adapted method of expressing knowledge in computer systems due to the fact that they assert both formal structuring and convenient flexibility. They have been proved to be useful in variety of applications spreading from weather forecasting to knowledge sharing and reusing it in information systems ([7]).

Ontologies by themselves can be treated as a method of expressing a decomposition of assumed part of reality that is going to be modelled using an ontology. This decomposition include a description of elementary objects taken from universe of discourse along with their inner characteristics and relations that may occur between them. Moreover, these elementary components (denoted as concepts) can be treated as a method of creating an abstract categorisation of real-world objects, allowing their convenient classification.

Such open-ended approach, despite obvious advantages, implicates a problem of heterogeneity of ontologies. There is no formal method that could assure a consistency between two independently created ontologies on the level of their definitions or their content ([19]). Imagine the situation illustrated on Figure 1.

Lets assume the existence of two independent computer systems utilising dedicated knowledge bases $KB$ and $KB'$, which incorporate ontologies $O$ (referenced

D. Hwang et al. (Eds.): ICCCI 2014, LNAI 8733, pp. 394–403, 2014.

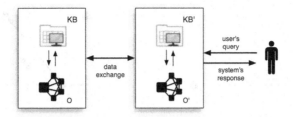

**Fig. 1.** Ontology integration use-case diagram

further as *a source ontology*) and *O'* (*a target ontology*) as a backbone. Further-more, lets assume that an end user of the knowledge base *KB'* sends a query concerning some topic of interests, but the proper answer for such request is not present in *KB'*, but in *KB*. The obvious necessity to select elements from the source ontology that correspond to elements targeted in the query sent to *O* naturally appears.

This issue, called ontology alignment, is a topic of variety of publications ([4], [18]) that describe different approaches to this problem. In our previous articles ([15]) we have created a novel solution of this task, based on analysing varying semantics of attributes assigned to concepts, serving as a foundation for finding mappings between every element that can be found in ontologies. The basic assumption of our methodology was a remark that an attribute acquire different meanings when assigned to different concepts - for example, an attribute *address* express different information when being a part of a concept *Building* and when included in a description of a concept *Webpage*. In [17] we have proved the eventual correctness of our ides utilising commonly accepted *OAEI* ([19]) evaluation datasets.

In this paper we want to concentrate on an aspect of managing changes that can be introduced in ontologies. In modern, web-based, distributed environments it cannot be expected that ontologies will not change over time. It is a common situation even in standard relational databases that their structure evolve and adapt to newly established requirements, especially when the amount of data that needs to be managed constantly increase ([20]). Moreover, assuming that some initial alignment between two ontologies has been designated, it is obvious that somewhere in time this mapping will no longer be valid.

The biggest drawback of any ontology mapping method is the fact that con-sidered task is very complex and time consuming, so is it necessary to repeat such procedure when only small changes have been applied to ontologies? Is it possible to revalidate only its selected components or introduce new matches? It is therefore critical to provide both flexible method of tracking changes applied to ontologies and a method of updating existing alignments. Consider a situation illustrated on Figure 2.

Lets assume that in some initial moment of time (denoted on a diagram as $t_0$) two ontologies $O$ and $O'$ have been aligned and such base alignment has been validated. After some time, in known moments, denoted as $t_1, t_2, t_3, t_4$ and $t_5$,

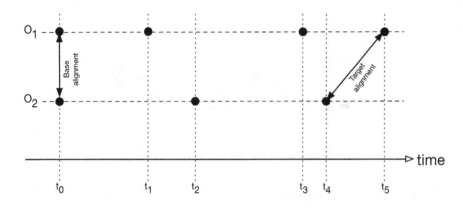

**Fig. 2.** Evolution of ontologies in time

considered ontologies have been updated, which can include modifying sets of their concepts, descriptions of these concepts' structures, relations between them or their instances. The natural question arises - what is the most efficient way of selecting initial mappings that are potentially no longer valid and revalidating them along with introducing new mappings that can be established between ontologies $O$ and $O'$ due to applied changes?

Therefore, the considered problem of updating initial ontology alignment can be described as follows: *For given two ontologies $O$ and $O'$ and a base alignment between them $M^{O,O'}$, one should determine a set of rules of modifying components of $M^{O,O'}$ concerning elements of ontologies O and O' that have been changed over time.*

The reminder of the following paper is as follows. In Section 2 we briefly describe current state of the art of ontology evolution. In part 3 we give basic notions and definitions used throughout the paper. Section 4 contains a proposal of a framework of utilising ontology logs that describe changes in ontologies over time in updating existing mappings between ontologies. A short summary and an overview of our upcoming research is given in Section 5.

## 2   Related Works

Throughout the recent years the topic of ontology alignment has gained great interest. During the most prominent evaluation campaign organised by *Ontology Alignment Evaluation Initiative* a number of systems have submitted their results ([19]). Implemented methods spread from basic comparison of ontologies on lexical level (for example, using variety of string similarity metrics) to more complex and sophisticated frameworks that include utilisation of external knowledge bases, Markov networks or machine learning ([4], [9], [18]).

On the other hand, the topic concerning managing ontology changes is still a current and opened problem ([13]). Increasing complexity and scope of ontologies rises a problem of providing a methodology for consistent collaborative ontology development ([12]). At a basic level it is a similar problem to managing changes in a source code of complex computer systems created by a group of programmers, but due to ontologies' characteristic feature the eventual solution needs to not only compare ontologies on a lexical level (for example, by tracking changes in OWL files), but also to compare and present structural changes, that implicates evolving semantics.

An interesting approach to the considered issues can be found in [5] where change history management framework for evolving ontologies has been introduced. Authors address several subproblems related to ontology management in time such as ontology versioning, tracking change's provenance, consistency assertion, recovery procedures, change representation and visualisation of ontology evolution. Experimental results showing accurate and consistent outcomes have also been provided and broadly described.

What is worth emphasising is the fact that considered issue of managing changes in ontologies over time is a multidisciplinary problem. It includes heterogeneity resolution, validating inner knowledge expressed using managed ontologies, analysing changes in such knowledge and how it evolve over time or providing convenient users' tools that support collaborative ontology creation ([8]).

Nevertheless, to our knowledge there is no research done in the field of analysing not only ontology evolution, but also the way it influence initial mappings that have been designated between evolving ontologies. Several solutions beside concentrating on providing reliable ontology mappings, also address a problem of calculation complexity minimisation, therefore attempting to decrease a time necessary to designate valid alignments between ontologies ([2], [3], [6]), but none of them address the issue of managing an evolution of these alignments.

The opened questions, that still need to be answered, relate to ontology alignment invalidations, detecting changes that need to be applied to initial mappings of ontologies, resolving potential inconsistencies that may occur and, last but not least, avoiding the necessity of relaunching the whole ontology mappings procedure.

## 3   Basic Notions

We define ontology as a triple:

$$O = (C, R, I) \tag{1}$$

where $C$ is a finite set of concepts, $R$ is a finite set of relations between concepts $R = \{r_1, r_2, ..., r_n\}$, $n \in N$, $r_i \subset C \times C$ for $i \in [1, n]$ and $I$ is a finite set of instances. We also assume a set $A$ of all possible attributes and a set $V$ of their valid valuations such that $V = \bigcup_{a \in A} V_a$, where $V_a$. By a pair $(A, V)$ we denote a *real world* that will be expressed using ontologies.

Every concept $c$ from the set $C$ is defined as a triple:

$$c = (Id^c, A^c, V^c) \tag{2}$$

where $Id^c$ is a unique identificator, $A^c$ is a set of its attributes and $V^c$ is a set of domains of these attributes defined as $V^c = \bigcup_{a \in A^c} V_a$, where $V_a$ is a domain of a particular attribute $a$.

An instance $i$ from the set $I$ is defined as:

$$i = (id, A_i, v_i) \tag{3}$$

where by $id$ we denote its identificator, by $A_i$ a set of assigned attributes and by $v_i$ a function $v_i : A_i \rightarrow \bigcup_{a \in A_i} V_a$ that is used to assign specific values from the set $V_a$ to particular attributes from the set $A_i$.

A member $i = (id, v_i, A_i)$ of the set $I$ is an instance of a concept $c = (Id^c, A^c, V^c)$ if $A^c \subseteq A_i$ and $\forall_{a \in A_i \cap A^c} v_i(a) \in V^c$. By $Ins(O, c)$ we denote a set of instances of a concept $c$ within ontology $O$.

In order to express meaning that attributes obtain while being included within concepts we assume a set $D_A$ of their atomic descriptions (e.g. $year\_of\_birth$). By $L_s^A$ we denote a sublanguage of the sentence calculus constructed with members of $D_A$ and elementary logic operators. Semantics of attributes is given by a function:

$$S_A : A \times C \rightarrow L_s^A \tag{4}$$

This approach gives the possibility to formally describe (using logic sentences) varying roles that attributes get when they participate in different structures of different concepts. For example, an attribute *Phone* behaves differently within a concept *Employee* and a concept *Hardware*. Utilising such approach gave us a possibility to define formal criteria for identifying equivalency, generalisation and contradiction between attributes ([15]).

Moreover, we assume a set $D_R$ with atomic descriptions of relations and $L_s^R$ denoting another sublanguage of the sentence calculus. We use it to define semantics of relations from the set $R$:

$$S_{R,O} : R \rightarrow L_s^R \tag{5}$$

Hence, we have provided a set of criteria for relationships between relations including equivalency, generalisation and contradiction ([16]).

For convenience, by $Rel(c_1, c_2)$ we denote a set of directed relations between two concepts $c_1, c_2 \in C$. Formally it can be defined as: $Rel(c_1, c_2) = \{r \in R | (c_1, c_2) \in r\}$.

In order to track, aforementioned in the first section of the paper, changes that can be applied to ontologies in time we introduce the notion of ontology log defined below.

**Definition 1.** *A log of an ontology $O$, denoted as $J_O$, is a set containing tuples of the form $\langle timestamp, e_s, e_t \rangle$ where timestamp represents a moment of time in which particular change took place, $e_s$ is an element from the ontology $O$ before change and $e_t$ is the same element after change. Due to the fact that changes in ontologies may include not only their modifications, but also removing some of* *their components or including new ones, both $e_s$ and $e_t$ can acquire the value $\phi$, which is used to represent the following:*

- $\langle timestamp,\ \phi,\ e_t \rangle$ *denotes the fact that at a certain moment of time a new element $e_t$ has been included into the ontology O.*
- $\langle timestamp,\ e_s,\ \phi \rangle$ *denotes the fact that at a certain moment of time an $e_s$ has been removed from the ontology O.*

Entries in ontology log can express certain changes that can be introduced in ontologies. Below we give a few examples of such modifications and how they could be defined within a log:

- change on a level of concepts' structures:
  $\langle 2014.05.10,\ A^c = \{a, b, c\},\ A^c = \{a, c, e, f\} \rangle$
- adding a new concept:
  $\langle 2014.02.22,\ \phi,\ c = ('Person', \{full_name\}, \{A - Za - z\}) \rangle$
- change on a level of attributes' semantics:
  $\langle 2014.04.12,\ S_A(c, a) = x \vee y,\ S_A(c, a) = x \vee z \rangle$
- change on a level of instance valuations:
  $\langle 2014.04.12,\ v_i(name) =' John',\ v_i(name) =' Joe' \rangle$
- change on a level of concepts' relations
  $\langle 2013.11.23,\ Rel(c_1, c_2 = \{r_1, r_2\},\ Rel(c_1, c_2) = \{r_1, r_3\}) \rangle$

By $TL$ we denote the timeline of changes included in ontologies. We define it as an ordered set $TL = \{t_i \mid i \in N\}$ of moments of time. It contains the element $t_0$ which represents the starting point in which ontology changes began to be tracked in dedicated logs. Obviously, for any ontology $O$ the following condition is met: $\{timestamp \mid \langle timestamp,\ e_s,\ e_t \rangle \in J_O\} \subseteq TL$. Having such structures we can introduce the notion of *a state of an ontology.*

**Definition 2.** *A state of the ontology O, denoted as $O_n$, represents an $(A, V)$-based ontology O after applying changes taken from its log $J_O$ with timestamps earlier than $t_n \in TL$.*

Assuming described base definitions, our alignment framework consists of four functions $\lambda_A, \lambda_C, \lambda_R$ and $\lambda_I$ ([17]):

- $\lambda_A : A^c \times A^{c'} \to [0, 1]$ representing the degree to which an attribute $a \in A^c$ can be aligned to an attribute $a' \in A^{c'}$.
- $\lambda_C : C \times C' \to [0, 1]$ representing the degree to which a concept $c \in C$ can be aligned into a concept $c' \in C'$.
- $\lambda_R : R \times R' \to [0, 1]$ representing the degree to which a relation $r \in R$ can be aligned into a relation $r' \in R'$.
- $\lambda_I : I \times I' \to [0, 1]$ representing the degree to which an instance $i \in I$ can be aligned into an instance $i' \in I'$.

Due to the fact that we build described functions on top of processing of semantics of attributes and we distinguish possible relationships between them (utilising the analysis of logic sentences assigned to them when they are included within different concepts) we have noticed that it is frequently easier to align detailed knowledge into general one, rather than in the other direction. Therefore, all of the above functions that calculate the degree to which it is possible to align selected element from the source ontology into the selected element taken from the target ontology are not symmetrical ([17]).

A structure of an alignment denoted as $M^{O,O'}$ of two $(A, V)$-based ontologies $O$ and $O'$ can be expressed with three sets $M_C^{O,O'}$, $M_R^{O,O'}$ and $M_I^{O,O'}$. It is defined as:

$$M_C^{O,O'} = \{(c, c', \lambda_C(c, c')) \mid c \in C \land c' \in C' \land \lambda_C(c, c') \geq T_C\}$$
$$M_R^{O,O'} = \{(r, r', \lambda_R(r, r')) \mid r \in R \land r' \in R' \land \lambda_R(r, r') \geq T_R\} \qquad (6)$$
$$M_I^{O,O'} = \{(i, i', \lambda_I(i, i')) \mid i \in I \land i' \in I' \land \lambda_I(i, i') \geq T_I\}$$

where $T_C$, $T_R$ and $T_I$ are user-defined thresholds.

Note that $M^{O,O'}$ does not include a set representing matches of single attributes. According to the base definition of an ontology (Equation 1) and concepts' structure (Equation 2) attributes acquire useful meanings only when included in certain concepts. By itself they do not carry any kind of knowledge. Moreover, aligned ontologies are all based on the same real world $(A, V)$, which contains a set of raw attributes possible to by utilised. Hence, there is no necessity or potential gain in mapping them.

In the next section we will present a framework that allows to designate changes that need to be applied in a base alignment $\langle M_C^{O_0,O_0'}, M_R^{O_0,O_0'}, M_I^{O_0,O_0'} \rangle$ between ontologies $O_0$ and $O_0'$ in the moment $t_0$ in order to update it into the valid alignment $\langle M_C^{O_n,O_n'}, M_R^{O_n,O_n'}, M_I^{O_n,O_n'} \rangle$ between ontologies $O_n$ and $O_n'$ in the moment $t_n$.

## 4   Revalidating Initial Ontology Alignment by Using Alignment Triggers

As it has been addressed in previous sections, it is important to identify situations expressed in ontology log that imply the necessity of re-aligning ontologies. It is obvious that if such decision is made, it must not refer to the whole ontology, but only some of its parts. Therefore, we propose a concept for building triggers for the automatic alignment process. An alignment trigger is understood as a rule of a form "if $A$ then $B$", where $A$ is a condition and $B$ is an action. Condition $A$ should be related with a concrete level (of concept, attribute or relationship). Its satisfaction should inform about the significant change on a given level. Action $B$, in turn, should cause running the alignment process on the this level and all those below it. Between triggers there can be some relationships owing to which one can optimise the process for running them in an re-alignment process.

We can assume that the condition $A$ can be associated with entries in ontologies' logs. Hence, action $B$ can be associated with modifications that need to be applied to initial initial ontology alignment $M^{O_0, O_0'}$. In our framework the general form of such modification rule is as follows:

$$\langle ADD/REMOVE/RECALCULATE, mapping \rangle \tag{7}$$

Statements $ADD/REMOVE/RECALCULATE$ refer to actions of adding new mappings, removing invalid ones and revalidating existing assertions in the initial alignment. $mapping$ is an element of the form corresponding to sets $M_C^{O_0, O_0'}$, $M_R^{O_0, O_0'}$ or $M_I^{O_0, O_0'}$.

The algorithm for revalidating the initial ontology alignment $M^{O_0, O_0'}$, by designating a set of rules of its modifications using aforementioned triggers, is given below:

---

**Algorithm 1.** Revalidating initial ontology alignment

---

**Require:** $O_0$, $O_0'$, $M^{O_0, O_0'}$, $J_{O_0}$, $J_{O_0'}$

**Ensure:** $R_M$ - a set of rules of updating $M^{O_0, O_0'}$

1: $J = J_{O_0} \cup J_{O_0'}$
2: sort the set $J$ by timestamps of its elements
3: **for all** $\langle timestamp, e_s, e_t \rangle \in J$ **do**
4:     **if** $e_s = \phi$ **then**
5:         add a set of matching *add* rules to $R_M$
6:     **end if**
7:     **if** $e_t = \phi$ **then**
8:         add a set of matching *removal* rules to $R_M$
9:     **end if**
10:     add a set of matching *recalculation* rules to $R_M$
11: **end for**
12: Return $R_M$

---

In order to match certain rules and rules related to them, the algorithm needs to identify a type of change that took place in the ontology. The main idea is tightly connected to expressivity levels of elements in ontologies and how modifications done on these certain levels entail necessary changes of the initial alignment. For example, altering a some concept's set of attributes should not only implicate revalidating mappings of this concept, but also mappings of relations that connect it with other concepts (due to the fact that a modification of concept's structure can imply that certain relations will no longer hold).

Finding matching rules that need to be added to the resulting set of Algorithm 1 can be described as a process of looking through a set of aforementioned alignment triggers. These triggers can be associated with types of log entries and what kind of base alignment's updates they entail.

Lets consider a log entry: $\langle 2014.05.13, A^c = \{a, b, c\}, A^c = \{a, b\}\rangle$. Such assertion can initialise the following trigger:

$$
\begin{aligned}
&\langle *, A^c, A^c \rangle \rightarrow \\
&\{\langle RECALCULATE, (c, c', \lambda_C(c, c'))\rangle \mid (c, c', \lambda_C(c, c')) \in M_C^{O_0, O_0'}\} \cup \\
&\{\langle RECALCULATE, (r, r', \lambda_R(r, r'))\rangle \mid (r, r', \lambda_R(r, r')) \in M_R^{O_0, O_0'} \wedge \\
&(r \vee r' \in (Rel(c, c') \cup Rel(c', c))\}
\end{aligned}
\tag{8}
$$

where $*$ represents any valid timestamp.

The action presented above is being implied by a matching log entry concerning a change in a structure of a concept $c$. It entails that Algorithm 1 need to add to its result a set of rules that will force recalculating the degree to which concept $c$ can be aligned to any other concept $c'$ with which it has been initially aligned. Additionally, mappings of relations that connected modified concept $c$ with other concepts will also be recalculated. Eventually, existing matches will be preserved or removed when invalidated in the light of the criterion from Equation 6.

## 5    Future Works and Summary

In this paper we have presented our proposal of a method of tracking changes in ontologies over time and reevaluating their initial alignment. We have developed a formal framework based on a notion of ontology log - a structure containing precise descriptions of modifications that have been introduced in ontologies. Eventually they are used as preconditions for initialising alignment triggers that describe certain modifications which assert that the initial mappings between ontologies will remain valid.

Due to the limited space available for this paper we are able to provide only one example for an alignment trigger. In our upcoming publications we will concentrate on providing a complete set of such triggers along with an attempt to using them in practical application related to expressing a process of software development using ontologies.

We also want to research the issue of providing formal metrics that could be used to measure the efficiency of updates applied to base alignment and incorporate temporal logics to illustrate a timeliness of knowledge and how certain information changed over time ([11]). Moreover we plan to implement our ideas, by extending previously created web-based ontology editor (that incorporates in a novel way a schemaless database engine) and some of the modern solutions in the field [1].

# References

1. Cao, S.T., Nguyen, L.A., Szalas, A.: WORL: a nonmonotonic rule language for the semantic web. Vietnam Journal of Science 1(2), 57–69 (2014)
2. Chua, W.W.K., Kim, J.-J.: BOAT: Automatic alignment of biomedical ontologies using term informativeness and candidate selection. log of Biomedical Informatics 45, 337–349 (2012)
3. Duong, T.H., Jo, G.S., Jung, J.J., Nguyen, N.T.: Complexity analysis of ontology integration methodologies: a comparative study. log of Universal Computer Science 15, 877–897 (2009)
4. Euzenat, J., Shvaiko, P.: Ontology Matching, 2nd edn. Springer, Heidelberg (2013)
5. Khattak, A.M., Latif, K., Lee, S.: Change management in evolving web ontologies. Knowledge-Based Systems 37, 1–18 (2013)
6. Jiménez-Ruiz, E., Grau, B.C., Horrocks, I.: LogMap and LogMapLt results for OAEI 2012. Ontology Matching (2013)
7. Fensel, D.A.: Ontologies: A Silver Bullet for Knowledge Management and Electronic Commerce, 2nd edn. (2010)
8. Flouris, G., Manakanats, D., Kondylaiks, H., Plexousakis, D., Antoniou, G.: Ontology change: classification and survey. Knowl. Eng. Rev. 23 (2008)
9. Ma, Y., Lu, K., Zhang, Y., Jin, B.: Measuring ontology information by rules based transformation. Knowledge-Based Systems 50, 234–245 (2013)
10. Nguyen, N.T.: Advanced Methods for Inconsistent Knowledge Management (Advanced Information and Knowledge Processing). Springer (2008)
11. Nguyen, V., Nguyen, N.T.: A Method for Temporal Knowledge Integration Using Indeterminate Model of Time. Cybernetics and Systems 44(2-3), 222–244 (2013)
12. Noy, N.F., Kunnatur, S., Klein, M., Musen, M.A.: Tracking Changes During Ontology Evolution. In: McIlraith, S.A., Plexousakis, D., van Harmelen, F. (eds.) ISWC 2004. LNCS, vol. 3298, pp. 259–273. Springer, Heidelberg (2004)
13. Pastuszak, J., Orłowski, C.: Model of rules for IT organization evolution. In: Nguyen, N.T. (ed.) Transactions on Computational Collective Intelligence IX. LNCS, vol. 7770, pp. 55–78. Springer, Heidelberg (2013)
14. Plessers, P., De Troyer, O., Casteleyn, S.: Understanding ontology evolution: A change detection approach. Web Semantics: Science, Services and Agents on the World Wide Web 5, 39–49 (2007)
15. Pietranik, M., Nguyen, N.T.: A Method for Ontology Alignment Based on Semantics of Attributes. Cybernetics and Systems Cybernetics and Systems 43(4), 319–339 (2012)
16. Pietranik, M., Nguyen, N.T.: Ontology Relation Alignment Based on Attribute Semantics. In: Nguyen, N.-T., Hoang, K., Jędrzejowicz, P. (eds.) ICCCI 2012, Part II. LNCS, vol. 7654, pp. 49–58. Springer, Heidelberg (2012)
17. Pietranik, M., Nguyen, N.T.: A multi-attribute based framework for ontology aligning, Neurocomputing (June 28, 2014) ISSN 0925-2312, doi: 10.1016/j.neucom.2014.03.067
18. Sabou, M., d'Aquin, M., Motta, E.: Exploring the semantic web as background knowledge for ontology matching. log on Data Semantics XI, 156–190 (2008)
19. Shvaiko, P., Euzenat, J.: Ontology Matching: State of the Art and Future Challenges. IEEE Trans. Knowl. Data Eng. 25(1), 158–176 (2013)
20. Vossen, G.: Big data as the new enabler in business and other intelligence. Vietnam Journal of Science 1(2), 57–69 (2014)

# Rule-Based Reasoning System
# for OWL 2 RL Ontologies

Jaroslaw Bak and Czeslaw Jedrzejek

Institute of Control and Information Engineering,
Poznan University of Technology,
M. Sklodowskiej-Curie Sqr. 5, 60-965 Poznan, Poland
{jaroslaw.bak,czeslaw.jedrzejek}@put.poznan.pl

**Abstract.** In this paper we present a method of transforming OWL 2 ontologies into a set of rules which can be used in a forward chaining rule engine. We use HermiT reasoner to perform the TBox reasoning and to produce classified form of an ontology. The ontology is automatically transformed into a set of Abstract Syntax of Rules and Facts. Then, it can be transformed into any forward chaining reasoning engine. We present an implementation of our method using two engines: Jess and Drools. We evaluate our approach by performing the ABox reasoning on the number of benchmark ontologies. Additionally, we compare obtained results with inferences provided by the HermiT reasoner. The evaluation shows that we can perform the ABox reasoning with considerably better performance than HermiT. We describe the details of our approach as well as future research and development.

## 1 Introduction

In the last decade, the use of ontologies in information systems has become more and more popular in various fields, such as web technologies, database integration, multi agent systems, natural language processing, etc. One of the most popular way to express an ontology is to use the Web Ontology Language (OWL) [12]. It is based on description logics (DLs) which are a family of knowledge representation languages.

In order to utilize all features that an ontology provides we need to apply a reasoning engine. However, we can use different engines with ontologies expressed in different OWL 2 Profiles [10] (as well as in different fragments of OWL 1.1[1], eg. Horn-$\mathcal{SHIQ}$). For instance, for an ontology within the OWL 2 EL profile we can use the HermiT[2] reasoner; but for an ontology within the OWL 2 QL profile, which expressive power is quite limited, we can use the REQUIEM[3] reasoner. As a result it is important to choose the right reasoner for a given ontology in order to obtain the best possible results in reasoning or query answering.

In this work we focus on ontology-based reasoning using a standard forward chaining rule engine. Thus, we mainly concentrate on the OWL 2 RL profile.

---

[1] http://www.w3.org/Submission/owl11-overview/

[2] http://www.hermit-reasoner.com/

[3] http://www.cs.ox.ac.uk/isg/tools/Requiem/

D. Hwang et al. (Eds.): ICCCI 2014, LNAI 8733, pp. 404–413, 2014.

However, the presented methodology can handle ontologies with expressivity beyond OWL 2 RL. This is enabled by employing HermiT to perform the TBox reasoning (with the terminological part of an ontology). As a result we can apply a rule-based reasoning engine to perform the ABox reasoning (with the assertional part of the ontology). It is in accordance with the idea behind OWL 2 RL - a requirement of scalable reasoning without the significant loss of the expressive power. In that case, a relatively lightweight ontology can be applied to perform inferences over a large number of instances.

In this paper we present a reasoning tool which is able to perform ontology-based reasoning using a standard forward chaining rule engine. The paper makes the following contributions:

- we present a transformation method of an OWL 2 ontology into a set of rules and a set of facts (if an ontology contains ABox),
- we propose Abstract Syntax of Rules and Facts (ASRF),
- we provide a reasoning schema compatible with our methodology,
- we describe an implementation of our approach using two forward chaining rule engines: Jess [4] and Drools[4],
- we evaluate our methodology by performing experiments using OWL 2 compatible ontologies and the number of reasoning engines.

The remainder of this paper is organized as follows. Firstly, we introduce the background and motivation of our work. Then, we describe our approach of an OWL 2 ontology transformation into two sets of rules and facts, respectively. Next, we provide our Abstract Syntax of Rules and Facts. Later, we present the implementation details as well as experiments. Finally, we describe the related work and we present the conclusions as well as future directions of our research.

## 2  Background and Motivation

Rule-based approaches to ontology-based reasoning achieve significant gains in reasoning complexity [15]. However, the current specification of the OWL 2 RL Profile provides the number of predefined entailment rules as a starting point for practical implementation with rule-based systems. These rules, called the OWL 2 RL/RDF[5] rules are based on universally quantified first-order implications over RDF[6] triples which are represented as ternary predicate $T$ with three elements: the subject, the predicate and the object. Moreover, OWL 2 RL/RDF rules follow the OWL 2 RDF-Based Semantics[7] which is the semantics of OWL 2 Full (which is known to be computationally undecidable with regard to consistency and entailment checking [2]). Nevertheless, if an OWL 2 RL ontology satisfies Theorem PR1 in [10] it follows OWL 2 Direct Semantics[8] which is the

---

[4] http://www.jboss.org/drools
[5] http://www.w3.org/2007/OWL/wiki/Profiles#OWL_2_RL
[6] http://www.w3.org/TR/rdf-primer/
[7] http://www.w3.org/TR/owl2-rdf-based-semantics/
[8] http://www.w3.org/TR/owl2-direct-semantics/

typical semantics for the OWL 2 RL Profile (description logic-based semantics). This characterization is one of the most confusing thing in the OWL 2 specification. As a result, an implementation of a reasoning engine which follows the OWL 2 Direct Semantics requires to satisfy preconditions in Theorem PR1. However, the assertional entailments obtained from a rule-based reasoning engine using OWL 2 RL/RDF rules over an OWL 2 RL ontology follow the Direct Semantics [14]. It means if we want to apply OWL 2 RL/RDF rules, we need to perform the TBox entailments using different reasoning engine. Though, we can raise the abstraction level (from triple-based rule representation) and instead represent the input OWL 2 RL ontology using an axiom-based data structure as shown in [11]. However, the main difference between both semantics is that the RDF-Based Semantics can be applied to arbitrary RDF graphs [7] which do not respect the various restrictions of the OWL 2 syntax. As a result one needs to decide which semantics is required in an application and then follow it during the implementation.

From the practical point of view, usually the terminological part of an ontology is rarely modified in contrast to the assertional part. In that case we can separate the TBox from the ABox. As a result we are able to apply different reasoning schemes and engines. Moreover, we need to perform the TBox reasoning only once (or every time it changes) using e.g. some DL reasoner and then we can perform the ABox reasoning using e.g. a rule-based engine each time when new individual assertions occur. Therefore, we can follow the OWL 2 RDF-Based Semantics in both reasoning engines. Such an approach is presented in [9], where the Pellet[9] engine is used with a rule-based system of Jena[10].

It is worth noticing the presented work is devoted to the development of the Rule-based Query Answering and Reasoning system (RuQAR). However, we present only the reasoning features while the query answering part remain to be carried out in the next release. Thus, the main idea of our work is to provide not only an OWL 2 RL reasoning framework but also a scalable query answering tool in which data is stored in a relational database.

## 3    OWL 2 RL Ontology Transformation

Application of a rule-based reasoning engine to an ontology-based reasoning requires a transformation method of an ontology into a set of rules. Since we mainly focus on the OWL 2 RL Profile, we split the reasoning process into two sub-processes: the TBox reasoning and the ABox reasoning. According to this we developed a methodology of transforming an OWL 2 ontology into a set of rules and a set of facts. In that case we can execute the TBox reasoning and the ABox reasoning separately. Moreover, as we want to perform a rule-based reasoning with different engines we propose Abstract Syntax of Rules and Facts (ASRF), thus enabling easy translation of an OWL 2 ontology into the language of a reasoning engine.

---

[9] http://clarkparsia.com/pellet/
[10] http://jena.apache.org/

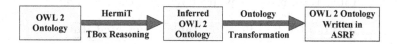

**Fig. 1.** OWL 2 ontology transformation schema

The transformation schema of an OWL 2 ontology into a set of rules and a set of facts expressed in ASRF is presented in Figure 1. Firstly, an OWL 2 ontology is loaded into the HermiT engine. We assume, that the ontology is consistent. Then, the TBox reasoning is executed. As a result we obtain a new classified version of the ontology (new TBox). Finally, the ontology is transformed into two sets: a set of rules and a set of facts (if it contains the assertional part). Both are expressed in the ASRF notation. In that way we separate the TBox part (set of rules) from the ABox part (set of facts). Thus, we can perform ABox reasoning with a forward chaining rule engine.

The transformation of an inferred TBox into a set of rules is performed in accordance with the OWL 2 RL/RDF rules. Generated rules contain two kinds of rules: equality rules and ontology-dependant rules (so-called the ABox rules). Equality rules are directly taken from Table 4 of the OWL 2 RL Profile. These rules are expressed in ASRF and they axiomatize the semantics of equality. They are ontology-independent in contrast to the ABox rules.

The transformation of an ontology, after classification performed by HermiT, into a set of ASRF rules is executed in the following way. For each supported rule (see Table 1 for the details) and the corresponding OWL 2 RL axiom we create a rule reflecting the expression in a given ontology. In other words, it means that rather than transforming the semantics of the OWL 2 RL language into rules we create rules according to this semantics and a given ontology. For instance, if we have an ObjectProperty *hasSibling* which is defined as a SymmetricObjectProperty we create a rule which reflects that when an instance of this property occurs, a symmetric instance should also occur (the following shortcuts are made: $S$ for Subject, $P$ for Predicate and $O$ for Object):

$$\begin{aligned} If \quad & (Triple \ (S \ ?x) \ (P \ "hasSibling") \ (O \ ?y)) \\ Then \quad & (Triple \ (S \ ?y) \ (P \ "hasSibling") \ (O \ ?x)) \end{aligned} \tag{1}$$

Rule (1) follows the semantics of *prp-symp* rule from Table 5 in the OWL 2 RL Profile specification. For each OWL 2 RL axiom which occurs in a given ontology we generate semantically equivalent rule containing a direct reference to the ontology. The generated rule is an instantiated version of the corresponding OWL 2 RL/RDF rule for a particular TBox. Generated rules can be perceived as ontology instance related rules (instantiated rules or ABox rules). These rules are ontology-dependant because they express the semantics of a given ontology and are intended for reasoning with the facts. As a result we provide a set of rules which can be directly applied in a forward chaining engine after the translation from ASRF notation to the engine's language. Such an approach provides an execution of reasoning task directly with the assertional part. It has a positive influence on reasoning efficiency since the semantics of the TBox part

**Table 1.** Currently supported OWL 2 RL entailment rules

| OWL 2 RL Specification Table | Supported Rules |
|---|---|
| Table 4. The Semantics of Equality | eq-sym, eq-trans, eq-rep-p eq-rep-s, eq-rep-o |
| Table 5. The Semantics of Axioms about Properties | prp-dom, prp-rng, prp-fp, prp-ifp, prp-symp, prp-trp, prp-eqp1, prp-spo1, prp-eqp2, prp-inv1, prp-inv2 |
| Table 6. The Semantics of Classes | cls-int1, cls-int2, cls-uni, cls-svf1, cls-svf2, cls-avf, cls-hv1, cls-hv2, cls-maxc2 |
| Table 7. The Semantic of Class Axioms | cax-sco, cax-eqc1, cax-eqc2 |

is directly represented by the generated ASRF rules. Moreover, an additional positive impact comes from the fact that the number of conditions in the body of each rule is smaller than in the corresponding OWL 2 RL/RDF rule.

Table 1 shows currently supported rules by our implementation. This set comes from the specification of OWL 2 [10]. However, this set is smaller than the original one. We decided to use the simplest subset of OWL 2 RL/RDF rules which is easily implementable in any reasoning engine. Moreover, we excluded each rule which is a "constraint" rule (e.g. *cls-nothing2* from Table 6 in the OWL 2 RL Profile) and each rule which does not have an impact on the ABox reasoning (e.g. all rules from Table 9 in the OWL 2 RL Profile). However, some rules remain to be implemented, e.g. *cls-maxqc3* from Table 6.

Presented transformation method may produce more entailments during reasoning than those represented by OWL 2 RL/RDF rules. It is caused by the fact that we apply the TBox reasoning with a DL-based reasoner. However, it depends on the expressivity of a given ontology. Nevertheless, the application of our method to ontology beyond the OWL 2 RL Profile will not produce the same entailments as derived by an appropriate DL-based reasoner. In this case, the reasoning with rules generated by our methodology is sound but not complete. We observed such an issue in our evaluation with the LUBM benchmark where all results produced by our method were within entailments derived by HermiT. However, HermiT produced more results which is correct since the expressivity of LUBM is beyond OWL 2 RL.

## 4   Abstract Syntax of Rules and Facts

As we mentioned in previous section the TBox reasoning is performed with the HermiT engine. Then, an ontology is automatically transformed in Abstract

Syntax of Rules and Facts. The main purpose of developing such a syntax is to rise an abstraction level providing more universal representation of rules and facts (assertional part of a knowledge base). As a result the application of ASRF expressions requires mapping schema between the ASRF notion and the language of a selected reasoning engine.

We applied the Extended Backus-Naur Form (EBNF) [13] notation as a technique to express our Abstract Syntax of Rules and Facts. This context-free grammar is presented in Figure 2. Non-terminal symbols are inside brackets ($<$ and $>$) while other symbols are the terminal ones.

The ASRF syntax is a first-order logic-based notation. Each fact is an atom which consists of a set of terms. Each term is a variable (preceded by '?') or a constant. Furthermore, each term is one of the following types: *Subject*, *Predicate*, *Object* or *Argument*. Similarly, each atom is one of the following types: *Triple* or *Comparison* ($\leq, \neq$, etc.). Each rule consists of the body $B$ (IF part) and the head $H$ (THEN part) of a rule ($B \rightarrow H$). Both elements contain atoms. Variables are universally quantified. Moreover, we can use additional operators like 'or' statement to express disjunction (only in the body of a rule) which is in accordance with the OWL 2 RL Profile. Both the body and the head can contain constants and/or variables in their atoms. In contrast, it is not allowed in the facts representation. By allowing to use comparisons we support SWRL Built-ins that can be employed in order to compare values.

The default meaning of the head of each rule is to assert (infer) new triple (fact). In order for a rule to be applied, all the conditional elements that occur in the body must hold. For instance, rule (1) follows the ASRF syntax as well as example facts (2) and (3). Fact (3) is inferred by applying rule (1) to fact (2).

$$(Triple\ (S\ ``Person1")(P\ ``hasSibling")(O\ ``Person2")) \tag{2}$$

$$(Triple\ (S\ ``Person2")(P\ ``hasSibling")(O\ ``Person1")) \tag{3}$$

Our ASRF syntax is similar to the syntaxes of well-known rule languages like Jess or Clips[11]. However, it is less powerful and is limited to expressions available in the OWL 2 RL Profile. For instance, we can not infer about inconsistencies in a knowledge base.

## 5   Implementation and Experiments

RuQAR implements our method of transforming OWL 2 ontologies into a set of rules and a set of facts expressed in the ASRF syntax. The tool is developed in Java and allows to perform ABox reasoning with two state-of-the-art rule engines: Jess and Drools. RuQAR is implemented as a library which can be included in applications requiring efficient ABox reasoning. RuQAR uses the OWL API [6] to handle ontology files as well as to extract the logical axioms from the ontology. We use Drools in version 5.5 and Jess in version 7.1.

---

[11] http://clipsrules.sourceforge.net/

```
<Rule>              ::=   If <ConditionalAtom>+
                          Then <Atom>+

<ConditionalAtom>   ::=   <Atom> |
                          <Logic-operator> <ConditionalAtom> |
                          <Comparison>

<Atom>              ::=   ( Triple ( Subject <Argument> )
                                   ( Predicate <Constant> )
                                   ( Object <Argument> ) ) |
                          ( Triple <Term>+ )

<Logic-operator>    ::=   AND | OR | NOT

<Term>              ::=   ( <TermType> <Argument> )

<TermType>          ::=   Subject | Predicate | Object

<Comparison>        ::=   ( <Argument> <Comparator> <Argument> )

<Argument>          ::=   <Constant> | <Variable>

<Comparator>        ::=   equal | not equal | greater than |
                          greater than or equal | less than |
                          less than or equal | different from

<Fact>              ::=   ( Triple ( Subject <Constant> )
                                   ( Predicate <Constant> )
                                   ( Object <Constant> ) )

<Constant>          ::=   A finite sequence of characters.

<Variable>          ::=   A finite sequence of characters
                          without white spaces preceded by '?' sign.
```

**Fig. 2.** Abstract Syntax of Rules and Facts in EBNF

We evaluated RuQAR using test ontologies taken from the KAON2 website[12]: Vicodi[13] - an ontology about European history, Semintec[14] - an ontology about financial domain and LUBM[15] - an ontology benchmark about organizational structures of universities. We used different datasets of each ontology (Semintec_0, Semintec_1, etc.) where the higher number means bigger ABox set.

Evaluation schema for each ontology was the following. Firstly, we performed the TBox reasoning using HermiT. Then, the classified ontology was loaded into an engine and the ABox reasoning was executed. In each case we recorded the reasoning time and counted the resulting ABox size. We performed the ABox reasoning with the following engines: Jess, Drools and HermiT. We verified that the reasoners produced identical results (a similar empirical approach is applied in [3] and [11] in order to compare their OWL 2 RL reasoners with Pellet/RacerPro and HermiT, respectively). However, HermiT provided more reasoning results in the LUBM case. It is correct, since only Vicodi is within the OWL 2 RL Profile. However, this is the main cause of extremely large differences of reasoning times in comparison to Jess and Drools. Nevertheless, all results inferred by Jess and Drools were among the results produced by HermiT. In each case we obtained better performance in ABox reasoning with Jess/Drools than with HermiT. For instance, for the Semintec_4.owl ontology, appropriate times for Jess, Drools and HermiT were the following (results were identical): over 3 seconds, over 5 seconds and over 16 seconds, respectively. As we can see from Figure 3 Jess performed better than Drools while HermiT was always on the third place. Obtained results confirm that our method increases the ABox reasoning in comparison to the DL-based reasoner.

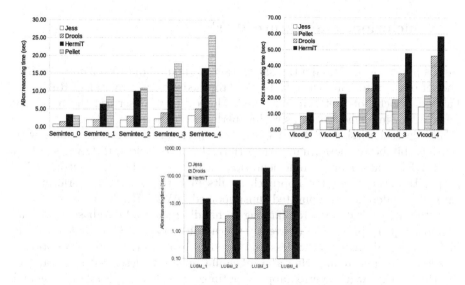

**Fig. 3.** The ABox reasoning times of the tested ontologies

# 6   Related Work

The most closely related work is an approach applied in DLEJena [9]. However, we do not restrict ourselves to one reasoning tool (Jena versus Jess and Drools). Furthermore, we apply slightly different transformation method - we do not use template rules to produce instantiated rules but Java-coded generation. Such an approach do not produce redundant instantiated rules as in [9]. Moreover, in this approach the entailment rules are generated at runtime while in our methodology ABox rules are generated before the reasoning process.

A pair of OWL 2 RL reasoners is presented in [11] where Drools and Jess are used to infer with rules directly representing the semantics of the OWL 2 RL Profile. Both aforementioned approaches follow the Direct Semantics which is the same semantics provided by RuQAR.

Another approach [5] provides OWL 2 RL reasoning and it is based on partial-indexing for optimising scalable rule-based materialisation using the set of template rules similar to DLEJena.

Scalable OWL 2 RL reasoner was presented in [8] where the inference engine is implemented inside the Oracle database system. This work introduces novel techniques for parallel processing as well as special optimisations of computing *owl:sameAs* relationships.

In [3] a method for storing asserted and inferred knowledge in a relational database is presented. Moreover, they also propose a novel database-driven forward chaining method which allows to perform scalable reasoning over OWL 2 RL ontologies with large ABoxes.

# 7   Conclusions and Future Work

In this paper we presented a transformation method of an OWL 2 ontology into one set of rules and one set of facts. We proposed Abstract Syntax of Rules and Facts in which both sets are expressed. Moreover, we described the reasoning schema, our implementation as well as performed experiments.

The current version of RuQAR is able to perform the ABox reasoning with considerably better performance than HermiT. Nevertheless, RuQAR is not an OWL 2 RL conformant[16] implementation since it cannot handle arbitrary RDF graph. However, presented approach results in better reasoning performance, since some inferences are omitted (unsupported OWL 2 RL/RDF rules).

In the next release we are planning to handle relational database as well as optimized query processing (currently we can only use query methods available in forward chaining engines: Jess and Drools). Moreover, we plan to optimize reasoning process by applying and extending methods described in [3] and in [1]. Due to the ASRF syntax, applied optimizations will be usable in Jess and Drools, and in other forward chaining rule engines.

To the best of our knowledge presented work is the first implementation of the OWL 2 RL reasoning in Drools and Jess (except the work presented in [11] that implements directly the semantics of OWL 2 RL) which can be applied in any application requiring efficient ABox reasoning.

We also plan to perform tests with the latest versions of Jess and Drools, 8.0 and 6.0, respectively. It will be useful to check if the reasoning efficiency is increased in the newer versions. As a result in Drools we will be able to compare two different algorithms: ReteOO (Drools 5.5) and PHREAK (Drools 6.0).

**Acknowledgments.** This work has been funded by the Polish National Science Centre (decision no. DEC-2011/03/N/ST6/01602) and DS 04/45/DS-PB/0105 grant.

---

[16] http://www.w3.org/TR/owl2-conformance/

# References

1. Bak, J., Brzykcy, G., Jedrzejek, C.: Extended rules in knowledge-based data access. In: Palmirani, M. (ed.) RuleML 2011. LNCS, vol. 7018, pp. 112–127. Springer, Heidelberg (2011)
2. Boris, M.: On the properties of metamodeling in owl. Journal of Logic and Computation 17(4), 617 (2007)
3. Faruqui, R.U., MacCaull, W.: OwlOntDB: A scalable reasoning system for OWL 2 RL ontologies with large ABoxes. In: Weber, J., Perseil, I. (eds.) FHIES 2012. LNCS, vol. 7789, pp. 105–123. Springer, Heidelberg (2013)
4. Hill, E.F.: Jess in Action: Java Rule-Based Systems. Manning Publications Co., Greenwich (2003)
5. Hogan, A., Pan, J.Z., Polleres, A., Decker, S.: SAOR: Template rule optimisations for distributed reasoning over 1 billion linked data triples. In: Patel-Schneider, P.F., Pan, Y., Hitzler, P., Mika, P., Zhang, L., Pan, J.Z., Horrocks, I., Glimm, B. (eds.) ISWC 2010, Part I. LNCS, vol. 6496, pp. 337–353. Springer, Heidelberg (2010)
6. Matthew Horridge and Sean Bechhofer. The owl api: A java api for working with owl 2 ontologies. In: OWLED (2009)
7. Horrocks, I., Patel-Schneider, P.F.: Knowledge representation and reasoning on the semantic web: Owl. pp. 365–398 (2010)
8. Kolovski, V., Wu, Z., Eadon, G.: Optimizing enterprise-scale OWL 2 RL reasoning in a relational database system. In: Patel-Schneider, P.F., Pan, Y., Hitzler, P., Mika, P., Zhang, L., Pan, J.Z., Horrocks, I., Glimm, B. (eds.) ISWC 2010, Part I. LNCS, vol. 6496, pp. 436–452. Springer, Heidelberg (2010)
9. Meditskos, G., Bassiliades, N.: Dlejena: A practical forward-chaining owl 2 rl reasoner combining jena and pellet. J. Web Sem. 8(1), 89–94 (2010)
10. Motik, B., Grau, B.C., Horrocks, I., Wu, Z., Fokoue, A., Lutz, C.: Owl 2 web ontology language profiles. In: W3C Recommendation, 2nd edn. (2012)
11. O'Connor, M.J., Das, A.: A pair of owl 2 rl reasoners. In: Klinov, P., Horridge, M. (eds.) OWLED. CEUR Workshop Proceedings, vol. 849. CEUR-WS.org (2012)
12. Patel-Schneider, P.F., Horrocks, I.: Owl 1.1 web ontology language (2006), http://www.w3.org/Submission/owl11-overview/
13. Pattis, R.E.: Ebnf: A notation to describe syntax (1980)
14. Polleres, A., Hogan, A., Delbru, R., Umbrich, J.: RDFS and OWL reasoning for linked data. In: Rudolph, S., Gottlob, G., Horrocks, I., van Harmelen, F. (eds.) Reasoning Weg 2013. LNCS, vol. 8067, pp. 91–149. Springer, Heidelberg (2013)
15. Volz, R.: Web ontology reasoning with logic databases. PhD thesis (2004)

# A Consensus-Based Method for Solving Concept-Level Conflict in Ontology Integration

Trung Van Nguyen[1] and Hanh Huu Hoang[2]

[1] College of Sciences, Hue University
77 Nguyen Hue Street, Hue City, Vietnam
nvtrung@hueuni.edu.vn
[2] Hue University
3 Le Loi Street, Hue City, Vietnam
hhhanh@hueuni.edu.vn

**Abstract.** Ontology reuse has been an important factor in developing shared knowledge in Semantic Web. The ontology reuse enables knowledge sharing more easily between intelligent ontology-based systems. However, this cannot completely reduce conflict potentials in ontology integration. This paper presents a method based on the consensus theory and a evaluation function of similarity measure between concepts which is used for a proposed algorithm for ontology integration at the concept level.

**Keywords:** consensus theory, ontology integration, concept level, similarity distance.

## 1 Introduction

The development of Semantic Web enables ontologies of organisations and personal usages being created heterogeneously. Ontology reuse has been a key factor in ontol-ogy development in terms of sharing knowledge in Semantic Web. The ontology re-use enables knowledge sharing more easily between intelligent ontological systems. However, this cannot completely reduce conflict potentials in ontology integration. For instance, Fig. 1 and Fig. 2 show that four experts from ontology can differently describe the concept of "course". The question is: how can we integrate these ontologies? or, how should we deal with this kind of inconsistency between ontologies?

There are different approaches for ontology integration or more specifically ontology matching. This paper proposes a consensus theory based method to resolve this kind of problem.

Our approach will be detailed and structured in this paper as follows: Section 2 formulates the problems of ontology integration and some drawbacks of current approaches for this problem. Meanwhile, Section 3 presents basic concepts of consensus theory and definitions of semantic distances. We define the distance space based on similarity measure between concepts in reference. Based on this, we propose an ontology integration algorithm for ontology conflicts at concept level. The paper concluded with discussions in Section 4.

D. Hwang et al. (Eds.): ICCCI 2014, LNAI 8733, pp. 414–423, 2014.

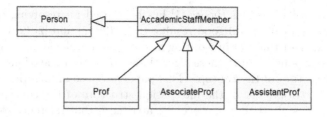

**Fig. 1.** Excerpt of referenced ontology $O_{REF-TREE}$

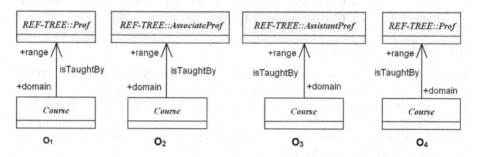

**Fig. 2.** Excerpt of ontologies of 4 different experts

## 2 Problems of Ontology Integration

A definition for ontologies integration has proposed in [7] is: "For given ontologies $O_1, O_2, \ldots, O_n$ one should determine one ontology $O^*$ which best represents them". The key problem is we have to solve conflicts or inconsistencies between entities in the ontologies. One classify inconsistencies between ontologies into the following levels:

- Inconsistency on the instance level: There are several instances with the same name having different descriptions in different ontologies.
- Inconsistency on the concept level: There are several concepts with the same name having different structures in different ontologies.
- Inconsistency on the relation level: Between the same two concepts there are inconsistent relations in different ontologies.

In the last 10 years, the problem of ontology integration has been a challenging issue which attracts several efforts of ontology research community. Authors in [3,4] presented some recently mechanisms for ontology integrations. In our point of view, current approaches have drawbacks. String comparison on entities' names, or even on terms extracted from entities' descriptions (by using NLP techniques) is not sufficient for completely evaluation of the similarity of those two entities. The reason lied on the synonym and homonym problems. Based on ontology graph structure, we can have better results, but we would face the complexity of the problem, especially in ontology with a big number of entities.

In recent years, consensus theory [1] has been used for resolving conflicts in ontologies and gained several good results: In [7] (2007) and [2] (2011), authors have proposed algorithms for integrating inconsistent ontologies on concept level. However, in our point of view, these algorithms only focus on attribute list of the integraton structure of the concept. This paper propose an algorithm not only generate the attribute list of the integration structure of the concept but also calculate domains of attributes, especially attributes that reference to concepts.

## 3    Resolving Inconsistency between Ontologies at Concept Level Using Consensus Theory

### 3.1    Consensus Theory

Consensus theory [1] is an appropriate tool to build collective intelligence. Several results of consensus theory using for knowledge integration have been proposed in [7]. In this section, we show some basic notations which are directly used in the problem of ontology integration.

By $\mathbf{U}$ we denote a finite set of objects representing possible values for a knowledge state. We also denote:

- $\mathbf{\Pi}_k(\mathbf{U})$ is the set of all $k$-element subsets (with repetitions) of set $\mathbf{U}$ ($k \in \mathbb{N}$, set of natural numbers).
- $\mathbf{\Pi}(\mathbf{U}) = \bigcup_{k \in \mathbb{N}} \mathbf{\Pi}_k(\mathbf{U})$ is the set of all nonempty subsets with repetitions of set $\mathbf{U}$. An element in $\mathbf{\Pi}(\mathbf{U})$ is called as a conflict profile.

**Definition 1 - Distance function.**
A distance function $d : \mathbf{U} \times \mathbf{U} \to [0, 1]$ is defined so that it has these following features:

1. Nonnegative: $\forall x, y \in \mathbf{U} : d(x, y) \leq 0$ ,
2. Reflexive: $\forall x, y \in \mathbf{U} : d(x, y) = 0 \Leftrightarrow x = y$ ,
3. Symmetrical: $\forall x, y \in \mathbf{U} : d(x, y) = d(y, x)$ .

We call the space $(\mathbf{U}, d)$ which is defined in the above way is a distance space. With an $\mathsf{X} \in \Pi(\mathbf{U})$, we denote:

- $d(x, \mathsf{X}) = \sum_{y \in \mathsf{X}} d(x, y)$, with $x \in \mathbf{U}$ .
- $d_{t\_mean}(\mathsf{X}) = \frac{1}{M(M+1)} \sum_{x, y \in \mathsf{X}} d(x, y)$, with $M = |\mathsf{X}|$ .
- $d_{min}(\mathsf{X}) = min_{x \in \mathbf{U}} \, d(x, \mathsf{X})$ .

**Definition 2 - Consensus function.**
By a consensus function in space $(\mathbf{U}, d)$, we mean a function

$$C \colon \Pi(\mathbf{U}) \to 2^{\mathbf{U}} .$$

For a conflict profile $\mathsf{X} \in \Pi(\mathbf{U})$, the set $C(\mathsf{X})$ is called the representation of $\mathsf{X}$, and an element in $C(\mathsf{X})$ is called a consensus of profile $\mathsf{X}$. $C(\mathsf{X})$ is a normal set (without repetitions).

Consensus functions need to satisfy some postulates [7] in order to elect the "proper" representation(s) from a conflict profile. The mostly used consensus function are $\mathcal{O}_1$-functions. The functions $C(\mathsf{X}), \mathsf{X} \in \Pi(\mathbf{U})$, of this kind satisfy the so-called $\mathcal{O}_1$-postulate [7]:

$$\left(x \in C(\mathsf{X})\right) \Rightarrow \left(d(x, \mathsf{X}) = min_{y \in \mathbf{U}} \, d(y, \mathsf{X})\right) .$$

**Definition 3 - Criteria for Consensus Susceptibility.**
Not from any conflict profile we can choose a consensus solution in general and $\mathcal{O}_1$-consensus in specifically. We say that, profile $\mathsf{X}$ is susceptible to consensus in relation to postulate $\mathcal{O}_1$ iff:

$$d_{t\_mean}(\mathsf{X}) \geq d_{min}(\mathsf{X}) .$$

## 3.2    Integrate Inconsistent Ontologies on the Concept Level Using $\mathcal{O}_1$-function

**Definition 4 - Ontology.**
An ontology is a quadruple $\langle \mathbf{C}, \mathbf{I}, \mathbf{R}, \mathbf{Z} \rangle$, where:

- $\mathbf{C}$ is a set of concepts (classes).
- $\mathbf{I}$ is a set of instances of concepts.
- $\mathbf{R}$ is a set of binary relations defined on $\mathbf{C}$.
- $\mathbf{Z}$ is a set of axioms which are formulae of first-order logic and can be interpreted as integrity constraints or relationships between instances and concepts, and which cannot be expressed by the relations in set $\mathbf{R}$, nor as relationships between relations included in $\mathbf{R}$.

**Definition 5 - The real world.**
Let $\mathbf{A}$ is a finite set of attributes. Each attribute $a \in \mathbf{A}$ has a domain $\mathbf{V}_a$. Let $\mathbf{V} = \bigcup_{a \in \mathbf{A}} \mathbf{V}_a$, we call $(\mathbf{A}, \mathbf{V})$ as a real world. A domain ontology that refers to the real world $(\mathbf{A}, \mathbf{V})$ is called $(\mathbf{A}, \mathbf{V})$-based.

**Definition 6 - Structure of a concept.**
A concept in an $(\mathbf{A}, \mathbf{V})$-based ontology is defined as a triple $(c, \mathbf{A}^c, \mathbf{V}^c)$, where:

- $c$ is the unique name of the concept,
- $\mathbf{A}^c \subseteq \mathbf{A}$ is a set of attributes describing the concept,
- $\mathbf{V}^c = \bigcup_{a \in \mathbf{A}^c} \mathbf{V}_a$ is the domain of attributes $(\mathbf{V}^c \subseteq \mathbf{V})$ .

The pair $(\mathbf{A}^c, \mathbf{V}^c)$ is called the structure of concept $c$ .

**Definition 7 - Relations between attributes.**
Two attributes $a, b$ in structure of a concept can have following relations:

- Equivalence: $a$ is equivalence to $b$, denoted as $a \leftrightarrow b$, if $a$ and $b$ reflect the same feature for instances of the concept. For example, $occupation \leftrightarrow job$ .
- Generalization: $a$ is more general than $b$, denoted as, $a \to b$, if information given by property $a$ including information given by property $b$. For example: $dayOfBirth \to age$ .
- Contradiction: $a$ is contradictory with $b$, denoted as $a \downarrow b$, if their domains are the same two-element set and values of them for the same instance are contradictory. For example: $isFree \downarrow isLent$, where $\mathbf{V}_{isFree} = \mathbf{V}_{isLent} = \{true, false\}$ which can be used to describe instances in the $Book$ concept whether its instances' property $isFree$ changed to $isLent$ .

**Definition 8 - The ontology integration problem on the concept level.**
Let $O_1, O_2, \ldots, O_n$, $(n \in \mathbb{N})$ are $(\mathbf{A}, \mathbf{V})$-based ontologies. Let the same concept $c$ belong to $O_i$ is $(c, \mathbf{A}^i, \mathbf{V}^i)$, $i = 1, 2, \ldots, n$. From the profile $\mathsf{X} = \{(\mathbf{A}^i, \mathbf{V}^i) : i = 1, 2, \ldots, n\}$, we have to determine the pair $(\mathbf{A}^*, \mathbf{V}^*)$ which presents the best structure for the concept $c$.

**Postulates for Optimised Integration $(\mathbf{A}^*, \mathbf{V}^*)$** Inspired by [7], we formulate the following postulates for determination of pair $(\mathbf{A}^*, \mathbf{V}^*)$ .

P1. For $a, b \in \mathbf{A} = \bigcup_{i=1}^{n} \mathbf{A}^i$ and $a \leftrightarrow b$ , all occurrences of $a$ in all sets $\mathbf{A}^i$ may be replaced by attribute $b$ or vice versa.

P2. If in any set $\mathbf{A}^i$ attributes $a$ and $b$ appear simultaneously and $a \to b$ then attribute $b$ may be removed.

P3. For $a, b \in \mathbf{A} = \bigcup_{i=1}^{n} \mathbf{A}^i$ and $a \downarrow b$ , occurrences of $a$ in all sets $\mathbf{A}^i$ may be replaced by attribute $b$ or vice versa.

P4. Occurrence of an attribute in set $\mathbf{A}^*$ should be dependent only on the appearances of this attributes in sets $\mathbf{A}^i$.

P5. An attribute $a$ appears in set $\mathbf{A}^*$ if it appears in at least half of sets $\mathbf{A}^i$.

P6. Set $\mathbf{A}^*$ is equal to $\mathbf{A}$ after applying postulates P1-P3.

P7. For each attribute $a \in \mathbf{A}^*$, its domain $\mathbf{V}_a^*$ is determined so that:

$$d_a(\mathbf{V}_a^*, \mathsf{X}_a) = min\{d_a(\mathbf{V}_a, \mathsf{X}_a) : \mathbf{V}_a \in \mathbf{U}_a\} ,$$

where:

- $\mathsf{X}_a$ is the conflict profile which is formulated from domains $\mathbf{V}_a^i, i = 1, \ldots, n$.
- $\mathbf{U}_a$ is the universe set, contains all possible values for $\mathbf{V}_a$.
- $d_a$ is the distance function between elements in $\mathbf{U}_a$.

Postulates P1-P6 are adapted to ones in [7]. We propose the P7 postulate to gain the result of consensus theory. More specifically, we use the $\mathcal{O}_1$ function to determine the optimal domain for the attribute $a \in \mathbf{A}^*$. It is important to formulate distance space $(\mathbf{U}_a, d_a)$ for using $\mathcal{O}_1$ function to find the consensus domain.

The important issue is about appropriately defining space distance $(\mathbf{U}_a, d_a)$ to compute the optimised solution for the ranges of properties in integration set.

In the remain part of this paper, we are going to describe the way how to formulate the space distance and integration algorithm according to these postulates.

**Formulate the Distance Function between Two Concepts** There are several ways to measure the similarity between concepts in an ontology. In this paper, we use idea of [5]. According to these authors, we allocate weight values to the edges between concepts:

$$w(parent, child) = 1 + \frac{1}{2^{depth(child)}}$$

where, $depth(child)$ present the depth of concept $child$ from the root concept in ontology hierarchy. The so-called semantic distance between concepts can be determined using **Algorithm 1** [5]:

---

**Algorithm 1.** Calculate semantic distance between concepts

---

**Input**: Concepts $c_1, c_2$ in ontology hierarchy.
**Output**: Semantic distance between $c_1, c_2$, denoted as $Sem\_Dis(c_1, c_2)$
**begin**
    **if** *($c_1, c_2$ is the same concept)* **then**
    |   $Sem\_Dis(c_1, c_2) := 0$;
    **else if** *(there exists the direct path relation between $c_1$ and $c_2$)* **then**
    |   $Sem\_Dis(c_1, c_2) := w(c_1, c_2)$;
    **else if** *(there exists the indirect path relations between $c_1$ and $c_2$)* **then**
        Determine shortestPath$(c_1, c_2)$ is the shortest path from $c_1$ to $c_2$ in
        the ontology hierarchy;
        $Sem\_Dis(c_1, c_2) := \sum_{(c_i, c_j) \in shortestPath(c_1, c_2)} w(c_i, c_j)$;
    **else**
        Determine $cpp$ = the nearest common parent concept of the two
        concepts $c_1, c_2$;
        $Sem\_Dis(c_1, c_2) :=$
        $min\{Sem\_Dis(c_1, cpp)\} + min\{Sem\_Dis(c_2, cpp)\}$;
    **end**
**end**

---

We can see clearly that, the $Sem\_Dis$ function is not normalized, i.e. its values may be out of $[0, 1]$. We can normalise it and formulate a distance space $(\mathbf{U}, d)$ from the ontology hierarchy like this:

- $\mathbf{U}$: the set of concepts in the ontology hierarchy.
- $d$: $\mathbf{U} \times \mathbf{U} \rightarrow [0, 1]$
   $d(c_1, c_2) \mapsto 1 - \frac{1}{Sem\_Dis(c_1, c_2) + 1}$

**Algorithm for Determining the Optimal Integration Based on the Consensus** Based on postulates that are presented in previous section, we propose an algorithm for determining integration structure for concept $c$ from element ontologies $O_1, O_2, \ldots, O_n$ (**Algorithm 2**).

---

**Algorithm 2.** Determine the optimal integration structure for concept

---

Input:
- Conflict profile $X = \{(\mathbf{A}^i, \mathbf{V}^i),\ i = 1, \ldots, n\}$, where $(\mathbf{A}^i, \mathbf{V}^i)$ is structure of concept $c$ in ontology $O_i$.
- Ontology hierarchy $O_{REF-TREE}$ that use for reference. $C_{REF-TREE}$ is the set of concepts in the ontology $O_{REF-TREE}$.
- Distance space $(\mathbf{U}, d)$ that formulate for ontology $O_{REF-TREE}$ as in previous section $(\mathbf{U} = C_{REF-TREE})$.

Output: Pair $(\mathbf{A}^*, \mathbf{V}^*)$ present the best structure of concept $c$.
begin

Step 1 | Set $\mathbf{A}^* := \bigcup_{i=1}^n \mathbf{A}^i$;

Step 2 | foreach $a, b \in \mathbf{A}^*$ do
    if $((a \leftrightarrow b)$ and $(a$ does not occur in relationships with other attributes from $\mathbf{A}^*$ $))$ then
     | Set $\mathbf{A}^* := \mathbf{A}^* \setminus \{a\}$;
    end
    if $((a \rightarrow b)$ and $(b$ does not occur in relationships with other attributes from $\mathbf{A}^*$ $))$ then
     | Set $\mathbf{A}^* := \mathbf{A}^* \setminus \{b\}$;
    end
    if $((a \downarrow b)$ and $(b$ does not occur in relationships with other attributes from $\mathbf{A}^*$ $))$ then
     | Set $\mathbf{A}^* := \mathbf{A}^* \setminus \{b\}$;
    end
  end

Step 3 | foreach $a \in \mathbf{A}^*$ do
    if (the number of occurrences of $a$ in pairs $(\mathbf{A}^i, \mathbf{V}^i)$ is smaller than $\frac{n}{2}$) then
     | Set $\mathbf{A}^* := \mathbf{A}^* \setminus \{a\}$;
    else
     Set $X_a := \{\mathbf{V}_1, \mathbf{V}_2, \ldots, \mathbf{V}_k\}$ where $\mathbf{V}_j$ is the domain of attribute $a$ in pairs $(\mathbf{A}^i, \mathbf{V}^i)$
     and $\mathbf{V}_j \in C_{REF-TREE},\ j = 1, \ldots, k,\ i = 1, \ldots, n$;
     if ($X_a$ is susceptible to consensus in relation to postulate $\mathcal{O}_1$) then
      Determine $\mathbf{V}_a^*$ as $\mathcal{O}_1$ consensus in distance space $(\mathbf{U}, d)$:
        $d(\mathbf{V}_a^*, X_a) = min\{d(\mathbf{V}_a, X_a)\ :\ \mathbf{V}_a \in \mathbf{U}\}$ ;
      Set $\mathbf{V}_a^*$ as domain of attribute $a$ in $\mathbf{A}^*$;
     else
      | Set $\mathbf{A}^* := \mathbf{A}^* \setminus \{a\}$;
     end
    end
  end

Step 4 | foreach $a \in \mathbf{A}^*$ do
    if (there is a relationship $a \leftrightarrow b$ or $a \rightarrow b$ or $a \downarrow b$) then
     | $\mathbf{A}^* := \mathbf{A}^* \cup \{b\}$
    end
  end
end

---

We have remarks for this algorithm:

- The complexity of the algorithm is $O(m^3)$ where $m = \left| \bigcup_{i=1}^n \mathbf{A}^i \right|$ ($m$ is the number of different attributes from sets $\mathbf{A}^i,\ i = 1, \ldots, n$).
- The algorithm only show way to determine optimal attributes for ones has domain in the ontology hierarchy $O_{REF-TREE}$. For attributes which have other kinds of domain (such as range of numbers) we can still use consensus theory to determine optimal domain (solution in [6] is an example).
- The algorithm determines consensus structure for concept $c$ in both component: attributes and their domains. We can make some modifications in step 3 of the algorithm to get some interesting results:
  - Get $\mathbf{V}_a^*$ from $X_a$ rather than from $\mathbf{U}_a$, i.e: $d_a(\mathbf{V}_a^*, X_a) = min\{d_a(\mathbf{V}_a, X_a)$: $\mathbf{V}_a \in X_a\}$. We can use this condition in cases that $C_{REF-TREE}$ contains a large of concepts so that the algorithm has less time to run.

- If $X_a$ is not susceptible in relation to $\mathcal{O}_1$, we can choose $\mathbf{V}_a^*$ as the common parent concept of concepts in $X_a$ rather than remove attribute $a$ from $\mathbf{A}^*$.

We consider a small example for our algorithm: Let $(\mathbf{A}, \mathbf{V})$ is a real world where:

- $\mathbf{A} = \{cid, isTaughtBy, isFinish, isActive, sched, tkb\}$ [1]
- $\mathbf{V}_{cid} = [1, 1000]$
- $\mathbf{V}_{isTaughtBy} = \{AscProf, Prof, AssiProf, AcademicStaffMember\}$
- $\mathbf{V}_{isFinish} = \{Yes, No\}$
- $\mathbf{V}_{isActive} = \{Yes, No\}$
- $\mathbf{V}_{sched} = \{Mon, Tue, Wed, Thurs, Fri, Sat, Sun\}$
- $\mathbf{V}_{tkb} = \{2, 3, 4, 5, 6, 7, 8\}$

Relationships between the attributes: $\{tkb \leftrightarrow sched, isFinish \downarrow isActive\}$

Concepts of ontologies which are reference to ontology $\mathcal{O}_{REF-TREE}$ (Fig. 1). First of all, we formulate distance space $(\mathbf{U}, d)$: Weight of edges in ontology $\mathcal{O}_{REF-TREE}$:

- $w[Person, AcademicSM] = 1 + 1/2 = 1.5$
- $w[AcademicSM, AscProf] = 1 + 1/2^2 = 1.25$
- $w[AcademicSM, Prof] = 1 + 1/2^2 = 1.25$
- $w[AcademicSM, AssiProf] = 1 + 1/2^2 = 1.25$

**Table 1.** Structures of concept *Course* from 5 ontologies

| Ontology | Structure of concept *Course* |
|---|---|
| $O_1$ | $\{(cid, [1, 1000]), (isActive, V_{isActive}), (sched, V_{sched}), (isTaughtBy, AssiProf)\}$ |
| $O_2$ | $\{(cid, [1, 1000]), (isFinish, V_{isFinish})\}$ |
| $O_3$ | $\{(isActive, V_{isActive}), (tkb, V_{isFinish}), (cid, [1, 1000])\}$ |
| $O_4$ | $\{(cid, [1, 1000]), (isTaughtBy, Prof)\}$ |
| $O_5$ | $\{(cid, [1, 1000]), (isTaughtBy, AssiProf)\}$ |

Result for step by step adapt to the algorithm:

- Step 1: $\mathbf{A}^* = \{cid, isActive, sched, isTaughtBy, isFinish, tkb\}$.
- Step 2: Remove 2 attributes $isFinish$ and $tkb$ from $\mathbf{A}^*$. After this step, we have: $\mathbf{A}^* = \{cid, isActive, sched, isTaughtBy\}$.
- Step 3:
- Consider $cid$: This attribute occurs 4 times in sets $\mathbf{A}^i$. By [6] we set its domain $\mathbf{V}_{cid}^* = [1, 1000]$.

---

[1] *tkb* is an acronym for *"thoi khoa bieu"* in Vietnamese, which is equals to *"schedule"* in English.

- Similarly, consider $isActive$: This attribute occurs 3 times ($> 5/2$), and its domain is $V_{isActive}^* = \{Yes, No\}$.
- Consider $sched$: It occurs 2 times ($< 5/2$). So we remove it from $\mathbf{A}^*$. Now, we have: $\mathbf{A}^* = \{cid, isActive, isTaughtBy\}$ .
- Consider $isTaughtBy$. It occurs 3 times ($> 5/2$) in sets $\mathbf{A}^i$ and, the domains reference to $O_{REF-TREE}$. Set $\mathsf{X}_{isTaughtBy} = \{2*AssiProf, Prof\}$ (it means that, the concept $AssiProf$ occurs 2 times in the profile). We easily get these folowing results:
  * $d(Person, Prof) = 0.73$
  * $d(AcademicSM, AscProf) = d(AcademicSM, Prof)$
    $= d(AcademicSM, AssiProf) = 0.56$
  * $d(Prof, AssiProf) = 0.71$
  * $d(Person, \mathsf{X}) = 0.55$
  * $d(Prof, \mathsf{X}) = 0.36$
  * $d(AssiProf, \mathsf{X}) = 0.18$
  * $d(AssocProf, \mathsf{X}) = d_{t-mean}(\mathsf{X}_{isTaughtBy}) = 0.238$
  * $d_{min}(\mathsf{X}_{isTaughtBy}) = min\{d(Person, \mathsf{X}), d(AcademicStaffmember, \mathsf{X}),$
    $d(Prof, \mathsf{X}), d(AssiProf, \mathsf{X}), d(AscProf, \mathsf{X})\} = 0.18 = d(AssiProf, \mathsf{X})$

  We have $d_{t-mean}(\mathsf{X}_{isTaughtBy}) \geq d_{min}(\mathsf{X}_{isTaughtBy})$. So profile $\mathsf{X}_{isTaughtBy}$ is susceptible in relation to postulate $\mathcal{O}_1$. And the domain of $isTaughtBy$ is $V_{isTaughtBy}^* = AssiProf$.

  − Step 4: Add the attribute $isFinish$ back to $\mathbf{A}^*$.

  Finally, we have the structure of the concept $course$ is:

$$(\mathbf{A}^*, \mathbf{V}^*) = \{(cid, [1, 1000]), (isActive, \{Yes, No\}), (isFinish, \{Yes, No\}),$$
$$(isTaughtBy, AssiProf)\}$$

## 4    Conclusion

In this paper, we use a way to formulate distance space from a similarity measure of concepts and use it for develop an algorithm determining integration of ontologies which are conflict on concept level. The paper showed that consensus theory is an appropriate and effective way for ontology integration problem.

In the future, we would like to analyse the opportunities of using other consensus functions for determining consensus integration. We also would like to apply the approach of this paper for other level of conflict in ontologies.

## References

1. Barthélemy, J.-P., Janowitz, M.F.: A formal theory of consensus. SIAM Journal on Discrete Mathematics 4(3), 305–322 (1991)
2. Duong, T.H., Nguyen, N.T., Kozierkiewicz-Hetmanska, A., Jo, G.: Fuzzy ontology integration using consensus to solve conflicts on concept level. In: ACIIDS Posters, pp. 33–42 (2011)
3. Euzenat, J., Shvaiko, P.: Ontology Matching, Second Edition. 2, pp. 1–511. Springer, Heidelberg (2013)

4. Fareh, M., Boussaid, O., Chalal, R., Mezzi, M., Nadji, K.: Merging ontology by semantic enrichment and combining similarity measures. International Journal of Metadata, Semantics and Ontologies 8(1), 65–74 (2013)
5. Ge, J., Qiu, Y.: Concept similarity matching based on semantic distance. In: Fourth International Conference on Semantics, Knowledge and Grid, SKG 2008, pp. 380–383. IEEE (2008)
6. Nguyen, N.T.: Representation choice methods as the tool for solving uncertainty in distributed temporal database systems with indeterminate valid time. In: Monostori, L., Váncza, J., Ali, M. (eds.) IEA/AIE 2001. LNCS (LNAI), vol. 2070, pp. 445–454. Springer, Heidelberg (2001)
7. Nguyen, N.T.: Advanced methods for inconsistent knowledge management. Springer (2007)

# Betweenness versus Linerank

Balázs Kósa*, Márton Balassi, Péter Englert, and Attila Kiss

Eötvös Loránd University,
Pázmány Péter sétány 1/C, 1117 Budapest, Hungary
{balhal,bamrabi,enpraai,kiss}@inf.elte.hu

**Abstract.** In our paper we compare two centrality measures of networks, namely betweenness and Linerank. Betweenness is a popular, widely used measure, however, its computation is prohibitively expensive for large networks, which strongly limits its applicability in practice. On the other hand, the calculation of Linerank remains manageable even for graphs of billion nodes, therefore it was offered as a substitute of betweenness in [4]. Nevertheless, to the best of our knowledge the relationship between the two measures has never been seriously examined. As a first step of our experiments we calculate the Pearson's and Spearman's correlation coefficients for both the node and edge variants of these measures. In the case of the edges the correlation is varying but tends to be rather low. Our tests with the Girvan-Newman algorithm for detecting clusters in networks [7] also underlie that edge betweenness cannot be substituted with edge Linerank in practice. The results for the node variants are more promising. The correlation coefficients are close to 1 almost in all cases. Notwithstanding, in the practical application in which the robustness of social and web graphs to node removal is examined node betweenness still outperforms node Linerank, which shows that even in this case the substitution still remains a problematic issue. Beside these investigations we also clarify how Linerank should be computed on undirected graphs.

**Keywords:** big data, networks, centrality measures, betweenness, Linerank.

## 1 Introduction

In a network centrality measures indicate the importance, interestingness of the nodes and the edges and they play a crucial role in many solutions to practical problems e.g. who are the most influential opinion-shapers in a community, which web pages contain the most relevant information about a certain topic [8] or which nodes should be deleted from a network in order to make the system to fall to pieces [2]. In [4] centrality measures are divided into three families. The first group is constituted by the *degree related measures*, the second group

---

* This work was partially supported by the European Union and the European Social Fund through project FuturICT.hu (grant no.: TAMOP-4.2.2.C-11/1/KONV-2012-0013)

D. Hwang et al. (Eds.): ICCCI 2014, LNAI 8733, pp. 424–433, 2014.

consists of the *diameter related measures*, while the third group contains the *flow based measures*. We focus on the last group in our paper. Here, flow refers to the amount of information that may pass through a node or an edge. The most important member of this group, betweenness centrality, was proposed by Freeman[1]. For a given node $v$, it measures the ratio of those shortest paths that go through $v$. Formally, $v^{bet} = \sum_{u,w} \frac{b_{u,v,w}}{b_{u,w}}$, where $b_{u,w}$ and $b_{u,v,w}$ respectively denote the number of the shortest paths between nodes $u$, $w$ and the number of those shortest paths from the previous ones that pass through $v$. The definition of this measure on edges can be formulated in a similar way.

Unfortunately, the computation of the exact values of betweenness is prohibitively expensive for large networks. For the 'node-variant' the best known algorithms work in time $O(nm)$, where $n$ denotes the number of nodes, while $m$ the number of edges in a graph [4]. For this reason several attempts have been made to estimate the value of betweenness by using a carefully selected sample. As an orthogonal direction in [4] a new flow based centrality measure, Linerank, was introduced whose computation remains practically manageable even for graphs of billion nodes. As its name suggests the definition of Linerank was greatly inspired by Pagerank [8]. Roughly speaking, in the first step the original graph is transformed into the corresponding *line graph* on which the Pagerank values of the nodes are calculated. Since in a line graph the nodes represent the edges of the original graph by accomplishing the previous step one gains values measuring the importance of edges in a similar way as Pagerank measures the importance of nodes. However, we want to emphasize that in [4] this measure on edges has not been introduced, Linerank has been only defined on nodes. Our results below confirm that this was a wise decision indeed. Nevertheless, in what follows we will refer to this measure as *edge Linerank*. In order to obtain a measure on nodes the previous scores of the incident edges of a node should be aggregated. The details will be given in Section 2.

In [4], however, the relationship between the newly proposed measure and the older variant was not investigated thoroughly. In our paper we compare node and edge betweenness with node and edge Linerank respectively from different aspects. First, the Pearson's and Spearman's correlation coefficients are calculated both on real world and generated graphs. It turns out that the correlation between node Linerank and betweenness is higher than 0.9 almost in all cases, whereas for the edge versions it ranges from 0.2 to 0.7. These results suggest that node Linerank is a very promising candidate for substituting node betweenness, while this interchangeability is far more questionable for the edge variants.

After these initial results we study two practical applications of the betweenness measure and examine whether it can be substituted with Linerank without significantly worsening the performance of these methods. Firstly, we consider the Girvan-Newman community detection algorithm [7]. Here, edges are removed from the graph according to the decreasing order of their betweenness values. However, after the removal of the edge with the highest betweenness score the

---

[1] Strictly speaking, Anthonisse introduced this measure earlier than Freeman in a technical report, however, this work has never been published [7].

betweenness values of the remaining edges should be recalculated in each step. Sooner or later the graph falls to pieces and the resulting components are to be considered as communities. Of course, later these clusters may also go to pieces. The hierarchy of communities is depicted by means of a dendrogram. Each level of this tree represents a possible clustering. In the last step the one with the highest modularity is chosen to be the final solution.

In our experiment instead of betweenness we calculated the Linerank value of the edges. In the comparison we used the same random benchmark graphs as in the calculation of the correlation coefficients. This model [5] generates graphs with communities, whose sizes vary according to a power law distribution with exponent $\beta$. The degree distribution is also assumed to be power law with exponent $\gamma$. Beside these parameters one can specify a mixing parameter $\mu$ s.t. each node shares a fraction $1 - \mu$ of its edges with the nodes of its cluster and a fraction $\mu$ with the other nodes of the graph. The number of nodes is also given as a parameter. In order to evaluate the performances of the betweenness and Linerank versions of the Girvan-Newman algorithm we applied *normalized mutual information*, since it is a widely used measure for testing the effectiveness of network clustering algorithms [3]. The results clearly show that the betweenness version significantly outperforms the Linerank version. On the one hand, this is not surprising since we have already observed that the correlation between these two measures is varying and it is never too strong. On the other hand, in their original paper Girvan and Newman tried three different variants of the betweenness measure and they found that the quality of the clusterings was not affected noticeably by the choice of the centrality measure. Our analysis reveals that this is no longer the case in the case of Linerank.

Secondly, we repeated the experiments of Boldi et al. in which they examined which nodes have the strongest impact in determining the structure of a network [2]. Or, in other words, which node-removal order influences this structure the most. They considered several centrality measures including Pagerank, harmonic centrality and betweenness. They removed the nodes in decreasing order according to these measures. Contrast to the Girvan-Newman algorithm however, in this case the order of the removal was fixed in the first step, which means that the aforementioned values were not recalculated after each deletion. The authors reported that in several cases betweenness outperformed the rest of the candidates. In our research instead of taking into account several centrality measures we focused solely on node betweenness and Linerank. Unlike in the previous case the difference between the performance of these two measures was unnoticeable for the generated benchmark graphs. However, in the case of real world graph networks betweenness outperformed Linerank again. This indicates that in practice one should still be careful when node betweenness is to be substituted with node Linerank.

The paper is organized as follows. In Section 2 the computation of Linerank is explained in more detail. Next, in Section 3 the results of our experiments are delineated. In Section 3.1 the Pearson's and Spearman's correlation coefficients are calculated. Then, in Section 3.2 edge betweenness is compared to edge

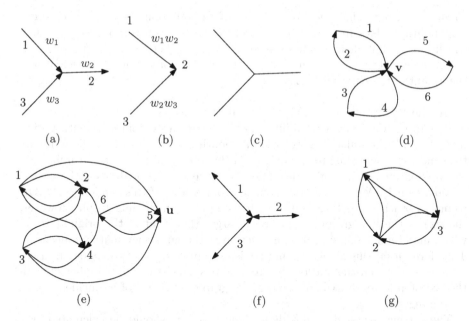

**Fig. 1.** (a) A graph $G$ with weights. (b) $L(G)$. (c) An undirected graph $H$. (d) $\tilde{H}$. (e) $L(\tilde{H})$. (f) The result of substituting the undirected edges of $H$ with bidirected edges. (g) The line graph of the graph on (f).

Linerank by using the Girvan-Newman algorithm. Afterwards, in Section 3.3 the node variants are considered in order to determine the node removal order in networks and then to assess the influence of these removal orders. Finally, in Section 4 we conclude by summarizing our work.

## 2   The Computation of Linerank

As it has already been outlined in the introduction for a graph $G$ Linerank is calculated by first constructing the line graph of $G$ denoted by $L(G)$. In a line graph each edge of the original graph is represented by a node. Let $G$ be a directed graph and let $e_1 = (u_1, v_1)$, $e_2 = (u_2, v_2)$ be edges of $G$. In $L(G)$ there is an edge from the node representing $e_1$ to the node representing $e_2$, if and only if $v_1$ coincides with $u_2$, i.e., the target node of $e_1$ is the same as the source node of $e_2$. Furthermore, if $e_1$, $e_2$ are weighted edges with weights $w_1$ and $w_2$ respectively, then the weight of the previous edge in the line graph is defined to be product of $w_1$ and $w_2$. An example can be seen in Fig. 1. (a) and (b). In what follows we consider such graphs, whose neigbouring edges have the same weights, i.e., if a node has $k$ outgoing edges, then each of these edges has weight $\frac{1}{k}$. We call these graphs *uniformly weighted*.

**Proposition 1.** *Let $G$ be an arbitrary uniformly weighted graph and let $v$ be an arbitrary node of $L(G)$. Then, each outgoing edge of $v$ has the same weight.*

*Proof.* First, note that the target nodes of the outgoing edges of $v$ represent neighbouring edges in the original graph, i.e., their source node is the same, which means that their weights are equal to each other. In $L(G)$ the weight of the outgoing edges of $v$ is computed as the product of the previous weight and the weight of the edge represented by $v$.

On the line graph a random walker at the current step either moves to a neighbouring node with probability $\beta$, or jumps to a random node with probability $1 - \beta$. If the walker moves to a neighbouring node, then she decides among the candidates according to the weights of the joining edges. Thus, for uniformly weighted graphs each neighbouring node has the same probability to be visited (Proposition 1). We seek the stationary probabilities of this random walk. Or, to put in other words Pagerank is to be computed on the line graph. However, the size of the line graph can be much larger than that of the original graph which may render the explicit construction of the adjacency matrix unfeasible. Therefore, in [4] this adjacency matrix is decomposed into two sparse matrices by means of which the stationary probabilities can be computed efficiently. In the last step for each node of the original graph the scores of its incident edges are aggregated.

The original paper does not detail how Linerank should be calculated over undirected graphs. It is tempting to substitute each undirected edge with two oppositely directed edges. However, this approach would result in completely useless Linerank values. To be specific for graph $G$ denote $\tilde{G}$ the result of the previous construction. Then the following statement can be proven.

**Proposition 2.** *Let $G$ be an undirected graph. For each node $u$ of $L(\tilde{G})$ the outdegree of each of the in-neighbours of $u$ is the same as the indegree of $u$. (An in-neighbour is defined to be the source node of an ingoing edge of $u$.)*

*Proof.* First, note that if the indegree of a node $v$ in $\tilde{G}$ is $k$, then the outdegree of $v$ is also $k$, which is a straightforward consequence of the definition of $\tilde{G}$. The statement obviously follows from this observation, since in this case the indegree of the representative of an outgoing edge – denote it $u$ – of $v$ in $L(\tilde{G})$ is also $k$. What is more, the outdegree of each of the in-neighbours of $u$ is also $k$ as they correspond to the ingoing edges of $v$. Consider an example in Fig. 1. (c)-(e).

**Corollary 1.** *For an arbitrary undirected, uniformly weighted graph $G$ the stationary probabilities are the same for each node of $L(\tilde{G})$.*

*Proof.* The statement is a straightforward consequence of Propositions 1 and 2.

Thus, instead of adding extra edges the line graph should be constructed as if the original undirected edges were bidirected. An example can be found in Fig. 1. (c), (f)-(g). Our experiments have shown that this construction avoids the preceding anomaly. What is more, it also saves a considerable amount of memory space.

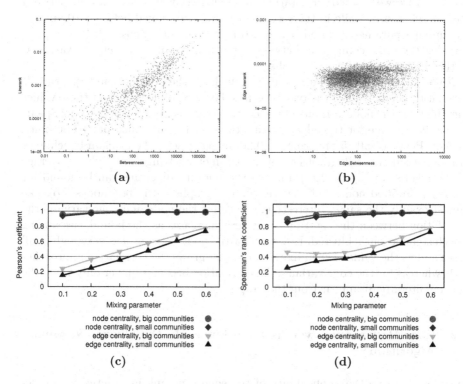

**Fig. 2.** (a) Node betweenness and Linerank values for the polblogs dataset. (b) Edge betweenness and Linerank for the same dataset. (c) Pearson's correlation coefficients for the benchmark graphs with 5000 nodes. (d) Spearman's correlation coefficients for the same set of benchmark graphs.

## 3    Experiments

### 3.1    The Correlation between Linerank and Betweenness

As a first step in our investigation we calculated the betweenness and Linerank values of the nodes and the edges of real world graphs. In all cases the scatter plots suggested a strong correlation between the node Linerank and betweenness values, whereas for the edge variants the relationship remained somewhat blurred. As a typical example in Figure 2 (a) and (b) we have included the plots belonging to the polblogs dataset [1], a directed network of hyperlinks between weblogs on US politics recorded in 2005 with 1490 nodes and 19090 edges. Beside the aforementioned real world graphs we also conducted the same experiment on random graphs described in the introduction. Owing to the costly computation of the exact betweenness values we used graphs of rather smaller sizes, namely with 1000 and 5000 nodes. The $\gamma$ exponent of the power law distribution of the degrees was set to 2, while the $\beta$ exponent of the power law distribution of the sizes of the clusters was chosen to be 1. We further distinguished two cases. In the first case we worked with rather large clusters whose size ranged between 20 and 100 nodes, while in the second case this size ranged between 10 and 50. The mixing parameter varied between 0.1 and 0.6 with steps of 0.1. The plots revealed the same connection between the Linerank and betweenness values as in the case of the real world graphs.

Next, to quantify this relationship we calculated the Pearson's and Spearman's correlation coefficients of the two measures. For the polblogs dataset these values were high for the node variants of the centrality measures: Pearson's: 0.82 Spearman's: 0.89; while for the edge variants the correlation turned out to be much weaker: Pearson's: 0.15 Spearman's: 0.26. In the case of the random benchmark graphs we generated 10 graphs for every parameter settings and took the average of the results. In Figure 2 (c) and (d) the relevant diagrams can be found for graphs with 5000 nodes. The curves for the graphs with 1000 nodes look like almost exactly the same. Interestingly, both for the node and edge variants the correlation between the centrality measures increases as the boundaries among the clusters becomes blurred. However, for the node versions it is very high in all cases and as the value of the mixing parameter grows the correlation approaches to 1, whereas for the edge variants the correlation becomes higher only when the clusters literally disappear from the graph.

## 3.2    Edge Betweenness versus Edge Linerank, the Girvan-Newman Algorithm

In order to assess the applicability of the edge Linerank in practice we implemented two versions of the well-known Girvan-Newman algorithm [7] for detecting communities in a network. The first version uses edge betweenness for finding the next edge to remove from the graph as in the original paper, whereas the second applies edge Linerank for this purpose. To compare the performance of the two variants we employed *normalized mutual information* [7], which is a frequently used measure for testing community detection algorithms.

In our experiments we used the same set of random benchmark graphs as in the previous case. Again, we generated 10 graphs with every parameter setting and we took the average of the normalized mutual information values. The behaviour on graphs with 1000 and 5000 nodes was indistinguishable, therefore we only included the diagrams related to the graphs with 5000 nodes. As the plots in Fig. 3. (a) and (b) clearly show the betweenness version of the Girvan-Newman algorithm significantly outperforms the Linerank version. Indeed, the scores of the latter are extremely low, which indicates the unusability of this method in practice. Of course, one may anticipate this result from the observations of the previous subsection, however, the former experiments only revealed that the correlation between the edge Linerank and betweenness values was rather low especially when the graphs contained quite definite clusters, but they did not foretell the superiority of betweenness. What is more, although these results suggest the inapplicability of edge Linerank for detecting clusters, since the correlation in the case of more scattered graphs was higher, the measure may still

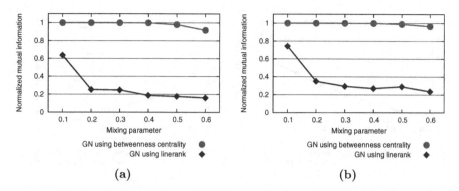

**Fig. 3.** (a) Effectiveness of GN algorithm versions on the benchmark graphs with larger clusters. (b) The same information for the benchmark graphs with smaller clusters.

prove to be useful in certain scenarios, where the presence of clusters is not so remarkable.

### 3.3    Node Betweenness versus Node Linerank, the Robustness of Networks to Node Removal

After testing the applicability of edge Linerank we tried to find out to what degree node betweenness can be substituted with node Linerank in a practical application. For this purpose using node Linerank and betweenness we conducted the experiment of Boldi et al. again in which they tested to what extent the node removals can disrupt the structure of the web and social networks [2]. More precisely, in the course of node removal $\vartheta m$ edges are deleted, where $m$ denotes the number of edges and $0 \leq \vartheta \leq 1$. In the first step one defines an order among the nodes by using a measure and then considering the nodes in decreasing order starts to remove their incident edges. As soon as the number of the deleted edges becomes greater than or equal to $\vartheta m$, the process stops. The authors were interested in how the node removal orders based on different measures influence the fraction of reachable pairs as $\vartheta$ increases. They also wanted to assess the divergence between the distance distributions of the old and new graphs. They tried several different approaches to measure these changes and they have found that the *relative harmonic-diameter change* reflects the differences the best.

In our own experiments again we used both generated and real world graphs. However, in this case we increased the number of nodes of the random graphs to 10000 and 50000. Accordingly, the sizes of the clusters also were also set higher. For the larger clusters these values ranged between 40 and 200, whereas for the smaller clusters between 20 and 100. The rest of the parameters remained the same. As the diagrams in Fig. 4. (a) and (b) show the results are indistinguishable

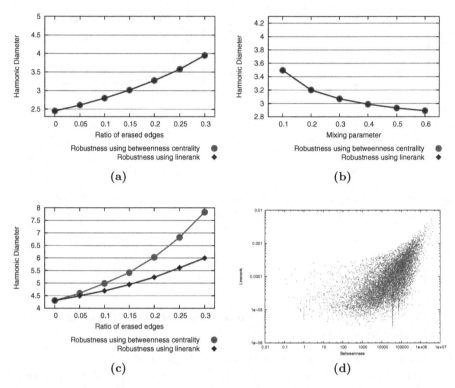

**Fig. 4.** (a) The relative harmonic-diameter change for the benchmark graphs with increasing ratio of the deleted edges (the mixing parameter is fixed). (b) The same data with varying mixing parameters (the ratio of the deleted edges is fixed). (c) The relative harmonic-diameter change for the CA-AstroPh network. (d) The scatter plot of the node betweenness and Linerank values for the CA-AstroPh network.

for node Linerank and betweenness. We only plotted the data belonging to the benchmarks graphs with 50000 nodes and larger clusters, however, the rest of the diagrams look exactly the same. Neither the changes of the mixing parameter nor the increase in the ratio of deleted edges influences this behaviour.

Nonetheless, in the case of real world graphs the scenario is somewhat less straightforward. As one can see in Fig. 4. (c) for the CA-AstroPh dataset [6], which is the collaboration network from the e-print ArXiv in the Astro Physics category (nodes: 18772, edges: 198110), the difference between the relative harmonic-diameter change is more significant. On the other hand, as the scatter plot in Fig. 4. (d) suggests the correlation between node Linerank and betweenness is still high. We experienced the same phenomenon for several real world graphs, which indicates that although the correspondence between the two measures seems to be strong, in practice one should still be careful, when node betweenness is to be substituted with node Linerank.

# 4   Conclusions

In our paper we compared two flow based centrality measures betweenness and Linerank. We have found that in the case of edges the correlation between these measures varies but tends to be rather low. Our experiments with the Girvan-Newman algorithm also underlined that edge betweenness cannot be substituted with edge Linerank in practice. The results for the node variants are more promising. In our tests both Pearson's and Spearman's correlation coefficients were close to 1 in most of the cases. For the generated benchmark graphs this strong correspondence persisted in the practical application in which we examined the robustness of social and web graphs to node removal. However, for real world graphs, although the correlation seemingly remained high, node betweenness outperformed node Linerank. This which shows that even in this case the substitution of the former with the latter remains problematic. Beside these investigations we have also clarified how Linerank should be computed on undirected graphs.

# References

1. Adamic, L.A., Glance, N.: The political blogosphere and the 2004 u.s. election: Divided they blog. In: Proceedings of the 3rd International Workshop on Link Discovery, LinkKDD 2005, pp. 36–43. ACM (2005)
2. Boldi, P., Rosa, M., Vigna, S.: Robustness of social and web graphs to node removal. Social Netw. Analys. Mining 3(4), 829–842 (2013)
3. Fortunato, S., Lancichinetti, A.: Community detection algorithms: A comparative analysis: Invited presentation, extended abstract. In: Proceedings of the Fourth International ICST Conference on Performance Evaluation Methodologies and Tools, VALUETOOLS 2009, pp. 27:1–27:2. ICST (Institute for Computer Sciences, Social-Informatics and Telecommunications Engineering) (2009)
4. Kang, U., Papadimitriou, S., Sun, J., Tong, H.: Centralities in large networks: Algorithms and observations. In: SDM, pp. 119–130. SIAM / Omni Press (2011)
5. Lancichinetti, A., Fortunato, S., Radicchi, F.: Benchmark graphs for testing community detection algorithms. Phys. Rev. E 78(4) (2008)
6. Leskovec, J., Kleinberg, J., Faloutsos, C.: Graph evolution: Densification and shrinking diameters. ACM Trans. Knowl. Discov. Data 1(1) (2007)
7. Newman, M.E.J., Girvan, M.: Finding and evaluating community structure in networks. Phys. Rev. E, 69(2) (2004)
8. Page, L., Brin, S., Motwani, R., Winograd, T.: The pagerank citation ranking: Bringing order to the web. In: Proceedings of the 7th International World Wide Web Conference, pp. 161–172 (1998)

# On Decomposing Integration Tasks
# for Hierarchical Structures

Marcin Maleszka

Institute of Informatics, Wroclaw University of Technology, St. Wyspianskiego 27,
50-370 Wroclaw, Poland
marcin.maleszka@pwr.edu.pl

**Abstract.** Hierarchical structures have became a common data structure in modern applications, thus the needs to process them efficiently has increased. In this paper we provide a full description of a mathematical model of complex tree integration, whicha allows the representation of various types of hierarchical structures and integration tasks. We use this model to show that it is possible to decompose some integration tasks by splitting them into subtasks. Thanks to this procedure it is possible to solve large integration tasks faster, without the need to develop new, less computationally complex, algorithms. Decomposition is an important step towards developing methods of multi-stage integration of hierarchical structures.

## 1 Introduction

In recent years hierarchical and graph-based data structures have become more common in both theoretical and practical applications. Many papers use thesauri and ontologies, while applications store data in XML format. With the spread of hierarchical structures, the need for more efficient tools for processing them has increased significantly. In this paper we show that it possible to decompose some integration tasks for trees into sub-tasks, thus increasing the speed without developing brand new tools.

In our previous work we developed both a mathematical model of hierarchical structures that facilitates integration [10] and various properties [11] and algorithms for such [12]. Recently we have found the model lacking and had to modify it [9]. The general outline remains unchanged: we introduce the notion of criteria that describe the integration of terms of input structures in relation to output structures (i.e. what relation the result has to the input). This allows to precisely define the aim of some integration task. It also allows mathematical analysis of the integration task.

In this paper we elaborate on our previous hypothesis, that it is possible to decompose integration tasks for some criteria [11]. We understand decomposition as 1) diving the set of inputs 2) integating the subsets 3) integrating the results of subtasks into the final result. This allows decreasing the processing time without any modification to the integration algorithm (computational complexity of the

D. Hwang et al. (Eds.): ICCCI 2014, LNAI 8733, pp. 434–443, 2014.
© Springer International Publishing Switzerland 2014

algorithm remains the same). Decomposition is an important step towards a more important area of research - multi-stage integration.

This paper is organized as follows: in Section 2 we present similar research into creating mathematical models of tree integration; in Section 3 we define the model of complex tree integration that allows the further analysis; in Section 4 we analyze the decomposition properties of some criteria; we conclude this paper with some final remarks and future work aspects in Section 5.

## 2    Related Works

Criteria-based approach to integrating hierarchical data structures was researched multiple times in the literature, mostly in relation to facilitating XML document processing. The most interesting approach may be attributed to Passi and Madria [8,13] who propose general criteria for integration, based on previous research in database area. Out of the three main criteria they propose, in this paper we make use of completeness and correctness (integrated tree should include and correctly represent all concepts from input schemas and the integrated schema should represent the union of input schema domains) and minimality (if an element occurs multiple times in different input trees, it should occur only once in the integrated tree). The understandability criterion proposed by these authors cannot be adapted to the mathematical model proposed in this paper, as the authors provide them more in form of general guidelines for designing integration algorithms.

A formal research into hierarchical structure integration that is very similar to the one discussed in this paper was done in a series of works [6,7,14]. They analyze both technical criteria as a means to classify integration algorithms (element vs. structure, language vs. constraint, etc.) and criteria describing the produced output (e.g. schema similarity as a measure of the relation between the number of correctly mapped elements in input trees to the total number of elements, precision and recall based on information retrieval measures as the relation between the number of correctly mapped nodes to all mapped nodes, and the number of nodes correctly mapped by the automatic systems to the number of nodes mapped by the user). Much of our earliest mathematical models was based on this research.

Mathematical models of integration tasks were first analyzed in [1] and all later research related to it. It is one of first research papers on tree integration. It proposed the median approach, that is finding a single solution to multiconsensus function that averages the input trees the most (the sum of distances to all input trees is minimal). As this problem was later proved NP-hard [3], multiple approximate algorithms were proposed to solve it [2,4,5], each adding more mathematical apparatus the area.

## 3    Integration Model

Due to numerous restrictions of existing data structures we were forced to define the Complex Tree to best represent the integration process. Our first definition from [10] is lacking when it comes to fully describing the decomposition properties of trees, so in our most recent work [9] we propose a new approach.

First we define a so-called dendrite of a tree:

**Definition 1. *Dendrite*** *is a rooted tree* $D = (W, E)$, *where:*

- $W$ *is a finite set of nodes,*
- $E$ *is a set of edges, where* $E \subseteq W \times W$.

We also require the following symbols:

Let $A$ be a set of attributes. By $l \in A$ we denote a marked attribute called label (labels are not identifiers of nodes). Let $V_a$ be the domain of an attribute $a \in A$. Set $V = \bigcup_{a \in A} V_a$ is the set of values of all attributes.

Due to this notations one can say that $(A, V)$ is some representation of real world.

Let $T$ be a finite set of node types. Type is some characteristic o node – all nodes of the same type represent objects of the same class and have the same attributes.

Using this notations we define the complex tree as follows:

**Definition 2. *Complex tree*** *is a five* $CT = (W, E, T_{CT}, A_{CT}, V_{CT})$, *where:*

- $(W, E)$ *is the dendrite of the tree,*
- $T_{CT} : W \to T$ *is a function that assigns each node a single type,*
- $A_{CT} : W \to 2^A$, *where* $\forall_{w \in W} l \in A_{CT}(w)$ *is a function that assigns each node attributes; at least one attribute – $l$ – is assigned to each node,*
- $V_{CT} : W \times A \to V$ *is a partial function which assigns each node and each attribute a value such that:*

$$\forall_{w \in W} \forall_{a \in A_{CT}(w)} V_{CT}(w, a) \in V_a$$

According to this definition, a node $w \in W$ that represents a real world object is denoted as $w = (T_{CT}(w), A_{CT}(w), V_{CT}(w))$. An object is represented by its label $V_{CT}(w, l)$, type $T_{CT}(w)$ and set of attributes $A_{CT}(w)$ and their values $V_{CT}(w, a), a \in A_{CT}(w)$.

We use the symbol $\overline{CT}$ to represent the set of all possible complex trees representing whole or part of the real world.

### 3.1   Integration Task

In this subsection we define the integration task, integration function and the task of finding a set representative. These definitions are not equivalent – one may use them to find either a single solution or a whole set of solutions.

To formally represent the integration task we require additional notations. Let $\Pi_k(B)$ be the set of all $k$-element subsets with repetitions of set $B$ and $2^B$ be the powerset of $B$. Let also:

$$\Pi(B) = \bigcup_{k \in \mathbf{N}} \Pi_k(B) \tag{1}$$

Therefore $\Pi(B)$ is the set of all non-empty finite subsets with repetitions of $B$. The integration function is defined as follows:

**Definition 3. *Integration function* for complex trees is a function $I$ of a form:**

$$I : \Pi(\overline{CT}) \to 2^{\overline{CT}}$$

such that for each complex tree $CT^* \in I(\Pi(\overline{CT}))$ one or more criteria $K_j, j \in \{1, \ldots, 16\}$ are met.

The set of complex trees resulting from integration is denoted as $\mathbf{CT^*}$.

Some example of criteria are described in the next subsection. Each tree may be described by normalized measures $M_j(CT^* | CT_1, CT_2, \ldots, CT_N)$. Critrion $K_j$ is met if $M_j = 1$.

We may define the following general integration task of complex trees:

**Definition 4. *Complex tree integration task*:**
For a given set $N$ of complex trees $\mathbf{CT} = \{CT_1, CT_2, \ldots, CT_N\}$ find such set of complex trees $\mathbf{CT^*}$, which elements meet one or more criteria $K_j$, $j \in \{1, \ldots, 16\}$.

If the aim of integration is to find a single solution, we use the following:

**Definition 5. *Task of finding a representative of complex trees set*:**
For a given set $N$ of complex trees $\mathbf{CT} = \{CT_1, CT_2, \ldots, CT_N\}$ find such complex tree $CT^*$, which **best** represents $\mathbf{CT}$.

We understand the word *best* as meeting the chosen group of criteria.
In several cases we also require the definition of node equivalence:

**Definition 6. *Nodes $w$ from complex tree $CT$ and $w'$ from complex tree $CT'$ are called *equivalent*, if:**

$$T_{CT}(w) = T_{CT'}(w'),$$

$$A_{CT}(w) = A_{CT'}(w'),$$

$$V_{CT}(w) = V_{CT'}(w').$$

Two possible situations may occur during integration:

- There exist equivalent nodes in a single tree. This is the multi-set interpretation of integration task.
- There are no equivalent nodes in a single tree. This is the set interpretation of integration task.

## 3.2   Criteria Examples

In this section we define two criteria that we use in this paper. The full list of criteria developed for this model may be found in [10][9].

Criteria of the completeness group are required to ensure that elements are not lost during integration. As nodes are the basic part of the tree, the structure completeness is one of the most important criteria for integration.

**Definition 7.** *Structure completeness is the measure of the ration of number of nodes in the input trees to the number of equivalent nodes in output tree.*

$$C_S(CT^*|CT_1,\ldots,CT_N) = \frac{1}{card(W_1 \cup \ldots \cup W_N)} \sum_{w \in W_1 \cup \ldots \cup W_N} m_W(w, W^*),$$

(2)

*where the characteristic function of $W^*$:*

$$m_W(w, W^*) = \begin{cases} 1 \text{ if } w \in W^*, \\ 0 \text{ if } w \notin W^*. \end{cases}$$

The values of structure completeness measure may be in the range $[0, 1]$. Structure completeness is equal 1, if $W^*$ is the sum of sets $W_1$, $W_2$, $\ldots$ and $W_N$. Structure completeness is equal 0, if $W^*$ does not have common elements with $V_1$, $V_2$, $\ldots$ or $V_N$.

For this measure, *structure completeness criterion is met if the structure completeness measure is maximum (equal 1).*

Minimality is a criterion we defined based on actual usage requirements in applications. It was observed that experts tend to compare the input and output trees based on visual characteristics, e.q. size and depth of the tree. The criterion and its measures are used to formalize this approach.

**Definition 8.** *Minimality is a measure of th ration of the output tree size to the input trees size.*

We understand the *tree size* as its various characteristics. Thus minimality may be calculated in multiple ways. Here we present two variants: number of nodes and tree depth:

$$M_W(CT^*|CT_1,\ldots,CT_N) = \min\left\{1, \frac{card(W_1) + card(W_2) + \ldots + card(W_N)}{card(W^*)}\right\}$$

(3)

$$M_D(CT^*|CT_1,\ldots,CT_N) =$$
$$= \min\left\{1, \frac{max_{w \in W_1} depth(w) + \ldots + max_{w \in W_N} depth(w)}{max_{w \in W^*} depth(w)}\right\} \quad (4)$$

Minimality is equal 1, when size of $CT^*$ is no larger than the sum of sizes of $CT_1$, $CT_2$, $\ldots$ and $CT_N$ – for abovementioned variants, the cardinality of set $W^*$ is not larger than the sum of cardinalities of $W_1$, $W_2$, $\ldots$ and $W_N$, and the maximum depth of a node in $W^*$ is no larger than the sum of maximum depth of nodes in $W_1$, maximum depth of node in $W_2$, ... and maximum depth of node in $W_N$. Minimality is close to 0 (in real world situations never equal 0), when size of $CT^*$ is much larger than the sum of sizes of $CT_1$, $CT_2$, ... and $CT_N$ – in abovementioned cases, when cardinality of set $W^*$ is much larger than the sum of cardinalities of sets $W_1$, $W_2$, ... and $W_N$, or when the maximum depth of node in $W^*$ is much larger than the sum of depths of deepest nodes in $W_1$, $W_2$, ... and $W_N$.

For this measure, *minimality criterion is met if the given minimality measure variant is maximum (equal 1).*

## 4   Decomposition of Integration Tasks

The criteria defined in the previous section allow decomposition of integration task. We have proposed a generalized outline of decomposition in hierarchical structures integration in our previous work [11]. The decomposition is understood as a following process:

1. dividing the integration task into subtasks with smaller set of inputs;
2. determining the results of integration in subtasks;
3. integrating the results of subtasks into a final solution.

Decomposition allows reduction of computation time for algorithms with high computational complexity.

**Definition 9.** *Integration task may be decomposed, if it satisfies the following requirement:*

– *for even N:*

$$I(CT_1, CT_2, \ldots, CT_N) = I\big(I(CT_1, CT_2), I(CT_3, CT_4), \ldots, I(CT_{N-1}, CT_N)\big) \tag{5}$$

– *for odd N:*

$$I(CT_1, CT_2, \ldots, CT_N) =$$
$$= I\big(I(CT_1, CT_2), I(CT_3, CT_4), \ldots, I(CT_{N-2}, CT_{N-1}), CT_N\big) \tag{6}$$

**Theorem 1.** *Each integration task for N complex trees that satisfies the structure completeness criterion may be decomposed.*

*Proof.* The proof is performed for two cases, when $N$ is even and when $N$ is odd:
*1) For even N:*
The integration process requires structure completenes measure equal 1, which means that:

$$\forall_{w \in W_1 \cup W_2 \cup \ldots \cup W_N} : w \in W^*. \tag{7}$$

Therefore after integration is finished $W^* = W_1 \cup W_2 \cup \ldots \cup W_N$.

As we consider no other criteria:

$$CT^* = I(CT_1, CT_2, \ldots, CT_N) \Leftrightarrow W^* = W_1 \cup W_2 \cup \ldots \cup W_N. \tag{8}$$

For sets of nodes the following equivalence is true:

$$W_1 \cup W_2 \cup \ldots \cup W_N = (W_1 \cup W_2) \cup (W_3 \cup W_4) \cup \ldots \cup (W_{N-1} \cup W_N). \tag{9}$$

Based on (8) the sum of sets $W_1' = W_1 \cup W_2$ is the set of nodes in the complex tree that results from integration $CT_1' = I(CT_1, CT_2)$. Similarly $W_2' = W_3 \cup W_4$ is the set of nodes in $CT_2' = I(CT_3, CT_4)$, and so on. Therefore:

$$W_1 \cup W_2 \cup \ldots \cup W_N = (W_1 \cup W_2) \cup (W_3 \cup W_4) \cup \ldots \cup (W_{N-1} \cup W_N) = W_1' \cup W_2' \cup \ldots \cup W_{\frac{N}{2}}'. \tag{10}$$

allows the following:

$$I(CT_1, CT_2, \ldots, CT_N) = I(CT_1', CT_2', \ldots, CT_{\frac{N}{2}}'). \tag{11}$$

$CT_i'$ (for $i = 1, \ldots, \frac{N}{2}$) are the results of integrating pairs of trees, thus:

$$I(CT_1, CT_2, \ldots, CT_N) = I(CT_1', CT_2', \ldots, CT_{\frac{N}{2}}') =$$
$$= I(I(CT_1, CT_2), I(CT_3, CT_4), \ldots, I(CT_{N-1}, CT_N)). \tag{12}$$

*2) For odd N:*
The proof if similar as for even $N$ until the substitution (10). For odd $N$ it has the following form:

$$W_1 \cup W_2 \cup \ldots \cup W_N =$$
$$= (W_1 \cup W_2) \cup (W_3 \cup W_4) \cup \ldots \cup (W_{N-2} \cup W_{N-1}) \cup W_N = W_1' \cup W_2' \cup \ldots \cup W_{\frac{N}{2}}' \cup W_N. \tag{13}$$

This allows the following notation:

$$I(CT_1, CT_2, \ldots, CT_N) = I(CT_1', CT_2', \ldots, CT_{\frac{N-1}{2}}', CT_N). \tag{14}$$

$CT_i'$ (for $i = 1, \ldots, \frac{N}{2}$) are results of integrating pairs of trees, thus:

$$I(CT_1, CT_2, \ldots, CT_N) = I(CT_1', CT_2', \ldots, CT_{\frac{N-1}{2}}', CT_N) =$$
$$= I(I(CT_1, CT_2), I(CT_3, CT_4), \ldots, I(CT_{N-2}, CT_{N-1}), CT_N). \tag{15}$$

**Theorem 2.** *Each integration task of N complex trees satisfying the cardinality variant of minimality ($M_W$) may be decomposed.*

*Proof.* The proof is performed for two cases, when $N$ is even and when $N$ is odd:
*1) For even N:*

The cardinality variant of minimality $(M_W)$ is equal 1 if:

$$card(W^*) \leq card(W_1) + card(W_2) + \ldots + card(W_N). \tag{16}$$

Assume that the result of integrating two complex trees $CT_1' = I(CT_1, CT_2)$ has the set of nodes $W_1'$. Minimality criterion is met, so:

$$card(W_1') \leq card(W_1) + card(W_2). \tag{17}$$

Similar inequality occurs to other pairs of trees. These inequalities may be grouped as follows:

$$card(W_1') + card(W_2') + \ldots + card(W_{\frac{N}{2}}') \leq card(W_1) + card(W_2) + \ldots + card(W_N). \tag{18}$$

The result of integrating complex trees $CT_1'$, $CT_2'$, ..., $CT_{\frac{N}{2}}'$ is some element $\widetilde{CT}$. The set of nodes of this tree is denoted as $\widetilde{W}$. Therefore:

$$card(\widetilde{W}) \leq card(W_1') + card(W_2') + \ldots + card(W_{\frac{N}{2}}'), \tag{19}$$

which together with (18) gives the following inequalities:

$$card(\widetilde{W}) \leq card(W_1') + card(W_2') + \ldots + card(W_{\frac{N}{2}}') \leq$$
$$\leq card(W_1) + card(W_2) + \ldots + card(W_N). \tag{20}$$

Therefore we have:

$$card(\widetilde{W}) \leq card(W_1) + card(W_2) + \ldots + card(W_N). \tag{21}$$

This means that the minimality criterion is met and we may denote:

$$I(CT_1, CT_2, \ldots, CT_N) = I\big(I(CT_1, CT_2), I(CT_3, CT_4), \ldots, I(CT_{N-2}, CT_{N-1})\big). \tag{22}$$

*2) For odd N.*

The proof is similar as for even $N$ until the inequality (18). For odd $N$ it has the following form:

$$card(W_1') + card(W_2') + \ldots + card(W_{\frac{N-1}{2}}') + card(W_N) \leq$$
$$\leq card(W_1) + card(W_2) + \ldots + card(W_N) \tag{23}$$

As for even $N$, the result of integrating complex trees $CT_1'$, $CT_2'$, ..., $CT_{\frac{N-1}{2}}'$ and $CT_N$ is some element $\widetilde{CT}$. Its set of nodes is denoted as $\widetilde{W}$. Thus:

$$card(\widetilde{W}) \leq card(W_1') + card(W_2') + \ldots + card(W_{\frac{N-1}{2}}') + card(W_N), \tag{24}$$

together with (23) gives the following inequality:

$$card(\widetilde{W}) \leq card(W_1') + card(W_2') + \ldots + card(W_{\frac{N-1}{2}}') + card(W_N) \leq$$
$$\leq card(W_1) + card(W_2) + \ldots + card(W_N). \tag{25}$$

Therefore:

$$card(\widetilde{W}) \leq card(W_1) + card(W_2) + \ldots + card(W_N). \qquad (26)$$

This means that the minimality criterion is met and we may denote:

$$I(CT_1, CT_2, \ldots, CT_N) =$$
$$= I\bigl(I(CT_1, CT_2), I(CT_3, CT_4), \ldots, I(CT_{N-2}, CT_{N-1}), CT_N\bigr). \quad (27)$$

## 5   Conclusions

In this paper we define the most complete definition of complex tree and the integration task for it. This allows to perform complex analysis of the integration process. One of properties of integration that we determined thanks to using the mathematical model is the possibility of decomposing the integration task.

Decomposition is the act of splitting a single integration task into multiple tasks taking place in a hierarchy of its own. Thus we may understand decomposition as one of possible aspects of multi-stage integration. Multi-stage integration is an important area that many researchers have indicated needs to be developed. Here we provide a small step towards developing such approaches.

This research will be used as a starting point for author's further work into multi-stage integration. With the research presented in this paper we may consider multi-stage integration as follows. The integration task is divided into subtasks. Each pair of subtasks is integrated into a partial result. Each pair of results is then integrated in turn, until only a single result remains. It may be also possible to determine other important characteristics of integration task, based on the proposed complete mathematical model.

**Acknowledgement.** This research was co-financed by a Ministry of Higher Education and Science grant no. B30012/I32.

## References

1. Barthelemy, J.P., McMorris, F.R.: The median procedure for n-trees. Journal of Classification 3, 329–334 (1986)
2. Barthelemy, J.P.: Thresholded consensus for n-trees. Journal of Classification 5, 229–236 (1988)
3. Bordewich, M., Semple, C.: On the computational complexity of the rooted subtree prune and regraft distance. Annals of Combinatorics 8(4), 409–423 (2005)
4. Cole, R., Farach-Colton, M., Hariharan, R., Przytycka, T., Thorup, T.: An o(nlog n) algorithm for the maximum agreement subtree problem for binary trees. SIAM J. Comput. 30(5), 1385–1404 (2000)
5. Day, W.H.E.: Optimal algorithms for comparing trees with labeled leaves. Journal of Classification 2, 7–28 (1985)
6. Do, H.H.: Schema Matching and Mapping-based Data Integration. Ph.D. Thesis, University of Leipzig (2006)

7. Do, H.-H., Melnik, S., Rahm, E.: Comparison of Schema Matching Evaluations. In: Chaudhri, A.B., Jeckle, M., Rahm, E., Unland, R. (eds.) NODe-WS 2002. LNCS, vol. 2593, pp. 221–237. Springer, Heidelberg (2003)
8. Madria, S., Passi, K., Bhowmick, S.: An XML Schema integration and query mechanism system. Data and Knowledge Engineering 65, 266–303 (2008)
9. Maleszka, M.: Knowledge Generalization during Hierarchical Structures Integration. In: Nguyen, N.T., Attachoo, B., Trawiński, B., Somboonviwat, K. (eds.) ACI-IDS 2014, Part I. LNCS, vol. 8397, pp. 242–250. Springer, Heidelberg (2014)
10. Maleszka, M., Nguyen, N.T.: A Method for Complex Hierarchical Data Integration. Cybernetics and Systems 42(5), 358–378 (2011)
11. Maleszka, M., Nguyen, N.T.: Some Properties of Complex Tree Integration Criteria. In: Jędrzejowicz, P., Nguyen, N.T., Hoang, K. (eds.) ICCCI 2011, Part II. LNCS, vol. 6923, pp. 1–9. Springer, Heidelberg (2011)
12. Maleszka, M., Mianowska, B., Nguyen, N.T.: A method for collaborative recommendation using knowledge integration tools and hierarchical structure of user profiles. Knowledge-Based Systems 47, 1–13 (2013)
13. Passi, K., Lane, L., Madria, S., Sakamuri, B.C., Mohania, M., Bhowmick, S.: A Model for XML Schema Integration. In: Bauknecht, K., Tjoa, A.M., Quirchmayr, G. (eds.) EC-Web 2002. LNCS, vol. 2455, pp. 193–202. Springer, Heidelberg (2002)
14. Rahm, E., Bernstein, P.A.: A survey of approaches to automatic schema matching. The VLDB Journal 10, 334–350 (2001)

# A Web-Based Multi-Criteria Decision Making Tool for Software Requirements Prioritization

Philip Achimugu, Ali Selamat[*], and Roliana Ibrahim

UTM-IRDA Digital Media Centre,
K-Economy Research Alliance & Faculty of Computing, Universiti Teknologi Malaysia,
Johor Bahru, 81310, Johor, Malaysia
check4philo@gmail.com, {aselamat,roliana}@utm.my

**Abstract.** Multiple-criteria decision making (MCDM) is widely used in ranking choices from a set of available alternatives with respect to multiple criteria. To analytically rank requirements under various criteria, we propose a tool called requirements prioritizer (RP) which has the capacity of keeping records of project stakeholders with their relative weights against each requirement, utilized by the system to compute an ordered list of prioritized requirements. The proposed approach offers a novel way of involving stakeholders in the entire decision making process irrespective of their numbers in an automated fashion. In this proposed approach, the relative weights assigned by each stakeholder are normalized and aggregated. The output of the system consists of prioritized requirements with an automatically generated graph showing the relative values of requirements across project stakeholders in a chronological order.

**Keywords:** MCDM, software, requirements, prioritization, tool.

## 1 Introduction

Requirements prioritization can be considered to be a multi-criteria decision making (MCDM) process [1]. It is a technique used to rank items or requirements from a pool based on some pre-defined criteria. This ranking has to do with the pair wise comparisons of requirements in order to determine their relative importance by help of a weight scale. MCDM is widely used in many domains to solve decision making problems. Software engineering is one domain where the application MCDM seems to be inevitable. During software development, requirements are elicited, analyzed and modeled before implementation. Prioritizing software requirements enhances the chances of developing quality systems with preferential requirements of stakeholders. Furthermore, it helps stakeholders or developers balance or determine the possibility of implementing specified requirements with respect to available resources such as skilled programmers, time and budget among others. Consequently, disagreements, conflicts or breaches in contract are avoided. This research presents a scalable

---

[*] Corresponding author.

D. Hwang et al. (Eds.): ICCCI 2014, LNAI 8733, pp. 444–453, 2014.

web-based tool known as Requirements Prioritizer (RP) to assist stakeholders specify and prioritize requirements from any geographical location at runtime.

The rest of the paper is organized as follows: in section 2, we discuss the related works and identify limitations of existing prioritization techniques. In section 3, we briefly introduce the proposed architecture and evaluation results. Section 4 concludes the research and suggests an area for future work.

## 2     Related Work

Researchers and practitioners have proposed various prioritization techniques in recent years. These techniques have been validated using experimental studies [2], case studies [3], empirical studies [4, 5] and literature studies [6]. Existing prioritization techniques are frequently categorized under two main concepts. These are: techniques that are applied to small number of requirements and techniques that are applied to large numbers of requirements. Techniques like round-the-group prioritization, multi-voting system, pair-wise analysis, weighted criteria analysis and the Quality Function Deployment approach falls under the former while techniques like MoSCoW, Binary Priority List, Planning Game and Wiegers's matrix techniques falls under the latter. However, considering the paradigm shift in software development processes, supporting tools are needed to cater for large number of requirements.

In a related development, Berander and colleagues [7] categorized existing techniques into two main classes, that is, (1) techniques which allows stakeholders to allot weights to various requirements (2) techniques which permit negotiations among stakeholders where priorities are arrived at through consensus. Techniques which apply to the first class are Analytical Hierarchy Process (AHP), Cumulative Voting, Numerical Assignment, Planning Game and Wieger's method. An example of a technique in the second class is Win-Win and Multi-criteria Preference Analysis Requirements Negotiation (MPARN) approaches.

AHP has gain tremendous attention among practitioners and seems to be the most popular technique. AHP is a typical multiple criteria decision-making technique that has been adopted for prioritizing software requirements [8-10]. However, the major drawbacks of AHP are scalability and rank reversals. The former describes the inability of the technique to perform well as requirements increases while the latter has to do with the inability of AHP to reflect or update new ranks whenever a new requirement is added or deleted from the list. Notwithstanding, in terms of reliability of prioritization results, AHP is considered to be the best. Bubblesort and Case base rank has also received attention in literature. But the problem faced by AHP also applies to Bubblesort and Case base rank as well.

So many prioritization techniques exist but supporting tool that is required to allow practitioners exercise or perform this exercise is unavailable. Concise analysis of some selected techniques can be found in [11, 12]. In this research, we proposed and implement a generic technique that can help developers determine preferential requirements of stakeholders through a weighting scale so as to plan for software releases.

# 3     Proposed Architecture and Evaluation

The proposed architecture for the RP is presented in this section. This approach has the merit of carrying all relevant stakeholders along in the entire decision making process. The proposed architecture consists of five major subsystems which are: user interface, model management, database management, prioritization management, and output management (Figure 1). However, the various components that constitute the architecture are described below:

## 3.1     User Interface

The user interface enables stakeholders to visually and interactively register themselves, and rank requirements. There are two major types of user devices: PC and mobile terminal devices (e.g., personal digital assistants and smart phones). Since requirements prioritization is considered as a multiple criteria decision making (MCDM) process, there is need to develop a scalable and interactive decision support system that is capable of comparing requirements relatively in order to determine an ordered list of prioritized requirements.

The mathematical model underpinning the proposed architecture is described as follows:

Let $X$ comprise of specified requirements with distinct attributes of $\sigma$-functionalities, where $N$ are all the number requirements of $X$. Prioritization $g$ defined on the measurable space $(X, N)$ is a set function $g : N \rightarrow [0,1]$ which satisfies the following properties:

$$g(\phi) = 0, g(X) = 1 \tag{1}$$

But for requirements $R_1, R_2, R_3, ... R_n$; the prioritization equation will be:

$$R_1 \subseteq R_2 \subseteq R_3, ... R_n \in N \rightarrow [0,1] \tag{2}$$

From the above definition, $X, N, g$ are said to be the parameters used to measure or determine the relative weights of requirements. This process is monotonic. Consequently, the monotonicity condition is obtained as:

$$g(R_1 \cup R_n) \geq \max\{g(R_1), g(R_n)\} ; g(R_1 \cap R_n) \leq \min\{g(R_1), g(R_n)\} \tag{3}$$

In the case where $g(R_1 \cup R_n) \geq \max\{g(R_1), g(R_n)\}$, the prioritization function $g$ attempts to determine the total number of requirements being prioritized and if $g(R_1 \cap R_n) \leq \min\{g(R_1), g(R_n)\}$, the prioritization function attempts to compute the relative weights of requirements provided by the relevant stakeholders. However, for updating rank status when requirements evolve, we have:

$$\int h(X) g(X) = \vee (h(R_1) \wedge g(R_n)) \tag{4}$$

Where $h(X)$ is a linear combination of an attribute function such that $R_1 \subset R_2 \subset ... \subset R_n \in X$.

Therefore, for the requirement set, the global weights of requirements are measured by:

$$W_i = \sum_{i=1}^{n} R_i \tag{5}$$

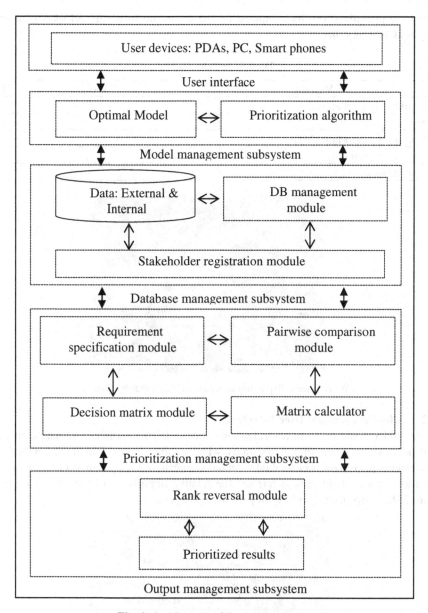

**Fig. 1.** Architecture of the proposed system

### 3.1.1    PC User Interface
The RP provides a familiar and consistent graphic user interface (Figure 2). With this interface, stakeholders can specify and rank requirements. Furthermore, the user interface displays the section that allows stakeholders to pair wisely compare requirements and rate them based on a 5-point scale shown in Table 1 as well as

display the prioritized results computed by the algorithm in the model management subsystem, among others.

**Table 1.** Weight scale

| Variables | Weight |
|---|---|
| Extremely Important (EI) | 5 |
| Very Important (VI) | 4 |
| Moderately Important (MI) | 3 |
| Fairly Important (FI) | 2 |
| Least Important (LI) | 1 |

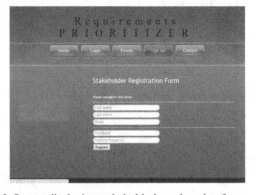

**Fig. 2.** Screen displaying stakeholder's registration form

### 3.1.2   Mobile Terminal User Interface

A stakeholder could perform prioritization exercise from a mobile terminal, such as smart phones or personal digital assistants, which enables them communicate with the requirements prioritizer. The mobile terminal connects to the system via the internet, thus allowing stakeholders to determine the relative importance of the requirements. Figure 3 shows the user interface for pair-wise comparisons.

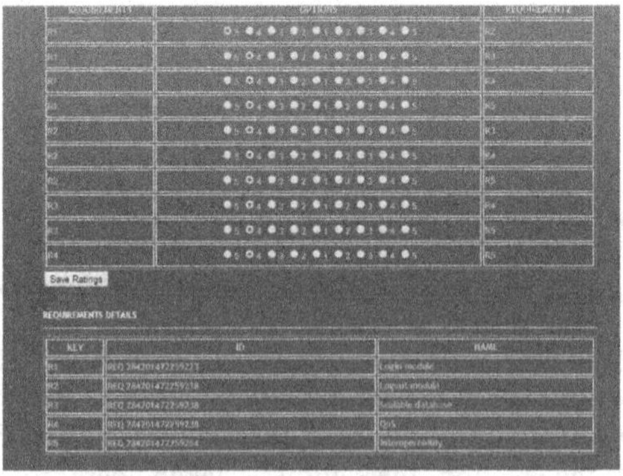

**Fig. 3.** Pairwise comparisons

## 3.2    Model Management Subsystem

This subsystem has two components: optimal model and prioritization algorithm. The optimal model contains the constraints, frameworks, theories, formulas, assumptions, and definitions, among other details, which are needed to prioritize requirements. One of the aims of the optimal model is to compute ranks that precisely tallies with the original weights allotted by the stakeholders. The algorithm governing the calculation of prioritized requirements consists of the under-listed steps:

*Step 1:* Given a prioritization event E with Requirements $R_1$, $R_2$, $R_3$,..., $R_n$ and Stakeholders $S_1$, $S_2$, $S_3$,..., $S_m$, a pairwise comparison of requirements is created for each stakeholder as follows:

| $R_1$ | $R_2$ | | $R_2$ | $R_3$ | ....... | $R_{n-1}$ | $R_n$ |
|---|---|---|---|---|---|---|---|
| $R_1$ | $R_3$ | | $R_2$ | $R_4$ | | | |
| $R_1$ | $R_4$ | | $R_2$ | $R_5$ | | | |
| $R_1$ | $R_5$ | | $R_2$ | R... | | | |
| $R_1$ | R... | | $R_2$ | $R_n$ | | | |
| $R_1$ | $R_n$ | | | | | | |
| | | | | | | | |

*Step 2:* For each stakeholder $S_1$ to $S_m$, rank of each pairwise comparison for requirements is given as:

$$rank_{j.S_i} \qquad (6)$$

Where, $1 \leq j \leq NoOfComparisons$ and $1 \leq i \leq m$

*Step 3:* Find the sum of the ranks of each of the pairwise comparisons, for all the stakeholders, thus:

$$rankSum_j = \sum_{i=1}^{m} rank_{j.S_i} \qquad (7)$$

*Step 4:* From Equation 2, the matrix of rankings is generated.

*Step 5:* Having derived the matrix of the rankings, the square of the matrix is calculated.

*Step 6:* The sum of the each of the rows of the matrix is calculated and the result is a (n x 1) matrix called the Eigenvector, which represents the individual collective ranking of all the requirements.

*Step 7:* The Eigenvector is normalised; the sum of the all the values in the Eigenvector is calculated and is used to divide each of the values in the Eigenvector. This allows all the values to be on a scale of 1 and the sum of all the values to tend towards 1.

Figure 4 shows the process of validating requirements and stakeholders in the model management system by an administrator. The registered stakeholders and elicited requirements become valid and ready for the exercise only when the administrator of the system has confirmed registration by adding them to the console. Once this is done, the entire number of stakeholders for a particular project and elicited requirements are displayed or pair wise comparisons.

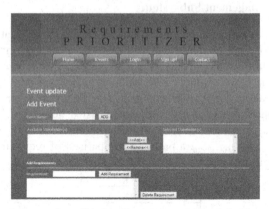

**Fig. 4.** Screen for validating requirements and stakeholders

## 3.3    Database Management Subsystem

The database system was developed in the Microsoft SQL server. The database organizes internal and external data as well as the information required for the system to function. Most of the internal data are stakeholder's weights. External data consist of the elicited requirements and stakeholder's registration information. Each requirement is allotted a unique identification number to avoid redundancy. This identification number is a combination of the word Event and the date and time of creation. The database management module stores the data in the database, retrieves data and controls the database.

## 3.4    Prioritization Management Subsystem

This subsystem consists of four basic modules which are: requirements specification, pairwise comparison, decision matrix and matrix calculation modules. Pseudo code 1 is used to calculate the decision matrix weights in order to compute the final ranks of requirements across all project stakeholders.

```
1. Set  weight  as  the  array  (matrix)  of  all  stakeholder-
   prioritized events.
2. Set No. as the number of requirements.
3. Calculate the square of (weight) as weightSquare.
4. Calculate the Eigen vector and set as finalVector.
5. Convert the list of requirements to array as arrReq.
6. Rearrange finalVector and arrReq based on the weights of each
   of the requirements in the finalVector.
```

**Pseudo code 1:** Calculating global weights of requirements

A good requirements prioritization technique is one which involves all the relevant stakeholders, provides them the flexibility of assessing and ranking the requirements by means of subjective or relative weights and aggregates weights to generate reliable prioritized results. All the elicited requirements are stored into the database system (Microsoft SQL server) in our case.

## 3.5    Output Management Subsystem

The output of the system is generated as soon as all the stakeholders have completed the weighting exercise. Therefore, this subsystem is responsible for displaying the prioritized results. It also deals with the problem of rank reversals (requirements change or evolution) inherent in existing techniques. This causes requirements to be included or deleted from the list. When this happens, the RP will automatically update ranks and re-compute the new ranks based on the new stakeholder's weight (Pseudo code 2). Lastly, our system is intelligent enough to generate prioritized elements in chronological order based on the final results and automatically generate a chart reflecting the prioritized requirements (Figure 5).

```
For each added or deleted requirement,
Add the value to the list of rating.


UpdateQuery Method

Background: For every user added to an event, the stakeholder's id
is added to the table of stakeholder's ratings in the database with
the value 0. The value 0 indicates that the stakeholder has not
rated the event.
So this method builds the query for the update.

    1.  Initialize the query statement as details with update
        statement.
    2.  Initial string variable as rating.
    3.  For each rating made by the user,
            a.  Add the rating to string variable rating.
    4.  Add rating to details.
    5.  Add WHERE clause to specify the stakeholder's id and the
        event id.
```

**Pseudo code 2:** Rank reversals

The system was designed in a way that stakeholders can specify and rank requirements from any geographical location. The system supports flexibility because requirements could be elicited remotely via emails or social networks before prioritization begins. The interface developed can integrate with modern information and communication technologies, principally the Internet. Determining the weights of stakeholder's requirements is achieved by synthesizing priorities over all levels obtained by varying numbers of requirements. The output of this system will help developers in determining where to invest more efforts in order to enhance the delivery of good quality software that meets user's requirements.

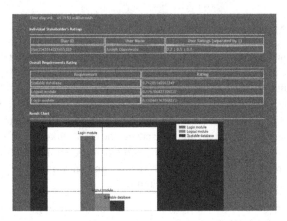

**Fig. 5.** Output of the decision support system

# 4    Conclusion/Future Work

The aim of this research was to identify the limitations of existing prioritization techniques with the aim of improving on them. It was eventually discovered that existing techniques actually suffer from mainly scalability problems, large disparity or disagreement between ranked weights, rank reversals, as well as unreliable results. These were addressed at one point or the other during the course of undertaking this research. The method utilized in this research consisted of intelligent algorithms implemented with C# and MicrosoftSQL server 2012. Various algorithms and models were formulated in order to enhance the usability of the proposed approach. The developed tool was designed and implemented in a way to cater for as much requirements and stakeholders available for any software project; it is easy to use with friendly user interface, reduced computational complexities and has addressed rank reversals issues. For the future work, we hope to validate the tool in a real-life setting with large numbers of stakeholders and requirements alike. Finally, the developed tool is able to classify ranked requirements in chronological order with an accompanied graph to visualize the prioritized results at a glance. For dependency issues, requirements are thoroughly analyzed to track redundant, conflicting, independent and dependent requirements before the prioritization process begins.

**Acknowledgement.** This work is supported by the Research Management Centre (RMC) at the Universiti Teknologi Malaysia under Research University Grant (Q.J130000.2510.03H02), the Ministry of Science, Technology & Innovations Malaysia under Science Fund (R.J130000.7909.4S062) and the Ministry of Higher Education (MOHE) Under Exploratory Research Grant Scheme (R.J130000.7828.4L051).

# References

1. Perini, A., Angelo, S., Paolo, A.: A machine learning approach to software requirements prioritization. IEEE Transactions on Software Engineering 39(4), 445–461 (2013)
2. Karlsson, L., Thelin, T., Regnell, B., Berander, P., Wohlin, C.: Pair-wise comparisons versus planning game partitioning – experiments on requirements prioritisation techniques. Empirical Software Engineering 12(1), 3–33 (2007)
3. Karlsson, J.: Software requirements prioritizing. In: Proceedings of 2nd International Conference on Requirements Engineering, pp. 110–116 (1996)
4. Lehtola, L., Kauppinen, M.: Empirical evaluation of two requirements prioritization methods in Product Development Projects. In: Proc. European Software Process Improvement Conference, Trondheim, Norway, pp. 161–170 (2004)
5. Perini, A., Ricca, F., Susi, A.: Tool-supported requirements prioritization: comparing the AHP and CBRank methods. Information and Software Technology 51(6), 1021–1032 (2009)
6. Avesani, P., Bazzanella, C., Perini, A., Susi, A.: Facing scalability issues in requirements prioritization with machine learning techniques. In: Proceedings of 13th IEEE International Conference on Requirements Engineering, Paris, France, pp. 297–306. IEEE Computer Society (2005)
7. Berander, P., Andrews, A.: Requirements prioritization. In: A. Aurum, C. Wohlin (Eds.), Engineering and Managing Software Requirements. Springer (2005)
8. Karlsson, J., Wohlin, C., Regnell, B.: An evaluation of methods for prioritizing software requirements. Information and Software Technology 39, 14–15 (1998)
9. Karlsson, J., Ryan, K.: A cost-value approach for prioritizing requirements. IEEE Software 14(5), 939–947 (1997)
10. Voola, P., Vinaya Babu, A.: Interval evidential reasoning algorithm for requirements prioritization. In: Satapathy, S.C., Avadhani, P.S., Abraham, A. (eds.) Proceedings of the InConINDIA 2012. AISC, vol. 132, pp. 915–922. Springer, Heidelberg (2012)
11. Khari, M., Nikunj, K.: Comparison of six prioritization techniques for software requirements. Journal of Global Research in Computer Science 4(1), 38–43 (2013)
12. Achimugu, P., Selamat, A., Ibrahim, R., Mahrin, M.N.: A systematic literature review of software requirements prioritization research. Information and Software Technology 56(6), 568–585 (2014)

# The Selection of Multicriteria Method Based on Unstructured Decision Problem Description

Jarosław Wątróbski, Jarosław Jankowski, and Zbigniew Piotrowski

Faculty of Computer Science and Information Technology,
West Pomeranian University of Technology Szczecin, Poland
{jwatrobski,jjankowski,zpiotrowski}@wi.zut.edu.pl

**Abstract.** Decision support processes and methods require applying numerous mathematical transformations, including one of the developed processes of multicriteria analysis. The core of most existing processes is usually one of the multicriteria decision aid methods (MCDA). The paper presents research focused on identifying which factors of a decision situation are significant for selecting a multicriteria method. The identified factors were analyzed with data-mining methods. Conclusions contain an outline of factors of decision situations that support MCDA methods to support decisions in particular situations.

**Keywords:** decision support, multicriteria methods, MCDA.

## 1 Introduction

Decision support is a vague domain in which mathematics, sociology, and technical science overlap. The last category includes applying algorithms in software tools [13][16] with interfaces making it easier to perform data processing and analysis. They can also exceed those basic functions by driving a dialog with a decision maker, leading to the expression of a description of a decision situation. Entering data about decision alternatives can include much more than just data entry of values of respective criteria [21]. Data mining, text mining, and Web mining methods and software agents technology enable actively searching for scenarios that fulfill prerequisites to include them in a set of decision alternatives [7][19]. While different methods can be applied for solving a decision problem, their usage is dependent on the structure and characteristics of the problem [21]. Wrongly applied methods can deliver results that do not meet decision-maker preferences [26]. While several approaches can be used to solve this problem based on preferences, characteristics of input parameters or number of criteria or decision makers [10.11], it is difficult to adjust methods to specifics of a problem, due to a lack of clear rules. The presented paper proposes identification of evidence of selection methods for decision-making purposes, on the basis of selected factors and the context of the decision situation. An attempt was made to discover decision rules for the selection of the method best suited for the multicriteria decision-making situations' described environmental information. The work is presented as follows: after the literature review and problem

D. Hwang et al. (Eds.): ICCCI 2014, LNAI 8733, pp. 454–465, 2014.

statement, the conceptual framework is presented with assumptions for the proposed approach. In the next stage, a rules database is created for several decision schemes and illustrated using decision trees. After that, conclusions and final remarks are presented.

## 2    Literature Review

Among many decision taxonomies, from this article's standpoint, the most significant is basic division into single and multicriteria decisions. Single-criterion decisions cover finding the most desired value for a defined criterion of judging scenarios and analysis allowing to find an "optimal" scenario that is expected in various methods, e.g., operational research. More interesting from a scientific standpoint, however more difficult to apply in practical terms, is the multicriteria decision approach. A multicriteria decision occurs when a goal of a decision maker includes finding best values for two or more independent criteria. Identifying subsequent practical needs for aiding multicriteria decisions, a group of methods and algorithms was developed, which covers ranking the decision alternatives and selecting the best. Depending on its characteristics, a decision can be assigned to one of the reference decision problematics [1]: $\alpha$ - choice problematic  based on finding a subset of set containing the best solution, $\beta$ - sorting problematic related to splitting alternatives into predefined categories, $\gamma$ - ordering problematic and building a ranking of decision alternatives from the best one to the worst one. A decision problem is more widely defined by B. Roy [22] as a representation of an element of a global decision, which accounts for the progress state of a decision problem and can be analyzed independently and serve as a reference point for decision aiding. The literature distinguishes several types of decision alternatives, with respect to real alternatives (for a complete and finished project that can be implemented) and fiction alternatives (describing an idealistic project, not fully developed or even imagined). With respect to completeness of the analyzed decision situation, available actions can be categorized as global alternatives and partial alternatives. The main difficulty in applying multicriteria methods (MCDA) lies in considering multiple points of view when grading decision alternatives (choice alternatives), i.e., multicriteria judgement [8]. Subsequent levels can be distinguished in multicriteria decision analysis like defining a subject of a decision and scope of participation of an analyst, analyzing consequences of a decision and designing decision criteria, modeling global preferences, selecting research procedures.

Formulating a multicriteria problem involves decision situations where a finite set of decision alternatives (actions), $A$ is graded according to $n$ criteria $g_1, g_2, \ldots, g_n$ constituting a criteria family $G=\{1,2,\ldots,n\}$. Without any loss of generality, it can be assumed that the higher the value of criterion function $g_i(a)$, the better alternative $a \in A$ is, considering the criterion $g_i$ for all $i \in G$.

The following describes the work on selecting a method covering the main research areas. The first area is data collection that is used by experts (analysts) in solving a given decision situation. Science literature was used as a reference data source because scientific publications usually give a mathematical rationale for selecting a multicriteria method [25]. The second research area was atomizing premises of a description of a decision situation in a way that is possible to translate each determinant into particular characteristics of a multicriteria method. The subject of an aided decision situation determines the selection of a decision algorithm . An analyst supporting a decision maker (DM) defines a structure based on the nature of a decision, which involves considering a domain of a decision that is covered by a decision [22]. Determining the nature of a decision requires an analyst to use his knowledge of the subject area. The subject of a supported decision determines which decision method should be chosen. An analyst supporting a DM defines a structure based on a decision, which in this paper is considered as a selected part of the reality that is covered by a decision. An analyst is required to use his knowledge gained from experience, in order to determine the nature of a decision. He should understand the specific language of a domain where a decision is made. The outcome of interaction between an analyst and a DM is a selection of one of the aforementioned problematics. One of the factors qualifying a problem into one of the problematics is resource availability, as well as possibility to implement a decision only once (usually due to the high cost of this action). In real life, it is the main factor for judging a decision situation belonging to problematics $\alpha$ or $\gamma$. A set of guidelines for selecting a multicriteria method according to available inputs and desired outputs was considered as an entry point of the analysis. The mentioned guidelines were described by Guitouni in [10,11]. While fitting MCDA methods to various decision problems, both practical and theoretical, some weaknesses of those guidelines surfaced. The first and most important weakness (identified in [11]) is the assumption of a reliable identification of a decision situation by a DM and stakeholders. Unfortunately, such a convenient scenario rarely occurs. Hence, more factors need to be considered. The next step of the research was analyzing the domain areas of decision problems. While performing the research, a set of environmental factors was outlined, which describes the alignment of a multicriteria method to a decision situation. A sample decision problem fully outlining the complexity of decisions made in modern organizations is the selection of a location for a new facility of an enterprise. The considered problem is multicriterial, where criteria cover various and distinct domain areas. Moreover, a decision has a significant impact on an organization, where success in implementing a decision can impact a company's position in the long term. The review showed that making a framework for adjusting the decision support method can increase the usefulness of the multicriteria method and better describe structuralized problems. In the next section, the conceptual framework for this purpose is presented.

## 3     Conceptual Framework and Proposed Approach

The starting point for the construction of a framework is presented in [11] approach to extract a set of values that affect the modeling preferences. In this paper, the presented approach was expanded and included in the description of the decision situation set, a broader category that includes the influence on the course of the entire decision-making process. Defined in this way, a set of metadata shows a pattern of decision making as defined by the formula:

$$\Phi = \{W, P, Q, V, U, K\} \tag{1}$$

Where the values of vectors $P=\{p_1, p_2,..., p_n\}$ and $Q=\{q_1, q_2,..., q_n\}$ define the support conditions respectively for relations $P$ and $Q$ and are referred to as the threshold preferences $p$ and indiscernibility threshold $q$. Set $W=\{W_1,W_2,...,W_n\}$ contains values $w_i$ or subsets of values $W_i$ that define the absolute and relative validity criterion $C_i$. In the case of determining the relative weights, the value $w_{ij}$ of the subset $W_i=\{w_{i1},w_{i2},...,w_{in}\}$ are defined on a Likert scale [6] and when defining the single-element subset $W_i=\{w_i\}$, the value can be crisp or fuzzy depending on the specification of the decision maker. The set $U=\{u_1,u_2,...,u_n\}$ is a set of utility functions for the attribute corresponding to the criterion $C_i$. Set $V=\{v_1,v_2,...,v_n\}$ includes veto values defined by a decision maker, which define the criterion for the rejection of the decision-making preferences option under consideration due to the significant difference in the individual single criterion. In a situation where $p_i=q_i=0$ for the criterion $C_i$ a sectional scale is required and preference is described by the operator " $>$" on the scale used. $K$ is the description of the problem domain serving the decision to select the method of practical applications, which confirmed matching the selected method.

In the first stage a model of the decision problem which takes into account the environmental characteristics and context of the decision situation is proposed. The decision problem is defined as four ordered elements according to the formula:

$$(C, \Phi, A, \Psi) \tag{2}$$

Where $C$ is a set of criteria, $\Phi$ represents the "context" based on metadata, $A$ is a set of potential decision variants shown as model data of the variants, and $\Psi$ is a collection of methods considered in solving the problem of decision making. The intention of the decision maker is to select the option that best meets their preference for a specific set of criteria. Further consideration was adopted as a solution to the problem of decision making to maximize the outcome of the transformation $F$ designating the degree of fulfillment of the criteria selected by the successive variants of decision making. This takes into account the assumptions of the shape of the decision-making process selected by $\Phi$, as shown in the equation:

$$G(a_p) = \max F(C(A), \Phi) \tag{3}$$

Where $a_p$ is the most preferred option selected from a set of decision-making variants $A$ and $G(a_p)$ a performance variant $a_p$ denoted also as an assessment of the fulfillment of criteria $C$. The set $\Phi$ is a set of characteristics describing decisions by the decision maker, such as agreement on compensation between the criteria, which will also be referred to as metadata decision situation. According to the definitions introduced above, the problem can be expressed as the problem of finding such a transformation $F$ which will choose the decision-making variant delivering maximum value to the decision maker, considering a set of criteria $C$, in a manner consistent with the characteristics described by the set $\Phi$, of internal characteristics of metadata of the decision-making process. To demonstrate the achievement of the objective, a definition was formalized, which combined a multicriteria selection problem to which decision-making situations are classified satisfying the following assumptions:

1. $max(C_1,...,C_k)$, $min(C_{k+1},C_n)$, where $n \geq 2$, and $(C_1,...,C_k)$ is a set of profit criteria and $(C_{k+1},C_n)$ is a set of cost criteria *(assumptions of multicriteria situation)*;

2. $C \subset \Psi \vee \Psi \equiv \{\Psi_1, \Psi_2,..., \Psi_i\}$ where $C$ is the set of all criteria, $\Psi$ is an area of reality, which is defined as a decision problem, and $\Psi_i$ is a specific subset of fields (for one area of the organization), $i=1,...,n$, $l \leq n$ (assumption of variety of sources describing the decision situation);

3. $D(A) \subset D_1 \cup D_2 \cup ... \cup D_l$ where $D(A)$ is the area of decision variants, $D_i$ is the source of the description of phenomena $i=1,...,n$, $l \leq n$ (assumption describes the diversity of variants of decision making);

4. $\delta A_{ij} << \delta C_i(A_j)$, where $\delta A_{ij}$ is the change in value of the variant $A_j$ relative to criterion $C_i$, $\delta C_i(A_j)$ is the change in the performance criterion $C_i$ for the variant $A_j$ (assumption of a significant impact of small changes in the value of individual attributes).

This framework is based on an analysis of the contexts of applications of multicriteria obtained from the scientific literature. It is assumed that the decision maker gathered from the sets $C$ and $\Phi$ are the basis for the selection shown in the formula (3), the conversion of $F$. For the purpose of simplifying the research, a detailed analysis of the literature was carried out [1, 4, 5, 27, 9, 14, 18, 20, 15, 28, 23, 24, 3, 12, 2]. This made it possible to extract a subset of methods $\Psi'$ commonly used in decision situations based on:

$$\Psi' = \{A_H, T_P, E_3, P_{T1}, P_{T2}, fP_T, fT_P, fK_O\} \qquad (4)$$

Where $A_H$ - AHP, $T_P$ - TOPSIS, $E_3$ - ELECTRE III, $P_{T1}$ - PROMETHEE I $P_{T2}$ - PROMETHEE I/II, , $fP_T$ - Fuzzy PROMETHEE, $fT_P$ - Fuzzy TOPSIS, $fK_O$ - Fuzzy Kuo. It can be observed that subset $\Psi'$ contains methods belonging to the methods' category that represent different approaches to calculation-assisted decisions. Consideration of methods belonging to different categories allowed the identification of mathematical operations that reflect the anticipated effect of the implementation of the calculation of the MCDA. Accepted categories methods are: methods based on the intercriterial relationship $\Psi_r$ (4), methods based on the aggregation of performance attributes to a single-criterion utility $\Psi_u$ and fuzzy methods $f\Psi$: $\Psi_r\{E_3, P_{T1}, P_{T2}\}$, $\Psi_u = \{A_H, T_P,\}$, $f\Psi = \{fP_T, fT_P, fK_O\}$.

## 4     Building a Rules Database Using the Context of the Decision Situation

Determination of the selection methods for a multicriteria decision-making problem is specified by characteristics of the decision area and the related decision-making situation. The decision maker selects subjects from a dictionary of terms constituting the domain, $D(K)$ of $K$ that describe the domain of the decision problem. The set $K$ includes a set of parameters representing the context in the decision-making situations listed below: $K1$ - item-category decisions with domain from the set $\{0, 1, 2, 3, 4, 5, 6, 7, 8, 9\}$; $K2$ - area of decision with domain from the set $\{0,1,2,3\}$; $K3$ - potentially conflicting criteria with domain from the set $\{0,1\}$; $K4$ - description of the decision making with domain from the set $\{0,1\}$; K5 - cardinality of attributes with domain from the set $\{0,1,2,3\}$; K6 - amount of decision variants with domain from the set $\{0,1,2,3\}$; K7 - degree of knowledge of the decision situation with domain from the set $\{0,1,2,3,4,5\}$; K8 - coverage of the intended decision with domain from the set $\{0,1,2,3\}$; K9 - purpose of decision with domain from the set $\{0,1\}$.

The analysis of the relationship between the description, K, of the decision situation and the choice of method, $\Psi$ performed using data-mining techniques. Solutions developed by the authors (using methods they selected) were considered as reference solutions for decision situations in the contexts and decision trees respectively. In particular, the modified parameters included the maximal number of branches and variance within the set $\{2,3,4\}$. Furthermore, we applied three decision-tree-split methods in the experiment namely: $\chi^2$ method, entropy minimization method and Gini reduction method. In the last step a set of MCDA methods was divided into subcategories of utility-value methods and outranking methods and to separately perform analyses for crisp and fuzzy methods. The created decision trees are dependent on one (identifying only a multicriteria approach) to five determining factors. The error rate that was found was within a range from 0.08 for a correctly fitted method of a subgroup of methods to 0.57 which shows that parameters selected for such a scenario are not suitable in this case. The operation was carried out for the designation of the distribution of the three methods: $\chi2$ at the significance level of 0.4, entropy minimization, reduction Gini designated as M1, M2, M3. A total of 63 decision trees were determined. Coefficients determined the number of correctly classified samples (the leaf) in relation to the value assigned to the leaf. The results are shown in Table 1.

**Table 1.** Number of correctly classified samples in leaves

| Name | M1 | M2 | M3 |
|------|----|----|----|
| Minimum observations in a leaf | 2 | 2 | 2 |
| Maximum observations in a leaf | 5 | 5 | 5 |
| Number of branch node | 2 | 3 | 4 |
| The maximum depth of the tree | 10 | 10 | 10 |
| Number of rules on node | 5 | 5 | 5 |
| The significance level for X2 | 0.3 | 0.3 | 0.3 |

The values of the correct qualifications for all trees induced, with the different settings and in accordance with the methodology set out above, are presented in Table 2.

The next steps presented in Table 2 correspond to the decision tree construction carried out for different amounts of leaf node (2,3,4) and for the following transformations: 1. $\Psi' \rightarrow \Psi$ , 2. $\Psi' \rightarrow \{ \Psi'_u, \Psi'_r \}$, 3. $\Psi' \rightarrow \{ f\Psi', \neg\Psi' \}$, 4. $\neg f\Psi' \rightarrow \neg f\Psi'$, 5. $\neg f\Psi' \rightarrow \{ \Psi'_u, \Psi'_r \} \cap \neg f\Psi'$, 6. $f\Psi' \rightarrow f\Psi'$ and 7. $\neg f\Psi' \rightarrow \{ \Psi'_u, \Psi'_r \} \cap f\Psi'$. A graphical analysis of the relationship between the description of the decision situation and the selection method was performed by using a multicriteria decision tree. Methods have been designated as follows: $P\_T$ - PROMETHEE I/II, $T\_P$ - TOPSIS, $A\_H$ - AHP, $E\_3$ - ELECTRE III, $FP\_T$ - Fuzzy PROMETHEE, $FT\_P$ - Fuzzy

460     J. Wątróbski, J. Jankowski, and Z. Piotrowski

TOPSIS, $FK\_O$ - Fuzzy Kuo. Figure 1 provides an indication of which decision tree fits the situation to the method of multicriteria decision making for six of the seven methods included in the collection $\psi'$. The accuracy of classification was at 61%.

**Table 2.** Values of classification for all trees induced

| Leafs | | 2 | | 3 | | 4 | |
|---|---|---|---|---|---|---|---|
| 1 | $\chi^2$ | 0.42 | K3, K6 | 0.42 | K3, K6 | 0.42 | K3, K6 |
| | Entropy | 0.61 | K1, K6, K7, K8 | 0.61 | K1, K2, K6, K7 | 0.55 | K1, K6, K7 |
| | Gini | 0.64 | K1, K6. K7, K8 | 0.61 | K1, K6. K7 | 0.55 | K1, K6, K7 |
| 2 | $\chi^2$ | 0.76 | K3, K6 | 0.76 | K3, K6 | 0.76 | K3, K6 |
| | Entropy | 0.35 | K1, K2, K7, K8 | 0.91 | K1, K2, K7, K8 | 0.91 | K1, K2, K4, K6, K8 |
| | Gini | 0.88 | K1, K2, K6, K8 | 0.88 | K1, K2, K6, K8 | 0.88 | K1, K2, K6, K8 |
| 3 | $\chi^2$ | 0.79 | K1 | 0.76 | K1 | 0.79 | K1 |
| | Entropy | 0.88 | K1, K2, K7 | 0.88 | K1, K6, K7, K8 | 0.85 | K1, K7 |
| | Gini | 0.88 | K1, K2, K7 | 0.85 | K1, K7 | 0.85 | K1, K7 |
| 4 | $\chi^2$ | 0.57 | K5, K6 | 0.57 | K5, K6 | 0.57 | K5,K6 |
| | Entropy | 0.71 | K2, K6, K7 | 0.62 | K6, K7 | 0.57 | K6, K7 |
| | Gini | 0.66 | K5, K6 | 0.62 | K6, K7 | 0.62 | K6, K7 |
| 5 | $\chi^2$ | 0.76 | K7, K8 | 0.81 | K7, K8 | 0.81 | K7, K8 |
| | Entropy | 0.90 | K1, K6, K7 | 0.90 | K1, K5 | 0.86 | K1, K5 |
| | Gini | 0.90 | K1, K6, K7 | 0.90 | K1, K5 | 0.86 | K1, K5 |
| 6 | $\chi^2$ | 0.75 | K7, K9 | 0.75 | K7, K9 | 0.75 | K7, K9 |
| | Entropy | 0.75 | K7, K8 | 0.75 | K7, K8 | 0.75 | K7, K8 |
| | Gini | 0.75 | K2, K7 | 0.75 | K7, K8 | 0.75 | K7, K8 |
| 7 | $\chi^2$ | 0.92 | K1 | 0.92 | K1 | 0.92 | K1 |
| | Entropy | 0.92 | K1 | 0.92 | K1 | 0.92 | K1 |
| | Gini | 0.92 | K1 | 0.92 | K1 | 0.92 | K1 |

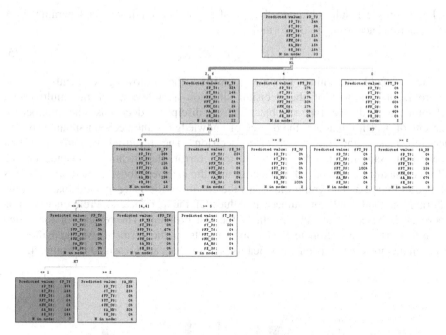

**Fig. 1.** Decision tree to choose the multicriteria method

Figure 2 provides an indication of how the decision-tree decision situation fit into the category of multicriteria methods for fuzzy or sharp data. The correctness of the classification is 88%.

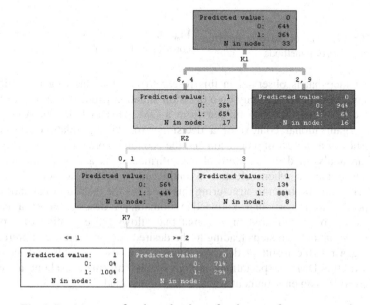

**Fig. 2.** Decision tree for the selection of a sharp or fuzzy approach

Based on the results obtained the set of significant environmental parameters $K$ was generated according to the formula:

$$K = \{K_1, K_2, K_6, K_7, K_8\} \tag{5}$$

The selected reasoning scheme identifies the following elements of the environmental context of the decision situation as relevant for the multicriteria selection method: the type-of-object-chosen discipline decisions, the number of variants of decision making, the degree of knowledge of the decision situation by the decision maker and the reach of a decision. The resulting wrongly qualified coefficients samples in relation to the total (error classification) depends on the allowed maximum number of leaf nodes and the selected division method, which is plotted in Figure 3. The graph illustrates the condition for using methods based on the information load, and not numerical values of the parameters. Preference of this approach is based on the input data of the test problem. The results for the tested combinations of induction test data and outputs in the form of averaged three of division methods for division classification error values are presented in Figure 4.

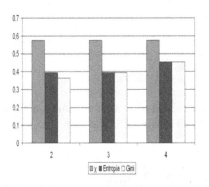

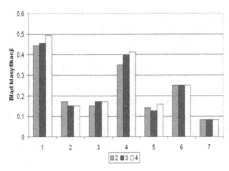

**Fig. 3.** The error rate of classification for multicriteria selection methods

**Fig. 4.** Average classification error for the constructed decision trees

The first fact that is observed in this research was that selecting a multicriteria method is not possible based only on a decision situation itself. It is a result of some intangible data, which are also hidden from a DM. This includes knowledge of a global or organizational context of a decision. An existing relation between the characteristics of a subject of a decision and the economic, practical, or psychological scope of an analyzed decision were also confirmed. It is a task of an analyst to consider those factors when selecting a multicriteria method to solve a decision situation. The main problem in structuring this aspect is the inability to quantify many factors specific to a particular knowledge domain. A description of a decision situation needs to be delivered in a format that allows the execution of a process according to some defined steps leading to the desired outcome. Hence, a prerequisite of obtaining a reliable result of decision support is to use methods and techniques aligned with the DM's expectations and available information about a decision problem, as well as an environment where a decision is made.

Future work assumes the extension of the proposed approach with other multicriteria methods and areas of reality in order to cover the situation of decision making not solved by the proposed algorithm. With the inclusion of more areas, special attention should be paid to the selection of the factors contained in the set K. It should be noted that there may be categories of decision-making situations that are poorly presented in the scientific literature or public data sets. In such a situation, an independent structure of the knowledge base with the help of domain experts should be considered.

# 5  Summary

The research of decision situation context is a complex and difficult to define task. Every decision takes place in a specific context, and it is necessary to select key characteristics and structures important for selecting the decision support method. Selecting the research procedure should take into account the parameters associated with the decision situation. The approach presented in this paper is one of the possible patterns of selection methods according to various aspects of the decision. The performed process of multivariate analysis of decision support methods took into account the set of possible values of the preferential information, linear compensation between the criteria of binary relations, manner of accounting, and shape of the function of the performance and characteristics of the veto. The proposed methods, based on data mining and prediction using decision trees, resulted in the rule-based system that determines the correctness of the proposed multicriteria method. The quality of decisions is the result of four elements based on the definition of the subject decision, the consequences of the implementation of decision variants, modeling global preferences, and selection procedures. Future direction of research in this area should take into account not only the correctness of methodical aspects, but also the quality of the recommendations. Selection of the proper method and procedures is essential for results, as it is the initial part of the process and determines the course of the analysis of the decision, which is often closely associated with the selected method.

# References

1. Araz, C., Ozkarahan, I.: Supplier evaluation and management system for strategic sourcing based on a new multicriteria sorting procedure. International Journal of Production Economics 106, 585–606 (2007)
2. Augusto, M., Lisboa, J., Yasin, M., Figueira, J.R.: Benchmarking in a multiple criteria performance context: An application and a conceptual framework. European Journal of Operational Research 184, 244–254 (2008)
3. Augusto, M.R., Lisboa, J.O., Yasin, M., Figueira, J.R.: Benchmarking in a multiple criteria performance context: An application and a conceptual framework. European Journal of Operational Research 184, 244–254 (2008)
4. Brans, J.P., Vincke, P.: A preference ranking organisation method. Management Science 31, 647–656 (1985)

5. Brans, J.P., Vincke, P., Mareschal, B.: How to select and how to rank projects: The PROMETHEE method. European Journal of Operational Research 24, 228–238 (1986)
6. Donegan, H.A., Dodd, F.J., McMaster, T.B.M.: A New Approach to AHP Decision-Making, Journal of the Royal Statistical Society. Series D (The Statistician) 41(3), 295–302 (1992)
7. Dong, C., Loo, G.: Flexible Web-Based Decision Support System Generator (FWDSSG) utilising software agents. In: Proceedings of the 12th International Workshop on Database and Expert Systems Applications Dexa 2000, pp. 0892–0910. IEEE Computer Society (2001)
8. Fortemps, P., Greco, S., Słowiński, R.: Multicriteria Choice and Ranking Using Decision Rules Induced from Rough Approximation of Graded Preference Relations. In: Tsumoto, S., Słowiński, R., Komorowski, J., Grzymała-Busse, J.W. (eds.) RSCTC 2004. LNCS (LNAI), vol. 3066, pp. 510–522. Springer, Heidelberg (2004)
9. Goumas, M., Lygerou, V.: An extension of the PROMETHEE method for decision making in fuzzy environment: Ranking of alternative energy exploitation projects. European Journal of Operational Research 123, 606–613 (2000)
10. Guitouni, A., Martel, J.M.: Tentative guidelines to help choosing an appropriate MCDA method. European Journal of Operational Research 109, 501–521 (1998)
11. Guitouni, A., Martel, J.M., Vincke, P.: A Framework to Choose a Discrete Multicriterion Aggregation Procedure. Defence Research Establishment Valcatier, DREV (1998)
12. Hokkanen, J., Salminen, P.: ELECTRE III and IV Decision Aids in an Environmental Problem. Journal of Multi-Criteria Decision Analysis 6, 215–226 (1997)
13. MCDA Software Package,
   http://www.cs.put.poznan.pl/ewgmcda/index.php/software
14. Kangas, A., Kangas, J., Pykaelaeinen, J.: Outranking Methods As Tools in Strategic Natural Resources Planning. Silva Fennica 35, 215–227 (2001)
15. Li, D.F.: Compromise ratio method for fuzzy multi-attribute group decision making. Applied Soft Computing 7, 807–817 (2007)
16. MPERIA Project Report, Comparison of Multi – Criteria Decision Analytical Software Searching for ideas for developing a new EIA- specific multi - criteria software Jyri Mustajoki Mika Marttunen Finnish Environment Institute (February 19, 2013)
17. Munda, G.: Multiple Criteria Decision Analysis and Sustainable Development, Multiple Criteria Decision Analysis: State of the Art Surveys. Springer, New York (2005)
18. Ozerol, G., Karasakal, E.: A Parallel between Regret Theory and Outranking Methods for Multicriteria Decision Making Under Imprecise Information. Theory and Decision 65(1), 45–70 (2008)
19. Pistolesi, G.: MicroDEMON: A Decision-making Intelligent Assistant for Mobile Business. Intelligent. In: Gupta, J.N.D., Forgionne, G.A., Manuel Mora, T. (eds.) Decision-Making Support Systems Decision Engineering, pp. 237–254. Springer, Heidelberg (2006)
20. Rao, R., Davim, J.: A decision-making framework model for material selection using a combined multiple attribute decision-making method. The International Journal of Advanced Manufacturing Technology 35, 751–760 (2008)
21. Roy, B.: Paradigms and challenges. In: Figueira, J., Greco, S., Ehrgott, M. (eds.) Multiple Criteria Decision Analysis. State of the Art Surveys, Springer Science and Business Media, Inc. (2005)
22. Roy, B.: The outranking approach and the foundations of ELECTRE methods. Theory and Decision 31, 49–73 (1991)

23. Saaty, T.L.: How to make a decision: The analytic hierarchy process. European Journal of Operational Research 48, 9–26 (1990)
24. Saaty, T.L.: The Analytic Hierarchy and Analytic Network Processes for the Measurement of Intangible Criteria and for Decision-Making, Multiple Criteria Decision Analysis: State of the Art Surveys (2005)
25. Wang, X., Triantaphyllou, E.: Ranking irregularities when evaluating alternatives by using some ELECTRE methods. Omega 36, 45–63 (2008)
26. Wang, X., Triantaphyllou, E.: Ranking irregularities when evaluating alternatives by using some ELECTRE methods. Omega 36, 45–63 (2008)
27. Wang, J.J., Yang, D.L.: Using a hybrid multi-criteria decision aid method for information systems outsourcing. Computers & Operations Research 34, 3691–3700 (2007)
28. Wei, C.C., Chien, C.F., Wang, M.J.J.: An AHP-based approach to ERP system selection. International Journal of Production Economics 96, 47–62 (2005)

# Multi-criteria Utility Mining
# Using Maximum Constraints

Guo-Cheng Lan[1], Tzung-Pei Hong[2,3], and Yu-Te Chao[2]

[1] Industrial Technology Research Institute,
Computational Intelligence Technology Center, Hsinchu, Taiwan
[2] Department of Computer Science and Information Engineering,
National University of Kaohsiung, Kaohsiung, Taiwan
[3] Department of Computer Science and Engineering,
National Sun Yat-Sen University, Kaohsiung, Taiwan
rrfoheiay@gmail.com, tphong@nuk.edu.tw, ny152_david@hotmail.com

**Abstract.** Most of the existing studies in utility mining use a single minimum utility threshold to determine whether an item is a high utility item. This way is, however, hard to reflect the nature of items. This work thus presents another viewpoint about defining the minimum utilities of itemsets. The maximum constraint is adopted, which is well explained in the text and suitable to some mining domains when items have different utility values. In addition, an effective two-phase mining approach is proposed to cope with the problem of multi-criteria utility mining under maximum constraints. The experimental results show the performance of the proposed approach.

**Keywords:** Data mining, utility mining, maximum constraint, multiple thresholds.

## 1    Introduction

Currently, utility mining has been one of important research topics due to its utility evaluation. The reason for this is that utility mining considers both the quantities and profits of items in a transaction database to evaluate the actual utility of an item in that database [8]. In the existing studies related to utility mining [3][6][8], all items in the utility-based framework are treated uniformly since a single minimum utility threshold is used as the utility requirement for the items. However, a single minimum utility is not easily used to reflect the natures of the items. For example, in retailing business, the profit of "LCD TV" is obviously higher than that of "Milk". As this example notes, only a utility requirement is not easily used to reflect the importance of the two items.

To address this, Lan *et al.* [4] then presented a new issue, namely utility mining with multiple minimum utilities using minimum constraints, which agreed the users to assign different minimum utility requirements for items by the significance of the items, such as profit or cost. For example, assume the minimum utilities of the two items, *A* and *B*, are 20% and 40%, respectively, and then the minimum utility of their

D. Hwang et al. (Eds.): ICCCI 2014, LNAI 8733, pp. 466–471, 2014.

superset $\{AB\}$ is 20%. However, when the minimum utility value of an itemset is defined as the minimum utility among the items in that itemset, the itemset may be a high utility itemset, but some items included that itemset may be not high utility items. In this case, it is doubtable whether the itemset is worth considering. For example, if the utility ratio of item $B$ is 30%, and is less than its minimum utility 40%, then its superset $\{AB\}$ should not be worth considering. As this notes, it is thus reasonable in some sense that the actual utilities of all items in an interesting itemset must be larger than or equal to the maximum of the minimum utilities of the items contained in it.

Due to the above reason, this work provides another viewpoint about defining the minimum utilities of itemsets under the maximum constraint when items have different minimum utilities. The maximum constraint is used, which is well explained and may be suitable to some domains, and also the number of unnecessary utility itemsets can be effectively reduced by the maximum constraint when compared to the minimum constraints. To our best knowledge, this is the first work on mining high utility itemsets with the consideration of the multi-criteria using maximum constraints in utility mining topic. Finally, the experimental results show the proposed approach has good performance in execution efficiency.

The remaining parts of this paper are organized as follows. The related works are reviewed in Section 2. The problem to be solved and its definitions are described in Section 3. The proposed approach is stated in Section 4. The experimental results are showed in Section 5. Finally, the conclusions are stated in Section 6.

## 2    Review of Related Works

In practical applications, items may have different criteria to assess their importance [7]. That is, the individual support requirement of each item should be different. To address this problem, Liu *et al.* presented a new issue, namely association-rule mining with multiple minimum supports [5], which agreed the users to assign different minimum requirements for items by the significance of the items, such as profit or cost. In Liu *et al.*'s study [5], they designed a minimum constraint, which was that the minimum value of the minimum supports of all items in an itemset was regarded as the minimum support of that itemset, to determine the minimum support of an itemset. Different from Liu *et al.*'s study [5], Wang *et al.* [7] then presented a bin-oriented, non-uniform support constraint, which allowed the minimum support value of an itemset to be any function of the minimum supports of items contained in the itemset.

On the other hand, the utility function in Yao *et al.*'s study [8] considered not only quantities of items in a transaction but also the profits of the items in a set of transactions to measure the utility of an item. By using a transaction dataset and a utility table together, the discovered itemset is able to better match a user's expectations than if found by considering only the transaction dataset itself. Compared to association-rule mining [1][5][7], the actual utility of an itemset can be effectively evaluated by the utility function in utility mining.

Based on the utility function, however, the downward-closure property in association rule mining cannot be kept in the problem of utility mining. To address this, Liu *et al.* then proposed a two-phase approach (*TP*), which was developed the upper-bound model to avoid information losing in mining [6]. Afterward, most of existing approaches were based on the framework of the *TP* algorithm to cope with various applications, such as on-shelf utility mining [3], and so on.

However, traditional utility mining [1] only adopts a single minimum utility threshold to determine whether or not an item is a high utility item in a database. But, in practical application, items may have different criteria to assess their importance [7]. To address this problem, Lan *et al.* then presented a new issue, namely utility mining with multiple minimum utilities using minimum constraints, which agreed the users to assign different minimum utility requirements for items by the significance of the items, such as profit or cost. As mentioned previously, however it is reasonable in some sense that the actual utilities of all items in an interesting itemset must be larger than or equal to the maximum of the minimum utilities of the items contained in it.

## 3    Problem Statement and Definitions

In this section, some terms are given to illustrate the multi-criteria utility mining problem with maximum constraints. Assume there are ten transactions in a quantitative transaction database (*QDB*), as shown in Table 1, and there are six items in Table 1, respectively denoted from *A* to *F*. In addition, the value attached to each item in the corresponding slot is the sold quantity in a transaction. The profits of the six items are 3, 10, 1, 6, 5 and 2, and their minimum utility criteria are 0.20, 0.40, 0.25, 0.15, 0.20 and 0.15, respectively.

**Table 1.** The ten transactions in this example

| TID | A | B | C | D | E | F |
|---|---|---|---|---|---|---|
| $Trans_1$ | 1 | 0 | 2 | 1 | 1 | 1 |
| $Trans_2$ | 0 | 1 | 25 | 0 | 0 | 0 |
| $Trans_3$ | 0 | 0 | 0 | 0 | 2 | 1 |
| $Trans_4$ | 0 | 1 | 12 | 0 | 0 | 0 |
| $Trans_5$ | 2 | 0 | 8 | 0 | 2 | 0 |
| $Trans_6$ | 0 | 0 | 4 | 1 | 0 | 1 |
| $Trans_7$ | 0 | 0 | 2 | 1 | 0 | 0 |
| $Trans_8$ | 3 | 2 | 0 | 0 | 2 | 3 |
| $Trans_9$ | 2 | 0 | 0 | 1 | 0 | 0 |
| $Trans_{10}$ | 0 | 0 | 4 | 0 | 2 | 0 |

In this study, an itemset *X* is a subset of the items *I*, $X \subseteq I$. If $|X| = r$, the set *X* is called an *r*-itemset. $I = \{i_1, i_2, ..., i_n\}$ is a set of items may appear in the transaction. In addition, a transaction (*Trans*) consists of a set items purchased with their quantities. A quantitative database *QDB* is then composed of a set of transactions. That is, *QDB*

= {$Trans_1$, $Trans_2$, ..., $Trans_y$, ..., $Trans_z$}, where $Trans_y$ is the $y$-th transaction in $QDB$. Based on Yao et al.'s utility function, the utility $u_{yi}$ of an item $i$ in $Trans_y$ is the external utility $s_i$ multiplied by the quantity $q_{zj}$ of $i$ in $Trans_y$, and the utility $u_{yX}$ of an itemset $X$ in $Trans_y$ is the summation of the utilities of all items in $X$ in $Trans_y$. Furthermore, the actual utility $au_X$ of $X$ in a database $D$ is the summation of the utilities of $X$ in the transactions including $X$ of $QDB$. For example, the utility of {$BC$} in $Trans_2$ can be calculated as $1*10 + 25*1$, which is 35, and then the actual utility $au_{\{BC\}}$ of {$BC$} in Table 1 can be calculated as $35 + 22$, which is 57.

Further, the actual utility ratio $aur_X$ of an itemset $X$ is the summation of the utilities of $X$ in the transactions including $X$ in $QDB$ over the summation of the transaction utilities of all transactions in $QDB$. For example, in Table 1, the summation of the transaction utilities of all transactions is 202 ($= 18 + 35 + 12 + 22 + 24 + 12 + 8 + 45 + 12 + 14$). Since the actual utility $au_{\{BC\}}$ of {$BC$} in Table 1 is 57, the actual utility ratio $au_{\{BC\}}$ can be calculated as 57/202, which is about 0.2822.

Finally, let $\lambda_i$ be the predefined individual minimum utility threshold of an item $i$. Note that here a maximum constraint is used to select the maximum value of minimum utilities of all items in $X$ as the minimum utility threshold $\lambda_X$ of $X$. Hence, an itemset $X$ is called a high utility itemset ($HU$) if $au_X \geqq \lambda_X$. For example in Table 1, since the minimum utility threshold of {$BC$} is 0.4, {$BC$} is not a $HU$ under the maximum constraint due to its actual utility ratio ($= 0.2822$).

However, the downward-closure property cannot be kept in the problem of this work. For example, the actual utility ratio and minimum utility of {$A$} are 0.1188 ($= 24/202$) and 0.2, so that {$A$} is not a high utility itemset, but its superset {$AE$} is. To avoid information losing in mining, the existing transaction-utility upper-bound ($TUUB$) model [6] is used to achieve this goal. The main concept of $TUUB$ is that the transaction utility of a transaction is used as the upper-bound of any subsets in that transaction, and the transaction-utility upper-bound ratio $tuubr_X$ of an itemset $X$ in $QDB$ is the summation of the transaction utilities of the transactions including $X$ in $QDB$ over the summation of transaction utilities of all transactions in $QDB$. If $tuubr_X \geqq \lambda_X$, the itemset $X$ is called a high transaction-utility upper-bound itemset ($HTUUB$). For example, in Table 1, since the itemset {$A$} appears in the four transactions, $Trans_1$, $Trans_5$, $Trans_8$, and $Trans_9$, and their transaction utilities are 18, 24, 45, and 12. Then, $tuubr_{\{A\}}$ of {$A$} is 0.4900 ($= (18 + 24 + 45 + 12) / 202$). Accordingly, {$A$} is a $HTUUB$ due to its minimum utility threshold ($= 0.2$).

## 4     The Proposed Mining Algorithm

The execution process of the two-phase multi-criteria approach using maximum constraints (abbreviated as $TPM_{max}$) is then stated as follows.

INPUT: A set of items, each with a profit value and an individual minimum utility threshold, a quantitative transaction database $QDB$, in which each transaction includes a subset of items with quantities.

OUTPUT: A final set of high utility itemsets ($HUs$) satisfying their individual minimum utilities under the maximum constraints.

**Phase 1: Finding high transaction-utility upper-bound itemsets**

STEP 1: For each $Trans_y$ in $QDB$, find the utility $u_{yz}$ of each item $i_z$ in $Trans_y$.

STEP 2: For each item $i$ in $QDB$, calculate the transaction-utility upper-bound $tuub_i$ of $i$.

STEP 3: For each item $i$ in $QDB$, if the transaction-utility upper-bound $tuub_i$ of the item $i$ is larger than or equal to the corresponding minimum utility threshold $\lambda_i$ of the item $i$, put it in the set of $HTUUB_1$.

STEP 4: Set $r = 1$, where $r$ represents the number of items in the current set of candidate utility $r$-itemsets $(C_r)$ to be processed.

STEP 5: Generate from the set $HTUUB_r$ the candidate set $C_{r+1}$, in which all the $r$-sub-itemsets of each candidate must be contained in the set of $HTUUB_r$.

STEP 6: For each candidate $(r+1)$-itemset $X$ in the set $C_{r+1}$, find the transaction-utility upper-bound $tuub_X$ of $X$ in $QDB$.

STEP 7: For each candidate utility $(r+1)$-itemset $X$ in $C_{r+1}$, do the following substeps.

   (a) Find the maximum value $\lambda_X$ of the minimum utility thresholds of all items in $X$ as the minimum utility threshold of $X$, $\lambda_X$.

   (b) Check whether the transaction-utility upper-bound $tuub_X$ of $X$ is larger than or equal to the minimum utility threshold $\lambda_X$. If it is, put it in set $HTUUB_{r+1}$.

STEP 8: If $HTUUB_{r+1}$ is null, do STEP 9; otherwise, set $r = r + 1$ and repeat STEPs 5 to 8.

**Phase 2: Finding high utility itemsets $(HUs)$ satisfying their minimum utilities**

STEP 9: Scan the database $QDB$ once to the actual utility $au_X$ of $X$ in all $HTUUB$ sets.

STEP 10: For each itemset $X$ in all $HTUUB$ sets, do the following substeps.

   (a) Find the maximum value $\lambda_X$ among the minimum utility thresholds of all items in $X$ as the minimum utility threshold of $X$, $\lambda_X$.

   (b) Check whether the actual utility $au_X$ of $X$ is larger than or equal to the minimum utility threshold $\lambda_X$. If it is, put it in set $HU_{r+1}$.

STEP 11: For each itemset $X$ in $HU$ set, check whether its each subset is also the member of the $HU$ set. If yes, keep it in the $HU$ set; otherwise, remove $X$.

STEP 12: Output the set of high utility itemsets satisfying their own criteria, $HUs$.

# 5     Experimental Evaluation

In this section, a series of experiments were conducted to show the performance of the proposed $TPM_{max}$. In the experiments, the public $IBM$ data generator [2] was used to produce the synthetic data "T10I4N4KD200K". Figure 1 showed the performance of the proposed $TPM_{max}$ and the traditional $TP$ [6] under various $\lambda_{min}$. Here the symbol $\lambda_{min}$ represented the minimum value of minimum utilities of all items in databases, and $\lambda_{min}$ was regarded as the minimum utility threshold in traditional utility mining.

As shown in the figure, it could be clearly observed that the execution efficiency of the proposed $TPM_{max}$ is faster than that of the traditional $TP$ under various thresholds. The main reason is that the maximum constraints could be effectively used to reduce a large number of unnecessary itemsets in mining when compared with the $TP$. Hence, the proposed viewpoint using maximum constraints might be a proper framework when items had different minimum utilities.

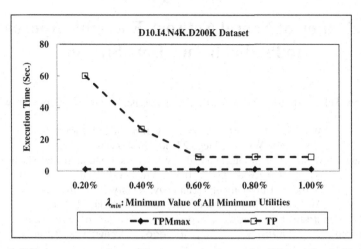

**Fig. 1.** Efficiency comparison of the two approaches under different thresholds $\lambda_{min}$

# 6    Conclusion

This work has presented a new research issue, namely multi-criteria utility mining with the maximum constraints, to define the individual minimum utility of an itemset in a database when items have different criteria. In particular, this work also presents the two-phase approach to cope with the problems of multi-criteria utility mining. The experimental results show that proposed $TPM_{max}$ has good performance on efficiency.

# References

1. Agrawal, R., Srikant, R.: Fast Algorithm for Mining Association Rules. In: International Conference on Very Large Data Bases, pp. 487–499 (1994)
2. IBM Quest Data Mining Project, Quest synthetic data generation code, http://www.almaden.ibm.com/cs/quest/syndata.html
3. Lan, G.C., Hong, T.P., Tseng, V.S.: Discovery of High Utility Itemsets from On-Shelf Time Periods of Products. Expert Systems with Applications 38(5), 5851–5857 (2011)
4. Lan, G.C., Hong, T.P., Chao, Y.T.: Multi-Criteria Utility Mining Using Minimum Constraints. In: The 27th International Conference on Industrial, Engineering & Other Applications of Applied Intelligent Systems (2014)
5. Liu, B., Hsu, W., Ma, Y.: Mining Association Rules with Multiple Minimum Supports. In: International Conference on Knowledge Discovery and Data Mining, pp. 337–341 (1999)
6. Liu, Y., Liao, W.K., Choudhary, A.: A Fast High Utility Itemsets Mining Algorithm. In: International Workshop on Utility-based Data Mining, pp. 90–99 (2005)
7. Wang, K., He, Y., Han, J.: Mining Frequent Itemsets Using Support Constraints. In: The 26th International Conference on Very Large Data Bases, pp. 43–52 (2000)
8. Yao, H., Hamilton, H.J., Butz, C.J.: A Foundational Approach to Mining Itemset Utilities from Databases. In: The 4th SIAM International Conference on Data Mining, pp. 482–486 (2004)

# Evaluation of Neural Network Ensemble Approach to Predict from a Data Stream

Zbigniew Telec[1], Bogdan Trawiński[1], Tadeusz Lasota[2], and Grzegorz Trawiński[3]

[1] Wrocław University of Technology, Institute of Informatics,
Wybrzeże Wyspiańskiego 27, 50-370 Wrocław, Poland
[2] Wrocław University of Environmental and Life Sciences, Dept. of Spatial Management,
ul. Norwida 25/27, 50-375 Wrocław, Poland
[3] Wrocław University of Technology, Faculty of Electronics,
Wybrzeże S. Wyspiańskiego 27, 50-370 Wrocław, Poland
{bogdan.trawinski,zbigniew.telec}@pwr.edu.pl,
tadeusz.lasota@up.wroc.pl, grzegorz.trawinsky@gmail.com

**Abstract.** We have recently worked out a method for building reliable predictive models from a data stream of real estate transactions which applies the ensembles of genetic fuzzy systems and neural networks. The method consists in building models over the chunks of a data stream determined by a sliding time window and enlarging gradually an ensemble by models generated in the course of time. The aged models are utilized to compose ensembles and their output is updated with trend functions reflecting the changes of prices in the market. In the paper we present the next series of extensive experiments to evaluate our method with the ensembles of artificial neural networks. We examine the impact of the number of aged models used to compose an ensemble on the accuracy and the influence of the degree of polynomial trend functions employed to modify the results on the performance of neural network ensembles. The experimental results were analysed using statistical approach embracing nonparametric tests followed by post-hoc procedures designed for multiple $N{\times}N$ comparisons.

**Keywords:** artificial neural networks, data stream, sliding windows, ensembles, trend functions, property valuation.

## 1 Introduction

Numerous strategies and techniques for mining data streams have been devised during the last decade. Processing data streams presents a big challenge because it requires considering memory limitations, short processing times, and single scans of incoming data. Gaber in his overview paper categorizes them into four main groups: two-phase techniques, Hoeffding bound-based, symbolic approximation-based, and granularity-based ones [1]. Much effort is devoted to the issue of concept drift which occurs when data distributions and definitions of target classes change over time [2], [3], [4]. Comprehensive reviews of ensemble based methods for handling concept drift in data streams can be found in [5], [6].

D. Hwang et al. (Eds.): ICCCI 2014, LNAI 8733, pp. 472–482, 2014.

For a few years we have been working out and testing methods for generating regression models to assist with real estate appraisal based on fuzzy and neural approaches: i.e. genetic fuzzy systems and artificial neural networks as both single models [7] and ensembles built using various resampling techniques [8], [9], [10], [11], [12], [13]. An especially good performance revealed evolving fuzzy models applied to cadastral data [14], [15]. Evolving fuzzy systems are appropriate for modelling the dynamics of real estate market because they can be systematically updated on demand based on new incoming samples and the data of property sales ordered by the transaction date can be treated as a data stream.

In this paper we present the results of our further study on the method to predict from a data stream of real estate sales transactions based on ensembles of regression models [16], [17], [18]. The goal of research reported in this paper is to apply artificial neural networks (*ANN*) to our method, namely general linear model, multilayer perceptron, and radial basis function neural networks. Having prepared a new real-world dataset we investigated the impact of the number of aged models used to compose an ensemble on the accuracy and the influence of degree of polynomial trend functions applied to modify the results on the performance of single models and ensembles. The scope of extensive experiments was enough to conduct advanced statistical analysis of results obtained including nonparametric tests followed by post-hoc procedures devised for multiple $N \times N$ comparisons.

## 2 Ensemble Approach to Predict from a Data Stream

Our ensemble approach to predict from a data stream lies in systematic building models over chunks of data and utilizing aged models to compose ensembles. The output produced by component models is corrected by means of trend functions reflecting the changes of prices in the market over time. The outline our approach to is illustrated in Fig. 1. The data stream is partitioned into data chunks of a constant length $t_c$. The sliding window, which length is a multiple of a data chunk, delineates training sets; in Fig. 1 it is double the chunk. We consider a point of time $t_0$ at which the current model was built over data that came in between time $t_0-2t_c$ and $t_0$. The models created earlier that have aged gradually are utilized to compose an ensemble so that the current test set is applied to each component model. However, in order to compensate ageing, their output produced for the current test set is updated using trend functions determined over all data since the beginning of the stream; we denote them as *BegTrends*. As the functions to model the trends of price changes the polynomials of the degree from one to five were employed: *Ti(t)*, where *i* stands for the degree. The method of updating the prices of premises with the trends is based on the difference between a price and a trend value in a given time point. More detailed description of the approach presented in the paper can be found in [19].

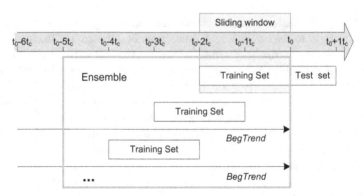

**Fig. 1.** Outline of ensemble approach to predict from a data stream

# 3 Experimental Setup

The experiments were conducted with our system implemented in Matlab. The system was designed to carry out research into machine learning algorithms using various resampling methods and constructing and evaluating ensemble models for regression problems. We have extended our system to include functions of building ensembles over a data stream. The trends are modelled using the Matlab function *polyfit*.

Real-world dataset used in experiments was drawn from an unrefined dataset containing above 100 000 records referring to residential premises transactions accomplished in one Polish big city with the population of 640 000 within 14 years from 1998 to 2011. In this period the majority of transactions were made with non-market prices when the council was selling flats to their current tenants on preferential terms. First of all, transactional records referring to residential premises sold at market prices were selected. Then, the dataset was confined to sales transaction data of residential premises (apartments) where the land was leased on terms of perpetual usufruct. The other transactions of premises with the ownership of the land were omitted due to the conviction of professional appraisers stating that the land ownership and lease affect substantially the prices of apartments and therefore they should be used separately for sales comparison valuation methods. The final dataset counted 9795 samples. Due to the fact we possessed the exact date of each transaction we were able to order all instances in the dataset by time, so that it can be regarded as a data stream. Four following attributes were pointed out as main price drivers by professional appraisers: usable area of a flat (*Area*), age of a building construction (*Age*), number of storeys in the building (Storeys), the distance of the building from the city centre (*Centre*), in turn, price of premises (*Price*) was the output variable.

Following parameters of our experiments were determined. As single models, general linear model (*GLM*), multilayer perceptron (*MLP*) and radial basis function neural networks (*RBF*) were built using *glm*, *mlp*, and *rbf* Matlab functions, respectively. In each function the number of inputs and outputs was set to four and one, respectively. In *glm* and *mlp* linear output unit activation functions and in *rbf* a radially symmetric Gaussian function as hidden unit activation function were used.

The functions *mlp* and *rbf* were run with three neurons in a hidden layer. The number of epochs to learn each network was equal to 100. As the performance measure the root mean square error (*RMSE*) was used.

The results of evaluating experiments were considered within six years 2005-2010, marked with grey shades in Fig. 2. This period was chosen because after Poland entered the European Union (EU) in 2004 a rise of real estate prices could be observed. Moreover, the prices of residential premises were increasing rapidly during the worldwide real estate bubble. In turn the period after the bubble burst and during the global financial crisis was characterized by unstable real estate market and great fluctuations of prices due to nervous behaviour of both buyers and sellers. We build the ensembles at the beginning of each quarter and so obtained 24 observation points. The trend of premises price changes over six years from 2005 to 2010, shown in Fig. 3, can be modelled by the polynomial function of degrees three and four.

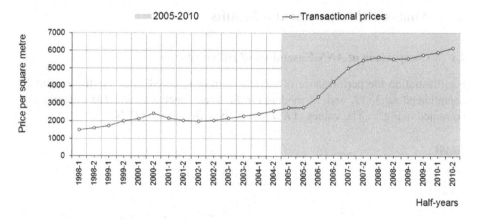

**Fig. 2.** Change trend of average transactional prices per square metre over time

Based on the results of our previous study, we determined following parameters of our experiments including two phases:

*1) Generating single ANN models*
- Set the length of the sliding window to 12 months, $t_w = 12$.
- Set the starting point of the sliding window, i.e. its right edge, to 2000-01-01 and the terminating point to 2010-12-01.
- Set the shift of the sliding window to 1 month, $t_s = 1$.
- Move the window from starting point to terminating point with the step $t_s = 1$.
- At each stage generate a *ANN* from scratch over a training set delineated by the window. In total 108 single models were built for each ANN.

*2) Building ANN ensembles*
- Select a period to investigate the real estate market in Poland, i.e. 2005-2010.
- At the beginning of each quarter ($t_0$) build ensembles composed of 3, 9, 12, 15, 21, and 24 ageing *ANNs*. An ensemble is created in the way described in Section 2 with the shift equal to one month, $t_s = 1$.

- Take test sets actual for each $t_0$ over a period of 3 months, $t_t = 3$.
- Compute the output of individual *ANNs* and update it using trend functions of degree from one to five determined for *BegTrends*.
- As the aggregation function of ensembles use the arithmetic mean.

The analysis of the results was performed using statistical methodology including nonparametric tests followed by post-hoc procedures designed especially for multiple $N \times N$ comparisons [20], [21], [22]. The routine starts with the nonparametric Friedman test, which detect the presence of differences among all algorithms compared. After the null-hypotheses have been rejected the post-hoc procedures are applied in order to point out the particular pairs of algorithms which produce differences. For $N \times N$ comparisons nonparametric Nemenyi's, Holm's, Shaffer's, and Bergmann-Hommel's procedures are employed.

# 4    Analysis of Experimental Results

## 4.1    Comparison of *ANN* Ensembles of Different Size

For illustration the performance of models of three selected sizes, i.e. the ensembles comprehending 3, 12, and 24 models for *T4* trend functions for *GLM* ensembles is presented in Fig. 3. The values of *RMSE* are given in thousand PLN.

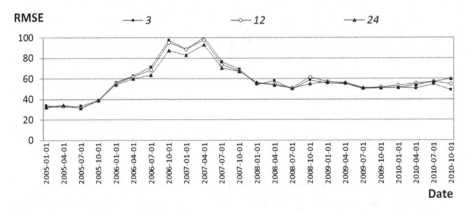

**Fig. 3.** Performance of *GLM* ensembles of different size for correction with *T4* trend functions

The Friedman tests performed in respect of *RMSE* showed that there were significant differences among ensembles in majority of cases. Average ranks of individual models for polynomial trend functions of degrees from *T1* to *T5* and with no correction *(noT)* produced by the tests for *GLM*, *MLP* and *RBF* ensembles are shown in Tables 1, 2, and 3, respectively. The tests indicating significant differences among models at the level of 0.05 are marked with italics in the tables and the lower rank value the better model.

**Table 1.** Average rank positions of *GLM* ensembles of size 3 to 24 by Friedman tests

| Trend | p-value | 1st | 2nd | 3rd | 4th | 5th | 6th |
|---|---|---|---|---|---|---|---|
| *noT* | *0.000* | *3 (1.29)* | *9 (2.29)* | *12 (3.00)* | *15 (3.67)* | *21 (4.75)* | *24 (6.00)* |
| *BegT1* | *0.000* | *3 (1.54)* | *9 (2.25)* | *12 (3.00)* | *15 (3.71)* | *21 (4.71)* | *24 (5.79)* |
| *BegT2* | *0.000* | *3 (1.42)* | *9 (2.21)* | *12 (3.04)* | *15 (4.00)* | *21 (4.92)* | *24 (5.42)* |
| BegT3 | 0.899 | 9 (3.25) | 3 (3.29) | 12 (3.46) | 24 (3.50) | 21 (3.71) | 15 (3.79) |
| *BegT4* | *0.049* | *24 (2.63)* | *21 (3.13)* | *15 (3.54)* | *3 (3.54)* | *12 (3.96)* | *9 (4.21)* |
| BegT5 | 0.077 | 24 (2.75) | 21 (3.08) | 15 (3.46) | 12 (3.71) | 3 (3.71) | 9 (4.29) |

**Table 2.** Average rank positions of *MLP* ensembles of size 3 to 24 by Friedman tests

| Trend | p-value | 1st | 2nd | 3rd | 4th | 5th | 6th |
|---|---|---|---|---|---|---|---|
| *noT* | *0.000* | *3 (2.00)* | *9 (2.46)* | *12 (2.92)* | *15 (3.50)* | *21 (4.50)* | *24 (5.63)* |
| BegT1 | 0.404 | 15 (3.04) | 21 (3.13) | 24 (3.42) | 12 (3.58) | 9 (3.79) | 3 (4.04) |
| *BegT2* | *0.000* | *24 (1.71)* | *21 (2.38)* | *15 (3.21)* | *12 (4.17)* | *9 (4.63)* | *3 (4.92)* |
| *BegT3* | *0.000* | *24 (1.00)* | *21 (2.00)* | *15 (3.04)* | *12 (4.08)* | *9 (5.00)* | *3 (5.88)* |
| *BegT4* | *0.000* | *24 (1.25)* | *21 (2.42)* | *15 (3.42)* | *12 (4.04)* | *9 (4.83)* | *3 (5.04)* |
| *BegT5* | *0.000* | *24 (1.25)* | *21 (2.46)* | *15 (3.42)* | *12 (4.13)* | *9 (4.83)* | *3 (4.92)* |

**Table 3.** Average rank positions of *RBF* ensembles of size 3 to 24 by Friedman tests

| Trend | p-value | 1st | 2nd | 3rd | 4th | 5th | 6th |
|---|---|---|---|---|---|---|---|
| *noT* | *0.000* | *3 (1.88)* | *9 (2.46)* | *12 (2.96)* | *15 (3.75)* | *21 (4.58)* | *24 (5.38)* |
| BegT1 | 0.257 | 15 (3.04) | 12 (3.04) | 21 (3.46) | 9 (3.46) | 24 (3.88) | 3 (4.13) |
| *BegT2* | *0.000* | *24 (1.83)* | *21 (2.67)* | *15 (3.38)* | *12 (3.96)* | *9 (4.38)* | *3 (4.79)* |
| *BegT3* | *0.000* | *24 (1.00)* | *21 (2.04)* | *15 (3.04)* | *12 (3.96)* | *9 (5.04)* | *3 (5.92)* |
| *BegT4* | *0.000* | *24 (1.25)* | *21 (2.38)* | *15 (3.38)* | *12 (4.17)* | *9 (4.75)* | *3 (5.08)* |
| *BegT5* | *0.000* | *24 (1.29)* | *21 (2.38)* | *15 (3.42)* | *12 (4.13)* | *9 (4.75)* | *3 (5.04)* |

Due to limited space we do not present detailed results of post-hoc procedures. Following main observations could be done based on the results of Shaffer's and Bergmann-Hommel's post-hoc procedures. The greater number of models with corrected outputs with *T4* and *T5*, and additionally with *T2* and *T3* in the case of *MLP* and *RBF*, in an ensemble the better performance. However, for *MLP* and *RBF* the differences were statistically significant, but it was not the case for *GLM*. For no correction and also with *T1, T2* for *GLM*, the models revealed reverse behaviour.

## 4.2    Comparison of GFS Ensembles Using Trend Functions of Different Degrees

For illustration the performance of models with corrected output using *T1 and T3* trend functions and without output correction (*noT*) of size equal to 24 for *MLP* ensembles is presented in Fig. 4. The values of *RMSE* are given in thousand PLN.

The Friedman tests performed in respect of *RMSE* showed that there were significant differences among ensembles in each case. Average ranks of individual models for ensemble sizes from 3 to 24 produced by the tests for *GLM*, *MLP* and *RBF* ensembles are shown in Tables 4, 5, and 6, respectively. The tests indicating significant differences among models at the level of 0.05 are marked with italics in the tables and the lower rank value the better model.

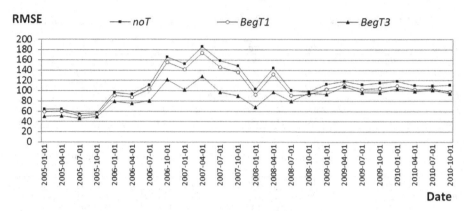

**Fig. 4.** Performance of *MLP* ensembles with correction using trend functions for Size=24

Due to limited space we do not present detailed results of post-hoc procedures. Following main observations could be done based on the results of Shaffer's and Bergmann-Hommel's post-hoc procedures: The best performance provided correction with trend functions *T3, T4,* and *T5* and additionally with *T2* in the case of *MLP* and *RBF*. However, only the differences between the aforementioned models and the models with no correction were statistically significant. The statistically significant differences were also detected for ensembles of size 24 and 21 in the following cases: *T4* vs *T2, T5* vs *T2* for *GLM* as well as *T3* vs *T1, T4* vs *T1* for both *MLP* and *RBF*.

**Table 4.** Average rank positions of *GLM* ensembles determined by Friedman test

| Size | p-value | 1st | 2nd | 3rd | 4th | 5th | 6th |
|------|---------|-----|-----|-----|-----|-----|-----|
| 3 | 0.000 | T3 (2.38) | T4 (3.04) | T5 (3.29) | T2 (3.38) | T1 (4.04) | noT (4.88) |
| 9 | 0.005 | T3 (2.83) | T4 (2.88) | T5 (3.13) | T2 (3.71) | T1 (3.79) | noT (4.67) |
| 12 | 0.003 | T4 (2.83) | T3 (2.83) | T5 (3.00) | T1 (3.83) | T2 (3.88) | noT (4.63) |
| 15 | 0.000 | T4 (2.75) | T3 (2.83) | T5 (2.83) | T1 (3.88) | T2 (4.04) | noT (4.67) |
| 21 | 0.000 | T4 (2.54) | T5 (2.54) | T3 (3.04) | T1 (3.79) | T2 (4.21) | noT (4.88) |
| 24 | 0.000 | T4 (2.38) | T5 (2.38) | T3 (3.21) | T2 (3.67) | T1 (4.25) | noT (5.13) |

**Table 5.** Average rank positions of *MLP* ensembles determined by Friedman test

| Size | p-value | 1st | 2nd | 3rd | 4th | 5th | 6th |
|------|---------|-----|-----|-----|-----|-----|-----|
| 3 | 0.000 | T3 (1.88) | T4 (2.96) | T2 (3.00) | T5 (3.46) | T1 (4.29) | noT (5.42) |
| 9 | 0.005 | T3 (2.83) | T4 (2.88) | T5 (3.13) | T2 (3.71) | T1 (3.79) | noT (4.67) |
| 12 | 0.003 | T4 (2.83) | T3 (2.83) | T5 (3.00) | T1 (3.83) | T2 (3.88) | noT (4.63) |
| 15 | 0.000 | T4 (2.75) | T5 (2.83) | T3 (2.83) | T2 (3.88) | T1 (4.04) | noT (4.67) |
| 21 | 0.000 | T3 (2.13) | T4 (2.75) | T2 (3.04) | T5 (3.13) | T1 (4.33) | noT (5.63) |
| 24 | 0.000 | T3 (2.04) | T4 (2.71) | T5 (3.04) | T2 (3.08) | T1 (4.38) | noT (5.75) |

**Table 6.** Average rank positions of *RBF* ensembles with determined by Friedman test

| Size | p-value | 1st | 2nd | 3rd | 4th | 5th | 6th |
|------|---------|-----|-----|-----|-----|-----|-----|
| 3 | 0.000 | T3 (2.08) | T4 (3.00) | T2 (3.13) | T5 (3.42) | T1 (4.17) | noT (5.21) |
| 9 | 0.000 | T3 (2.17) | T4 (2.96) | T2 (3.13) | T5 (3.29) | T1 (4.21) | noT (5.25) |
| 12 | 0.000 | T3 (2.17) | T4 (2.92) | T2 (3.13) | T5 (3.29) | T1 (4.21) | noT (5.29) |
| 15 | 0.000 | T3 (2.17) | T4 (2.88) | T2 (3.13) | T5 (3.25) | T1 (4.17) | noT (5.42) |
| 21 | 0.000 | T3 (2.21) | T4 (2.79) | T2 (3.08) | T5 (3.08) | T1 (4.21) | noT (5.63) |
| 24 | 0.000 | T3 (2.21) | T4 (2.75) | T5 (2.96) | T2 (3.08) | T1 (4.29) | noT (5.71) |

## 4.3   Comparison of ANN Ensembles with Other Approaches

The ensembles built using *GLM*, *MLP*, and *RBF* networks were compared with two other methods in terms of accuracy. For comparison we took the results of our previous investigations into evolving fuzzy systems and ensembles constructed by means of genetic fuzzy systems (*GFS*) conducted over the same datasets [23], [24]. For statistical tests we selected the ensembles providing the best performance, i.e. the ones comprising 24 component models which output was corrected using polynomial functions of degree 3 or 4. The evolving fuzzy models were generated using the *Flexfis* algorithm [25]. The *RMSE* of examined methods was computed in the same 24 observation time points as reported in section 3. The Friedman test performed in respect of *RMSE* values showed that there were significant differences among models. Average ranks of compared methods produced by the test are shown in Table 4, where the lower rank value the better model. Adjusted p-values for Nemenyi's, Holm's, Shaffer's, and Bergmann-Hommel's post-hoc procedures for N×N comparisons for all possible pairs of algorithms are shown in Table 5. The p-values indicating the statistically significant differences between given pairs of algorithms are marked with italics. The significance level considered for the null hypothesis rejection was 0.05. Following main observations could be done: *Flexfis*, *GLM-T4*, and *GFS-T4* models revealed significantly better performance than *MLP-T3* and *RBF-T3* ones. At the same time no significant differences among *Flexfis*, *GLM-T4*, and *GFS-T4* models could be observed.

**Table 7.** Average rank positions of compared methods determined by Friedman test

| p-value | 1st | 2nd | 3rd | 4th | 5th |
|---|---|---|---|---|---|
| 0.000 | Flexfis (1.67) | GLM-T4 (1.79) | GFS-T4 (2.63) | MLP-T3 (4.46) | RBF-T3 (4.46) |

**Table 8.** Adjusted p-values for N×N comparisons for all 10 hypotheses

| Method vs Method | pNeme | pHolm | pShaf | pBerg |
|---|---|---|---|---|
| MLP-T3 vs Flexfis | 9.58E-09 | 9.58E-09 | 9.58E-09 | 9.58E-09 |
| RBF-T3 vs Flexfis | 9.58E-09 | 9.58E-09 | 9.58E-09 | 9.58E-09 |
| GLM-T4 vs MLP-T3 | 5.15E-08 | 4.12E-08 | 3.09E-08 | 3.09E-08 |
| GLM-T4 vs RBF-T3 | 5.15E-08 | 4.12E-08 | 3.09E-08 | 3.09E-08 |
| MLP-T3 vs GFS-T4 | 5.90E-04 | 3.54E-04 | 3.54E-04 | 2.36E-04 |
| RBF-T3 vs GFS-T4 | 5.90E-04 | 3.54E-04 | 3.54E-04 | 2.36E-04 |
| Flexfis vs GFS-T4 | 0.357638 | 0.143055 | 0.143055 | 0.143055 |
| GLM-T4 vs GFS-T4 | 0.678892 | 0.203667 | 0.203667 | 0.143055 |
| GLM-T4 vs Flexfis | 1.000000 | 1.000000 | 1.000000 | 1.000000 |
| MLP-T3 vs RBF-T3 | 1.000000 | 1.000000 | 1.000000 | 1.000000 |

## 5   Conclusions and Future Work

Our further investigation into the method to predict from a data stream of real estate sales transactions based on ensembles of regression models is reported in the paper. The core of our approach is incremental expanding an ensemble by models built from scratch over successive chunks of a data stream determined by a sliding window. In order to compensate ageing the output produced by individual component models for the current test dataset is updated using trend functions which reflect the changes of the market. In our research we employed artificial neural networks as the base machine learning algorithms and the trends were modelled over data that came in from the beginning of a stream.

The experiments aimed at examining the impact of the number of aged models used to compose an ensemble on the accuracy and the influence of degree of polynomial trend functions applied to modify the results on the accuracy of single models and ensembles. The data driven models, considered in the paper, were generated using real-world data of sales transactions taken from a cadastral system and a public registry of real estate transactions. The whole dataset was ordered by transaction date forming a sort of a data stream. The comparative experiments consisted in generating ensembles of *ANN* models for 24 points of time within the period of six years using the sliding window one year long which delineated training sets. The predictive accuracy of *ANN* ensembles for different variants of ensemble sizes and polynomial trend functions, was compared using nonparametric tests of statistical significance. The *ANN* ensembles were compared also with evolving fuzzy systems and ensembles constructed by means of genetic fuzzy systems.

The results proved the usefulness of ensemble approach incorporating the correction of individual component model output. For the majority of cases the bigger ensembles encompassing 21 and 24 *ANN* models produced more accurate predictions than the smaller ensembles. Moreover, they outperformed significantly the single models. As for correcting the output of component models, the need to apply trend functions to update the results provided by ageing models is indisputable. However, the selection the most suitable trend function in terms of the polynomial degree has not been definitely resolved. In majority of cases the trend functions of higher degree, i.e. three and four provided better accuracy. However, the differences were not statistically significant. Therefore, further study is needed into the selection of correcting functions dynamically depending on the nature of price changes.

**Acknowledgments.** This paper was partially supported by the "Młoda Kadra" funds of the Wrocław University of Technology. Many thanks also to Edwin Lughofer for granting us his FLEXFIS algorithm code for experiments.

# References

1. Gaber, M.M.: Advances in data stream mining. Wiley Interdisciplinary Reviews: Data Mining and Knowledge Discovery 2(1), 79–85 (2012)
2. Brzeziński, D., Stefanowski, J.: Reacting to Different Types of Concept Drift: The Accuracy Updated Ensemble Algorithm. IEEE Transactions on Neural Networks and Learning Systems 25(1), 81–94 (2014)
3. Sobolewski, P., Woźniak, M.: Concept Drift Detection and Model Selection with Simulated Recurrence and Ensembles of Statistical Detectors. Journal for Universal Computer Science 19(4), 462–483 (2013)
4. Tsymbal, A.: The problem of concept drift: Definitions and related work. Technical Report. Department of Computer Science, Trinity College, Dublin (2004)

5. Kuncheva, L.I.: Classifier Ensembles for Changing Environments. In: Roli, F., Kittler, J., Windeatt, T. (eds.) MCS 2004. LNCS, vol. 3077, pp. 1–15. Springer, Heidelberg (2004)
6. Minku, L.L., White, A.P., Yao, X.: The Impact of Diversity on Online Ensemble Learning in the Presence of Concept Drift. IEEE Transactions on Knowledge and Data Engineering 22(5), 730–742 (2010)
7. Król, D., Lasota, T., Trawiński, B., Trawiński, K.: Comparison of Mamdani and TSK Fuzzy Models for Real Estate Appraisal. In: Apolloni, B., Howlett, R.J., Jain, L. (eds.) KES 2007, Part III. LNCS (LNAI), vol. 4694, pp. 1008–1015. Springer, Heidelberg (2007)
8. Lasota, T., Telec, Z., Trawiński, B., Trawiński, K.: Exploration of Bagging Ensembles Comprising Genetic Fuzzy Models to Assist with Real Estate Appraisals. In: Corchado, E., Yin, H. (eds.) IDEAL 2009. LNCS, vol. 5788, pp. 554–561. Springer, Heidelberg (2009)
9. Lasota, T., Telec, Z., Trawiński, B., Trawiński, K.: A Multi-agent System to Assist with Real Estate Appraisals Using Bagging Ensembles. In: Nguyen, N.T., Kowalczyk, R., Chen, S.-M. (eds.) ICCCI 2009. LNCS, vol. 5796, pp. 813–824. Springer, Heidelberg (2009)
10. Graczyk, M., Lasota, T., Trawiński, B., Trawiński, K.: Comparison of Bagging, Boosting and Stacking Ensembles Applied to Real Estate Appraisal. In: Nguyen, N.T., Le, M.T., Świątek, J. (eds.) Intelligent Information and Database Systems. LNCS, vol. 5991, pp. 340–350. Springer, Heidelberg (2010)
11. Krzystanek, M., Lasota, T., Telec, Z., Trawiński, B.: Analysis of Bagging Ensembles of Fuzzy Models for Premises Valuation. In: Nguyen, N.T., Le, M.T., Świątek, J. (eds.) Intelligent Information and Database Systems. LNCS, vol. 5991, pp. 330–339. Springer, Heidelberg (2010)
12. Kempa, O., Lasota, T., Telec, Z., Trawiński, B.: Investigation of bagging ensembles of genetic neural networks and fuzzy systems for real estate appraisal. In: Nguyen, N.T., Kim, C.-G., Janiak, A. (eds.) ACIIDS 2011, Part II. LNCS, vol. 6592, pp. 323–332. Springer, Heidelberg (2011)
13. Lasota, T., Telec, Z., Trawiński, G., Trawiński, B.: Empirical Comparison of Resampling Methods Using Genetic Fuzzy Systems for a Regression Problem. In: Yin, H., Wang, W., Rayward-Smith, V. (eds.) IDEAL 2011. LNCS, vol. 6936, pp. 17–24. Springer, Heidelberg (2011)
14. Lasota, T., Telec, Z., Trawiński, B., Trawiński, K.: Investigation of the eTS Evolving Fuzzy Systems Applied to Real Estate Appraisal. Journal of Multiple-Valued Logic and Soft Computing 17(2-3), 229–253 (2011)
15. Lughofer, E., Trawiński, B., Trawiński, K., Kempa, O., Lasota, T.: On Employing Fuzzy Modeling Algorithms for the Valuation of Residential Premises. Information Sciences 181, 5123–5142 (2011)
16. Trawiński, B., Lasota, T., Smętek, M., Trawiński, G.: An Attempt to Employ Genetic Fuzzy Systems to Predict from a Data Stream of Premises Transactions. In: Hüllermeier, E., Link, S., Fober, T., Seeger, B. (eds.) SUM 2012. LNCS, vol. 7520, pp. 127–140. Springer, Heidelberg (2012)
17. Trawiński, B., Lasota, T., Smętek, M., Trawiński, G.: Weighting Component Models by Predicting from Data Streams Using Ensembles of Genetic Fuzzy Systems. In: Larsen, H.L., Martin-Bautista, M.J., Vila, M.A., Andreasen, T., Christiansen, H. (eds.) FQAS 2013. LNCS, vol. 8132, pp. 567–578. Springer, Heidelberg (2013)
18. Telec, Z., Lasota, T., Trawiński, B., Trawiński, G.: An Analysis of Change Trends by Predicting from a Data Stream Using Neural Networks. In: Larsen, H.L., Martin-Bautista, M.J., Vila, M.A., Andreasen, T., Christiansen, H. (eds.) FQAS 2013. LNCS, vol. 8132, pp. 589–600. Springer, Heidelberg (2013)

19. Trawiński, B.: Evolutionary Fuzzy System Ensemble Approach to Model Real Estate Market based on Data Stream Exploration. Journal of Universal Computer Science 19(4), 539–562 (2013)
20. Demšar, J.: Statistical comparisons of classifiers over multiple data sets. Journal of Machine Learning Research 7, 1–30 (2006)
21. García, S., Herrera, F.: An Extension on "Statistical Comparisons of Classifiers over Multiple Data Sets" for all Pairwise Comparisons. Journal of Machine Learning Research 9, 2677–2694 (2008)
22. Trawiński, B., Smętek, M., Telec, Z., Lasota, T.: Nonparametric Statistical Analysis for Multiple Comparison of Machine Learning Regression Algorithms. International Journal of Applied Mathematics and Computer Science 22(4), 867–881 (2012)
23. Telec, Z., Trawiński, B., Lasota, T., Trawiński, K.: Comparison of Evolving Fuzzy Systems with an Ensemble Approach to Predict from a Data Stream. In: Bădică, C., Nguyen, N.T., Brezovan, M. (eds.) ICCCI 2013. LNCS, vol. 8083, pp. 377–387. Springer, Heidelberg (2013)
24. Trawiński, B., Smętek, M., Lasota, T., Trawiński, G.: Evaluation of Fuzzy System Ensemble Approach to Predict from a Data Stream. In: Nguyen, N.T., Attachoo, B., Trawiński, B., Somboonviwat, K. (eds.) ACIIDS 2014, Part II. LNCS, vol. 8398, pp. 137–146. Springer, Heidelberg (2014)
25. Lughofer, E.: FLEXFIS: A robust incremental learning approach for evolving TS fuzzy models. IEEE Transactions on Fuzzy Systems 16(6), 1393–1410 (2008)

# Some Novel Improvements for MDL-Based
# Semi-supervised Classification of Time Series

Vo Thanh Vinh[1] and Duong Tuan Anh[2]

[1] Faculty of Information Technology, Ton Duc Thang University, Viet Nam
[2] Faculty of Computer Science & Engineering,
Ho Chi Minh City University of Technology, Viet Nam
`vtvinh@it.tdt.edu.vn, dtanh@cse.hcmut.edu.vn`

**Abstract.** In this paper, we propose two novel improvements for semi-supervised classification of time series: an improvement technique for Minimum Description Length-based stopping criterion and a refinement step to make the classifier more accurate. Our first improvement applies the non-linear alignment between two time series when we compute Reduced Description Length of one time series exploiting the information from the other. The second improvement is a post-processing step that aims to identify the class boundary between positive and negative instances accurately. Experimental results show that our two improvements can construct more accurate semi-supervised time series classifiers.

**Keywords:** Time series, semi-supervised classification, stopping criterion, MDL principle, X-Means.

# 1    Introduction

In time series data mining, classification is a crucial problem which has attracted lots of researches in the last decade. However, most of the current methods assume that the training set contains a great number of positive/labeled data. Such an assumption is unrealistic in the real world where we have a small set of labeled data, in addition to abundant unlabeled data. In such circumstances, semi-supervised classification is a suitable paradigm.

Semi-supervised classification (SSC) method will train itself by trying to expand the set of labeled data with the most similar unlabeled data until reaching a stopping criterion. Though several semi-supervised approaches have been proposed, only a few could be used for time series data, due to its special characteristic within.

Most of the time series SSC methods have to suggest a good stopping criterion. The SSC approach for time series proposed by Wei et al. in 2006 [8] uses a stopping criterion which is based on the minimal nearest neighbor distance, but this criterion can not work correctly in some situations. Ratanamahatana and Wanichsan, in 2008 [6], proposed a stopping criterion for SSC of time series which is based on the historical distances between candidate instances from the set of unlabeled instances to the initial positive instances. The most well-known stopping criterion so far is the one

D. Hwang et al. (Eds.): ICCCI 2014, LNAI 8733, pp. 483–493, 2014.

using Minimum Description Length (MDL) proposed by Begum et al., 2013 [1]. Even though this newest state-of-the-art stopping criterion gives a breakthrough for SSC of time series, it is still not effective to be used in some situations where time series may have some distortion along the time axis and the computation of Reduced Description Length for them becomes so rigid that the stopping point for the classifier can not be found precisely.

In this work, we propose two novel improvements for SSC of time series: an improvement technique for MDL-based stopping criterion and a refinement step to make the classifier more accurate. Our first improvement applies the non-linear alignment between two time series when we compute Reduced Description Length of one time series exploiting the information from the other. The second improvement is a post-processing step that aims to identify the class boundary between positive and negative instances accurately. Experimental results show that our two improvements can construct more accurate semi-supervised time series classifiers.

The rest of this paper is organized as follows. Section 2 reviews some background. Section 3 gives details of the two proposed improvements, followed by a set of experiments in Section 4. Section 5 concludes the work and gives suggestions for future work.

## 2     Background

In this section, we introduce briefly the framework of semi-supervised time series classification, dynamic time warping distance and MDL-based stopping criterion.

### 2.1     Semi-supervised Classification of Time Series

SSC technique can help build better classifiers in situations where we have a small set of labeled data, in addition to abundant unlabeled data. The main ideas of SSC of time series are summarized as follows. Given a set $P$ of positive instances and a set $N$ of unlabeled instances, the algorithm iterates the following two steps:

Step 1: We find the nearest neighbor of any instance of our training set from the unlabeled instances.

Step 2: This nearest neighbor instance, along with its newly acquired positive label, will be added into the training set.

Note that the above algorithm has to be coupled with the ability to stop adding instances at the correct time. This important issue will be addressed later. The algorithm for SSC of time series is given as follows:

**Self_Training_Classifier ($P, N$)**
    // $P$: Positive/Labeled set and $N$: Negative/Unlabeled set
**while** (the stopping criterion)
        nearest_obj = One_Nearest_Neighbor ($P, N$)
        $P = P \cup$ {nearest_obj}
        $N = N -$ {nearest_obj}
**end**

## 2.2    Dynamic Time Warping Distance

One problem with time series data is the distortion in the time axis, making Euclidean distance unsuitable. However, this problem can be effectively addressed by Dynamic Time Warping (DTW), a distance measure that allows non-linear alignment between the two time series to accommodate sequences that are similar but out of phase [2]. Given two time series $Q$ and $C$ which have length $n$ and $m$ respectively: $Q = q_1, q_2,\ldots, q_n$ and $C = c_1, c_2\ldots, c_m$.

DTW is a dynamic programming technique which calculates all possible warping paths between two time series for finding minimum distance. To calculate DTW between the two above time series, firstly we construct a matrix $D$ with size $m \times n$. Every element in matrix $D$ is cumulative distance defined as:

$$\gamma(i, j) = d(i, j) + \min\{ \gamma(i\text{-}1, j), \gamma(i, j\text{-}1), \gamma(i\text{-}1, j\text{-}1)\}$$

where $\gamma(i, j)$ is $(i, j)$ element of matrix that is a summation between $d(i, j) = (q_i\text{-} c_j)^2$, a square distance of $q_i$ and $c_j$, and the minimum cumulative distance of three adjacent elements to $(i, j)$.

Next, we choose the optimal warping path which has minimum cumulative distance defined as:

$$DTW(Q,C) = \min \sum_{k=1}^{K} w_k$$

where $w_k$ is $(i, j)$ at $k^{th}$ element of the warping path, and $K$ is the length of the warping path.

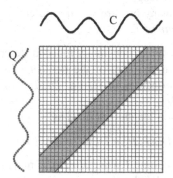

**Fig. 1.** DTW with Sakoe-Chiba band

In addition, for a more accurate distance measure, some global constraints were suggested to DTW. A well-known constraint is Sakoe-Chiba band [7], shown in Fig. 1. The Sakoe-Chiba band constrains the indices of the warping path $w_k = (i, j)_k$ such that $j - r \leq i \leq j + r$, where $r$ is a term defining the allowed range of warping, for a given point in a sequence. Due to space limitation, interested readers may refer to ([3], [7]) for more detail about DTW.

Due to evident advantages of DTW for time series data, our proposed work does incorporate DTW distance measure into our algorithm.

## 2.3     Stopping Criterion Based on MDL Principle

In this subsection, we review the state-of-the-art stopping criterion for SSC of time series which is based on the Minimum Description Length principle. This stopping criterion was proposed by Begum et al, in 2013 [1]. This is the first work to address SSC of time series using MDL.

**Definition 1.** *Discrete Normalization Function*: A discrete function *Dis_Norm* is the function to normalize a real-value subsequence $T$ into b-bit discrete value of range $[1, 2^b]$. It is defined as below:

$$Dis\_Norm(T) = round[(T - min)/(max - min)]*(2^b - 1) + 1$$

where *min* and *max* are the minimum and maximum value in $T$ respectively.

After casting the original real-valued data to discrete values, we are interested in determining how many bits are needed to store a particular time series $T$. It is called the *Description Length* of $T$.

**Definition 2.** *Description Length*: A description length $DL$ of a time series $T$ is the total number of bits required to represent it.

$$DL(T) = w*\log_2 c$$

where $w$ is the length of $T$ and $c$ is the cardinality (the number of values we discretize the time series).

**Definition 3.** *Hypothesis*: A hypothesis $H$ is a subsequence used to encode one or more subsequences of the same length.

We are interested in how many bits are required to encode $T$ given $H$. It is called the *Reduced Description Length* of $T$.

**Definition 4.** *Reduced Description Length*: A reduced description length of a time series $T$ given hypothesis $H$ is the sum of the number of bits required in order to encode $T$ exploiting the information in $H$. i.e. $DL(T|H)$, and the number of bits required for $H$ itself, i.e. $DL(H)$.

Thus, the reduced description length is defined as:

$$DL(T, H) = DL(H) + DL(T|H)$$

One simple approach of encoding $T$ using $H$ is to store a *difference vector* between $T$ and $H$. Therefore: $DL(T|H) = DL(T - H)$.

Example: Given $A$ and $H$, two time series of length 20:
$A = 1\ 2\ 4\ 4\ 5\ 6\ 8\ 8\ 9\ 10\ 11\ 13\ 13\ 14\ 17\ 16\ 17\ 18\ 19\ 19$
$H = 1\ 2\ 3\ 4\ 5\ 6\ 7\ 8\ 9\ 10\ 11\ 12\ 13\ 14\ 15\ 16\ 17\ 18\ 19\ 20$

Without encoding, the bit requirement to store $A$ and $H$ is $2*20*\log_2 20 = 173$ bits. The difference vector $A' = |A - H| = 0\ 0\ 1\ 0\ 0\ 0\ 1\ 0\ 0\ 0\ 0\ 1\ 0\ 0\ 2\ 0\ 0\ 0\ 0\ 1$. And in the difference vector, there are 5 mismatches. The bit requirement is now just $20*\log_2 20 + 5*(\log_2 20 + \lceil \log_2 20 \rceil) = 134$ bits, which brings out a good data compression.

Assume that there exists only a single positive instance as the initial training set [1]. The SSC procedure using MDL-based stopping criterion can be outlined as follows.

First, it selects the seed positive instance as hypothesis. It selects the nearest neighbor of any of the instance(s) in the labeled training set from the unlabeled dataset. It encodes this instance in terms of the hypothesis and keeps the rest of the dataset uncompressed. Then it computes the reduced description length of the whole dataset. If it can achieve a data compression then the instance in question is a true positive. It continues to test to see if unlabeled instances can be added to the positive pool by this data compression criterion. Once the SSC module starts including instances dissimilar to the hypothesis, it no longer achieves data compression and the first occurrence of such an instance is the point where the SSC module should *stop*.

Even though this stopping criterion is the best one for SSC of time series so far, it is still not effective to be used in some situations where time series may have some distortion along the time axis and the way of computing Difference Vector for them becomes so rigid that the stopping point for the classifier can not be found precisely.

## 3     The Proposed Method

This work aims to improve the MDL-based stopping criterion and at the same time improve the accuracy of the classifier. We devise an improvement technique for the MDL-based stopping criterion and propose a refinement step to make the classifier more accurate.

### 3.1     New Stopping Criterion Based on MDL Principle

The original MDL-based stopping criterion is really simple, which finds mismatch points by one-to-one alignment between two time series and then calculates Reduced Description Length using the number of mismatch points. In fact, it is hard to find bit saves in this method because the time series may have some distortion in the time axis and a lot of mismatches will be found and there are not many bit saves.

We propose a more flexible technique for finding mismatch points. Instead of linear alignment, we use a non-linear alignment when finding mismatch points. This method attempts to find an optimal matching between two time series for determining as fewer mismatch points as possible.

The principle of our proposed method is in the same spirit of the main characteristic of Dynamic Time Warping (DTW). Therefore, we can modify the algorithm of computing DTW distance between two time series in order to include the finding of mismatch points between them. Fig. 2 shows our proposed mismatch count algorithm based on the calculation of DTW distance. There are two phases in this algorithm. At first phase, we calculate the DTW distance. The second phase goes backward along the found warping path and finds the number of mismatch points. In addition, at first phase, we use Sakoe-Chiba band constraint (through the user-specified parameter $r$) for limiting the meaningless warping paths between the two time series.

### 3.2     Refinement Step

In this work, we include to the framework of semi-supervised time series classification algorithm given in Subsection 2.1 a process called *Refinement*. The aim of this process is to check again the training set and modify it in order to obtain a more accuracy classifier. This process is based on the finding of *ambiguous labeled instances*, and these ambiguous instances will be classified again using the confident true labeled instances. The refinement process is iterated until the training set becomes stable, i.e. the training set before and after a refinement iteration are the same.

| | **mismatch_count = Count_Mismatch (x, y, r)**<br>// x: Time series,  y: Time series,  r: Sakoe-Chiba band constraint |
|---|---|
| | // Phase 1: Calculate DTW with Sakoe-Chiba band constraint |
| 1 | matrix[1,1] = square(x[1] − y[1]) |
| 2 | **for** i = 2 **to** length(y) **do** |
| 3 |   matrix[1, i] = matrix[1, i-1] + square(x[1] − y[i]) |
| 4 | **end** |
| 5 | **for** i = 2 **to** length(x) **do** |
| 6 |   matrix[i, 1] = matrix[i -1, 1] + square (x[i] − y[1]) |
| 7 | **end** |
| 8 | **for** i = 2 **to** length(x) **do** |
| 9 |   **for** j = 2 **to** length(y) **do** |
| 10 |     **if** \|i − j\| <= r **then** |
| 11 |       min_val = MIN(matrix[i -1, j], matrix[i, j -1], matrix[i-1, j - 1]); |
| 12 |       matrix[i, j] = min_val + square(x[i] - y[j]) |
| 13 |     **else** |
| 14 |       matrix[i, j] = +INFINITY |
| 15 |     **end** |
| 16 |   **end** |
| 17 | **end** |
| | // Phase 2: Finding minimum number of mismatch points |
| 18 | i = length(x); j = length(y) |
| 19 | mismatch_count = 0 |
| 20 | **if** x[i] != y[j] **then** |
| 21 |   mismatch_count = mismatch_count + 1 |
| 22 | **end** |
| 23 | **while** i > 1 OR j > 1 **do** |
| 24 |   value = matrix[i, j] − square (x[i] − y[j]) |
| 25 |   **if** i > 1 AND j > 1 AND value = matrix[i-1, j - 1] **then** |
| 26 |     i = i − 1; j = j - 1 |
| 27 |   **else if** j > 1 AND value = matrix[i, j - 1] **then** |
| 28 |     j = j - 1 |
| 29 |   **else if** i > 1 AND value = matrix[i- 1, j] **then** |
| 30 |     i = i - 1 |
| 31 |   **end** |
| 32 |   **if** x[i] != y[j] **then** |
| 33 |     mismatch_count = mismatch_count + 1 |
| 34 |   **end** |
| 35 | **end** |

**Fig. 2.** Mismatch-count algorithm between two time series with Sakoe-Chiba band constraint

Fig. 3 shows our proposed refinement algorithm. In this algorithm, *AMBI* is the set of ambiguous labeled instances, *P* is the positive set and *N* is the negative set. The set *AMBI* consists of the instances which are near the positive and negative boundary. This algorithm classifies the instances in *AMBI* basing on the current *P* and *N*. The process of detecting *AMBI* and classifying the instances in *P* is repeated until *P* and *N* are unchanged. Finally, the instances in *AMBI* that can not be labeled will be classified the last time.

The ambiguous instance detection process is done under the following rules:

1. The instances in *P* which were classified as positive by SSC but their nearest neighbors are in the negative set *N*, then their nearest neighbors and themselves are ambiguous.

2. The instances in *N* which were classified as negative by SSC but their nearest neighbors are in the positive set *P*, then their nearest neighbors and themselves are ambiguous.

3. The instances which were classified as positive by *X-means-classifier* (explained later) but are classified as negative by SSC, these are considered ambiguous.

The process of classifying instances in *AMBI* is done using One-Nearest- Neighbor (1-NN) in which the instance in *AMBI* which is nearest to *P* or *N* will be labeled first.

In this work, we propose a method called *X-means-Classifier* that can be used as SSC method for time series. This is a clustering-based approach which applies X-means algorithm, an extended variant of k-means which was proposed by Pelleg and Moore in 2000 [5]. One outstanding feature of X-means is that it can automatically estimate the suitable number of clusters during the clustering process. The SSC method based on X-means consists of the following steps. First, we use X-means to cluster the training set (including positive and unlabeled instances). Then, if there exists one cluster which contains the positive instance, all the instances in it will be classified as positive instances, and all the rest are classified as negative. X-means-Classifier will be used to initialize the *AMBI* in the Refinement process (Line 1 in the algorithm in Fig. 3).

| | **Refinement (P, N)**<br>// *P*: positive/labeled set (output of Improved MDL method)<br>// *N*: negative/unlabeled set (output of Improved MDL method) |
|---|---|
| 1 | *AMBI* = Find ambiguous instances in *P* and *N* |
| 2 | *P = P − AMBI; N = N − AMBI* |
| 3 | **repeat** |
| 4 | Classify *AMBI* by new training set *P* and *N* and then add each classified instance to *P* and *N*. |
| 5 | *AMBI* = Find ambiguous instances in *P* and *N* |
| 6 | *P = P − AMBI;  N = N − AMBI* |
| 7 | **until** (*P* and *N* are unchanged) |
| 8 | Classify *AMBI* by new training set *P* and *N* and then add each classified instance to *P* and *N*. |

**Fig. 3.** The outline of Refinement process in SSC

# 4    Experimental Evaluation

We implemented our proposed method and previous methods with Matlab 2012 and conducted the experiments on the Intel Core i7-740QM 1.73 GHz, 4GB RAM PC. After the experiments, we evaluate the classifier by measuring the precision, recall and F-measure of the retrieval. The precision is the ratio between the correctly classified positive test data and the total number of test instances classified as positive. The recall is the ratio between the correctly classified positive test data and the total number of all positive instances in the test dataset. An F-measure is the ratio defined by the formula: $F = 2*p*r/(p + r)$ where $p$ is precision and $r$ is recall. In general, the higher the F-measure is, the better the classifier.

Our experiments were conducted over the datasets from UCR Time Series Data Mining archive [4]. Details of these datasets are shown in Table 1. Besides, MIT-BIH Supraventricular Arrhythmia Database and St. Petersburg Arrhythmia Database that are used to compare the stopping criteria are downloaded from ([9]).

**Table 1.** Datasets used in the evaluation experiments

| Datasets | Number of classes | Size of Dataset | Time series Length |
|---|---|---|---|
| Yoga | 2 | 300 | 426 |
| Words Synonyms | 25 | 267 | 270 |
| Two Patterns | 4 | 1000 | 128 |
| MedicalImages | 10 | 381 | 99 |
| Synthetic Control | 6 | 300 | 60 |
| TwoLeadECG | 2 | 23 | 82 |
| Gun-Point | 7 | 50 | 150 |
| Fish | 7 | 175 | 463 |
| Lightming-2 | 2 | 60 | 637 |
| Symbols | 6 | 25 | 398 |

## 4.1    Comparing two MDL-Based Stopping Criteria

We perform a comparison between our improvement technique and the previous MDL-based stopping criteria [1] on four dasets. Due to space limitation, here we show the experimental results on the three datasets: MIT-BIH Supraventricular Arrhythmia Database, St. Peterburg Arrhythmia Database and Gun Point Training Set in Fig. 4, Fig. 5 and Fig. 6, respectively.

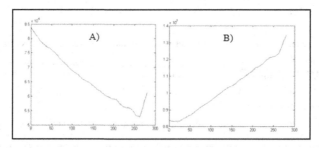

**Fig. 4.** A) Stopping point by our MDL (Proposed Method) at iteration 262 (Nearly perfect). B) Stopping point by MDL (Previous Method) at iteration 10 (too early).

From Figs. 4, 5, and 6, we can see that our improvement technique suggests a better stopping point in most of the datasets. Detecting a good stopping point is very crucial in SSC of time series. We attribute this desirable advantage of our improvement technique to the flexible way of determining mismatches between two time series when computing Reduced Description Length of one time series exploiting the information in the other.

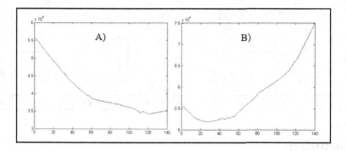

**Fig. 5.** A) Stopping point by our MDL (Proposed Method) at iteration 121 (Nearly perfect). B) Stopping point by MDL (Previous Method) at iteration 28 (too early)

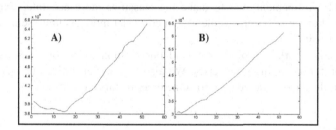

**Fig. 6.** A) Stopping point by our MDL (Proposed Method) at iteration 15 (Nearly perfect). B) Stopping point by MDL at iteration 3 (too early).

## 4.2    Effects of Refinement Step

Now we compare SSC by our new MDL-based stopping criterion with and without Refinement step. Table 2 reports the experimental results (precision, recall and F-measure) of this comparison. The results show that our proposed Refinement step brings out better performance in all the datasets. In most of datasets, the performance of the proposed method is better, for example, on Two-Paterns F = 81.4%, on Synthetic-Control F = 98.99%, on TwoLeadECG    F = 85.714%, on Fish F = 92.683%, on Symbol F = 94.118%. Specially, on the Synthetic-Control dataset, SSC without Refinement gives F = 14.815%, while with Refinement, F-measure reaches to 98.99%, a perfect result. These experimental results show that the Refinement step in SSC can improve the accuracy of the classifier remarkably.

**Table 2.** Experiment results with and without Refinement

| Datasets | Without Refinement | | | With Refinement | | |
|---|---|---|---|---|---|---|
| | Precision | Recall | F-measure | Precision | Recall | F-measure |
| Yoga | 0.64 | 0.35036 | 0.45283 | 0.57609 | 0.38686 | **0.46288** |
| WordsSynonyms | 0.94737 | 0.3 | 0.4557 | 0.625 | 0.41667 | **0.5** |
| Two Patterns | 1.0 | 0.41328 | 0.58486 | 1.0 | 0.68635 | **0.814** |
| MedicalImages | 0.57276 | 0.91133 | 0.70342 | 0.56587 | 0.93103 | **0.70391** |
| Synthetic Control | 1.0 | 0.08 | 0.14815 | 1.0 | 0.98 | **0.9899** |
| TwoLeadECG | 0.88889 | 0.66667 | 0.7619 | 0.75 | 1.0 | **0.85714** |
| Gun-Point | 0.93333 | 0.58333 | 0.71795 | 1.0 | 0.625 | **0.76923** |
| Fish | 0.94737 | 0.81818 | 0.87805 | 1.0 | 0.86364 | **0.92683** |
| Lightning-2 | 0.7619 | 0.4 | 0.52459 | 0.6875 | 0.55 | **0.61111** |
| Symbols | 1.0 | 0.75 | 0.85714 | 0.88889 | 1.0 | **0.94118** |

# 5    Conclusions

In this paper, we have proposed two novel improvements for semi-supervised classification of time series: an improvement technique for MDL-based stopping criterion and a refinement step to make the classifier more accurate. Experimental results show that our two improvements can construct more accurate semi-supervised time series classifiers.

As for future work, we plan to extend our method to perform semi-supervised classification for streaming time series. We also plan to generalize our method to the case of multiple classes and adapt it to some other distance measures.

# References

1. Begum, N., Hu, B., Rakthanmanon, T., Keogh, E.: Towards a Minimum Description Length Based Stopping Criterion for Semi-Supervised Time Series Classification. In: Proc. of IEEE 14th Int. Conf. on Information Reuse and Integration (IRI), San Francisco, CA, August 14-16, pp. 333–340 (2013)
2. Berndt, D., Clifford, J.: Using Dynamic Time Warping to Find Patterns in Time Series. In: Proceedings of AAAI Workshop on Knowledge Discovery in Databases, KDD 1994, Seattle, Washington, USA, pp. 359–370 (1994)
3. Keogh, E., Ratanamahatana, C.A.: Exact Indexing of Dynamic Time Warping. Knowledge and Information Systems 7, 358–386 (2005)
4. Keogh, E.: The UCR Time Series Classification/Clustering Homepage (2008), http://www.cs.ucr.edu/~eamonn/time_series_data/
5. Pelleg, D., Moore, A.: X-means: Extending k-means with Efficient Estimation of the Number of Clusters. In: Proc of ICML (2000)
6. Ratanamahatana, C.A., Wanichsan, D.: Stopping Criterion Selection for Efficient Semi-Supervised Time Series Classification. In: Lee, R.Y. (ed.) Software Engineering, Artificial Intelligence, Networking and Parallel/Distributed Computing. SCI, vol. 149, pp. 1–14. Springer, Heidelberg (2008)

7. Sakoe, H., Chiba, S.: Dynamic Programming Algorithm Optimization for Spoken Word Recognition. IEEE Trans. Acoustics, Speech, and Signal Proc. ASSP-26, 43–49 (1978)
8. Wei, L., Keogh, E.: Semi-Supervised Time Series Classification. In: Proceedings 12th ACM SIGKDD International Conference on Knowledge Discovery and Data Mining (2006)
9. Website on Time Series Classification:
   http://www.cs.ucr.edu/~nbegu001/SSL_myMDL.htm

# A Novel Method for Mining Class Association Rules with Itemset Constraints

Dang Nguyen[1,2], Bay Vo[1,2], and Bac Le[3]

[1] Division of Data Science and [2] Faculty of Information Technology,
Ton Duc Thang University, Ho Chi Minh City, Viet Nam
`nguyenphamhaidang@outlook.com, vodinhbay@tdt.edu.vn`
[3] Computer Science Department, University of Science, VNU - Ho Chi Minh, Viet Nam
`lhbac@fit.hcmus.edu.vn`

**Abstract.** Mining class association rules with itemset constraints is very popular in mining medical datasets. For example, when classifying which populations are at high risk for the HIV infection, epidemiologists often concentrate on rules which include demographic information such as sex, age, and marital status in the rule antecedents. However, two existing methods, post-processing and pre-processing, require much time and effort. In this paper, we propose a lattice-based approach for efficiently mining class association rules with itemset constraints. We first build a lattice structure to store all frequent itemsets. We then use paternity relations among nodes to discover rules satisfying the constraint without re-building the lattice. The experimental results show that our proposed method outperforms other methods in the mining time.

**Keywords:** Associative classification, Class association rule, Data mining, Useful rules, Lattice.

## 1   Introduction

Mining class association rules (CARs) with itemset constraints is one of variations of mining rules including mining association rules [1], mining CARs [2], mining sequential rules [3]. There are two basic methods to discover CARs with itemset constraints: post-processing and pre-processing methods. The post-processing method first mines the complete set of CARs by using an algorithm such as CBA [4], ECR-CARM [5], CAR-Miner [6], or PMCAR [2] and then filters out the ones which do not satisfy the itemset constraint in the post-processing step. One example of this strategy is CAR-Miner-Post [7]. This kind of strategy is very inefficient because all CARs have to be generated and often a large number of candidates must be tested in the last step. Another strategy, pre-processing, first filters out records which do not contain the constrained itemset in the pre-processing step, mines all CARs, and then obtains the constrained CARs in the post-processing step. In [7], the authors proposed SC-CAR-Miner, an algorithm integrates the itemset constraint into the actual mining process to generate only the CARs which satisfy the constraint. Since this strategy can use the properties of the constraint much more effectively, its execution time is much lower than those of the post- and pre-processing strategies.

D. Hwang et al. (Eds.): ICCCI 2014, LNAI 8733, pp. 494–503, 2014.

In practice, the itemset constraint is often changed by end-users. Consequently, the available methods must be executed again whenever the constraint is changed. Thus, the response cannot be immediately returned to the end-users. Under this context, the present study proposes an efficient method for mining CARs with itemset constraints which are in the form of the presence of specific items in the rule antecedents and are frequently changed by end-users.

The main contributions of this paper are stated as follows. Firstly, we develop a lattice structure to store all frequent itemsets (Section 4.1). Secondly, we provide a theorem for quickly checking the paternity relation between two nodes in the lattice (Section 4.2). Finally, we propose an efficient algorithm for mining CARs with the itemset constraint based on the lattice (Section 4.2).

## 2    Some Concepts and Notations

Let $D$ be a dataset with $n$ attributes $\{A_1, A_2, ..., A_n\}$ and $|D|$ records (objects) and let $C = \{c_1, c_2, ..., c_k\}$ be a list of class labels. A specific value of an attribute $A_i$ and class $C$ are denoted by lower-case letters $a_i$ and $c_j$, respectively. An item is described as an attribute and a specific value for that attribute, denoted by $\langle (A_i, a_{im}) \rangle$. An itemset is a set of items and a *Constraint_Itemset* is a specific itemset considered by end-users. A class association rule $r$ has the form *itemset* $\rightarrow c_j$, where $c_j \in C$ is a class label. A rule $r$ satisfies the itemset constraint if its antecedent (*itemset*) contains the *Constraint_Itemset*. The actual occurrence $ActOcc(r)$ of rule $r$ in $D$ is the number of objects in $D$ which match $r$'s antecedent. The support of rule $r$, denoted by $Sup(r)$, is the number of objects in $D$ which match $r$'s antecedent and are labeled with $r$'s class. The confidence of rule $r$, denoted by $Conf(r)$, is defined as:

$$Conf(r) = \frac{Sup(r)}{ActOcc(r)}$$

**Table 1.** Example of a dataset

| OID | A | B | C | Class |
|-----|-----|-----|-----|-------|
| 1 | a1 | b1 | c1 | 1 |
| 2 | a1 | b2 | c1 | 2 |
| 3 | a2 | b2 | c1 | 2 |
| 4 | a3 | b3 | c1 | 1 |
| 5 | a3 | b1 | c2 | 2 |
| 6 | a3 | b3 | c1 | 1 |
| 7 | a1 | b3 | c2 | 1 |
| 8 | a2 | b2 | c2 | 2 |

A sample dataset is shown in Table 1 where *OID* is an object identifier. It contains eight objects, three attributes, and two classes (1 and 2). For example, consider rule $r:\{(A,a1)\} \rightarrow 1$. We have $ActOcc(r) = 3$ and $Sup(r) = 2$ since there are three objects with $A = a1$, in which two objects have the same class 1. In addition, we have

$$Conf(r) = \frac{Sup(r)}{ActOcc(r)} = \frac{2}{3}.$$

## 3     Related Work

In this section, we review some algorithms for mining frequent itemsets with itemset constraints.

Since the introduction of mining frequent itemsets with itemset constraints [8], various strategies have been proposed. They can be classified into three main categories: post-processing, pre-processing, and constrained itemset filtering. Post-processing methods firstly mine frequent itemsets, and then check them against the itemset constraint. Two examples are Apriori+ [9] and FP-Growth+ [10]. Pre-processing methods firstly restrict the source dataset to records which contain constrained itemsets, and then find frequent itemsets on the filtered dataset. Examples include MCFPTree [10] and Pre-CAP [11]. Constrained itemset filtering methods integrate itemset constraints into the actual mining process to generate only frequent itemsets which satisfy the constraint. CAP [9] and MFS_DoubleCons [12] belong to this strategy.

However, three approaches for mining frequent itemsets with itemset constraints cannot be applied for mining CARs with itemset constraints since they do not generate constrained rules directly. They succeed in mining frequent itemsets with itemset constraints only. Obviously, the problem of mining class association rules with itemset constraints studied in this paper is totally different from the problem of mining frequent itemsets with itemset constraints. Hence, it requires a different strategy for mining.

## 4     Mining CARs with Itemset Constraints

### 4.1     Lattice Structure

In this section, we describe the lattice structure used to store all frequent itemsets in the dataset.

**Definition 1:** Let $X$ be a $k$-itemset. The children itemsets generated from $X$ based on the equivalence class concept are:

$$childrenEC(X) = \{XA \mid XA \text{ is a } (k+1)\text{-itemset}, A \notin X \text{ and } A \neq \varnothing\}$$

**Definition 2:** Let $X$ be a $k$-itemset. The children itemsets generated from $X$ based on the lattice concept are:

$$childrenL(X) = \{BX \mid BX \text{ is a } (k+1)\text{-itemset}, BX \notin childrenEC(X) \text{ and } X \subset BX\}$$

**Definition 3:** Each node in the lattice is a tuple as follows:

$$\langle id, itemset, (Obidset_1, ..., Obidset_k), pos, total, traverse, childrenEC, childrenL \rangle$$

Where:

1. $id$ : A positive integer storing the identity of the node

2. $(Obidset_1, ..., Obidset_k)$ : A list of $Obidsets$ in which each $Obidset_i$ is a set of object identifiers that contain the $itemset$ and class $c_i$ ($k$ is the number of classes)

3. $pos$ : A positive integer storing the position of the class with the maximum cardinality of $Obidset_i$, i.e., $pos = \text{argmax}_{i \in [1,k]} \{|Obidset_i|\}$

4. $total$ : A positive integer which stores the sum of cardinality of all $Obidset_i$, i.e.,

$$total = \sum_{i=1}^{k} |Obidset_i|$$

5. $traverse$ : A flag which indicates whether or not the node already generated a rule

6. $childrenEC$ : A list of children nodes generated from $itemset$ based on the equivalence class definition

7. $childrenL$ : A list of children nodes generated from $itemset$ based on the lattice definition

In the lattice, the $itemset$ is converted into form $att \times values$ for easily programming, where:

(a) $att$ : A positive integer which represents a list of attributes

(b) $values$ : A list of values, each of which is contained in one attribute in $att$

**Example 1:** In Table 1, itemset $X = \langle (A, a2) \rangle$ is contained in two objects 3 and 8, both of which belong to class 2. Thus, the node which contains itemset $X$ has the form $1 \times a2(\varnothing, \underline{38})$ in which $att=1$, $values=a2$, $Obidset_1 = \varnothing$ (i.e. no objects contain both itemset $X$ and class 1), $Obidset_2 = \{3,8\}$ (or $Obidset_2 = 38$ for short) (i.e. two objects 3 and 8 contains both itemset $X$ and class 2), $pos=2$ (a line under position 2 of list $Obidset_i$), and $total=2$. $pos$ is 2 because the cardinality of $Obidset$ for class 2 is maximum (2 versus 0).

Compared to the MECR-tree [6] and the SCR-tree [7], the lattice structure has a significant advantage that we do not need to re-build the lattice when the itemset constraint is changed. Based on the generated lattice, we use paternity relations among nodes to discover class association rules satisfying the itemset constraint. This noticeably improves the response time.

## 4.2    Proposed Algorithm

In this section, we firstly introduce some theorems for quickly determining the support of an infrequent itemset and the paternity relation between two nodes. Then, we present an efficient algorithm called L-CAR-Miner for mining CARs with the itemset constraint based on the lattice structure.

**Theorem 1 [7]:** Given two nodes $X$ and $Y$, if $X.att = Y.att$ and $X.values \neq Y.values$, then $X.Obidset_i \cap Y.Obidset_i = \varnothing$ ($\forall i \in [1, k]$).

Theorem 1 implies that if two nodes $X$ and $Y$ have the same attributes, it is not necessary to combine them as node $XY$ because $Sup(XY) = 0$.

**Theorem 2 [13]:** Let $XA$ and $XB$ be two nodes of $k$-itemset. If $\forall XB \in X.childrenEC$ and $XA$ is generated before $XB$ in lattice, $\neg \exists Y \in XB.childrenEC \cup XB.childrenL$ so that $Y \in XA.childrenL$.

Theorem 2 implies that finding children nodes which belong to $XA.childrenL$ is easily performed in two loops and one if statement as follows: (i) let $Y \in X.childrenL$, (ii) let $YZ \in Y.childrenEC$, and (iii) if $XA \subset YZ$, then $YZ \in XA.childrenL$. This theorem allows the proposed algorithm to be better than the one mentioned in [14] because it eliminates a large number of candidates.

**Theorem 3:** Let $Y \in X.childrenL$, $YZ \in Y.childrenEC$, and $XA \in X.childrenEC$. If $A = Z$, then $XA \subset YZ$.

**Proof:** Regarding Definition 2, $Y$ has the form $BX$ where $BX$ is a $(k+1)$-itemset. It can be inferred that $YZ$ has the form $BXZ$. Thus, if $A = Z$, then $XA \subset BXA = YZ$ ∎

Using Theorem 3 allows us to check quickly the condition (iii) $XA \subset YZ$ in Theorem 2. Instead of checking $XA$ with all subsets of $YZ$, we need to check only the condition $A = Z$.

Based on the proposed lattice structure and three theorems, the algorithm L-CAR-Miner is developed for efficiently mining CARs with the itemset constraint. The proposed algorithm is briefly described as follows. Firstly, the lattice structure is built to store frequent itemsets (showed in Figure 1). Secondly, when the itemset constraint is changed, the sub-lattice which begins at the itemset constraint is traversed to generate all CARs satisfying the constraint (showed in Figure 2).

The algorithm firstly finds the root node of the lattice ($L_r$) which contains frequent 1-itemsets at the first level (Line 1). Procedure BUILD-LATTICE is then recursively called with the parameters $L_r$ and $minSup$ to build the lattice structure (Lines 2-19). The function of procedure UPDATE-LATTICE($X$, $XA$) is to update the paternity relations of a node with its children nodes (Lines 20-23) while the function of procedure TRAVERSE-LATTICE is to find all rules which satisfy the itemset constraint (Lines 24-30). Procedure GENERATE-RULE generates a rule from a node (Lines 31-33).

We apply the proposed algorithm to the example dataset in Table 1 with $minSup = 25\%$ and $minConf = 50\%$ to illustrate its basic ideas. Firstly, L-CAR-Miner finds all frequent 1-itemsets. The result after this first step is $L_r = \{1 \times a1(\underline{17}, 2), 1 \times a2(\varnothing, \underline{38}),$ $1 \times a3(\underline{46}, 5), 2 \times b2(\varnothing, \underline{238}), 2 \times b3(\underline{467}, \varnothing), 4 \times c1(\underline{146}, 23), 4 \times c2(7, \underline{58})\}$. The algorithm then calls procedure BUILD-LATTICE with two parameters $L_r$ and $minSup$ to build the lattice structure. When a node in the lattice is generated, its paternity relations with its children nodes are updated through procedure UPDATE-LATTICE. The lattice structure constructed by the proposed algorithm is shown in Figure 3. Note that the solid and dashed lines represent *childrenEC* and *childrenL*, respectively.

---

**Input:** Dataset $D$ and minimum support *minSup*
**Output:** Lattice $L$ containing all frequent itemsets in $D$
**Procedure:**

1.  Let $L_r$ be the root of the lattice. $L_r$ includes a set of nodes where each node contains a frequent 1-itemset and has the form:

$$\langle id, itemset, (Obidset_1, ..., Obidset_k), pos, total, false, \{\ \}, \{\ \}\rangle$$

**BUILD-LATTICE($L_r$, *minSup*)**

2.  for all $l_x \in L_r$.children do

3.      $P_i = \varnothing$;

4.      for all $l_y \in L_r$.children, with $y > x$ do

5.          if $l_y.att \neq l_x.att$ then // using Theorem 1

6.              assign an incremental integer to $O.id$;

7.              $O.att = l_x.att \cup l_y.att$; // using bitwise operation

8.              $O.values = l_x.values \cup l_y.values$;

9.              $O.Obidset_i = l_x.Obidset_i \cap l_y.Obidset_i$; // $\forall i \in [1, k]$

10.             $O.pos = \text{argmax}_{i \in [1,k]} \{|O.Obidset_i|\}$;

11.             $O.total = \sum_{i=1}^{k} |O.Obidset_i|$;

12.             $O.traverse = false$;

13.             if $|O.Obidset_{O.pos}| \geq minSup$ then // $O$ satisfies *minSup*

14.                 add $O.id$ to $l_x.childrenEC$;

15.                 add $O.id$ to $l_y.childrenL$;

16.                 UPDATE-LATTICE($l_x$, $O$);

17.                 $P_i = P_i \cup O$;

18.                 add $O$ to lattice $L$;

19.     BUILD-LATTICE($P_i$, *minSup*);

**UPDATE-LATTICE($X$, $XA$)**

20. for each child node $Y$ in $X.childrenL$ do // using Theorem 2

21.     for each child node $YZ$ in $Y.childrenEC$ do

22.         if $A = Z$ then // using Theorem 3

23.             add $YZ.id$ to $XA.childrenL$;

---

**Fig. 1.** Building the lattice structure

**Input:** Lattice $L$, minimum confidence *minConf*, and the itemset constraint *Constraint_Itemset*

**Output:** All CARs satisfying *minSup*, *minConf*, and *Constraint_Itemset*

**Procedure:**

**TRAVERSE-LATTICE**($l$, *minConf*)

24. if $l.traverse = false$ then
25.     GENERATE-RULE($l$, *minConf*);
26.     $l.traverse = true$ ;
27.     for each child node $X$ in $l.childrenEC$ do
28.        TRAVERSE-LATTICE( $X$ , *minConf*);
29.     for each child node $Y$ in $l.childrenL$ do
30.        TRAVERSE-LATTICE( $Y$ , *minConf*);

**GENERATE-RULE**($l$, *minConf*)

31. $conf = \left| l.Obidset_{l.pos} \right| / l.total$ ;
32. if $conf \geq minConf$ then
33.     CARs=CARs $\cup \left\{ l.itemset \rightarrow c_{pos} \left( \left| l.Obidset_{l.pos} \right|, conf \right) \right\}$ ;

**Fig. 2.** Traversing the sub-lattice to generate constrained CARs

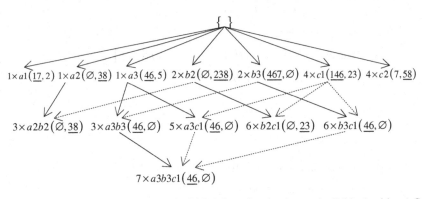

**Fig. 3.** Lattice structure generated by L-CAR-Miner for the dataset in Table 1 with *minSup* = 25% and *minConf* = 50%

To generate all CARs satisfying the itemset constraint and *minConf*, the algorithm uses procedure TRAVERSE-LATTICE to traverse the sub-lattice which begins at the constraint. For example, consider *Constraint_Itemset* = $\langle (B, b3) \rangle$. The algorithm first looks up node $2 \times b3 (\underline{467}, \varnothing)$. Because this node did not generate a rule yet, the algorithm uses procedure GENERARTE-RULE to generate rule $2 \times b3 \rightarrow 1$ (i.e. "if B = b3 then Class = 1"). Then the algorithm considers node $6 \times b3c1 (\underline{46}, \varnothing)$ and generates rule $6 \times b3c1 \rightarrow 1$ (i.e. "if B = b3 and C = c1 then Class = 1"). Similarly, rules $7 \times a3b3c1 \rightarrow 1$ and $3 \times a3b3 \rightarrow 1$ are generated from node $7 \times a3b3c1 (\underline{46}, \varnothing)$ and node $3 \times a3b3(\underline{46},\varnothing)$, respectively. Note that the algorithm takes into account node $7 \times a3b3c1(\underline{46},\varnothing)$ after generating a rule from node $3 \times a3b3(\underline{46},\varnothing)$ through the *childrenEC*. However, the rule is not generated because it is already done before through the *childrenL* of node $6 \times b3c1(\underline{46},\varnothing)$.

## 5 Experiments

All experiments were conducted on a computer with an Intel Core i7-2637M CPU at 1.7 GHz and 4 GB of RAM, running Windows 7 Enterprise (64-bit) SP1. The experimental datasets were obtained from the UCI Machine Learning Repository[1]. The algorithms were coded in C# using Microsoft Visual Studio .NET Premium 2013 with .NET Framework 4.5.50938. Characteristics of experimental datasets are described in Table 2. The table shows the number of attributes (including the class attribute), the number of class labels, the number of distinctive values (i.e. the total number of distinct values in all attributes), and the number of objects (or records) in each dataset. Note that *minConf* = 50% was used for all experiments.

**Table 2.** Characteristics of the experimental datasets

| Dataset | #attributes | #classes | #distinctive values | #objects |
|---|---|---|---|---|
| German | 21 | 2 | 1,077 | 1,000 |
| Lymph | 19 | 4 | 63 | 148 |
| Chess | 37 | 2 | 76 | 3,196 |
| Connect-4 | 43 | 3 | 130 | 67,557 |

To show the efficiency of L-CAR-Miner, we compared its execution time with those of Pre-CAR-Miner (a pre-processing method) and SC-CAR-Miner [7]. Since SC-CAR-Miner always outperforms CAR-Miner-Post [7], we do not include CAR-Miner-Post in the comparison. The results are shown in Figures 4 and 5. Note that Pre, SC, and L are runtimes of Pre-CAR-Miner, SC-CAR-Miner, and L-CAR-Miner, respectively.

To initialize the *Constraint_Itemset*, we define the selectivity of a constraint as the ratio of the number of items selected to be the constraint against the total number of items. Thus, a constraint with 0% selectivity means no items, while a constraint with 100% selectivity is the one selecting all the items (distinctive values) in the dataset. Note that the runtime of L-CAR-Miner includes both execution times of two phases, lattice building and mining.

The results show that Pre-CAR-Miner performed worst for all experimental datasets. It is slower than SC-CAR-Miner and L-CAR-Miner because it must generate all CARs and check a huge number of candidates. L-CAR-Miner is always the fastest of all tested algorithms. For example, consider the Chess dataset with *minSup* = 50% and selectivity = 90%. The runtime of L-CAR-Miner was 3.807(s) while Pre-CAR-Miner was 40.357(s) and SC-CAR-Miner was 17.468(s).

[1] http://mlearn.ics.uci.edu

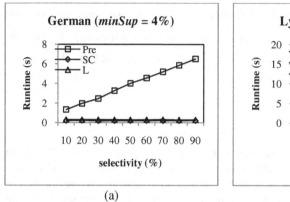

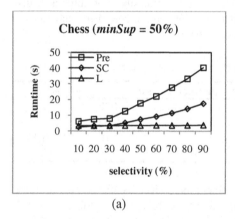
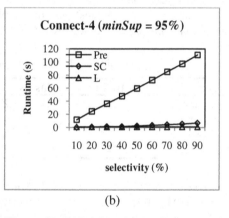

**Fig. 4.** Runtimes of Pre-CAR-Miner, SC-CAR-Miner, and L-CAR-Miner for the German (a) and Lymph (b) datasets with the constraint selectivity

**Fig. 5.** Runtimes of Pre-CAR-Miner, SC-CAR-Miner, and L-CAR-Miner for the Chess (a) and Connect-4 (b) datasets with the constraint selectivity

# 6    Conclusions

This paper proposes an efficient approach for mining CARs with itemset constraints. Unlike the post-processing and pre-processing, the proposed approach does not need to re-build its data structure when the constraint is changed. The framework of the proposed algorithm is based on the lattice structure and three theorems for quickly pruning infrequent itemsets and determining the paternity relation between two nodes.

To validate the efficiency of the proposed algorithm, a series of experiments was conducted on four popular datasets, namely German, Lymph, Chess, and Connect-4. The experimental results show that the proposed method is superior to existing methods.

However, when the minimum support value is very low, the cost for storing the lattice is very high. This can cause the memory leakage. We will study the solution for reducing the memory consumption in the future.

**Acknowledgment.** This work was supported by Vietnam's National Foundation for Science and Technology Development (NAFOSTED) under Grant Number 102.01-2012.17.

# References

1. Vo, B., Le, B.: Mining traditional association rules using frequent itemsets lattice. In: The 39th International Conference on Computers & Industrial Engineering (CIE 2009), pp. 1401–1406. IEEE (2009)
2. Nguyen, D., Vo, B., Le, B.: Efficient Strategies for Parallel Mining Class Association Rules. Expert Systems with Applications 41, 4716–4729 (2014)
3. Van, T.-T., Vo, B., Le, B.: IMSR_PreTree: an improved algorithm for mining sequential rules based on the prefix-tree. Vietnam Journal of Computer Science 1, 97–105 (2014)
4. Liu, B., Hsu, W., Ma, Y.: Integrating classification and association rule mining. In: The 4th International Conference on Knowledge Discovery and Data Mining (KDD 1998), pp. 80–86 (1998)
5. Vo, B., Le, B.: A novel classification algorithm based on association rules mining. In: Richards, D., Kang, B.-H. (eds.) PKAW 2008. LNCS, vol. 5465, pp. 61–75. Springer, Heidelberg (2009)
6. Nguyen, L.T., Vo, B., Hong, T.-P., Thanh, H.C.: CAR-Miner: An efficient algorithm for mining class-association rules. Expert Systems with Applications 40, 2305–2311 (2013)
7. Nguyen, D., Vo, B.: Mining class-association rules with constraints. In: Van Huynh, N., Denoeux, T., Tran, D.H., Le, A.C., Pham, B.S. (eds.) KSE 2013, Part II. AISC, vol. 245, pp. 323–334. Springer, Heidelberg (2014)
8. Srikant, R., Vu, Q., Agrawal, R.: Mining association rules with item constraints. In: The 3rd International Conference on Knowledge Discovery and Data Mining (KDD 1997), pp. 67–73 (1997)
9. Ng, R.T., Lakshmanan, L.V.S., Han, J., Pang, A.: Exploratory mining and pruning optimizations of constrained associations rules. In: ACM SIGMOD International Conference on Management of Data, pp. 13–24. ACM (1998)
10. Lin, W.-Y., Huang, K.-W., Wu, C.-A.: MCFPTree: An FP-tree-based algorithm for multi-constraint patterns discovery. International Journal of Business Intelligence and Data Mining 5, 231–246 (2010)
11. Nguyen, D., Truong, T., Vo, B.: Mining Frequent Patterns containing HIV-Positive in HIV Voluntary Counseling and Testing data. ICIC Express Letters 8, 541–546 (2014)
12. Duong, H., Truong, T., Vo, B.: An efficient method for mining frequent itemsets with double constraints. Engineering Applications of Artificial Intelligence 27, 148–154 (2014)
13. Vo, B., Le, T., Hong, T.-P., Le, B.: An effective approach for maintenance of pre-large-based frequent-itemset lattice in incremental mining. Applied Intelligence (in press, 2014)
14. Nguyen, L.T., Vo, B., Hong, T.-P., Thanh, H.C.: Classification based on association rules: A lattice-based approach. Expert Systems with Applications 37, 11357–11366 (2012)

# A *PWF* Smoothing Algorithm for K-Sensitive Stream Mining Technologies over Sliding Windows

Ling Wang[1,*], Zhao Yang Qu[1], Tie Hua Zhou[2], Xiu Ming Yu[2], and Keun Ho Ryu[2]

[1] Department of Computer Science, School of Electrical & Computer Engineering,
Northeast Dianli University, Jilin, China
smile2867ling@gmail.com, qzywww@mail.nedu.edu.cn
[2] Database/Bioinformatics Laboratory, School of Electrical & Computer Engineering,
Chungbuk National University, Chungbuk, Korea
{thzhou,yuxiuming,khryu}@dblab.chungbuk.ac.kr

**Abstract.** The development of Streaming Mining technologies as a hotspot entered the limelight, which is more effectively to avoid big data and distributed streams mining problems. Especially for the *IoT* and Ubiquitous Computing may interact with the real world's humans and physical objects in a sensory manner. They require quantitative guarantees regarding the precision of approximate answers and support distributed processing of high-volume, fast, and variety streams. Recent works on mining Top-*k* synopsis processing over data streams is that utilize all the data between a particular point of landmark and the current time for mining. Actually, the landmark and parameter *k* are two more important factors to obtain high-quality approximate results. Therefore, we proposed a Proper-Wavelet Function (*PWF*) algorithm to smooth the approximate approach, in order to reduce *k*-effect to the final approximate results. Finally, we demonstrate the effectiveness of our algorithm in achieving high-quality *k*-nearest neighbors mining results with applying wider proper *k* values.

**Keywords:** Top-*k* synopsis processing, stream mining, sliding windows, *PWF* smoothing algorithm, distributed processing.

## 1 Introduction

In many real-world applications, data stream are usually collected in a sensory manner such as sensor network, ubiquitous sensor network, online e-commerce data analysis, weather forecasting and smart grid. Particularity, requirements for a continuous, fast, high-volume, adaptable, costly streaming data, in an approximate analysis is a requirement for a fast response to users on forward predicates, such as some famous Data Stream Management Systems (*DSMS*) [1, 2]. Approximate query processing [3, 4 and 5] has emerged as a cost-effective solution to mitigate the issue of dealing with huge data volumes and stringent response-time requirements of

---

* Corresponding author.

D. Hwang et al. (Eds.): ICCCI 2014, LNAI 8733, pp. 504–514, 2014.
© Springer International Publishing Switzerland 2014

today's decision-support systems. Actually, recent work on querying data streams has focused on systems where newly arriving data is processed and continuously streamed to the users in real time. Among all kinds of sketching techniques, the wavelet-based approaches [6, 7, and 8] have received the most research attention due to the property of dimensionality reduction and the simplicity of transforming the data cells. Wavelets provide a mathematical tool for the hierarchical decomposition of functions, with a long history of successful applications in signal and image processing. *Haar* wavelets have been used extensively in synopsis construction multiple applications in image analysis, signal processing, and streaming databases. Zhu et al. considered burst detection using a summary structure called SWT, which is a shifted-wavelet tree based on the Haar wavelet transform [9]. Teng et al. applied the *Haar* wavelet transform concept to discover the frequent temporal patterns of data streams [10]. Recent work has demonstrated the effectiveness of the wavelet decomposition in reducing large amounts of data to a compact set of wavelet coefficients (termed "wavelet synopsis") that can be used to provide fast and reasonably accurate approximate answers to queries. In fact, errors can vary widely and unpredictably, even for identical queries on nearly-identical values in distinct parts of the data.

Continuous $k$-NN queries aims at retrieving the similarity between two or more streams is more challenging since the query object, the data, or both may change over time [7, 11, 12 and 13]. The Euclidean distance has become the most widely used similarity measure to retrieve important patterns, and note that a DSMS usually monitors a massive number of data streams. Under this circumstance, in addition to the simple similarity search between two streams, discovering k-NN queries usually cost much higher complexity, for example, *RFN* [14], *QEPs* [15] and rank based $k$-NN [16]. We study the problem of similarity search and continuous nearest neighbor discovery within an arbitrary range from the wavelet synopsis over distributed streams, where both of the query sequence and the data sequences are changed over time. Recent works on mining Top-k synopsis processing over data streams is that utilize all the data between a particular point of landmark and the current time for mining. The landmark usually refers to the time or tuple-number when the sliding window starts, and the interval of the most recent landmarks could decide the basic-size of sliding windows. Actually, the landmark and parameter $k$ are two more important factors to obtain high-quality approximate results. Therefore, we proposed a Proper-Wavelet Function (*PWF*) algorithm to smooth the approximate approach, in order to reduce $k$-effect to the final results. Finally, we demonstrate the effectiveness of our algorithm in achieving high-quality k-nearest neighbors mining results with applying wider proper $k$ values (local Top-k synopsis processing).

## 2    Overall Approach and Motivation

Many applications of wavelet transforms consider representing the input in terms of high level coefficients and broader characteristics of the original data, and those are typically referred to as a synopsis. These synopses are used subsequently in learning, classification, aggregation, event detection. In a multiple streams discussion, the traditional wavelet transform to process in limited memory are not easy to hold all values to do the real time analysis. One way to mine useful wavelet coefficients instead of exist data stored in synopses is to do further processing, and enables unbiased approximate query answers with low error guarantees on the accuracy of individual answers [18]. The other way is that uses statistical functions to measure the central tendency of data. In this paper, we tested several kinds of different statistical functions, and finally proposed our PWF algorithm to smooth the approximate approach in a real-time way. Here, "smoothing" is means that we found a Proper-Wavelet Function would help to afford a wider range of $k$ values, furthermore to reduce the $k$-effect (local Top-$k$ synopsis processing) in $k$-nearest neighbors' queries (see Fig.1.). There are two different kinds of value $k$, one is local Top-$k$ synopsis ($k_1$) and the other is k-nearest neighbor queries ($k_2$). They are both have a big impact on space usage and calculation precision for approximation approach. If $k_1$ is too small to select far fewer itemsets leads to low calculation precision, otherwise almost all the itemsets could be selected and greatly increase the load space of the real time system. At the same time, $k_2$ is also has the same effect for the final approximation results. Because $k_1$ is processed in the first phase and it decided the local size of each sliding windows in our design, and it is truly more important than $k_2$ after testing. Therefore, in this paper, we only focus on $k$-sensitive Top-$k$ synopsis processing to discuss the approximated calculation.

Our second contribution is that achieves high-quality $k$-nearest neighbors mining results with applying wider proper $k_1$ values. That's because our *PWF* function lets $k_1$ quickly to start a smooth increasing of identified itemsets' types. $k_1$ is usually hard to control and to set up by personal opinion because of a variable value to decide the basic-size of sliding windows. Moreover, it can directly decide the recognition of identified itemsets' types, and it is a big effect of the accurate of results. We presented an adaptive way to predict proper-k value from small to large until smooth increasing starts. This predict proper-k value is relatively smaller than the usual values, that means more calculate space usage could be saved and effectively support the precision of the approximate calculations.

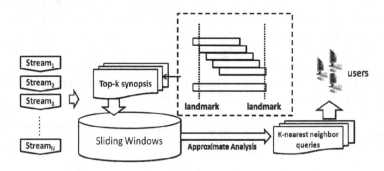

**Fig. 1.** Overall approach

# 3    Proper-Wavelet Function and Central Tendency Measuring

Nowadays, since the data streams evolve continuously without limit, it is impractical to store complete details for each stream. Instead, the queries are usually processed from the limited memory in which the behaviors of the data streams are summarized. Among all kinds of sketching techniques, the wavelet-based approaches have received the most attention dues to the property of dimensionality reduction and the simplicity of transforming the data cells. A class of algorithms for stream processing focuses on the recent past of data streams by applying sliding windows. Many of wavelet transforms consider representing the input in terms of the high level coefficients and broader characteristics of the original data, typically referred to as synopses.

## 3.1    Parameters and Function Definition

However, these methods also cannot get away from arbitrary judgment of two important values: How closed to destination query point (extended boundary- $E_b$)? How many Top-k objects for each local synopsis ($k_l$)? Therefore, our proposed PWF algorithm connects these two parameters were bound together by a mutual suppression agreement. First, consider an object $O$ that is involved only in queries that request a bound $E_b$ on the value of $O$ alone. Then it suffices to fix the bound width of $O$ to be the smallest of the precision constraints: $W_o = \min(E_b)$ for queries $Q_m$ with $S_m = \{O\}$. Second, in order to obtain the most suitable value of $E_b$, the most frequency itemset $F_{max}$ and the total stream types $T_t$ are have very closely linked with it. With the continuous inflow of data flow, we set the $a = F_{max} / T_t$ to control $E_b$ and the number of identified stream data types by using the Top-$k$ mining, it could help approximate results to keep a good balance until $a$ is greater or equal to the 0.5.

The other way is that uses statistical functions to measure the central tendency of data. In this paper, we tested several kinds of different statistical functions, and finally proposed our PWF algorithm to smooth the approximate approach in a real-time way. The general way to measure the central tendency of data, Suppose that we have some attribute $X$, like *salary*, which has been recorded for a set of objects. Let $x_1, x_2, \ldots, x_N$ be the set of $N$ observed values or observations for $X$. Here, these values may also be referred to as the data set (for $X$). If we were to plot the observations for salary, where would most of the values fall? This gives us an idea of the central tendency of the data. Measures of central tendency include the *mean, median, mode,* and *midrange*. The most common and effective numeric measure of the "center" of a set of data is the (arithmetic) mean. Let $x_1, x_2, \ldots, x_N$ be a set of $N$ values or observations, such as for some numeric attribute $X$, like salary. The mean of this set of values is

$$\bar{x} = \frac{\sum_{i=1}^{N} x_i}{N} = \frac{x_1 + x_2 + \ldots + x_N}{N} \tag{3.1}$$

This corresponds to the built-in aggregate function, average (*avg()* in SQL), provided in relational database systems.

Most of the central tendency measuring is easy to retrieve, but exactly only focus on the "center" is too one-side. Therefore, we proposed a novel Proper-Wavelet Function to combine wavelet transform with central tendency methods to derive a new computing equation: Function $(PWF) = (F_{max}/F_{min}, mean)$ . $F_{max}$ is the highest frequency in each streaming line, and $F_{min}$ is the lowest one. In fact, we have tested several kinds of different functions focus on these three characters, such as $|(F_{max}/F_{min})- mean|$, $((F_{max} + F_{min} /2), mean)$, $(F_{max} -F_{min} )$ and so on. Our conclusion is that Function $(PWF)$ makes the $k$-effect and $E_b$ value of the fastest convergence, and let the streaming types smoothly increasing.

## 3.2    PWF Algorithm

In our proposed real time processing flow, k most similarity object streams that would be stored in a combination synopsis as shown in the Fig.2. The original data into a new representation of pair $(m, n)$ to present PWF function parameters, and they would be stored in a sub-window of each synopsis as the most representative factors. Here, $\{S_1, S_2, S_3 ...\}$ represent different kinds of streams, such as Stream 1, Stream 2, etc. Here, the pair $(m, n)$ is equals to $(PWF) = (F_{max}/F_{min}, mean)$. After PWF copulating of each streaming lines, a Top-k list based on the parameter $E_b$ value could be formed and removed to the final synopsis. Finally, give a line express the update objects for supporting real time analysis in sliding windows and a statistical computing over the synopsis to obtain the approximate results.

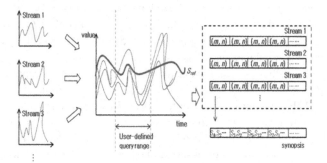

**Fig. 2.** *PWF* function transforms phases

In a stream application, a new mechanism is needed to support continuous queries over real-time data that is continuously updated from the environment. For effective processing of stream data, new data structures, techniques, and algorithms are needed. Our proposed PWF algorithm contains three key points to reduce the $k$-effect for both of Top-$k$ and $k$-nearest neighbors processing. First, the size of sliding window is a very important and very big effect value for approximate results. It's very difficult to maintain the Top-$k$ lists for each streaming datasets when the size of sliding window is too large. Otherwise, it would be take more space usage to calculate. Therefore, we track the size of landmark use parameter $a$ ,until it equals to 0.5. Then, it can help us to get a relatively safe value of sliding windows' size. Second, how many itmesets could be identified is decided by parameter $E_b$, and it could be calculated under the previous value control. Third, function $(PWF)$ transforms the original datasets into paired sets $(V_{pair})$ of each stream over each sub-window, and the goal of continuous $k$-NN queries is to find the $k$ streams among all these paired transforms with the highest similarities to the query stream compared to that of other streams in the user-defined time range. In our testing, it reduces $k$-cost wasted because of $a$, and the function $(PWF)$ is the fastest way to gather the value $E_b$ which ensures the itemsets within a stable balance.

The detailed PWF algorithm is shown in the following:

```
Input: s, data streams flow;
Output: Syn,Top-k lists of stream IDs.
For α = Fₘₐₓ / Tₜ, E_b = i×j;
  if α < 0.5, then s++,j++;
    if 1≤j≤10, then E_b = i×j;
    else E_b = i×j, i = E_b +1, j=1;
  else if α≥0.5; return E_b, Fₘₐₓ;
    for calculate Fₘᵢₙ, mean;
    return Function (PWF);
        if V_pair < E_b then remove to the local synopsis;
        else delete from synopsis lists;
Retrun Syn.
```

## 4     Experiment and Evaluation

In this section, we describe a number of experiments run to evaluate the performance of *Haar* [18], *FH* [17] and our proposed *PWF*. We used the real data here where the daily average temperature data of 17 cities randomly selected around Asia, which were obtained from the temperature data archive of the University of Dayton (*www.engr.udayton.edu/weather/*). In the test, we consider the data from each city as a stream, each of which has 5581 data points and a total of 106039 data for this evaluation.

### 4.1     Landmark and $E_b$ Effect for *Haar*, *FH* and *PWF* Methods

In the previous discussion, landmarks decide the size of sliding windows as shown in this experiment 100, 200 and 500. And, $E_b$ is the extended boundary to the destination query points which could set to 1°C, 2°C, 5°C, 10°C, 20 °C, 30°C and 40°C. The first experiment examined the impacts of itemsets recognition of data categories exactly by these two parameters, and recognition rate contrast of *Haar*, *FH* and *PWF* with some real temperature datasets. The temperature information of 17 cities flows into memory as 17 different streams, and one of them is selected as a query stream. No special obvious difference on the category recognition rate of these three methods without any approximation processing, as shown in the Fig.3 and Fig.4. They ensure the identification accuracy for itemsets recognition after the transformation. However, the itemsets recognition rate is growing as the parameter $E_b$ increasing in Fig.3, until $E_b$

=40 almost all of the itemsets could be identified. But, it wasn't particularly sensitive to the size of the window, here says "landmarks" as shown in Fig.4. This experiment further proved that the itemsets recognition rate is more influenced by parameter $E_b$.

Although these three methods did not show the special difference on the itemsets recognition and they all ensure the identification accuracy after transformation, *FH* and *PWF* to deal with the problem of space waste have outstanding performance as shown in Fig.5. *FH* and *PWF* methods only take up less than 1% space usage, however, *Haar* is far more than them and take up over 20% space usage at most of the time. They are affected by the parameter $E_b$ of sensitive increasing. However, occupying space usage is reduced as the growth of the window sizes on the contrary, as shown in the Fig.6. Actually, it would directly affect the precision of the final results that it's not better for the greater window size. The following discussion contrast among them will prove this judgment, and show that *Syn-stream* (Synopsis stream lists which are selected from local Top-k processing) recognition rate is a more important judgment condition.

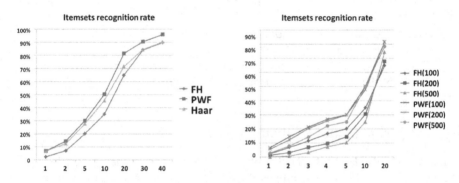

**Fig. 3.** Itemsets recognition rates contrast of different $E_b$

**Fig. 4.** Itemsets recognition rates contrast of different Landmarks

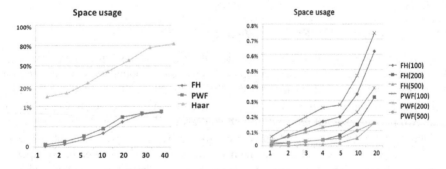

**Fig. 5.** Space usage comparison of Haar, FH and PWF

**Fig. 6.** Space usage comparison of different Landmarks

## 4.2    Smoothing Operations

In the previous discussion, there are not significant differences between *FH* and *PWF* methods. However, a strange phenomenon that there is a steady growth with the *Syn-streams* recognition rate of almost 60% as shown in Fig.7 and Fig.8. It's not hard to find that 60% is a balance vector for these original datasets, whatever uses any kinds of approximate methods. The important thing is that who can reach or find this balance vector earlier, and it is a key to determine the final result accuracy through the actual test. In fact, our proposed *PWF* is not only could quickly calculate the balance vector, but also smoothing and maintain this status with different sizes of window sizes (*landmark*) and $E_b$. For example, the value of parameter is testing from lower to higher as shown in Fig.7 (from 1 to 20). *FH (100)* would reach the balance vector 60% until $E_b$ is equals to 2, and the status to keep 3 intervals. However, PWF has been reached the key value until $E_b$ is equals to 1, and the status to keep 4 intervals. Moreover, *FH (200)* reach the balance until $E_b$ is equals to 4, and the status to keep 3 intervals. However, *PWF* has been reached the balance until $E_b$ is equals to 2, and the status to keep 5 intervals, as shown in Fig.7 and Fig.8. In this way, our new proposed *PWF* is much better than *FH* method, the details will show in Table 1, Table 2 and Table 3.

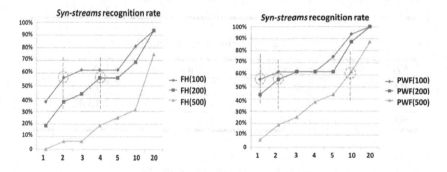

**Fig. 7.** Syn-streams recognition rate for *FH*    **Fig. 8.** Syn-streams recognition rate for *PWF*

A smoothing operation is the area within the scope of the safe $E_b$ value, and the smooth interval to keep wider. However, if the identified value is suddenly increased a lot, that's because many noisy steams were extended, such as the $E_b$ equals to 5. In the approximate analyses, the original data have been screening via mining preprocessing and only stored valuable local results in the synopsis, then remove those of invaluable data to the secondary storage. Therefore, the most difficult thing is that how to use these reduced data to calculate the final correct approximate results. The following analysis on the precision of the approximate results will prove this opinion.

## 4.3     Approximate Results

In fact, the correct result for the 4-NN query is {5, 2, 11 and 13}, and most of the final approximate results are keeping right except traditional method *Haar*, as shown in Table 1, 2, and 3. Stream 12 is a noisy stream dataset is supplied by *Haar*, but Stream 2 because of high error rate without attention, as shown in Table 4. Here says error rate of original datasets maybe generated via testing error of the machines, missing data, or some other mistakes lead to these data have been lost. Most of these streams datasets are valuable and the error rate is below 0.5%, except Stream 2 and 9 that give a warning that almost 50% dataset are not available. However, even Stream 2 has only 61% data are valuable, it also could get a fair calculated because of synopsis local processing. Our proposed methods could solve this problem using real time synopsis processing such as *FH* and *PWF*, and the traditional solution *Haar* cannot do this. After testing, our proposed *PWF* is not only could quickly calculate the balance vector, but also smoothing and keep this status for longer with different sizes of window sizes (*landmark*) and $E_b$. Therefore, *PWF* is the best way to deal with the stream approximate processing in real time.

**Table 1.** Approximate results for *Haar*

| *Haar* | $E_b$ (expect) | Status | 4-NN |
|--------|----------------|--------|------|
| - | 1 | - | 5,11,13,**12** |

**Table 2.** Approximate results for *FH*

| *FH* | $E_b$ (expect) | Status | 4-NN |
|------|----------------|--------|------|
| 100 | 2~5 | 3 | 5,**2**,11,13 |
| 200 | 4~7 | 3 | 5,**2**,13,11 |
| 500 | 17~18 | 1 | 5 |

**Table 3.** Approximate results for *PWF*

| *PWF* | $E_b$ (expect) | Status | 4-NN |
|-------|----------------|--------|------|
| 100 | 1~4 | 4 | 5,**2**,11,13 |
| 200 | 2~7 | 5 | 5,13,**2**,11 |
| 500 | 10~13 | 3 | 13,5,**2** |

**Table 4.** Error rates for each original stream

| *Streams* | 1 | **2** | 3 | 4 | 5 | 6 | 7 | 8 |
|-----------|------|----------|-------|------|-------|-------|-------|-------|
| error | 9.28% | **39.17%** | 0.22% | 0.7% | 0.25% | 0.29% | 0.34% | 0.25% |
| 9 | 10 | 11 | **12** | 13 | 14 | 15 | 16 | Query |
| 42.5% | 0.29% | 0.48% | **0.32%** | 0.45% | 0.2% | 0.22% | 0.22% | 0.23% |

# 5  Conclusion

In this paper, we proposed PWF algorithm to smooth the approximate approach in a real time way. Here, "smoothing" is means that we found a *Proper-Wavelet Function* would help to afford a wider range of $k$ values, furthermore to reduce the $k$-effect in $k$-nearest neighbors' queries. After real testing with several kinds of different statistical functions and wavelet transform methods, landmarks and $E_b$ are two key values to have a big impact on space usage and calculation precision for approximation approach. Finally, our experimental results demonstrate the accuracy and efficiency of our approximate technique *PWF* algorithm to allow continuous $k$-neighbor neighbors queries on the large, fast and various incoming stream database.

**Acknowledgments.** This work was supported by the National Natural Science Foundation of China (No.51077010) and by the Jilin provincial department of science and technology (No.20120338).

# References

1. Arasu, A., Babcock, B., Babu, S., Cieslewica, J., Datar, M., Ito, K., Motwani, R., Srivastava, U., Widom, J.: STREAM: the Stanford Data Stream Management System. Technical Report, Stanford University (2004)
2. Botan, I., Derakhshan, R., Dindar, N., Haas, L., Miller, R.J., Tatbul, N.: SECRET: a model for analysis of the execution semantics of stream processing systems. In: Very Large Data Base, pp. 232–243. VLDB Press (2010)
3. Jin, C., Yi, K., Chen, L., Yu, J.X., Lin, X.: Sliding-window Top-k Queries on Uncertain Streams. In: Very Large Data Base, pp. 301–312. VLDB Press, New Zealand (2008)
4. Teng, W.G., Chen, M.S., Yu, P.S.: Resource-aware Mining with Variable Granularities in Data Steams. In: 2004 Fourth SIAM International Conference on Data Mining, pp. 527–531. SIAM Press, Lake Buena Vista (2004)
5. Tong, Y., Chen, L., Cheng, Y., Yu, P.S.: Mining Frequent Itemsets over Uncertain Databases, pp. 1650–1661. VLDB Press, Turkey (2012)
6. Guha, S., Harb, B.: Wavelet synopsis for data streams: minimizing non-euclidean error. In: 2005 Eleventh ACM SIGKDD International Conference on Knowledge Discovery in Data Mining, pp. 88–97. ACM Press, Chicago (2005)
7. Hung, H.P., Chen, M.S.: Efficient range-constrained similarity search on wavelet synopses over multiple streams. In: 2006 Fifteenth ACM International Conference on Information and Knowledge Management, pp. 327–336. ACM Press, Arlington (2006)
8. Sacharidis, D.: Constructing Optimal Wavelet Synopses. In: Grust, T., Höpfner, H., Illarramendi, A., Jablonski, S., Fischer, F., Müller, S., Patranjan, P.-L., Sattler, K.-U., Spiliopoulou, M., Wijsen, J. (eds.) EDBT 2006. LNCS, vol. 4254, pp. 97–104. Springer, Heidelberg (2006)
9. Zhu, Y., Shasha, D.: Efficient Elastic Burst Detection in Data Streams. In: 2003 Ninth ACM SIGKDD International Conference on Knowledge Discovery and Data Mining, pp. 336–345. ACM Press, Washington (2003)
10. Teng, W.G., Chen, M.S., Yu, P.S.: Resource-aware mining with variable granularities in data streams. In: 2004 Fourth SIAM International Conference on Data Mining, pp. 527–531. SIAM Press, Lake Buena Vista (2004)

514    L. Wang et al.

11. Tao, Y., Yi, K., Sheng, C., Kalnis, P.: Quality and Efficiency in High Dimensional Nearest Neighbor Search. In: 2009 ACM SIGMOD International Conference on Management of Data, pp. 563–576. ACM Press, Providence (2009)
12. Sharifzadeh, M., Shahabi, C.: VoR-Tree: R-trees with Voronoi Diagrams for Efficient Processing of Spatial Nearest Neighbor Queries, pp.1231–1242.VLDB Press (2010)
13. Wang, L., Zhou, T.H., Kim, K.A., Cha, E.J., Ryu, K.H.: Adaptive Approximation-based Streaming Skylines for Similarity Search Query. J. Software Engineering and Its Applications, 113–118 (2012)
14. Yao, B., Li, F., Kumar, P.: Reverse Furthest Neighbors in Spatial Database, pp. 664–675. IEEE Press, Shanghai (2009)
15. Aly, A.M., Aref, W.G., Ouzzani, M.: Spatial Queries with Two kNN Predicates, vol. 5(11), pp. 1100–1111. VLDB Press (2012)
16. Zhang, Y., Lin, X., Zhu, G., Zhang, W., Lin, Q.: Efficient Rank based kNN Query Processing over Uncertain Data, pp. 28–39. Long Beach (2010)
17. Wang, L., Zhou, T.H., Shon, H.S., Lee, Y.K., Ryu, K.H.: Extract and Maintain the Most Helpful Wavelet Coefficients for Continuous k-Nearest Neighbor Queries in Stream Processing, pp. 358–363. Springer-Verlag Press, Changsha (2010)
18. Garofalakis, M., Gibbons, P.B.: Wavelet synopses with error guarantees. In: 2002 ACM SIGMOD International Conference on Management of Data, pp. 476–487. ACM Press, Madison (2002)

# Subsume Concept in Erasable Itemset Mining

Giang Nguyen[1], Tuong Le[2,3], Bay Vo[2,3], Bac Le[4], and Phi-Cuong Trinh[5]

[1] University of Technology, Ho Chi Minh City, Vietnam
nh.giang@hutech.edu.vn
[2] Division of Data Science, Ton Duc Thang University, Ho Chi Minh City, Vietnam
[3] Faculty of Information Technology, Ton Duc Thang University, Ho Chi Minh City, Vietnam
{lecungtuong,vodinhbay}@tdt.edu.vn
[4] Faculty of Information Technology, University of Science, VNU Ho Chi Minh City, Vietnam
lhbac@fit.hcmus.edu.vn
[5] Ton Duc Thang University, Ho Chi Minh City, Vietnam
trinhphicuong@tdt.edu.vn

**Abstract.** In recent year, erasable itemset mining is an interesting problem in supply chain optimization problem. In the previous works, we presented dPidset structure, a very effective structure for mining erasable itemsets. The dPidset structure improves the preferment compared with the previous structures. However, the mining time is still large. Therefore, in this paper, we propose a new approach using the subsume concept for mining effectively erasable itemsets. The subsume concept helps early determine information of a large number of erasable itemsets without usual computational cost. The experiment was conducted to show the effectiveness of using subsume concept in the mining erasable itemsets process.

**Keywords:** data mining, erasable itemset, subsume concept.

## 1 Introduction

Pattern mining [1-2, 7, 12-16] is an essential task in data mining. In recent years, an interesting variation of pattern mining, the problem of mining erasable itemsets [3-6, 8-10] was first presented in 2009. A factory produces many products created from a number of items. Each product brings an income to the factory. A financial resource is required to buy and store all items. However, in a financial crisis situation this factory has not enough money to purchase all necessary components as usual. This problem is to find the itemsets which can best be erased so as to minimize the loss to the factory's gain. Managers can then utilize the knowledge of these erasable itemsets to make a new production plan. There are many methods to mine erasable itemsets in recent years such as: META [5], VME [6], MERIT [4], dMERIT+ [8] and MEI [9]. In which, MERIT and dMERIT+ use NC_Set and dNC_Set generated from PPC-tree. Besides, MEI uses dPidset structure. According to the experimental results in [9], MEI is new and the most notable algorithm. Although there are many algorithms for mining erasable itemsets, the mining time and memory usage are still quite large.

D. Hwang et al. (Eds.): ICCCI 2014, LNAI 8733, pp. 515–523, 2014.
© Springer International Publishing Switzerland 2014

Therefore, in this paper, we propose an improvement algorithm using subsume concept for mining effectively erasable itemsets. The experimental results show that sMEI algorithm outperforms MEI algorithm in terms of the mining time.

The rest of the paper is organized as follows. Section 2 presents basic concepts, and then the new method for mining erasable itemsets using subsume concept is proposed in section 3. Experimental result is presented in section 4. The paper concluded in section 5.

# 2    Basic Concepts

## 2.1    Erasable Itemset Mining

Let $I = \{i_1, i_2, ..., i_m\}$ be a set of all items which is the components of products. A product dataset $DB = \{P_1, P_2, ..., P_n\}$, where $P_i$ is a product. Each product presented in the form of $\langle Items, Val \rangle$, where *Items* are the items and *Val* is the profit that the factory obtains by selling this product. A set $X \subseteq I$ is also called an itemset. An example dataset in Table 1 will be used throughout this article.

**Table 1.** An example dataset $(DB_E)$

| Product | Items | Val ($) |
|---------|-------|---------|
| $P_1$ | a, b | 1,000 |
| $P_2$ | a, b, c | 200 |
| $P_3$ | c, e | 150 |
| $P_4$ | b, d, e, f | 50 |
| $P_5$ | d, e | 100 |
| $P_6$ | d, e, f, h | 200 |

**Definition 1 (Gain of an itemset).** Let $X (\subseteq I)$ be an itemset. The gain of $X$ is determined as:

$$g(X) = \sum_{\{P_k| X \cap P_k.Items \neq \varnothing\}} P_k.Val \qquad (1)$$

**Definition 2.** Given a threshold $\xi$ and a product dataset $DB$. Let $T = \sum_{P_k \in DB} P_k.Val$ be the total profit of the factory. An itemset $X$ is erasable if:

$$g(X) \leq T \times \xi \qquad (2)$$

The problem of mining erasable itemsets is to find all erasable itemsets which have gain $g(X)$ less than $T \times \xi$ in the dataset.

## 2.2    Pidset Structure

**Definition 3 (Index of gain).** Let $P_i$ be a product in *DB*. An array is defined as the index of gain:

$$G[i] = P_i.Val \tag{3}$$

**Definition 4 (pidset).** The pidset of an itemsets $X$ is denoted as follows:

$$p(X) = \bigcup_{A \in X} p(A) \tag{4}$$

where $A$ is an item in itemset $X$ and $p(A)$ is the pidset of item $A$, i.e., the set of product identifiers which includes $A$.

**Definition 5 (Gain of an itemset based on pidset).** The gain of an itemset $X$ denoted by $g(X)$ is computed as follows:

$$g(X) = \sum_{P_k \in p(X)} G[k] \tag{5}$$

where $G[k]$ is the element at position $k$ of $G$.

**Theorem 1.** Let $XA$ and $XB$ be two itemsets with the same prefix $X$ and $p(XA)$ and $p(XB)$ are pidsets of $XA$ and $XB$, respectively. The pidset of $XAB$ is computed as follows:

$$p(XAB) = p(XB) \cup p(XA) \tag{6}$$

*Example 1.* For $DB_E$, $p(ab) = \{1, 2, 4\}$ and $p(ac) = \{1, 2, 3\}$. According to Theorem 1, $p(abc) = p(acb) = p(ab) \cup p(ac) = \{1, 2, 4\} \cup \{1, 2, 3\} = \{1, 2, 3, 4\}$.

## 2.3    dPidset Structure

**Definition 6 (dPidset).** The dPidset of pidsets $p(XA)$ and $p(XB)$, denoted as $dP(XAB)$, is defined as follows:

$$dP(XAB) = p(XB) \setminus p(XA) \tag{7}$$

According to Definition 6, the dPidset of $p(XA)$ and $p(XB)$ is the product identifiers which only exist on $p(XB)$.

*Example 2.* We have $p(ab) = \{1, 2, 4\}$ and $p(ac) = \{1, 2, 3\}$. Based on Definition 6, $dP(abc) = p(ac) \setminus p(ab) = \{1, 2, 3\} \setminus \{1, 2, 4\} = \{3\}$. Note that reversing the order of $ab$ and $ac$ will get a different result. Consequently, $dP(acb) = p(ab) \setminus p(ac) = \{4\}$.

**Theorem 2.** Let $XA$ and $XB$ be two itemsets and $dP(XA)$ and $dP(XB)$ are the dPidsets of $XA$ and $XB$, respectively. The dPidset of $XAB$ is computed as follows:

$$dP(XAB) = dP(XB) \setminus dP(XA) \tag{8}$$

*Example 3.* Based on $DB_E$, $p(a) = \{1, 2\}$, $p(b) = \{1, 2, 4\}$ and $p(c) = \{2, 3\}$. According to Definition 6, $dP(ac) = p(c) \setminus p(a) = \{3\}$ and $dP(ab) = p(b) \setminus p(a) = \{4\}$. Based on Theorem 4, $dP(abc) = dP(ac) \setminus dP(ab) = \{3\} \setminus \{4\} = \{3\}$. In *Exam.* 2 and 3, $dP(abc) = \{3\}$. So, these examples verify Theorem 2.

**Theorem 3.** The gain of $XAB$ is determined based on that of $XA$ as follows:

$$g(XAB) = g(XA) + \sum_{P_k \in dP(XAB)} G[k] \tag{9}$$

where $g(XA)$ is the gain of $X$ and $G[k]$ is the element at position $k$ of $G$.

*Example 4.* We have $p(a) = \{1, 2\}$, $p(b) = \{1, 2, 4\}$ and $p(c) = \{2, 3\}$, so $g(a) = 1200$, $g(b) = 1250$ and $g(c) = 350$. According to *Example 3*, $dP(ac) = \{3\}$ and $dP(ab) = \{4\}$ so $g(ac) = g(a) + \sum_{P_k \in dP(ac)} G[k] = 1200 + 150 = 1350$ and $g(ab) = g(a) + \sum_{P_k \in dP(ab)} G[k] = 1200 + 50 = 1250$. Besides, $dP(abc) = \{3\}$ so $g(abc) = g(ab) + \sum_{P_k \in dP(abc)} G[k] = 1250 + 150 = 1400$.

# 3     Mining Erasable Itemsets Using Subsume Concept

## 3.1     Subsume Index Concept

**Definition 7 [11].** The subsume of a erasable 1-itemset, $A$, denoted by $S(A)$ is defined as follows:

$$S(A) = \{B \in I_1 \mid p(A) \subseteq p(B)\} \tag{10}$$

*Example 5.* We have $p(a) = \{1, 2\}$ and $p(b) = \{1, 2, 4\}$. Because $p(a) \subseteq p(b)$, thus $a \in S(b)$.

**Theorem 4.** Let the subsume of an item $A$ be $\{a_1, a_2, \ldots, a_m\}$. The gain of each of the $2^m$-1 nonempty subsets of $\{a_1, a_2, \ldots, a_m\}$ is equal to the gain of $A$.

*Example 6.* According to $DB$, we have $S(e) = \{d, f, h\}$. Therefore $2^m - 1$ nonempty subsets of $S(e)$ is $\{d, f, h, df, dh, fh\}$. Based on Theorem 4, the gain of $2^m - 1$ itemset which are combined $2^m - 1$ nonempty subsets of $S(e)$ with $e$ is equal to $g(e)$. In this case, we have the gain of $\{ed, ef, eh, edf, edh, efh\}$ is 500 dollar.

**Theorem 5.** Let $A, B, C \in I_1$ be three items. If $A \in S(B)$ and $B \in S(C)$ then $A \in S(C)$.

*Proof.* We have $A \in S(B)$ and $B \in S(C)$ therefore $p(B) \subseteq p(A)$ and $p(C) \subseteq p(B)$. So $p(C) \subseteq p(A)$ and thus this theorems is proven.

## 3.2    sMEI Algorithm

```
Input: product dataset DB and threshold ξ
Output: E_result, the set of all EIs
1.Determine T, G, and erasable 1-itemsets with their pid-
sets (E₁) in only one time scanning DB
2.Sort E₁ in descending order of their pidsets size
3.Find_Subsume(E₁)
4.Put E₁ to E_result
5.Expand_E(E₁)

procedure Find_Subsume(E₁)
1.for i ← |E₁| - 1  to 1 do
2.  for j ← i+1 to |E₁| - 1 do
3.    if j ∈ E₁[i].Subsumes then continue
4.    if E₁[i].pidset ⊇ E₁[j].pidset then
5.        add E₁[j]'s index, j, to E₁[i].Subsumes
6.        add E₁[j].Subsumes to E₁[i].Subsumes //using Theo-
rem 5

Procedure Expand_E(E_v)
1.for k ← 0 to |E_v| - 2
2.  E_next ← ∅
3.  if |E_v[k].Subsumes| > 0 then
4.    for each s generated from all elements of
E_v[k].Subsumes // 2^{m-1} nonempty subsets generated from its
subsume
5.        add ⟨s ∪ E_v[k], g(E_v[i])⟩ to E_result
6.  for j ←(k+1) to |E_v| - 1 do
7.    if(j∈E_v[k].Subsumes)
8.      continue
9.    E.Items = E_v[k].Items ∪ E_v[j].Items
10.   E.pidset ← E_v[k].pidset \ E_v[j].pidset
11.   E.gain = E_v[k].gain + Σ_{k∈E.pidset} G[k]
12.   if E.gain ≤ T × ξ then
13.       E_next ← E and E_result ← E
14. Expand_E(E_next)
```

Fig. 1. The sMEI algorithm

## 3.3    An Illustrated Example

Firstly, sMEI scans $DB_E$ to find erasable 1-itemsets with their pidset with $\xi = 30\%$ (Fig. 2).

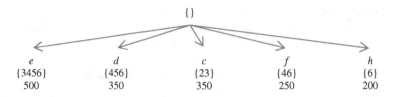

**Fig. 2.** Erasable 1-itemsets with their pidset for $DB_E$ with $\xi = 30\%$

Secondly, the algorithm finds the subsume index associated with erasable 1-itemsets (Table 2).

**Table 2.** Subsume index associated with erasable 1-itemsets

| Erasable 1-itemset | Subsume |
|:---:|:---:|
| e | d, f, h |
| d | f, h |
| c | |
| f | f |
| h | |

Thirdly, the algorithm calls the Expand_E for mining all erasable itemsets. If the erasable 1-itemsets, A, has the subsume then all itemsets which combined A and $2^{m-1}$ nonempty subsets of its subsume are erasable without calculating their pidsets and their gains. Fig. 3 shows all erasable itemsets for $DB_E$ with $\xi = 30\%$. In Fig. 3, the algorithm does not compute and store the pidsets of the nodes {ed, ef, eh, edf, edh, efh, edfh, ef, dh, dfh, fh}. Therefore, using the subsume concept reduces the runtime of erasable itemset mining.

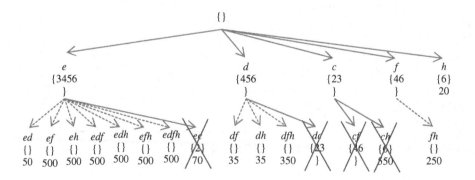

**Fig. 3.** Erasable itemsets for $DB_E$ with $\xi = 30\%$

# 4    Experimental Results

All experiments presented in this section were performed on a laptop with an Intel Core i3-3110M 2.4-GHz CPU and 4 GB of RAM. All the programs were coded in C# and .Net Framework Version 4.5.50709.

The experiments are conducted on datasets such as Chess and Mushroom[1]. To make these datasets look like product datasets, a column was added to store the profit of products. To generate values for this column, a function denoted by $N(100, 50)$, for which the mean value is 100 and the variance is 50, was created. The features of these datasets are shown in Table 3.

**Table 3.** Features of datasets used in experiments

| Dataset[2] | # of Products | # of Items |
|---|---|---|
| Chess | 3,196 | 76 |
| Mushroom | 8,124 | 120 |

Table 4 shows the number of subsume associated with erasable 1-itemsets and the number of erasable 1-itemsets. Note that an erasable 1-itemsets can have one or more subsumes. So the number of subsume associated with erasable 1-itemsets can be greater than the number of erasable 1-itemsets. The sMEI algorithm using subsume concept is more effective than MEI algorithm for the datasets having a greater number of subsume associated with erasable 1-itemsets.

**Table 4.** The numbers of subsume associted with erasable 1-itemsets

| Dataset | Threshold | # of erasable 1-itemset | # of subsume associated with erasable 1-itemset |
|---|---|---|---|
| Chess | 30 | 26 | 10 |
| | 40 | 35 | 18 |
| | 50 | 38 | 23 |
| Mushroom | 1 | 23 | 18 |
| | 2 | 30 | 18 |
| | 3 | 37 | 42 |

Figs. 4-5 report the number of nodes with subsume and the total nodes in various thresholds for Chess and Mushroom datasets. A larger number of nodes do not need to determine their pidsets and their gain. Therefore, the mining time is reduced.

---

[1] Downloaded from http://fimi.cs.helsinki.fi/data/
[2] These datasets are available at http://sdrv.ms/14eshVm

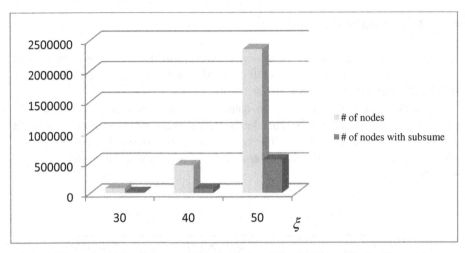

**Fig. 4.** The nodes with subsume and the total nodes of the sMEI's results for Chess dataset

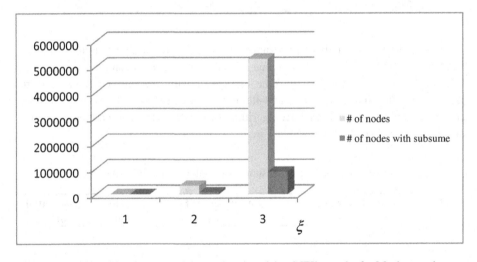

**Fig. 5.** The nodes with subsume and the total nodes of the sMEI's results for Mushroom dataset

Table 5 report the runtime of MEI and sMEI algorithms. These Figs show that sMEI is little faster than MEI.

**Table 5.** The runtime of MEI and sMEI algorithms for Chess and Mushroom datasets

| Dataset | Threshold | MEI (seconds) | sMEI (seconds) |
|---|---|---|---|
| | 30 | 0.194 | 0.197 |
| Chess | 40 | 0.994 | 0.950 |
| | 50 | 4.670 | 4.530 |
| | 1 | 0.080 | 0.069 |
| Mushroom | 2 | 0.690 | 0.690 |
| | 3 | 10.600 | 10.200 |

# 5     Conclusions and Future Work

This paper proposed sMEI, an efficient algorithm for mining erasable itemsets. First the subsume concept in erasable itemset mining was introduced. The subsume concept helps early determine information of a large number of erasable itemsets without usual computational cost. The experimental results show that sMEI algorithm outperforms MEI algorithm for mining erasable itemsets. In future work, we will focus in mining erasable closed/maximal itemsets.

# References

1. Agrawal, R., Srikant, R.: Fast algorithms for mining association rules. In: VLDB 1994, pp. 487–499 (1994)
2. Agrawal, R., Imielinski, T., Swami, A.: Mining association rules between set of items in large databases. In: SIGMOD 1993, pp. 207–216 (1993)
3. Deng, Z.H.: Mining top-rank-$k$ erasable itemsets by PID_lists. International Journal of Intelligent Systems 28(4), 366–379 (2013)
4. Deng, Z.H., Xu, X.R.: Fast mining erasable itemsets using NC_sets. Expert Systems with Applications 39(4), 4453–4463 (2012)
5. Deng, Z., Fang, G., Wang, Z., Xu, X.: Mining erasable itemsets. In: ICMLC 2009, pp. 67–73 (2009)
6. Deng, Z., Xu, X.: An efficient algorithm for mining erasable itemsets. In: Cao, L., Feng, Y., Zhong, J. (eds.) ADMA 2010, Part I. LNCS, vol. 6440, pp. 214–225. Springer, Heidelberg (2010)
7. Han, J., Pei, J., Yin, Y.: Mining frequent patterns without candidate generation. In: SIGMOD 2000, pp. 1–12 (2003)
8. Le, T., Vo, B., Coenen, F.: An efficient algorithm for mining erasable itemsets using the difference of NC-Sets. In: IEEE SMC 2013, Manchester, UK, pp. 2270–2274 (2013)
9. Le, T., Vo, B.: MEI: an efficient algorithm for mining erasable itemsets. Engineering Applications of Artificial Intelligence 27, 155–166 (2014)
10. Nguyen, G., Le, T., Vo, B., Le, B.: A new approach for mining top-rank-$k$ erasable itemsets. In: Nguyen, N.T., Attachoo, B., Trawiński, B., Somboonviwat, K. (eds.) ACIIDS 2014, Part I. LNCS, vol. 8397, pp. 73–82. Springer, Heidelberg (2014)
11. Song, W., Yang, B., Xu, Z.: Index-BitTableFI: An improved algorithm for mining frequent itemsets. Knowledge-Based Systems 21, 507–513 (2008)
12. Van, T.-T., Vo, B., Le, B.: IMSR_PreTree: an improved algorithm for mining sequential rules based on the prefix-tree. Vietnam J. Computer Science 1(2), 97–105 (2014)
13. Vo, B., Coenen, F., Le, T., Hong, T.-P.: A hybrid approach for mining frequent itemsets. In: IEEE SMC 2013, Manchester, UK, pp. 4647–4651 (2013)
14. Vo, B., Le, T., Coenen, F., Hong, T.-P.: Mining frequent itemsets using the N-list and subsume concepts. International Journal of Machine Learning and Cybernetics, http://dx.doi.org/10.1007/s13042-014-0252-2
15. Zaki, M.J.: Scalable algorithms for association mining. IEEE Transactions on Knowledge and Data Engineering 12(3), 372–390 (2000)
16. Zaki, M.J., Gouda, K.: Fast vertical mining using diffsets. In: SIGKDD 2003, pp. 326–335 (2003)

# Analyzing the Behavior and Text Posted by Users to Extract Knowledge

Soumaya Cherichi and Rim Faiz

LARODEC, IHEC Carthage
University of Carthage
Carthage Presidency, Tunisia
soumayacherichi@gmail.com,
Rim.Faiz@ihec.rnu.tn

**Abstract.** With the explosion of Web 2.0 platforms such as blogs, discussion forums, andsocial networks, Internet users can express their feelings and share information among themselves. This behavior leads to an accumulation of an enormousamount of information.Among these platforms are so-called microblogs. Microblogging(e.g. Twitter1), as a new form of online communication in whichusers talk about their daily lives, publish opinions or share information by short posts, hasbecome one of the most popular social networking services today, which makes it potentially alarge information base attracting increasing attention of researchers in the field of knowledgediscovery and data mining.Several works have proposed tools for tweets search, but, this area is still not well exploited. Our work consists of examining the role and impact of social networks, in particular microblogs, on public opinion. We aim to analyze the behavior and text posted by users to extract knowledge that reflect the interests and opinions of a population.This gave us the idea to offer new tool more developed that uses new features such as audience and RetweetRank for ranking relevant tweets. We investigate the impact of these criteria on the search's results for relevant information. Finally, we propose a new metric to improve the results of the searches in microblogs. More accurately, we propose a research model that combines content relevance, tweet relevance and author relevance. Each type of relevance is characterized by a set of criteria such as audience to assess the relevance of the author, OOV (Out Of Vobulary) to measure the relevance of content and others. To evaluate our model, we built a knowledge management system. We used a collection of subjective tweets talking about Tunisian actualities in 2012.

**Keywords:** microblogs, relevantinformation, analyzingtextposted, knowledge management system.

## 1    Introduction

In the current era, People are becoming more communicative through expansion of services and multi-platform applications, i.e., the so called Web 2.0 which establishes social and collaborative backgrounds. They commonly use various means including Blogs to share the diaries, RSS feeds to follow the latest information of their interest and Computer Mediated Chat (CMC) applications to hold bidirectional communications. Microblogging is one of the most recent products of CMC, in which users talk

D. Hwang et al. (Eds.): ICCCI 2014, LNAI 8733, pp. 524–533, 2014.
© Springer International Publishing Switzerland 2014

about their daily lives, publish opinions or share information by short posts. It was first known as Tumblelogs on April 12, 2005, and then came into greater use by the year 2006 and 2007, when such services as Tumblr and Twitter arose. The problem that we face is how to find this information and transform data collections into new knowledge, understandable, useful and interesting in the context where it is located. Information retrieval systems solve one of the biggest problems of knowledge management (KM): quickly finding useful information within massive data stores and ranking the results by relevance.

Recent years have revealed the accession of interactive media, which gave birth to a huge volume of data in blogs and micro-blogs more precisely. These micro-blogs attract more and more users due to the ease and the speed of information sharing es-pecially in real time.

Twitter has played a role in important events, but the service also allows people to communicate among a relatively small social circle, and a sizeable part of Twitter's success is because of this function.

Indeed a micro-blog is a stream of text that is written by an author. It is composed by regular and short updates that are presented to readers in reverse chronological order called time-line.

While micro-blogging services are becoming more famous, the methods for organiz-ing and providing access to data are also improving. Micro-bloggers as well as send-ing tweets are looking for the last updates according to their interests. Finding the most relevant tweets to a topic depends on the criteria of micro-blogs.

Unlike other micro-blogging service, Twitter is positioned by the social relationship of subscription. And since the association is led, it allows users to express their inter-est in the items of another micro-bloggers. The social network of Twitter is not lim-ited to bloggers and subscription relationships; it also includes all the contributors and data that interact in both contexts of use and publication of articles. We have analyzed the micro-blogging service Twitter and we have identified the main criteria of Twit-ter.

But the question arises what is the impact of each feature on the quality of results?

Our work consists in searching a new metric of features' impact on the search re-sults' quality. Several criteria have been proposed in the literature [1] and [2], but there are still other criteria that have not been exploited as audience which could be the size of the potential audience for a message: What is the maximum number of people who could have been exposed to a message?

We gathered the features on three groups: those related to content, those related to tweet and those related to the author. We used the coefficient of correlation with human judgment to define our score. For processing the content of tweets, we intend to use resources and linguistic methods Our experimental result uses a corpus of thousand subjective tweets which are neither answers nor retweets, and we also collected a corpus of human judgments to find the correlation coefficient.

The remainder of this paper is organized as follows. In section 2, we give an overview of related works. In section 3 we present an Analysis of short texts using NLP. In Section 4, we discuss experiments and obtained results. Finally, section 5 concludes this paper and outlines future work.

## 2     Related Works

A micro-blogging service is at once a communication mean and a collaboration sys-tem that allows sharing and disseminating text messages. In comparison with other social networks on the Web (for example Facebook, Myspace), the microblogs arti-cles are particularly short and submitted in real time to report a recent event. At the time of this writing, several micro-blogging services exist. In this paper, we will focus on the micro-blogging service Twitter which is the most popular and widely used. Twitter is characterized from similar sites by certain features and functionalities. An important characteristic is the presence of social relationships subscription. This di-rectional rela-tionship allows users to express their interest on the publications of a particular blogger. Twitter is distinguished from similar websites by some key fea-tures. The main one consists on the following social relationship. This directed asso-ciation enables users to express their interest in other micro-bloggers' posts, called tweets, which doesn't exceed 140 characters. Moreover, Twitter is marked by the retweet feature which gives users the ability to forward an interesting tweet to their followers.

Several works have focused on the analysis of data posted on microblogs, particu-larly in Twitter. [3] and [4] propose approaches for sentiment classification of Twitter mes-sages i.e. determine whether tweets express a positive, negative or neutral feel-ing. Positive and negative polarities correspond respectively to a favorable and unfa-vora-ble opinion. To solve this task the authors have used natural language processing and machine learning techniques.

Many studies have found that there is a high correlation between the information posted on the web and actual results. [5] have used tweets to analyze awareness and anxiety levels of Tokyo habitants the events of earthquakes tsunami and states of nuclear emergencies in Japan in 2011. [6] have presented a method to measure the prevalence of H1N1 disease in the population of United Kingdom. They sought in the tweets the symptoms related to the disease. The obtained results were compared with real results from the Health Protection Agency. [7] also analyzed the tweets to predict public opinion and then compared the results with surveys.

We find that most approaches for information retrieval in micro-blogs don't take into account all the features to narrow the search. In fact, each feature has a unique impact on the other ones. Based on this observation and to improve the results of research, we will try to overcome these limitations by measuring the impact of these criteria. We will propose a measurement metric impact criteria for improving out-comes re-search. The search for tweets is an information retrieval task ad-hoc whose objective is to select the items relevant micro-blogs in response to a query. The defini-tion of relevance in the search for tweets is not limited to textual similarity but also takes account of social interactions in the network. In this context, the relevance of the items depends also on the tweets' technical specificities and the importance of the author.

Regarding the relevance of content, several studies have used Okapi BM25 algorithm [8], other studies like work of [9] have added new features such as tweets' quality ie the tweet that contains the least amount of Out of vocabulary (OOV) is considered as the most informative one. Also Duan et al, consider that the longer the tweet, the bet-ter amount of information it contains.

Our work consists of examining the role and impact of social networks, in particular microblogs, on public opinion. We aim to analyze the behavior and text posted by users to extract knowledge that reflect the interests and opinions of a population. We introduce in this paper our approach for tweet search that integrates different criteria namely the social authority of micro-bloggers, the content relevance, the tweeting features as well as the hashtag's presence. We present in the next section the main features of our criteria.

# 3    Analysis of Short Texts Using NLP

Data analysis of social networks has become a major trend in the field of natural language processing. Thus, large communities NLP gave its fair share to the analysis of data microblogs. In recent years, major conferences have created workshops for data analysis in social networks. Several studies concerned with the analysis of short texts do not aim only to determine the polarity of the messages, but to use the messages to detect events or predict results.

Among the most important tasks for a ranking system tweet is the selection of features set. We offer three types of features to rank tweets:

— Content features refer to those features which describe the content relevance between queries and tweets.
— Tweet features refer to those features which represent the particular characteristics of tweets, as OOV and hashtags in tweet.
— Author features refer to those features which represent the authority of authors of the tweets in Twitter.

## 3.1    Content Relevance Features

The criterion "Content" refers to the thematic relevance traditionally calculated by IR systems standards. The thematic relevance is generally measured by one of several IR models. One of the models reference Information Retrieval IR is the probabilistic model [10] with the weighting scheme BM25 as matching request document function. For this reason, we have adopted this model for the calculation of the thematic relevance. Of course, it is made possible to calculate using any other IR model. BM25 is a search function based "bag of words", it allows us to organize all documents based on the occurrences of the query terms given in the documents. (cf section 2).

We used four content relevance features:

1. Relevance(T,Q): we used OKAPI BM25 score which measures the content relevance between the query Q and tweet T.

$$TF - IDF_{(w,Ti)} = TF_{(w,Ti)}.IDF_{(w,Ti)}$$
$$= TF_{w,Ti}((\log_2 * \frac{N}{DF_w}) + 1)$$

Knowing that: w is a term in the query Q and Ti is the tweet i.

2. Popularity(Ti,Tj,Q): with i and j in n and i≠j : it used to calculate the popularity of a tweet from the corpus. It measures the similarity between the

tweets in the context of the tweet's topic. We used cosine similarity, according to a study done by  Sarwar et al.(2001) cosine similarity is the most efficient similarity measure ,in addition, it is not sensitive to the size of each tweet:

$$Cosine(Ti,Tj) = \frac{\sum\limits_{w\in(Ti\cap Tj)} TFIDF_{w,Ti} * TFIDF_{w,Tj}}{\sqrt{\sum\limits_{w\in Ti}(TFIDF_{w,Ti})^2 * \sum\limits_{w\in Tj}(TFIDF_{w,Tj})^2}}$$

Knowing that w is a term in the query Q, Ti is tweet i, Tj is tweet j, i and j in n and  i≠j.

3.  Length of tweet (Lg(Ti,Q)): Length is measured by the number of characters that a tweet contains. It is said that more the tweet is long, more it contains information.

4.  Out of Vocabulary (OOV(Ti)): This feature is used to roughly approximate the language quality of tweets. Words out of vocabulary in Twitter include spelling errors and named entities. This feature aims to measure the quality language of tweet as follows:

$$Quality(T) = 1 - \frac{Number of OOV(Ti)}{Lg(Ti)}$$

The more number of out of vocabulary is small the more quality of tweet is better.

## 3.2    Tweet Relevance Features

We note that the thematic relevance depends solely on the item and query. Each tweet has many technical features, and each feature form selection criteria that we have exploited.

1.  Retweet (Ti,Q): is defined as the number of times a tweet is retweeted. In a rational manner, the most retweeted tweets are most relevant. Retweets are forwarding of corresponding original tweets, sometimes with comments of retweeters. According to Duan et al. (2010), they are supposed to contain no more information than the original tweets.

2.  Reply(Ti): An @reply is any update posted by clicking the "Reply" button on a Tweet, it will always begin with @username. This feature aims to calculate the number of reply to a tweet. Ultimately tweets that have received the most response are more relevant.

3.  Favor(Ti): this feature aims to calculate the number of times a tweet is classified as a favorite. According to [13], if a message is considered by many followers as a favorite, it means that it is relevant.

4.  Hashtag Count(Ti):The # symbol, called a hashtag, is used to mark keywords or topics in a Tweet. It was created organically by Twitter users as a way to categorize messages. This feature aims to calculate the number of hashtags in tweet.

5.  Urlcount(Ti):Twitter allows users to include URL as a supplement in their tweets. This feature aims to estimates the number of times that the URL appears in the tweet corpus. According to [11][14] and [15], tweets containing URLs are more in-formative.

### 3.3    Author Relevance Features

Each blogger has specific characteristics such as number of follower and number of mention Accordind to [15,16], users who have more followers and have been mentioned in more tweets, listed in more lists and retweeted by more important users are thought to be more authoritative.

1. Tweet Count(a):this feature represents the number of tweet posted by the author
2. Mention Count (author): A mention is any Twitter update that contains "@username" anywhere in the body of the Tweet, this means that @replies are also considered mentions. This feature aims to calculate the number of times an author is mentioned.
3. Follower(a):this feature represents the number of follower to the author
4. Following(a): this feature represents the number of subscriptions of the author (a) to other authors
5. Expertise(a): this feature was found by conducting a survey that asks people to rate the expertise of the blogger from 0 to 10.
6. RetweetRank (a): Retweet Rank looks up all recent retweets, number of followers, friends and lists of a user. It then compares these numbers with those of other users' and assigns a rank. Retweet Rank tracks both RTs posted using the Retweet button and other RTs (ReTweets) (e.g. RT @username).This feature is an indicator of how a blogger is influential on twitter.
7. TwitterPageRank(a): this feature represents the rank of author of the total twitter users using PageRank Algorithm
8. Audience (a): is the size of the potential audience for a message. What is the maximum number of people who could have been exposed to a message?

## 4    Metric Measure of the Impact of Criteria to Improve Search Results

We introduce a research model that combines tweets relevant content, the specificities of tweets and the authority of bloggers. This model considers the specificities of tweets and the authority of bloggers as important factors which contribute to the relevance of the results. The search for tweets is a task of information retrieval whose goal is to select the relevant sections in response to a user's request. To present an accurate list of articles, our model combines a score of content's relevance, a score of author's authority and a score of tweets' specificities. The objective of this combination is to provide a list of tweets that cover the subject of the request and are posted by major bloggers. After normalizing the feature scores, these three scores are combined linearly using the following formula:

$$Score(Ti, Q) = scoreContent(Ti, Q)$$
$$+\beta * scoreTweet(Ti, Q)$$
$$+\gamma * scoreAuthor(Ti, Q)$$

With score (Ti,Q) on [0, 2] and $\beta+\gamma=1$.

Where Ti and Q represent respectively, tweet and request. $\beta$ and $\gamma$ on [0,1] are a weighting parameter[12]. Scorecontent (Ti, Q) is the normalized score of the relevance of content. Scoretweet is the normalized score of the specificity of the tweet Ti and ScoreAuthor (a, Ti) is the normalized score of the importance of the author a corresponds to the blogger who published the tweet Ti.

We note that:

1. Scorecontent(Ti,Q)=Relevance(T,Q) + Lg(Ti) + Popularity(Ti,Tj,Q) + Quality(Ti);
2. ScoreTweet(Ti,Q)= Url count(Ti)+ Hashtag Count(Ti) + Retweet(Ti) + Reply(Ti) + Favor(Ti);
3. ScoreAuthor(a,Q)= TwitterPageRank(a) +Audience(a)+ Tweet Count(a) + Mention Count(a) + Expertise(a) + RetweetRank(a) + Follower(a) + Following(a).

### 4.1     Experimental Evaluation

We conducted a series of preliminary experiments on a collection of articles from Twitter, in order to evaluate the performance of our model.

**Description of the Collection.** With the absence of a standard framework for evaluating information retrieval in micro-blogs, we collected a set of articles and queries. Our concern is that the database size is small. We describe in the following collection of articles and the approach for collecting relevance judgments.

**Search Engine TWEETRIM.** We built a search engine that we have called ``TWEETRIM", which allows to calculate all scores and display the most relevant tweets according to these score. It has as input a query composed of three keywords and as output a set of relevant tweets relative to the query.

**Tweet Set.** We built a collection of articles, metadata about relationships subscription and reply. This corpus is collected manually ie a thousand blogs and thousands of tweets have been browsed. This collection contained a total of 3000 tweets published by 50 active Tunisian bloggers who are interested on the Tunisian news, we chose the period of March 4, 2012 until June 4, 2012.

**Queries and Relevance Judgments.** To perform queries and to collect the human judgement of relevance followed the following steps:

1. We collected 1000 queries on recent actualities in Tunisia from users,
2. then, we used the system that we have built which allows us to view the 10 results are especially relevant according to the score of the content,
3. and then, we asked 450 users to judge the 10 first results of each query.

We suppose that the content relevance already exists and we will improve our search result by varying our two other scores ScoreTweet and ScoreAuthor. We calculate the correlation coefficient between our scores and the corpus, which allowed us to find our weighting coefficients $\beta$ and $\gamma$.

## 4.2    Results

**Estimation of Weights.** We make a comparison within the values the values of correlation coefficients and from these results, we observe that the best correlation coefficient between βScoreTweet+γScoreAuthor with human judgment score = 0, 0,24462 when β = 0,4 and thus γ= 0,6..

**Evaluation of Our Model.** We compare, in Figure 2, the values of correlation's coefficients obtained by Tweet Features and Author Features with the parameters β, γ values respectively (1.0) and (0.1) obtained by experiments and the third configuration with β=0.4 and γ=0.6.

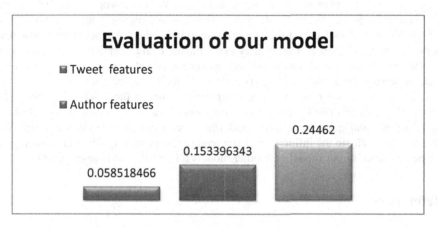

**Fig. 1.** Comparing correlation coefficients

We notice that the performance of the last 2 configurations are very close with a slight advantage for the combination ``Tweet Features & Author Features'' on the model based only on the specificities of the tweet and the importance of the author. We conclude that Author features have more impact on the search's results then Tweet features.

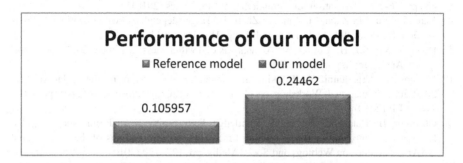

**Fig. 2.** Comparing our model with reference model

The reference model combines only the features linearly without weighting. This model gave us the correlation coefficient equal to 0.10 and our model gave us the correlation coefficient of 0,24462. Can clearly be seen a 43% improvement in the satisfaction of our human judgment.

## 5    Conclusion

Research conducted under the auspices of knowledge management varies greatly in direction and scope. There are several approaches that have been proposed which are based on the features. Therefore the choice of characteristics is important to obtain a satisfactory result and close to the human judgment. We have proposed in this paper a new metric for Social Research on twitter. This has to integrate relevance of content, the specificities of tweets and the author's importance where we incorporate new features such as the audience. The primary experimental evaluation that we conducted on a collection of articles of Twitter shows the measurement that we propose allows a better assessing the impact of bloggers and tweets' technical specificities.

Looking ahead, we plan to conduct experiments under the Micro-blog Text REtrieval Conference (TREC) evaluation framework that will include a collection of many articles and queries for larger and whose relevance judgments are social. We also need to evaluate the influence of each feature independently. We plan to compare the performance of our model with other models for social searching of tweets.

## References

1. Ben Jabeur, L., Tamine, L., Boughanem, M.: Un modèle de recherche d'information sociale dans les microblogs: cas de Twitter. In: Conférence sur les Modèles et l'Analyse des Réseaux: Approches Mathématiques et Informatique (MARAMI 2011), Grenoble (2011)
2. Cha, M., Haddadi, H., Benevenuto Krishna, F., Gummadi, P.: Measuring User Influence in Twitter: The Million Follower Fallacy. In: Proceedings of the Fourth International AAAI Con- ference on Weblogs and Social Media, ICWSM 2010 (2010)
3. Barbosa, L., Feng, J.: Robust sentiment detection on Twitter from biased and noisy data. In: Proceedings of the 23rd International Conference on Computational Linguistics: Posters, pp. 36–44. Association for Computational Linguistics (2010)
4. Jiang, L., Yu, M., Zhou, M., Liu, X., Zhao, T.: Target-dependent Twitter sentiment classification. In: Proc. 49th ACL: HLT, vol. 1, pp. 151–160 (2011)
5. Doan, S., Vo, B.K.H., Collier, N.: An analysis of Twitter messages in the 2011 Tohoearthquake. Arxiv preprint arXiv:1109.1618 (2011)
6. Lampos, V., Cristianini, N.: Tracking the u pandemic by monitoring the social web. In: 2010 2nd International Workshop on Cognitive Information Processing (CIP), pp. 411–416. IEEE (2010)
7. OConnor, B., Balasubramanyan, R., Routledge, B.R., Smith, N.A.: From tweets to polls: Linking text sentiment to public opinion time series. In: Proceedings of the International AAAI Conference on Weblogs and Social Media, pp. 122–129 (2010)
8. Robertson, S., Walker, S., Hancock-Beaulieu, M.: Okapi at TREC-7: Automatic Ad Hoc, Filtering, VLC and Interactive. In: Text REtrieval Conference TREC, pp. 199–210 (1998)

9. Duan, Y., Jiang, L., Qin, T., et al.: An empirical study on learning to rank of tweets. In: COLING Proceedings of the 23rd International Conference on Computational Linguistics Proceedings of the Conference, Beijing, China, August 23-27, pp. 295–303. Tsinghua University Press (2010)

10. Jones, S., Walker, K., Robertson, S.: A probabilistic model of information retrieval: Development and comparative experiments. Information Processing & Management 36(6), 779–808 (2000)

11. Damak, F., Pinel-Sauvagnat, K., Cabanac, G.: Recherche de microblogs: quels critères pour raffiner les résultats des moteurs usuels de RI? In: Conférence en Recherche d'Information et Applications 2012 CORIA 2012, Bordeaux, pp. 371–328 (2012)

12. Akermi, I., And Faiz, R.: Hybrid method for computing word-pair similarity based on web content. In: Proceedings of the International Conference on Web Intelligence, Mining and Semantics, WIMS 2012. ACM, New York (2012)

13. Cherichi, S., Faiz, R.: Recherche d'information pertinente dans les microblogs: Mesure métrique de l'impact des critères pour améliorer les résultats de la recherche. In: Conférence Internationale sur l'Extraction et la Gestion des Connaissances – Maghreb, Hammamet, Tunisie, EGC-M (2012)

14. Cherichi, S., And Faiz, R.: New metric measure for the improvement of search results in mi- croblogs. In: Proc. of the International Conference on Web Intelligence, Mining and Semantics (WIMS 2013). ACM, New York (2013)

15. Cherichi, S., And Faiz, R.: Relevant Information Discovery in Microblogs: New metric measure for the improvement of search results in microblogs. In: Proc. of INSTICC International Conference on Knowledge Discovery and Information Retrieval (KDIR 2013), Vilamoura, Portugal, September 19-22, SciTePress (2013)

16. Cherichi, S., Faiz, R.: Relevant information management in microblogs. In: International Conference on Knowledge Management, Information and Knowledge Systems (KMIKS 2013), Hammamet, Tunisia (April 2013)

# Common-Knowledge and Cooperation Management II S4n-Knowledge Model Case

Takashi Matsuhisa*

Department of Natural Science, Ibaraki National College of Technology
Nakane 866, Hitachinaka-shi, Ibaraki 312-8508, Japan
mathisa@ge.ibaraki-ct.ac.jp

**Abstract.** Issues of moral hazard and adverse selection abound in each and every contract where one has a self interest and information that the other party does not possess, and there is still need for more information on how you handle a party to a contract with more information than you. This paper re-examines the issue in the framework of a principal-agent model under uncertainty. We highlight epistemic conditions for a possible resolution of the moral hazard between the principal and the agents with **S4n**-knowledge, and we show that if the principalr and agents commonly know each agent's belief on the others' efforts, then all effort levels such that the expected marginal costs actually coincide for them can be characterised as the critical points of the refunded proportional rate function. This implies our recommendation that, for removing out such moral hazard in the principal-agents cooperation, the principal and agents should commonly know their beliefs on the others' effort levels.

**Keywords:** Belief, Common-knowledge, Conjecture, Effort level, Expected marginal cost, **S4n**-knowledge model, Moral hazard, Principal-agent model under uncertainty.
**AMS 2000 Mathematics Subject Classification**: Primary 91A35.
**Journal of Economic Literature Classification**: C62, C78.

## 1 Introduction

Issues of moral hazard and adverse selection abound in each and every contract where one has a self interest and information that the other party does not possess. While this is a fertile research area, there is still need for more information on how you handle a party to a contract with more information than you. The Global Financial Crisis is an epitome of the moral hazard: managers and employees as agents and shareholders as principals. In fact, still perplexes

* Research Fellow, Mathematical Research Institute of BUSAIKU-BUHI Foundation for Scientific Research. Mito-shi, Ibaraki 310-0033, Japan E-mail: takashimatsuhisa.mri.bsbh@gmail.com. He was partially supported by JSPS Grant in aids for Scientific Research (C) No. 23540175.

D. Hwang et al. (Eds.): ICCCI 2014, LNAI 8733, pp. 534–543, 2014.

bankers and shareholders alike: shareholders are still having problems with how they can handle their agents, while on the other hand insurers and bankers are struggling to structure products that will reduce the impact of moral hazard. Such moral hazards are the bottlenecks in buyer supplier cooperation, and the buyer-supplier management is another epitome.

The first formal analysis of the principal-agent relationship and the phenomena of moral hazard was made by Arrow [1]. The many sided moral hazard can arise when there are many agents that affect gross returns and their individual actions are not observed by each other, and especially, the treatment of the principal-agent model with many sided moral hazard was given by Holmstrom [3]. He formalized it as the issue in a partnership model whether there exist any sharing rules that both balances the budget and under which an efficient action is a Nash equilibrium. Holmstrom [3] and Williams and Radner [8] respectively analyzed the conditions for existing the sharing rule such that some actions profile satisfies the first-order conditions for an equilibrium.

Recently, Matsuhisa [4] and Matsuhisa and Jiang [6] adopted a new approach to the many sided moral hazard from the epistemic model point of view developed by Aumann [2] and his followers in game theory, they analyzed the moral hazard as the disagreement on expected marginal costs between the principal and agents in an extended model of principal and agents under uncertainty. He gave a necessity condition that the moral hazard will not be appeared; that is, under some technical assumptions, the principal and agent model under uncertainty disappears the moral hazard if the principal and agents could share fully information on their expected marginal costs in the following two cases: first they commonly known the marginal expected costs(Matsuhisa [4]), and secondly they communicate the costs as long run (Matsuhisa and Jiang [6]). However, in the papers they assume the existence of decision function consistent to the technical assumptions, and it has not been guaranteed.

This paper aims to remedy the defect. We re-examine a buyer-supplier cooperation with moral hazard as a problem of the principal-agent relationship. We present an extended principal-agent model under uncertainty, and we highlight hidden conditions for a possible resolution of the moral hazard between the buyer and the suppliers. For removing out such moral hazard in the buyer-supplier cooperation, our recommendation is that the principal and the agents should commonly know their beliefs on the others' effort:

Let us consider that there are the buyer (as principal) and suppliers (as agents) more than two: The buyer manufactures the productions made of parts supplied by the suppliers with paying their costs, and he/she gets a profit by selling the productions. Assume that the buyer and all suppliers aim to maximize each gross return independently. The moral hazard arises that there is not the sharing rule so that the buyer makes a contract with every supplier such that the total amount of all profits is refunded to each supplier in proportion to the supplier's contribution to the productions, i.e.; the expected costs are not equal between the buyer and the suppliers. To investigate the phenomenon in detail we shall extend the principal-agent model with complete information to the principal-agent model with incomplete information.

Now we assume each agent as well as the principal, $k$, has the below two abilities on knowledge:

**T**  each $k$ cannot know something when it does not occurs;
**4**  each $k$ knows that he/she knows something.

This structure is induced from a *reflexive and transitive* binary relation associated with the multi-modal logic **S4n**.

We focus on the situation that the principal and the agents interact each other from sharing information on the below:

**DF.**  Both the buyer and suppliers have the same decision function;
**PR.**  The refunded proportional rates to suppliers are functions of each supplier's effort level;
**CK.**  Both the buyer and suppliers commonly known each agent's belief on the others' efforts;

In this line we can show:

**Theorem 1.** *Under the above conditions* **DF, PR, CK** *all effort levels such that the expected marginal costs actually coincide for buyer and suppliers can be characterised as the critical points of the refunded proportional rate function. Consequently, if the refunded proportional rate is constant then all marginal costs have to coincide each other; i.e., there is no moral hazard.*

The same result has been obtained for **S5n**-knowledge model by Matsuhisa [5].

The paper is organized as follows: Section 2 reviews the moral hazard in the classical principal-agent model (i.e.: the principal-agent model with complete information) following Matsuhisa [4]. Section 3 recalls the formal model of knowledge and common-knowledge, and presents the principal-agent model under uncertainty, and an illustrative example of our contract design problem in the principal-agent model under uncertainty. Section 4 states main theorem formally with the proof. Finally we conclude remarks.

## 2  Moral Hazard[1]

Let us consider the principal-agents model as follows: There are the principal $P$ and $n$ agents $\{1, 2, \cdots, k, \cdots, n\}$ $(n \geq 1)$ in a firm. The principal makes a profit by selling the productions made by the agents. He/she makes a contract with each agent $k$ that the total amount of all profits is refunded each agent $k$ in proportion to the agent's contribution to the firm.

Let $e_k$ denote the measuring managerial effort for $k$'s productive activities, called $k$'s *effort level* or simply $k$'s *effort* with $e_k \in \mathbb{R}_+$. Let $I_k(x_k)$ be a real valued continuously differentiable function on $\mathbb{R}_+$. $I_k(x_k)$ is interpreted as the

---

[1] Section 2 in Matsuhisa [4].

profit by $k$'s effort $e_k$ selling the productions made by the agent $k$ with the cost $c(e_k)$. Here we assume $I_k'(x_k) \geq 0$ and the cost function $c(\cdot)$ is a real valued continuously differentiable function on $\mathbb{R}_+$. Let $I_P$ be the total amount of all profits: $I_P(x) = I_P(x_1, x_2, \cdots, x_k, \cdots, x_n) = \sum_{k=1}^{n} I_k(x_k)$.

The principal $P$ cannot observe these efforts $e_k$, and shall view it as a random variable $\mathbf{e}_k$ on a probability space $(\Omega, \mu)$; i.e., $\mathbf{e}_k$ is a $\mu$-measurable function from $\Omega$ to $\mathbb{R}_+$. We introduce the *ex-post* expectation: $\mathrm{Exp}[I_P(e)] := \sum_{\xi \in \Omega} I_P(\mathbf{e}(\xi)) \mu(\xi)$ and $\mathrm{Exp}[I_k(e_k)] := \sum_{\xi \in \Omega} I_k(\mathbf{e}_k(\xi)) \mu(\xi)$. The optimal plan for the principal then solves the following problem:

$$\mathrm{Max}_{e=(e_1, e_2, \cdots, e_k, \cdots, e_n)} \{\mathrm{Exp}[I_P(e)] - \sum_{k=1}^{n} \mathrm{Exp}[c(e_k)]\}.$$

Let $W_k(e_k)$ be the total amount of the refund to agent $k$: $W_k(e_k) = r_k I_P(e)$, with $\sum_{k=1}^{n} r_k = 1, 0 \leq r_k \leq 1$, where $r_k$ denotes the proportional rate representing $k$'s contribution to the firm. The optimal plan for each agent also solves the problem: For every $k = 1, 2, \cdots, n$,

$$\mathrm{Max}_{e_k} \{\mathrm{Exp}[W_k(e_k)] - \mathrm{Exp}[c(e_k)]\} \quad \text{subject to} \quad \sum_{k=1}^{n} r_k = 1, 0 \leq r_k \leq 1.$$

We assume that $r_k$ is independent of $e_k$, and the necessity conditions for critical points are as follows: For each agent $k = 1, 2, \cdots, n$, we obtain $\frac{\partial}{\partial e_k} \mathrm{Exp}[I_k(e_k)] - \mathrm{Exp}[c'(e_k)] = 0$ and $r_k \frac{\partial}{\partial e_k} \mathrm{Exp}[I_k(e_k)] - \mathrm{Exp}[c'(e_k)] = 0$. in contradiction to $0 \lneqq r_k \lneqq 1$ because $c'(e_k) = \frac{\partial}{\partial e_k} \mathrm{Exp}[I_k(e_k)] = r_k \frac{\partial}{\partial e_k} \mathrm{Exp}[I_k(e_k)]$

This contradictory situation is called a **moral hazard** in the principal-agents model; i.e., there is no equilibrium effort level as a solution of the contract design problem.

## 3   The Model

Let $N$ be a set of finitely many agents and let $k$ denote an agent and $P$ the principal. The specification is that $\bar{N} = \{P, 1, 2, \cdots, k, \cdots, n\}$ consists of the principal $P$ and the agents $N = \{1, 2, \cdots, k, \cdots, n\}$ in a firm. A state-space $\Omega$ is a non-empty *finite* set, whose members are called *states*. An *event* is a subset of the state-space. We denote by $2^{\Omega}$ the field of all subsets of it. An event $E$ is said to occur in a state $\omega$ if $\omega \in E$.

### 3.1   Information and Knowledge

By *RT-information* structure[2] we mean $\langle \Omega, (\Pi_i)_{i \in \bar{N}} \rangle$ in which $\Pi_i : \Omega \to 2^{\Omega}$ satisfies the two postulates: For each $i \in \bar{N}$ and for any $\omega \in \Omega$,

**Ref.**   $\omega \in \Pi_i(\omega)$;   **Trn.**   $\xi \in \Pi_i(\omega)$ implies $\Pi_i(\xi) \subseteq \Pi_i(\omega)$.

---

[2] RT-information stands for reflexive and transitive information.

The set $\Pi_i(\omega)$ will be interpreted as the set of all the states of nature that $i$ knows to be possible at $\omega$, or as the set of the states that $i$ cannot distinguish from $\omega$. We call $\Pi_i(\omega)$ $i$'s *information set* at $\omega$.

**Definition 1.** The **S4n**-*knowledge structure* is a tuple $\langle \Omega, (\Pi_i)_{i \in \bar{N}}, (K_i)_{i \in \bar{N}} \rangle$ that consists of a partition information structure $\langle \Omega, (\Pi_i)_{i \in \bar{N}} \rangle$ and a class of $i$'s *knowledge operator* $K_i : 2^\Omega \to 2^\Omega$ defined by $K_i E = \{ \omega \mid \Pi_i(\omega) \subseteq E \}$.

The event $K_i E$ will be interpreted as the set of states of nature for which $i$ knows $E$ to be possible. We record the properties of $i$'s knowledge operator: For every $E, F$ of $2^\Omega$,

**N.**  $K_i \Omega = \Omega$;  **K**  $K_i(E \cap F) = K_i E \cap K_i F$;  **T**  $K_i E \subseteq E$;
**4.**  $K_i E \subseteq K_i(K_i E)$;

## 3.2  Common-Knowledge

Let $S$ be a subset in $\bar{N}$. The *mutual knowledge operator* among a coalition $S$ is the operator $K_S : 2^\Omega \to 2^\Omega$ defined by the intersection of all individual knowledge: $K_S F = \cap_{i \in S} K_i F$, which interpretation is that everyone knows $E$.

**Definition 2.** The *common-knowledge operator* among $S$ is the operator $K_C^S : 2^\Omega \to 2^\Omega$ defined by $K_C^S F = \cap_{n \in \mathbb{N}} (K_S)^n F$.

An event $E$ is *common-knowledge* among $S$ at $\omega \in \Omega$ if $\omega \in K_C^S E$. The intended interpretations are as follows: $K_C^S E$ is the event that 'every agent in $S$ knows $E$' and "every agent in $S$ knows that 'every agent in $S$ knows $E$'," and "'every agent in $S$ knows that "everyone knows that 'every agent in $S$ knows $E$'," ."'

## 3.3  Principal-Agent Model under Uncertainty

Let us reconsider the principal-agent model and let notations and assumptions be the same as in the above section. We shall introduce the extended principal-agents model.

**Definition 3.** By a *principal-agent model under uncertainty* we mean s structure

$$\mathcal{M} = \langle \bar{N}, (\Omega, \mu), (\mathbf{e}_k)_{k \in N}, (I_k)_{k \in \bar{N}}, I_P, (r_k)_{k \in N}, (c_k)_{k \in N}, (\Pi_k)_{k \in N} \rangle$$

in which

1. $\bar{N} = \{ \mathrm{P}, 1, 2, \cdots, k, \cdots, n \}$ where $P$ is the principal and each $k$ is an agent, $(\Omega, \mu)$ is a probability space;
2. $\mathbf{e}_k$ is a random variable on $\Omega$ into $\mathbb{R}_+$ with $\mathbf{e}_k(\omega)$ a real variable in $\mathbb{R}_+$;
3. $I_k(x_k)$ is an agent $k$'s profit function with $I_k(e_k)$ the profit by his/her effort $e_k$, which is sufficiently many differentiable on $\mathbb{R}_+$ with $I_k' \geq 0$, and $I_P(x) = I_P(x_1, x_2, \cdots, x_n) = \sum_{k=1}^n I_k(x_k)$ is the profit function of the firm (the total amount of all the agents' profits);

4. $r_k$ is a proportional rate function in the contract, which is sufficiently many differentiable and weakly increasing on $\mathbb{R}_+$ with $0 < r_k \leqq 1$ for $k = 1, 2, \cdots, n$;

5. $c_k$ is the cost function for agent $k$, which is sufficiently many differentiable on $\mathbb{R}_+$ with $I'_k \geqq 0$ with $c_k(e_k)$ interpreted as the cost of $k$ for effort level $e_k$;

6. $(\Pi_k)_{k \in \bar{N}}$ is a non-partition information structure satisfying the two postulates **Ref** and **Trn**.

For each $e_k \in \mathbb{R}_+$ and $e = (e_1, e_2, \cdots, e_k, \cdots, e_n) \in \mathbb{R}_+^n$ let us denote by $[e_k]$ the event of $k$'s effort $[e_k] = \{\xi \in \Omega | \mathbf{e}_k(\xi) = e_k\}$ and by $[e]$ the event of total efforts $[e] = \cap_{k \in N}[e_k]$. For any non-empty subset $S$ of $\bar{N}$, we will denote $[e_S] = \cap_{k \in S}[e_k]$, and $[e_{-k}] = \cap_{l \in N \setminus \{k\}}[e_l]$

From now on we assume a principal-agent model under uncertainty is not with *complete information*, which means $\Pi_i(\omega) \neq \{\omega\}$ for some $\omega \in \Omega$..

## 3.4 Bayesian Approach

According to this we have to assume that each agent $k$ know his/her own effort $e_k$ but $k$ cannot know the others' effort $e_k$, and also the principal $P$ cannot know efforts for any agents. The former assumption can be formulated as

**KE.** $[e_k] \subseteq K_k([e_k])$ for every $e_k \in E_k$.

The later assumption means that the principal cannot have the exact knowledge on the agents' effort levels $e$ and also each agent cannot have the exact knowledge on the others's effort $e_{-k} \in \mathbb{R}_+^{n-1}$.

## 3.5 Belief and Conjecture

Following the interpretations we have to introduce the notion of *belief on the others' effort level*: By the principal $P$'s belief on the agents efforts $e$ we mean a probability $q_P(e)$ of $e$, and by an agent $k$'s *belief* on the other agents effort $e_{-k}$ we mean a probability $q_k(e_{-k})$ of $e_{-k}$. The *conjecture* $\mathbf{q}_P(e; \omega)$ of the principal $P$ for the agents' effort $e_k \in mathbbR$ ($k \in N$) is defined by $\mathbf{q}_P(e; \omega) = \mu([e] | \Pi_P(\omega))$, and the *conjecture* $\mathbf{q}_k(e_{-k}; \omega)$ of agent $k$ for the other agents' effort $e_{-k} \in \mathbb{R}_+^{n-1}$ is $\mathbf{q}_k(e_{-k}; \omega) = \mu([e_{-k}] | \Pi_k(\omega))$. By the event of $P$'s belief on the agents efforts $e$, we mean $[q_P(e)] := \{\xi \in \Omega \mid \mathbf{q}_P(e; \omega) = q_P(e)\}$, and by the event of $k$'s belief on the other agents efforts $e_{-k}$ we mean $[q_k(e_{-k})] := \{\xi \in \Omega \mid \mathbf{q}_k(e_{-k}; \omega) = q_k(e_{-k})\}$ It should be noted by **KE** that $\mathbf{q}_k(e_{-k}; \omega) = \mathbf{q}_k(e; \omega)$ and so $[\mathbf{q}_k(e_{-k}; \omega)] = [\mathbf{q}_k(e; \omega)]$.

## 3.6 Interim Expectation

By the *interim expectation* (or simply *expectation* ) of $I_P$ we mean

$$\text{Exp}[I_P(e) | \Pi_P](\omega) := \sum_{\xi \in [e]} I_P(\mathbf{e}_1(\xi), \mathbf{e}_2(\xi), \cdots, \mathbf{e}_n(\xi)) \mu(\xi | \Pi_P(\omega))$$

$$= I_P(e) \mathbf{q}_P(e; \omega)$$

and by the *interim expectation* (or simply *expectation* ) of $I_k$ we mean

$$\mathrm{Exp}[I_k(e_k)|\Pi_k](\omega) := I_k(e_k)\mathbf{q}_k(e_{-k};\omega) = \sum_{\xi\in[e_{-k}]} I_k(e_k(\xi))\mu(\xi|\Pi_k(\omega))$$

and the interim expectation of agent $k$'s income $W_k$ is

$$\mathrm{Exp}[W_k(e_k)|\Pi_k](\omega) := r_k(e_k)\mathrm{Exp}[I_P(e_k,\mathbf{e}_{-k})|\Pi_k(\omega)]$$
$$= \sum_{e_{-k}\in E_k} r_k(e_k)I_P(e_k,e_{-k})\mathbf{q}_k(e_{-k};\omega)$$

## 3.7   Contract Design Problem

We will treat the maximisation problems as for the optimal plans for the principal and agents: To find out effort levels $e = (e_1, e_2, \cdots, e_k, \cdots, e_n) \in \prod_{k\in N} E_k$ such that, subject to $\sum_{k=1}^n r_k = 1, 0 < r_k \le 1$,

**PE.**   $\mathrm{Max}_{e=(e_k)_{k=1,2,\cdots,n}}\{\mathrm{Exp}[I_P(e)|\Pi_P(\omega)] - \sum_{k=1}^n \mathrm{Exp}[c_k(e_k)]\}$;

**AE.**   $\mathrm{Max}_{e_k}\{\mathrm{Exp}[W_k(e_k)|\Pi_k(\omega)] - \mathrm{Exp}[c_k(e_k)]\}$.

*Example 1.* Let us consider the principal-agent model under uncertainty as follows;

- $\bar{N} = \{P, 1, 2\}$:
- $\Omega = \{\omega_1, \omega_2, \omega_3, \omega_4\}$, with each state is interpreted as the two types of effort levels $\{H, L\}$ for agent 1 and effort levels $\{h, l\}$ for agent 2 given by the table 1:
- The information partition $(\Pi_i)_{i=P,1,2}$ are:
  - $\mu$ is the equal probability measure on $2^\Omega$; i.e., $\mu(\omega) = \frac{1}{4}$:
  - The partitions $(\Pi_i)_{i=P,1,2}$ on $\Omega$:
    - The partition $\Pi_P$ on $\Omega$: $\Pi_P(\omega) = \Omega$.
    - The partition $\Pi_1$ on $\Omega$:

    $$\Pi_1(\omega) = \{\omega_1, \omega_2\} \text{ for } \omega = \omega_1, \omega_2, \quad \Pi_1(\omega) = \{\omega_3, \omega_4\} \text{ for } \omega = \omega_3, \omega_4.$$

    - The partition $\Pi_2$ on $\Omega$:

    $$\Pi_2(\omega) = \{\omega_1, \omega_3\} \text{ for } \omega = \omega_1, \omega_3, \quad \Pi_2(\omega) = \{\omega_2, \omega_4\} \text{ for } \omega = \omega_2, \omega_4.$$

**Table 1.** Types and Variables

| | h | l |
|---|---|---|
| H | $\omega_1$ | $\omega_2$ |
| L | $\omega_3$ | $\omega_4$ |

| $\mathbf{e}_*$ | $y_h$ | $y_t$ |
|---|---|---|
| $x_h$ | $\omega_1$ | $\omega_2$ |
| $x_l$ | $\omega_3$ | $\omega_4$ |

- $\mathbf{e}_i : \Omega \to \mathbb{R}_+$ with $\mathbf{e}_i(\omega)$ a real variable is defined by

$\mathbf{e}_1(\omega) = x_{\mathrm{h}}$ for $\omega = \omega_i (i = 1, 2), \mathbf{e}_1(\omega) = x_{\mathrm{l}}$ for $\omega = \omega_i (i = 3, 4)$ with $x_{\mathrm{h}} \geq x_{\mathrm{l}}$,

$\mathbf{e}_2(\omega) = y_{\mathrm{h}}$ for $\omega = \omega_i (i = 1, 3), \mathbf{e}_2(\omega) = y_{\mathrm{l}}$ for $\omega = \omega_i (i = 2, 4)$ with $y_{\mathrm{h}} \geq y_{\mathrm{l}}$,

with $[\mathbf{e}_1(\omega)] = \Pi_1(\omega), [\mathbf{e}_2(\omega)] = \Pi_2(\omega)$.
This means that agent 1's effort at $\omega_1, \omega_2$ is higher that the effort at $\omega_3, \omega_4$, and agent 2's effort at $\omega_1, \omega_3$ is higher that the effort at $\omega_2, \omega_4$;

- $I_1(x)$ and $I_2(y)$ are profit functions and $I_P(x, y) = I_1(x_1) + I_2(x_2)$ is the total amount of the profits;

Under the situation we obtain that $\mathrm{E}[W_1|\Pi_1](\omega)$ $= r_1(x_{\mathrm{h}})I_P(x_{\mathrm{h}}, y_j)(\omega = \omega_1, \omega_2, j = \mathrm{h, l}), \mathrm{E}[W_1|\Pi_1](\omega) = r_1(x_{\mathrm{l}})I_P(x_{\mathrm{l}}, y_j)(\omega = \omega_3, \omega_4, j = \mathrm{h, l}), \mathrm{E}[W_2|\Pi_2](\omega) = r_2(y_{\mathrm{h}})I_P(x_i, y_{\mathrm{h}})(\omega = \omega_1, \omega_3, i = \mathrm{h, l}), \mathrm{E}[W_2|\Pi_2](\omega) = r_2(y_{\mathrm{l}})I_P(x_i, y_{\mathrm{l}})(\omega = \omega_2, \omega_4, i = \mathrm{h, l})$. and $\mathrm{E}[I_P|\Pi_P](\omega) = \frac{1}{4}\sum_{i,j=\mathrm{h,l}}(I_1(x_i) + I_2(y_j))$. Then we can observe that there are no moral hazard if $r_1(e) = r_2(e) \equiv \frac{1}{2}$, hence any effort level can be a solution of the above contract problem **PE** and **AE**.

## 4    Main Theorem

For the beliefs $q_P, q_k$ of the principal $P$ and agent $k$, we refer the conditions:

**BCK.**    $\cap_{k \in N}[q_k] \cap K_C^N([q_P]) \neq \emptyset$

The interpretations of **BCK** is that all the agents commonly know the principal's belief $q_P$ at some state where all agents actually have their beliefs $(q_k)_{k \in N}$. Under the circumstances we can now restate Theorem 1 as follows:

**Theorem 2.** *In the principal-agent model under uncertainty with **KE**, assume that the principal $P$ and each agent $k$ have actually beliefs $q_P, q_k$. If all agents commonly know the principal's belief $q_P$ then every effort level $e_k(k \in N)$ as solutions of the the contract design problem **PE**, **AE** must be a critical point of $r_k$ for every $k \in N$; i.e.: $r'_k(e_k) = 0$ if **BCK** is true. In this case the proportional rate $r_k$ is determined by the principal belief: $r_k(e_k) = q_P(e_k)$.*

Before proceeding with the proof, we notice:

*Remark 1.* Theorem 2 can explain the resolution of moral hazard of Example 1: In fact, since $K_C^{\{1,2\}}([\mathbf{q}_\mathrm{P}(e(\omega); \omega)]) = \Omega$, we can see that $[\mathbf{q}_\mathrm{P}(e(\omega); \omega)]$ is common-knowledge everywhere among agents 1, 2, and further, $r_1$ and $r_2$ are the constant with $r_1(e_1(\omega)) = \mathbf{q}_\mathrm{P}(e_1(\omega); \omega) = \frac{1}{2}, r_2(e_2(\omega)) = \mathbf{q}_\mathrm{P}(e_2(\omega); \omega) = \frac{1}{2}$. Hence the resolution of moral hazard in Example 1 can be described by Theorem 2.

We shall turn to the proof of Theorem 2:

**Critical Points Condition.** Partially differentiating the expressions in the parentheses of the problems **PE** and **AE** with respect to $x_k$ yields the necessity condition for critical points for every $k \in N$: From **PE** we have,

$$\text{Exp}[I_k'(e_k)|\Pi_P(\omega)] = I_k'(e_k)\mathbf{q}_P(e;\omega) = \text{Exp}[c_k'(e_k)]; \tag{1}$$

and from **AE** the condition is also that, subject to $0 < r_k < 1$ and $\sum_{k \in N} r_k = 1$,

$$r_k'(e_k)\text{Exp}[I_k(e_k)|\Pi_k(\omega)] + r_k(e_k)\text{Exp}[I_k'(e_k)|\Pi_k(\omega)] = \text{Exp}[c_k'(e_k)] \tag{2}$$

The below proposition plays another central role to prove Theorem 2:

**Proposition 1 (Decomposition theorem).** *Suppose that the principal $P$ and each agent $k$ have beliefs $q_P, q_k$ with **BCK** in the principal-agent model under uncertainty with **KE**. Then we obtain that for every $k \in N$, $q_P(e) = q_k(e_{-k})q_P(e_k)$ for any $e = (e_1, e_2, \cdots, e_k, \cdots, e_n) \in \mathbb{R}_+^n$.*

*Proof.* Let $M$ denote $M = K_C^N([q_P])$, which is non-empty by **BCK**, and so take $\omega \in M$. For each $k \in N$ and for each $e = (e_1, e_2, \cdots, e_k, \cdots, e_n) \in \prod_{k \in N} E_k$, we set $H_k = [e_k] \cap M$, and we can easily observe that both $M$ or $H_k$ satisfy the below properties: An event $H$ is called $\Pi_k$-*invariant* if $\Pi_k(\xi) \subseteq H$ for any $\xi \in H$,

(i)   $\omega \in L \subseteq [\mathbf{q}_P(e;\omega)] \subseteq [\mathbf{q}_k(e;\omega)] = [\mathbf{q}_k(e_{-k};\omega)]$ for $L = M$ or $H_k$;
(ii)  $M$ is $\Pi_P$-invariant, and $L$ is $\Pi_k$-invariant.

Therefore, on considering $X = [e]$ it follows from Fundamental lemma in Matsuhisa and Kamiyama [7] for $L = H_k$ that $\mu([e_{-k}]|H_k) = \mu([e_{-k}] \mid \Pi_k(\omega))$, and so $\mu([e_{-k}]|H_k) = \mathbf{q}_k(e_{-k};\omega)$. Dividing by $\mu(M)$ yields that $\mu([e]|M) = \mathbf{q}_k(e_{-k};\omega)\mu([e_k]|M)$. On considering $X = [e]$ it follows from Fundamental lemma as above for $L = M$ that $\mu([e]|M) = \mu([e] \mid \Pi_P(\omega))$. Hence we can observe $\mu([e]|M) = \mathbf{q}_P(e;\omega)$. and thus $\mu([e_k]|M) = \mathbf{q}_P(e_k;\omega)$. It follows by the above three equations that $\mathbf{q}_P(e;\omega) = \mathbf{q}_k(e_{-k};\omega)\mathbf{q}_P(e_k;\omega)$. Noting $\omega \in [q_P(e)] \subseteq [q_k(e_{-k})]$, we have shown that $q_P(e) = q_k(e_{-k})q_P(e_k)$.

**Proof of Theorem 2.** Notations and assumptions are the same in Proposition 1. On viewing Eqs.(1), (2), it can be easily observed that the former part of Theorem 2 follows from Proposition 1. Especially, if $r_k$ is a constant function, then $r_k' = 0$, and so the latter part also follows immediately.   □

# 5   Concluding Remarks

This paper advocates a new approach to treat a moral hazard problem in principal-agent model focused on the beliefs of effort levels from the epistemic point of view. To highlight the structure of sharing private information on their conjectures about effort levels by common-knowledge between principal and agents is capable of helping us to make progress in 'problematic' of classical principal-agent models. In particular, common-knowledge on the conjectures play crucial role in removing out the moral hazard in the classical principal-agent model. In fact, for removing out the moral hazard in the buyer-supplier cooperation management we will recommend that they (the buyer and suppliers) should share fully information on only their conjectures on the others' effort levels but not expected marginal costs by making common-knowledge on the efforts.

It well ends this paper in our appraisals on our result and the assumptions of the principal-agent model under uncertainty. The above recommendation is not so fresh to us: In cooperative organization system, any concerned agents will, tacitly or consciously, share their information. We well know this is the first step towards to succeed our cooperation. The point in this paper is to clearfy what kind of information we have to share: It is the information on belief of the others' efforts not but their expected marginal costs.

What happened in the extended principal-agents model without the assumptions [CK] ? Common-knowledge on effort levels does not make a sufficient condition for resolution of the moral hazard: When CK does not hold, we can construct the typical examples, one of which the moral hazard still remain and the other of which it disappears. The former is given by modifying Example 1 in replacing the principal's information partition $\Pi_P$ with $\Pi_P(\omega) = \{\omega\}$. The latter is done by modifying Example 1 in replacing $\Pi_P$ with $\Pi_P(\omega_i) = \{\omega_1, \omega_4\}$ for $i = 1, 4$ and $\Pi_P(\omega_j) = \{\omega_2, \omega_3\}$ for $i = 2, 3$.

We have not also analyzed any other applications of the above presented model than the moral hazard. Can we treat the Global Financial Crisis by our framework? How about the extended agreement issues on TPP (Trans-Pacific Strategic Economic Partnership)? These are interesting and important and are our next agenda.

# References

1. Arrow, K.J.: Uncertanty and welfare economics of medical care. American Econ. Review 53, 941–973 (1963)
2. Aumann, R.J.: Agreeing to disagree. Ann. of Statistics 4, 1236–1239 (1976)
3. Holmstrom, B.: Moral Hazard in Teams. Bell J. of Econ. 13, 324–340 (1982)
4. Matsuhisa, T.: Moral hazard resolved by common-knowledge in principal-agent model. Int. J. of Intelligent Information and Database Systems 6(3), 220–229 (2012)
5. Matsuhisa, T.: Common-Knowledge and Cooperation Management I – S5n-knowledge model case – Working paper 2014. In: An extended abstract will be appeared in Proc. of CENet 2014. LNEE (2014)
6. Matsuhisa, T., Jiang, D.-Y.: Moral hazard resolved in communication Network. World J. of Social Sciences 1(3), 100–115 (2011)
7. Matsuhisa, T., Kamiyama, K.: Lattice structure of knowledge and agreeing to disagree. J. of Math. Econ. 27(4), 389–410 (2011)
8. Williams, S., Radner, R.: Efficiency in partnerships when the joint output is uncertain, Discussion Paper No.76, Kellog School of Management, Northwestern University (1989), http://kellogg.northwestern.edu/research/math/papers/760.pdf

# Modelling Mediator Assistance in Joint Decision Making Processes Involving Mutual Empathic Understanding

Rob Duell and Jan Treur

VU University Amsterdam, ASR Group,
De Boelelaan 1081,
1081 HV, Amsterdam, The Netherlands
r.duell@vu.nl, treur@cs.vu.nl

**Abstract.** In this paper an agent model for mediation in joint decision-making processes is presented for establishing mutual empathic understanding. Elicitation of affective states is an important criterion of empathy. In unassisted joint decision-making it can be difficult to recognise whether empathic responses are the result of experiencing the other individual's affective state, or whether these affective states are at least partly blended with own states that would also have developed in individual decision-making. The mediator agent assists two individual social agents in establishing and expressing empathy, as a means to develop solidly grounded joint decisions.

**Keywords:** mediation, empathy, joint decision making.

## 1 Introduction

Reaching a solidly grounded joint decision is a complex process involving multiple interacting individuals arriving at a common choice. The complexity of this process encompasses the mutual tuning of individual intentions and emotions, and the development of mutual empathic understanding between the interacting individuals. Ultimately, empathic understanding can be acknowledged between the interacting individuals by both verbal and nonverbal responses. One of the criteria for empathy concerns the elicitation of an individual's affective state upon observation or imagination of another individual's affective state (according to, e.g., [4,11]). It is difficult to establish the occurrence of mutual empathic understanding during an unassisted joint decision making process: are the verbal and nonverbal responses genuinely the result of experiencing the other individual's affective state, or the result of affective states that would also have developed in individual decision making? In [5] an analysis is made on the possible outcomes of joint decision processes. Based on this, this paper analyses, models and simulates mediator assistance for establishing mutual empathic understanding. This paper contributes a model of a mediator agent that supports in discerning between the development of genuine empathic understanding and

D. Hwang et al. (Eds.): ICCCI 2014, LNAI 8733, pp. 544–553, 2014.

situations with blended affective states. The mediator assists in the joint decision making by introducing the asymmetry in the interaction that is necessary to test for the elicitation criterion. In this way, the occurrence of true empathic understanding can be established in a conscious fashion. Acknowledgement of this form of understanding develops solidly grounded joint decisions.

## 2    Criteria of Empathy

In [12] the four criteria of empathy are summarized, as formulated in [4,11]:

1. Presence of an affective state in a person
2. Isomorphism of the person's own and the other person's affective state
3. Elicitation of the person's affective state upon observation or imagination of the other person's affective state
4. Knowledge of the person that the other person's affective state is the source of the person's own affective state

As described in [12], the social agent model models empathic responses for emotions and action tendencies. In reaching a joint decision, these empathic responses may be used in establishing mutual empathic understanding.

Assuming true faithful nonverbal and verbal expression these empathic responses indeed indicate the presence of an affective state, covering criterion 1. Also, under the same assumption, comparing empathic responses between interacting agents may indicate isomorphism of their respective emotional and tendency states, thus covering criterion 2. Criteria 3 and 4 however cannot be directly established by just looking at the empathic responses, but require more insight in the way that these empathic responses (could have) come into existence.

The elicitation criterion 3 actually means that one of the interacting agents expresses emotional and/or tendency states, followed by development (and subsequent expression) of the corresponding state(s) in the other agent. In other words, in covering criterion 3, a difference should be made between scenarios in which an agent develops emotional and/or tendency states independently of other agents, and scenarios where an agent could have developed these same states also strictly under the influence of other agents. In the latter case, the elicitation criterion is satisfied. Criterion 4 requires conformance to criterion 3, and can further be established by observing verbal empathic responses for emotional and tendency states.

To establish the influence of another agent, an agent should ignore the world stimulus and completely focus on the other agent. In that situation all four criteria can be tested for (nonverbal and verbal) empathic responses. For establishing mutual empathic understanding, this situation can be repeated for the other agent.

## 3   The Social Agent Model Used

In [12] a social agent model for joint decision making is presented addressing the role of mutually acknowledged emphatic understanding in decision making. This neurologically inspired cognitive agent model uses the following principles: mirroring (see also [9]), internal simulation (see also [6]) and emotion-related-valuing (see also [3]). Interacting social agents may develop mutual empathic understanding (see also [4]), which may be shown (nonverbal) and acknowledged (verbal). For further details, see [12].

## 4   The Extended Social Agent Model

This Section describes the adaptations to the social agent model to enable detecting true empathy between two social agents, as summarized in Table 1.

The adaptations concern the introduction of the states ES(p) and WS(p), and the introduction of an additional connection from the sensor state for context, to the sensory representation state for the stimulus. The ES(p) state is an externally observable execution state that maintains a high activation level when the social agent, using body language, expresses any preparation activity. The WS(p) state is a world state that is used by the mediator agent to communicate a pause in which the social agents can reduce their preparation activity. This reduction is realized by the connected SS(p) sensor state, that suppresses the social agents sensory representation for feeling. The additional connection, from the sensor state for context to the sensory representation for the stimulus, is necessary to make the agent use information from the other agents for its activation of preparation for action. In order to detect true empathy, the agent should not develop the preparation for action purely on its own.

It should be noted that the sensory representation state for context remains connected to the self-ownership states, as in the original social agent model. Also, all other connections and parameters to states SR(s) and SR(b) remain unchanged.

**Table 1.** Social agent model additions

| From state | Connection | | To state | $\tau$ | $\sigma$ |
|---|---|---|---|---|---|
| SS(context) | $\omega_{22c}$ | 1 | SR(s) | 1.5 | 20 |
| PS(a) | $\omega_{9a}$ | 1 | ES(p) | 0 | 20 |
| PS(b) | $\omega_{9b}$ | 1 | | | |
| WS(p) | $\omega_{1W}$ | 1 | SS(p) | - | - |
| SS(p) | $\omega_{24b}$ | -1 | SR(s) | 0.7 | 4 |

# 5   The Mediator Agent Model

For the mediator agent a dynamical systems perspective is adopted as advocated, for example, in [1,10]. The quantitative aspects are modeled in a mathematical manner as in [1,10,13]. This fits in the domain of small continuous-time recurrent neural networks as advocated by [2] and inspired by e.g., [7,8]; see also [13].

This Section describes the dynamical systems model for the mediator agent. The mediator model supports multiple main (major) episodes with different foci, aimed at detecting the presence of true empathy between the two social agents A and B. Between each two of these major episodes, a special (minor) pause episode enables the agents to come to rest before starting the next major episode. For each major episode, the mediator model goes trough the following pattern of dependencies:

(1) All activation levels of states that may be relevant for ending a specific episode are combined into a single belief on the progress of the episode. As described earlier, the aim for episodes is the detection of true empathy between the two social agents. As soon as empathy is acknowledged, the episode ends. In case no empathy develops, the mediator uses a timeout for each episode. The timeout is restarted after each pause, and dedicated triggers are developed for each separate episode. For instance, in the mediator agent model (Fig. 1), relevant states for ending an episode (X) are the communication on feeling and/or intention for an agent Z, WS(Z,e,b) and WS(Z,s,a,e) respectively, and a belief on the timeout trigger for this episode, belief(timeout(episode(X))). These states are combined to form a belief state on the progress of the episode: belief(progress(episode(X))). Also, the belief on the progress is safeguarded against reacting to communication on feeling and/or intention before entering the episode. In the model this is realized by the state belief(monitoring(episode(X))), that suppresses the belief on progress during the pre-episode timeframe.

(2) Each belief state on the progress of the episode uses an inhibiting connection, which causes the end of its related episode. In the mediator agent model (Fig. 1) the state belief(progress(episode(X))) suppresses the state desire(episode(X)), for ending episode(X). The desire(episode(X)) state maintains a high activation level as long as episode X has not yet been completed. To realize this, desire(episode(X)) uses a positive feedback-loop (and a related combination function) for balancing the inhibiting connection from state belief(progress(episode(X))). This feedback-loop is necessary because otherwise zero suppression (an inhibiting connection emanating from a state with a zero-valued activation level) can have an unwanted downward effect on a connected state with a non-zero (positive) activation level.

(3) Each (major and minor) episode suppresses its directly subsequent (major and minor) episode(s). For example, in a two-episode scenario, the first episode is represented in the mediator agent model (Fig. 1) by intention(episode(1)) and the second episode by intention(episode(2)). In this case, intention(episode(2)), is suppressed by inhibiting connections from the preceding major episode, intention(episode(1)), and by the minor pause episode, intention(pause). As soon as previous suppressions fall away and the second episode has not yet been

completed, the intention(episode(2)) state achieves a high activation level through the connection with its generator state desire(episode(2)).

(4) Each major episode determines the focus for either or both social agents. The major episodes are represented in the model by states intention(episode(X)). Ultimately, an active episode determines the focus for either or both social agents. For example, the focus for the first episode is agent A, the second episode focuses on agent B, and a third episode on both agents A and B. In the first episode the mediator agent communicates this to agent A as a context with the state with label EC(A,c). In the second episode the focus is communicated to agent B through EC(B,c). The focus for the third episode is communicated by high activation levels for both EC(A,c) and EC(B,c).

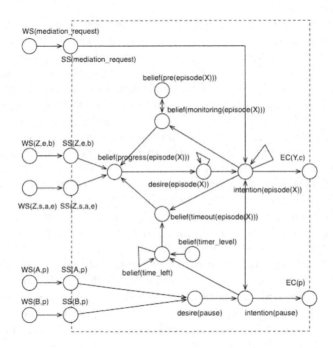

**Fig. 1.** Overview of the mediator agent model

As mentioned earlier, between each of the major episodes represented by intention(episode(X)), a special (minor) pause episode enables the agents to come to rest before starting the next major episode. This pause episode is indicated in the model by intention(pause). It should be noted that the extended social agent model, through body language, expresses any preparation activity by an observable execution state ES(p). This ES(p) state maintains a high activation level when the social agent internally prepares for an action or an emotion expression. The mediator agent observes the ES(p) states of social agents A and B through world states WS(A,p) and WS(B,p) respectively. As both social agents

**Table 2.** Instantiation of Mediator Agent model for three episodes

| Episode (X) | Focus (Y) | Empathy (Z) |
|:-----------:|:---------:|:-----------:|
| 1 | A | B |
| 2 | B | A |
| 3 | A, B | - |

**Table 3.** State properties used in the mediator agent model

| Notation | Description |
|:--------:|:------------|
| WS(mediation_request) | world state for communication of a mediation request |
| WS(A, s, a, e) | world state for communication by agent B of ownership for agent A of action a with effect e and stimulus s |
| WS(A, e, b) | world state for communication by agent B of ownership for agent A of emotion indicated by body state b and effect e |
| WS(A, p) | world state for body state of agent A expressing preparation activity p |
| SS(W) | sensor state for world state W |
| belief(progress(episode(X))) | belief state for ending episode with sequence number X |
| belief(monitoring(episode(X))) | belief state for suppression of belief on progress for episode X, before episode with sequence number X occurs |
| belief(pre(episode(X))) | belief state indicating the pre-episode timeframe for episode with sequence number X |
| belief(timeout(episode(X))) | belief state for timeout trigger for episode with sequence number X |
| belief(time_left) | belief state about the time left |
| belief(timer_level) | belief state about the limit level; time-left decreases until it reaches the timer-level |
| desire(episode(X)) | desire state for episode with sequence number X; maintains a high activation level as long as episode X has not yet been completed |
| desire(pause) | desire state for pause episode |
| intention(episode(X)) | intention state for episode with sequence number X |
| intention(pause) | intention state for pause episode |
| EC(A, c) | communication to agent A of context c with focus on agent A |
| EC(B, c) | communication to agent B of context c with focus on agent B |
| EC(p) | communication to agent A and B of pause episode |

A and B should come to rest before starting a next (major) episode, these world states are combined into a single state desire(pause). The desire(pause) state maintains a high activation level as soon as either of the social agents indicates any preparation activity, usually from the beginning of a major episode until the agent comes to rest again. Each major episode intention(episode(X)) suppresses the pause episode, intention(pause); after these suppressions fall away, intention(pause) takes over as long as its desire(pause) state maintains a high

activation level. The mediator agent communicates pause episodes to the social
agents by a high activation level for EC(p).

The state properties used in the mediator agent model are summarized in
Table 3. The connections between state properties (the arrows in Fig. 1) have
weights, as indicated in Table 4.

**Table 4.** Overview of the connections, weights, and settings used for (initial activation) level, threshold $\tau$ and steepness $\sigma$ parameters in the mediator agent model.

| From state | Weight | To state | Level | $\tau$ | $\sigma$ |
|---|---|---|---|---|---|
| - | - | WS(mediation_request) | 1 | - | - |
| WS(W) | $\omega_{0W}$ | SS(W) | 0 | - | - |
| belief(progress(episode(X))) | $\omega_{11X}$ | desire(episode(X)) | 1 | 0.5 | 20 |
| desire(episode(X)) | $\omega_{12X}$ | | | | |
| SS(A,p) | $\omega_{21}$ | desire(pause) | 0 | 0 | 20 |
| SS(B,p) | $\omega_{22}$ | | | | |
| SS(mediation_request) | $\omega_{31X}$ | intention(episode(X)) | 0 | 1.9 | 80 |
| intention(episode(X'<X)) | $\omega_{32X'X}$ | | | | |
| intention(pause) | $\omega_{33X}$ | | | | |
| desire(episode(X)) | $\omega_{34X}$ | | | | |
| desire(pause) | $\omega_{41}$ | intention(pause) | 0 | 0.1 | 20 |
| intention(episode(X)) | $\omega_{42X}$ | | | | |
| SS(Z,s,a,e) | $\omega_{51ZX}$ | belief(progress(episode(X))) | 0 | 0.5 | 20 |
| SS(Z,e,b) | $\omega_{52ZX}$ | | | | |
| belief(monitoring(episode(X))) | $\omega_{53X}$ | | | | |
| belief(timeout(episode(X))) | $\omega_{54X}$ | | | | |
| intention(episode(X)) | $\omega_{61X}$ | belief(monitoring(episode(X))) | 1 | 0.2 | 20 |
| belief(pre(episode(X))) | $\omega_{62X}$ | | | | |
| belief(monitoring(episode(X))) | $\omega_{63X}$ | belief(pre(episode(X))) | 1 | - | - |
| intention(episode(X)) | $\omega_{71X}$ | belief(timeout(episode(X))) | 0 | 0.3 | 10 |
| belief(time_left) | $\omega_{72X}$ | | | | |
| belief(timer_level) | $\omega_{73}$ | belief(time_left) | 1 | 0.05 | 2 |
| intention(pause) | $\omega_{74}$ | | | | |
| belief(time_left) | $\omega_{75}$ | | | | |
| intention(episode(X)) | $\omega_{8XY}$ | EC(Y,c) | 0 | 0.5 | 20 |
| intention(pause) | $\omega_9$ | EC(p) | 0 | 0.5 | 20 |

# 6   Agent Interaction

As discussed earlier, because of the elicitation criterion, empathy is in principle
an asymmetrical situation between two interacting individuals. In our configuration, two social agents A and B represent these two interacting individuals.
In addition, a mediator agent introduces the asymmetry that is necessary for

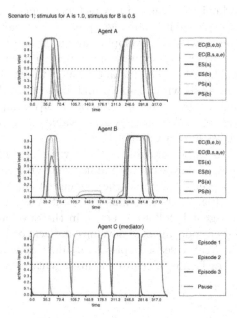

**Fig. 2.** Scenario 1: agent B depends on agent A; the mediator agent applies three episodes, see also Table 2. In each graph the vertical axis shows activation level, the horizontal axis shows time. Top: agent A, middle: agent B, bottom: mediator agent.

establishing empathy between these two social agents. The mediator agent introduces episodes in which one of the social agents is established as the focus of the interaction, and the other social agent is tested for communicating acknowledgement of emotion and action tendency as part of an empathic response. For establishing the concept of mutual empathic understanding, at least two of these episodes are necessary. After these two asymmetrical episodes, a third episode may re-establish symmetry for normal agent interaction. In order to create independent episodes, the mediator may introduce short pauses in which the social agents can come to rest by reducing their preparation activity.

The interactions between the agents, flowing through body and world states, comprise: (1) expression of feeling and intention, (2) verbal empathic responses for ownership of feeling and intention, (3) expression of preparation activity for feeling and/or intention, (4) verbal communication of episode focus, and (5) verbal invitation for suppression of preparation activity.

## 7   Simulation Results

The mediator model supports multiple episodes. In order to establish true empathy between two social agents, at least two independent episodes are necessary. In each of these two episodes, one of the social agents is selected as being in focus, and the other agent is tested for developing true empathy with regard to the agent in focus. A third episode is added in which both agents are in focus simultaneously. In this third episode it is no longer possible to establish true empathy, as both agents may be capable of developing intention and feeling without any specific agent as focus, whereas they may or may not process influences of each other during their development. The instantiation of the mediator agent model for these three episodes is shown in Table 2 and Table 4. In the simulation, the following weights from Table 4 have a value of -1: $\omega_{11X}$, $\omega_{12X}$, $\omega_{32XX}$, $\omega_{33X}$, $\omega_{42X}$, $\omega_{53X}$, $\omega_{61X}$, and $\omega_{72X}$. Weight $\omega_{71X}$ has a value of 0.4 and $\omega_{73}$ has a value of 0.05. All other weights have a value of 1.

The stimulus plays an important role in the development of intention and, in relation to that, feeling. A low stimulus implies that an agent depends on the other agent for action preparation and the associated emotional response and feeling. In Scenario 1 (Fig. 2) the stimulus for agent A is high and for agent B low. This means that agent B follows agent A in the development of intention and feeling. In the first episode, agent B develops and acknowledges full empathy. As agent B is in focus for the second episode, its development is insufficient to express intention and/or feeling. Because of this, agent A does not develop any intention or feeling.

## 8   Discussion

This paper presents a dynamical systems model for assisting social agents in reaching a joint decision involving mutually acknowledged empathic understanding. The model addresses the criteria of empathic understanding by introducing episodes in which one of the social agents is selected as being in focus, and the other agent is tested for developing genuine empathy. Acknowledged empathy, expressing empathic understanding of how the other agent feels about a considered option, is an important factor in solidly grounded joint decisions (see also [5,12]).

The main contribution of this model is that the mediator agent addresses all of the four criteria as formulated in [4,11] for establishing empathy. This paper takes the approach that in order to adhere to all four criteria for empathy it is not sufficient to just register nonverbal and verbal expressions; the mediator agent has to define a focus in order to gain insight in the way that these empathic responses come into existence. The presented mediator model operationalizes this focus by (verbally) creating a context in which one of the interacting agents is able to ignore the world stimulus so that this agent can fully focus on the other agent. The simulations show that the model does what it is conceived to do, in that mutually acknowledged empathy can be established in accordance with the four criteria.

The computational architecture as used in this paper does not mean that the results are dependent on the specific design and implementation aspects for the social agents. The mediator agent depends only on external aspects that are assumed to exist in human interaction, such as the expression of intention and emotion, and communication.

The computational model contributed in this paper may provide a basis to further explore the development of support for joint decision making processes, for example in the form of a mediation assistant agent. Such an assistant may provide analysis and give process advice in order to develop a joint decision, and take care that no escalating conflicts arise. This will be a direction of future research.

# References

1. Ashby, W.R.: Design for a Brain. Chapman and Hall, London (1952)
2. Beer, R.D.: On the dynamics of small continuous-time recurrent neural networks. Adaptive Behavior 3, 469–509 (1995)
3. Damasio, A.R.: Descartes' Error: Emotion, Reason and the Human Brain. Papermac, London (1994)
4. De Vignemont, F., Singer, T.: The empathic brain: how, when and why? Trends in Cogn. Sciences 10, 437–443 (2006)
5. Duell, R., Treur, J.: A Computational Analysis of Joint Decision Making Processes. In: Aberer, K., Flache, A., Jager, W., Liu, L., Tang, J., Guéret, C. (eds.) SocInfo 2012. LNCS, vol. 7710, pp. 292–308. Springer, Heidelberg (2012)
6. Hesslow, G.: Conscious thought as simulation of behaviour and perception. Trends Cogn. Sci. 6, 242–247 (2002)
7. Hopfield, J.J.: Neural networks and physical systems with emergent collective computational properties. Proc. Nat. Acad. Sci (USA) 79, 2554–2558 (1982)
8. Hopfield, J.J.: Neurons with graded response have collective computational properties like those of two-state neurons. Proc. Nat. Acad. Sci (USA) 81, 3088–3092 (1984)
9. Iacoboni, M.: Mirroring People: the New Science of How We Connect with Others. Farrar, Straus & Giroux, New York (2008)
10. Port, R.F., van Gelder, T.: Mind as motion: Explorations in the dynamics of cognition. MIT Press, Cambridge (1995)
11. Singer, T., Leiberg, S.: Sharing the Emotions of Others: The Neural Bases of Empathy. In: Gazzaniga, M.S. (ed.) The Cognitive Neurosciences, 4th edn., pp. 973–986. MIT Press (2009)
12. Treur, J.: Modelling Joint Decision Making Processes Involving Emotion-Related Valuing and Empathic Understanding. In: Kinny, D., Hsu, J.Y.-j., Governatori, G., Ghose, A.K. (eds.) PRIMA 2011. LNCS, vol. 7047, pp. 410–423. Springer, Heidelberg (2011)
13. Treur, J.: An Integrative Dynamical Systems Perspective on Emotions. Biologically Inspired Cognitive Architectures Journal 4, 27–40 (2013)

# Real-Time Head Pose Estimation Using Weighted Random Forests

Hyunduk Kim, Myoung-Kyu Sohn, Dong-Ju Kim, and Nuri Ryu

Dept. of Convergence, Daegu Gyeongbuk Institute of Science & Technology (DGIST),
50-1 Sang-Ri, Hyeongpung-Myeon, Dalseong-Gun, Daegu, 711-873, Korea

**Abstract.** In this paper we proposed to real-time head pose estimation based on weighted random forests. In order to make real-time and accurate classification, weighted random forests classifier, was employed. In the training process, we calculate accuracy estimation using preselected out-of-bag data. The accuracy estimation determine the weight vector in each tree, and improve the accuracy of classification when the testing process. Moreover, in order to make robust to illumination variance, binary pattern operators were used for preprocessing. Experiments on public databases show the advantages of this method over other algorithm in terms of accuracy and illumination invariance.

**Keywords:** Head pose estimation, Random Forests, Real time, Illumination invariant.

## 1    Introduction

Recently, estimation of the human head pose has become an intriguing and actively addressed research topic, inspired by the increasing demands of many human-head-related computer vision applications such as human-computing interfaces (HCI), driver surveillance systems, entertainment systems, and so on. Over the last decade, great effort has been dedicated to develop efficient, accurate, and robust algorithms for this problem. Nonetheless, most of the approaches suffer from the flaws in terms of efficiency, accuracy, or robustness against partial occlusion and the variations of head pose, illumination, and facial expression. Due to its relevance and to the challenges posed by the problem, there has been considerable effort in the computer vision community to develop fast and reliable algorithms for head pose estimation [1]. The several approaches to head pose estimation can be briefly divided into two categories: feature-based approaches and appearance-based approaches.

The feature-based approaches combine the location of facial features (e.g. eyes, mouth, and nose tip) and a geometrical face model to calculate precise angles of head orientation [2]. In general, these approaches can provide accurate estimation results for a limited range of poses. However, these approaches have difficulty dealing with low-resolution images due to invisible or undetectable facial points. Moreover, these approaches depend on the accurate detection of facial points. Hence, these approaches

D. Hwang et al. (Eds.): ICCCI 2014, LNAI 8733, pp. 554–562, 2014.

are typically more sensitive to occlusion than appearance-based methods, which use information from the entire facial region [3].

The appearance-based approaches discretize the head poses and learn a separate detector for each pose using machine learning techniques that determine the head poses from entire face images [3]. These approaches include multi-detector methods, manifold embedding methods, and non-linear regression methods. Generally, multi-detector methods train a series of head detectors each attuned to a specific pose and assign a discrete pose to the detector with the greatest support [1, 4]. Manifold embedding based methods seek low-dimensional manifolds that model the continuous variation in head pose. These methods are either linear or nonlinear approaches. The linear techniques have an advantage in that embedding can be performed by matrix multiplication; however, these techniques lack the representational ability of the non-linear techniques [1, 5]. Non-linear regression methods use nonlinear regression tools (e.g. Support Vector Regression, neural networks) to develop a functional mapping from the image or feature data to a head pose measurement. These approaches are very fast, work well in the near-field, and give some of the most accurate head pose estimates in practice. However, they are prone to error from poor head localization [1, 6].

In this paper, we propose a novel head pose estimation algorithm for gray-level images; this method consists of two techniques. First weighted random forests were employed. Second, in order overcome the problem caused by illumination variation on face, binary pattern operators were employed for preprocessing. The experimental results show the efficiency of the proposed algorithm. The remainder of this paper is organized as follows. Previous works regarding to the head pose estimation are introduced in Section 2. After that, the proposed weighted random forest is described in detail in Section 3. Experiments results and a discussion of those results are reported in Section 4. Finally, the paper concludes in Section 5.

## 2 Related Work

### 2.1 Feature-Based Approach

In the feature-based methods, the head pose is inferred from the extracted features, which include the common feature visible in all poses, the pose-dependent feature, and the discriminant feature together with the appearance information.

Vatahska et al. [7] use a face detector to roughly classify the pose as frontal, left, or right profile. After his, they detect the eyes and nose tip using AdaBoost classifiers, and the detections are fed into a neural network which estimates the head orientation. Whitehill et al. [8] present a discriminative approach to frame-by-frame head pose estimation. Their algorithm relies on the detection of the nose tip and both eyes, thereby limiting the recognizable poses to the ones where both eyes are visible. Yao and Cham [9] propose an efficient method that estimates the motion parameters of a human head from a video sequence by using a three-layer linear iterative process. Morency et al. [10] propose a probabilistic framework called Generalized Adaptive View-based Appearance Model integrating frame-by-frame head pose estimation, differential registration, and keyframe tracking.

## 2.2     Appearance-Based Approach

In the appearance-based methods, the entire face region is analyzed. The representative methods of this type include the manifold embedding method, the flexible-model-based method, and the machine-learning-based method. The performance of both kinds of methods may deteriorate as a consequence of feature occlusion and the variation of illumination, owing to the intrinsic shortcoming of 2D data. Generally, the appearance-based methods outperform the feature-based methods, because the latter rely on the error-prone facial feature extraction.

Balasubramanian et al. [11] propose the Biased Manifold Embedding (BME) framework, which uses the pose angle information of the face images to compute a biased neighborhood of each point in the feature space, before determining the low-dimensional embedding. Huang et al. [12] present Supervised Local Subspace Learning (SL2), a method that learns a local linear model from a sparse and non-uniformly sampled training set. SL2 learns a mixture of local tangent spaces that is robust to under-sampled regions, and due to its regularization properties it is also robust to over-fitting. Osadchy et al. [13] describe a method for simultaneously detecting faces and estimating their pose in real time. The method employs a convolutional network to map images of faces to points on a low-dimensional manifold parameterized by pose, and images of non-faces to points far away from that manifold.

# 3     Proposed Head Pose Estimation Algorithm

## 3.1     Random Forests Framework

The Random Forests (RF) algorithm consists of two steps; training and testing. The training step is basically to construct multi tree-shaped classifiers, which involves data induction, construction of tree-shaped structure, and optimization of parameters. In testing step, the intermediate outcomes generated by the trees are integrated to provide the final result.

A tree $T$ in a forest $F = \{T_i\}$ is built from the set of annotated samples $P = \{P_i = (I_i, c_i)\}$ randomly extracted from the training images, where $I_i$ and $c_i$ are the intensity of the training images and the annotated head pose class labels, respectively. Starting from the root, each tree is built recursively by assigning a binary test $\phi(I) \rightarrow \{0, 1\}$ to each non-leaf node. Such test sends each sample either to the left or right child, in this way the training samples $P$ arriving at the node are split into two sets, $PL(\phi)$ and $PR(\phi)$.

The best test $\phi*$ is chosen from a pool of randomly generated ones ($\{\phi\}$): all samples arriving at the node are evaluated by all tests in the pool and a predefined information gain of the split $IG(\phi)$ is maximized:

$$\phi^* = Arg \max_\phi IG(\phi).$$     (1)

The process continues with the left and the right child using the corresponding training sets $PL(\phi*)$ and $PR(\phi*)$ until a leaf is created when either the maximum tree depth is reached, or less than a minimum number of training samples are left .

## 3.2    Training Step

The essence of the RF is to model the problem of head pose estimation as a classification problem, and to build a non-linear mapping of a set of depth images into a set of head pose class label. The mapping is accomplished by a set of tree-shaped classifiers, which shows better performance and is less prone to over-fit than an individual classifier as reported in [14]. The whole procedure of the training is described as follows.

The construction of the tree-shaped classifiers is characterized by the rule that each tree is built independently and is arbitrarily added to the existing trees. In our words, all the trees are trained with the same parameters but on different training sets (=N). These sets are generated from the original training set using the bagging strategy. In this strategy, samples are sampled randomly from the entire data pool with replacement until enough samples are collected. That is, some sample data will occur more than once and some will be absent.

When the training set for the current tree is drawn by sampling using bagging strategy, some samples are left out, called out-of-bag data (N/3), which are used for determining weight of current tree after construction of the tree. In this strategy, the out-of-bag data is selected to be the internal evaluation data for the on-growing forest. This choice makes efficient use of the data pool in that it averts the need of assigning an exclusive subset of the data pool for validation; otherwise, datasets with shrunk size are used for training and testing purposes. Furthermore, since the set of out-of-bag data is different for each classifier, the diversity of data for prediction is improved, which result in a more comprehensive assessment of the current forest and thus better performance of the classifiers, as reported in [15].

To overcome the problem caused by illumination variation on face, various approaches have been introduced, such as preprocessing and illumination normalization techniques, illumination invariant feature extraction techniques, and 3D face modeling techniques. Among above mentioned approaches, local binary pattern (LBP) [16] has received increasing interest for face representation in general [17]. LBP was originally proposed for texture description [18], and has been widely exploited in many applications such as video retrieval, aerial image analysis, and visual inspection. In addition, centralized binary pattern (CBP) [19] and center-symmetric local binary pattern (CS-LBP) [20] introduced for facial expression recognition and image representation.

We employ the binary tests $\phi_{F_1,F_2,\tau}(I)$ as follow:

$$\phi_{F_1,F_2,\tau}(I) = \begin{cases} |F_1|^{-1} \sum_{q \in F_1} I(q) - |F_2|^{-1} \sum_{q \in F_2} I(q) > \tau, & \text{go to left child} \\ \text{otherwise,} & \text{go th right child} \end{cases}, \qquad (2)$$

where $I$ is the gray level, $F_1$ and $F_2$ are two asymmetric rectangles defined within the sample, and $\tau$ is a threshold. This binary test is the difference between the average values of two rectangular areas rather than single pixel differences in order to sensitive to noise.

During training, for each non-leaf node starting from the root, we generate a large pool of binary tests $\{\phi_k\}$ by randomly choosing parameters $F_1$, $F_2$, and $\tau$. For efficiency reason, the number of binary tests is determined depend on the depth of the tree. That is, the number of the binary test increases with increasing the depth of the tree. In this strategy, the samples are split roughly at the beginning levels, and are divided more finely at deeper levels. The binary test which maximizes a specific optimization function, which is called information gain, is picked.

Our information gain $IG(\phi)$ is defined as follows:

$$IG(\phi) = \sum\nolimits_{i \in \{L,R\}} (\mu_i - \mu)^2 - \sum\nolimits_{i \in \{L,R\}} \frac{n_i}{n_i + n_j} \left[ \sum_{j=1}^{n_i} (c_{ij} - \mu_i)^2 \right], \qquad (3)$$

where $n_i$ and $\mu_i$ are the number of samples and the mean of class at the child node $i$, respectively, $c_{ij}$ is the head pose class label of the $j$-th patch contained in child node $i$, and $\mu$ is the mean of class at the parent node. The information gain $IG(\phi)$ indicates the difference between the within variance and weighted between variance.

For each leaf, class distribution $p(c|T)$ and class label $c_{max}$ that received the majority of votes is stored. The distributions are estimated from the training samples that arrive at the leaf and are used for estimation the head pose.

After each tree has been trained, for each vector that included in out-of-bag data set, find the class that has got the majority of votes in the tree and compare it to the ground-truth class label. The weight vector of tree is computed by a ratio of the number of correct classified out-of-bag data to all the vectors in the out-of-bag data set.

## 3.3    Testing Step

Given a new gray image of a head, image is guided by the binary tests stored at the nodes. At each node of a tree, the stored binary test evaluates the input image, sending it either to the right of left child, all the way down until a leaf. Arriving at a leaf, a tree outputs the weight vector $w$, class distribution $p(c|T)$ and class label $c_{max}$. Because leaves with a low probability are not very informative and mainly add noise to the estimate, we discard all votes if $p(c_{max}|T)$ less than an empiric threshold $P_{max}$. The final class distribution is generated by weighted arithmetic averaging of each remained distribution of all trees as follows:

$$p(c_i \mid F) = \frac{1}{|F|} \sum_{t=1}^{|F|} w_t(c_i) p(c_i \mid T_t). \qquad (4)$$

Here, $w_t(c_i)$ means the weight correspond to $t$-th tree and $i$-th class. We choose $c_i$ as the final class of an input image if $p(c_i|F)$ has the maximum value.

# 4    Experiments

In order to evaluate the performance of our algorithm, we performed a 5-fold, subject-independent cross-validation based on the CMU Multi-PIE database, which contains more than 750,000 images of 337 people recorded in up to four sessions over the span of five months. Subject were imaged under 15 view points and 19 illumination conditions while displaying a range of facial expressions [14]. In our paper, first session, 249 person, 2 expressions (neutral and smile), 19 illuminations and 7 view points, which consist of $0°$, $\pm15°$, $\pm30°$, and $\pm45°$, were employed. All of these face images were cropped to 32×32. Fig. 1 shows an example of the CMU Multi-PIE databases. In Fig. 1, top row is neutral expression and bottom row is smile expression, and columns show the head pose classes. Fig. 2 shows sample images transformed by various binary pattern operators such as LBP, CBP, and CS-LBP, respectively.

**Fig. 1.** Example of CMU Multi-PIE databases

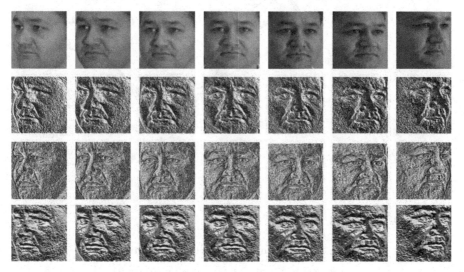

**Fig. 2.** Example of various binary pattern images

Training a forest involves the choice of several parameters. We always stop growing a tree when the depth reaches 15, or if there are less than 10 samples left for training. Moreover, we set other parameters as follows. The number of trees in the forest is 100; the number of training images for each tree is 15000; the number of out-of-bag data for each tree is 5000; the maximum number of binary tests is 2250, i.e., 150 different combinations of $F_1$ and $F_2$, each with 15 different threshold $\tau$; the maximum size of the sub-patches defining the areas $F_1$ and $F_2$ in the tests is half the size of the sample.

**Table 1.** Comparison of classification accuracies of different algorithm

| Algorithm | Raw | LBP | CBP | CS-LBP |
|-----------|-----|-----|-----|--------|
| RF | 80.0% | 80.6% | 81.7% | 83.4% |
| Weighted RF | 81.4% | 82.5% | 83.5% | 84.9% |

**Table 2.** Comparison of classification accuracies of different class

| Algorithm | Class1 | Class2 | Class3 | Class4 | Class5 | Class6 | Class7 |
|-----------|--------|--------|--------|--------|--------|--------|--------|
| Raw | 75.3% | 76.7% | 95.4% | 78.8% | 73.3% | 86.2% | 84.3% |
| LBP | 76.1% | 79.8% | 91.7% | 80.6% | 75.6% | 84.2% | 89.2% |
| CBP | 78.1% | 72.2% | 98.8% | 83.0% | 78.1% | 88.9% | 85.2% |
| CS-LBP | 79.4% | 80.4% | 94.0% | 80.4% | 76.4% | 83.2% | 90.4% |

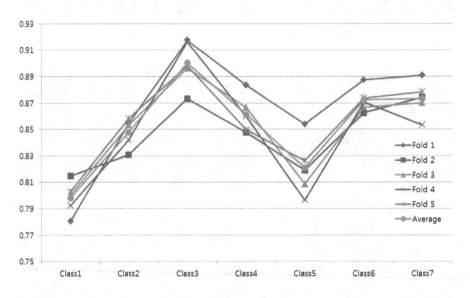

**Fig. 3.** The average of weights used in our experiment

Table 1 show the comparison results of the classification accuracies of different binary pattern between original Random Forests and weighted Random Forests. Because of illumination change, the result of the LBP, CBP, CS-LBP image were better than those of the raw image. As a result, the weighted Random Forests has performance better than that of original Random Forests about 1.4%, 1.9%, 1.8%, and 1.5%, correspond to raw, LBP, CBP and CS-LBP, respectively.

Fig. 3 show the average of weights used in our experiment and table 2 show the comparison of classification accuracies of different class with weighted Random Forests correspond to raw, LBP, CBP, CS-LBP, respectively. In the Fig.3, one can observe that the curves have similar trends for every fold. In particularly, the class 3 has the largest weight about 0.9 and the class 1 has smallest weight about 0.8. For these reasons, the trends of classification accuracies are similar to that of the average of weights. In particularly, class 3 has the highest classification accuracy and the class 1 has the lowest classification accuracy in the table 2.

# 5    Conclusion

In this paper we proposed to real-time head pose estimation based on weighted random forests. In order to make real-time and accurate classification, weighted random forests classifier was employed. In the training process, we calculate accuracy estimation using preselected out-of-bag data. The accuracy estimation determine the weight vector in each tree, and improve the accuracy of classification when the testing process. Moreover, in order to make robust to illumination variance, binary pattern operators were used for preprocessing. Experiments on public databases show the advantages of this method over other algorithm in terms of accuracy and illumination invariance.

**Acknowledgment.** This work was supported by the DGIST R&D Program of the Ministry of Education, Science and Technology of Korea (14-IT-03). It was also supported by Ministry of Culture, Sports and Tourism (MCST) and Korea Creative Content Agency (KOCCA) in the Culture Technology (CT) Research & Development Program (Immersive Game Contents CT Co-Research Center).

# References

1. Murphy-Chutorian, E., Trivedi, M.M.: Head pose estimation in computer vision: A survey. IEEE Transaction. on Pattern Analysis and Machine Intelligence 31(4), 607–626 (2009)
2. Gee, A.: Cipolla. R.: Determining the gaze of faces in images. Image and Vision Computing 12(10), 639–647 (1994)
3. Li, Y., Wang, S., Ding, X.: Person-independent head pose estimation based on random forest regression. In: 17th IEEE International Conference on Image Processing, pp. 1521–1524 (2010)
4. Huang, C., Ai, H., Li, Y., Lao, S.: High-performance rotation invariant multiview face detection. IEEE Transaction on Pattern Analysis and Machine Intelligence 29(4), 671–686 (2007)
5. Raytchev, B., Yoda, I., Sakaue, K.: Head pose estimation by nonlinear manifold learning. In: 17th IEEE International Conference on Pattern Recognition, pp. 462–466 (2004)

6.  Li, Y., Gong, S., Liddell, H.: Support vector regression and classification based multi-view face detection and recognition. In: 4th IEEE International Conference on Automatic Face and Gesture Recognition, pp. 300–305 (2000)
7.  Vatahska, T., Bennewitz, M., Behnke, S.: Feature-based head pose estimation from images. In: 7th IEEE-RAS International Conference on Humanoid Robots, pp. 330–335 (2007)
8.  Whitehill, J., Movellan, J.R.: A discriminative approach to frame-by-frame head pose tracking. In: 8th IEEE International Conference on Automatic Face and Gesture Recognition, pp. 1–7 (2008)
9.  Yao, J., Cham, W.K.: Efficient model-based linear head motion recovery from movies. In: IEEE International Conference on Computer Vision and Pattern Recognition, pp. 414–421 (2004)
10. Morency, L., Whitehill, J., Movellan, J.: Generalized adaptive view-based appearance model: Integrated framework for monocular head pose estimation. In: 8th IEEE International Conference on Automatic Face and Gesture Recognition, pp. 1–8 (2008)
11. Balasubramanian, V.N., Te, J., Panchanathan, S.: Biased manifold embedding: A framework for person-independent head pose estimation. In: IEEE International Conference on Computer Vision and Pattern Recognition, pp. 1–7 (2007)
12. Huang, D., Storer, M., De la Torre, F., Bischof, H.: Supervised local subspace learning for continuous head pose estimation. In: IEEE International Conference on Computer Vision and Pattern Recognition, pp. 2921–2928 (2011)
13. Osadchy, M., Cun, Y.L., Miller, M.L.: Synergistic face detection and pose estimation with energy-based models. Machine Learning Research 8, 1197–1215 (2007)
14. Breiman, L.: Random Forests. Machine learning 45(1), 5–32 (2001)
15. Ying, Y., Wang, H.: Dynamic random regression forests for real-time head pose estimation. Machine Vision and Applications 24(8), 1705–1719 (2013)
16. Ojala, T., Pietikainen, M., Maenpaa, T.: Multi-resolution Gray-scale and Rotation Invariant Texture Classification with Local Binary Patterns. IEEE Transaction on Pattern Analysis and Machine Intelligence 24(7), 971–987 (2002)
17. Ahonen, T., Hadid, A., Pietikainen, M.: Face Description with Local Binary Patterns: Application to Face Recognition. IEEE Transaction on Pattern Analysis and Machine Intelligence 28(12), 2037–2041 (2006)
18. Ojala, T., Pietikainen, M., Harwood, D.: A Comparative Study of Texture Measures with Classification based on Featured Distribution. Pattern Recognition 29(1), 51–59 (1996)
19. Fu, X., Wei, W.: Centralized Binary Patterns Embedded with Image Euclidean Distance for Facial Expression Recognition. In: 4th IEEE International Conference of Neural Computation, pp. 115–119 (2008)
20. Heikkilä, M., Pietikäinen, M., Schmid, C.: Description of Interest Regions with Center-Symmetric Local Binary Patterns. In: Kalra, P.K., Peleg, S. (eds.) ICVGIP 2006. LNCS, vol. 4338, pp. 58–69. Springer, Heidelberg (2006)

# An Integer Programming Approach for Two-Sided Matching with Indifferences

Naoki Ohta and Kazuhiro Kuwabara

College of Information Science and Engineering, Ritsumeikan University,
1-1-1 Noji Higashi, Kusatsu, Shiga 525-8577 Japan
n-ohta@fc.ritsumei.ac.jp, kuwabara@is.ritsumei.ac.jp

**Abstract.** To make use of the collective intelligence of many autonomous self-interested agents, it is important to form a team on which all the agents agree. Two-sided matching is one of the basic approaches to form a team that consists of agents from two disjoint agent groups. Traditional two-sided matching assumes that an agent has a totally ordered preference list of the agents it is to be paired with, but it is unrealistic to have a totally ordered list for a large-scale two-sided matching problem. In this paper, we propose an integer programming based approach to solve a two-sided matching program that allows indifferences in agents' preferences, and show how an objective function can be defined to find a matching that minimizes the maximum discontentedness of agents in one group.

**Keywords:** multi-agent system, two-sided matching, integer programming.

## 1 Introduction

Forming a team of autonomous agents is an important capability in a multi-agent system in order to make use of the collective capabilities of intelligent agents. Generally, an autonomous agent acts to maximize its utility. An agent will not form a team if forming a team does not increase the agent's utility.

Two-sided matching is one of the basic methods to form a team. In the two-sided matching problem, two disjoint agent groups exist. A member of one group is paired with one or multiple members of the other group. Each agent has a preference for certain agents of the other group to be paired with. A two-sided matching algorithm finds a matching that all agents agree with. Two-sided matching is used in a variety of fields in the real world, such as matching interns with hospitals [1,2,3,4]. We can expect that two-sided matching is used for resource allocation [5].

The Gale Shapley algorithm is one of the major algorithms used to solve the two-sided matching problem [1,4]. This algorithm assumes that an agent has a totally ordered preference list of the agents it is to be paired with. A matching found by this algorithm is known to be *stable*. Stable matching means that there is no pair of agents that collectively prefer another match.

D. Hwang et al. (Eds.): ICCCI 2014, LNAI 8733, pp. 563–572, 2014.

The two-sided matching problem can also be solved by Integer Programming (IP) [7,8]. By properly defining constraints in the IP, a stable matching can be calculated. Although the matching found by the Gale Shapley algorithm is known to result in the best matching for one group and the worst matching for the other group [6], various stable matchings can be obtained by using a different objective function when IP is used.

In traditional two-sided matching that uses the Gale Shapley algorithm, it is assumed that all agents have a totally ordered preference list of the agents in the other group. This assumption is somewhat unrealistic for a large scale two-sided matching. For example, let us suppose that there are 1000 job seekers and many companies that potentially offer jobs. When we consider this job matching problem as a two-side matching, it is very difficult for a company to determine the entire order of preferences for job seekers. To cope with this kind of issue, two-sided matching was extended to allow indifferences in the preference list of agents [9]. Three types of stability are identified for two-sided matching with indifferences [9].

In this paper, we propose an IP based algorithm for two-sided matching with indifferences. Three types of stable conditions [9] are encoded in the constraints of the IP. By defining a suitable objective function, we can find a matching such that the maximum of the discontentedness of agents in one group is minimized. The key idea of the proposed algorithm is based on the concept of "nucleolus" [10] in coalition game research.

The rest of this paper is organized as follows. First, we review the model of two-sided matching (Section 2). Next, we propose an algorithm of two-sided matching with indifferences that makes use of IP (Section 3). Finally, we present the conclusion of this paper (Section 4).

## 2   Two-Sided Matching

### 2.1   Traditional Two-Sided Matching

In this section, we illustrate traditional two-sided matching and a well-known algorithm called the Gale Shapley algorithm [1]. There are two finite and disjoint sets of agents $L$ and $R$. In two-sided matching, a member of $L$ is paired with one or multiple members of $R$, and a member of $R$ is paired with one or multiple members of $L$. Each agent $p$ ($p \in L \cup R$) has capacity $n_p$. $n_p$ is a natural number and represents the number of agents that can be matched from the other group. Therefore, $\forall l \in L$, $n_l \leq |R|$ and $\forall r \in R$, $n_r \leq |L|$ hold.

Every member of $L$ has preferences over $R$, and every member of $R$ has preferences over $L$.

**Definition 1 (preference).** *The preference over $X$ is a total order on $X$.*

Assume that there is an agent $p$ whose preference is denoted by $\succ_p$. If $a \succ_p b$ holds, $p$ prefers agent $a$ to agent $b$.

**Definition 2 (matching).** *Given a two-sided matching $(L, R)$, $(L, R)$'s matching $m$ is a function, that satisfies the following conditions:*

- *Its domain is $R \cup L$.*
- *$\forall r \in R$, $m(r) \subseteq L$ holds.*
- *$\forall l \in L$, $m(l) \subseteq R$ holds.*
- *$\forall p \in L \cup R$, $|m(p)| \le n_p$.*
- *$\forall q \in m(p)$, $p \in m(q)$ holds.*

Two-sided matching where $\forall p \in L \cup R$, $n_p = 1$ holds is called *one-to-one* matching, and two-sided matching where $\forall l \in L$, $n_l = 1$ or $\forall r \in R$, $n_r = 1$ holds, is called *many-to-one* matching.

There are a number of algorithms to solve two-sided matching problems. These algorithms find a matching with a good property such as *stable* matching.

**Definition 3 (the lowest pair).** *Given an agent $p$ $(p \in L \cup R)$ and matching $m$, the lowest pair $m_w(p)$ is an agent that satisfies the following conditions:*

- *$\forall q \in m(p) \setminus \{m_w(p)\}$, $q \succ_p m_w(p)$*

**Definition 4 (blocking pair).** *For matching $m$ of a given two-sided matching $(L, R)$, the blocking pair $(l, r)(l \in L$, $r \in R)$ is a pair of agents that satisfies the following conditions:*

- *$|m(l)| < n_l$ or $r \succ_l m_w(l)$ holds.*
- *$|m(r)| < n_r$ or $l \succ_r m_w(r)$ holds.*

If a blocking pair $(l, r)$ exists for a matching $m$, $l$ and $r$ have an incentive to collectively deviate from the matching $m$. Therefore, a matching with a blocking pair is not stable.

**Definition 5 (stable matching).** *Given a two-sided matching $(L, R)$, a matching $m$ is stable when there is no blocking pair of $m$.*

The Gale Shapley algorithm is a well-known algorithm for calculating stable matchings.

**Definition 6 (Gale Shapley Algorithm).** *The Gale Shapley algorithm calculates stable matchings as follows: (We call this algorithm "L proposed." If transposed $L$ to $R$ and $R$ to $L$, we call it "R proposed.")*

- *Given matching $m$ where $\forall p \in R \cup L$, $m(p) = \emptyset$*
- *While there is an agent $l \in L$ who satisfies $|m(l)| < n_l$ and has not proposed all members of $R$, execute the following procedure:*
  - *$l$ proposes agent $r$ that satisfies $\forall r' \in R' \setminus \{r\}$, $r \succ_l r'$ where $R'$ is a set of agents who have not been proposed by $l$.*
  - *$r$ replies as follows:*
    - *if $|m(r)| < n_r$ holds, $r$ accepts. $l$ joins $m(r)$ and $r$ joins $m(l)$.*
    - *if $|m(r)| = n_r$ and $m_w(r) \succ_r l$ hold, $r$ does not accept.*

&ast; *if $|m(r)| = n_r$ and $l \succ_r m_w(r)$ hold, $r$ accepts $l$'s proposal, and drops $m_w(r)$. $m_w(r)$ defects from $m(r)$,$r$ defects from $m(m_w(r))$, $l$ joins $m(r)$, and $r$ joins $m(l)$.*

− *If there is no agent who is not paired, and who has not proposed all members of $R$, this algorithm returns the matching $m$.*

The Gale Shapley algorithm finds a stable matching that is best for members of the proposing group [6].

These characteristics of the matching are desirable in such cases as assigning students to professors' seminars in a university, where the preferences of the students are usually placed before the professors', but for the cases where the two groups are on an equal basis, other characteristics are wanted.

IP can find a stable matching that has other characteristics by defining a proper objective function. Before discussing the objective function, we show how a two-sided matching problem can be represented in the IP framework. There are $|L| \times |R|$ variables $x(l,r)(\forall l \in L, \forall r \in R)$ $x(l,r) \in \{0,1\}$ holds. If $x(l,r)$ is 0, $l \notin m(r)$ and $r \notin m(l)$ hold. If $x(l,r)$ is 1, $l \in m(r)$ and $r \in m(l)$ hold.

The constraints that are needed to find a stable matching in one-to-one matching can be defined as follows [7].

**Definition 7 (constraints for stable matchings).** *If variables $x(l,r)(\forall l \in L, \forall r \in R)$ satisfy the following conditions, the matching is stable.*

− *$\forall l \in L$, $\sum_{r \in R} x(l,r) = 1$ holds.*
− *$\forall r \in R$, $\sum_{l \in L} x(l,r) = 1$ holds.*
− *$\forall l \in L$, $\forall r \in R$, $\sum_{l'(l' \in L,\ l \succ_r l')} x(l',r) + \sum_{r'(r' \in R,\ r \succ_l r')} x(l,r') \leq 0$ holds.*

Let us consider an example of an objective function. Assume that we define a function $r(p,q)$ to denote that $q$ is the $r(p,q)$-th best agent for agent $p$. When we define an objective function as minimizing $\sum_{l \in L, r \in R} r(l,r) \times r(r,l) \times x(l,r)$, a stable matching that maximizes the sum of all agents' satisfaction can be found.

We can find matchings with various characteristics by using the IP. However, the IP has a drawback. The IP is more complex than the Gale Shapley algorithm, and solving the IP problem is NP-hard in general. However, there are IP solvers such as IBM ILOG CPLEX, which solve the IP programs efficiently. By using these IP solvers, we can find matchings within a reasonable time.

## 2.2   Two-Sided Matching with Indifferences

In some applications, especially those with a large number of agents, the assumption that all agents have totally ordered preferences is unrealistic. Two-sided matching was extended to allow an agent to have indifferent preferences about agents in the other group [9].

**Definition 8 (indifferent preference).** *If there is a set of agents $X$, the indifferent preference over $X$ is a total order on $SX$, which is a partition of $X$.*

Assume that there is an agent $p$ whose preference $p$'s are represented as $\succ'_p$. $A \succ'_p B$ means that $p$ prefers a member of $A$ to a member of $B$. We will write it also as $a \succ''_p b$ if $a \in A$ and $b \in B$ holds. In addition, if two agents $p'$ and $p''$ are members of one group ($A$ or $B$), $p$ likes $p'$ as good as $p''$. In the following, we write an agent's preference as $(\{a, b\}, \{c, d, e\}, \{f, g\})$, which means the following:

- $A = \{a, b\}$, $B = \{c, d, e\}$, $C = \{f, g\}$.
- Assume that there are two agents $p, p'$ which are members of one group. The agent $p$ likes $p'$ as good as $p''$. (For example, the agent likes $c$ as good as $e$).
- The agent prefers a member of A to a member of B, a member of B to a member of C, and a member of A to a member of C.

Assume that there is an agent $p$, and that $p$ has an indifferent preference over $X$.

For two-sided matching with indifferences, three types of stable conditions are proposed [9].

**Definition 9 (weakly stable).** *If a pair of agents $(l, r)$ does not exist that satisfies the following conditions for a given matching $m$, $m$ is said to be* weakly stable.

- $r \notin m(l)$ and $l \notin m(r)$ hold.
- $|m(l)| < n_l$ or $m(l)$ has an agent $r'$ that $r \succ''_l r'$ holds.
- $|m(r)| < n_r$ or $m(r)$ has an agent $l'$ that $l \succ''_r l'$ holds.

**Definition 10 (strongly stable).** *If a pair of agents $(l, r)$ does not exist that satisfies the following conditions for a given matching $m$, $m$ is said to be* strongly stable.

- $r \notin m(l)$ and $l \notin m(r)$ hold.
- One of the following conditions is satisfied.
  - The following conditions are satisfied.
    * $|m(l)| < n_l$ or $m(l)$ has an agent $r'$ such that $r \succ''_l r'$ holds.
    * $|m(r)| < n_r$ holds, $m(l)$ has an agent $r'$ such that $r \succ''_l r'$ holds, or $l$ likes $r$ as good as $r'$.
  - The following conditions are satisfied.
    * $|m(l)| < n_l$ holds, $m(r)$ has an agent $l'$ such that $l \succ''_r l'$ holds, or $r$ likes $l$ as good as $l'$.
    * $|m(r)| < n_r$ or $m(l)$ has an agent $r'$ such that $r \succ''_l r'$ holds.

**Definition 11 (super stable).** *If a pair of agent $(l, r)$ does not exist for a given matching $m$, $m$ is said to be* super stable.

- $r \notin m(l)$ and $l \notin m(r)$ hold.
- $|m(l)| < n_l$ holds, $m(r)$ has an agent $l'$ that $l \succ''_r l'$ holds, or $r$ likes $l$ as good as $l'$.
- $|m(r)| < n_r$ holds, $m(l)$ has an agent $r'$ that $r \succ''_l r'$ holds, or $l$ likes $r$ as good as $r'$.

If we introduce an artificial ordering in the preferences with indifferences to make it totally ordered (for example, ties are randomly broken), we can apply the traditional two-sided matching algorithm such as the Gale Shapley algorithm to the modified two-sided matching problem. The resultant matching is weakly stable in the original two-sided matching problem in the above sense.

## 3    Integer Programming for Two-Sided Matching with Indifferences

### 3.1    Constraints for Stable Matching

To calculate a *stable* matching using IP, we need to define proper constraints. As mentioned in Section 2.1, a matching is expressed by $|L| \times |R|$ variables $x(l,r)$. We add new $|L|+|R|$ variables $e(p) \in \{0,1\}$ $(p \in X \cup Y)$ to represent the constraints for stable matchings in a many-to-many two-sided matching problem. Here, $e(p) = 1$ means that $m(p) < n_p$ holds (that is, agent $p$ still can accept more agent(s) from the other group), and $e(p) = 0$ means that $m(p) = n_p$ holds.

For a two-sided matching with indifferences, three types of stable matchings exist: weakly stable, strongly stable, and super stable [9]. First, we describe the constraints to find a weakly stable matching.

**Definition 12 (constraints for weakly stable matchings).** *If the following linear constraints are satisfied, the matching is weakly stable.*

(1) $\forall l \in L, \sum_{r \in R} x(l,r) \leq n_l$

(2) $\forall r \in R, \sum_{l \in L} x(l,r) \leq n_r$

(3) $\forall l \in L, \sum_{r \in R} x(l,r) + |R| \times e(l) \geq n_l$

(4) $\forall r \in R, \sum_{l \in L} x(l,r) + |L| \times e(r) \geq n_r$

(5) $\forall l \in L, \forall r \in R, -x(l,r) + e(l) + e(r) < 2$

(6) $\forall l \in L, \forall r \in R, \forall l' \in L \setminus \{l\}$ such that $l \succ''_r l'$ holds, $-x(l,r) + x(l',r) + e(l) < 2$

(7) $\forall l \in L, \forall r \in R, \forall r' \in R \setminus \{r\}$ such that $r \succ''_l r'$ holds, $-x(l,r) + x(l,r') + e(r) < 2$

(8) $\forall l \in L, \forall r \in R, \forall l' \in L \setminus \{l\}, \forall r' \in R \setminus \{r\}$ such that $l \succ''_r l'$ and $r \succ''_l r'$ holds, $-x(l,r) + x(l',r) + x(l,r') < 2$

**Theorem 1.** *A matching is weakly stable if and only if it satisfies the constraints described in definition 12.*

*Proof.* Constraints (1) and (2) require that $\forall p \in L \cup R, |m(p)| \leq n_q$ holds. Thus, if and only if variables $x(l,r)$ satisfy constraints (1) and (2), the matching satisfies the conditions in definition 2.

Constraints (3) and (4) describe the constraints regarding newly added variables $e(p)$, which require that $e(p) = 1$ $(p \in R \cup L)$ holds when $m(p) < n_p$.

In the following, we prove that the matching is weakly stable if $(l,r)$ satisfies the constraints (5), (6), (7), and (8) which are constraints for a pair of agents $l \in L$ and $r \in R$, $(l,r)$.

Constraint (5) requires that $x(l,r) = 0$, $e(l) = 1$, and $e(r) = 1$ do not hold at the same time. This means that $(l,r)$ does not satisfy the condition of: $x(l,r) = 0$, $|m(l)| < n_l$, and $|m(r)| < n_r$. If and only if constraint (5) is not satisfied, then $(l,r)$ satisfies the first condition, the first part of the second condition and the first part of the third condition in definition 9.

Constraint (6) requires that $x(l,r) = 0$, $x(l',r) = 1$, and $e(l) = 1$ hold when $l \succ''_r l'$. In other words, $(l,r)$ and $l'(l \succ''_r l')$ does not satisfy the condition of: $x(l,r) = 0$, $x(l',r) = 1$, and $|m(l)| < n_l$. If and only if constraint (6) is not satisfied, then $(l,r)$ satisfies the first condition, the first part of the second condition, and the second part of the third condition in definition 9.

Similarly, if and only if constraint (7) is not satisfied, $(l,r)$ satisfies the first condition, the first part of the second condition, and the second part of the third condition, and if and only if constraint (8) is not satisfied, then $(l,r)$ satisfies the first condition, the second part of the second condition, and the second part of the third condition.

Therefore, if and only if the matching satisfies the conditions in definition 12, is it weakly stable. ∎

To calculate a *strongly stable* matching, we need to change the conditions in definition 12 as follows:

- constraint (6): change "$l \succ''_r l'$ holds" to "$l \succ''_r l'$ holds or $r$ likes $l$ as good as $l'$".
- constraint (7): change "$r \succ''_l r'$ holds" to "$r \succ''_l r'$ holds or $l$ likes $r$ as good as $r'$".
- constraint (8) : change "$l \succ''_r l'$ and $r \succ''_l r'$ holds," to "either (i) or (ii) is satisfied: (i) $r \succ''_l r'$ holds, or $l$ likes $r$ as good as $r'$ and $l \succ''_r l'$ holds; (ii) $l \succ''_r l'$ holds, or $r$ likes $l$ as good as $l'$ and $r \succ''_l r'$ holds."

In addition, if we change the conditions in definition 12 as follows, a *super stable* matching can be found.

- constraints (6) and (7): same as above.
- constraint (8): change "$l \succ''_r l'$ and $r \succ''_l r'$ holds," to "$l \succ''_r l'$ or $r$ likes $l$ as good as $l'$, and $r \succ''_l r'$ or $l$ likes $r$ as good as $r'$"

*Example 1.* Let us consider the following two-sided matching problem with indifferences, $(L, R)$.

- $L = \{l_1, l_2, l_3\}$, $R = \{r_1, r_2\}$
- Agents' preferences: $P(l_1) = (\{r_2\}, \{r_1\})$, $P(l_2) = (\{r_1\}, \{r_2\})$, $P(l_3) = (\{r_1, r_2\})$, $P(r_1) = (\{l_1, l_2\}, \{l_3\})$, $P(r_2) = (\{l_3\}, \{l_1, l_2\})$
- $n_{l_1} = 1$, $n_{l_2} = 1$, $n_{l_3} = 2$, $n_{r_1} = 1$, $n_{r_2} = 2$

The constraints to obtain a weakly stable matching of this two-sided matching are given as follows:

- $x(l_1, r_1) + x(l_1, r_2) \leq 1$, $x(l_2, r_1) + x(l_2, r_2) \leq 1$, $x(l_3, r_1) + x(l_3, r_2) \leq 2$
- $x(l_1, r_1) + x(l_2, r_1) + x(l_3, r_1) \leq 1$, $x(l_1, r_2) + x(l_2, r_2) + x(l_3, r_2) \leq 2$

- $x(l_1, r_1) + x(l_1, r_2) + 2e(l_1) \geq 1$, $x(l_2, r_1) + x(l_2, r_2) + 2e(l_2) \geq 1$, $x(l_3, r_1) +$
  $x(l_3, r_2) + 2e(l_3) \geq 2$
- $x(l_1, r_1) + x(l_2, r_1) + x(l_3, r_1) + 3e(r_1) \geq 1$, $x(l_1, r_2) + x(l_2, r_2) + x(l_3, r_2) +$
  $3e(r_2) \geq 2$,
- $-x(l_1, r_1) + e(l_1) + e(r_1) < 2$, $-x(l_1, r_2) + e(l_1) + e(r_2) < 2$, $-x(l_2, r_1) +$
  $e(l_2) + e(r_1) < 2$, $-x(l_2, r_2) + e(l_2) + e(r_2) < 2$, $-x(l_3, r_1) + e(l_3) + e(r_1) < 2$,
  $-x(l_3, r_2) + e(l_3) + e(r_2) < 2$
- $-x(l_1, r_1) + x(l_3, r_1) + e(l_1) < 2$, $-x(l_2, r_1) + x(l_3, r_1) + e(l_2) < 2$, $-x(l_3, r_2) +$
  $x(l_1, r_2) + e(l_3) < 2$, $-x(l_3, r_2) + x(l_2, r_2) + e(l_3) < 2$
- $-x(l_1, r_2) + x(l_1, r_1) + e(r_2) < 2$, $-x(l_2, r_1) + x(l_2, r_2) + e(r_1) < 2$
- $-x(l_2, r_1) + x(l_3, r_1) + x(l_2, r_2) < 2$

## 3.2  Objective Function

Next, we consider an objective function. The matching found by using the Gale Shapley algorithm in a traditional two-sided matching problem is the *best* matching to the proposing group in the sense that there is no stable matching in which any agent in the proposing group can be paired with an agent more preferable in the other group. However, in two-sided matching with indifferences, such a matching does not necessarily exist.

To solve this issue, we propose a new definition of the *best* matching that is based on the concept of "nucleolus" in cooperative game theory [10], and we also call such a matching *nucleolus*.

The worse agents pair, the more are agents discontented. Assume there are a group of agents $L$ and a group of matchings $M$. Intuitively the nucleolus for $L$ in $M$ minimizes the discontentedness of the agent who has the worst discontentedness in $L$ in $M$

Nucleolus matching can formally be defined as follows (In this paper, we define nucleolus for $L$. If transposed $L$ to $R$ and $R$ to $L$, we define nucleolus for $R$). Assume that there are a set of agents $L$ and a set of matchings $M$. First, let us define $rank(l, r)$ to mean that agent $l \in L$ likes agent $r \in R$ as the $rank(l, r)$-th best. If agent $l$ prefers being paired with any agent to being paired with no agent (In this paper, we assume that all agents satisfy this condition), $rank(l, \emptyset) = |R| + 1$ holds. For example, if agent $l$'s preference $P(l)$ is given as $P(l) = (\{r_1, r_3\}, \{r_2\})$, $rank(l, r_1) = 1$, $rank(l, r_2) = 3$, $rank(l, r_3) = 1$, and $rank(l, \emptyset) = |R| + 1 = 4$ hold.

Next, we define a rank vector of agent $l$ for matching $m$ as $y_{l,m}$. $y_{l,m}$ is calculated as follows. First, a set of $rank(l, r)$ where $r \in m(l)$ is obtained. Second, if $|m(l)| < n_l$ holds, we add $rank(l, \emptyset)$ to this set $n_l - m(l)$ times. Then, this set is sorted in descending order and set to $y_{l,m}$. Note that $|y_{l,m}| = n_l$. Continuing the example above, if $n_l = 3$, $m(l) = \{r_3\}$, $y_{l,m} = \{4, 4, 1\}$.

Next, we define a rank vector of $L$ for matching $m$ as $Y_{L,m}$, which is a sorted list in the descending order of the union of all the rank vector of agents in $L$ ($\bigcup_{l \in L} y_{l,m}$). Note that the size of $Y_{L,m}$ is $\sum_{l \in L} n_l$, which is denoted by $n_L$.

For example, assume that there are two agents in $L$ ($L = \{l_1, l_2\}$). For matching $m$, if $y_{l_1,m}$ is given as $\{5, 3, 1\}$ and $y_{l_2,m}$ is given as $\{3, 2\}$, $Y_{L,m}$ is $\{5, 3, 3, 2, 1\}$.

Since different types of stable matchings (such as weakly stable, strongly stable, and super stable) are defined, a nucleolus is potentially different for each type of stable matching.

**Definition 13 (nucleolus).** *Matching $m$ is a nucleolus for a set of agents $L$ in a group of matchings $M$ if the following conditions are satisfied for any given matching $m'$ $(\in M)$.*

- *There exists an integer $i$ $(0 \leq i \leq \sum_{l \in L} n_l)$ that satisfies the followings:*
  - *$\forall 1 \leq j \leq i$, $Y_{L,m}(j) = Y_{L,m'}(j)$*
  - *if $i \neq n_L$, $Y_{L,m}(i+1) < Y_{L,m'}(i+1)$*
  *where $Y_{L,m}(i)$ means the $i$-th element of $Y_{L,m}$.*

In other words, the nucleolus for $L$ in $M$ has the lexicographically minimum of rank vector of $L$ in $M$. Since the lexicographical ordering is transitive, there is always one or more nucleoli in stable matchings.

We can calculate the nucleolus for $L$ to use the following objective function.

**Definition 14 (objective function for the nucleolus).** *If variables $x(y, z)$ maximize the following function, the matching expressed by the variables is a nucleolus for $L$ in a group of matchings that satisfy the constraints for stable matching.*

$$\sum_{l \in L,\ r \in R} ((n_L + 1)^{|R+1|} - (n_L + 1)^{rank(l,r)}) \times x(l, r)$$

**Theorem 2.** *A matching expressed by the variables in definition 14 is the nucleolus.*

Due to space limitations, we only briefly sketch the theorem's proof below. Assume that there are two matchings $m$, $m'$ and $m$ is *better* than $m'$. Additionally, we assume that the 1st $\sim$ $i$-th element of $Y_{L,m}$ is equal to that of $Y_{L,m'}$, and the $i+1$-th element of $Y_{L,m}$ is less than that of $Y_{L,m'}$. In this case, $m$'s objective function is minimized where the $i+2$-th $\sim$ last element of $Y_{L,m}$ is equal to the $i$-th element of $Y_{L,m}$. $m'$'s objective function is maximized where $i+2$-th $\sim$ last element of $Y_{L,m'}$ is 1. $m$'s objective function is more than $m'$'s objective function when $m$'s objective function is minimized and $m'$'s objective function is maximized. Therefore, if $m$ is *better* than $m'$, $m$'s objective function is more than that of $m'$. The matching which maximizes the objective function is the nucleolus.

*Example 2.* The objective function to calculate a nucleolus of the two-sided matching in example 1 is given as follows.

$$100x(l_1, r_1) + 120x(l_1, r_2) + 120x(l_2, r_1) + 100x(l_2, r_2) + 120x(l_3, r_1) + 120x(l_3, r_2)$$

The nucleolus minimizes the maximum discontentedness of agents in one group. In this sense, satisfying this condition means that agents regard the worst pair as important. In future, we should propose other objective functions, for example, where agents regard the best pair as important while minimizing the maximum discontentedness of agents in one group.

## 4  Conclusion

Two-sided matching is one of the main approaches to form a team in a multi-agent system. The matching is found based on the agents' preferences of the agents they are to be paired with. For a large scale multi-agent system, handling indifferences in the agents' preferences is important.

In this paper, we applied the IP to two-sided matching with indifferences. The constraints were defined to find three types of stable matchings. In addition, by defining an objective function properly, a stable matching with proper characteristics could be found. As an example, we showed an objective function that allowed us to obtain a matching that minimizes the maximum of agents' discontendedness in one group.

Future works include identifying constraints to find a matching that satisfies another character called "Pareto efficiency" [11] and extending the proposed algorithm to satisfy incentive compatibility.

## References

1. Gale, F., Shapley, L.S.: College Admissions and the Stability of Marriage. American Mathematical Monthly 69, 9–15 (1962)
2. Roth, A.E.: The Evolution of the Labor Market for Medical Interns and Residents: A Case Study in Game Theory. Journal of Political Economy 92, 991–1016 (1984)
3. Roth, A.E.: The National Residency Matching Program as a Labor Market. Journal of American Medical Association 275(13), 1054–1056 (1996)
4. Roth, A.E., Sotomayor, M.A.O.: Two-sided Matching: A study in Game-Theoretic Modeling and Analysis. Cambridge University Press (1990)
5. Haas, C., Kimbrough, S.O., Caton, S., Weinhardt, C.: Preference-Based Resource Allocation: Using Heuristics to Solve Two-Sided Matching Problems with Indifferences. In: Altmann, J., Vanmechelen, K., Rana, O.F. (eds.) GECON 2013. LNCS, vol. 8193, pp. 149–160. Springer, Heidelberg (2013)
6. McVitie, D., Wilson, L.B.: The Stable Marriage Problem. Communications of the ACM 14, 486–490 (1971)
7. Gusfield, D., Irving, R.W.: The Stable Marriage Problem: Structure and Algorithms. MIT Press, Cambridge (1989)
8. Vande Vate, J.H.: Linear Programming Brings Marital Bliss. Oper. Res. Lett. 8, 147–153 (1988)
9. Irving, R.W.: Stable Marriage and Indifference. Discrete Applied Mathematics Archive 48(3), 261–272 (1994)
10. Schmeidler, D.: The Nucleolus of a Characteristic Function Game. Journal of Applied Mathematics 17, 1163–1170 (1969)
11. Kamiyama, N.: A New Approach to the Pareto Stable Matching Problem, Mathematics of Operations Research (2013), http://dx.doi.org/10.1287/moor.2013.0627 (published online in Articles in Advance October 24, 2013)

# DC Programming and DCA for Nonnegative Matrix Factorization

Hoai An Le Thi[1], Tao Pham Dinh[2], and Xuan Thanh Vo[1]

[1] Laboratory of Theoretical and Applied Computer Science EA 3097
University of Lorraine, Ile de Saulcy, 57045 Metz, France
{hoai-an.le-thi,xuan-thanh.vo}@univ-lorraine.fr
[2] Laboratory of Mathematics, National Institute for Applied Sciences-Rouen
Avenue de l'Université- 76801 Saint-Etienne-du-Rouvray cedex, France
pham@insa-rouen.fr

**Abstract.** Techniques of matrix factorization or decomposition always play a central role in numerical analysis and statistics with many applications in real-world problems. Recently, the NMF dimension-reduction technique, popularized by Lee and Seung with their multiplicative update algorithm (an adapted gradient approach) has drawn much attention of researchers and practitioners. Since many of existing algorithms lack a firm theoretical foundation, and designing efficient scalable algorithms for NMF still is a challenging problem, we investigate DC programming and DCA for NMF.

**Keywords:** Nonnegative matrix factorization, Multiplicative update algorithm, DC programming, DCA.

## 1 Introduction

Nonnegative matrix factorization (NMF) is the problem of approximating a given nonnegative matrix by the product of two low-rank nonnegative matrices, i.e., given a matrix $A \in \mathbb{R}^{m \times n}$ and a positive integer $r < \min\{n, m\}$, one desires to compute two low-rank matrices $U \in \mathbb{R}_+^{m \times r}$ and $V \in \mathbb{R}_+^{n \times r}$ such that

$$A \approx UV^T = \sum_{j=1}^{r} U_{:j} V_{:j}^T. \tag{1}$$

This problem was first introduced in 1994 by Paatero and Tapper [13], and received a considerable interest after the works of Lee and Seung [6,7]. NMF has been successfully applied in many applications such as text mining [6,20,17,1], image processing [6,4], spectral data analysis [14,1], bioinformatics [11,2], recommendation system [21], non-stationary speech denoising [18], e.t.c.

Typically, the goodness of the approximation (1) is measured by the sum of the squares of the errors on the entries, which leads to the following optimization problem.

D. Hwang et al. (Eds.): ICCCI 2014, LNAI 8733, pp. 573–582, 2014.
© Springer International Publishing Switzerland 2014

$$\min \left\{ F(U, V) = \frac{1}{2} \|A - UV^T\|_F^2 : U \in \mathbb{R}_+^{m \times r}, \ V \in \mathbb{R}_+^{n \times r} \right\}. \qquad (2)$$

The problem (2) is a non-convex optimization with respect to variables $U$ and $V$, and it is shown that solving NMF is NP-hard [19]. Hence, one only hopes to find a local minimum in practice.

The most commonly used algorithm is multiplicative update (MU) algorithm proposed by Lee and Seung in their seminal papers [6,7]. But some issues related to MU's performance and problems with convergence [3,9,10] were reported. Lin [9] proposed a modified version of MU algorithm that fixed the convergence issue. But this new algorithm was reported to be slower than the original version ([9]). Despite the lack of convergence results, the multiplicative update method is still the most used algorithm for NMF because of its efficiency and simplicity.

In this paper, we investigate difference of convex (DC) programming and DC algorithm (DCA) for solving the NMF problem. To apply DCA, we will find a region where there is at least one solution of the problem (2), and represent the objective function $F$ as a DC function. DCA applied to the resulting DC program generates a solution that is guaranteed to holds necessary conditions of a local optimum. Different from most of other algorithms where the factors $U$ and $V$ are updated in an alternative manner, our method updates $U$ and $V$ simultaneously.

Throughout the paper, we use uppercase letters to denote matrices. For a matrix $X \in \mathbb{R}^{m \times n}$, the notation $X_{i:}$ (resp. $X_{:j}$) refers to the $i$th row (resp. $j$th column) of matrix $X$. The notation $\circ$ denotes the component-wise product of matrices. The matrix inner product of two matrices $X, Y$ is defined by $\langle X, Y \rangle = \text{trace}(X^T Y)$. The projection of a vector $x \in \mathbb{R}^n$ on a subset $\Omega \subset \mathbb{R}^n$, denoted by $P_\Omega(x)$, is the nearest element of $\Omega$ to $x$ w.r.t. Euclidean distance. If $\Omega = \mathbb{R}_+^n$ is the nonnegative orthant, the projection is denoted by $[x]_+$ and defined as $[x]_+ = \max(0, x)$ element-wisely.

The rest of this paper is organized as follows. The next section states some characteristics of the NMF problem. In section 3 we present DC programming and DCA for general DC programs, and show how to apply DCA to solve the NMF problem. Finally, the numerical experiments are presented in section 4 and section 5 concludes the paper.

## 2 Characteristics of the NMF Problem

The KKT optimality conditions (necessary conditions for local optimality) for the problem (2) are given as follows

$$U \geq 0, \ \nabla_U F(U, V) \geq 0, \ U \circ \nabla_U F(U, V) = 0, \qquad (3a)$$

$$V \geq 0, \ \nabla_V F(U, V) \geq 0, \ V \circ \nabla_V F(U, V) = 0, \qquad (3b)$$

where

$$\nabla_U F(U, V) = UV^T V - AV, \ \nabla_V F(U, V) = VU^T U - A^T U. \qquad (4)$$

**Definition 1.** *We call* $(U, V)$ *a stationary point of the NMF problem if and only if* $U$ *and* $V$ *satisfy the KKT conditions* (3).

The following result shows that finding solution for the problem (2) can be restricted in a bounded region.

**Theorem 1.** *The NMF problem* (2) *has a solution* $(U, V)$ *such that*

$$\|U\|_F \le \mathbf{b}, \quad \|V\|_F \le \mathbf{b}. \tag{5}$$

*where* $\mathbf{b} = (\sqrt{r}\|A\|_2)^{1/2}$.

*Proof.* Suppose that $U, V$ satisfy the KKT conditions. From (3), we have

$$0 = \sum_{ij}(U \circ \nabla_U F(U, V))_{ij} = \langle U, \nabla_U F(U, V) \rangle = \langle UV^T, UV^T - A \rangle.$$

Thus,

$$\|UV^T\|_F^2 = \langle A, UV^T \rangle \le \|A\|_2 \|UV^T\|_F \implies \|UV^T\|_F^2 \le \|A\|_2^2.$$

Moreover,

$$\|UV^T\|_F^2 = \|\sum_{i=1}^{r} U_{:i}V_{:i}^T\|_F^2 = \sum_{i,j=1}^{r} \langle U_{:i}, U_{:j} \rangle \langle V_{:i}, V_{:j} \rangle \ge \sum_{i=1}^{r} \|U_{:i}\|^2 \|V_{:i}\|^2.$$

By normalizing, we can always force $\|U_{:i}\| = \|V_{:i}\|$ for any $i = 1, \ldots, r$, and so $\|U\|_F = \|V\|_F$. Then, we have

$$\|A\|_2^2 \ge \sum_{i=1}^{r} \|U_{:i}\|^4 \ge \frac{1}{r}\left(\sum_{i=1}^{r} \|U_{:i}\|^2\right)^2 = \frac{1}{r}\|U\|_F^4 = \frac{1}{r}\|V\|_F^4.$$

This implies the conclusion. ∎

## 3   DCA for Solving the NMF Problem

### 3.1   Outline of DC Programming and DCA

A general DC program is that of the form:

$$\alpha = \inf\{F(x) := G(x) - H(x) \mid x \in \mathbb{R}^n\} \quad (P_{dc}),$$

where $G, H$ are lower semi-continuous proper convex functions on $\mathbb{R}^n$. Such a function $F$ is called a DC function, and $G - H$ a DC decomposition of $F$ while $G$ and $H$ are the DC components of $F$. Note that, the closed convex constraint $x \in C$ can be incorporated in the objective function of $(P_{dc})$ by using the indicator function on $C$ denoted by $\chi_C$ which is defined by $\chi_C(x) = 0$ if $x \in C$, and $+\infty$ otherwise.

For a convex function $\theta$, the subdifferential of $\theta$ at $x_0 \in \mathrm{dom}\theta := \{x \in \mathbb{R}^n : \theta(x_0) < +\infty\}$, denoted by $\partial\theta(x_0)$, is defined by

$$\partial\theta(x_0) := \{y \in \mathbb{R}^n : \theta(x) \geq \theta(x_0) + \langle x - x_0, y \rangle, \forall x \in \mathbb{R}^n\}.$$

A point $x^*$ is called a *critical point* of $G - H$, or a generalized Karush-Kuhn-Tucker point (KKT) of $(\mathrm{P}_{dc})$) if

$$\partial H(x^*) \cap \partial G(x^*) \neq \emptyset. \tag{6}$$

Based on local optimality conditions and duality in DC programming, the DCA consists in constructing two sequences $\{x^k\}$ and $\{y^k\}$ (candidates to be solutions of $(\mathrm{P}_{dc})$ and its dual problem respectively). Each iteration $k$ of DCA approximates the concave part $-H$ by its affine majorization (that corresponds to taking $y^k \in \partial H(x^k)$) and minimizes the resulting convex function.

**Generic DCA Scheme**
**Initialization:** Let $x^0 \in \mathbb{R}^n$ be an initial guess, $k \leftarrow 0$.
**Repeat**
- Calculate $y^k \in \partial H(x^k)$
- Calculate $x^{k+1} \in \arg\min\{G(x) - \langle x, y^k \rangle : x \in \mathbb{R}^n\}$    $(P_k)$
- $k \leftarrow k + 1$
**Until** convergence of $\{x^k\}$.

Convergences properties of DCA and its theoretical basic can be found in [15,8]. It is worth mentioning that

- DCA is a descent method *without linesearch*.
- If the optimal value $\alpha$ of problem $(\mathrm{P}_{dc})$ is finite and the infinite sequences $\{x^k\}$ and $\{y^k\}$ are bounded then every limit point $x^*$ of the sequences $\{x^k\}$ (resp. $\{x^k\}$) is a critical point of $G - H$.
- DCA has a *linear convergence* for general DC programs, and has a finite convergence for polyhedral DC programs.

A deeper insight into DCA has been described in [8]. For instant it is crucial to note the main feature of DCA: DCA is constructed from DC components and their conjugates but not the DC function $f$ itself which has infinitely many DC decompositions, and there are as many DCA as there are DC decompositions. Such decompositions play a critical role in determining the speed of convergence, stability, robustness, and globality of sought solutions. It is important to study various equivalent DC forms of a DC problem. This flexibility of DC programming and DCA is of particular interest from both a theoretical and an algorithmic point of view. For a complete study of DC programming and DCA the reader is referred to [15,8,16] and the references therein.

## 3.2    DCA for Solving the NMF Problem

Instead of solving the problem (2) on the whole space $U, V \geq 0$, we only solve on the subspace restricted by (5) and consider the problem

$$\min_{(U,V) \in \mathcal{S}_U \times \mathcal{S}_V} F(U,V) := \frac{1}{2}\|A - UV^T\|_F^2, \tag{7}$$

where $\mathcal{S}_U = \{U \in \mathbb{R}_+^{m \times r} : \|U\|_F \le \mathbf{b}\}$ and $\mathcal{S}_V = \{V \in \mathbb{R}_+^{n \times r} : \|V\|_F \le \mathbf{b}\}$.

For any $(U, V) \in \mathcal{S}_U \times \mathcal{S}_V$ and for any $H \in \mathbb{R}^{m \times r}$ and $K \in \mathbb{R}^{n \times r}$ we have

$$F(U+H, V+K) = F(U,V) + DF(U,V)[H,K] + \frac{1}{2} D^2 F(U,V)[H,K] + o(\|H\|_F^2 + \|K\|_F^2)$$

where

$$DF(U,V)[H,K] = \langle \nabla_U F(U,V), H \rangle + \langle \nabla_V F(U,V), K \rangle,$$

and

$$\begin{aligned}
D^2 F(U,V)[H,K] &= \langle V^T V, H^T H \rangle + 2\langle UV^T - A, HK^T \rangle + 2\langle UK^T, HV^T \rangle + \langle U^T U, K^T K \rangle \\
&\le \|V\|_F^2 \|H\|_F^2 + (2\|U\|_F \|V\|_F + \|A\|_F)(\|H\|_F^2 + \|K\|_F^2) + \|U\|_F^2 \|K\|_F^2 \\
&\le (1 + 3\sqrt{r})\|A\|_2 (\|H\|_F^2 + \|K\|_F^2).
\end{aligned}$$

Therefore, for $\rho \ge (1 + 3\sqrt{r})\|A\|_2$, the function

$$h(U,V) = \frac{\rho}{2}(\|U\|_F^2 + \|V\|_F^2) - F(U,V)$$

is convex on $\mathcal{S} = \mathcal{S}_U \times \mathcal{S}_V$.

Let $g(U,V) = \frac{\rho}{2}(\|U\|_F^2 + \|V\|_F^2)$, we have a DC decomposition $g - h$ of $F$ on $\mathcal{S}$ and corresponding DC program

$$\min \{g(U,V) - h(U,V) : (U,V) \in \mathcal{S}\}. \tag{8}$$

DCA applied to (8) consists of computing two sequences $\{(U^k, V^k)\}$ and $\{(\overline{U}^k, \overline{V}^k)\}$ with

$$\overline{U}^k = \nabla_U h(U^k, V^k) = \rho U^k - \nabla_U F(U^k, V^k), \tag{9}$$

$$\overline{V}^k = \nabla_V h(U^k, V^k) = \rho V^k - \nabla_V F(U^k, V^k), \tag{10}$$

$$(U^{k+1}, V^{k+1}) \in \arg\min \left\{ \frac{\rho}{2}(\|U\|_F^2 + \|V\|_F^2) - \langle \overline{U}^k, U \rangle - \langle \overline{V}^k, V \rangle : (U,V) \in \mathcal{S} \right\}.$$

Computing $(U^{k+1}, V^{k+1})$ can be split into two problems separately as follows

$$U^{k+1} \in \arg\min_{U \in \mathcal{S}_U} \left\{ \frac{\rho}{2}\|U\|_F^2 - \langle \overline{U}^k, U \rangle \right\} = \arg\min_{U \in \mathcal{S}_U} \left\{ \frac{1}{2}\|U - \frac{1}{\rho}\overline{U}^k\|_F^2 \right\},$$

$$V^{k+1} \in \arg\min_{V \in \mathcal{S}_V} \left\{ \frac{\rho}{2}\|V\|_F^2 - \langle \overline{V}^k, V \rangle \right\} = \arg\min_{V \in \mathcal{S}_V} \left\{ \frac{1}{2}\|V - \frac{1}{\rho}\overline{V}^k\|_F^2 \right\}.$$

Hence $U^{k+1}$ (resp. $V^{k+1}$) is the projection of the point $\frac{1}{\rho}\overline{U}^k$ (resp. $\frac{1}{\rho}\overline{V}^k$) onto $\mathcal{S}_U$ (resp. $\mathcal{S}_V$) and can be explicitly express as follows

$$U^{k+1} = P_{\mathcal{S}_U}\left(\frac{1}{\rho}\overline{U}^k\right) = \begin{cases} \frac{1}{\rho}[\overline{U}^k]_+ & , \text{if } \|[\overline{U}^k]_+\|_F \le \rho\mathbf{b} \\ \frac{\mathbf{b}}{\|[\overline{U}^k]_+\|_F}[\overline{U}^k]_+ & , \text{if } \|[\overline{U}^k]_+\|_F > \rho\mathbf{b} \end{cases}, \tag{11}$$

$$V^{k+1} = P_{\mathcal{S}_V}\left(\frac{1}{\rho}\overline{V}^k\right) = \begin{cases} \frac{1}{\rho}[\overline{V}^k]_+ & , \text{if } \|[\overline{V}^k]_+\|_F \le \rho\mathbf{b} \\ \frac{\mathbf{b}}{\|[\overline{V}^k]_+\|_F}[\overline{V}^k]_+ & , \text{if } \|[\overline{V}^k]_+\|_F > \rho\mathbf{b}. \end{cases} \tag{12}$$

DCA for solving the problem (7) is summarized in Algorithm 1. The following theorem shows that the solutions given by Algorithm 1 are really the solutions of the NMF problem (2).

---

**Algorithm 1. (DCA)**

---

Initialize $U^0, V^0$ satisfied (5), $k \leftarrow 0$.

**repeat**

  - Compute $(\overline{U}^k, \overline{V}^k)$ using (9),(10).
  - Compute $(U^{k+1}, V^{k+1})$ using (11),(12).
  - $k \leftarrow k + 1$.

**until** Stopping criterion is satisfied.

---

**Theorem 2.** *Every limit point generated by Algorithm 1 is a stationary point of the problem (2). Moreover, the sequence $\{U^k, V^k)\}$ generated by Algorithm 1 has at least one limit point.*

*Proof.* Suppose that $(U^*, V^*)$ is limit point generated by Algorithm 1. Then, by theory of DC programming and DCA, $(U^*, V^*)$ is a stationary point of (7) (and (8)) (since both $g$ and $h$ are continuously differentiable, a generalized KKT point is also an ordinary KKT point). Thus, there exist $\alpha, \beta \geq 0$ and $\mu \in \mathbb{R}_+^{m \times r}$, $\nu \in \mathbb{R}_+^{n \times r}$ such that

$$
\begin{cases}
U^* \geq 0, & V^* \geq 0, \\
\nabla_U F(U^*, V^*) + 2\alpha U^* = \mu, & \nabla_V F(U^*, V^*) + 2\beta V^* = \nu, \\
\mu \circ U^* = 0, & \nu \circ V^* = 0, \\
\|U^*\|_F^2 \leq \mathbf{b}^2, & \|V^*\|_F^2 \leq \mathbf{b}^2, \\
\alpha \cdot (\|U^*\|_F^2 - \mathbf{b}^2) = 0, & \beta \cdot (\|V^*\|_F^2 - \mathbf{b}^2) = 0.
\end{cases}
\tag{13}
$$

It is clear that if $\alpha = \beta = 0$ then $U^*$ and $V^*$ satisfy the KKT conditions (3). Now we assume $\alpha > 0$. Then $\|U^*\|_F^2 = \mathbf{b}^2$. From the first three properties in (13), we have

$$
2\alpha \|U^*\|_F^2 = -\langle U^*, \nabla_U F(U^*, V^*) \rangle = -\langle V^*, \nabla_V F(U^*, V^*) \rangle = 2\beta \|V^*\|_F^2.
$$

Combining this and the fifth property in (13), we deduce that $\beta = \alpha > 0$ and $\|V^*\|_F^2 = \|U^*\|_F^2 = \mathbf{b}^2$. For any $i = 1, \ldots, r$, from the first three properties in (13), we have

$$
2\alpha \|U_{:i}^*\|^2 = -\langle U_{:i}^*, (\nabla_U F(U^*, V^*))_{:i} \rangle = -\langle V_{:i}^*, (\nabla_V F(U^*, V^*))_{:i} \rangle = 2\beta \|V_{:i}^*\|^2.
$$

Therefore, $\|U_{:i}^*\|^2 = \|V_{:i}^*\|^2$ for any $i = 1, \ldots, r$. Then

$$
\|U^*(V^*)^T\|_F^2 = \|\sum_{i=1}^r U_{:i}^*(V_{:i}^*)^T\|_F^2 = \sum_{i,j=1}^r \langle U_{:i}^*, U_{:j}^* \rangle \langle V_{:i}^*, V_{:j}^* \rangle
$$

$$
\geq \sum_{i=1}^r \|U_{:i}^*\|^2 \|V_{:i}^*\|^2 = \sum_{i=1}^r \|U_{:i}^*\|^4 \geq \frac{1}{r} (\sum_{i=1}^r \|U_{:i}^*\|^2)^2.
$$

That is $\|U^*\|_F^2 \leq \sqrt{r}\|U^*(V^*)^T\|_F$.

Moreover, we have

$$\|U^*(V^*)^T\|_F^2 - \langle A, U^*(V^*)^T \rangle = \langle U^*, \nabla_U F(U^*, V^*) \rangle = -2\alpha\|U^*\|_F^2 < 0.$$

So

$$\|U^*(V^*)^T\|_F^2 < \langle A, U^*(V^*)^T \rangle \leq \|A\|_F\|U^*(V^*)^T\|_F.$$

$$\Rightarrow \|U^*\|_F^2 \leq \sqrt{r}\|U^*(V^*)^T\|_F < \sqrt{r}\|A\|_F = b^2.$$

We have a contradiction. Thus, $\alpha = \beta = 0$ and $(U^*, V^*)$ satisfy the KKT conditions (3). The last conclusion is trivial due to the fact that the sequence $\{U^k, V^k)\}$ is bounded.                                                                  ∎

*Comments on the proposed method:*

- DCA applied to (8) is very simple and inexpensive: each iteration of DCA consists of computations of the projection of points onto a nonnegative ball that all are explicitly computed.

- In DCA, $U^{k+1}$ and $V^{k+1}$ are separately but simultaneously computed. This can avoid getting stuck in a bad local solution as updating $U$ and $V$ alternatively.

- This method can be easily extended to other variants of NMF with more complicated constraints as long as the projection is still easy to compute.

## 4    Experiments

In this section we present experimental results to assess the performance of our algorithm as compared to the standard multiplicative update (MU) algorithm [7]. We implemented all algorithms in Matlab 7.7 on a Core(TM) i5-3360M $2 \times 2.80$ GHz computer with 4GB memory.

### 4.1    Experimental Setup

*Stopping criterion.* Let

$$\Delta_U = \min(U, \nabla_U F(U, V)), \quad \Delta_V = \min(V, \nabla_V F(U, V)).$$

Then, it is easy to see that the KKT conditions (3) are equivalent to $\Delta_U = 0$ and $\Delta_V = 0$ and so equivalent to

$$\Delta = \sqrt{\|\Delta_U\|_F^2 + \|\Delta_V\|_F^2} = 0.$$

The stopping criterion we use is

$$\Delta \leq \epsilon\Delta_0, \tag{14}$$

where $\Delta_0$ is value of $\Delta$ at the initial value $(U_0, V_0)$, and $\epsilon$ is a tolerance.

*Initialization.* We use the initialization strategy proposed in [5] as follows. Let $U \in \mathbb{R}_+^{m \times r}$ and $V \in \mathbb{R}_+^{n \times r}$ be random matrices and $D$ is a diagonal matrix such that $D_{ii} = \sqrt{\|V_{:i}\|_2 / \|U_{:i}\|_2}$, $(i = 1, \ldots, r)$. Compute the scaling factor $\alpha = \langle A, UV^T \rangle / \|UV^T\|_F^2$, then the initialization $(U_0, V_0)$ is determined by

$$U_0 = UD\sqrt{\alpha}, \quad V_0 = VD^{-1}\sqrt{\alpha}.$$

By this way, we see that $\|(U_0)_{:i}\|_2 = \|(V_0)_{:i}\|_2$, $\forall i = 1, \ldots, r$, and

$$\|U_0 V_0^T\|_F^2 = \langle A, U_0 V_0^T \rangle \Rightarrow \|U_0 V_0^T\|_F \le \|A\|_2.$$

Thus, $(U_0, V_0)$ is satisfied the conditions (5).

*Datasets.* We used both synthetic and real datasets for comparison. The synthetic datasets are created as follows ([12]). For each triple $(m, n, r)$, we randomly generated $U \in \mathbb{R}_+^{m \times r}$ and $V \in \mathbb{R}_+^{n \times r}$ with 40% sparsity. Then, we computed $A = UV^T$ and added Gaussian noise to each element where the standard deviation is 1% of the average magnitude of elements in $A$.

For real dataset, we used the CBCL face image database [6]. The database contains $n = 2429$ facial images, each consisting of $m = 19 \times 19$ pixels, and constituting an $m \times n$ matrix $A$.

## 4.2   Experimental Results

For each dataset, we ran all algorithms with the same 10 initializations created by the method in Sec. 6.2. The average results are summarized in Tables 1-2. The reported results include the computing time, the residue and the sparsity of the computed factors $U$ and $V$. The residue is computed as $\|A - UV^T\|_F$. While the sparsity of the factor $U$ (resp. $V$) is defined as the percentage of zero elements in the matrix $U$ (resp. $V$). Since the factors $U$ and $V$ generated by MU algorithm do not contain exact zero elements, we will regard very small elements $(< 10^{-10})$ as zeros. It is worth noting that the sparsity is a desired property of NMF algorithms since it is helpful in interpreting the results in many applications ([6]).

Observe from the numerical results, we see that

- The MU algorithm is slow and shows difficulty in meeting the stopping criterion. The MU algorithm only reaches the stopping criterion on the synthetic datasets with $r = 10$ and $r = 20$ at tolerance $\epsilon = 10^{-3}$, but it slower than DCA 9 times. In all remaining cases, the execution time of MU algorithm exceeds the limit. While DCA uses at most a half of the allowable time in all cases.

- On Table 1, with $r = 10$ and $r = 20$, at the tolerance $\epsilon = 10^{-3}$, the MU algorithm gives lower residue than DCA. While at the tolerance $\epsilon = 10^{-4}$, with a little more execution time, DCA gives lower residue than MU algorithm does. When $r = 30$, DCA outperforms MU algorithm on both execution time and residue.

- On Table 2, for the residue comparison, two algorithms are comparable but MU algorithm consumes much more time than DCA. DCA also produces sparser solutions than MU algorithm does, especially on the factor $U$.

**Table 1.** Experimental results on $500 \times 300$ synthetic datasets with two stopping tolerances. We report the average time (in seconds), and residual of 10 initializations. The residue is computed as $\|A - UV^T\|_F$. A limit time of 30 seconds is imposed for slowly converging cases. The better results are highlighted in bold font.

| $r$ | Algorithm | $\epsilon = 10^{-3}$ Time | Residue | $\epsilon = 10^{-4}$ Time | Residue |
|---|---|---|---|---|---|
| 10 | DCA | **0.533** | 0.838 | **0.716** | 0.735 |
|    | MU | 4.603 | **0.751** | 30.0 | 0.735 |
| 20 | DCA | **3.027** | 1.751 | **4.018** | **1.442** |
|    | MU | 29.0 | **1.484** | 30.0 | 1.482 |
| 30 | DCA | **9.508** | **3.989** | **13.799** | **2.121** |
|    | MU | 30.0 | 6.611 | 30.0 | 6.611 |

**Table 2.** Experimental results on CBCL database with the stopping tolerance $\epsilon = 10^{-3}$. We report the average time (in seconds), residue, and the sparsity of factors ($U$ and $V$) of 10 initializations. The residue is computed as $\|A - UV^T\|_F$. For $r = 10$ (resp. 20 and 30), a limit time of 50 (resp. 60 and 100) seconds is imposed for slowly converging cases. The better results are highlighted in bold font.

| $r$ | Algorithm | Time | Residue | Spar. U | Spar. V |
|---|---|---|---|---|---|
| 10 | DCA | **11.528** | 79.075 | **19.0** | **7.7** |
|    | MU | 50.0 | **78.881** | 14.6 | 7.3 |
| 20 | DCA | **30.070** | **63.51** | **30.7** | **10.7** |
|    | MU | 60.0 | 63.59 | 17.7 | 10.0 |
| 30 | DCA | **53.243** | 54.17 | **38.2** | **12.6** |
|    | MU | 100.0 | 54.17 | 21.1 | 11.8 |

## 5  Conclusion

In this paper, a new algorithm for computing nonnegative matrix factorization (NMF) based on DC programming and DCA has been proposed. The original problem has been recast as a DC program with a nice DC decomposition. The resulting DCA enjoys explicit computation and strong convergence property that are desirable from both practical and theoretical viewpoints. The experimental results show that the proposed algorithm is more efficient than the standard multiplicative update algorithm of Lee and Seung ([7]) in computing NMF.

The existing methods may fail to solve other variants of NMF with additional constraints and regularizations that arise from applications. A direction for future research would be to extend the proposed method to these problems.

## References

1. Berry, M., Browne, M., Langville, A., Pauca, P., Plemmons, R.: Algorithms and applications for approximate nonnegative matrix factorization. Computational Statistics and Data Analysis, 155–173 (2006)

2. Devarajan, K.: Nonnegative Matrix Factorization: An Analytical and Interpretive Tool in Computational Biology. PLoS Computational Biology 4(7) (2008)
3. Gonzalez, E.F., Zhang, Y.: Accelerating the Lee-Seung algorithm for non-negative matrix factorization. Tech Report, Department of Computational and Applied Mathematics, Rice University (2005)
4. Guillamet, D., Vitria, J.: a, Non-negative matrix factorization for face recognition. Topics in Artificial Intelligence, 336–344 (2002)
5. Ho, N.-D.: Nonnegative Matrix Factorization: Algorithms and Applications. PhD Thesis, University catholique de Louvain (2008)
6. Lee, D.D., Seung, H.S.: Learning the Parts of Objects by Nonnegative Matrix Factorization. Nature 401, 788–791 (1999)
7. Lee, D.D., Seung, H.S.: Algorithms for Non-negetive matrix factorization. In: Advances in Neural Information Processing Systems, vol. 13, pp. 556–562 (2001)
8. Le Thi, H.A., Pham Dinh, T.: The DC (difference of convex functions) Programming and DCA revisited with DC models of real world nonconvex optimization problems. Annals of Operations Research 133, 23–46 (2005)
9. Lin, C.-J.: On the convergence of multiplicative update algorithm for non-negative matrix factorization. IEEE Transactions on Neural Networks (2007)
10. Lin, C.-J.: Projected gradient methods for nonnegative matrix factorization. Neural Computation 19, 2756–2779 (2007)
11. Kim, H., Park, H.: Sparse non-negative matrix factorizations via alternating non-negativity constrained least squares for microarray data analysis. Bioinformatics 23, 1495–1502 (2007)
12. Kim, J., Park, H.: Toward faster nonnegative matrix factorization: A new algorithm and comparisons. In: Proceedings of the 8th IEEE ICDM, pp. 353–362 (2008)
13. Paatero, P., Tapper, U.: Positive matrix factorization: a non-negative factor model with optimal utilization of error estimates of data values. Environmetrics 5, 111–126 (1994)
14. Pauca, V.P., Piper, J., Plemmons, R.J.: Nonnegative Matrix Factorization for Spectral Data Analysis. Linear Algebra and its Applications 416, 29–47 (2006)
15. Pham Dinh, T., Le Thi, H.A.: Convex analysis approach to DC programming: Theory, algorithms and applications. Acta Math. Vietnamica 22(1), 289–357 (1997)
16. Pham Dinh, T., Le Thi, H.A.: Dc optimization algorithms for solving the trust region subproblem. SIAM J. Optimization 8, 476–505 (1998)
17. Shahnaz, F., Berry, M.W., Langville, A.N., Pauca, V.P., Plemmons, R.J.: Document clustering using nonnegative matrix factorization. Information Processing and Management 42, 373–386 (2006)
18. Schmidt, M.N., Larsen, J., Hsiao, F.T.: Wind noise reduction using non-negative sparse coding. In: IEEE Workshop on Machine Learning for Signal Processing, pp. 431–436 (2007)
19. Vavasis, S.A.: On the complexity of nonnegative matrix factorization. SIAM Journal on Optimization 20, 1364–1377 (2009)
20. Xu, W., Liu, X., Gong, Y.: Document clustering based on non-negative matrix factorization. In: Proceedings of the 26th Annual International ACM SIGIR Conference on Research and Development in Information Retrieval, pp. 267–273 (2003)
21. Zhang, S., Wang, W., Ford, J., Makedon, F.: Learning from Incomplete Ratings Using Non-negative Matrix Factorization. In: Proc. of the 6th SIAM Conference on Data mining, pp. 549–553 (2006)

# An Ant Colony Optimization Algorithm for an Automatic Categorization of Emails

Urszula Boryczka, Barbara Probierz, and Jan Kozak

Institute of Computer Science, University of Silesia, Będzińska 39,
41–200 Sosnowiec, Poland
{urszula.boryczka,barbara.probierz,jan.kozak}@us.edu.pl

**Abstract.** This article presents a new approach to an automatic categorization of email messages which is based on Ant Colony Optimization algorithms (ACO). The aim of this paper is to create an algorithm that would allow one to improve the classification of emails into folders (the email foldering problem) by using solutions that have been applied in Ant Colony algorithms, data mining and Social Network Analysis (SNA). The new algorithm which is proposed here has been tested on the publicly available Enron email data set. The obtained results confirm that this approach allows one to improve the accuracy with which new emails are assigned to particular folders based on an analysis of previous correspondence.

**Keywords:** Enron E-mail, Ant Colony Optimization, Social Network Analysis.

## 1 Introduction

People have been using emails on an almost daily basis since 1971, i.e. when Ray Tomlinson sent his first email message. Currently billions of emails are sent via the Internet every day, whereas more than 100 trillion of them are sent annually. It is estimated that the total number of email accounts worldwide amounting to 3.9 billion in 2013 will have increased to over 4.9 billion by the end of 2017. A typical user receives approx. 40-50 email messages every day. Some even get hundreds of emails a day, which is why email users devote a considerable part of their working time to reading and answering email messages. At the same time many emails that are sent to the users contain unnecessary information and should be filtered. As a result, there has recently been a growing interest in creating systems that would automatically help the users manage their emails.

Based on an analysis of this problem, the aim of the article has been established, i.e. to create an adaptive algorithm which will make it possible to improve the accuracy of classifying email messages. In fact, it is to better match new emails to particular folders (the email foldering problem). The proposed algorithm is based on a modified version of the ACDT algorithm which contains elements of social network analysis (or, more specifically, of communication network analysis). This algorithm was used in a previously prepared data set of

D. Hwang et al. (Eds.): ICCCI 2014, LNAI 8733, pp. 583–592, 2014.

emails which had been obtained from the public Enron email data set especially for this purpose.

This article is organized as follows. Section 1 comprises an introduction to the subject of this article. In section 2, social network analysis is presented. Section 3 describes categorization of email into folders and Enron e-mail dataset. Section 4 describes Ant Colony Optimization in Data Mining, especially Ant Colony Decision Tree approach. Section 5 focuses on the presented, new version of the ACO approach based on ACDT algorithm and social network analysis. Section 6 presents the experimental study that has been conducted to evaluate the performance of the proposed algorithm, taking into consideration Enron e-mail dataset. Finally, we conclude with general remarks on this work and a few directions for future research are pointed out.

## 2   Social Network Analysis

Social network analysis (SNA) plays an extremely important role in studies of data sets containing email messages. Most of all, SNA provides a specific perspective on an analysis because it does not focus on individual units or macrostructures but studies the connections between particular units or groups. A social network is usually represented as a graph. According to the mathematical definition, a graph is an ordered pair $G = (V, E)$, where $V$ denotes a finite set of a graph's vertices, and $E$ denotes a finite set of all two-element subsets of set V that are called edges, which link particular vertices such that:

$$E \subseteq \big\{ \{u, v\} : u, v \in V, u \neq v \big\}. \tag{1}$$

vertices represent objects in a graph whereas edges represent the relations between these objects. Depending on whether this relation is symmetrical, a graph which is used to describe a network can be directed or undirected.

Social network analysis has a wide range of applications. It is primarily used in large organizations and companies as a tool for supporting strategic human resource management or knowledge management in an organization. SNA supports a company's innovativeness and an analysis of business processes as well as training needs. Additionally, it is used in marketing research for creating a map of a social network of customers. However, social network analysis primarily allows managers to familiarize themselves with the informal structure of an organization and the flow of information within a company.

The first studies of social networks were conducted in 1923 by Jacob L. Moreno, who is regarded as one of the founders of social network analysis. SNA is a branch of sociology which deals with the quantitative assessment of the individual's role in a group or community by analyzing the network of connections between individuals. Moreno's 1934 book that is titled Who Shall Survive presents the first graphical representations of social networks as well as definitions of key terms that are used in an analysis of social networks and sociometric networks [16].

Many studies that were carried out as part of SNA were aimed at finding correlation between a network's social structure and efficiency [12]. At the beginning, social network analysis was conducted based on questionnaires that were filled out by hand by the participants [5]. However, research carried out by using email messages has become popular over time [1]. Some of the studies found that research teams were more creative if they had more social capital [11]. Social networks are also associated with discovering communication networks. The database which was used in the experiments that are presented in this article can be used to analyze this problem. G. C. Wilson and W. Banzhaf, among others, discussed such an approach, which they described in their article [18].

## 3  Categorization of Email into Folders

The classification of text is the main email management tool. It is a process of assigning each document $d_i$ from a given data set $D_t = \{(d_1, c_1), \ldots, (d_m, c_m)\}$ to one of the predefined classes based on a set of values of attributes that describe a given document. Therefore, one can find a mapping that assigns one class from set $C = \{c_1, \ldots, c_m\}$ to a given document $d_i$ which is represented by the vector of features $(a_1, \ldots, a_n)$. The mapping:

$$f : R^n \ni (a_1, \ldots, a_n) \to c_m \in C \tag{2}$$

is referred to as a classifier or a classification mapping. The aim of the categorization is correct allocation of e-mail messages in a folder by using an algorithm to generate a classifier based on the training set.

The classification of documents has a wide range of applications; primary among them is the filtering of spam. Another typical application of the classification of documents is in thematic catalogs which organize information by subject or which find discussion threads. Moreover, classification is also used to determine the importance of an email message by assigning the appropriate priority to it as well as to extract specific information items from a text, for example, about terrorist attacks, which leads to creating a concise data structure and not a set of documents, as is the case with searching for information on the Internet. The email foldering problem is a special case of the classification issue. It is about assigning emails to folders that are created or deleted by users over time. Such folders can be used for categorizing tasks to do, project groups or certain recipients. Email foldering is a complex problem because an automatic classification method can work for one user while for another it can lead to errors.

The first studies on methods of categorizing emails were carried out in the 1990s. D. Lewis introduced the model of concept learning for text classification systems, including systems for retrieving documents, automatic indexing and filtering electronic mail [15]. In their article [14] Kiritchenko and Matwin presented research which indicated that classification with SVM gives much better results than the naive Bayes classifier. In their article [17] M. Wang, Y. He and M. Jiang presented a categorization of email messages which was based on the information bottleneck (IB) and maximum entropy methods. The IB method was used in order to find key words, and then email subjects and address groups were used as features that were supplementary to email texts.The maximum entropy model was used to improve the classifier's accuracy. R. Bekkerman, A. McCallum and G. Huang [2] presented a case study of benchmark email foldering based on the example of two data sets of emails: Enron and SRI. They classified emails from seven email boxes into thematic folders based on four classifiers: Maximum Entropy (MaxEnt), Naive Bayes, Support Vector Machine (SVM) and Wide-Margin Winnow. Before running the training classifiers the data were cleaned and standardized. The folders that were deleted were those which contained a small number of messages as well as folders that had been found to be outdated, i.e. those which had been automatically created by the email application (e.g. the inbox, sent items, deleted items) as well as those which had been archived for all users within a certain organization and which could be found in the hierarchy of folders used by all of Enron's former employees. However, folders which had been archived individually by particular employees were kept.

The Enron email data set constitutes a set of data which were collected and prepared as part of the CALO Project (a Cognitive Assistant that Learns and Organizes). It contains more than 600,000 email messages which were sent or received by 158 senior employees of the Enron Corporation. The data set was taken over by the Federal Energy Regulatory Commission during an investigation that was carried out after the company's collapse and then it was made available to the public. A copy of the database was purchased by Leslie Kaelbling with the Massachusetts Institute of Technology (MIT), and then it turned out that there were serious problems associated with data integrity. As a result of work carried out by a team from SRI International, especially by Melinda Gervasio, the data were corrected and made available to other scientists for research purposes.

This database is considered to be one of the most valuable data sets because it consists of real email messages that are available to the public, which is usually problematic as far as other data sets are concerned due to data privacy. These emails are assigned to personal email accounts and divided into folders. There are no email attachments in the data set and certain messages were deleted because their duplicate copies could be found in other folders. The missing information was reconstructed, as far as possible, based on other information items; if it was not possible to identify the recipient the phrase no_address@enron.com was used.

The Enron data set is commonly used for the purpose of studies that deal with social network analysis, natural language processing and machine learning. The classification of emails can have many different applications; in particular, it can be used to filter emails based on the priority criteria associated with assigning emails to folders that have been created by a user, and to identify spam.

## 4   Ant Colony Optimization in Data Mining

Ant Colony Decision Trees (ACDT) algorithm [3] employs Ant Colony Optimization techniques [8] for constructing decision trees and decision forests. Ant

Colony Optimization is a branch of a newly developed form of artificial intelligence called swarm intelligence. Swarm intelligence is a form of emergent collective intelligence of groups of simple individuals: ants, termites or bees in which a form of indirect communication via pheromone was observed. Pheromone values encourage the ants following the path to build good solutions of the analyzed problem and the learning process occurring in this situation is called positive feedback or auto-catalysis.

In this paper we defined an ant algorithm to be a distracted system inspired by the observation of real ant colony behavior exploiting the stigmergic communication paradigm. The optimization algorithm in this paper was inspired by the previous works on Ant Systems (AS) and, in general, by the term – stigmergy. This phenomenon was first introduced by P. P. Grasse [13].

An essential step in this direction was the development of Ant System by Dorigo et al. [8], a new type of heuristic inspired by analogies to the foraging behavior of real ant colonies, which has proven to work successfully in a series of experimental studies. Diverse modifications of AS have been applied to many different types of discrete optimization problems and have produced very satisfactory results [7]. Recently, the approach has been extended by Dorigo et al. [6,9,10] to a full discrete optimization metaheuristic, called the Ant Colony Optimization (ACO) metaheuristic.

Ant Colony Optimization (ACO) approach has been successfully applied to many difficult combinatorial problems. Ant Colony Decision Trees (ACDT) algorithm is the first ACO adaptation to the task of constructing decision trees.

In each ACDT step an ant chooses an attribute and its value for splitting the objects in the current node of the constructed decision tree. The choice is made according to a heuristic function and pheromone values. The heuristic function is based on the Twoing criterion, which helps ants select an attribute-value pair which well divides the objects into two disjoint sets, i.e. with the intention that objects belonging to the same decision class should be put in the same subset. The best splitting is observed when objects are partitioned into the left and right subtrees such that objects belonging to the same decision class are in the same subtree. Pheromone values indicate the best way (connection) from the superior to the subordinate nodes – all possible combinations are taken into account.

As mentioned before, the value of the heuristic function is determined according to the splitting rule employed in CART approach, that is, in the algorithm proposed by Breiman et al. in 1984 [4]. The probability of choosing the appropriate split in the node is calculated according to a classical probability used in ACO:

$$p_{i,j} = \frac{\tau_{m,m_{L(i,j)}}(t) \cdot \eta_{i,j}^{\beta}}{\sum_{i}^{a} \sum_{j}^{b_i} \tau_{m,m_{L(i,j)}}(t) \cdot \eta_{i,j}^{\beta}} \tag{3}$$

where $\eta_{i,j}$ is a heuristic value for the split using the attribute $i$ and value $j$; $t$ is a step of the algorithm; $\tau_{m,m_{L(i,j)}}$ is an amount of pheromone currently available at step $t$ on the connection between nodes $m$ and $m_{L(i,j)}$ (it concerns the attribute $i$ and value $j$), and $\beta$ is the relative importance of the heuristic value.

The pheromone trail is updated by increasing pheromone levels on the edges connecting each tree node with its parent node (excepting the root):

$$\tau_{m,m_L}(t+1) = (1-\gamma) \cdot \tau_{m,m_L}(t) + Q(T) \tag{4}$$

where $Q(T)$ determines the evaluation function of decision tree (see equation (5)), and $\gamma$ is a parameter representing the evaporation rate, equal to 0.1.

The evaluation function for decision trees will be calculated according to the following equation:

$$Q(T) = \phi \cdot w(T) + \psi \cdot a(T,P), \tag{5}$$

where $w(T)$ is the size (number of nodes) of the decision tree $T$; $a(T,P)$ is the accuracy of the classification object from a training set $P$ by the tree $T$; and $\phi$ and $\psi$ are constants determining the relative importance of $w(T)$ and $a(T,P)$.

## 5    Proposed Algorithm

The proposed method entails using a modified version of the ACDT algorithm (which is described in Section 4) and transforming a data set of emails into a decision table which is understood as structure (6). For such a data set the ACDT algorithm was prepared; it contains elements of communication network analysis which entails analyzing the list of recipients. Let us assume that:

$$S = (U, A \cup \{dec\}), \tag{6}$$

where:

$U$ is a set of objects $U = \{u_1, \ldots, u_n\}$,
$A$ is a set of attributes $a_j : U \rightarrow V_j$,
$dec$ is a special attribute called the decision $dec : U \rightarrow \{1, \ldots, d\}$.

The decision table that has been prepared consists of the following attributes:

1. from - the sender;
2. word1 - the first word which is used in the subject of an email (with the exception of basic words and copulas); additionally, words which belong to the set of decision classes are supported;
3. word2 - the second word which is established similarly to word1;
4. word3 - the third word which is established similarly to word1 and word2;
5. cc - the Boolean value which indicates whether the person who has received an email was added as a recipient of a copy of an email (if not then it means that the person was the addressee of an email);
6. length - number of characters of the mail (with white spaces);
7. category - a decision class, i.e. a folder, to which an email message is assigned.

In a decision table there is one decision attribute, i.e. category, which defines the folder to which an email is to be assigned. The number of decision classes depends on the case that is being analyzed; it is provided in Tab. 1 for each data set.

**Table 1.** Parameters in data sets

|           | N. of objects | N. of class (email folders) | Number of attributes from | word1 | word2 | word3 | cc | length |
|-----------|------|-----|-----|------|------|------|---|------|
| beck-s    | 1971 | 101 | 390 | 527  | 670  | 549  | 2 | 1331 |
| farmer-d  | 3672 | 25  | 412 | 827  | 985  | 864  | 2 | 1679 |
| kaminski-v| 4477 | 41  | 821 | 1231 | 1304 | 1058 | 2 | 2461 |
| kitchen-l | 4015 | 46  | 597 | 1170 | 1207 | 996  | 2 | 2138 |
| lokay-m   | 2493 | 11  | 295 | 842  | 955  | 863  | 2 | 1654 |
| sanders-r | 1188 | 30  | 272 | 442  | 485  | 423  | 2 | 1033 |
| williams-w3| 2769| 18  | 196 | 523  | 597  | 540  | 2 | 1056 |

Decision tables were created for a data set which had been prepared in this way; these are described in Tab. 1. These data sets are very large (see Tab. 1) – they are composed of a large number of decision classes and have attributes with many values, mainly with continuous values. Therefore, Ant Colony Optimization algorithms were used to analyze this data set because they perform very well as far as such problems are concerned [3]. What is more, these algorithms can be further modified.

---

**Algorithm 1.** Pseudo code of the proposed algorithm

```
1  dataset = prepare_decision_tables(person)
2  pheromone = initialization_pheromone_trail();
3  for i=1 to number_of_iterations do
4      best_classifier = NULL;
5      for j=1 to number_of_ants do
6          new_classifier = build_classifier_ACDT(pheromone, dataset);
7          new_classifier = check_contacts_SNA(new_classifier, dataset);
8          assessment_of_the_quality_classifier(new_classifier);
9          if new_classifier is_higher_quality_than best_classifier then
10             best_classifier = new_classifier;
11         endIf
12     endFor
13     update_pheromone_trail(best_classifier, pheromone);
14 endFor
15 result = best_constructed_classifier;
```

---

Another application of the ACDT algorithm (which is based on a modification of this algorithm at the present stage) entails exploring the communication network between people if an email was sent to a group of persons, i.e. $cc = true$. The list of all recipients is analyzed, which has an influence on which decision class (email folder) a classifier will choose. This decision is also influenced by the preferences of the group of users who contact one another; therefore, if the users contacted one another with the same frequency then the emails they received were classified in the same way (uses the voting rule (7)). The way in which such algorithms work is presented based on the example of Alg. 1.

$$dUG(u) := \arg\max_c N_c(u), \tag{7}$$

where $dUG$ is a group of users, $c$ is a email folder; $N_c(u)$ is the number of votes for the email $u \in U$ classified into folder $c$, such that $N_c(u) := \#\{j : p_j(u) = c\}$, where $p_j$ is a j-th person.

## 6    Experiments

The proposed algorithm was implemented in C++. All computations were carried out on a computer with an Intel Core i5 2.27 GHz processor, 2.9 GB RAM, running on the Debian GNU/Linux operating system.

The experiments were repeated 30 times for each data set with the same standard parameter settings which were related to Ant Colony Optimization algorithms (which were adopted for the ACDT algorithm). Given the size of the data set, the number of generations of the Ant Colony Optimization algorithm was initially restricted to 30 for a population of 5 ants. The run-time of the proposed algorithm ranged, depending on the data set, between 7 and 400 seconds for one run of the algorithm. This is, however, a time during which a classifier is created whereas classification itself is carried out very quickly. The results for the other algorithms were cited based on the article titled [2].

**Table 2.** Compare all approaches in terms of classification accuracy

|             | Algorithms presented in [2] | | | | Proposed |
|             | MaxEnt | Naive Bayes | SVM | Wide-margin Winnow | algorithm |
|---|---|---|---|---|---|
| beck-s      | 0.558 | 0.320 | 0.564 | 0.499 | **0.583** |
| farmer-d    | 0.766 | 0.648 | 0.775 | 0.746 | **0.811** |
| kaminski-v  | 0.557 | 0.461 | 0.574 | 0.516 | **0.695** |
| kitchen-l   | 0.584 | 0.356 | 0.591 | 0.546 | **0.625** |
| lokay-m     | 0.836 | 0.750 | 0.827 | 0.818 | **0.888** |
| sanders-r   | 0.716 | 0.568 | 0.730 | 0.721 | **0.829** |
| williams-w3 | 0.944 | 0.922 | 0.946 | 0.945 | **0.962** |

The obtained results, which are presented in Tab. 2 and Fig. 1, indicate that there is a significant improvement in the classification of emails when the proposed algorithm is used. This is particularly important because the algorithms that are described in [2] required a thorough process of data cleaning. The proposed algorithm, at the present stage, does not require that much work be carried out when preparing a data set for research purposes, and its adaptability allows one to obtain stable results even for "uncleaned", real data sets. The proposed algorithm each time achieves increasingly better results for seven data sets that have been created for seven users (who were selected so as to make a comparison with other algorithms possible). As for three data sets (beck-s, farmer-d and williams-w3), the accuracy with which a folder is assigned to an email improved

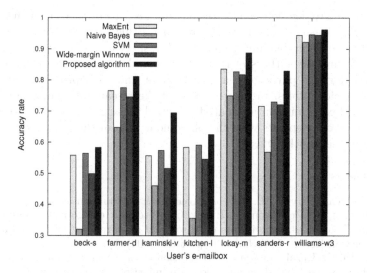

**Fig. 1.** The correctness of the proposed categorization method with respect to article [2]

by 2-3% in relation to the best of the other methods that were compared. For two other data sets (kitchen-1 i lokay-m) there was a large, 5-percent improvement, whereas for the kaminski-v i sanders-r there was a very large improvement, i.e. of more than 10%.

Other elements that are related to analyzing algorithms also need to be compared, i.e. those which could not be compared at this stage. Nonetheless, the classification stage itself is very similar for all the methods; therefore, potential differences may only result from the complex structure of the classifiers.

## 7    Conclusions

The proposed approach led to a significant improvement in the classification of emails into folders. The analysis itself of the decision table that had been prepared especially for this purpose allowed one to obtain satisfactory results when using the adaptive ACDT algorithm. The addition of elements of social network in the form of an analysis of communication between users made it possible to obtain much better results.

Based on the experiments that were carried out, it was confirmed that the accuracy of classification, i.e. the correctness of an automatic categorization of email messages, was considerably improved when Ant Colony Optimization algorithms were used. The aim of this article has been achieved. The observations that were made during the experiments suggest that the adaptability that results from using ACO algorithms may improve an analysis of the email foldering problem even more after employing social network analysis mechanisms.

The above-mentioned adaptation abilities of Ant Colony Optimization algorithms additionally enhance their operational abilities while, at the same time, data sets only have to be slightly cleaned. As for the Enron email data set, it is possible to create decision tables without the need to further clean them – the process which was carried out by the authors of this data set will suffice, which represents a considerable improvement with respect to the other methods.

In the future the authors of this article intend to adapt the social network mechanism for this purpose to a larger extent and to improve the process of creating decision tables. In future stages of the research, the incorporation of elements of text mining in an analysis of email message content and the direct coupling of these elements with the pheromone trail of the proposed algorithm should produce positive effects.

# References

1. Aral, S., Van Alstyne, M.: Network structure & information advantage (2007)
2. Bekkerman, R., McCallum, A., Huang, G.: Automatic categorization of email into folders: Benchmark experiments on enron and sri corpora. Center for Intelligent Information Retrieval, Technical Report IR (2004)
3. Boryczka, U., Kozak, J.: Ant colony decision trees – A new method for constructing decision trees based on ant colony optimization. In: Pan, J.-S., Chen, S.-M., Nguyen, N.T. (eds.) ICCCI 2010, Part I. LNCS, vol. 6421, pp. 373–382. Springer, Heidelberg (2010)
4. Breiman, L., Friedman, J.H., Olshen, R.A., Stone, C.J.: Classification and Regression Trees. Chapman & Hall, New York (1984)
5. Cummings, J.N., Cross, R.: Structural properties of work groups and their consequences for performance. Social Networks 25, 197–210 (2003)
6. Doerner, K.F., Merkle, D., Stützle, T.: Special issue on ant colony optimization. Swarm Intelligence 3(1), 1–2 (2009)
7. Dorigo, M., Di Caro, G., Gambardella, L.: Ant algorithms for distributed discrete optimization. Artif. Life 5(2), 137–172 (1999)
8. Dorigo, M., Stützle, T.: Ant Colony Optimization. MIT Press, Cambridge (2004)
9. Dorigo, M., Birattari, M., Blum, C., Clerc, M., Stützle, T., Winfield, A.F.T. (eds.): ANTS 2008. LNCS, vol. 5217. Springer, Heidelberg (2008)
10. Dorigo, M., Birattari, M., Stützle, T., Libre, U., Bruxelles, D., Roosevelt, A.F.D.: Ant colony optimization – artificial ants as a computational intelligence technique. IEEE Comput. Intell. Mag. 1, 28–39 (2006)
11. Gloor, P., Grippa, F., Putzke, J., Lassenius, C., Fuehres, H., Fischbach, K., Schoder, D.: Measuring social capital in creative teams through sociometric sensors. International Journal of Organisational Design and Engineering (2012)
12. Gloor, P.A.: Swarm Creativity: Competitive Advantage through Collaborative Innovation Networks. Oxford University Press, USA (2006)
13. Grasse, P.-P.: Termitologia, vol. II. Masson, Paris (1984)
14. Kiritchenko, S., Matwin, S.: Email classification with co-training. Tech. rep., University of Ottawa (2002)
15. Lewis, D.D.: Representation and Learning in Information Retrieval. Ph.D. thesis, Department of Computer Science, University of Massachusetts (1992)
16. Moreno, J.L.: Who Shall Survive? Foundations of Sociometry, Group Psychotherapy and Sociodrama. Beacon House, Beacon (1953)
17. Wang, M., He, Y., Jiang, M.: Text categorization of enron email corpus based on information bottleneck and maximal entropy (2010)
18. Wilson, G.C., Banzhaf, W.: Discovery of email communication networks from the enron corpus with a genetic algorithm using social network analysis (2009)

# Goal-Oriented Requirements
# for ACDT Algorithms

Jan Kozak and Urszula Boryczka

Institute of Computer Science, University of Silesia,
Będzińska 39, 41-200 Sosnowiec, Poland
{jan.kozak,urszula.boryczka}@us.edu.pl

**Abstract.** This paper is devoted to the new application of the ACDF approach. In this work we propose a new way of an virtual-ant performance evaluation. This approach concentrates on the decision tree construction using ant colony metaphor the goal of experiments is to show that decision trees construction may by oriented not only at accuracy measure. The proposed approach enables (depending on the decision tree quality measure) the decision tree construction with high value of accuracy, recall, precision, F-measure or Matthews correlation coefficient. It is possible due to use of nondeterministic, probabilistic approach - Ant Colony Optimization. The algorithm proposed was examined and the experimental study confirmed that the goal-oriented ACDT can create expected decision trees, accordance to the specified measures.

**Keywords:** Decision Tree, Ant Colony Optimization, Evaluation of Classification.

## 1   Introduction

The authors propose an approach entailing a goal-oriented ACDT algorithm which is aimed at evaluating classification based on different measures: recall, precision, F-measure or the Matthews correlation coefficient. This approach is possible as a result of changing an agent-ant's goal function (evaluation of a solution's quality) while maintaining the method of determining the value of the heuristic function. This gives a significant advantage to the ACDT algorithm over other, classical approaches which do not make it possible to adapt a classifier to a given measure when the division criteria are invariant. Consequently, one can construct trees with better recall, precision, F-measure or the Matthews correlation coefficient, depending on current needs, by using a different evaluation of classification as the goal function of an ant colony algorithm.

Data mining and machine learning have been the subject of increasing attention over the past 30 years. Ensemble methods, popular in machine learning and pattern recognition, are learning algorithms that construct a set of many individual classifiers, called base learners, and combine them to classify new data points or samples by taking a weighted or unweighted vote of their predictions. It is now well-known that ensembles are often much more accurate than the individual classifiers that make them up. The success of ensemble approaches on many benchmark data sets has raised considerable interest in understanding why such methods succeed and identifying circumstances in which they can be expected to produce good results.

Ant Colony Optimization is a branch of a newly developed form of artificial intelligence called swarm intelligence [5,7] . Swarm intelligence is a form of emergent collective intelligence of groups of simple individuals: ants, termites or bees in which a form of indirect communication via pheromone was observed. Pheromone values encourage the ants following the path to build good solutions of the analyzed problem and the learning process occurring in this situation is called positive feedback or auto-catalysis.

In this paper we defined an ant algorithm to be a multi–agent system inspired by the observation of real ant colony behavior exploiting the stigmergic communication paradigm. The optimization algorithm in this paper was inspired by the previous works on Ant Systems (AS) and, in general, by the term — stigmergy. This phenomenon was first introduced by P. P. Grasse [11].

An essential step in this direction was the development of Ant System by Dorigo et al. [7], a new type of heuristic inspired by analogies to the foraging behavior of real ant colonies, which has proven to work successfully in a series of experimental studies. Diverse modifications of AS have been applied to many different types of discrete optimization problems and have produced very satisfactory results [6,8].

ACO is not the only technology that is used in the decision tree construction. As far as we know, evolutionary computation was also used in this area. This achievement was published by Chai et al. [4]. In paper [10], the basic population was assembled by decision trees firstly constructed by C4.5 algorithm. The memetic algorithm, proposed by Kretowski [14] in decision tree induction is also worth mentioning. Whereas GP with Simulated Annealing was presented in [9]. On the other hand there are no articles in which there is classification measure other than classification accuracy.

This article is organised as follows. Section 1 comprises an introduction to the subject of this article. Section 2 and 3 describe decision trees and, Ant Colony Decisin Tree approach. Section 4 describes the Evaluation of Classification. Section 5 focuses on the new idea of the ACDT algorithm. Section 6 presents the experimental study that was conducted to evaluate the performance of the new approach to the ACDT algorithm by taking into consideration six data sets. Finally, we conclude with general remarks on this work, and a few directions for future research are pointed out.

## 2   Decision Trees

One of the most efficient and widely applied learning algorithms search the hypothesis (solution) space consisting of decision trees [15]. The term hypothesis is understood as a combination of attribute values which determine the way to undertake a specific decision. A decision tree learning algorithm searches the space of such trees by first considering trees that test only one attribute and making an immediate classification. Then they consider expanding the tree by replacing one of the leaves by a test of the second attribute. Various heuristics are applied to choose which test to include in each iteration and when to stop growing the tree. The evaluation function for decision trees will be calculated according to the following formula:

$$Q(T) = \phi \cdot w(T) + \psi \cdot a(T, S) \tag{1}$$

where: $w(T)$ – the size (number of nodes) of the decision tree $T$; $a(T, S)$ – the accuracy of the classification samples from a test set $S$ by the tree $T$; $\phi$ and $\psi$ – constants determining the relative importance of $w(T)$ and $a(T, S)$.

Constructing optimal binary decision trees is an NP–complete problem, where an optimal tree is one which minimizes the expected number of tests required for identification of the unknown samples, as shown by Hyafil et al. in [12]. Classification And Regression Tree (CART) approach was developed by Breiman et al. in 1984 [3] .

Twoing criterion, firstly proposed in CART, will search for two classes that will make up together more then 50% of the data. Twoing splitting rule maximizes the following change-of-impurity measure which implies the following maximization problem for nodes $m_l$, $m_r$:

$$\operatorname*{arg\,max}_{a_j \leq a_j^R, j=1,\ldots,M} \left( \frac{P_l P_r}{4} \left[ \sum_{k=1}^{K} |p(k|m_l) - p(k|m_r)| \right]^2 \right), \tag{2}$$

where: $p(k|m_l)$, $p(k|m_r)$ – the conditional probability of the class $k$ provided in node $m_l$, $m_r$; $P_l$, $P_r$ – the probability of transition samples into the left or right node $m_l$, $m_r$; $K$ – number of decision classes; $a_j$ – $j$-th variable, $a_j^R$ is the best splitting value of variable $a_j$.

## 3   Ant Colony Decision Trees Algorithm

Ant Colony Optimization (ACO) approach has been successfully applied to many difficult combinatorial problems. Ant Colony Decision Trees (ACDT) algorithm is the first ACO adaptation to the task of rule induction and constructing decision trees, but also rule induction approach – Ant-Miner [2,16].

In each ACDT step an ant chooses an attribute and its value for splitting the samples in the current node of the constructed decision tree. The choice is made according to a heuristic function and pheromone values. The heuristic function is based on the Twoing criterion (eq. (2)), which helps ants select an attribute-value pair which well divides the samples into two disjoint sets, i.e. with the intention that samples belonging to the same decision class should be put in the same subset. The best splitting is observed when similar number of samples exists in the left subtree and in the right subtree, and samples belonging to the same decision class are in the same subtree. Pheromone values indicate the best way (connection) from the superior to the subordinate nodes – all possible combinations are taken into account.

As mentioned before, the value of the heuristic function is determined according to the splitting rule employed in CART approach (see formula (2)). The probability of choosing the appropriate split in the node is calculated according to a classical probability used in ACO [8]:

$$p_{i,j} = \frac{\tau_{m,m_{L(i,j)}}(t) \cdot \eta_{i,j}^{\beta}}{\sum_i^a \sum_j^{b_i} \tau_{m,m_{L(i,j)}}(t) \cdot \eta_{i,j}^{\beta}}, \tag{3}$$

where:

$\eta_{i,j}$ – a heuristic value for the split using the attribute $i$ and value $j$,

$\tau_{m,m_{L(i,j)}}$ – an amount of pheromone currently available at time $t$ on the connection between nodes $m$ and $m_L$, (it concerns the attribute $i$ and value $j$),

$\beta$ – the relative importance with experimentally determined values 3, respectively.

The initial value of the pheromone trail is determined similarly to the Ant–Miner approach and depends on the number of attribute values. The pheromone trail is updated (4) by increasing pheromone levels on the edges connecting each tree node with its parent node:

$$\tau_{m,m_L}(t+1) = (1 - \gamma) \cdot \tau_{m,m_L}(t) + Q(T), \tag{4}$$

where $Q(T)$ is a quality of the decision tree (see formula (1)), and $\gamma$ is a parameter representing the evaporation rate, equal to 0.1.

## 4   Evaluation of Classification

The evaluation of classification quality is one of the problems associated with machine learning and it is crucial in determining whether a given classifier is of good or poor quality. There are, however, no classifiers that could be changed depending on the measure that is being used, or optimized for several measures while, when it comes to real-life problems, it is, for example, precision or the balancing of two different measures of evaluating classification [13] that may often turn out to be more important.

The text below presents selected and most popular measures of binary classification quality (for data sets with two decision classes) which can be determined based on a confusion matrix [17]. All of them are based on a confusion matrix, i.e. a table that is connected with classification (tab. 1), which makes it possible to better evaluate the quality of this classification. The confusion matrix contains information about an object's decision class as well as the class into which this object has been classified, based on which one can determine the values of accuracy, recall, precision, F-measure and the Matthews correlation coefficient.

It should be noted that the accuracy of classification is a measure that only shows how many objects have been correctly classified whereas, for example, precision makes it possible to assess the confidence with which one can assume that an object belonging to a given class will be correctly classified. This is of particular significance when it is more important that the objects belonging to a

**Table 1.** Confusion matrix

|  | Predicted positive | Predicted negative |
|---|---|---|
| Positive examples | True positive (TP) | False negative (FN) |
| Negative examples | False positive (FP) | True negative (TN) |

given class be correctly classified into this class – it is better to assign too many objects to a given class than to omit one of the relevant objects. Analogously, as for recall, it is information whether a given class contains objects that have been correctly assigned to it that is more important – it is better to omit one of the relevant objects than to incorrectly assign an irrelevant object to a class.

**Accuracy** constitutes one of the most popular classification evaluation measures. It should, however, be noted that this measure does not provide sufficient evaluation, for example, for data sets with a considerable diversity of decision classes, which was observed in [13], among others. The accuracy of classification describes the ratio of objects that have been correctly classified to all objects in a class.

$$ev_{acc}(T, S) = \frac{(TP + TN)}{(TP + TN + FP + FN)}. \tag{5}$$

**Recall** is a different, simple measure which is used for binary classification. It is determined based on equation (6), i.e. the ratio of objects that have been correctly classified into class $P$ to all objects that should have been classified into this class. Therefore, as for recall, it can be stated that it is better to incorrectly assign an object belonging to class $N$ to class $P$ than to incorrectly classify an object belonging to class $P$.

$$ev_{rec}(T, S) = \frac{TP}{(TP + FN)}. \tag{6}$$

**Precision** is a measure that evaluates a classifier based on an incorrect classification of objects belonging to class $N$ into class $P$. Here it is better to omit certain objects from class $P$ than to incorrectly assign objects belonging to class $N$ to class $P$. Precision is determined based on the ratio of objects that have been correctly classified into class $P$ to all objects that have been assigned to this class:

$$ev_{prec}(T, S) = \frac{TP}{(TP + FP)}. \tag{7}$$

**F-measure** (F1 score) constitutes an attempt to balance precision and recall. This is a frequently used measure for evaluating binary classification (two decision classes). In its simplest form it is determined based on the ratio of a double product of precision and recall to the sum of these two measures:

$$ev_{fm}(T, S) = 2 \cdot \frac{ev_{prec}(T, S) \cdot ev_{rec}(T, S)}{ev_{prec}(T, S) + ev_{rec}(T, S)}, \tag{8}$$

Therefore, after making appropriate substitutions the following equation is obtained:

$$ev_{fm}(T, S) = \frac{(TP + TP)}{(TP + TP + FP + FN)}. \qquad (9)$$

**Matthews correlation coefficient** is also used as an evaluation measure in binary classification. In contrast to the above-mentioned recall, precision and F-measure, it is determined based on the entire confusion matrix and calculated by using the following equation:

$$ev_{mcc}(T, S) = \frac{(TP \cdot TN - FP \cdot FN)}{\sqrt{(TP + FP) \cdot (TP + FN) \cdot (TN + FP) \cdot (TN + FN)}}. \qquad (10)$$

The Matthews correlation coefficient is considered to be a good measure which can be used when the disproportion between the cardinalities of decision classes is very large.

## 5   Goal-Oriented ACDT Algorithm

A new approach to the ACDT algorithm is based on changing the goal function of agent-ants by modifying the way of determining the evaluation function for estimating decision tree quality (eq. (1)) which, for example, influences pheromone trail updating (eq. (4)). This will cause solutions with a higher value of a given classification evaluation measure to be rewarded with the pheromone trail.

In the experiments that are described here equation (1) was reduced to this form:

$$Q(T) = \phi \cdot w(T) + \psi \cdot ev(T, S), \qquad (11)$$

where $ev(T, S)$ (which is also referred to as the goal function in this article) denotes a selected method of evaluating the classification of decision tree $T$ that was constructed by an agent-ant based on data set $S$. The value of $ev(T, S)$ is determined depending on the selected measure in the following way: accuracy - eq. (5), recall - eq. (6), precision - eq. (7), f-measure - eq. (9), The Matthews correlation coefficient - eq. (10).

## 6   Experiments

A variety of experiments was conducted to test the performance and behaviour of the proposed algorithm. In this section we will consider an experimental study (see Table 2, Figs. 1 and 2 – the best results are presented in bold) performed for the following adjustments. We performed 30 experiments for each data set. Each experiment included 200 generations with the population size of the ant colony equal to 15. In the presented approaches the twoing criterion as well as Error-Based Pruning procedure have been used. The parameters values employed in ACO are established in the way firstly presented in [1]: $q_0 = 0.3$, $\alpha = 3.0$, $\gamma = 0.1$, $\phi = 0.05$ and $\psi = 1.0$. The experiments were carried out on an Intel Core i5 2.27 GHz Computer with 2.9 GB RAM.

## 6.1 Data Sets

An evaluation of the performance behavior of ACDT was performed using 6 public-domain data sets from the UCI (University of California at Irvine) data set repository. Data sets are estimated by 10-fold cross-validation. We used an additional data set – a clean set. The results are tested on a clean set that has not been used to build the classifier.

The characteristics of data sets, which is important when carrying out experiments, is as presented below (the number of objects, the number of conditional attributes, the number of objects in class "0" and the number of objects in class "1"): *australian*, 690, 14, 307, 383; *bcw*, 699, 9, 458, 241; *heart*, 270, 13, 120, 150; *hepatitis*, 155, 19, 123, 32; *horse − colic*, 366, 22, 136, 230; *ttt*, 958, 9, 625, 332.

Roughly speaking, the ratio of objects in class "0" to all objects is similar for the *australian* and *heart* sets (44%) and for the *ttt* and *bcw* sets (65-66%). The number of objects in class "0" in relation to the size of the data set is the smallest for the *horse − colic* set (37%) and the largest for the *hepatitis* set (79%).

## 6.2 Results of Experiments

The aim of the experiments was to compare five versions of the ACDT algorithm which use the proposed measures for evaluating the quality of classification. The way of evaluating the quality of classification is crucial in selecting an agent-ant which is to be used to update the pheromone trail; a particular evaluation method also influences the new value of an updated pheromone. Also, the values of each of these measures were taken into account (irrespective of the current method of determining $ev(T, S)$), as well as the size of the decision tree, its height and the run-time of the algorithm, i.e. the time during which the classifier was created.

The obtained results (which are presented in Table 2) confirm that, in most cases, the maximum value for each of the measures is determined when the algorithm using a given measure (the measure in question) as its goal function is working. This is true irrespective of which data set is analyzed. In other words, the highest accuracy is obtained when measure (5), i.e. accuracy, is used in equation (1) as $ev(T, S)$, and the highest precision is obtained when it is precision (7) that is optimized, etc. This means that it is possible to maximize the value of a given measure by changing the way of determining $ev(T, S)$ - interestingly, this can often be achieved without a considerable worsening of accuracy (which here is treated as a measure that was originally used in the ACDT algorithm).

When comparing the results for the *australian* set one can notice that accuracy remains at a similar level while recall improves by nearly 10% in relation to the classical approach and by not less than 5% more than when the other measures are used. The situation is similar for precision (a 6-percent improvement). The disproportion was slightly smaller when F-measure was used. Very similar dependencies were found for the *hepatitis* set; however, here the differences in the accuracy of classification were larger, and the highest value of accuracy was obtained when F-measure was used. These results are similar to those obtained for the *heart* set, where, additionally, recall makes the accuracy of classification considerably worse. Larger differences in accuracy, but with the previously described observations still holding true, can be noticed for the remaining data sets: *bcw*, *horse − colic* i *ttt*, i.e. those in which the difference in the number of objects belonging to decision classes is at 33-35%.

**Table 2.** The obtained value of the test measures depending on the goal function – the average of all performance

| | Measure | australian | | | | | bcw | | | | |
|---|---|---|---|---|---|---|---|---|---|---|---|
| | | acc | rec | prec | F1 | MCC | acc | rec | prec | F1 | MCC |
| Goal function | acc | **0.846** | 0.849 | 0.816 | 0.828 | 0.691 | **0.937** | 0.932 | 0.973 | 0.950 | 0.861 |
| | rec | 0.835 | **0.932** | 0.761 | 0.834 | **0.692** | 0.909 | **0.956** | 0.914 | 0.934 | 0.793 |
| | prec | 0.838 | 0.744 | **0.878** | 0.801 | 0.677 | 0.918 | 0.897 | **0.979** | 0.935 | 0.832 |
| | F1 | 0.843 | 0.882 | 0.803 | **0.838** | 0.702 | 0.932 | 0.929 | 0.969 | **0.952** | 0.854 |
| | MCC | 0.833 | 0.838 | 0.804 | 0.817 | 0.670 | 0.935 | 0.932 | 0.973 | 0.949 | **0.863** |

| | Measure | heart | | | | | hepatitis | | | | |
|---|---|---|---|---|---|---|---|---|---|---|---|
| | | acc | rec | prec | F1 | MCC | acc | rec | prec | F1 | MCC |
| Goal function | acc | 0.767 | 0.717 | 0.762 | 0.728 | 0.537 | 0.789 | 0.891 | 0.852 | 0.868 | 0.390 |
| | rec | 0.707 | **0.798** | 0.654 | 0.710 | 0.443 | 0.788 | **0.966** | 0.806 | 0.878 | 0.318 |
| | prec | 0.751 | 0.563 | **0.849** | 0.657 | 0.514 | 0.775 | 0.839 | **0.874** | 0.852 | 0.403 |
| | F1 | **0.780** | 0.753 | 0.766 | **0.749** | 0.561 | **0.814** | 0.941 | 0.845 | **0.888** | **0.410** |
| | MCC | 0.776 | 0.727 | 0.777 | 0.736 | **0.567** | 0.789 | 0.889 | 0.858 | 0.867 | 0.404 |

| | Measure | horse-colic | | | | | ttt | | | | |
|---|---|---|---|---|---|---|---|---|---|---|---|
| | | acc | rec | prec | F1 | MCC | acc | rec | prec | F1 | MCC |
| Goal function | acc | **0.818** | 0.733 | 0.798 | **0.755** | **0.620** | 0.858 | 0.925 | 0.868 | 0.895 | 0.681 |
| | rec | 0.764 | **0.820** | 0.667 | 0.729 | 0.543 | 0.761 | **0.992** | 0.736 | 0.845 | 0.465 |
| | prec | 0.815 | 0.637 | **0.854** | 0.720 | 0.608 | 0.823 | 0.837 | **0.888** | 0.859 | 0.627 |
| | F1 | 0.809 | 0.726 | 0.786 | 0.745 | 0.603 | 0.858 | 0.922 | 0.869 | **0.899** | 0.679 |
| | MCC | 0.814 | 0.723 | 0.791 | 0.747 | 0.609 | **0.863** | 0.924 | 0.875 | 0.894 | **0.693** |

Abbrev.: acc – accuracy; rec – recall; prec – precision; F1 – F-measure; MCC – Matthews correlation coefficient

F-measure can be regarded as the most universal goal function; the results are usually correct for each of the measures when F-measure is used. This probably results from the way in which F-measure's values are determined, i.e. as described in Section 4. F-measure somewhat balances precision and recall. The size of trees that are constructed by using F-measure (Fig. 2) and the run-time of the algorithm (Fig. 1) do not also differ much from the average values obtained in the other cases.

Figure 1 presents a comparison between the number of nodes of the constructed trees in relation to the classical ACDT (with accuracy). Definitely the

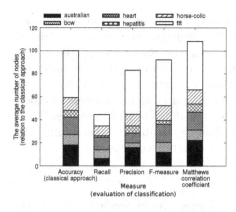

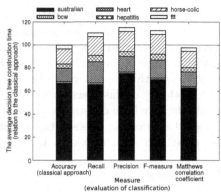

**Fig. 1.** The average number of nodes which was established based on all of the data sets in relation to the classical approach

**Fig. 2.** The average decision tree construction time which was established based on all of the data sets in relation to the classical approach

smallest decision trees are constructed when using recall as the goal function. Interestingly, even a large reduction in the number of nodes in a tree does not necessarily result in a large deterioration in accuracy in this case, and it often increases the value of recall.

Figure 2 presents a comparison between decision tree constructing times. The algorithm with accuracy and MCC as the goal function is the fastest, and it is slightly slower for recall, F-measure and precision. One can, however, state that the algorithm works in a similar way for each of the measures.

## 7 Conclusions

This article presents the ACDT algorithm with five different measures for evaluating the quality of classification which were used as the ACDT algorithm's goal function. For all of these approaches it has been observed that the change of the goal function leads to the maximization of a selected measure although the maximum value is sometimes found when a different measure is used. This possibility is of vital importance for data sets that are difficult to evaluate by using the accuracy of classification. The authors' proposal allows one to construct trees based on the division criteria that are related to accuracy and yet to maximize them in terms of another, selected measure. This makes it possible to construct not only an accurate but also, for example, precise classifier.

The results indicate that, for the ACDT algorithm, F-measure can be regarded as a better universal measure (i.e. which gives average values of results) than accuracy. Further research is also required to evaluate the goal function in terms of the sizes of constructed decision trees. The results of the experiments show that trees with a smaller number of nodes and of smaller height are constructed when the recall measure is used, and larger trees are constructed when the precision measure is used.

As for the research on evaluating the quality of classification in Ant Colony Optimization algorithms that are used for classification, the next step should be to use new measures in the process of constructing decision trees, i.e. the ACDF algorithm which is popular in this field. One can consider using different measures in such a way that each tree will be evaluated by a different measure, as a result of which the set of classifiers (in a decision forest) will contain classifiers (decision trees) that have been constructed based on several different goal functions. When combined with voting, this solution should allow one to obtain stable results with respect to each of the measures, i.e. to achieve a certain consensus.

# References

1. Boryczka, U., Kozak, J.: Ant colony decision trees – A new method for construct-ing decision trees based on ant colony optimization. In: Pan, J.-S., Chen, S.-M., Nguyen, N.T. (eds.) ICCCI 2010, Part I. LNCS, vol. 6421, pp. 373–382. Springer, Heidelberg (2010)
2. Boryczka, U., Kozak, J.: New Algorithms for Generation Decision Trees – Ant–Miner and Its Modifications. In: Abraham, A., et al. (eds.) Foundations of Comput. Intel. 6. SCI, vol. 206, pp. 229–264. Springer, Heidelberg (2009)
3. Breiman, L., Friedman, J.H., Olshen, R.A., Stone, C.J.: Classification and Regres-sion Trees. Chapman & Hall, New York (1984)
4. Chai, B.-B., Zhuang, X., Zhao, Y., Sklansky, J.: Binary linear decision tree with ge-netic algorithm. In: International Conference on Pattern Recognition, vol. 4 (1996)
5. Corne, D., Dorigo, M., Glover, F.: New Ideas in Optimization. McGraw–Hill, Cam-bridge (1999)
6. Dorigo, M., Di Caro, G.: New Ideas in Optimization. McGraw–Hill, London (1999)
7. Dorigo, M., Stützle, T.: Ant Colony Optimization. MIT Press, Cambridge (2004)
8. Dorigo, M., Birattari, M., Blum, C., Clerc, M., Stützle, T., Winfield, A.F.T. (eds.): ANTS 2008. LNCS, vol. 5217. Springer, Heidelberg (2008)
9. Folino, G., Pizzuti, C., Spezzano, G.: Genetic programming and simulated anneal-ing: A hybrid method to evolve decision trees. In: Poli, R., Banzhaf, W., Langdon, W.B., Miller, J., Nordin, P., Fogarty, T.C. (eds.) EuroGP 2000. LNCS, vol. 1802, pp. 294–303. Springer, Heidelberg (2000)
10. Fu, Z., Golden, B.L., Lele, S., Raghavan, S., Wasil, E.A.: Diversification for better classification trees. Computers & OR 33(11), 3185–3202 (2006)
11. Grasse, P.-P.: Termitologia, vol. II. Masson, Paris (1984)
12. Hyafil, L., Rivest, R.: Constructing optimal binary decision trees is NP–complete. Inf. Process. Lett. 5(1), 15–17 (1976)
13. Kozak, J., Boryczka, U.: Dynamic version of the ACDT/ACDF algorithm for H-bond data set analysis. In: Bădică, C., Nguyen, N.T., Brezovan, M. (eds.) ICCCI 2013. LNCS, vol. 8083, pp. 701–710. Springer, Heidelberg (2013)
14. Krętowski, M.: A memetic algorithm for global induction of decision trees. In: Geffert, V., Karhumäki, J., Bertoni, A., Preneel, B., Návrat, P., Bieliková, M. (eds.) SOFSEM 2008. LNCS, vol. 4910, pp. 531–540. Springer, Heidelberg (2008)
15. Murphy, O.J., McCraw, R.L.: Designing Storage Efficient Decision Trees. IEEE Transactions on Computers 40, 315–320 (1991)
16. Otero, F.E.B., Freitas, A.A., Johnson, C.G.: Handling continuous attributes in ant colony classification algorithms. In: CIDM, pp. 225–231 (2009)
17. Rokach, L., Maimon, O.: Data Mining With Decision Trees: Theory And Applica-tions. World Scientific Publishing (2008)

# Implementing Population-Based ACO

Rafał Skinderowicz

Institute of Computer Science, Silesia University, Sosnowiec, Poland
`rafal.skinderowicz@us.edu.pl`

**Abstract.** Population-based ant colony optimization (PACO) is one of
the most efficient ant colony optimization (ACO) algorithms. Its strength
results from a pheromone memory model in which pheromone values
are calculated based on a population of solutions. In each iteration an
iteration-best solution may enter the population depending on an update
strategy specified. When a solution enters or leaves the population the
corresponding pheromone trails are updated. The article shows that the
PACO pheromone memory model can be utilized to speed up the process
of selecting a new solution component by an ant. Depending on the values
of parameters, it allows for an implementation which is not only memory
efficient but also significantly faster than the standard approach.

**Keywords:** population based ant colony optimization, travelling salesman problem, pheromone memory.

## 1 Introduction

Ant Colony Optimization (ACO) is a nature inspired metaheuristic mainly used
to solve combinatorial optimization problems [1]. In the ACO a number of artificial ants iteratively construct solutions to the problem tackled. In each iteration
an ant extends its partial solution with a new component selected according to
the probability distribution based on the *heuristic information* and *pheromone
trails* corresponding to the available solution components. The heuristic information is usually *static* and given *a priori* and only the pheromone trails are
updated based on the ants decisions. The pheromone trails comprise a *pheromone
memory* shared by the ants. The pheromone memory allows the collective of ants
to learn and thus its use is essential for the good performance of the algorithm
[2]. A number of the ACO modifications were proposed of which many differ in
the usage of pheromone memory. A detailed survey of many ACO algorithms
can be found in [1].

In the population-based ACO algorithm, the pheromone values change according to a population of solutions. The population is updated in every iteration
using an *iteration best solution*. The iteration best solution may be added to
the population if it meets the criteria specified by the *update strategy* used. The
size of the population is limited, hence adding a solution may require removal
of one of the solutions from the population. An important feature of the PACO
is that the values (levels) of pheromone trails depend only on the current contents of the population. This fact allows for an implementation in which the

D. Hwang et al. (Eds.): ICCCI 2014, LNAI 8733, pp. 603–612, 2014.

values of pheromone trails are stored in a compact, *indirect form* opposed to a full pheromone matrix [8]. It is especially valuable for a parallel implementation, but we will show that it allows for an efficient serial implementation of the PACO.

The remainder of this paper is organized as follows. A brief description of the PACO algorithm is given in Section 2. Section 3 describes how the PACO pheromone memory model can be utilised to speed up the solution construction process. Section 4 presents results of the computational experiments conducted along with comments. Finally, in Section 5 we give conclusions and present ideas for future work.

## 2    Population-Based ACO

In this section we describe how the ants construct solutions in the PACO with the focus on the selection rule. Next we focus on the pheromone memory model used in the PACO along with the strategies used to update the pheromone trails.

### 2.1    Solution Construction Process

The solution construction process used in the PACO closely resembles the process present in the ant colony system (ACS) algorithm, which in turn differs only slightly from the one used in the ACO [2,3]. In the ACO and many ACO–related algorithms, including PACO, each ant constructs a solution to the problem tackled iteratively. An ant starts with a randomly chosen solution component (node) and in each iteration it appends a new component based on the *selection rule*. In the case of TSP, the ant starts at randomly selected city (node) and in every subsequent step selects one of the unvisited nodes until the solution is complete. Specifically, the ant located at the node $i$ decides which of the unvisited neighbour nodes (denoted by $\mathcal{S}_i$) to choose based on two factors. The first one is called *heuristic information*, $\eta_{ij}$, and depends on the problem ($j$ denotes an unvisited neighbour node). In the case of TSP, $\eta_{ij}$ is equal to the inverse distance between the cities $i$ and $j$. A value, $\tau_{ij}$, of the *pheromone trail* deposited by the ants on the edge $(i, j)$ is the second factor.

The decision process of an ant in the PACO consists of two steps. Firstly, a random number $r$ is drawn with uniform probability from $[0, 1]$. Secondly, $r$ is compared with the value of a parameter $q_0 \in [0, 1)$. If $r \leq q_0$, the ant chooses the node $j \in \mathcal{S}_i$ according to:

$$j = \arg \max_{k \in \mathcal{S}_i} \tau_{ik} \cdot \eta_{ik}^{\beta} , \qquad (1)$$

where $\beta$ is a parameter denoting the relative importance of the heuristic information. Obviously, the choice specified by (1) is *greedy*. If $r > q_0$ the next node is selected with the probability:

$$P(j|i) = \frac{\tau_{ij} \cdot \eta_{ij}^{\beta}}{\sum_{k \in S_i} \tau_{ik} \cdot \eta_{ik}^{\beta}} . \qquad (2)$$

The parameter $q_0$ affects the *exploitation* to *exploration* ratio. For the values of $q_0$ close to 1 the solution search process becomes mainly exploitative, i.e. in the most cases the ants select nodes connected with short edges and high concentration of the pheromone.

## 2.2   PACO Pheromone Memory Model

In order to understand how the PACO pheromone memory works it is convenient to compare it with the pheromone memory used in the Ant Colony System (ACS). In the ACS the values of the pheromone trails are stored in a *pheromone matrix* $[\tau_{ij}]$ of size $N \times N$, where $N$ is the number of nodes (equal to the size of the problem) [2]. At the beginning, all the values in the pheromone matrix are set to an initial value $\tau_0$. Each time an ant selects an edge $(i, j)$ some of the corresponding pheromone is evaporated resulting in a smaller probability of selecting it by the other ants. This process is called a *local pheromone update*. After all the ants have built complete solutions a *global pheromone update* is performed, which consists in increasing the values of the pheromone trails corresponding to the components of the current global-best solution. The global update increases the probability of building solutions close to the best-so-far. It is worth noting that the values in the pheromone matrix accumulate all the changes made by the ants since the start of the algorithm.

In contrast to the ACS, the values of the pheromone trails in the PACO depend on the contents of a *population* of solutions selected among the iteration-best solutions found so far. The elements of the population are stored in a *solution archive*, denoted by P, updated once per iteration according to the specified update strategy. The size of the archive, denoted by $K$, is limited, hence a solution has to be removed if addition of a new one would result in exceeding the limit. Each time a solution enters the archive a pheromone is increased on the edges corresponding to the elements of the solution. Analogically, each time a solution is removed from the archive the previously deposited pheromone is removed from the respective pheromone trails. Specifically, the general form of the PACO pheromone matrix is given by:

$$\tau_{ij} = \tau_0 + \Delta \cdot |\{\pi \in P \,|\, (i, j) \in \pi\}| \,, \tag{3}$$

where

$$\Delta = \frac{\tau_{\max} - \tau_0}{K} \,, \tag{4}$$

where $\tau_0 = 1/(N - 1)$ and $\tau_{\max}$ is a parameter. The updating of the archive corresponds to the global pheromone update in the ACO. No local pheromone update is performed in the PACO, what reduces the computation time required to update the pheromone from $O(N^2)$ to $O(N)$ compared with the ACO. It is worth noting, that all the values in the PACO pheromone memory can be calculated based on the *contents* of the archive.

Several population update strategies were proposed by Guntch [3,4]. In this paper we consider the three investigated in [5], namely: the *age-based* strategy, the *quality-based* strategy and the *elitist-based* strategy.

The age-based strategy works in a FIFO (First-In-First-Out) fashion. After each iteration of the PACO the iteration-best solution is appended to the solution archive. If the size of the archive exceeds the specified limit, $K$, the eldest solution is removed.

In the quality-based strategy after each iteration the iteration best solution is considered for addition to the archive. Its quality is compared with the quality of the worst solution in the archive and if its higher, then the new solution replaces the worst, otherwise the archive remains unchanged. Compared to the age-based strategy the quality-based strategy places stronger emphasis on the exploitation by focusing its search on the neighbourhood of the best quality solutions found so far.

The elitist-based strategy works similarly to the age-based strategy but with one exception – the archive always contains the best solution found so far, called the *elitist solution* (denoted by $e$). Each time a solution enters the archive it is compared with the elitist solution and if its quality is higher, then it replaces the elitist solution. Otherwise, the new solution is appended to the archive and the eldest of the remaining $K-1$ solutions is removed. Pheromone trails corresponding to the components of the elitist solution receive an amount of pheromone equal to $\omega_e \cdot \tau_{\max}$, where $\omega_e$ is a parameter. The pheromone trails for the rest of the solutions in the archive are updated to $\Delta \cdot (1 - \omega_e)/(K - 1)$.

## 3   Selection Rule Optimization

As it was mentioned, the pheromone memory model used in the PACO does not require to explicitly store the pheromone matrix, because the values can be computed based on the contents of the solution archive. Obviously, it allows to reduce the memory consumption compared with the full pheromone matrix, especially if the size of the archive is small, i.e. $K \ll N^2$. It is especially beneficial for the hardware implementation of the PACO on FPGA as showed by Scheuermann et. al [8]. Small size of the archive allowed to store all the solutions in the memory and parallelize the solution construction process. A small size of the archive allows also to efficiently transfer the contents of the pheromone memory between CPUs in the distributed memory architecture compared with the relatively long time required to transfer the full pheromone matrix [6].

The reduction of the memory consumption is not the only potential benefit of the PACO pheromone memory model. If $K$ is small the number of pheromone trails with the non-initial, $\tau_0$, value is also small. This fact can be used to speed up the process of selecting a new solution component. Lets assume that an ant located at the node $i$ is performing a greedy selection of a next node $j$ according to Eq. (1), then it has to consider the set $\mathcal{S}_i$ of the unvisited neighbours of $i$. The elements of the set $\mathcal{S}_i$ can be divided into two disjoint subsets: $\mathcal{S}_i^P$ and $\mathcal{S}_i^D$. The first set contains only the elements for which the corresponding edge belongs to at least one of the solutions in the archive, i.e. $\mathcal{S}_i^P = \{j | (i,j) \in \pi \wedge \pi \in P\}$. The second set contains the elements for which the value of the pheromone trail is equal to $\tau_0$ and can be defined as $\mathcal{S}_i^D = \mathcal{S}_i \setminus \mathcal{S}_i^P$. In order to find the next element we need to calculate the product of the pheromone value and the heuristic information for all elements in $\mathcal{S}_i^{\mathrm{P}}$ (in order to find the maximum). On the other hand, the pheromone trails for the elements in $\mathcal{S}_i^{\mathrm{D}}$ are all equal to $\tau_0$, hence we are interested in finding the element with the highest value of the heuristic information.

To what extent this observation can be useful depends on the problem tackled. In the case of the TSP (and related problems) the set of neighbours $\mathcal{S}_i$ considered by an ant located at the node $i$ is usually restricted to a so-called *candidate set* which consists of $cl$ nearest neighbours of a node [2]. If $cl \ll N$ (typically $cl \le 30$), where $N$ is the number of nodes, then the selection process becomes significantly faster compared with iterating over the full list of unvisited neighbours of the node. But even in this case it is possible to take an advantage of the fact that $\mathcal{S}_i = \mathcal{S}_i^{\mathrm{P}} \cup \mathcal{S}_i^{\mathrm{D}}$. In other words, we need to check all elements in $\mathcal{S}_i^{\mathrm{P}}$ but stop iterating over the elements of $\mathcal{S}_i^{\mathrm{D}}$ as soon as we find the first unvisited node, assuming that we access the elements in descending order of heuristic information ($\eta_{ij}$) value. Obviously, the difference should be noticeable if $K < cl$.

## 4   Experiments

In order to verify if the optimization described in Section 3 can result in speeding up the computations compared with the unmodified PACO several experiments were conducted on the TSP instances selected from the well-known TSPLIB repository [7]: *kroA100, tsp225, lin318, u574, rat783, pcb1173, pr2392* and *rl5915*. PACO requires to set a number of parameters. Most of the values were set as suggested in [5] and the rest are based on preliminary experiments. Specifically, the values were as follows: $m = 10$ ants, $\beta = 4$, $q_0 = (N - 50)/N$, $cl = 30$ (size of the candidate list), $\omega_e = 0.5$. The size of the solution archive $K$ was set to 8 (unless stated otherwise) what means that only 8 solutions were passed between successive iterations, while in every iteration $m = 10$ new solutions were constructed. All algorithms were run with a time limit as a stop condition, namely: 30 seconds for the instances: *kroA100, tsp225, lin318*; 60 seconds for the instances: *u574, rat783, pcb1173*; 120 seconds for the largest instances: *pr2392* and *rl5915*.

For every combination of the parameters values the computations were repeated 30 times. To minimize the influence of the implementation efficiency on the algorithms' performance both were implemented using the same framework and shared most of the code. The algorithms were implemented in C++ and compiled with the GCC v4.8.1 compiler with -Ofast switch. The computations were carried on a computer with Intel Xeon X5650 six-core 2.66GHz processor with 12MB cache, running under the control of Scientific Linux. A single processor core was utilised by the algorithm.

### 4.1   Relative Speedup Analysis

In the first part of the experiments the PACO with the modified (optimized) selection rule (denoted by FPACO) was compared with the unmodified PACO

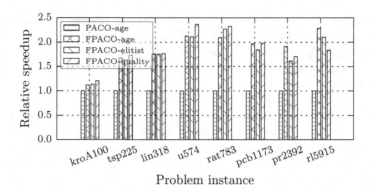

**Fig. 1.** Mean speedup of the PACO with the optimized selection rule (FPACO) with various population update strategies relative to the unmodified PACO with the *age-based* update strategy (no local search was used).

with the age-based strategy. No pheromone matrix was used in the FPACO, i.e. the pheromone values were calculated based on the contents of the solution archive.

The archive was simply a list of $K$ solutions, but an additional auxiliary array **A** of size $N \times K$ ($N$ being the size of the problem) was used to speed up the access to the pheromone values. Specifically, if a solution $X$ entered the archive, entries corresponding to the edges $(i, j) \in X$ were updated as follows. If an edge $(i, j)$ was not a part of any other solution in the archive a new entry $\langle j, \tau_0 + \Delta \rangle$ was put in a row $A_i$ (see Eq. (3) and Eq. (4)). Otherwise, the edge $(i, j)$ was already a part of $k$ ($1 \le k < K$) solutions and the corresponding entry $\langle j, \tau_0 + k\Delta \rangle$ was replaced with $\langle j, \tau_0 + (k+1)\Delta \rangle$. The auxiliary array speeds up the access to the pheromone values, especially if the same edge is a part of multiple solutions in the archive.

The unmodified PACO was run with a solution archive and a full pheromone matrix of size $N \times N$, were $N$ is the size of the problem. It is worth noting that the pheromone matrix allows for a faster access to a pheromone value than the auxiliary array, because the latter requires to check in the worst case all the entries in a respective row of **A**.

All algorithms were run for with a fixed time limit and the relative speedup was calculated based on the ratio of the mean number of iterations made to the mean number of iterations made by the PACO algorithm with the age-based update strategy. Figure 1 shows that the FPACO was significantly faster for all instances except for *kroA100* and regardless of the update strategy used. Thanks to the optimized selection rule the FPACO with the age-based update strategy was able to perform approximately twice the number of iterations for the largest instances than the unmodified PACO. Obviously, the larger number of iterations should result in solutions of better quality. Mean solution quality for the algorithms investigated is shown in Fig. 2. As can be seen, more iterations result in solutions closer to the optimum but not in all cases. The advantage is significant especially for the largest instances and algorithms with age-based

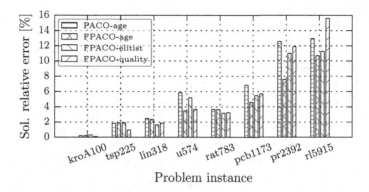

**Fig. 2.** Mean solution quality of the PACO with the optimized selection rule (FPACO) with three update strategies and the unmodified PACO with the *age-based* update strategy (no local search was used).

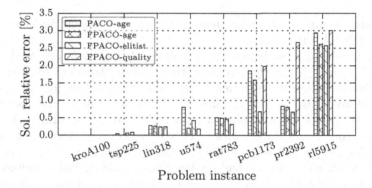

**Fig. 3.** Mean solution quality of the PACO with the optimized selection rule (FPACO) with three update strategies and the unmodified PACO with the *age-based* update strategy (no local search was used).

update strategy. Comparing the results for the FPACO with the age-based, the elitist-based and the quality-based update strategies, respectively, we can see that the age-based strategy is the most consistent in terms of quality and beats the other two for larger instances.

Generally, the quality of the solutions obtained was rather poor, especially for the largest instances. Certainly, better quality solutions could be obtained if the time limit was increased, but the ACO algorithms are usually combined with a local search heuristic what allows to significantly improve the quality of results [1]. In the second part of the experiments the algorithms were run with the same time limits but the *3-Opt local search* (LS) with don't look bits was used to improve every solution generated by the ants [11]. As can be seen in Fig. 3 the PACO with the local search obtained significantly better results, in most cases below 1% from the optimum. The time limit of 120 seconds was too

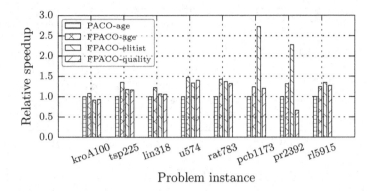

Fig. 4. Speedup of the PACO with the optimized selection rule (FPACO) with various population update strategies relative to the unmodified PACO with the *age-based* update strategy (3-Opt local search was used).

low for the instances with more than 1000 nodes and should be increased to obtain results of better quality.

Figure 4 shows the relative speedup for all the algorithms investigated. Again, the FPACO was significantly faster than the PACO, although the differences for the *age-based* update strategy were much lower than in the non-LS case, e.g. 1.24x vs 2.28x for the *rl5915* instance. The differences can be explained by the fact that the 3-Opt LS has complexity of $O(N^3)$ compared with $O(N^2)$ time required to construct a single solution. Time to perform the local search for each solution dominated the time required to build it and, in consequence, the speed up of the construction process had smaller effect on the algorithm runtime. The age-based strategy showed the most consistent behaviour when combined with LS, while the speedup of the elitist-based version varied greatly between the instances.

Table 1 presents detailed results for the PACO and the FPACO with and without LS applied. As can be seen, the FPACO achieved significantly better results in almost half of the cases, especially for the larger instances.

## 4.2   Archive Size Analysis

The number of solutions $K$ stored in the archive has a direct effect on the time required to select a next node when constructing a solution as described in Section 3. To check the influence of the archive size on the FPACO performance the algorithm was run with the age-based update strategy and various values of $K$: 2, 4, 8, 16, 32 and 64.

Figure 5 shows how the number of iterations made in a fixed time limit (60 sec.) and the quality of the solutions changed depending on the archive size, $K$. As can be seen, the results vary between the problem instances, but generally the number of iterations performed becomes lower with the increasing size of the solution archive. However, the decrease is slow due to high similarity of the

**Table 1.** Comparison of the mean number of iterations and mean solution error (relative to optimum) for the PACO and FPACO with the *age-based* update strategy. Values in bold refer to a significant difference to the other algorithm according to the *non-parametric Wilcoxon rank-sum test* (significance level $\alpha = 0.01$).

| Instance | PACO | | FPACO | | PACO + LS | | FPACO + LS | |
|---|---|---|---|---|---|---|---|---|
| | Iterations | Error [%] | Iterations | Error [%] | Iterations | Error [%] | Iterations | Error [%] |
| kroA100 | 69340 | 0,21 | 77606 | 0,21 | 33887 | 0 | 36548 | 0 |
| tsp225 | 33730 | 1,82 | 56618 | 1,89 | 19920 | 0,04 | 26923 | 0 |
| lin318 | 25506 | 2,49 | 44821 | 2,34 | 10958 | 0,28 | 13414 | 0,26 |
| u574 | 24431 | 5,85 | 51938 | **3,47** | 6455 | 0,81 | 9500 | **0,2** |
| rat783 | 18022 | 3,65 | 37693 | 3,64 | 11051 | 0,5 | 15807 | 0,49 |
| pcb1173 | 11310 | 6,84 | 22244 | **4,58** | 2536 | 1,86 | 3144 | **1,59** |
| pr2392 | 8445 | 12,57 | 16166 | **7,59** | 2936 | 0,83 | 3856 | 0,81 |
| rl5915 | 1708 | 12,96 | 3902 | **10,7** | 403 | 2,94 | 502 | **2,61** |

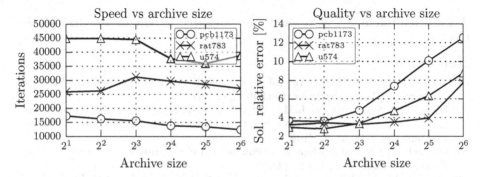

**Fig. 5.** Performance of the FPACO with the *age-based* update strategy in terms of the number of iterations performed (left) and the quality of the solutions obtained (right) vs the archive size for the instances *pcb1173*, *rat783* and *u574*.

solutions stored in the archive, i.e. it is likely that the solutions entering the archive share most of the edges or, even, have an identical structure as the ones already in the archive. Thanks to the auxiliary matrix **A** updating the pheromone value for the same edge multiple times has no effect on the pheromone lookup time and, in consequence, on the algorithm runtime. The larger archive size results also in worse quality solutions. This is an effect of the lower number of iterations performed and also negative effect of low diversity of the solutions in the archive. Summarizing, the small values of $K$ offer the best quality, at least for the *age-based* update strategy.

## 5    Summary

In the present paper we discussed how the pheromone memory used in the PACO can be exploited to obtain a time and memory efficient implementation. We proposed how the selection rule used by ants when constructing solutions can be modified to take into account the fact that relatively small (assuming small archive size) part of all possible pheromone trails have non-default ($\tau_0$) value at the same time. The PACO with the modified selection rule, denoted by FPACO, was tested on several instances of the TSP. The results showed that the FPACO was in many cases almost two times faster than the *unmodified* PACO when no local search was used, and about 25% faster when 3-Opt LS was used. It is worth noting that the FPACO consumed only a fraction of memory relative to the PACO because the pheromone matrix of size $N \times N$ was replaced with the auxiliary array of size $N \times K$, where $K$ is the size of the solution archive (equal to size of the population). For the value of $K = 8$ used in the experiments, the memory consumed by the pheromone memory was reduced from $O(N^2)$ to $O(N)$. The results confirm that PACO is a very effective metaheuristic. Future research should focus on its efficient parallelization, especially with the proposed selection rule optimization. It would also be worthwhile to compare it with the ACS with a selective pheromone memory proposed in [9,10].

**Acknowledgments.** This research was supported in part by PL-Grid Infrastructure.

# References

1. Dorigo, M., Birattari, M.: Ant colony optimization. In: Encyclopedia of Machine Learning, pp. 36–39. Springer, Heidelberg (2010)
2. Dorigo, M., Gambardella, L.M.: Ant colony system: a cooperative learning approach to the traveling salesman problem. IEEE Transactions on Evolutionary Computation 1(1), 53–66 (1997)
3. Guntsch, M.: Ant algorithms in stochastic and multi-criteria environments. PhD thesis, Karlsruhe, Univ., Diss. (2004)
4. Guntsch, M., Middendorf, M.: A population based approach for ACO. In: Cagnoni, S., Gottlieb, J., Hart, E., Middendorf, M., Raidl, G.R. (eds.) EvoWorkshops 2002. LNCS, vol. 2279, pp. 72–81. Springer, Heidelberg (2002)
5. Oliveira, S.M., Hussin, M.S., Stützle, T., Roli, A., Dorigo, M.: A detailed analysis of the population-based ant colony optimization algorithm for the TSP and the QAP. In: Proceedings of the 13th Annual Conference Companion on Genetic and Evolutionary Computation, pp. 13–14. ACM (2011)
6. Pedemonte, M., Nesmachnow, S., Cancela, H.: A survey on parallel ant colony optimization. Applied Soft Computing 11(8), 5181–5197 (2011)
7. Reinelt, G.: Tsplib95,
   http://www.iwr.uni-heidelberg.de/
   groups/comopt/-software/tsplib95/index.html
8. Scheuermann, B., So, K., Guntsch, M., Middendorf, M., Diessel, O., ElGindy, H., Schmeck, H.: Fpga implementation of population-based ant colony optimization. Applied Soft Computing 4(3), 303–322 (2004)
9. Skinderowicz, R.: Ant colony system with selective pheromone memory for TSP. In: Nguyen, N.-T., Hoang, K., Jędrzejowicz, P. (eds.) ICCCI 2012, Part II. LNCS, vol. 7654, pp. 483–492. Springer, Heidelberg (2012)
10. Skinderowicz, R.: Ant colony system with selective pheromone memory for SOP. In: Bădică, C., Nguyen, N.T., Brezovan, M. (eds.) ICCCI 2013. LNCS, vol. 8083, pp. 711–720. Springer, Heidelberg (2013)
11. Stützle, T., Dorigo, M.: ACO algorithms for the traveling salesman problem. In: Evolutionary Algorithms in Engineering and Computer Science, pp. 163–183 (1999)

# Finding Optimal Strategies in the Coordination Games

Przemyslaw Juszczuk

Institute of Computer Science,
University of Silesia, ul.Bedzinska 39, Sosnowiec, Poland
przemyslaw.juszczuk@us.edu.pl

**Abstract.** In this article we present a new algorithm which is capable to find optimal strategies in the coordination games. The coordination game refers to a large class of environments where there are multiple equilibria. We propose a approach based on the Differential Evolution where the fitness function is used to calculate the maximum deviation from the optimal strategy. The Differential Evolution (DE) is a simple and powerful optimization method, which is mainly applied to continuous problems. Thanks to the special operator of the adaptive mutation, it is possible to direct the searching process within the solution space. The approach used in this article is based on the probability of chosing the single pure strategy.

**Keywords:** Differential Evolution, Coordination Game, Optimal Strategy.

## 1 Introduction

This paper presents a new approach to find optimal strategies in the $n$-person coordination games in the normal form. These kind of games are also called the strategic form games in which every player has the finite set of the pure strategies and the payoff function mapping each strategy profile (i.e. each combination of strategies, one for each player) to a real number that captures preferences of the player over the possible outcomes of the game. We consider only games in which interests of the players are opposite and they do not cooperate. A basic feature of the competition is that the final outcome depends primarily on the combination of strategies selected by the adversaries. Many social or economical situations have in common that there are a number of decision makers (also called players) with various objectives involved, who for some reason do not cooperate and the outcome depends only on the actions chosen by the different decision makers. The simple game in the strategic form may be desribed as:

$$\Gamma = \langle N, \{A_i\}, M \rangle, i = 1, 2, ..., n \tag{1}$$

where:

- $N = \{1, 2, ..., n\}$ is the set of players;
- $\{A_i\}$ is the finite set of strategies for the $i$-th player with $m$-strategies;
- $M = \{\mu_1, \mu_2, ..., \mu_n\}$ is the finite set of the payoff functions.

D. Hwang et al. (Eds.): ICCCI 2014, LNAI 8733, pp. 613–622, 2014.
© Springer International Publishing Switzerland 2014

The strategy profile will be defined as follows: $a = (a_1, ..., a_n)$ for all players. Moreover:

$$a_{-i} = (a_1, ..., a_{i-1}, a_{i+1}, ..., a_n), \tag{2}$$

will be the strategy profile excluding the $i$-th player. The mixed strategy for the $i$-th player will be denoted as:

$$a_i = (P(a_{i_1}), P(a_{i_2}), ..., P(a_{i_m})), \tag{3}$$

where $P(a_{i_1})$ will be the probability of chosing the strategy 1 by the player $i$.

The main problem in all game theory is finding the Nash equilibrium. In general, the Nash equilibrium is a strategy profile such that no deviating player could achieve a payoff higher than the one that the specific profile gives him:

$$\forall_i, \forall_j \; \mu_i(a) \geq \mu_i(a_{i_j}, a_{-i}), \tag{4}$$

where $i$ is the $i$-th player, $j$ is the number of the strategies for given player, $\mu_i(a)$ is the payoff for the $i$-th player for the strategy profile $a$ and $\mu_i(a_{i_j}, a_{-i})$ is the payoff for the $i$-th player using strategy $j$ against the profile $a_{-i}$. The Nash equilibrium is often considered as the optimal choice. We can specify a mixed Nash equilibria which in the most of the situation are very difficult to find, (opposite to a pure Nash equilibria). In this article we consider a very popular coordination games in which there are many pure and mixed optimal strategies (the Nash equilibrium). Multiple equilibria are common in economic models and the cooperation game is example of a problem in which many optimal strategies leads to increased difficulty.

We will show an approach which allows to find a mixed optima which are far less popular than a pure optimal strategies. First we present a way to code the probability distribution over the set of pure strategies for all players - this will give us the genotype which will be decoded on the basis of the proposed fitness function. Decoded value - the fenotype of the individual will be equal to the distance between optimal mixed strategy and the actual strategy of the players.

The article is organized as follows: first we give some background related to the coordination games, next we describe a main assumptions related to these games, we shortly describe the Differential Evolution algorithm. In the section five we give a detailed description of the proposed approach. We end with experiments and conclusions.

## 2    Related Works

Finding the Nash equilibrium is not a trivial problem. There exist many mathematical methods which are capable to find optimal set of strategies. One of the state of the art algorithms is the famous Lemke Howson algorithm which is an approach similar to the simplex method [9]. The large number of algorithms are mainly focused on finding equilibria in the general types of the games. There exists algorithms based on the support enumeration [4,13], algorithms for the games with large number of active strategies [15]. A separate class of the algorithms is based on the computation of the pessimistic approximation of the Nash equilibrium. Such algorithms give $\epsilon = \frac{3}{4}$ [6], $\epsilon = \frac{1}{2}$ [3] and $\epsilon = 0.3393$ [20], where $\epsilon$ means the maximum deviation from the optimal strategy for the worst player. Of course all above methods focus on the finding the best approximation of the Nash equilibrium, but there are many interesting concepts in the game theory which are also very difficult problems. One of the most famous concepts in the coordination games is the focal point concept firstly described by the Thomas Schelling in the 1960 [16]. J. Mehta, Ch. Starmer and R. Sugden described the focal point concept in pure coordination games [11]. The coordination games are examples in which the main problem is to find the similar strategy for all involved players - so it is communication problem in coordination games described in [2]. One of the very popular methods used in the coordination games is so called fictious play firstly described in 1951 by the G. Brown [1]. Another approach assumes that there are multiple fictious decisions that precede the real choice [17]. On the other hand there is evidence, that some examples of the coordination games cannot be resolved with the fictious play approach [5]. The most articles related with the coordination games concern the theoretical approach in which there are only two players which are using pure strategies. We give example of the algorithm which is capable to find the approximate solution in mixed strategies for the $n$-players.

## 3   Coordination Games

Coordination game is the special case of the strategic form game in which $n$-players win only if they select the same pure strategy. The mathematical definition of the coordination game is the same, as the classical game in the strategic form. On the fig. 1 a pure coordination game example is given.

|  |  | Player 2 | |
|---|---|---|---|
|  |  | A | B |
|  | A | 10,10 | 0,0 |
| Player 1 | B | 0,0 | 10,10 |

**Fig. 1.** A pure coordination game

The above example is often used for the pure theoretical concept in which the solution exists only if players use the same strategy. The coordination games are the common class of games in which group of players must select the same option (strategy). These games have at least $n$ Nash equilibria in the pure strategies (where in the classical random game the pure Nash equilibrium may not exist at all). Such situation significantly affects the level of difficulty, because players do not know about prefered choice of the other players. There are two main approaches used for solve such games. First is based on the mentioned before the focal point (also called the Schelling Point [16]). Schelling asked subjects to choose independently and without communication where in New York City they would try to meet one another. Those who chose the same meeting location as their partner would receive a positive (hypothetical) payoff, equal to that of their partners and independent of the specific location. So the main idea in this approach is to give players some kind of a hypothetical markup which should help with making the decision. It is interesting, that without no particular reasons, players more often choose the strategy which is marked up in some way.

There exist approaches based on the so called fictitious play (examples of the articles related with the fictitious play were given in the previous section). Unfortunately in the fictious play often there is additional assumption in which before actual game there is a set of fictious moves which affect on the final decision of the players.

We will try to give an approximate algorithm based on the different approach in which main objective is to find the mixed optimal strategy. The mixed strategy is the probability distribution over the set of pure strategies, so players will choose the strategy from the set of active strategies. We will consider example coordination games generated with the GAMUT [12]. It is the standard test suite used in the game theory and it allows to generate different classes of games.

## 4    The Classical Differential Evolution

The Differential Evolution (DE) is a population based heuristic recently proposed by Storn and Price [18] and deeply studied by Jouni Lampinen, Ivan Zelinka [7,8,14], and others. Nowadays the DE has become one of the most frequently used evolutionary algorithms for solving the global optimization problems. It was used in many optimization problems: neural network train [10], filter design [19] and many more. The pseudocode of the general DE algorithm:

1. Create the initial population of genotypes $P_0 = \{X_{1,0}, X_{2,0}, ..., X_{n,0}\}$,
2. Set the generation number $g = 0$
3. Until the stop criterion is not met:
   (a) Compute the fitness function for every genotype in the population $\{f(X_{1,g}), f(X_{2,g}), ..., f(X_{n,g})\}$
   (b) Create the population of trial genotypes $V_g$ based on $P_g$
   (c) Make crossover of genotypes from the population $P_g$ and $V_g$ to create population $U_g$
   (d) Choose the genotypes with the highest fitness function from population $U_g$ and $P_g$ for the next population
   (e) $g = g + 1$

The most important part which is capable to effectively improve performance and accelerate the convergence of the algorithm is the mutation operator. Mutation is a process that adds randomly-generated increments to the selected genes. It is applied first to generate a trial vector, next this vector is used in the crossover procedure. Note that the mutation is successively subjected to each genotype within the population. The population generated after the mutation is called the trial population $V$. The most popular strategy denoted by abbreviation DE/rand/1/ generates the individual by adding the weighted difference of two points:

$$\forall_i, \forall_j \ V_{i,j} = P_{r_1,j} + F \cdot (P_{r_2,j} - P_{r_3,j}).$$

An individual $P_{r_1,j}$ is denoted as a target vector. $(P_{r_2} - P_{r_3})$ is the differential vector created from the two random individuals $P_{r_2}$ and $P_{r_3}$. Moreover $F$ is the mutation parameter. The crossover schema is used as the additional genetic operator which allows to combine individuals from the parent population $P$ and trial population $V$. A selection schema used in this algorithm guarantees that individuals promoted to the next generation are not worse than individuals in the actual generation.

$$\forall_i, P^*{}_i = \begin{cases} U_i \text{ when } f(U_i) \le f(P_i), \\ P_i \text{ in other case,} \end{cases}$$

where $P^*$ is the individual in the next generation, $f()$ is the fitness function. This selection schema is similar to the greedy algorithm, in which only the best solutions are promoted to the next stage of the algorithm. The core of the proposed solution is the fitness function, which will be described in the next section.

## 5 Proposed Approach

Our goal is to calculate approximate optimal strategies for a randomly generated coordination game for $n$ players. It is obvious that in any coordination game the mixed Nash equilibrium will not be the Pareto equilibrium. The Pareto equilibrium gives all players the maximal possible payoff. In this type of games, all mixed equilibria are dominated by a pure equilibria. It means that any pure Nash equilibrium will give players the higher payoff than the mixed equilibrium. On the other hand, in the $n$ players games approximate mixed Nash equilibrium has a much greater chance for selecting than pure Nash equlibrium. Our approach assumes, that we will focus only on the mixed strategy profile and take into account the probability, that will not guarantee high expected payoff for all players.

Genotype of a single individual includes probability distributions over the set of pure strategies for all players. A simple example with only two players and three pure strategies is shown on the fig. 2.

The main concept in the proposed fitness function calculation is dependency between optimal strategies of all players in the coordination game. In such case all payoffs are equal. In any other strategy profile payoffs are different. We add second condition: sum of all probabilities over the set of pure strategies for every

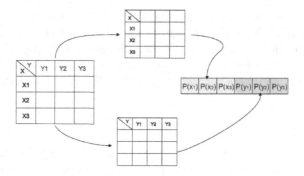

**Fig. 2.** The single individual genotype

player must be equal to 1.

$$f_1 = \sum_{i=1}^{n} \sum_{j=1}^{n} c \cdot |u_i - u_j|; i \neq j \tag{5}$$

where $u_i$ is the payoff for the $i$-th player, and $c$ is constant value. Additional probabilities dependency is given below:

$$f_2 = \sum_{i=1}^{n} |1 - \sum_{j=1}^{m} P(a_{ij})| \tag{6}$$

$f_1$ means the maximal deviation from the optimal strategy - in other words, the worst case scenario. $f_2$ means that in the beginning of the algorithm all genes in the individual genotype are values randomly selected from the range $\langle 0 : 1 \rangle$. Over the time each genotype should represent probabilities values over the set of strategies. Finally we add some penalty function for pure equilibria:

$$\forall_i, f_3 = \begin{cases} f_3 + c \text{ if } g[i] = 1, \\ f_3 \quad \text{ in other case,} \end{cases}$$

where $g[i]$ is the $i$-th gene in the genotype. Sum of all above functions is the fitness function:

$$f = f_1 + c \cdot f_2 + f_3.$$

$c$ constant in both above equations is used to add some penalty value. The fitness function equal to 0 is identified as global optimum.

## 6    Experiments

The main goal for this article is to show, that the Differential Evolution is capable to give a set of solutions in the $n$ players coordination games. There are many algorithms based for example on the fictious play in which an optimum in pure strategies is known, and the only problem is the way to select the strategy. These algorithms are mainly used for the games with 2-players and small number of strategies. In this article the main problem is the calculation of the mixed optimal equilibrium, which may be later used for example in the fictitious play. As it was mentioned before, we generated a coordination games on the basis of the GAMUT program. All generated games had solutions in the pure and mixed strategy profiles. The number of players ranged from 3 to 5 players. The number of pure strategies was always equal to number of players and all payoffs were normalized. We generated 10 different games for every size (3, 4 and 5 players games). The experiment was repeated 30 times. In the Differential Evolution algorithm we used the classical mutation schema described in the previous section with the mutation factor equal to 0.7, and the binomial crossover with the crossover parameter $CR$ value equal to 0.5. The population size was set to number of players times number of pure strategies.

First of all, we calculated the $\epsilon$ values for different games. $\epsilon$ in the Nash equilibrium calculation problem means the maximal deviation from the optimal strategy. In the other words $\epsilon$ is the distance from the global optimum. The results for the 3-players games may be seen on the fig. 3. At the boxplot we can see the minimum, maximum, average and median values for 10 different games. In every case the median value was lower than 0.1.

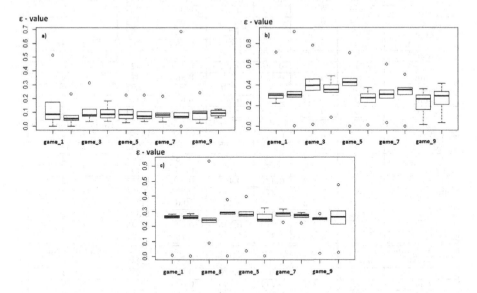

**Fig. 3.** $\epsilon$ values for different types of the coordination games. a) 3-players games, b) 4-players games, c) 5-players games

4-players games seem to be much more complicated. The genotype size in this case was 16 elements long and the median values for the $\epsilon$ value were in the range $[0.2, 0.4]$. For a few games there were results far more worse than $\epsilon$ equal 0.5, but those are just outliers and do not impact significantly on the overall results. Finally, the same boxplot for the 5-players games may be seen on the 3. These results are surprisingly good. They seem to be very similar to the results obtained for the 4-players games, but in this case $\epsilon$ values are more repetitive.

The second part of the experiments was to calculate the average, minimum, maximum, median and standard deviation payoff values for different size of the games. Our goal was to push towards the Pareto Nash equilibrium, so large payoff value was more desirable. In the tab. 1 we give all calculated values. As we can see, these values differ a lot from each other.

A short tab. 1 summary may be seen on the fig. 4. Interesting fact is that there were any additional functions used in the fitness function to, in some way, increase the payoffs. In the case of the 3-players games, results are very good. Payoffs are normalized, so the theoretical maximum payoff value is equal 1. Of course we cannot assume, that there exist an equilibrium with payoff equal 1 so

**Table 1.** Payoff values for different size of coordination games

| 3-players games | Minimum | 1-quartile | Median | 3-quartile | Maximum | average | Standard deviation |
|---|---|---|---|---|---|---|---|
| game1 | 0.449 | 0.596 | 0.680 | 0.717 | 0.901 | 0.668 | 0.151 |
| game2 | 0.004 | 0.346 | 0.416 | 0.627 | 0.927 | 0.465 | 0.237 |
| game3 | 0.077 | 0.455 | 0.561 | 0.721 | 0.839 | 0.554 | 0.231 |
| game4 | 0.302 | 0.387 | 0.541 | 0.675 | 0.955 | 0.566 | 0.218 |
| game5 | 0.582 | 0.632 | 0.647 | 0.713 | 0.902 | 0.678 | 0.093 |
| game6 | 0.016 | 0.408 | 0.436 | 0.492 | 0.651 | 0.427 | 0.166 |
| game7 | 0.258 | 0.381 | 0.439 | 0.550 | 0.803 | 0.475 | 0.139 |
| game8 | 0.054 | 0.642 | 0.667 | 0.736 | 1 | 0.691 | 0.219 |
| game9 | 0.235 | 0.341 | 0.395 | 0.664 | 0.674 | 0.413 | 0.124 |
| game10 | 0.352 | 0.390 | 0.457 | 0.632 | 0.864 | 0.494 | 0.152 |
| 4-players games | Minimum | 1-quartile | Median | 3-quartile | Maximum | Average | Standard deviation |
| game1 | 0.223 | 0.273 | 0.301 | 0.317 | 0.717 | 0.323 | 0.099 |
| game2 | 0.005 | 0.284 | 0.306 | 0.337 | 0.916 | 0.332 | 0.154 |
| game3 | 0.018 | 0.348 | 0.395 | 0.456 | 0.785 | 0.408 | 0.222 |
| game4 | 0.089 | 0.328 | 0.355 | 0.398 | 0.486 | 0.353 | 0.083 |
| game5 | 0.001 | 0.392 | 0.427 | 0.461 | 0.709 | 0.417 | 0.152 |
| game6 | 0.013 | 0.231 | 0.275 | 0.313 | 0.375 | 0.238 | 0.109 |
| game7 | 0.033 | 0.273 | 0.308 | 0.35 | 0.6 | 0.3 | 0.138 |
| game8 | 0.002 | 0.304 | 0.354 | 0.372 | 0.501 | 0.319 | 0.122 |
| game9 | 0.052 | 0.262 | 0.310 | 0.345 | 0.644 | 0.345 | 0.165 |
| game10 | 0.014 | 0.291 | 0.343 | 0.386 | 0.751 | 0.367 | 0.145 |
| 5-players games | Minimum | 1-quartile | Median | 3-quartile | Maximum | Average | Standard deviation |
| game1 | 0.009 | 0.254 | 0.262 | 0.271 | 0.281 | 0.246 | 0.066 |
| game2 | 0.003 | 0.251 | 0.260 | 0.270 | 0.283 | 0.232 | 0.083 |
| game3 | 0.089 | 0.227 | 0.242 | 0.253 | 0.634 | 0.249 | 0.119 |
| game4 | 0.003 | 0.280 | 0.290 | 0.294 | 0.376 | 0.271 | 0.098 |
| game5 | 0.037 | 0.268 | 0.277 | 0.296 | 0.398 | 0.266 | 0.095 |
| game6 | 0.003 | 0.233 | 0.243 | 0.279 | 0.323 | 0.212 | 0.113 |
| game7 | 0.225 | 0.267 | 0.282 | 0.292 | 0.314 | 0.279 | 0.025 |
| game8 | 0.222 | 0.259 | 0.272 | 0.281 | 0.292 | 0.266 | 0.022 |
| game9 | 0.017 | 0.244 | 0.249 | 0.256 | 0.284 | 0.231 | 0.076 |
| game10 | 0.162 | 0.215 | 0.284 | 0.313 | 0.643 | 0.267 | 0.074 |

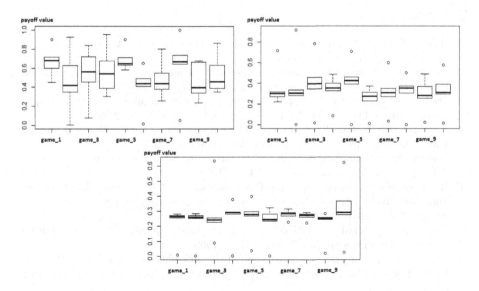

**Fig. 4.** 3-person games - payoffs

results above 0.5 are very good. On the other hand, the same results obtained for the 4 and 5-players games are much lower.

## 7   Conclusions and Future Work

The Differential Evolution is a universal method capable to solve different continuous and discrete optimization problems. In this article we presented the game theory problem, which can be easily transformed into the continuous function optimization problem. Appropriate coding leads to the simple genotype, where each gene is limited to the range $[0, 1]$. The best obtained results are very promising, although there is still much room for improvements. Unlike other mathematical optimization methods, the proposed approach is capable to give the set of solutions in the single algorithm run which is very desirable especially for the games with number of players greater than 2.

The Differential Evolution algorithm allows to obtain a very good results in the problem of finding optimal strategies in the coordination games, but we should remember, that it is only a part of the problem. It could be interesting, to answer the question, if the mixed equilibria in coordination games for $n$-players could be used as the alternative strategies opposite to classical pure strategy equilibria.

# References

1. Brown, G.W.: Iterative solution of games by fictitious play, Activity analysis of production and allocation (1951)
2. Cooper, R., Dejong, D., Forsythe, R., Ross, T.: Communication in Coordination Games. The Quarterly Journal of Economics 107(2), 739–771 (1992)
3. Daskalakis, C., Mehta, A., Papadimitriou, C.: A note on approximate Nash equilibria. Journal Theoretical Computer Science 410, 1581–1588 (2009)
4. Dickhaut, J., Kaplan, T.: A program for finding Nash equilibria, Working papers, University of Minnesota, Department of Economics (1991)
5. Foster, D.: On the Nonconvergence of Fictitious Play in Coordination Games. Games and Economic Behavior 25, 79–96 (1998)
6. Kontogiannis, S.C., Panagopoulou, P.N., Spirakis, P.G.: Polynomial algorithms for approximating Nash equilibria of bimatrix games. Journal Theoretical Computer Science 410, 1599–1606 (2009)
7. Lampinen, J., Zelinka, I.: Mixed variable non-linear optimization by differential evolution. In: Proceedings of Nostradamus (1999)
8. Lampinen, J., Zelinka, I.: On stagnation of the differential evolution algorithm. In: Proceedings of Mendel, 6th International Mendel Conference on Soft Computing (2000)
9. Lemke, C.E., Howson, J.T.: Equilibrium Points of Bimatrix Games. Society for Industrial and Applied Mathematics 12, 413–423 (1964)
10. Magoulas, G.D., Plagianakos, V.P., Vrahatis, M.N.: Neural Network-Based Colonoscopic Diagnosis using On-Line Learning and Differential Evolution. Applied Soft Computing 4, 369–379 (2004)
11. Mehta, J., Starmer, C., Sugden, R.: Focal Points in Pure Coordination Games An Experimenta Investigation. Theory and Decision 36(2), 163–185 (1994)
12. Nudelman, E., Wortman, J., Shoham, Y., Leyton-Brown, K.: Run the GAMUT: A Comprehensive Approach to Evaluating Game-Theoretic Algorithms. In: Proceedings of the Third International Joint Conference on Autonomous Agents and Multiagent Systems, vol. 2 (2004)
13. Porter, R., Nudelman, E., Shoham, Y.: Simple Search Methods for Finding a Nash Equilibrium. Games and Economic Behavior 63, 642–662 (2008)
14. Price, K., Storn, R., Lampinen, J.: Differential evolution: a practical approach to global optimization. Springer, Heidelberg (2005)
15. Sandholm, T., Gilpin, A., Conitzer, V.: Mixed-integer programming methods for finding Nash equilibria. In: Proceedings of the 20th National Conference on Artificial Intelligence, vol. 2, pp. 495–501 (2005)
16. Schelling, T.C.: The strategy of conflict, 1st edn. Harvard University Press, Cambridge (1960)
17. Sela, A., Herreiner, D.: Fictitious play in coordination games. International Journal of Game Theory 28, 189–197 (1999)
18. Storn, R., Price, K.: Differential evolution - a simple and efficient heuristic for global optimization over continuous spaces. Journal of Global Optimization 11(4), 341–359 (1997)
19. Storn, R.: Differential evolution design of an iir-Filter. In: IEEE International Conference on Evolutionary Computation, ICE 1996, pp. 268–273 (1996)
20. Tsaknakis, H., Spirakis, P.G.: An optimization approach for approximate nash equilibria. In: Deng, X., Graham, F.C. (eds.) WINE 2007. LNCS, vol. 4858, pp. 42–56. Springer, Heidelberg (2007)

# Cryptanalysis of Transposition Cipher Using Evolutionary Algorithms

Urszula Boryczka[2] and Kamil Dworak[1,2]

[1] Future Processing, Bojkowska 37A, 44-100 Gliwice, Poland,
[2] University of Silesia, Institute of Computer Science,
Będzińska. 39, 41-200 Sosnowiec, Poland
urszula.boryczka@us.edu.pl,
kamil.dworak@us.edu.pl
http://ii.us.edu.pl/en
http://www.future-processing.com

**Abstract.** This paper presents how techniques such as evolutionary algorithms (*EAs*) can optimize complex cryptanalysis processes. The main goal of this article is to introduce a special algorithm, which allows executing an effective cryptanalysis attack on a ciphertext encoded with a classic transposition cipher. In this type of cipher, the plaintext letters are modified by permutation. The most well-known problem, which is often solved with optimization techniques operating on a set of permutations, is the Travelling Salesman Problem (*TSP*). The mentioned algorithm uses a specially prepared function of assessment of the individuals with a set of genetic operators, used in the case of *TSP* problem.

**Keywords:** cryptanalysis, evolutionary algorithms, TSP, crossover, cryptology.

## 1  Introduction

Currently, restrictions based on the access control for chosen users or simple authentication, do not provide sufficient security. The danger of intercepting such data as logins and passwords, which leads to uncontrollable access, can be identified by software engineers or database administrators of a system. The technique that may be of use here is cryptography, that is, the science of encrypting and decrypting information. Its purpose is not only to obscure a message, but to transform it in such a way that it is readable only to the sender and recipient. With the rise of cryptography, cryptanalysis was born that objective is to decipher information without knowing the key and searching for any bugs and oversights in cryptographic systems. It is worth mentioning that there are Polish references in the cryptanalysis's history. The Enigma machine, used by Germans to send secret messages, was cracked by the Poles and then transferred to the British intelligence. Cryptanalysis processes are not the simplest and the fastest ones. Their operation time is long and their memory requisition is very high. In

D. Hwang et al. (Eds.): ICCCI 2014, LNAI 8733, pp. 623–632, 2014.

view of these criteria, any kind of optimization methods for these processes can be very useful. One of the best techniques to be mentioned is the use of *EA*.

Techniques based on artificial intelligence are gaining popularity in the computer security. Over the past few years, there have been many publications like Simulated Annealing or Particle Swarm Optimization which concern the application of various evolutionary in order to optimize currently applied ciphers (e.g. generating cases with sub-keys of very good cryptographic properties, also known as the *S-blocks*) or cryptographic systems (e.g. developing new, original cryptanalysis attacks, focusing directly on the given encryption algorithm). There is growing interest in the computer security community towards evolutionary computation (*EC*) techniques as a result of recent successes, but there is still a number of open problems in this field that should be addressed.

The next section of this paper is the presentation of basic information on *EAs* and a survey concerning the previous study on the use of certain evolutionary techniques in cryptanalysis. The third section contains basic information on cryptography and the transposition cipher. In the next section, the basics of cryptanalysis and frequency analysis are presented, followed by the example of an attack [1]. The fifth section contains the observations, results and test comparison of the presented attack [1] and its modifications. The last section is a summary of this study.

## 2   Popularity of Evolutionary Algorithms

In 1998, Andrew John Clark [4] presented positive results of his research on cryptanalysis of cyphertexts, based on classic ciphers supported by such techniques as Tabu Search, simulated annealing or *EA*. A thesis written by Bethany Delman presented many examples of using *EAs* in cryptanalysis [7]. Laskari I Meletiou and other scientists suggested the attack on the simplified version of *DES* with the use of the Particle Swarm Optimization method, which turned out successfully [13]. Toemh and Arumugam presented a modification of the cryptanalysis attack on a classical transposition cipher using *EAs* [18]. Wafaa A. Ghonaim based her dissertation on evolutionary techniques like Ant Colony Optimization, Artifical Neutral Networks or Particle Swarm Optimization in cryptanalysis on the example of the *DES* standard [8].

*EA* is a simple randomized algorithm generating a population of $P(t)$ individuals in the $t$ iteration of the algorithm [14]. The individual, represented by a special data structure, presents a single possible solution to the given problem [14]. Every solution is assessed by special fitness function [14]. Some units will be chosen by means of crossover and mutation to create the next population of possible solutions [9].

# 3   Information Safety – Cryptography

The main task of the cryptographer is to encode and decode some information (for example data located in the database or the content of a confidential document) using simple mathematical operations, like addition or transposition [12]. Cryptography obscures the message content, not information about its existence [16]. Even if the attacker obtains part of the data, he is not able to read it, because they are completely ineligible to him.

Encryption involves converting the text in such a way that is not readable. In most cases cryptography is carried out by using two keys, the encryption key *K1* and the *K2* decryption key. By using a suitable decryption key the person receiving the message is able to recover the original, legible information.

In this paper, a classic transposition cipher is used as example. Each plaintext sign is exactly one sign in the cryptogram. Marks remain the same, and only the order is modified. The message is written into the table with a width equal to the length of $\pi$ (Table 1). The table must be fully completed. If the message does not completely fill it in, additional random signs can be added.

**Table 1.** Transposition cipher - encryption process

| 5 | 2 | 3 | 7 | 1 | 6 | 4 |
|---|---|---|---|---|---|---|
| t | o |   | b | e |   | o |
| r |   | n | o | t |   | t |
| o |   | b | e |   | t | h |
| a | t |   | i | s |   | t |
| h | e |   | q | u | e | s |
| t | i | o | n | **e** | **a** | **d** |

Then we create the ciphertext. If the signs in the table are read according to the number of columns from top to bottom, a ciphertext is produced:

ET SUEO TEI NB OOTHTSDTROAHT T EABOEIQN

The drawn signs have been underlined. The process of decrypting the ciphertext is not complicated. It is based on an inverse operation. By knowing the key we can complement the content of the table by filling in the part of the ciphertext in the respective columns. We are then able to read the message by following the table from left to right.

# 4    Decrypting without a Key – Cryptanalysis

Cryptanalysis is the science of deciphering a plaintext without knowing the key [16]. In most cases, the main goal of the attacker is to obtain the decryption key [17]. One of basic concepts of cryptanalysis is frequency analysis. This involves counting the frequency of all signs in the selected ciphertext and then creating a ranking of the most common symbols [2]. A comparison is made between the distribution of the ciphertext symbol and comparable text. This way the attacker tries to infer the cryptogram equivalents of the signs. Frequency analysis does not mean only checking the individual signs. The notation $n$-gram is often used. It is based on the fact that we search for the prevalence of two-character, digram/bigrams, or three-character, trigrams, pieces of text and then compare the frequency of their occurrence in a natural language, such as English [11].

When using $EC$ in the field of cryptanalysis, the main issue is how to define a fitness used to find the best keys. In 1988 Andrew John Clark [4] presented an interesting method based on $n$-grams which was used to compute a fitness function's value. An attacker receives the encrypted message and the key's length by which it is encoded. The initial population is made by generating a random permutation of the key length. The fitness function value is calculated based on the following:

$$F_f = \beta \cdot \Sigma_{i,j \in A} |K^b_{(i,j)} - D^b_{(i,j)}| + \gamma \cdot \Sigma_{i,j,k \in A} |K^t_{(i,j,k)} - D^t_{(i,j,k)}| \tag{1}$$

where:
$A$ - the language alphabet,
$i$, $j$, $k$ - next indexes of the message,
$F_f$ - the fitness function,
$K$ - known language statistics,
$D$ - the frequency of occurrence in the decrypted text,
$b$, $t$ - the indices to denote bigram/trigram statistics,
$\beta, \gamma$ - weight parameters, determining the priority of the statistics (the condition $\beta + \gamma = 1$ must be fulfilled).

The unigrams statistics are completely omitted here. A transposition cipher has the same number of signs before and after the encryption process. Comparing these statistics does not make sense, thus it can be left out [4]. The value of the fitness function is based on the incidence of $n$-grams in the decrypted text. The total sum of bigrams and trigrams is subtracted from the known language statistics. In addition, there are two parameters, beta and gamma, allowing to assign the weight which determines the usefulness of this statistic, i.e. of $n$-gram. In some cases, calculating the bigrams may be sufficient and more effective. Trigrams increase the algorithm's complexity to $O(K^3) + O(n)$, where $K$ is the alphabet size and $n$ is the length of the message. The smaller value of the function suggest that the differences between the statistics and the decrypted text are insignificant. It can thus be concluded that the correct description key has smallest possible value of the function. Due to the additional random signs generated during the encryption process, obtaining the value 0 becomes practically impossible. The classical cryptosystems, nowadays entirely broken, had a general property where the real key producing the encryptions/decryptions was close to original ciphertext/plaintext. This significantly aided in defining the fitness functions. Modern ciphers do not exhibit this property. This is a major problem and the reason behind the relative lack of applications of these heuristic techniques into modern cryptosystems, where while testing a key that has 255 right bits out of 256 (99.6 correct). The resulting plaintext would then appear completely random due to a property that has been named the Avalanche Effect.

Article [1] presents an example of cryptanalysis attack on a classic transposition cipher based on *EAs*, hereinafter referred to as "the original attack". The same fitness function as the one used by Clark is presented here.

The steps of the presented operator are shown in Fig. 1.

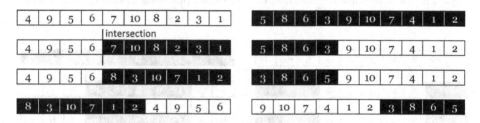

**Fig. 1.** Original attack's crossover [1]

The intersection point is chosen at random. In the case of the first parent, if the right part is cut off, then the second parent's left part is cut off as well. In the next step, the separated fragment is sorted in the order set out by the second parent, and then added to the potential offspring. Finally, the sorted fragment is moved to the other side of the offspring according to the intersection point. The mutation operator transfers the whole string one position to the left.

The details on the aforementioned algorithm are presented in the paper [1].

One of the most popular tasks connected to *EA* is the *TSP* problem. Its aim is to find an optimal route with visiting an $n$ number of cities only once so as to the total route will be possibly short [3]. The set of solutions in this task is a set of permutations of $n$ cities, where each permutation is the solution [14]. The size of the set is $n!$ [14].

The set of solutions of the transposition cipher, as in the *TSP* problem, is a set of key permutations used to decode the ciphertext. The size of this population is determined by the number of columns of the table mentioned in the previous section. Undoubtedly, we can apply genetic operators here, specially dedicated for permutation problems used to find the shortest route in the graph. The most popular types of crossover in the *TSP* problem are *PMX*, *OX* and *CX* [9].

The *PMX* crossover has been described by Goldberg and Lingle [10]. At first, it assumes a subroute between two randomly chosen intersection points should be outlined, which are next exchanged between the offsprings. In the next step, other genes not appearing in the offspring are exchanged, based on the outlined subroutes. The rest of the genes of an individual are filled with unused alleles.

Another method of crossover between chromosomes is the *CX* crossover [15]. A newly created individual comes from only one of the parents. The first offspring is created by taking only the gens of the first parent. Every point is outlined from the same parent based on the corresponding (according to the position) point in the second parent. This is how a cycle is made. When the point already existing in the offspring chromosome is reached, the cycle is closed and another unused point of the parent is taken.

The last type of crossover mentioned in this study is the *OX* crossover. It has been developed by Davis [5]. At the beginning, a subroute is outlined in every parent, which is automatically copied to the offspring. Next, starting from the second intersection point of one of the parents, gens of the second parent are copied in the same order, excluding the symbols appearing in the offspring. This crossover is presented in Fig. 2.

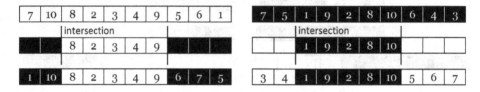

**Fig. 2.** Original *OX* crossover

It was decided to implement two recently presented modifications of the *OX* crossover operator [6], hereinafter referred to as *OX4* and *OX5* operators.

In the *OX4* modification, the intersection points of the parents are changed in a way that they could appear in various places in the chromosomes [6]. Additionally, the sizes of the newly outlined substrings can be of various lengths. The rest of the operations were untouched as in the *OX* crossover. The *OX4* crossover example is shown in the Fig. 3.

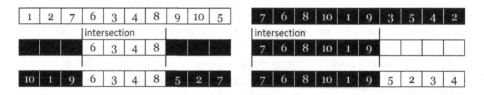

**Fig. 3.** Modification of *OX* - *OX4* crossover

The *OX5* crossover comes down to outlining 4 points (2 pairs) of intersections in chromosomes in a way, that two subroutes of various lengths are created [6] Then we proceed analogically with the classic the *OX* crossover.

The *PMX*, *CX* and *OX5* crossovers obtained poor results so they are completely omitted in this study.

## 5    Experiments and Results

This section presents a comparison of the original attack with a set of further modifications based on subsequent crossover operators presented earlier. All the parameters are shown in Table 2. Algorithms were implemented in C# programming language and tested on a PC with a processor of core frequency 3.4GHz and 4GB RAM memory. The tested texts contain around 4000 signs.

**Table 2.** Parameters of the evolutionary algorithms

| Number of Iterations | Unlimited (time limit to 5 minutes) |
|---|---|
| Crossover's Probability | 0.7 |
| Mutation's Probability | 0.25 |
| Selection's Type | roulette wheel method |
| $\beta, \gamma$ | 0.5 |

While testing the transposition's cipher cryptanalysis, the minimum value $(F_f(MIN))$ always coincided with the best population. This function counts the number of inconsistencies that arise with respect to the occurrence of digrams and trigrams between English and the decrypted message. It can be concluded that the smaller the value of the function for a particular individual, the more reliable the solution is. It was decided to select only the most interesting tests.

**Table 3.** Results obtained in the original attack [1]

| ID | N | $K_L$ | $F_f(MAX)$ | $F_f(MIN)$ | $F_f(Median)$ | Iteration | Time |
|---|---|---|---|---|---|---|---|
| 10 | 300 | 11 | 4160 | 3057 | 3660 | 9 | 00:28 |
| 14 | 400 | 13 | 4336 | 3054 | 3700 | 9 | 00:39 |
| 17 | 450 | 15 | 3919 | 3055 | 3444 | 16 | 01:15 |
| 24 | 600 | 17 | 3984 | 3049 | 3380 | 25 | 02:05 |
| 29 | 1000 | 19 | 3880 | 3058 | 3383 | 26 | 03:50 |

where:
*ID* - identity of a single test,
*N* - number of solutions in the population,
$K_L$ - key's length,
$F_f(MIN)/F_f(MAX)$ - the best/worst value in the population,
$F_f(Median)$ - median value adjustment of individuals in the population,
*Iteration* - number of the iteration which found the best solution,
*Time* - duration of the algorithm (for finding the correct key).

When analyzing the convergence graphs in Fig. 4 and results presented in Table 3 and Table 4 it can be easily seen that crossover operator originating from the *TSP* problem [5], has a positive influence on the speed and accuracy of the algorithm.

**Table 4.** Results obtained in the evolutionary attack with OX Crossover

| ID | N | $K_L$ | $F_f(MAX)$ | $F_f(MIN)$ | $F_f(Median)$ | Iteration | Time |
|----|-----|-----|------|------|------|----|-------|
| 10 | 300 | 11 | 4620 | 3057 | 3900 | 9 | 00:27 |
| 14 | 350 | 13 | 4587 | 3054 | 3720 | 11 | 00:32 |
| 17 | 400 | 15 | 4190 | 3055 | 3590 | 16 | 01:06 |
| 24 | 500 | 17 | 4025 | 3049 | 3237 | 21 | 01:36 |
| 29 | 600 | 19 | 3737 | 3058 | 3238 | 37 | 03:22 |

If we compare Table 3 and Table 4 we can notice that the *OX* crossover requires smaller population, so algorithm needs fewer individuals to find perfect solution. Comparing the convergence charts presented in Fig. 4 we can observe that the *OX* crossover attack quickly stabilized (in 12th iteration).

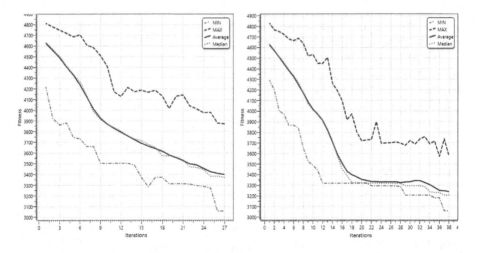

**Fig. 4.** Convergence's chart for test 29 for original attack (left) and the OX (right)

Later, it takes around 15 iterations to find valid decryption key using mutation. However, presented modification of original [1] attack works faster and uses 60-90% individuals of previously proposed attack (depends on the key's length).

**Table 5.** Results obtained in the evolutionary attack with the OX4 Crossover

| ID | N | $K_L$ | $F_f(MAX)$ | $F_f(MIN)$ | $F_f(Median)$ | Iteration | Time |
|----|-----|----|------|------|------|----|------|
| 19 | 150 | 11 | 4439 | 3057 | 3818 | 10 | 00:16 |
| 14 | 300 | 13 | 4491 | 3054 | 3705 | 13 | 00:33 |
| 17 | 400 | 15 | 4160 | 3055 | 3640 | 14 | 00:56 |
| 24 | 400 | 17 | 4360 | 3049 | 3680 | 18 | 01:12 |
| 29 | 400 | 19 | 4060 | 3058 | 3440 | 23 | 01:30 |

Results presented in Table 5, Table 4 and in Fig. 5 represent modification that uses the $OX4$ crossover. For each presented case, regardless of the length of the decryption key, only 400 individuals were required to find an ideal solution of the problem. Due to such a small population the proper decoding key could be found even 2-3 times faster than in the case of the original attack.

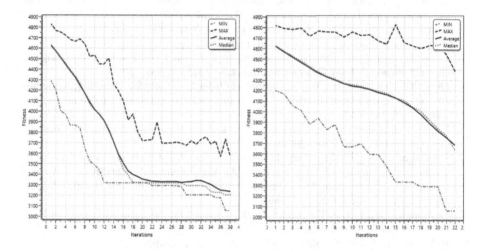

**Fig. 5.** Convergence's chart for test 29 for the OX (left) and the OX4 (right)

Analysing the above charts, we can easily notice that the $OX4$ attack obtained convergence during 2 - 3 iterations then immediately found a better solution.

# 6   Conclusion and Future Work

Experiments and results presented in the previous sections showed that *EAs* can be successfully used in classic cryptanalysis. It was proved that applying solutions commonly used in other kinds of problems such as the *TSP* problem, have positive influence on improving cryptanalytic processes.

One might wonder how a change of the crossover's probability will affect the algorithm or how attack will handle other types of crossovers like *EAX*, *PX* or

*EX.* Perhaps further studies devoted to their mutation could improve the quality and speed of the results that have been obtained so far.

Positive results achieved by using the presented algorithms closely connected to *TSP* justify the application of *EAs* in traditional cryptanalysis, which has a positive influence on optimizing the algorithm. *EAs* open new possibilities for further research in this field, which undoubtedly pave the way for works in this discipline of study.

# References

1. Boryczka, U., Dworak, K.: Genetic transformation techniques in cryptanalysis. In: Nguyen, N.T., Attachoo, B., Trawiński, B., Somboonviwat, K. (eds.) ACIIDS 2014, Part II. LNCS, vol. 8398, pp. 147–156. Springer, Heidelberg (2014)
2. Bauer, F.L.: Decrypted Secrets. Methods and Maxims of Cryptology. Springer, Heidelberg (2002)
3. Bryant, K.: Genetic Algorithms and the Traveling Salesman Problem. Departamenr of Mathematics, Harvey Mudd College (December 2000)
4. Clark, A.J.: Optimisation Heuristics for Cryptology. PhD thesis, Information Research Centre Faculty of Information Technology Queensland (1998)
5. Davis, L.: Applying Adaptive Algorithms to Epistatic Domains. In: Proceedings of the International Joint Conference on Artifical Intelligence (1985)
6. Deep, K., Mebrahtu, H.: New Variations of Order Crossover for Travelling Salesmap Problem. International Journal of Combinatorial Optimization Problems and Informatics 2(1) (2011)
7. Delman, B.: Genetic Algorithms in Cryptography. Rochester Institute of Technology. Rochester, New York (2004)
8. Ghonaim, W.A.: Evolutionary Computation in Cryptanalysis. Departament of Mathematics Faculty of Science (Girls). Al-Azhar University (2013)
9. Goldberg, D.E.: Genetic Algorithms in Search Optimization, and Machine Learning. Addison-Wesley, Boston (1989)
10. Goldberg, D.E., Lingle, R.: Alleles, Loci, and the TSP. In: Proceedings of the First International Conference on Genetic Algorithms (1985)
11. Kahn, D.: The Code-breakers. Scribner, New York (1996)
12. Kenan, K.: Cryptography in the databases. The Last Line of Defense. Addison Wesley Publishing Company (2005)
13. Laskari, E.C., Meletiou, G.C., Stamatiou, Y.C., Vrahatis, M.N.: Evolutionary computation based cryptanalysis: A first study. Nonlinear Analysis 63 (2005)
14. Michalewicz, Z.: Genetic Algorithms + Data Structures = Evolution Programs. Springer, London (1996)
15. Oliver, I.M., Smith, D.J., Holland, J.R.C.: A Study of Permutation Crossover Operators on the Traveling Salesman Problem. In: Proceedings of the Second International Conference on Genetic Algorithms on Genetic Algorithms and Their Application. Erlbaum Associates Inc., Hillsdale (1987)
16. Schneier, B.: Applied Cryptography: Protocols, Algorithms, and Source Code In C, 2nd edn. Wiley (1996)
17. Stinson, D.S.: Cryptography. Theory And Practice. Chapman & Hall/CRC Taylor & Francis Group, Boca Raton (2006)
18. Toemeh, R., Arumugam, S.: Breaking Transposition Cipher with Genetic Algorithm. Electronics and Electrical Engineering 7(79) (2007)

# Improved Video Scene Detection Using Player Detection Methods in Temporally Aggregated TV Sports News

Kazimierz Choroś

Institute of Informatics, Wrocław University of Technology,
Wybrzeże Wyspiańskiego 27, 50-370 Wrocław, Poland
kazimierz.choros@pwr.edu.pl

**Abstract.** Many strategies of content-based indexing have been proposed to recognize sports disciplines in sports news videos. It may be achieved by player scenes analyses leading to the detection of playing fields, of superimposed text like player or team names, identification of player faces, detection of lines typical for a given playing field and for a given sports discipline, recognition of player and audience emotions, and also detection of sports objects and clothing specific for a given sports category. The analysis of TV sports news usually starts by the automatic temporal segmentation of videos, recognition, and then classification of player shots and scenes reporting the sports events in different disciplines. Unfortunately, it happens that two (or even more) consecutive shots presenting two different sports events although events of the same discipline are detected as one shot. The strong similarity mainly of colour of playing fields makes it difficult to detect a cut. The paper examines the usefulness of player detection methods for the reduction of undetected cuts in temporally aggregated TV sports news videos leading to better detection of events in sports news. This approach has been tested in the Automatic Video Indexer AVI.

**Keywords:** content-based video indexing, TV sports news analyses, sports video categorization, temporal segmentation, undetected cuts, scene detection, temporal aggregation, news headlines, player detection, AVI Indexer.

## 1 Introduction

Sports videos are the most frequently viewed videos on the Web. Effective retrieval of video data cannot be achieved using standard text indexing and retrieval procedures. More sophisticated content-based video indexing and retrieval methods should be applied. The main goal of research and experiments with sports videos is to propose and develop automatic methods such as automatic detection or generation of highlights, video summarization and content annotation, player detection and tracking, action recognition, ball detection and tracking, kick detection such as penalty, free, and corner kick, replay detection, player number localization and recognition, text detection and recognition for player and game identification, detection of advertisement billboards and banners, authentic emotion detection of audience, and so on. Because of a huge commercial appeal sports videos became nowadays a dominant application area for video automatic indexing and retrieval.

D. Hwang et al. (Eds.): ICCCI 2014, LNAI 8733, pp. 633–643, 2014.

Many approaches proposed for automatic video indexing [1-3] are based on a video structure. A video is composed of different structural units such as: acts, episodes (sequences), scenes, camera shots and finally, single frames [4]. The most general unit is an act. So, a film is composed of one or more acts. Then, acts include one or more sequences, sequences comprise one or more scenes, and finally, scenes are built out of camera shots. The main structural, basic elements of a video are scenes and shots. A scene is defined as a group of consecutive shots sharing similar visual properties and having a semantic correlation – following the rule of unity of time, place, and action. A shot is usually defined as a continuous video acquisition with the same camera, so, it is a sequence of interrelated consecutive frames recorded contiguously and representing a continuous action in time and space. A shot change is when a video acquisition is done with another camera. The simplest and the most frequent way to perform a change between two shots is called a cut. In this case, the last frame of the first video sequence is directly followed by the first frame of the second video sequence. The process of the detection of shot changes present in the video sequences, the process of segmentation a sequence of video frames is called temporal segmentation of a video.

There are many proposed methods of temporal segmentation and they are still evolving and improving [1-3] because their high efficiency promote the efficacy of the next steps of content-based video indexing process. In our research on temporal aggregation [5] of TV sports news and video categorization very efficient temporal segmentation methods are desirable. The main purpose of sports news processing is to categorize sports video shots and scenes in TV sports news. This processing should take into account that shots in news are relatively short and they come from various sports. Furthermore, studio discussions, commentaries, interviews, charts, tables, announcements of future games, discussions of decisions of sports associations, etc., so non-sports parts are presented at the same broadcasted news. Due to the automatic categorization of sports events videos can be automatically indexed. The retrieval of individual sports news and sports highlights such as the best or actual games, tournaments, matches, contests, races, cups, etc., special player behaviours or actions like penalties, jumps, or race finishes, etc. in a desirable sports discipline becomes more effective.

The temporal aggregation method [5] detects and aggregates two kinds of shots: sequences of long studio shots unsuitable for content-based indexing and player scene shots adequate for sports categorization. A player scene is a scene presenting the sports game, i.e. a given scene was recorded on the sports fields such as playgrounds, tennis courts, sports hall, swimming polls, ski jumps, etc. All other non-player shots and scenes usually recorded in a TV studio such as commentaries, interviews, tables, announcements of future games, discussions of decision of sports associations, etc. are called studio shots or studio scenes. Studio shots are slightly useful for indexing and therefore can be rejected. It was observed that the studio scenes may be even two thirds of TV sports news. The temporal aggregation method leads to the rejection of non-player scenes before starting content analyses and then to the significant reduction of computing time and providing more effective analyses.

The tests have shown that applying the temporal aggregation method it was possible to reject about half of video material and despite this almost all sports scenes reported in TV sports news have been indexed. The temporal aggregation efficacy would be better if we could detect cuts also in the sequence of shots from different sports events but of the same sports discipline. It happens for example in headlines of TV sports news very useful for indexing purposes. Headlines are regarded as a summary of sports news [6, 7]. In headlines a series of several single shots of soccer games (mainly after the games of the UEFA Champions League or FIFA World Cup qualifications) is very often presented. Because of a strong similarity (colours, histograms, lightings etc.) of these scenes it happens that they are detected as one scene. The playing fields are very similar in these scenes but players differ, mainly player's clothing. This difference can be applied to improve the results of temporal segmentation and in consequence to improve scene detection.

The paper is organized as follows. The next section describes some related works in the area of automatic player detection methods. The principles of the temporal aggregation are outlined in the third section. The forth section presents the experimental results obtained in the AVI Indexer of the player detection in TV sports news videos. The object recall as well as pixel recall are also analyzed in this section. The improved temporal segmentation based on comparison of colour of player's clothing is proposed in the fifth section. The final conclusions and the future research work areas are discussed in the last sixth section.

## 2    Related Works

There has been much research carried out in the area of automatic recognition of video content and of visual information indexing and retrieval. Different solutions have been proposed for player or other objects detection. For example a scheme to detect and locate the players and the ball on the grass playfield in soccer videos has been proposed in [8]. A shape analysis-based approach was applied to identify the players and the ball from the roughly extracted foreground, which was obtained by a trained colour histogram-based playfield detector and connected component analysis.

Then a player segmentation methodology for videos of soccer games has been presented in [9], by applying such algorithms as dominant colour region detection, as well as referee and player identification. Many other works and tests have been performed with soccer video. In [10] a method is described to perform automatic multiple player detection, unsupervised labelling and efficient tracking in broadcast soccer videos. Player detection is achieved by combining the ability of dominant colour based background subtraction and a boosting detector with Haar features. Hundreds of player samples with the player detector have been collected, and learn codebook based player appearance model by unsupervised clustering algorithm. A player can be recognized as player of one of two teams, referee, or outlier. The learning procedure can be generalized to different videos without additional processes. Based on detection and labelling, multiple player tracking with Markov chain Monte Carlo (MCMC) data association was performed.

The methods presented in [11] aim at detecting and recognizing players on a sport-field, based on a distributed set of loosely synchronized cameras. Whereas, the system presented in [12] is based on the combination of histogram of oriented gradients (HOG) descriptors and linear support vector machine (SVM) classification. The proposed detection system incorporates a dominant colour based segmentation technique of a football playfield, a 3D playfield modelling algorithm based on Hough transform and player tracking algorithm. The goal of the method proposed in [13] was to identify regions that perform certain activities in a scene. Using a tracking method a classification of these regions to active or static was performed. Then in [14] a method has been presented to represent the player's action on the panoramic background. This method was effective for the sports video like diving, jumps, which the player performs his or her action in a large arena, and the camera motion mainly includes horizontal and vertical direction.

## 3    Scene Detection by Temporal Aggregation

The temporal aggregation method enables us to detect and to aggregate two kinds of shots: sequences of long studio shots unsuitable for content-based indexing and player scene shots adequate for sports categorization. It has two main advantages. First of all it detects player scenes, therefore the most informative parts of videos. Then, it significantly reduces video material analyzed in content-based indexing of TV sports news because it permits to limit indexing process only to player scenes. Globally, the length of all player scenes is significantly lower than the length of all studio shots.

The temporal aggregation is specified by three values: minimum shot length as well as lower and upper limits representing the length range for the most informative shots. The values of these parameters was determined basing on the exhaustive analyses of TV sports news and the analyses of their high-level structure.

Very short shots including single frames are relatively very frequent. Generally very short shots of one or several frames are detected in case of dissolve effects or they are simply wrong detections. The causes may be different. Most frequently it is due to very dynamic movements of players or of a camera during the game, as well as due to light flashes during the interviews. These extremely short shots resulting from temporal segmentation are joined with the next shot in a video. So, the first two steps of the temporal aggregation of shots also leads to the significant reduction of wrong cuts incorrectly detected during temporal segmentation.

As the result of the temporal aggregation of shots we receive the sequences of shots of the lengths between two a priori defined values. Furthermore, all sequences of shots are separated always by only one very long studio shot. In such a case every sequence of shots can be treated as a pseudo-scene (Fig. 1).

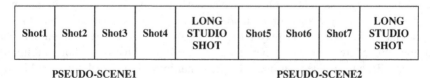

| Shot1 | Shot2 | Shot3 | Shot4 | LONG STUDIO SHOT | Shot5 | Shot6 | Shot7 | LONG STUDIO SHOT |
|---|---|---|---|---|---|---|---|---|

PSEUDO-SCENE1                           PSEUDO-SCENE2

**Fig. 1.** Pseudo-scene set as the result of the temporal aggregation of shots

Unfortunately, it happens that two shots are detected as only one shot (cut is not detected) most frequently if these shots of different scenes are of the same sports disciplines. The main difference between these two shots easy observed by humans is the colours of player's clothing. So, the player detection methods can be useful to solve this problem. Undetected cuts diminish the efficacy of the temporal aggregation because shots become longer, and therefore may be treated as non-informative part of a video after its temporal aggregation.

## 4    Detection of Player Fields and Players

The process of the detection of player fields and players and its usefulness for the improvement of scene detection has been tested in the AVI Indexer [15].

The first step of the procedure implemented was the detection of the area where foreground objects can be expected. This area in the case of sports videos is a player field such as soccer playing field or tennis court, etc. At the beginning the number of colours in a analyzed frame was reduced to only eight basic colours. Then the dominant colour was chosen. As the dominant colour the most frequent colour was taken. The area belonging to the playing field was determined by the pixel of the dominant colour. In this area the players have been detected in the next step.

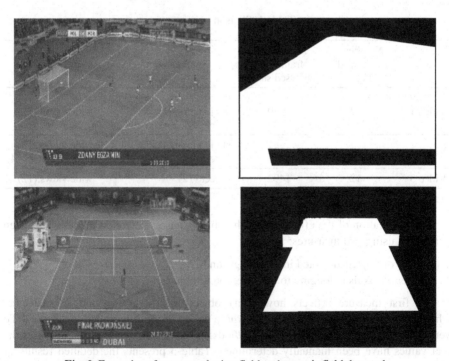

**Fig. 2.** Examples of a soccer playing field and a tennis field detected

Three methods of foreground objects have been implemented in the AVI Indexer:

- pixel blocs detection method (based on [16]),
- Gaussian mixture model based method (based on [17]),
- Bayes decision rule based method (based on [18, 19]).

Some additional problems had to be solved. These methods enabled us to detect not only foreground objects but also lines of playing fields. The lines were removed using the Hough transform [20]. Next, there have been great differences in the performance of these methods depending if the frame was recorded as a wide view or a close-up view. Our conclusion is that these two kinds of frames should be analyzed by different algorithms. In our research wide view frames have been more interesting because the undetected cuts are mainly observed between the wide view shots.

The tests have been performed with four TV sports news from the video data base used also in other experiments of the AVI Indexer project. Three of them have been treated as the training set and the forth sports news has been analyzed to check the usefulness of foreground object detection methods for further video analyses. All tested videos were segmented in the AVI Indexer and the numerous analyses were conducted to designate the best parameters for the TV sports news videos with special characteristics adequate for the video editing style as well as for specificity of sports videos. The characteristics of the training video set is presented in Table 1 and 2.

**Table 1.** Shots in training set

| | Number of analysed shots | Number of frames in analysed shots | Sport categories | Additional information |
|---|---|---|---|---|
| Video 1 | 7 | 930 | cross-country skiing (1), ski jumping (1), soccer (3), basketball (2) | wide views (7) |
| Video 2 | 17 | 2127 | ski jumping (1), soccer (14), hockey (2) | wide views (13), close-up views (4) |
| Video 3 | 12 | 1492 | soccer (12) | wide views (12) |

The verification of the efficiency of the objects/player detection methods has been performed using two measures:

- number of objects detected in the foreground,
- number of pixels belonging to the foreground objects.

The first measure reflects how many objects of the foreground are detected, whereas, the second one verifies if the whole object of the foreground is detected. To verify the efficiency of the detection methods and to calculate these two measures the real values have been manually determined. Table 3 presents the detailed results of object detection.

**Table 2.** Detailed data of shots in a testing video

| Frame number | Duration [mm:ss:ms] | Number of frames | Sport category | Additional information |
|---|---|---|---|---|
| 1 | 00:07:88 | 197 | hockey | close-up view of players |
| 2 | 00:04:96 | 124 | hockey | wide view, many players on the field |
| 3 | 00:06:00 | 150 | hockey | wide view, many players on the field |
| 4 | 00:07:00 | 175 | basketball | wide view, many players on the field |
| 5 | 00:04:24 | 106 | ski jumping | wide view, ski jump |
| 6 | 00:05:20 | 130 | tennis | wide view, two players |
| 7 | 00:03:20 | 80 | soccer | wide view, many players on the field |
| 8 | 00:04:88 | 122 | soccer | close-up view of players |
| 9 | 00:02:48 | 62 | soccer | close-up view of players |
| 10 | 00:04:00 | 100 | soccer | wide view, many players on the field |
| 11 | 00:03:48 | 87 | soccer | wide view, many players on the field |
| 12 | 00:06:44 | 161 | tennis | wide view, two players |
| 13 | 00:06:24 | 156 | tennis | wide view, two players |
| 14 | 00:11:80 | 295 | tennis | wide view, two players |

**Table 3.** Detailed results of object detections in the shots of the tested video

| Shot | RNO | RNP | Pixel blocs detection method | | | | Gaussian mixture model based method | | | | Bayes decision rule based method | | | |
|---|---|---|---|---|---|---|---|---|---|---|---|---|---|---|
| | | | NO | % | NP | % | NO | % | NP | % | NO | % | NP | % |
| 1 | 6 | 97732 | 1 | 16.67 | 28842 | 29.51 | 3 | 50.00 | 54926 | 56.20 | 3 | 50.00 | 60009 | 61.40 |
| 2 | 6 | 17323 | 3 | 50.00 | 7154 | 41.30 | 6 | 100.00 | 15954 | 92.10 | 6 | 100.00 | 15954 | 92.10 |
| 3 | 5 | 12813 | 2 | 40.00 | 4218 | 32.92 | 3 | 60.00 | 6835 | 53.34 | 3 | 60.00 | 7364 | 49.67 |
| 4 | 6 | 49059 | 0 | 0.00 | 0 | 0.00 | 0 | 0.00 | 0 | 0.00 | 0 | 0.00 | 0 | 0.00 |
| 5 | 1 | 9002 | 1 | 100.00 | 3593 | 39.91 | 1 | 100.00 | 6889 | 76.53 | 1 | 100.00 | 7138 | 79.29 |
| 6 | 2 | 1429 | 1 | 50.00 | 987 | 69.07 | 1 | 50.00 | 1058 | 74.04 | 1 | 50.00 | 1001 | 70.05 |
| 7 | 15 | 13927 | 6 | 40.00 | 5148 | 36.96 | 9 | 60.00 | 8991 | 64.56 | 6 | 40.00 | 5347 | 38.39 |
| 8 | 3 | 22913 | 1 | 33.33 | 12238 | 53.41 | 3 | 100.00 | 20858 | 91.03 | 3 | 100.00 | 20082 | 87.64 |
| 9 | 3 | 38656 | 1 | 33.33 | 13857 | 35.85 | 1 | 33.33 | 14936 | 38.64 | 1 | 33.33 | 27420 | 70.93 |
| 10 | 11 | 9251 | 5 | 45.45 | 4124 | 44.58 | 9 | 81.81 | 7733 | 83.59 | 10 | 90.91 | 8553 | 92.45 |
| 11 | 17 | 12192 | 6 | 35.29 | 4573 | 37.51 | 13 | 76.47 | 10006 | 82.07 | 11 | 64.71 | 7349 | 60.28 |
| 12 | 2 | 3609 | 1 | 50.00 | 1215 | 33.67 | 1 | 50.00 | 2558 | 70.88 | 1 | 50.00 | 2495 | 69.13 |
| 13 | 2 | 1512 | 1 | 50.00 | 638 | 42.20 | 2 | 100.00 | 1359 | 89.88 | 2 | 100.00 | 1297 | 85.78 |
| 14 | 2 | 2244 | 1 | 50.00 | 1046 | 46.61 | 2 | 100.00 | 2047 | 91.22 | 2 | 100.00 | 2008 | 89.48 |

where:      RNO – real number of objects,
            RNP – real number of pixels belonging to the objects,
            NO – number of objects detected,
            NP – number of detected pixels belonging to the objects.

Two of the tested methods: Gaussian and Bayes led to the very similar results
(Table 4). Two thirds of objects as well as two thirds of pixels have been detected.
Whereas, the third method based on pixel blocs enabled us to detect significantly
lower number of objects. The next tests will verify if this level of efficacy is sufficient
for better detection of scenes.

**Table 4.** Average recall of object detections

| Method | Object recall [%] | Pixel recall [%] |
|---|---|---|
| Pixel blocs method | 42.43 | 38.82 |
| Gaussian mixture model based method | 68.69 | 68.86 |
| Bayes decision rule based method | 67.07 | 67.61 |

## 5    Improved Temporal Segmentation Based on Detected Players

The methods of player detection can be useful for improving temporal segmentation
when there are undetected cuts between very similar shots of the same discipline
because of the great similarities of backgrounds (Figure 3). Then the detection of
foreground objects, in the case of sports videos the detection of players, and the
analyses of the colours of the foreground objects can reduce the number of undetected
video transitions.

**Fig. 3.** Two similar consecutive frames of two different shots from different sports events of
very similar colour histograms

The frames of a shot identified in headlines of TV sports news have been analysed and the objects/players have been detected in these frames using Gaussian foreground detection method. The results of this analysis for ten frames are presented in Table 5.

It can be easily observed that there is a significant change between the frame 5 and the frame 6. The numbers of players in both frames are the same or almost the same but their colours are completely different. It suggests that there is a cut between these two frames. So, despite the facts that two frames are very similar (their histograms are not significantly different), and further the number of player remains almost the same (in a dynamic shot this number obviously varies) but because the colours of detected players (the colours of their sports clothing) are different the shot analysed is composed of two shots of two different sports events.

The methods of foreground and background selection can be effectively used for the reduction of the number of undetected cuts. In consequence it leads to better detection of scenes in sports videos for example in headlines of TV sports news.

**Table 5.** Number of detected objects and their colours in consecutive 10 frames of a shot identified in headlines of sports news

| Frame number | Number of objects detected | Colours of detected objects | | | | | |
|---|---|---|---|---|---|---|---|
| | | black | blue | green | red | white | yellow |
| 1 | 9 | 1 | 2 | | 5 | | 1 |
| 2 | 8 | 1 | 2 | | 4 | | 1 |
| 3 | 10 | 1 | 3 | | 5 | | 1 |
| 4 | 11 | 1 | 4 | | 5 | | 1 |
| 5 | 11 | 1 | 4 | | 5 | | 1 |
| 6 | 9 | | | 5 | | 4 | |
| 7 | 9 | | | 5 | | 4 | |
| 8 | 9 | | | 5 | | 4 | |
| 9 | 10 | | | 6 | | 4 | |
| 10 | 10 | | | 6 | | 4 | |

# 6    Final Conclusion and Further Studies

Object detection methods with additional colour detection of player's clothing are useful for the improvement of temporal segmentation. Particularly that not all objects in an every frame must be detected. The verification of the efficiency of the objects/player detection methods has been performed using two measures: number of objects detected in the foreground and number of pixels belonging to the foreground objects. The methods which ensure two thirds of the recall of object detection enables us to detect cuts between very similar frames of two different shots from two different sports events. These scenes of different sports events are distinguishable if the players are detected and compared as televiewers do watching the sports broadcast.

The results of tests performed in the AVI Indexer have confirmed that the detection of foreground objects is also useful for the reduction of undetected cuts and then for the improvement of scene detection in the temporal aggregation process.

In further research the tests on more reach video material will be performed, mainly with other than soccer shots. Then, the detection of other specific elements in sports shots of sports news is planed. Finally, new computing techniques will be still developed leading to new functions implemented in the Automatic Video Indexer.

# References

1. Hu, W., Xie, N., Li, L., Zeng, X., Maybank, S.: A survey on visual content-based video indexing and retrieval. IEEE Transactions on Systems, Man, and Cybernetics, Part C: Applications and Reviews 41(6), 797–819 (2011)
2. Del Fabro, M., Böszörmenyi, L.: State-of-the-art and future challenges in video scene detection: a survey. Multimedia Systems 19(5), 427–454 (2013)
3. Asghar, M.N., Hussain, F., Manton, R.: Video indexing: a survey. International Journal of Computer and Information Technology 3(1), 148–169 (2014)
4. Choroś, K.: Video structure analysis for content-based indexing and categorisation of TV sports news. International Journal of Intelligent Information and Database Systems 6(5), 451–465 (2012)
5. Choroś, K.: Temporal aggregation of video shots in TV sports news for detection and categorization of player scenes. In: Bădică, C., Nguyen, N.T., Brezovan, M. (eds.) ICCCI 2013. LNCS, vol. 8083, pp. 487–497. Springer, Heidelberg (2013)
6. Choroś, K.: Headlines usefulness for content-based indexing of TV sports news. In: Zgrzywa, A., Choroś, K., Siemiński, A., et al. (eds.) Multimedia and Internet Systems: Theory and Practice. AISC, vol. 183, pp. 65–76. Springer, Heidelberg (2013)
7. Choroś, K.: Automatic detection of headlines in temporally aggregated TV sports news videos. In: Proceedings of the 8th International Symposium on Image and Signal Processing and Analysis ISPA 2013, pp. 140–145 (2013)
8. Huang, Y., Llach, J., Bhagavathy, S.: Players and ball detection in soccer videos based on color segmentation and shape analysis. In: Sebe, N., Liu, Y., Zhuang, Y.-t., Huang, T.S. (eds.) MCAM 2007. LNCS, vol. 4577, pp. 416–425. Springer, Heidelberg (2007)
9. Nunez, J.R., Facon, J., Brito Junior, A.: Soccer video segmentation: referee and player detection. In: Proceedings of the 15th International Conference on Systems, Signals and Image Processing IWSSIP 2008, pp. 279–282 (2008)
10. Liu, J., Tong, X., Li, W., Wang, T., Zhang, Y., Wang, H.: Automatic player detection, labeling and tracking in broadcast soccer video. Pattern Recognition Letters 30(2), 103–113 (2009)
11. Delannay, D., Danhier, N., De Vleeschouwer, C.: Detection and recognition of sports(wo)men from multiple views. In: Proceedings of the Third ACM/IEEE International Conference on Distributed Smart Cameras ICDSC 2009, pp. 1–7 (2009)
12. Maćkowiak, S., Konieczny, J., Kurc, M., Maćkowiak, P.: A complex system for football player detection in broadcasted video. In: Proceedings of the International Conference on Signals and Electronic Systems ICSES 2010, pp. 119–122 (2010)
13. Mentzelopoulos, M., Psarrou, A., Angelopoulou, A., García-Rodríguez, J.: Active foreground region extraction and tracking for sports video annotation. Neural Processing Letters 37(1), 33–46 (2013)

14. Zhang, J., Qiu, J., Wang, X., Wu, L.: Representation of the player action in sport videos. In: Proceedings of the Signal and Information Processing Association Annual Summit and Conference APSIPA 2013, pp. 1–4 (2013)
15. Choroś, K.: Video Structure Analysis and Content-Based Indexing in the Automatic Video Indexer AVI. In: Nguyen, N.T., Zgrzywa, A., Czyżewski, A., et al. (eds.) Advances in Multimedia and Network Information System Technologies. AISC, vol. 80, pp. 79–90. Springer, Heidelberg (2010)
16. Nawaz, M., Fatah, O.A., Comas, J., Aggoun, A.: Extracting foreground in video sequence using segmentation based on motion, contrast and luminance. In: Proceedings of the International Symposium on Broadband Multimedia Systems and Broadcasting BMSB 2012, pp. 1–3 (2012)
17. Li, L., Huang, W., Gu, I.Y., Tian, Q.: Foreground object detection from videos containing complex background. In: Proceedings of the 11th ACM International Conference on Multimedia, pp. 2–10 (2003)
18. Zivkovic, Z.: Improved adaptive Gaussian mixture model for background subtraction. In: Proceedings of the 17th International Conference on Pattern Recognition ICPR 2004, vol. 2, pp. 28–31 (2004)
19. Ying-hong, L., Hong-fang, T., Yan, Z.: An improved Gaussian mixture background model with real-time adjustment of learning rate. In: Proceedings of the International Conference on Information Networking and Automation ICINA 2010, vol. 1, pp. 512–515 (2010)
20. Fernandes, L.A., Oliveira, M.M.: Real-time line detection through an improved Hough transform voting scheme. Pattern Recognition 41(1), 299–314 (2008)

# An Overlapped Motion Compensated Approach for Video Deinterlacing

Shaunak Ganguly[1], Shaumik Ganguly[1], and Maria Trocan[2]

[1] Amity School of Engineering and Technology,
AUUP, Sec-125, Noida, India
[2] Institut Superieur d'Electronique de Paris,
28 rue Notre Dame des Champs, Paris, France
maria.trocan@isep.fr

**Abstract.** In this paper a block-based motion compensated, contour-preserving deinterlacing method is proposed. It classifies the frame texture according to its contours content and adapts the motion estimation and interpolation in order to ensure high quality image reconstruction. As frame reconstruction is block-based, overlapped motion compensation with adaptive low-pass filters is employed in order to avoid blocking artifacts. The experimental results show significant improvement of the proposed method over classical motion compensated and adaptive deinterlacing techniques.

## 1 Introduction

Interlaced video has been used for long time due to its satisfactory trade-off between frame rate and bandwidth capacity [1]. However, modern day display devices have a fixed resolution display unit, thereby making it impossible to display the interlaced video without further treatment of the video, as the interlaced video format has only a half vertical resolution. It is important to scale the interlaced fields into frames of a given resolution by applying some interpolation in order to fill-in the missing pixels in the resized frame. Resolution synthesis is based on a stochastic model explicitly reflecting the fact that pixels fall into different classes such as edges of different orientation and smooth textures [2].

In motion compensated interpolation, the missing lines are interpolated along the estimated motion trajectory. For example, Dynamic Predictive Search Algorithm (DPSA) is a fast motion estimation (ME) algorithm with high performance efficiency and low computational complexity. It shows that temporally and spatially adjacent macro-blocks are not just statically correlated, but also dynamic alterations in their motion content are highly coherent [3]. Hence, block based motion compensation is implemented. In block motion compensation (BMC), the frames are partitioned in non-overlapping blocks of pixels. Each block is predicted from a block of equal size in the reference frame. The blocks are not transformed in any way apart from being shifted to the position of the predicted block, encountering thus blocking artifacts.

D. Hwang et al. (Eds.): ICCCI 2014, LNAI 8733, pp. 644–652, 2014.

Overlapped block-based motion-compensation (OBMC) is a good solution to these problems because it not only increases prediction accuracy but also avoids blocking artifacts [4]. When using OBMC, blocks are typically twice as big in each dimension and overlap quadrant-wise with all 8 neighbouring blocks. Thus, each pixel belongs to 4 blocks. In such a scheme, there are 4 predictions for each pixel which are summed up to a weighted mean. Overlapped block motion compensation is derived as a linear estimator of each pixel intensity, given that the only motion information available to the decoder is a set of block-based vectors. OBMC predicts the current frame of a sequence by repositioning overlapping blocks of pixels from the previous frame, each weighted by some smooth window.

In addition to motion compensation, to enhance the reconstruction even more, a substantial solution would be to utilize the change in brightness of the image. Optical flow (OF) is the distribution of apparent velocities of movement of brightness patterns in an image. Optical flow can arise from relative motion of objects and the viewer. Consequently, optical flow can give important information about the spatial arrangement of the objects viewed and the rate of change of this arrangement. Discontinuities in the optical flow can help in segmenting images into regions that correspond to different objects. The relationship between the optical flow in the image plane and the velocities of objects in the three dimensional world is not necessarily obvious. We perceive motion when a changing picture is projected onto a stationary screen, whereas motion in terms of optical flow is predicted based on change in brightness [5], [6].

In this paper we propose an adaptive OBMC-based deinterlacing method; the field is partitioned into blocks, and depending on their edge-content, OBMC-based on block-based motion estimation (in case of smooth, non textured areas) and OBMC-based on optical-flow ME (in case of textured areas) are deployed. This way, the contours accuracy is preserved while reducing the overall motion-compensation possible-resulting artifacts.

The paper is organized as follows: Section 2 describes the proposed solution for smooth video deinterlacing. Some experimental results obtained with the proposed method for different video sequences are presented in Section 3. Finally, conclusions are drawn in Section 4.

## 2    Overlapped Motion-Compensated Deinterlacing

The flow chart of the proposed algorithm is depicted in Figure 1. As the field interpolation is based on motion estimation, the temporal correlation between the fields is firstly checked as described in [7]. If the fields are correlated (absence of a cut), bidirectional motion information will be estimated, otherwise unidirectional, backward or forward ME will be performed. However, if both the past and future frames have no correlation with respect to the current frame, basic spatial interpolation is employed [7].

Further, the current field is passed through a Canny egde-detector and further partitioned into blocks of fixed size, each block being categorized as high or low textured depending on the density of detected edges. High textured blocks are then interpolated using in a motion compensated manner, using an optical flow predictor, whereas, block-based motion estimation is used for low-textured blocks. Thereafter, overlapped block motion compensation is applied in order to alleviate the possible blocking artifacts. In the followings, we propose to detail each processing step of the proposed deinterlacing method.

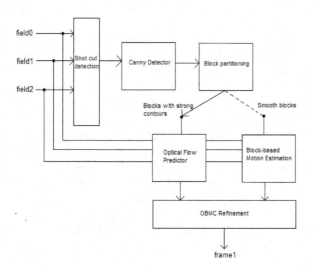

**Fig. 1.** Flow-chart diagram of proposed deinterlacing algorithm

For each block $b_n$ of size 'BxB' belonging to the currrent field $f_n$, its number of edges is derived as in eq.(1), by counting the amount of pixels on contours in the collocated block $c_n$ in the mask field, obtained with the Canny filter:

$$CE_{b_n} = \Sigma_{k=0}^{B^2} c_{n^k} \tag{1}$$

where, $CE_{b_n}$ is the number of identified edges in block $b_n$. The block $b_n$ is classified as highly textured if $CE_{b_n}$ is significant with respect to the blocksize $B^2$, i.e.:

$$CE_{b_n} > T_b \tag{2}$$

($T_b$ is a threshold depending on $B^2$), or smooth, if eq.(2) does not hold. If the number of contours is significant (as in eq. (2)) optical flow based motion estimation is implemented, otherwise we use block-based estimation.

Motion vectors (MV) are obtained for each block on the backward and/ or forward directions for the current field, based on whether a shot-cut is detected, and applying either OF-based estimation proposed by Liu in [5], or simple block-based ME.

We assume that the motion trajectory is linear, so the obtained forward motion vectors (MVs) are split into backward (MVB) and forward (MVF) motion vector fields for the current field $f_n$. As a block in $f_n$ could have zero or more than one MVs passing through, the corresponding $MV_n$ for the block $b_n \in f_n$ are obtained by the minimization of the euclidian distance between $b_n$'s center, $(y_{n,0}, x_{n,0})$, and the passing vectors MVs. In our minimization, we consider only the MVs obtained for the blocks in the neighbourhood of the collocated block $b_{n-1}$ in the left field $f_{n-1}$ (thus a total of nine MVs, obtained for $b_{n-1}$ and the blocks adjacent to $b_{n-1} \in f_{n-1}$, as these MVs are supposed to be the most correlated to the one in the current block, e.g., belonging to the same motion object).

If the motion vector MV corresponding to the collocated block $b_{n-1} \in f_{n-1}$ lies on the line:

$$\frac{y - y_{n-1,0}}{MV_y} = \frac{x - x_{n-1,0}}{MV_x} \tag{3}$$

where $(y_{n-1,0}, x_{n-1,0})$ is the center of $b_{n-1}$ and $MV_x$, respectively $MV_y$, measures the displacement along the $x$, respectively $y$ axis, the distances from the center $(y_{n,0}, x_{n,0})$ of the current block $b_n$ to the MVs lines are obtained as:

$$D_{k \in \{1,...,9\}} = \frac{|MV_{k,x}y_{n,0} - MV_{k,y}x_{n,0} + MV_{k,y}x_{n-1,k} - MV_{k,x}y_{n-1,k}|}{\sqrt{MV_{k,x}^2 + MV_{k,y}^2}}. \tag{4}$$

$MV_n$ is the closest motion vector to the current block $b_n$, if its corresponding distance to the center of $b_n$, $(y_{n,0}, x_{n,0})$, is minimal, i.e.: $D_n = min(D_{k \in \{1,...,9\}})$. Hence, $MV_n$ is generated for each block, containing the motion-estimation in the x and y directions for every pixel.

## 2.1 Overlapped Motion Compensation

OBMC is derived as the linear estimator of each pixels intensity, given that the motion information available is only a set of block-based motion vectors [4]. For the OBMC prediction of the current block, eight motion vectors of neighboring blocks are considered, namely the MV corresponding to the up, down, left, right, upleft, upright, downleft and downright blocks, in addition to the current block motion vector. OBMC predicts the current frame of a sequence by repositioning overlapping blocks of pixels from the previous frame, each weighted by some smooth window. Instead of restricting the MV to consist of a single vector, we define the MV to include motion vectors from blocks in some neighborhood of the particular pixel. If we assume that both the image and the block motion field are realizations of a stationary process, the position of neighboring blocks (relative to the current block) and the best weights to associate with vectors from those blocks should only depend on the relative pixel position within the block and not on the absolute pixel position. Like the standard block-based algorithms, we limit the complexity by constraining the weights to be independent of the transmitted data [4], [8].

The predicted frame using block-based as well as OF estimations are both refined through OBMC, using a weighted window. We have considered several low-pass windowing functions, i.e. Hamming, Hanning, Blackman and Raised Cosine (one weighting function is visually depicted in Figs. 2).

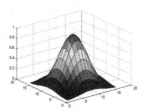

**Fig. 2.** Hamming weighting window

It should be noted that any other low-pass filtering weighting matrices can be used; in our experimental setup we have considered only the classical low-pass windows. The motion-compensated block to be further used in the deinterlaced frame is obtained as average of the backward and forward block-predictions. For a predicted block $b_n$, it is refined through the given weighted matrix of one the mentioned window, as follows:

$$
\begin{aligned}
rb_n = (&w_{centre} * b_n(i + MVy, j + MVx) + w_{up} * b_n(i + MVy + B, j + MVx) + \\
&w_{down} * b_n(i + MVy - B, j + MVx) + w_{left} * b_n(i + MVy, j + MVx + B) + \\
&w_{right} * b_n(i + MVy, j + MVx - B) + w_{upleft} * b_n(i + MVy + B, j + MVx + B) + \\
&w_{upright} * b_n(i + MVy + B, j + MVx - B) + \\
&w_{downleft} * b_n(i + MVy - B, j + MVx + B) + \\
&w_{downright} * b_n(i + MVy - B, j + MVx - B))/W.
\end{aligned}
\tag{5}
$$

where, $rb_n$ is the refined prediction of the block, MVy and MVx are motion vectors obtained in the y and x directions for each pixel, B is the block size, $w_{centre,up,right,left,down,upright,upleft,downright,downleft}$ correspond to the weight-matrix obtained for the respective portion of the window w used and W is the summation of all the weight-matrices (normalization factor).

Finally, the deinterlaced block is found by a contour-based interpolation as:

$$
b_n(i, j) = \frac{b_n(i - 1, j + x_0) + b_n(i + 1, j - x_0) + (k * rb_n(i, j))}{k + 2}
\tag{6}
$$

where $k$ is a weight for the interpolation and $x_0$ is obtained by edge-line minimization [7]. The exact value of $k$ will be empirically found through experiments, such that it maximizes the deinterlacing quality for all considered test sequences.

# 3   Experimental Results

To objectively and comprehensively present the performance of the proposed deinterlacing approach, our method has been tested on several CIF-352 × 288 (Foreman, Hall,Mobile, Stefan and News) and QCIF-176 × 144 (Carphoneand Salesman) video sequences, which have been chosen for their different texture content and motion dynamics. The selected video sequences were originally in progressive format. In order to generate interlaced content, the even lines of the even frames and the odd lines of the odd frames were removed. This way, objective quality measurements could be done, using the original sequences, i.e. progressive frames as references.

In our experimental framework we have used $4 \times 4(B = 4)$ and $8 \times 8(B = 8)$ pixel blocks for a $16 \times 16(S = 16)$ search motion estimation window. The OBMC weighting matrices used are the ones mentioned in Section 2.1. The weight $k$ for the MC interpolation is set to 62, value which has been empirically proved to maximize the deinterlacing quality through all tested sequences.

The tests were run on 50 frames for each sequence. The deinterlacing performance of our method is presented in terms of peak signal to-noise ratio (PSNR, in Table I) computed on the luminance component. The average PSNR results, for the different weighting windows in Section 2.1, are compared in Table II to Vertical Average (VA), Edge Line Average (ELA), Temporal Field Average (TFA), Adaptive Motion Estimation (AME) and Motion-Compensated Deinterlacing (MCD), which are the most common implementations in deinterlacing systems. Moreover, the efficiency of the proposed algorithm is compared to the work in [9], denoted by EPMC, [10] denoted by SMCD and the methods proposed in [11], [12] and [13] (these latter results are reported as in the corresponding references, NC denoting the non-communicated ones).

For visually showing the results of the proposed method, two deinterlaced frames are illustrated in Figs. 3 and 4.

As in the proposed method we have used block based motion-estimation for smooth surfaces and optical flow based algorithm for regions with strong contours, the threshold $T_b$ in eq. (2) is set to 8 for a block-size $B = 4$, and $T_b = 32$ for $B = 8$.

For a search window of $16 \times 16$ pixels, we have compared results for varying the ME block size from $4 \times 4$ to $8 \times 8$ in Table I and measured the computation time in Table III. Not only the computation time of having a block size of $8 \times 8$ is significantly less than for a $4 \times 4$-block ME, but also the obtained PSNR results are almost the same. In other words, by changing the block size from $4 \times 4$ to $8 \times 8$, we have reduced the computation time significantly without compromising with the quality of the reconstructed video sequence.

It is important to note that the reduction of computation time is not so significant for the two sequences in QCIF format. The resolution of QCIF is $176 \times 144$ pixels and after interlacing only half vertical resolution is left. For a search window of $16 \times 16$, it is not possible to accommodate a block size of $8 \times 8$. Therefore, we sampled the interlaced frame horizontally, and then used a block size of $4 \times 4$ for motion estimation and then scaling it by two, resulting

**Table 1.** Comparision of PSNR for different windows

| | Hamming | | Hanning | | Blackman | | Raised cosine | |
|---|---|---|---|---|---|---|---|---|
| | 4x4 | 8x8 | 4x4 | 8x8 | 4x4 | 8x8 | 4x4 | 8x8 |
| *Foreman(cif)* | 41.81 | 42.07 | 41.82 | 42.11 | 41.82 | **42.41** | 41.80 | 42.11 |
| *Hall(cif)* | 42.31 | 42.72 | 42.32 | **42.87** | 42.33 | 42.86 | 42.31 | 42.86 |
| *Mobile(cif)* | 33.46 | 35.18 | 33.47 | 35.22 | 33.46 | **35.26** | 33.46 | 35.24 |
| *Stefan(cif)* | 35.79 | 35.90 | 35.80 | 35.94 | 35.81 | **35.97** | 35.79 | 35.93 |
| *News(cif)* | 43.48 | **44.31** | 43.43 | 44.27 | 43.38 | 44.18 | 43.41 | 44.28 |
| *Carphone(qcif)* | 42.14 | 44 66 | 42.11 | 44.66 | 42.11 | **44.69** | 42.14 | 42.68 |
| *Salesman(qcif)* | 47.32 | **50.44** | 47.32 | **50.44** | 47.31 | **50.44** | 46.32 | **50.44** |

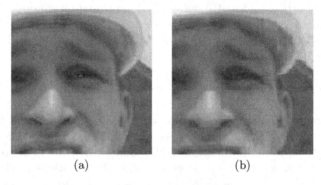

(a)                              (b)

**Fig. 3.** Deinterlacing result for a region within the $15^{th}$ frame of "Foreman" sequence with 8x8-block Blackman window: (a) original, (b) deinterlacing result

**Table 2.** PSNR Results

| | Foreman | Hall | Mobile | Stefan | News | Carphone | Salesman |
|---|---|---|---|---|---|---|---|
| *VA* | 32.15 | 28.26 | 25.38 | 27.30 | 34.64 | 32.17 | 31.52 |
| *ELA* | 33.14 | 30.74 | 23.47 | 26.04 | 32.19 | 32.33 | 30.51 |
| *TFA* | 34.08 | 37.47 | 27.96 | 26.83 | 41.06 | 37.39 | 45.22 |
| *AME* | 33.19 | 27.27 | 20.95 | 23.84 | 27.36 | 29.63 | 28.24 |
| *MCD* | 35.42 | 34.23 | 25.26 | 27.32 | 35.49 | 33.55 | 33.16 |
| *EPMC(S1)* | 37.09 | 39.27 | 31.54 | 30.02 | 41.63 | 37.53 | 45.61 |
| *EPMC(S2)* | 37.18 | 39.08 | 30.56 | 30.11 | 39.44 | 37.55 | 42.28 |
| [16] | 33.77 | NC | 27.66 | 28.79 | NC | NC | NC |
| [17] | NC | NC | NC | 24.59 | NC | NC | NC |
| [18] | 33.93 | 38.79 | 24.67 | 26.88 | NC | NC | NC |
| *SMCD(S1)* | 37.52 | 39.71 | 30.41 | 31.77 | 41.85 | 37.59 | 45.95 |
| *SMCD(S2)* | 37.63 | 39.86 | 30.58 | 31.82 | 42.00 | 37.74 | 45.09 |
| *OBMCOF4x4* | 41.82 | 42.33 | 33.47 | 35.82 | 43.49 | 42.15 | 47.32 |
| *OBMCOF8x8* | **42.07** | **42.72** | **35.18** | **35.90** | **44.31** | **44.66** | **50.44** |

**Table 3.** Comparison of computation time

(average time in seconds/frame for all OBMC interpolation, for a given sequence)

|                  | 4x4   | 8x8   |
|------------------|-------|-------|
| $Foreman(cif)$   | 0.881 | 0.356 |
| $Hall(cif)$      | 0.886 | 0.351 |
| $Mobile(cif)$    | 0.879 | 0.349 |
| $Stefan(cif)$    | 0.941 | 0.352 |
| $News(cif)$      | 0.874 | 0.353 |
| $Carphone(qcif)$ | 0.188 | 0.179 |
| $Salesman(qcif)$ | 0.186 | 0.183 |

(a)                              (b)

**Fig. 4.** Deinterlacing result for a region within the $20^{th}$ frame of "Mobile" sequence with 8x8-block Blackman window: (a) original, (b) deinterlacing result

in an $8 \times 8$ reference block size while passing it through the Canny detector to differentiate the method to be adopted. Hence, the computation time is almost the same as that obtained for a block size of $4 \times 4$.

## 4    Conclusion

In this paper, a combination of block-based interpolation, along with optical flow estimation and overlapped block-compensation for video deinterlacing is proposed. The deinterlacer adapts the interpolation approach in function of the texture content of an area: for smooth regions, simple block-based ME is implemented, whereas for highly textured areas optical flow estimation is performed. By deinterlacing on contours direction with the OF approach, the strong edges are not affected by the artifacts that a block-based ME might introduce. To further improve the interpolation results, the compensation is performed in an overlapped manner, using different weighting windows for the OBMC. The obtained results clearly show that the proposed deinterlacing method surpasses all other techniques in terms of quality of reconstructed images. Furthermore, the proposed method acknowledges the possibility of improving image quality and simultaneously reducing execution time.

# References

1. Haan, G.D., Bellers, E.B.: Deinterlacing - An overview. Proceedings of the IEEE 86(9), 1839–1857 (1998)
2. Atkins, C.B.: Optical Image Scaling using Pixel Classification. In: International Conference on Image Processing (2001)
3. Abdoli, B.: A dynamic predictive search algorithm for fast block-based motion estimation, Theses and Dissertations (2012)
4. Orchard, M., Sullivan, G.: Overlapped block Motion Compensation: an estimation-theoretic approach. IEEE Image Processing (1994)
5. Liu, C.: Beyond Pixels: Exploring new Representations and applications for Motion Analysis, Doctoral Thesis. MIT (May 2009)
6. Horn, B.K.P., Schunck, B.G.: Determining optical flow. Artificial Intelligence 17, 185–203 (1981)
7. Trocan, M., Mikovicova, B., Zhanguzin, D.: An Adaptive Motion Compensated Approach for Video Deinterlacing. Multimedia Tools and Applications 61(3), 819–837 (2011)
8. Watanabe, H., Singhal, S.: Windowed Motion Compensation. Visual Communications and Image Processing (1991)
9. Zhanguzin, D., Trocan, M., Mikovicova, B.: An edge-preserving motion-compensated approach for video deinterlacing. In: IEEE/IET/BCS 3rd International Workshop on Future Multimedia Networking (June 2010)
10. Trocan, M., Mikovicova, B.: Smooth Motion Compensated Video Deinterlacing. In: 7th International Symposium on Image and Signal Processing and Analysis, ISPA (September 2011)
11. Chen, Y., Tai, S.: True motion- compensated de-interlacing algorithm. IEEE Transactions on Circuits and Systems for Video Technology 19 (October. 2009)
12. Wang, S.-B., Chang, T.-S.: Adaptive de-interlacing with robust overlapped block motion compensation. IEEE Transactions on Circuits and Systems for Video Technology 18(10), 1437–1440 (2008)
13. Lee, G.G., Wang, M.J., Li, H.T., Lin, H.Y.: A motion-adaptive deinterlacer via hybrid motion detection and edge-pattern recognition. EURASIP Journal on Image and Video Processing (2008)

# Enhancing Collaborative Filtering Using Semantic Relations in Data

Manuel Pozo, Raja Chiky, and Zakia Kazi-Aoul

Institut Supérieur d'Eléctronique de Paris, LISITE Lab,
28, rue Notre-Dame-des-Champs. 75006 Paris, France
{manuel.pozo,raja.chiky,zakia.kazi}@isep.fr
http://www.isep.fr

**Abstract.** Recommender Systems (RS) pre-select and filter information according to the needs and preferences of the user. Users express their interest in items by giving their opinion (explicit data) and navigating through the webpages (implicit data). In order to personalize users experience, recommender systems exploit this data by offering the items that the user could be more interested in. However, most of the RS do not deal with domain independency and scalability. In this paper, we propose a scalable and reliable recommender system based on semantic data and Matrix Factorization. The former increases the recommendations quality and domain independency. The latter offers scalability by distributing treatments over several machines. Consequently, our proposition offers quality in user's personalization in interchangeable item's environments, but also alleviates the system by balancing load among distributed machines.

**Keywords:** collaborative filtering, distributed systems, recommender system, semantic web technologies.

## 1 Introduction

The amount of information in the web has greatly increased in the past decade. This phenomenon has promoted the advance of Recommender Systems (RS) research area. The aim of these systems is to provide personalized recommendations. They help users by suggesting useful items to them, usually dealing with enormous amounts of data.

Typically, in order to create a top K items (K most relevant items) that should be presented first to the user, Recommender Systems study the interaction between users and items. For instance, users may rate items (such as films, books, etc.) using a 0-5 stars scale (explicit feedback), or user might just create a navigational path clicking links or purchasing items (implicit feedback). Hence, Recommender Systems exploit these feedbacks to set up recommendations. In the literature, Recommender Systems have been usually classified into two basic types: Content-based CB and Collaborative Filtering CF [1]. The former focus on the characteristics of the items in order to determine similarities between

D. Hwang et al. (Eds.): ICCCI 2014, LNAI 8733, pp. 653–662, 2014.
© Springer International Publishing Switzerland 2014

them, and finally recommend similar items to the one the user liked in the past. The latter groups users according to their preferences profile and recommend items that people from the same group have already liked [2]. However, the correct exploitation of the data and the recommendation accuracy are the trend challenges of RS. On the one hand, seeking more relations between users and items may increase the recommendation quality. To this task, the RS might use semantic information about items, which enhance the data representation and help to find out the underlying reasons for which a user may or may not be interested in a particular item [3]. On the other hand, as the quantity and the variety of information constantly increase, the scalability and the domain generality of the system are two important aspects to alleviate the time processing and the heterogeneity of data.

In this paper, we propose a semantic collaborative filtering recommender system in order to improve the recommendations quality and preserve the domain genericity. In addition, we use an easily distributable collaborative filtering technique based on the well known ALS algorithm [4] to ensure the system scalability.

This article is structured as follows: in section 2, we present related work. In section 3, we explain our general approach. In section 4, we expose the experimentation phase and the comparisons. Finally, we discuss in section 5 about the results and the future work.

## 2   Related Work

Recommender Systems (RS) use users feedback to predict interest in items. Nowadays, the current challenges for Recommender Systems are to improve their recommendations quality, their items domain genericity (i.e the domain independency) and their scalability. In order to address these issues, a trend topic is the use of semantic knowledge. For instance, [5] and [6] propose Content Based (CB) recommender systems based on items domain descriptions, such as ontologies. The approaches associate items and keywords that characterize them, creating a bag of words for each item. Hence, the items similarity looks for common keywords in bags. Recently, [7] proposed a recommender system architecture that facilitates the integration of heterogeneous data. The system creates a semantic graph representation of data in order to weight the relations between the nodes of the graph. The system explores the graph and chooses the best weights for predictions tasks.

Nevertheless, some authors have already demonstrated the effectiveness of more extended Collaborative Filtering (CF) techniques [2]. For example, [8] proposed a CF method that took the implicit users feedback into account. Authors used Alternating Least Squares (ALS) CF technique that is based on Matrix Factorization techniques. The ALS method can be easily distributed among machines, and thus, enhances the scalability of the system. [9] focus on the relevance of items in the ranking, rather than in the items ratings prediction using a technique similar to ALS. [10] proposes another approach that improves CF techniques using item-item similarity based on items description in Wikipedia.

In cases where sparsity is too high, the system guess artificial ratings for new items regarding the last user ratings on similar items.

Differently, but also aiming to improve CF, [11] suggests a three-layer representation for data: users, interests and items. For a user, an interest is a characteristic that an item must have. For an item, an interest is one of its attributes. Thus, authors construct a correlation matrix graph containing users hidden interests. Other approaches encourage the use of multi-criteria feedback to improve recommendations quality [12][13][14]. In order to explain an overall rating in items, authors also analyze these item attributes feedbacks and also predict their rating.

In [15] and [16], authors proposed an approach for reducing dynamically the big number of item features needed for the recommendations. Both approaches construct an items-attributes matrix that they reduce by using SVD algorithm, alleviating sparsity and noise.

However, most of the approaches lack distribution environment and domain genericity [3][17][11]. Thus, we propose a general semantic CF method based on ALS [8][9] and an ontology semantic technology. ALS allows the distribution of the computation among several machines using Matrix Factorization techniques, whereas ontologies add information representation and domain independency to the system.

## 3   General Approach

The proposed architecture includes a collaborative filtering engine that relies on domain knowledge about the items. This knowledge is represented using an ontology that will enhance the recommendation quality. The architecture is represented in figure 1 and is composed of two main modules: a semantic module and a recommendation module.

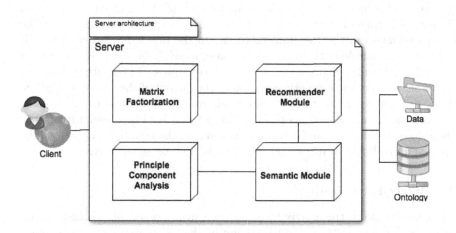

**Fig. 1.** Global architecture of our approach

The semantic module exploits the domain ontology to define the relations between items and their attributes. Because the number of those attributes could be huge, we apply a dimension reduction technique based on **Principal Component Analysis** (PCA) in order to select the most representative ones (preanalysis phase). The PCA also provides weights for attributes to give them more or less importance in the recommendation process. The recommendation module is based on matrix factorization method. Both modules will be described in the following subsections.

## 3.1   Semantic Module

A user could be interested in an item because of one or more of its attributes. For example, a user might appreciate a set of movies only because they have in common his favorite actor. Therefore, the purpose of this module is to take into account the attributes that compose items to better serve the user. For that, its relies on a domain ontology (i.e. an ontology that describes the domain for which we want to implement a recommendation system).

To transform traditional ratings into "semantic rating", we focus first on the number of occurrence of attributes that were rated by a user. This occurrence that we call **occurrence frequency** or **coincidence** is the number of times the attribute values of the items rated by the user are repeated. The second step is to calculate the semantic value ($sv$) based on the occurrence frequency. The equation used is presented in (1):

$$sv = r + E[r] * \frac{\left| \sum_{i=1}^{F} c_i * w_i \right|}{N} \tag{1}$$

Where r is the value of the initial rating, E[r] is the average of the user ratings, F is the total number of attributes, $c_i$ is the occurrence frequency of the attribute i in the set of items that have been rated by the user, $w_i$ is the weight calculated from the PCA phase, and N is the total number of items rated by the user. This equation takes into account positive and/or negative ratings. It can be used at two levels in the Recommendation. On the one hand, we can apply it to all ratings available in the original database, which helps to explain the users interest for particular characteristics of the rated items. On the other hand, we can choose to apply the semantic equation to the output of the recommendation. Indeed, suppose the recommendation module returns the result as top K items for a user, with an estimation of ratings of this top K. These user ratings will be transformed into a semantic rating according to the equation (1) and will be reordered accordingly in top K'. K' can be less than or equal to K.

## 3.2   Matrix Factorization

As its name suggests, the matrix factorization consists on decomposing a matrix into two or several matrices, which results in the same original matrices after their multiplication.

This method allows discovering latent or hidden relationships between a user and the items. For instance, we may suppose that two users have highly rated a movie because they like its actors/actress on it, or because it is an action movie, which may be their favorite movie genre. Thus, if we are able to discover these hidden-reasons, we may be capable of predicting the interest on each item, because the associated characteristics of preferences may correspond with the ones that the items contain.

The Matrix Factorization is closely related to Singular Value Decomposition (SVD), which is widely used to identify hidden relation factors in information retrieval. Collaborative Filtering method may adapt this technique in order to work on the typical sparse rating matrix.

The Matrix Factorization models the interaction between users and items by applying a scalar product of two vectors representing the latent features in a space of dimension f. As a consequence, each item $i$ is associated to one vector $q_i \in$, as well as each user $u$ is associated to a vector $p_u$. The dot product of both vectors represents an estimation of the ratings that the user may give to the item.

$$r'_{ui} = q_i^T p_u$$

Hence, the challenge is to obtain all these vectors $q_i$ and $p_u$; by solving the equation (2)

$$min_{q*,p*} \sum_{u,i \subset K} (r_{ui} - q_i^T p_u)^2 + \lambda(\| q_i \|^2 + \| p_u \|^2) \tag{2}$$

Where $K$ is the set of pairs $(u, i)$ in which the ratings of $r_{u,i}$ are available, and $\lambda$ is a regularization parameter that allows controlling the learning model [4].

**Alternating Least Squares (ALS).** To solve equation (2), [18] proposes an approach called Alternating Least Squares (ALS). In this equation, the vectors $q_i$ and $p_u$ are unknown, and thus the equation is not convergent. However, if we are able to fix one of them, the optimization problem becomes square, and we are able to optimally solve the equation. Thus, the ALS techniques fix in each iteration $q_i$ or $p_u$: when $p_u$ is fixed, the system computes $q_i$ by least squares, and vice versa. This fact allows an optimal convergence in the iteration of the equation. This method is highly interesting because it is easily distributable among multiple machines. In fact, the system computes each $q_i$ independently of the other factors, as well as for each $p_u$.

## 4    Experimentation

### 4.1    Experimental Dataset

In order to implement the semantic module, we need a dataset and an items domain description that defines the elements. We choose MovieLens, a movies dataset, which contains a good number of items (films) and users ratings.

**Table 1.** Experimentation: Dataset

| Users | Films | Ratings | Features |
|-------|-------|---------|----------|
| 100 | 1232 | 11019 | 9 |

In addition, we need the relative information about the attributes of the items, which is not provided in the MovieLens dataset. To alleviate this lack, we use an ontology in order to associate automatically the movies with their attributes. Filling a dataset might be a hard work because we have to deal with each item, its properties and its attributes [19]. As a consequence, it may become an obstacle for the experimentation [20]. For this task, we use the IMDb database that represents the relations described in the ontology. Thus, the use of both sources (IMDb and MovieLens) facilitates the integration of the semantic module, which needs items (MovieLens) and their attributes (IMDb). Note that the titles might have different representations, and thus, we needed a clean up phase in order to merge both sources.

Moreover, the ratings dataset in MovieLens and the associated attributes from IMDb are too big and this does not help to apply the semantic equation, which may require important processing time. In order to test our approach, we reduce the number of users to the 100 users who have more rated movies and we focus on only 9 attributes (actor or actress, color, editor, director, genre, language, producer, writer and productions year). Table 1 summaries information about our experimental dataset.

### 4.2 Semantic Recommender System

As far as we know, there is no existing recommender system that handles or facilitates the usage of the web semantic technologies. In addition, such a system might be really complex to implement in order to validate our approach. Thus, we decided to add a semantic layer to an existing recommender system. Moreover, our objective is to implement a collaborative filtering algorithm based on Matrix Factorization in order to be able to analyze huge volume of data. This algorithm is implemented in numerous recommender systems libraries. Hence, we choose a system that better fit our work environment and our constraints: the system has to use a collaborative filtering technique and be easily distributed for handling scalability issues. For our experimentations, we use Myrrix [21]. It is a natural evolution of the widely extended Apache Mahout libraries [22]. We consider Myrrix as a black box: it takes a dataset, analyzes it and executes the recommendation technique. To interact with the ontologies, we use the Jena library [23].

We test the semantic recommender system using two approaches:

- **Semantic Top K**: In this first approach, we aim to apply the semantic layer on the output data. Typically, recommender systems provide a list of K items ordered by user's preference. In this case, the semantic module re-

orders these items and gives back a new list in an other relevance order. This
approach allows better recommendations in a reduced execution time.
  – **Semantic Dataset**: In this second approach, we apply the semantic layer on
    the input of the system. A pre-analysis is done on the dataset before it goes
    into the recommender system. This approach allows finding out new items
    that were not taken into account a priori. However, the semantic module
    needs to analyze the whole dataset increasing the process time in comparison
    with the non-semantic analysis.

In the next section, we present how these approaches have modified the recommendations, and the quality of the obtained results.

### 4.3   Results

We tested our approach using the recommender system Mahout/Myrrix and
adding the semantic layer either at the input or the output. From the experimental dataset, we chose the user who has rated more items (the user with id
13 and 636 ratings). Then, we took out 60 ratings of this user (all of them have
been rated with a 5 over 5). Next, we ask to the system a list of 60 items for this
concrete user. The table 2 presents the top 10 items (over the 60 items) with the
higher prediction scores in the three approaches (non-semantic, semantic dataset
and semantic top K).

**Semantic Output: Semantic Top K.** The semantic top K approach works
over the 60 films asked to Mahout/Myrrix. The semantic algorithm re-evaluates
the predictions associated to each film and gives them back in a new order. The
table 2 shows up the modification in the top K. In concrete, we can see that the

**Table 2.** Experimentation: Top-10 results for a request of the user 13 of MovieLens

| Non-Sem. | | Sem.Ratings | | Sem.Top K | |
|---|---|---|---|---|---|
| ID | Prediction | ID | Prediction | ID | Prediction |
| 56 | 0.720 | 127 | 0.74 | 121 | 1.12 |
| 127 | 0.690 | 56 | 0.73 | 56 | 1.06 |
| 121 | 0.610 | 121 | 0.63 | 237 | 1.049 |
| 135 | 0.600 | 100 | 0.60 | 202 | 1.03 |
| 100 | 0.570 | 135 | 0.58 | 127 | 1.00 |
| 50 | 0.560 | 50 | 0.579 | 13 | 0.91 |
| 234 | 0.550 | 234 | 0.572 | 423 | 0.902 |
| 204 | 0.519 | 237 | 0.53 | 993 | 0.90 |
| 181 | 0.517 | 202 | 0.518 | 161 | 0.894 |
| 237 | 0.500 | 181 | 0.514 | 50 | 0.883 |

film 121 is now in the most preferable, when before it was the third one. The film 237 has gone from the 10th place to the 3rd one. Moreover, note that the half of items in the semantic top K items is absent in the non-semantic top 10; yet, they are still in the list of 60 extracted films.

**Semantic Input: Semantic Dataset.** In this approach, the semantic layer is applied on to the whole dataset before Mahout/Myrrix analyzes it. The results in the table 2 present some small differences against the non-semantic recommendations. The similarities are more important than semantic Top K, in particular the first top 7 recommended films are the same but in different order. However, we also see the emergence of new films (202 film in position 9).

In order to study in more details the impact of the semantic layer on the results, we performed some statistical comparison techniques. We use 3 evaluation methods, all of them are implemented in the Mahout/Myrrix libraries.

- The Area Under Curve (AUC) represents the probability to give a higher rating to a relevant random item than an irrelevant item. In this measure, the higher value is the better one.
- The Precision and Recall (PAR) measures the relevance of the items in the top K recommended list. It combines the fraction of all recommended items that are relevant (precision) and the fraction of all relevant items that were recommended (recall). In this case, the higher value is the better one.
- Estimated Strength (ES) measures the quality and reliability of the results. In this measure, the lower value is the better one.

The table 3 presents the obtained results using or not the semantic layer in the Top-K, and thus it may compare the approaches. The non-semantic dataset contains 0-5 scaled ratings whereas the semantic dataset ratings are between 0 and 10. In the row Semantic (0-5), the ratings have been scaled to 0-5 for a quick comparison with the non-semantic dataset.

**Table 3.** Comparisons: different scale datasets.

| Dataset | AUC | PAR | ES |
|---|---|---|---|
| Non-Semantic (0-5 ratings) | 0.6233 | 0.0692 | 0.929 |
| Semantic (0-5 ratings) | 0.6448 | 0.0728 | 0.937 |
| Semantic (0-10 ratings) | 0.7945 | 0.0329 | 0.1651 |

From this table, we can say that:

- Concerning semantic Dataset (0-10), we got good results for the AUC and the ES measures in comparison to the non-semantic approach.
- Semantic Dataset (0-5): In this case, we observe good results in the AUC and PAR measures. Moreover, the ES values in both semantic and non-semantic approach are very close.

Finally, note that these results are preliminary, and can be improved. Indeed, a closer analysis of the data set would improve the weightings applied to the semantic equation (equation (1)). Further, adding attributes and more users and feedbacks will improve the semantic equation, and therefore the quality of recommendations.

## 5 Conclusion

Recommender systems face a substantial challenge when dealing with huge amount of data. In this paper, our main goal was to help solving this problem by describing an easy development of a distributed recommender system. This was possible by using a powerful algorithm of Collaborative Filtering called ALS which can be easily distributed among several machines. In addition, we propose a semantic equation that allows to reorder the top K suggested items to the user according to his preferences (based on underlying reasons for which a user may or may not be interested in a particular item). The proposed approach in this paper has demonstrated an efficiency on the recommendation of films by using the MovieLens dataset and a film ontology, which was filled from IMDb data source. We decided to choose this domain because the MovieLens dataset is public and highly available.

Nevertheless, the design of the approach is very independent of the application domain. Our system may be used as a black box, where we can connect a database to introduce ratings as well as the application domain ontology. Our future work will focus on two main aspects: (1) To reproduce similar experimentation in different domain datasets in order to prove the genericity of our system, (2) To improve the semantic-layer in order to get better recommendation results and to reduce the time execution of the system.

## References

1. Kantor, P.B., Ricci, F., Rokach, L., Shapira, B.: Recommender systems handbook (2011)
2. Su, X., Khoshgoftaar, T.M.: A survey of collaborative filtering techniques. In: Advances in artificial intelligence 2009, vol. 4 (2009)
3. Peis, E., del Castillo, J.M., Delgado-López, J.: Semantic recommender systems. analysis of the state of the topic. Hipertext. net 6, 1–5 (2008)
4. Koren, Y., Bell, R., Volinsky, C.: Matrix factorization techniques for recommender systems. Computer 42(8), 30–37 (2009)
5. Mabroukeh, N.R., Ezeife, C.I.: Ontology-based web recommendation from tags. In: 2011 IEEE 27th International Conference on Data Engineering Workshops (ICDEW), pp. 206–211. IEEE (2011)
6. Werner, D., Cruz, C., Nicolle, C.: Ontology-based recommender system of economic articles. arXiv preprint arXiv:1301.4781 (2013)
7. Plumbaum, T., Lommatzsch, A., De Luca, E.W., Albayrak, S.: SERUM: Collecting semantic user behavior for improved news recommendations. In: Ardissono, L., Kuflik, T. (eds.) UMAP Workshops 2011. LNCS, vol. 7138, pp. 402–405. Springer, Heidelberg (2012)

8. Hu, Y., Koren, Y., Volinsky, C.: Collaborative filtering for implicit feedback datasets. In: Eighth IEEE International Conference on Data Mining, ICDM 2008, 263–272. IEEE (2008)
9. Takács, G., Tikk, D.: Alternating least squares for personalized ranking. In: Proceedings of the Sixth ACM Conference on Recommender Systems, pp. 83–90. ACM (2012)
10. Katz, G., Ofek, N., Shapira, B., Rokach, L., Shani, G.: Using wikipedia to boost collaborative filtering techniques. In: Proceedings of the Fifth ACM Conference on Recommender Systems, pp. 285–288. ACM (2011)
11. Liu, Q., Chen, E., Xiong, H., Ding, C.H., Chen, J.: Enhancing collaborative filtering by user interest expansion via personalized ranking. IEEE Transactions on Systems, Man, and Cybernetics, Part B: Cybernetics 42(1), 218–233 (2012)
12. Lakiotaki, K., Tsafarakis, S., Matsatsinis, N.: Uta-rec: a recommender system based on multiple criteria analysis. In: Proceedings of the 2008 ACM Conference on Recommender Systems, pp. 219–226. ACM (2008)
13. Mikeli, A., Apostolou, D., Despotis, D.: A multi-criteria recommendation method for interval scaled ratings. In: 2013 IEEE/WIC/ACM International Joint Conferences on Web Intelligence (WI) and Intelligent Agent Technologies (IAT), vol. 3, pp. 9–12 (2013)
14. Mikeli, A., Sotiros, D., Apostolou, D., Despotis, D.: A multi-criteria recommender system incorporating intensity of preferences. In: 2013 Fourth International Conference on Information, Intelligence, Systems and Applications (IISA), pp. 1–6 (2013)
15. Mobasher, B., Jin, X., Zhou, Y.: Semantically enhanced collaborative filtering on the web. In: Berendt, B., Hotho, A., Mladenič, D., van Someren, M., Spiliopoulou, M., Stumme, G. (eds.) EWMF 2003. LNCS (LNAI), vol. 3209, pp. 57–76. Springer, Heidelberg (2004)
16. Szwabe, A., Ciesielczyk, M., Janasiewicz, T.: Semantically enhanced collaborative filtering based on RSVD. In: Jędrzejowicz, P., Nguyen, N.T., Hoang, K. (eds.) ICCCI 2011, Part II. LNCS, vol. 6923, pp. 10–19. Springer, Heidelberg (2011)
17. Kapusuzoglu, H., Olguducu, S.: A relational recommender system based on domain ontology. In: 2011 International Conference on Emerging Intelligent Data and Web Technologies (EIDWT), pp. 36–41. IEEE (2011)
18. Koren, Y., Bell, R.: Advances in collaborative filtering. In: Recommender Systems Handbook, pp. 145–186. Springer, Heidelberg (2011)
19. Bouza., A.: Mo the movie ontology (2010), http://www.movieontology.org/
20. Avancha, S., Kallurkar, S., Kamdar, T.: Design of ontology for the internet movie database, imdb (2010)
21. Owen, S.: Myrrix (2013), http://www.myrrix.com
22. Owen, S., Anil, R., Dunning, T., Friedman, E.: Mahout in Action. First edn. Manning Publications Co., Manning Publications Co., 20 Baldwin Road, PO Box 261, Shelter Island, NY 11964 (2011)
23. McBride, B.: Jena: A semantic web toolkit. IEEE Internet Computing 6(6), 55–59 (2002)

# Security Incident Detection Using Multidimensional Analysis of the Web Server Log Files

Grzegorz Kołaczek and Tomasz Kuzemko

Wroclaw University of Technology,
Wybrzeze Wyspianskiego 27 str. 50-370 Wroclaw, Poland
Grzegorz.Kolaczek@pwr.edu.pl

**Abstract.** The paper presents the results of the research related to security analysis of web servers. The presented method uses the web server log files to determine the type of the attack against the web server. The web server log files are collections of text strings describing users' requests, so one of the most important part of the work was to propose the method of conversion informative part of the requests, to numerical values to make possible further automatic processing. The vector of values obtained as the result of web server log file processing is used as the input to Self-Organizing Map (SOM) network. Finally, the SOM network has been trained to detect SQL injections and brute force password guessing attack. The method has been validated using the data obtained from a real data center.

**Keywords:** web server, security, log files, intrusion detection.

## 1    Introduction

The security breaches are very frequent events in Internet and they are not limited only to big enterprises and the most popular web servers. Each day, there are hundreds of new attacks which are performed using newly discovered vulnerabilities and new types of malware and hacking tools. Because the hacking tools are widely available and also there are many tools which automate the computer system exploitation every Internet user should be prepared not only to protect himself/herself but also it is important to detect if the applied security countermeasures have not been broken [1]. One of the most widely known and discussed example of the degree of the threats related to web servers may be a heartbleed vulnerability discovered recently [2].

The diversity of the threats that may impact users' data security makes that classical protective mechanisms as firewalls, antivirus systems etc. must be combined with intrusion detection and prevention systems (IDS/IPS). Because of rapid changes in protection systems as well as in attack methods, the companies developing IDS/IPS solutions started to combine typical signature based solutions with methods related to soft computing (e.g neural networks, Support Vector Machines, etc.) [3].

The main aim of the research presented in this paper was to develop a new method for improving web servers security. The basic element of the proposed method is

D. Hwang et al. (Eds.): ICCCI 2014, LNAI 8733, pp. 663–672, 2014.
© Springer International Publishing Switzerland 2014

based on artificial neuron network which processes the complex data sets characterizing web server user's behavior. The data sets are derived from the web server status logs and the neural network type is Self Organizing Map (SOM).

## 2    Related Works

There are several works which present the applicability of SOM networks to solve the problem of security breaches in computer networks [4] [5] [6] [7] [8]. In most cases the Self Organizing Map (SOM) networks are used to process the numerical data sets. This means that there is no need to preprocess data before the SOM can be applied. For example, in [6], the network traffic analyzer has been presented. For each ISO/OSI network layer a separate SOM network has been trained. The final decision about security related event detection comes as a result of fusion of data provided for each ISO/OSI layer by dedicated SOM network separately. The similar, multilayer SOM approach has been presented in papers [5] [7][11]. In this approach, the authors used a lot of SOM network trained to determine the similarity for each specialized groups of features. These SOM networks constitute the first layer of the intrusion detection model. Then, the SOM from the second layer combines the data from the first stage of data processing and produces the final decision about system security. Next element which has been derived from the previous researches and applied in the approach which has been presented this paper, is the sliding window for SOM network. However, to the authors' best knowledge there are no publications considering the application of SOM networks to the web server log analysis and especially there is no defined method to transform web server logs into the numerical values which can be processed by SOM network.

The rest of the paper is organized as follows. The next section presents the background of the log analysis problem and defines the detection method of attacks against web server. After that, the experimental evaluation of the proposed method has been presented. The last section contains conclusions and describes future works.

## 3    Server Log Analysis

The main directions in research related to web servers security incident detection focus on the network traffic and server log analysis [1] [6] [9]. As each HTTP request to the web server can be recorded in server's access log file it becomes a natural source of the information about the server's healthiness as well security. The correct user request as well as the invalid or related to attacks against server will be recorded at access log. Then, the logs can be analyzed and symptoms of the security incidents can be recognized. Due to log file size it cannot be analyzed thoroughly by system administrator. So, some additional tools supporting the system administrator should be provided. The final solution must be a compromise between computational complexity, detection precision and speed. The proposed method is offline processing of the recorded user activity available in web server log files. This assumption gives greater possibility for data preprocessing optimization. It also is more flexible and easy to be applied in real web server environment.

The main steps of the web server incident detection method are presented in Fig. 1.

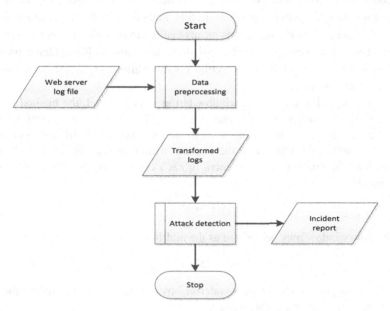

**Fig. 1.** Steps of Web Server incidents detection

Incident detection is performed in two main phases. The first one is web server log file processing. This phase is responsible for:

- sorting the access logs data by IP address and time stamp
- features selection
- transformation of selected features into numerical values
- encoding sequence number (in learning phase only)
- encoding the number of event/attack class (in learning class only)
- saving the data into CSV file

Next phase is respectively training the SOM network (in learning phase only) or attack detection in preprocessed log file. The network is trained in supervised mode so it is required that the data have information about class events. The learning phase is performed one time before the detection of the security incidents in web server log files. Important advantage of the proposed solution is that there is no association with specific types of attacks. The system using the information about the event class learns to recognize the characteristics of a given class of incidents. So, it is possible to use the approach to detect new and unknown attacks. The only element is required is extractor of characteristic attack features which will be able to recognize the information sufficient to distinguish new type of attack from other classes of traffic.

## 3.1   Data Preprocessing

The original web server log files are processed in two step procedure. The first step is log records aggregation. The predefined FIFO buffer has been used as a the sliding-window for data aggregation algorithm. In this method, only the information about a sequence of events is maintained. This means that the information about the time gaps between successive events is lost. Nevertheless, according to [4] hidden representation of time allows to get better detection results while using SOM networks than explicit time representation.

As data collected in the sliding window buffer is aggregated, the method must be proposed to set the particular traffic class for the buffer (in the training phase). This is due to the fact that the entire window is treated as one sample of data. Let all the classes contained in the window at the moment will be marked by a set $S$. The function $v$ assigns the number of occurrences of each class of the records in the current time window.

$$\underset{s \in S}{v} : S \to \mathbb{N} \tag{1}$$

Then the time window may be defined as the multiset:

$$M = \langle S, v \rangle \tag{2}$$

The class assigned to the window is defined by $m$, which is set by finding the class which is the most frequent in this window.

$$m : \underset{m \in S}{max}(v(m)) \tag{3}$$

For example, in the window containing the following sequence of traffic classes: (*normal, normal, sqlinjection,sqlinjection,normal*), the window class will be set to *normal*.

Categorical variables are variables that can assume values only from a limited set. Some of the data in web server logs are categorical so there must be algorithm defined to transform categorical variables into numerical values. One of the frequently used approaches is to assign individual values of categorical variables to consecutive integers. This method does not work, however when it is necessary to determine distance between two values. For example, the calculated Euclidean distance will be larger for extreme values than the neighboring. An alternative method of representation of categorical variables in numerical form is to convert them to binary form [10]. With such representation calculated Euclidean distance between all acceptable values that will be constant. The example of representation of the binary categorical variable "HTTP method" for the value of GET and HEAD are presented in the **Table 1**.

**Table 1.** Categorical variable binary representation

| Value | GET | POST | HEAD | OPTIONS |
|-------|-----|------|------|---------|
| GET   | 1   | 0    | 0    | 0       |
| HEAD  | 0   | 0    | 1    | 0       |

Finally, log file should be preprocessed to extract the most informative part of the record. The general idea of HTTP request/response interpretation has been presented in Fig. 2. The text string representing particular request is divided into a few separate parts which are interpreted by so called "extraction modules".

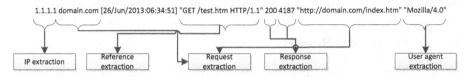

**Fig. 2.** Log record interpretation

The most important operations performed during features extraction phase include: identification and numerical representation of the continent related to IP address, representation of HTTP method and version as a binary vector, calculation of the URL request depth, calculation of the no alphanumerical characters number in the request, hash value calculation from the file extension, change of user-agent name into corresponding ID value, etc. After this phase the vector of numerical values describing the recorded in log file request is passed to the next module.

## 3.2 Network Training

To be able to use the system for the detection of attacks, it is necessary to train SOM network on the training data containing information about the class of the event. This phase is crucial, because the system is able to identify only those attacks that learned to recognize.

The training data must possess information about the class of the event. For example, normal traffic can be assigned to the class of *normal*, while attacks classes with names which correspond names of attacks like: *sqlinjection*, *pathtraversal* and so on. Label with name of the class should be added to each record of access logs used to train the system. It should apply the following format [CLASS], where CLASS is the name of the class to which is assigned to the record. Label in this form should be appended to the end of each row.

The detection system is multi-layered SOM network. Multilayer network consists of multiple layers of standard Kohonen network. Each of them is trained to determine the similarity of the analyzed record to the characteristics of a single attack category. At the input of the network is presented a single data sample. It was isolated three groups of features: *Request*, *Response*, and *Referer*. For each group there is a separate Kohonen network layer.

## 3.3 Attacks Detection

Trained Kohonen network system is ready to perform the classification of previously unknown records of access logs. For this purpose, it is necessary to convert logs into a metrics number. The neuron whose weight vector is most similar to the input is called the best matching unit (BMU). BMU of the sample is calculated by finding the node with the minimum distance to it. Distance sample to a node is calculated as the sum of the partial-weighted distances for each layers. Weight for each layer is defined as a parameter before network training. The classification is made on the basis of class BMU given sample. In the training phase to each node is assigned one class. This information is used in step of anomaly detection to determine previously unknown class of the sample. After finding the BMU for a given sample the information on the BMU class is determined. The class name is returned as the final decision on the classification.

## 4    Experimental Results

The full set of access log which has been used during experimental method evaluation, contains more than 600 thousand records from the server of one hosting company. Each record contains one task per client and HTTP server response. During the observation period, clients' requests coming from more than 100 thousands different IP addresses have been recorded. For the experiment subset comprising the first 10000 records was selected. As a result of the classification procedure performed by the expert two different types of attacks have been identified: SQL injection and brute force password scan. Finally, together with the normal requests the three classes of network traffic has been defined.

**Table 2.** Default configuration of the algorithm

| Parameter name | Value |
|---|---|
| Features | request, response, refer, user-agent, geoip |
| FIFO-size | 5 |
| Rows | 7 |
| Columns | 7 |
| Gridtype | Hexagonal |
| Weight | 1 |
| Training-ratio | 0.5 |
| Numer of iterations | 50 |

The implemented method using SOM networks allows setting the values of the following parameters in the detection algorithm: set of log features defining the SOM dimensions, the sliding window size, number of rows and columns of SOM network, gridtype, weights of features, training ration and number of iterations. The default values of these parameters have been presented in

In the first performed experiment the correlation of the algorithm parameters values and the detection precision (eq. 4) has been investigated.

$$precision = \frac{\sum diag(M_{conf})}{\sum(M_{conf})} \tag{4}$$

where $M_{conf}$ is a confusion matrix.

The results for the default values of algorithm parameters has been presented in Table 3 and Fig. 3.

**Table 3.** Confusion matrix for the default parameters algorithm

|  | Bruteforce | Normal | Sqlinjection |
|---|---|---|---|
| Bruteforce | 41 | 47 | 0 |
| Normal | 27 | 4955 | 1 |
| Sqlinjection | 0 | 15 | 58 |
| Precision | 98,25% | | |

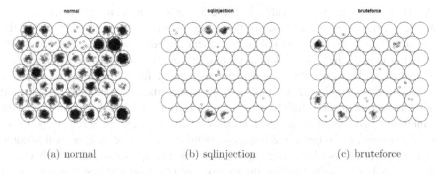

(a) normal                (b) sqlinjection                (c) bruteforce

**Fig. 3.** Mapping for each class of attack to the default SOM parameter values

Next experiment was intended to check the impact of feature selection on detection precision. This parameter determines which features describing HTTP requests and responses are taken into account when training SOM network and during classification. The experiment concerned the following combinations of features set values

**Table 4.** .Cofusion matrix for futures={request}

|  | Bruteforce | Normal | Sqlinjection |
|---|---|---|---|
| Bruteforce | 60 | 28 | 0 |
| Normal | 3 | 4962 | 18 |
| Sqlinjection | 0 | 20 | 53 |
| Precision | 98,66% | | |

**Table 5.** Confusion matrix for futures={request,response}

|  | Bruteforce | Normal | Sqlinjection |
|---|---|---|---|
| Bruteforce | 0 | 88 | 0 |
| Normal | 0 | 4968 | 15 |
| Sqlinjection | 0 | 29 | 44 |
| Precision | 97,43% | | |

**Table 6.** Confusion matrix for futures={request,response,refer}

|  | Bruteforce | Normal | Sqlinjection |
|---|---|---|---|
| **Bruteforce** | 55 | 33 | 0 |
| **Normal** | 19 | 4922 | 19 |
| **Sqlinjection** | 0 | 25 | 48 |
| **Precision** | | 97,69% | |

As the results presented in Table 4, Table 5, Table 6 show, more features used to generate metrics does not necessarily turns into better precision. Paradoxically, the best results were achieved using only one element set *feature=request*. The results are even better than in the case of default parameter values, using all five available features extractors. However, the main element responsible for the highest level of precision is extremely good classification of brute force attack. Meanwhile, sqlinjection attack has been identified slightly worse than in the case of default parameter values. An interesting results of brute force attack classification appear when using set of features consisting of *request* and *response*. In this case, no events appears that could have been classified as a brute force attack. All attack samples were incorrectly classified as normal. The main result of the experiment is the observation that a group of features selected for classification must be chosen individually in the context of a particular attack that we want to detect with the greatest accuracy. Alternatively, one can achieve relatively good classification precision of all attack types by selecting all available features.

The third experiment investigated the influence of the fifo-size for the classification precision of attacks against web servers.

**Table 7.** Confusion matrix for fifo-size=1

|  | Bruteforce | Normal | Sqlinjection |
|---|---|---|---|
| **Bruteforce** | 0 | 85 | 0 |
| **Normal** | 0 | 4997 | 0 |
| **Sqlinjection** | 0 | 64 | 0 |
| **Precision** | | 97,10% | |

**Table 8.** Confusion matrix for fifo-size=10

|  | Bruteforce | Normal | Sqlinjection |
|---|---|---|---|
| **Bruteforce** | 43 | 53 | 0 |
| **Normal** | 24 | 4932 | 14 |
| **Sqlinjection** | 0 | 13 | 62 |
| **Precision** | | 97,98% | |

**Table 9.** Confusion matrix for fifo-size=20

|              | Bruteforce | Normal | Sqlinjection |
|--------------|:----------:|:------:|:------------:|
| **Bruteforce**   | 39         | 52     | 0            |
| **Normal**       | 26         | 4937   | 14           |
| **Sqlinjection** | 0          | 20     | 48           |
| **Precision**    | 97,69%     |        |              |

From the obtained results of experiments with different parameter values fifo-size noticeable is the influence of this parameter on the accuracy of classification. First of all, when the sliding window is off (set to 1) none attack was not detected. The best total result was ever achieved with the default parameter values (fifo-size = 5). Brute force attacks are detected with slightly better precision for longer sliding window values . In turn, the attack *sqlinjection* is best recognized at the value fifo-size = 10.

On the basis of the experiment results can be concluded that the optimal length of the queue depends on the kind of attack we anticipate with the best precision. One should also remember that the selection of longer sliding window increases the probability that the information about a particular attack related to only a single request can be lost. This is due to the way in which traffic class is determined for the whole window.

## 5  Conclusions

Application of soft computing methods to recognition of attack against web server extends the possibilities of the security incident detection. Also previously unknown attacks can be detected only if they are described by similar set of features. The proposed method uses SOM networks for event type classification where events are defined by the records in web server log files. The approach defines also the method of transformation of log files records into corresponding numerical vectors which can be processed by SOM network.

Finally, the several experiments on real data sets have been performed. The results of the experiments are promising for security incidents detection precision. The average detection precision for the proposed method is about 98%.

Further work will focus on improving the rate of attacks detection. Despite the achievement of high values of precision indicator, the number of correctly detected brute force attacks are relatively low. This is widely known problem in computer attack detection where the number of normal events is much greater than the number of events related to attacks. The next steps will be dedicated to the improvement of learning phase to provide the better brute force attack recognition.

## References

1. Multi-agent platform for security level evaluation of information and communication services. Grzegorz, Kołaczek. Springer, Berlin
2. Egeber, P.: Background on Heartbleed (2014)

3. Gudkov, O.: Calculation Algorithm for Network Flow Parameters Entropy in Anomaly Detection. Kaspersky Lab (2012), http://www.kaspersky.com/images/Oleg%20Gudkov.pdf
4. Lichodzijewski, P., et al.: Host-based intrusion detection using self-organizing maps. In: Neural Networks, pp. 1714–1719 (2002)
5. Heywood, M.I.: Dynamic intrusion detection using self-organizing maps (2002)
6. Rhodes, C.: Multiple self-organizing maps for intrusion detection. In: 23rd National Information Systems Security Conference (2000)
7. Stevanovic, D., Vlajic, N.: Detection of malicious and non-malicious website visitors using unsupervised neural network learning. Applied Soft Computing 13(1), 698–708 (2013)
8. Łukasz, B., Katarzyna, N., Michał, A., Grzegorz, K.: SOM-based system for anomaly detection in network traffic. Wroclaw University of Technology, Wroclaw (2013)
9. Kolaczek, G., Juszczyszyn, K.: Traffic pattern analysis for distributed anomaly detection. In: Wyrzykowski, R., Dongarra, J., Karczewski, K., Waśniewski, J. (eds.) PPAM 2011, Part II. LNCS, vol. 7204, pp. 648–657. Springer, Heidelberg (2012)
10. Singh, N., Jain, A., Raw, R.S., Raman, R.: Detection of Web-Based Attacks by Analyzing Web Server Log Files. In: Mohapatra, D.P., Patnaik, S. (eds.) Intelligent Computing, Networking, and Informatics. AISC, vol. 243, pp. 101–109. Springer, Heidelberg (2014)
11. Budka, K.C., Deshpande, J.G., Thottan, M.: Network Security. In: Communication Networks for Smart Grids, pp. 209–225. Springer, London (2014)

# Analysis of Differences between Expected and Observed Probability of Accesses to Web Pages

Jozef Kapusta, Michal Munk, and Martin Drlík

Constantine the Philosopher University in Nitra, Tr. A. Hlinku 1, Nitra 949 74, Slovakia
{jkapusta,mmunk,mdrlik}@ukf.sk

**Abstract.** The paper introduces an alternative method for website analysis that combines two web mining research fields - discovering of web users' behaviour patterns as well as discovering knowledge from the website structure. The main objective of the paper is to identify the web pages, in which the value of importance of these web pages, estimated by the website developers, does not correspond to the actual perception of these web pages by the visitors. The paper presents a case study, which used the proposed method of the identification suspicious web pages using the analysis of expected and observed probabilities of accesses to the web pages. The expected probabilities were calculated using the PageRank method and observed probabilities were obtained from the web server log file. The observed and expected data were compared using the residual analysis. The obtained results can be successfully used for the identification of potential problems with the structure of the observed website.

**Keywords:** web usage mining, web structure mining, PageRank, support, observed, visit rate, expected visit rate.

## 1    Introduction

The aim of the web portal designers or developers is to provide information to users in a clear and understandable form. Information on web pages is interconnected by the hyperlinks. The website developer of designer affects visitors' behaviour by references. He/she indicates the importance of information displayed on web pages through the references. Probably more references head to more important web pages, which are directly accessed from the home page or are referred from other important web pages. Web pages are mostly understood as an information resource for users. They can also provide information in the opposite direction. The website providers can obtain the amount of information about their users or about users' behaviours, needs or interests.

The goal of this paper is to point out the connection between the estimated importance of web pages (obtained by methods of web structure mining) and visitors' actual perception of the importance of individual pages (obtained by methods of web usage mining – a part of web log mining). We can use possible differences between the expected probability and observed probability of accesses to the individual portal web pages to identify suspicious pages. Suspicious pages are defined as pages that are not ordered correctly in the hypertext portal structure.

D. Hwang et al. (Eds.): ICCCI 2014, LNAI 8733, pp. 673–683, 2014.
© Springer International Publishing Switzerland 2014

By knowledge discovery from web page structure (web structure mining, WSM) [1], we focused on the analysis of the quality and importance of web pages based on references among web pages. Determination of web page importance is based on the idea that the degree to which we can rely on the web page quality is transferred by references among web pages. If the web page is referred to by other important pages, the references on that page also become important. By the web usage mining (WUM), we start from the fact, that a user usually posts a large amount of information to the server during his visit of the web page. The most web servers automatically save this information in the form of log file records.

## 2     Related Work

The authors tried to combine web structure mining, web content mining and web usage mining methods in several studies. The combination of methods and techniques of these research fields could help to solve some typical types of web structure analysis issues.

The authors of similar experiments examined several other methods of web page quality estimation, which plays a crucial role in the contemporary web searching engines.

Usually, the estimation of web page quality was assured by the PR or TrustRank algorithms. However, low quality, unreliable data or spam stored in the hypertext structure caused less effective estimation of the web page quality.

Liu et al. [2] further utilized learning algorithms for web page quality estimation based on the content factors of examined web pages (the web page length, the count of referred hyperlinks).

Chua and Chan [3] dealt with the analysis of selected properties of examined web pages. They combined the content, structure and character of hyperlinks of web pages for the web page classification. They wanted to define web page classifiers for thematically oriented search engines more precisely. Each analysed web page was represented by a set of properties related to its content, structure and hyperlinks. Finally, they stated 14 properties of web pages, which they used consequently as the input of machine learning algorithms.

Jacob et al. [4] designed an algorithm WITCH (Webspam Identification Through Content and Hyperlinks), which combined a web structure and web content mining methods for the purpose of spam detection. We also found a similar approach in other experiments [5-7].

Lorentzen [8] found only a few studies, which combine two sub-fields of web mining. The most of the research has focused on a combination of the usage and content mining methods, but he also mentioned some examples of structure mining, which could be said to be web mining's equivalent to link analysis. For example, the Markov chain-based Site Rank and Popularity Rank combine structure and usage mining with a co-citation-based algorithm for the automatic generation of hierarchical sitemaps for websites, or for the automatic exploration of the topical structure of a given academic subject, based on the HITS algorithm, semantic clustering, co-link analysis and social network analysis.

Usually, as we wrote previously, the estimation of the web page quality was assured by the PR, HITS or TrustRank algorithms. However, low quality, unreliable data or spam stored in the hypertext structure caused less effective estimation of the web page quality [9, 10]. Jain at al. [11] provided a detailed review of PR algorithms in Web Mining, their limitations and a new method for indexing web pages.

Ahmadi-Abkenari [12] introduced web page importance metric of LogRank that works based on analysis on server level clickstream data set. The application of this metric means the importance of each page is based on the observation period of log data and independent from the downloaded portion of the Web. Agichtein et al. and Meiss et al. [13, 14] used the traffic data to validate the PageRank random surfing model. Su et al. [15] proposed and experimentally evaluated a novel approach for personalized page ranking and recommendation by integrating association mining and PageRank.

## 3    Data Pre-processing

We mentioned previously, that we tried to connect web usage mining and web structure mining methods for the purpose of identification the differences between expected and observed accesses to the web pages. We describe the details of the proposed approach in this section.

### 3.1    Data Pre-processing for PR Calculation

We developed the crawler, which went through and analysed web pages. The crawler began on the home page and read all hyperlinks on the examined web page. If the crawler found hyperlinks to the unattended web pages, it added them to the queue.

The crawler had created a site map which we have utilized later in the PR calculation of individual pages. The web crawler implemented the method, which has been operating in several steps:

- URL selection from the queue.
- An analysis of the content of selected web page for the purpose of finding new URL references.
- New URL references added to the queue.

The crawler was simple because it scanned only the hyperlinks between web pages. We consider this as the main limitation of the proposed method, because the crawler did not regard the actual position of the hyperlinks within the web page layout, which has a strong influence on the probability of being accessed by a website visitor.

Consequently, we calculated PR for different web pages. Brin and Page [16], the authors of the PR, introduced the Random Surfer Model. This model assumed that the user had clicked on the hyperlinks, he/she had never returned, and they have started on other random web pages.

PR of the web page $i$ ($PR(i)$) is defined as

$$PR(i) = (1 - d) + d \sum_{(j,i) \in E} \frac{PR(j)}{O_j}$$ (1)

where $d$ is Dumping Factor ($0 \leq d < 1$), $E$ is a set of oriented edges, i.e., hyperlinks between web pages, $j$ is a web page with a reference on the web page $i$, and $O_j$ is the count of hyperlinks referred to the other web pages from the web page $j$.

Therefore, the authors proposed Dumping Factor $d$ $\langle 0, 1 \rangle$. Dumping factor represented the probability that the random surfer would continue on the next web page. The value 1- $d$ meant the probability that the user would start on the new web page. The typical value of the variable $d$ is usually 0.85 [17]. We can iterate this calculation until the value of $Pr(i)$ begins to converge to the limit value [16].

## 3.2    Data Pre-processing for Visit Rate Finding of Individual Web Pages

We used the log files of the university web site. We removed unnecessary records and accesses of crawlers from the log file. The final log file had 573020 records over a period of three weeks. We prepared data at some levels:

- Data with session identification using standard time threshold (STT) [18-20]. The session identification method using standard time threshold (time-window) represents the most common method. Using this method, each time we had found subsequent records about the web page requests where the time of the web page displaying had been higher than explicitly selected time, we divided the user visits into several sessions. We used STT = 10 minutes.
- Data with path completion [21, 22]. The reconstruction of the web site visitors' activities represents another issue for WUM. This technique does not belong to the session identification approaches. It is usually the next step of the data pre-processing phase [23, 24]. The main aim of this stage is to identify any significant accesses to the web page which are not recorded in the log file.
  For example, if the user returns to the web page during the same session, the second attempt will probably display the cached version of the web page from the Internet browser. The path completion technique provides an acceptable solution to these problems. It assumes that we can add the missing records to the web server log file using the site map or eventually using the value of the variable *referrer* stored in the log file [25, 26].

The observed visit rate of the web pages were found by the WUM method and represented by the value of variable *support*. The variable *support* is defined as *support(X) = P(X)*. In other words, item $X$ has support $s$ if $s\%$ of transactions contain $X$, i.e. the variable *support* means the frequency of occurrence of given set of items in the database. It represents the probability of visiting a particular web page in identified sequences (sessions).

# 4    Analysis of Differences between Expected and Observed Probability of Accesses to Web Pages

We could see input data of web mining from two different views:

- Data, which depends on the web page developers, this data represents the input of the web structure mining.
- Data, which depends on the web page visitors, this data represents the input of the web usage mining.

Of course, we could dispute that the visitors' behaviour was determined by the structure and the content of the web page and vice versa, but we have had the primary origin of data in mind.

These two groups of data are interdependent. The web page developers should create web pages that reflect the needs of their visitors. We proved the dependence of values PR and real visit rate expressed as values of the variable *support*.

We analysed the dependence of expected values of the web page accesses on real values. The value of PR for individual web pages represented the probability of being visited by a random visitor. At the same time, PR expressed the importance of the web page from the web developer's perspective. If the web page had been important, the developer created more references directed to this web page than to other, less important, web pages. We compared the importance of the web page given by the developer with the real importance of this web page from the visitor's point of view.

We will describe one method of finding the differences between the expected and observed probability of accesses to the individual web pages of the website and analyse the obtained results.

# 5    Residual Analysis

We supposed that we had a log file from a web server with users' sessions identified using the STT method. Simultaneously, we calculated PR for each web page on the examined web site. Finally, we calculated values of the variable *support*. We only used the web pages with a value of variable *support* greater than 0.5 in the residual analysis.

We considered the comparable values for the comparison of expected (PR) and observed (support) visit rate. The variable *support* was from the interval 0-100 and represented the probability of visiting a particular web page in the identified sessions.

We should be aware that the values of PR of individual web pages created the probability distribution of the visits together. Therefore, the sum of PR should be 1. We transformed it into relative values for that reason.

We found inspiration in the residual analysis. The main idea of this method assumes that

$$Data = prediction\ using\ model\ (function) + residual\ value.$$

If we subtracted the values obtained from the model (expected values) from the observed values, we would have got errors (residual values). We could analyse the residual values for the purpose of the model appraisal.

The selection residues $e_i$ are defined as

$$e_i = y_i - \hat{y}_i \ , \tag{2}$$

where $\hat{y}_i$ are expected values predicted by the model and $y_i$ are observed values.

The residual analysis serves for the purpose of the model validity verifying and its improving because it helps to find out the relationships, which the model did not consider. For example, we can use the residual analysis for the regression model stability verifying, i.e., we can identify the incorrectness of the selected model using the correlated chart of residues and independent variable.

The values of variable *support* represented the observed values in the described experiment. The values of the variable PR represented the expected values. As we mentioned earlier, the main objective of the residual analysis was the identification of the outliers. We could visualize the residues in the charts of defined cases (Fig. 1).

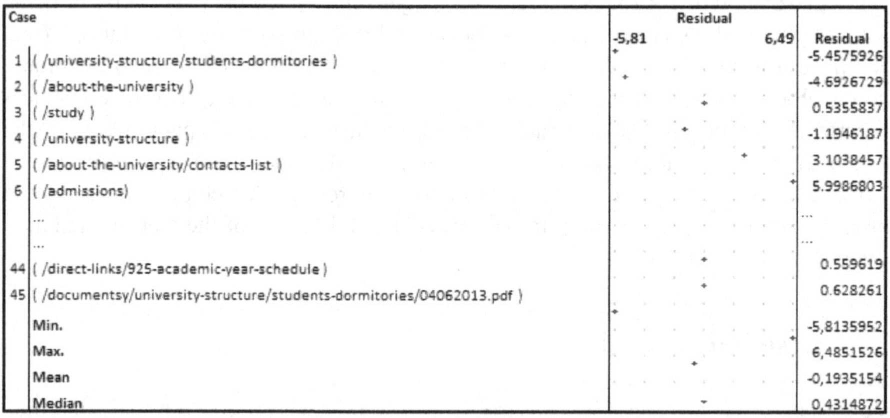

| Case | | Residual | | Residual |
|---|---|---|---|---|
| | | -5,81 | 6,49 | |
| 1 | ( /university-structure/students-dormitories ) | | | -5.4575926 |
| 2 | ( /about-the-university ) | | | -4.6926729 |
| 3 | ( /study ) | | | 0.5355837 |
| 4 | ( /university-structure ) | | | -1.1946187 |
| 5 | ( /about-the-university/contacts-list ) | | | 3.1038457 |
| 6 | ( /admissions) | | | 5.9986803 |
| | ... | | | ... |
| | ... | | | |
| 44 | ( /direct-links/925-academic-year-schedule ) | | | 0.559619 |
| 45 | ( /documentsy/university-structure/students-dormitories/04062013.pdf ) | | | 0.628261 |
| | Min. | | | -5,8135952 |
| | Max. | | | 6,4851526 |
| | Mean | | | -0,1935154 |
| | Median | | | 0,4314872 |

**Fig. 1.** Chart of residues example

The chart visualizes calculated values of residues, minimum, maximum, mean and median for each case. We did not consider the home page of the website in this case study because the value of the variable *support* for this page was equal to 53.51. The average value of the variable *support* was 3.09. Therefore, the value of the main page would have distorted the residues of other web pages in the charts.

We created a chart with expected, observed and residual values for better understanding (Fig. 2). The individual web pages ordered by the PR represent the x axis. We can see the web page identifier in the chart of residues. We could see from this chart that the expected values of PR and observed values of the variable *support* were different. The residuum had to be equal zero in the ideal case, i.e., the expected and observed values should be the same. It implies, the structure of the references to a given web page created by the web developers would be better if the value of the residuum is closer to the x axis.

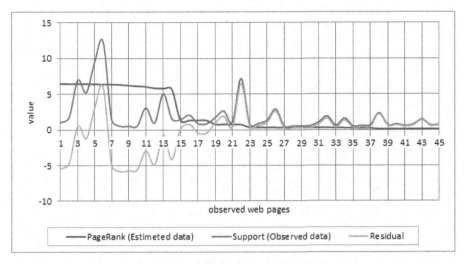

**Fig. 2.** Chart with expected, observed and residual values

# 6    Residual Outliers Identification

The identification of outliers is the objective of the residual analysis. The outlier identifies potentially "suspicious" web pages. In this case, the outliers identified the web pages where the structure of the web site (the intention of the developers, creators) did not reflect the real behaviour of the visitors. We created the chart of residues for the outliers' identification (Fig. 3). Considering the theory of residues we could have identified the outliers using rule ±2σ. It means that we considered the cases which were out of the interval

*Average of differences ± 2 standard deviation of differences.*

We calculated the following boundary values for the selected web pages of the website: -5.696351; 5.30932. The chart (Fig. 3) visualizes the identification of outliers. As we can see, we identified four "suspicious" web pages.
We identified two web pages with the residuum greater than +2σ, which were underestimated by web developers. Even though the web pages had few references from other web pages, the visit rate of these web pages was high. It could have been caused by the seasonal importance of the web pages' content. On the other hand, it is clear that these web pages should have had references from more relevant web pages on the web site.

The main menu of the examined web site caused the main problem of the described experiment. The main menu was available on each web page of the web site. It means that there were always the direct references to the main parts of the web site from all web pages. Therefore, there occurred the evident difference between the value of PR of web pages available directly from the main menu and other web pages.

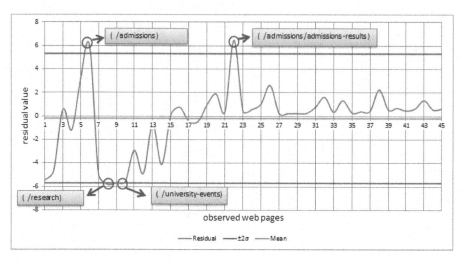

**Fig. 3.** Visualization the identification of outliers

At the same time, we identified two "suspicious" web pages with the residuum greater than -2σ. These web pages were overestimated by the web site developers. Even though the web pages had many references from other web pages, these web pages obtained a small visit rate.

# 7    Discussion

We identified the suspicious web pages on the basis of the rule ±2σ. We analysed 45 web pages and found four suspicious pages. It is questionable if the number of the suspicious web pages is sufficient. If we identify a greater number of suspicious web pages we could assess the boundary values using the quartile interval($Q_I$ − $1,5Q; Q_{III}$ + $1,5Q$). Figure 4 (Fig. 4) depicts the visualization of the suspicious web pages using the mentioned boundary values.

We utilize the results of the residual analysis to recommend website structure changes. In the described case, we should create references (hyperlinks) on the main page or add other items to the main menu for all pages where the value of residuum was greater than the boundary value. On the other hand, we should change the structure of the references on web pages where the value of residuum was lower than the boundary value.

Web usage mining methods examine the behaviour of visitors on the website. In general, we obtained the set of rules using these methods. We could have evaluated each useful rule subjectively if the rule were in accordance with the idea of the website developers.

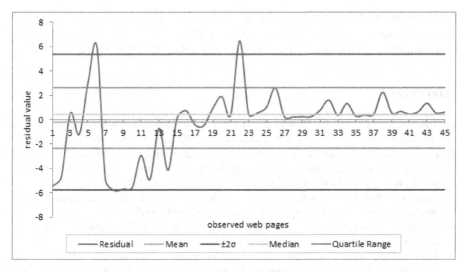

**Fig. 4.** Visualization boundary values of residual analysis

Consequently, we might identify problematic parts of the examined web site. We might compare the intention of the web site developers with the real visits of the web-site using a combination of web structure and web usage mining methods. Moreover, we might emphasize the potential differences using residual analysis.

The proposed method also has some limitations. We have already mentioned the limited behaviour of the crawler, which was used for the PR calculation of individual pages. It is necessary to take into account also other characteristics of the hyperlinks to improve the obtained results, i.e., their position on a given web page, detection of unwanted clicks, etc.

## 8    Conclusions

We paid attention to the new web structure mining method in this paper. We chose the algorithm for PR calculation for estimations of importance and quality of the individual web page. The quality of a given web page depends on the number and quality of the web pages which refer to it. We selected the PR method because this method expresses the probability of visiting of given web page by a random visitor.

We presented the case study of the proposed method of identification "suspicious" web pages in the last chapter. Following the conclusions of the previous experiments, we assumed that the expected visit rate would correlate with the real visit-rate.

We utilized the potential advantages of joining web structure and web usage mining methods in the residual analysis. We tried to identify the potential problems with the structure of the web site. Whereas the sequence rules analysis can only uncover the potential problems of web pages with higher visit rate, the proposed method of residual analysis can also detect the web pages with a low visit rate.

**Acknowledgements.** This paper is published with the financial support of the project of Scientific Grant Agency (VEGA), project number VEGA 1/0392/13 and Cultural and Educational Grant Agency (KEGA), project number 067UKF-4/2012.

# References

1. Srivastava, J., Cooley, R., Deshpande, M., Tan, P.-N.: Web usage mining: discovery and applications of usage patterns from Web data. SIGKDD Explor. Newsl. 1, 12–23 (2000)
2. Liu, Y., Zhang, M., Cen, R., Ru, L., Ma, S.: Data cleansing for web information retrieval using query independent features. Journal of the American Society for Information Science and Technology 58, 1884–1898 (2007)
3. Chau, M., Chen, H.: A machine learning approach to web page filtering using content and structure analysis. Decision Support Systems 44, 482–494 (2008)
4. Jacob, A., Olivier, C., Carlos, C.: WITCH: a new approach to Web spam detection. Yahoo! Research Report No. YR-2008-001 (2008)
5. Castillo, C., Donato, D., Gionis, A., Murdock, V., Silvestri, F.: Know your neighbors: web spam detection using the web topology. In: Proceedings of the 30th Annual International ACM SIGIR Conference on Research and Development in Information Retrieval, pp. 423–430. ACM, Amsterdam (2007)
6. Gan, Q., Suel, T.: Improving web spam classifiers using link structure. In: Proceedings of the 3rd International Workshop on Adversarial Information Retrieval on the Web, pp. 17–20. ACM, Banff (2007)
7. Ntoulas, A., Najork, M., Manasse, M., Fetterly, D.: Detecting spam web pages through content analysis. In: Proceedings of the 15th International Conference on World Wide Web (WWW), Edinburgh, pp. 83–92 (2006)
8. Lorentzen, D.G.: Webometrics benefitting from web mining? An investigation of methods and applications of two research fields. Scientometrics 99, 409–445 (2014)
9. Lili, Y., Yingbin, W., Zhanji, G., Yizhuo, C.: Research on Page Rank and Hyperlink-Induced Topic Search in Web Structure Mining. In: Conference Research on Page Rank and Hyperlink-Induced Topic Search in Web Structure Mining, pp. 1–4 (2011)
10. Wu, G., Wei, Y.: Arnoldi versus GMRES for computing pageRank: A theoretical contribution to google's pageRank problem. ACM Trans. Inf. Syst. 28, 1–28 (2010)
11. Jain, A., Sharma, R., Dixit, G., Tomar, V.: Page Ranking Algorithms in Web Mining, Limitations of Existing Methods and a New Method for Indexing Web Pages. In: Proceedings of the 2013 International Conference on Communication Systems and Network Technologies, pp. 640–645. IEEE Computer Society (2013)
12. Ahmadi-Abkenari, F., Selamat, A.: A Clickstream Based Web Page Importance Metric for Customized Search Engines. In: Nguyen, N.T. (ed.) Transactions on Computational Collective Intelligence XII. LNCS, vol. 8240, pp. 21–41. Springer, Heidelberg (2013)
13. Agichtein, E., Brill, E., Dumais, S.: Improving web search ranking by incorporating user behavior information. In: Proceedings of the 29th Annual International ACM SIGIR Conference on Research and Development in Information Retrieval, pp. 19–26. ACM, Seattle (2006)
14. Meiss, M.R., Menczer, F., Fortunato, S., Flammini, A., Vespignani, A.: Ranking web sites with real user traffic. In: Proceedings of the 2008 International Conference on Web Search and Data Mining, pp. 65–76. ACM, Palo Alto (2008)

15. Su, J.-H., Wang, B.-W., Tseng, V.S.: Effective Ranking and Recommendation on Web Page Retrieval by Integrating Association Mining and PageRank. In: Proceedings of the 2008 IEEE/WIC/ACM International Conference on Web Intelligence and Intelligent Agent Technology, vol. 3, pp. 455–458. IEEE Computer Society (2008)
16. Brin, S., Page, L.: The Anatomy of a Large-Scale Hypertextual Web Search Engine. Computer Networks 107–117 (1998)
17. Page, L., Brin, S., Motwani, R., Winograd, T.: The PageRank Citation Ranking: Bringing Order to the Web. Technical report. Technical report. Standford Digital, Standford (1998)
18. Cooley, R., Mobasher, B., Srivastava, J.: Data Preparation for Mining World Wide Web Browsing Patterns. Knowledge and Information System 1 (1999)
19. Catledge, L.D., Pitkow, J.E.: Characterizing browsing strategies in the World-Wide Web. Comput. Netw. ISDN Syst. 27, 1065–1073 (1995)
20. Pirolli, P., Pitkow, J., Rao, R.: Silk from a sow's ear: Extracting usable structures from the Web. In: Conference Silk From a Sow's Ear: Extracting Usable Structures from the Web (1996)
21. Dhawan, S., Lathwal, M.: Study of Preprocessing Methods in Web Server Logs. International Journal of Advanced Research in Computer Science and Software Engineering 3, 430–433 (2013)
22. Li, Y., Feng, B., Mao, Q.: Research on Path Completion Technique in Web Usage Mining. In: Proceedings of the 2008 International Symposium on Computer Science and Computational Technology, vol. 1, pp. 554–559. IEEE Computer Society (2008)
23. Gong, W., Baohui, T.: A New Path Filling Method on Data Preprocessing in Web Mining. In: Conference A New Path Filling Method on Data Preprocessing in Web Mining, pp. 1033–1035 (2012)
24. Klocoková, D.: Integration of heuristics elements in the web-based environment: Experimental evaluation and usage analysis. Procedia - Social and Behavioral Sciences 15,1010–1014 (2011)
25. Chitraa, V., Davamani, A.S.: An Efficient Path Completion Technique for web log mining. In: IEEE International Conference on Computational Intelligence and Computing Research (2010)
26. Zhang, C., Zhuang, L.: New Path Filling Method on Data Preprocessing in Web Mining. Proceedings of Computer and Information Science 1, 112–115 (2008)

# Method of Criteria Selection and Weights Calculation in the Process of Web Projects Evaluation

Paweł Ziemba, Mateusz Piwowarski, Jarosław Jankowski, and Jarosław Wątróbski

West Pomeranian University of Technology, Szczecin,
Faculty of Computer Science and Information Technology,
Żołnierska 49, 71-210 Szczecin, Poland
{pziemba,mpiwowarski,jjankowski,jwatrobski}@wi.zut.edu.pl

**Abstract.** The article outlines the issues of website quality assessment, reduction of website assessment criteria and factors that users employ in the assessment of websites. The research presents a selection procedure concerning significant choice criteria and revealing undisclosed user preferences based on the website quality assessment models. The formulated procedure utilizes feature selection methods derived from machine learning. Results concerning undisclosed preferences were verified through a comparison with those declared by website users.

**Keywords:** feature selection, Cohen kappa, CART, mean absolute deviation.

## 1 Introduction

Almost 634 million web pages operate in the world [1] with more than 2.4 billion users [2]. Some systems have a global reach, others local. News portals are one of the most popular types of websites and their popularity is strictly linked with quality. This hypothesis is present in publications concerning website assessment, whereby customers must be satisfied with their experience in using the website or they will not return. Thus, the assessment of website quality has become a priority for companies [4] and is affecting users' loyalty and usage frequency [24]. The importance of website quality assessment is reflected in other research which states that effective evaluation of websites has become a point of concern for practitioners and researchers [5]. Evaluation is an aspect of website development and operation that can contribute to maximizing the exploitation of invested resources [6]. Assessment of website quality, including the most popular types of websites, i.e. news portals and social platforms is challenging task in the most stages of online projects. An effort should be made to ensure that the website is reliable and reflects users' expectations. This fact is supported in practice by the ongoing development of methods of presenting information as well as information accessibility and development of new functions tailored to meet the current needs of users. Proposed in this paper approach is based on determining the criteria that implicitly influence users in their evaluation of the quality of web-

D. Hwang et al. (Eds.): ICCCI 2014, LNAI 8733, pp. 684–693, 2014.

sites. This procedure allows usage of weights that reflect their relevance and enables criteria reduction with low impact on the assessment.

## 2     Literature Review and Proposed Approach

Various models of website assessment varying in their uses of quality description criteria, the number of criteria, assessment scales and methods require a model which focuses on the information value of researched sites. According to the analysis, the following models have the highest usability in assessment of the news portals: eQual [7], Ahn [8], SiteQual [9], Web Portal Site Quality [10] and Website Evaluation Questionnaire [11]. The statement was influenced by the fact that these models elaborately treat the issue of presented information quality. One of the limitations of these models is the method of obtaining weights criteria during evaluation of services, as they are usually defined on the basis of declarative approach. Meanwhile, the criteria for determining weights based on surveys and explicit declared user preferences can generate errors in the study [12]. Declared user preferences may differ from actual preferences and from preferences acquired by analytical systems [15]. This means that users in the evaluation can be guided by criteria other than those declared [13][14]. In addition, permanent use in assessing the quality of all the criteria derived from these models would be very confusing, because there are a total of more than sixty of them. Proposed procedure is characterized by the possibility of generalization for applying it to the selection criteria relevant when assessing various types of sites. Presented method for determining the selection criteria and their weights is based on the assumption that surveyed multiple criteria evaluation is not accurate. The evaluation is only accurate if the websites' ratings, calculated as weighted averages of assessment criteria, correspond with the overall evaluations of these websites. The second assumption states that there is a subset of the criteria, which differentiates website quality to a considerable degree. There are also implicit values of weights to be used in multi-criteria assessment services which can give a solution similar to the overall assessment of service determined by the respondents. Based on the assumptions formulated in the research, heuristics inspired by algorithms used in the construction of feature selection, methods of machine learning were applied. According to this approach for finding the subset of classification criteria, there is a model in the sense of machine learning, characterized by a low degree of conflict cases, training model which enables us to a large extent to determine the overall assessment of the service on the basis of the criterion ratings. Therefore, building models of classifiers using subsets of criteria, which can be chosen subsets, may provide solutions close to optimum.

## 3     Procedure of Selecting Website Evaluation Criteria

The aim of the proposed procedure is to analyze the data obtained using feature selection procedures. The objects are to be analyzed with sets of marks awarded by each of the respondents for each service relative to successive criteria and the final marks. Each object consists, therefore, of a set of features, which are criteria for evaluation and descriptions of the class to which the object belongs in the form of global assessment. Feature selection methods examined the influence of individual characteristics on assigning an object to a specific class. An independent assessment of the characteristics of the use of the general characteristics of the data is carried out. Here the correlation coefficients between the values and characteristics belonging to a specific class can be used. These methods, as opposed to wrappers, choose characteristics regardless of the results of the classification, and the classifier is used only to verify the set of characteristics [27]. The use of filters eliminates the impact of the quality of the classifier to select features. In addition, the method was used to examine further features independently of each other, resulting in a ranking of the full set of features together with the numerical value of the significance of each feature. During the processes the following methods were used: ReliefF [27], Significance Attribute [27], Symmetrical Uncertainty [27] and individual indicators of Hellwig information capacity [27]. For ReliefF method individual characteristics can be applied for nearest neighbors, and sampling was performed on all objects. In this step, the obtained rankings were based on significant criteria features.

After the rankings and subsets of features, testing should be performed using the methods of classification. A decision trees classifier namely the advanced classification tree CART was deployed for this purpose. The Gini measure was used as a criterion for the distribution node in the tree. The minimum allowable cardinality of the node was set at five objects and the stop parameter for trimming the tree was misclassification error [27]. Also used was the estimation of the *a priori* probability of belonging to particular classes of objects [27]. Estimating *a priori* helped to improve the model classifier due to the fact that the frequencies of particular classes of decision-making were different. They were close to normal distribution, so the use of the *a priori* estimate was justified. In order to obtain stable results of the classification the 10 fold cross validation was used [27]. The classification of objects into one of seven classes represented specific assessment of the overall service. Each of the ranking criteria was iteratively eliminated in accordance with an important feature of the test rankings and other features used for classification. On the basis of the classification, a true positive rate and Cohen's kappa were determined. Cohen's kappa coefficient describes the compatibility between the expected and predicted objects belonging to the classes of decision-making. An important advantage of Cohen's Kappa coefficient, compared to a ratio of relevance classifier, is that it corrects a random compatibility classification [27]. Moreover, in the literature linguistic interpretations of the extent of compliance can be found, specified by the numeric value of Cohen's kappa coefficient :$K_C \in [0.0, 0.2]$: slight; $K_C \in (0.2, 0.4]$: fair; $K_C \in (0.4, 0.6]$: moderate; $K_C \in (0.6, 0.8]$: substantial; $K_C \in (0.8, 1]$: almost perfect [28].

The next stage of the study was to search for suboptimal subsets of criteria that will get the results of the evaluation criterion as close as possible to the results of the assessment of general services. For this purpose, based on the ratings of well-known news services given by respondents for each website included in the survey, the average overall score was calculated according to the formula (1):

$$G_s = \frac{\sum_{i=1}^{n} s_i}{L_{max}} / n * 100\% \tag{1}$$

$s_i$ – overall assessment of the service assigned by the *i*-th user,
$n$ – the number of users participating in the survey,
$L_{max}$ – evaluating the maximum value of the scale (in this case, seven).

For each obtained subset of criterias the average standardized assessment services were also calculated, using criterial evaluation and weight according to the formula (2):

$$O_s = \frac{\sum_{j=1}^{m}\sum_{i=1}^{n}(w_j * k_{ij})}{\sum_{j=1}^{m}(w_j * L_{max})} / n * 100\% \tag{2}$$

$k_{ij}$ – evaluation of service terms of $j$-th criterion, assigned by the $i$-th user,
$w_j$ – weight of the $j$-th criterion,
$m$ – number of criteria, in terms of evaluating service.

Because of different numbers of criteria resulted in obtaining the services of a different number of points, the score for each subset of criteria has been normalized to the range [0-1]. If a selected subset of the assessment criteria accurately reflects the quality of Internet service, it is between the assessment and the assessment of the overall service relationship or criterion GS ≈ OS. A comparison of standard GS and OS values obtained for the different subsets was performed using a mean absolute deviation measure of [27], according to the formula (3):

$$MAD = \frac{\sum_{i=1}^{n}\|O_i - G_i\|}{n} \tag{3}$$

$n$ - number of respondents' websites,
$G_i$ - average overall rating of the $i$-th website,
$O_i$ - average the criterion of the $i$-th website.

The final selection of a subset of criteria was based on the results of the classification and co-factors, and Cohen kappa values of mean absolute deviation. The steps of the method of selection criteria for the evaluation are shown in Figure 1.

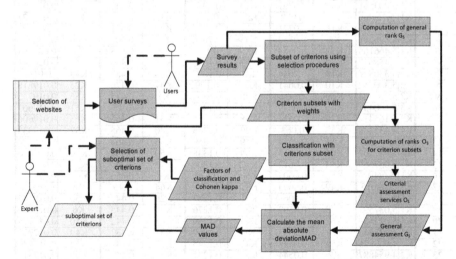

**Fig. 1.** The steps of the method of selection criteria for assessing the quality

# 4     Research Results

The first stage of the research was to gather user surveys for the assessment of the quality of the set of news portals. The survey included questions that correspond to each evaluation criteria using functioning models such as eQual, Ahn, SiteQual, WPSQ and WEQ based on top sites from national ranking [3]. In addition, the survey included a question regarding an overall rating of each service. For comparison purposes, users were also asked about their preferences explicitly declared as the weights of the criteria. Therefore, the users gave answers about how they evaluate each of the websites in terms of a total of 67 evaluation criteria; how important it is for them to have each of the criteria satisfied and how to generally evaluate each of the sites. The study used a seven-step Likert scale. Due to the very large number of questions, filling out the survey in one session would be very tedious and could result in unreliable responses to questions contained in it. Therefore, the survey was divided into three parts and accessed by the users at weekly intervals. The study collected 133 questionnaires, containing criteria evaluations and overall assessments of each site; therefore, base 532 objects were available. After collecting the questionnaires the results were analyzed using feature selection procedures. This made it possible to get rankings and relevance of the criteria contained in Table 1 with columns C for criterion and S for significance.

**Table 1.** Rankings criteria obtained using feature selection procedures

| Feature | Users | | ReliefF | | Symmetrical Uncertainty | | Significance Attribute | | Hellwig's Method | |
|---|---|---|---|---|---|---|---|---|---|---|
| | C | S | C | S | C | S | C | S | C | S |
| 1 | K1 | 5,804511 | K45 | 0,0797 | K16 | 0,1912 | K16 | 0,541 | K16 | 0,025877 |
| 2 | K1 | 5,75188 | K7 | 0,078 | K52 | 0,173 | K19 | 0,533 | K5 | 0,023821 |
| 3 | K1 | 5,578947 | K16 | 0,0771 | K5 | 0,1727 | K52 | 0,511 | K45 | 0,022993 |
| 4 | K9 | 5,496241 | K5 | 0,0759 | K45 | 0,1696 | K5 | 0,509 | K40 | 0,022245 |
| 5 | K4 | 5,466165 | K53 | 0,0653 | K19 | 0,166 | K53 | 0,503 | K7 | 0,021007 |
| 6 | K2 | 5,458647 | K32 | 0,0647 | K53 | 0,1628 | K55 | 0,488 | K53 | 0,020241 |
| 7 | K1 | 5,330827 | K19 | 0,0644 | K7 | 0,1627 | K63 | 0,485 | K19 | 0,0199 |
| 8 | K6 | 5,300752 | K10 | 0,0641 | K55 | 0,159 | K47 | 0,476 | K25 | 0,019818 |
| 9 | K2 | 5,225564 | K66 | 0,064 | K40 | 0,1562 | K7 | 0,475 | K55 | 0,019725 |
| 10 | K3 | 5,180451 | K40 | 0,0639 | K63 | 0,1443 | K32 | 0,47 | K51 | 0,019667 |
| 11 | K2 | 5,165414 | K29 | 0,0634 | K58 | 0,1369 | K51 | 0,466 | K63 | 0,019606 |
| 12 | K4 | 5,135338 | K55 | 0,063 | K29 | 0,1365 | K40 | 0,465 | K37 | 0,019333 |
| 13 | K4 | 5,120301 | K12 | 0,0619 | K49 | 0,1359 | K45 | 0,464 | K9 | 0,018878 |
| 14 | K6 | 5,082707 | K64 | 0,0615 | K64 | 0,1338 | K50 | 0,461 | K29 | 0,018706 |
| 15 | K1 | 5,067669 | K52 | 0,0612 | K12 | 0,1306 | K42 | 0,46 | K58 | 0,018601 |
| 16 | K1 | 5,06015 | K50 | 0,0605 | K50 | 0,129 | K9 | 0,456 | K64 | 0,018477 |
| 17 | K3 | 5,06015 | K49 | 0,0603 | K32 | 0,129 | K64 | 0,45 | K8 | 0,01831 |
| 18 | K5 | 5,045113 | K6 | 0,06 | K51 | 0,1283 | K29 | 0,447 | K52 | 0,018145 |
| 19 | K1 | 5,037594 | K58 | 0,0592 | K6 | 0,1276 | K49 | 0,445 | K6 | 0,018096 |
| 20 | K4 | 5,022556 | K63 | 0,059 | K47 | 0,1269 | K66 | 0,442 | K10 | 0,017794 |
| 21 | K4 | 4,977444 | K9 | 0,0563 | K8 | 0,1269 | K58 | 0,437 | K49 | 0,017776 |
| 22 | K4 | 4,969925 | K51 | 0,0547 | K42 | 0,1255 | K57 | 0,437 | K12 | 0,017739 |
| 23 | K2 | 4,954887 | K2 | 0,0544 | K37 | 0,1231 | K37 | 0,437 | K66 | 0,017317 |
| 24 | K5 | 4,954887 | K8 | 0,0537 | K9 | 0,1194 | K6 | 0,436 | K50 | 0,017225 |
| 25 | K4 | 4,93985 | K48 | 0,0523 | K66 | 0,1179 | K12 | 0,435 | K47 | 0,017055 |
| 26 | K5 | 4,917293 | K14 | 0,0522 | K59 | 0,117 | K59 | 0,435 | K28 | 0,016391 |
| 27 | K1 | 4,909774 | K44 | 0,0521 | K57 | 0,1149 | K62 | 0,435 | K17 | 0,016333 |

**Table 1.** *(continued)*

| Feature | Users | | ReliefF | | Symmetrical Uncertainty | | Significance Attribute | | Hellwig's Method | |
|---|---|---|---|---|---|---|---|---|---|---|
| | C | S | C | S | C | S | C | S | C | S |
| 28 | K5 | 4,887218 | K17 | 0,0521 | K62 | 0,1148 | K8 | 0,424 | K32 | 0,015936 |
| 29 | K4 | 4,842105 | K47 | 0,0521 | K17 | 0,1131 | K25 | 0,411 | K60 | 0,015473 |
| 30 | K4 | 4,834586 | K46 | 0,0514 | K10 | 0,1098 | K61 | 0,411 | K62 | 0,015401 |
| 31 | K3 | 4,819549 | K56 | 0,0512 | K14 | 0,1085 | K28 | 0,408 | K46 | 0,01526 |
| 32 | K4 | 4,81203 | K62 | 0,05 | K2 | 0,1085 | K17 | 0,408 | K15 | 0,015172 |
| 33 | K5 | 4,789474 | K28 | 0,0492 | K27 | 0,1072 | K56 | 0,406 | K14 | 0,015158 |
| 34 | K2 | 4,781955 | K31 | 0,048 | K44 | 0,102 | K60 | 0,402 | K2 | 0,015004 |
| 35 | K3 | 4,759398 | K41 | 0,0469 | K3 | 0,0999 | K2 | 0,396 | K27 | 0,015001 |
| 36 | K4 | 4,729323 | K42 | 0,0441 | K25 | 0,0979 | K46 | 0,396 | K56 | 0,014751 |
| 37 | K7 | 4,714286 | K37 | 0,0437 | K41 | 0,0977 | K44 | 0,394 | K26 | 0,014742 |
| 38 | K3 | 4,714286 | K11 | 0,0428 | K60 | 0,0969 | K14 | 0,394 | K41 | 0,014598 |
| 39 | K2 | 4,699248 | K3 | 0,0427 | K15 | 0,0966 | K15 | 0,388 | K3 | 0,014129 |
| 40 | K2 | 4,661654 | K15 | 0,0422 | K56 | 0,0958 | K27 | 0,387 | K44 | 0,01411 |
| 41 | K5 | 4,639098 | K27 | 0,0408 | K46 | 0,0948 | K10 | 0,381 | K48 | 0,013936 |
| 42 | K6 | 4,639098 | K43 | 0,04 | K28 | 0,0936 | K26 | 0,378 | K42 | 0,013731 |
| 43 | K1 | 4,616541 | K24 | 0,0395 | K20 | 0,0897 | K54 | 0,376 | K59 | 0,013439 |
| 44 | K3 | 4,586466 | K25 | 0,0393 | K26 | 0,0876 | K41 | 0,371 | K11 | 0,012663 |
| 45 | K3 | 4,571429 | K1 | 0,0393 | K61 | 0,087 | K18 | 0,367 | K24 | 0,012466 |
| 46 | K5 | 4,541353 | K33 | 0,0387 | K18 | 0,0828 | K3 | 0,367 | K54 | 0,012324 |
| 47 | K5 | 4,488722 | K57 | 0,0385 | K54 | 0,0812 | K43 | 0,362 | K18 | 0,012031 |
| 48 | K6 | 4,458647 | K59 | 0,0384 | K43 | 0,0811 | K11 | 0,359 | K43 | 0,011867 |
| 49 | K5 | 4,443609 | K54 | 0,0384 | K24 | 0,077 | K23 | 0,342 | K33 | 0,011342 |
| 50 | K8 | 4,43609 | K4 | 0,0382 | K11 | 0,0756 | K31 | 0,34 | K31 | 0,011248 |
| 51 | K3 | 4,421053 | K65 | 0,0371 | K33 | 0,0706 | K13 | 0,336 | K30 | 0,011238 |
| 52 | K5 | 4,37594 | K67 | 0,0365 | K23 | 0,0694 | K20 | 0,335 | K57 | 0,011169 |
| 53 | K2 | 4,345865 | K60 | 0,0359 | K31 | 0,0692 | K48 | 0,334 | K67 | 0,011106 |
| 54 | K3 | 4,330827 | K39 | 0,0346 | K30 | 0,0692 | K30 | 0,333 | K23 | 0,011037 |
| 55 | K3 | 4,315789 | K30 | 0,0339 | K48 | 0,0644 | K24 | 0,326 | K61 | 0,010818 |
| 56 | K6 | 4,315789 | K26 | 0,0332 | K4 | 0,0613 | K65 | 0,324 | K22 | 0,010709 |
| 57 | K5 | 4,300752 | K18 | 0,0331 | K35 | 0,0598 | K22 | 0,321 | K34 | 0,010469 |
| 58 | K1 | 4,270677 | K61 | 0,0329 | K65 | 0,0593 | K33 | 0,316 | K20 | 0,010468 |
| 59 | K1 | 4,24812 | K23 | 0,0329 | K34 | 0,0578 | K35 | 0,309 | K1 | 0,009537 |
| 60 | K6 | 4,172932 | K20 | 0,0328 | K22 | 0,057 | K34 | 0,303 | K65 | 0,009511 |
| 61 | K2 | 4,165414 | K36 | 0,032 | K36 | 0,0542 | K39 | 0,289 | K35 | 0,008776 |
| 62 | K6 | 4,120301 | K21 | 0,0284 | K39 | 0,054 | K36 | 0,286 | K4 | 0,008497 |
| 63 | K6 | 4 | K13 | 0,0267 | K67 | 0,0525 | K67 | 0,278 | K39 | 0,008287 |
| 64 | K3 | 3,796992 | K35 | 0,0258 | K13 | 0,051 | K4 | 0,278 | K13 | 0,008234 |
| 65 | K6 | 3,511278 | K34 | 0,0224 | K1 | 0,0443 | K21 | 0,269 | K36 | 0,008027 |
| 66 | K2 | 3,278195 | K22 | 0,0224 | K21 | 0,0381 | K1 | 0,248 | K21 | 0,00752 |
| 67 | K2 | 3,218045 | K38 | 0,0215 | K38 | 0 | K38 | 0 | K38 | 0,003743 |

When analyzing Table 1, regularity can be noticed and all the procedures for the selection of features for the least important criterion considered K38. Similarly, in all the rankings, among others, low-ranking features include: K21, K36, K34, K35 and K39. In turn, some of the most important criteria in each of the rankings are K16, K5, K19, K45 and K53. These results differ significantly from the ranking obtained on the basis of users' expressed responses in surveys, such as the features K16 and K63, which the users indicated are not very important compared to the rest of the features. Additionally, feature K13 which, according to the users, is considered one of the most important features, is positioned low in other rankings. There are also marked differences between the rankings created using various procedures for the selection of features. For example, the K52 feature was considered very important by the procedure Symmetrical Uncertainty and Attribute Significance, while in the ReliefF and Hellwig rankings it occupies lower positions. After creating rankings, the characteristics were tested using decision trees CART. For each subset of criteria a true positive rate for each class of decision-making and decision-making for all classes together was set. Moreover, for subsets, based on the confusion matrix, Cohen's kappa value was determined (KC). For the KC the standard error was set and on this basis, the confidence interval was 0.99 [26, pp. 543-547]. The confusion matrix for the classification carried out using the full set of 67 criteria was characterized by a pointer value KC = 0.738. Table 2 contains the value of the lower limit of the confidence interval of Cohen's kappa (KCmin) and the true positive rate for the worst classified class (TPRMIN). Subsets of criteria satisfying the conditions TRPMIN > 50 % and KCmin > 0.6 in Table 2 are marked with a pattern.

**Table 2.** KCmin and $TPR_{MIN}$ for subsets of criteria

| Features | Users | | ReliefF | | Symmetric uncertainty | | Attribute significance | | Hellwig method | |
|---|---|---|---|---|---|---|---|---|---|---|
| | $K_{Cmin}$ | $TPR_{MIN[\%]}$ | $K_{Cmin}$ | $TPR_{MIN[\%]}$ | $K_{Cmin}$ | $TPR_{MIN[\%]}$ | $K_{Cmin}$ | $TPR_{MIN[\%]}$ | $K_{Cmin}$ | $TPR_{MIN[\%]}$ |
| 15 | 0,537 | 51,85 | 0,570 | 57,14 | 0,589 | 42,86 | 0,563 | 37,14 | 0,528 | 17,14 |
| 16 | 0,542 | 45,71 | 0,590 | 60,00 | 0,578 | 42,86 | 0,589 | 37,14 | 0,589 | 37,14 |
| 17 | 0,528 | 62,96 | 0,592 | 60,00 | 0,581 | 42,86 | 0,588 | 51,43 | 0,582 | 37,14 |
| 18 | 0,545 | 57,14 | 0,625 | 60,00 | 0,575 | 42,86 | 0,620 | 37,14 | 0,582 | 37,14 |
| 19 | 0,551 | 62,96 | 0,626 | 60,00 | 0,595 | 51,43 | 0,629 | 54,29 | 0,603 | 45,71 |
| 20 | 0,557 | 62,96 | 0,613 | 54,29 | 0,608 | 54,29 | 0,596 | 57,14 | 0,603 | 45,71 |
| 21 | 0,550 | 48,15 | 0,610 | 54,29 | 0,608 | 54,29 | 0,596 | 57,14 | 0,608 | 51,43 |
| 22 | 0,540 | 48,15 | 0,611 | 54,29 | 0,625 | 54,29 | 0,586 | 62,86 | 0,607 | 51,43 |
| 23 | 0,558 | 60,00 | 0,615 | 54,29 | 0,633 | 54,29 | 0,589 | 62,86 | 0,622 | 54,29 |
| 24 | 0,558 | 60,00 | 0,616 | 57,14 | 0,636 | 54,29 | 0,636 | 62,86 | 0,624 | 54,29 |
| 25 | 0,585 | 51,43 | 0,620 | 57,14 | 0,643 | 57,14 | 0,634 | 62,86 | 0,632 | 57,14 |
| 26 | 0,599 | 51,43 | 0,623 | 57,14 | 0,643 | 57,14 | 0,629 | 62,86 | 0,633 | 51,43 |
| 27 | 0,592 | 51,43 | 0,626 | 57,14 | 0,634 | 62,86 | 0,619 | 60,00 | 0,633 | 51,43 |
| 28 | 0,574 | 40,00 | 0,635 | 57,14 | 0,626 | 56,47 | 0,619 | 60,00 | 0,635 | 51,43 |
| 29 | 0,620 | 62,86 | 0,640 | 57,14 | 0,623 | 56,47 | 0,619 | 60,00 | 0,637 | 51,43 |
| 30 | 0,620 | 62,86 | 0,640 | 57,14 | 0,623 | 56,47 | 0,619 | 60,00 | 0,634 | 62,86 |
| 67 | 0,682 | 60,00 | 0,682 | 60,00 | 0,682 | 60,00 | 0,682 | 60,00 | 0,682 | 60,00 |

Results presented in the Table 2 show that the subsets of criteria created by the feature selection procedures allow for a more correct classification than the plurality of subsets of the same criteria, created based on the ranking of the users. Hence, it can be concluded that each of the procedures for the selection of features allows users to better specify the criteria that determine the quality of websites. The next stage of the study consisted in assigning weights to criteria in Table 1 and calculating the mean absolute deviation according to equation (3). For MAD values for a subset of stances from 15 to 30, the criteria are shown in Table 3. Table 3 shows retained TRPMIN subsets which satisfy the conditions of> 50% and KCmin> 0.6, derived from Table 2.

**Table 3.** Values of mean absolute deviation for criteria subsets

| Features | 15 | 16 | 17 | 18 | 19 | 20 | 21 | 22 | 23 | 24 | 25 | 26 | 27 | 28 | 29 | 30 | 67 |
|---|---|---|---|---|---|---|---|---|---|---|---|---|---|---|---|---|---|
| Users | 2,41 | 2,51 | 2,59 | 2,94 | 2,57 | 2,62 | 2,53 | 2,67 | 2,69 | 2,66 | 2,68 | 2,61 | 2,67 | 2,65 | 2,6 | 2,8 | 2,09 |
| ReliefF | 2,02 | 1,97 | 1,92 | 1,82 | 1,86 | 1,86 | 1,94 | 1,96 | 1,85 | 1,83 | 1,88 | 1,92 | 1,95 | 1,9 | 1,91 | 1,91 | 1,85 |
| Sym. Uncert. | 1,81 | 1,77 | 1,83 | 1,86 | 1,78 | 1,81 | 1,79 | 1,79 | 1,71 | 1,77 | 1,77 | 1,81 | 1,87 | 1,89 | 1,84 | 1,89 | 1,85 |
| Sign. Attribute | 1,73 | 1,84 | 1,86 | 1,98 | 1,94 | 1,92 | 1,96 | 2,02 | 1,92 | 1,84 | 1,87 | 1,92 | 1,95 | 1,93 | 1,78 | 1,77 | 1,9 |
| Hellwig's Method | 1,5 | 1,47 | 1,5 | 1,52 | 1,45 | 1,56 | 1,54 | 1,59 | 1,59 | 1,57 | 1,6 | 1,54 | 1,5 | 1,6 | 1,54 | 1,68 | 1,81 |

When analyzing Table 3 it should be noted that the mean absolute deviation obtained for a subset of the criteria established on the basis of users is much higher than for other subsets. Moreover, for all subsets created using feature selection procedures, which meet the conditions TRPMIN > 50 % and KCmin > 0.6, the value of MAD is lower than the value obtained for the full set of features with weights which are the average weights given by users (users' 67 criteria).

# 5    Discussion

According to the linguistic interpretation, there was substantial agreement between the observed and predicted values of the classifier. After taking into account the confidence level 0.99 and the calculation of the standard error for the indicator KC = 0.738 the confidence interval [0.682, 0.794] was obtained. After examining the ratio KC for the full set of criteria, it is assumed that the reduction criteria for the lower limit of the confidence interval (determined in the same manner as for the full set of criteria) should reach a value higher than 0.6. Based on the interpretation of the linguistic values of KC, this means that consistency between observed and predicted values of the classifier, with a probability of 99 %, is maintained at a substantial level. In addition, it is assumed that the true positive rate for the worst classified class (TPRMIN) should reach a value of over 50 %. For a subset of the criteria obtained by use of the feature selection method, these values were achievable for a subset of up to about 20 criteria. In turn, for subsets based on the weights assigned by the user criteria, the parameters of such subsets number approximately 30 criteria. Table 2 shows sets of criteria ranging from 15 to 30 criteria. This analysis confirms the theory which states that, depending on which feature selection procedure is used, better identification of the relevant quality criteria and their weights can be found than from the users themselves. Based on the value of MAD, TPRMIN and KCmin selected subset of criteria characterizing the corresponding values of these parameters and the small cardinality criteria. The selected subset of a 21-piece set of criteria was created using individual indicators information capacity Hellwig (Method Hellwig, 21 criteria). This set complies with terms of the coefficients TPRMIN and KCmin, and is further characterized by a small number of the criteria used and one of the lowest values of the received MAD. Among other subsets that meet the established conditions, a slightly lower value of MAD is characterized by only a 27-piece set of criteria

692    P. Ziemba et al.

created using the Hellwig method. A much lower value of the mean absolute deviation, however, does not compensate for a greater multiplicity of evaluation criteria.

## 6    Summary

Proposed method for selection criteria that uses machine learning methods allowed the prediction of a suboptimal subset of criteria for assessing the quality of information services. This subset of the full set of criteria and their weights specified by the users allowed for a more precise multicriteria assessment of the information services. During the selection criteria it was also shown that declarations of the users' opinions regarding how important the criteria are do not correspond to the actual weights of the criteria. In other words, users were subconsciously guided by other criteria than those explicitly declared in their evaluation of the quality of websites. It can therefore be concluded that the developed method determines the implicit preferences that guide the users, assessing the true quality of the Internet service. Future work includes additional analysis related to using internal system parameters based on automated measurements.    Although for the indicated method of procedure only information services were presented as an example, it can adjusted to the designation criteria and their weights for the evaluation of websites of other type.

## References

1.  Netcraft, http://news.netcraft.com/archives/2012/12/04/december-2012-web-server-survey.html
2.  Internet World Stats, http://www.internetworldstats.com/stats.htm
3.  Gemius, http://www.gemius.pl/pl/aktualnosci/2014-02-17/01
4.  Kim, S., Stoel, L.: Dimensional hierarchy of retail website quality. Information & Management 41, 619–633 (2004)
5.  Chiou, W.C., Lin, C.C., Perng, C.: A strategic framework for website evaluation based on a review of the literature from 1995-2006. Information & Management 47, 282–290 (2010)
6.  Grigoroudis, E., Litos, C., Moustakis, V.A., Politis, Y., Tsironis, L.: The assessment of user-perceived web quality: Application of a satisfaction benchmarking approach. European Journal of Operational Research 187, 1346–1357 (2008)
7.  Barnes, S.J., Vidgen, R.: The eQual Approach to the Assessment of E-Commerce Quality: A Longitudinal Study of Internet Bookstories. In: Suh, W. (ed.) Web Engineering: Principles and Techniques, pp. 161–181. Idea Group Publishing (2005)
8.  Ahn, T., Ryu, S., Han, I.: The impact of Web quality and playfulness on user acceptance of online retailing. Information & Management 44, 263–275 (2007)
9.  Webb, H.W., Webb, L.A.: SiteQual: an integrated measure of Web site quality. Journal of Enterprise Information Management 17, 430–440 (2004)
10. Yang, Z., Cai, S., Zhou, Z., Zhou, N.: Development and validation of an instrument to measure user perceived service quality of information presenting Web Portals. Information & Management 42, 575–589 (2005)
11. Elling, S., Lentz, L., de Jong, M., van den Bergh, H.: Measuring the quality of governmental websites in a controlled versus an online setting with the 'Website Evaluation Questionnaire'. Government Information Quarterly 29, 383–393 (2012)

12. Zenebe, A., Zhou, L., Norcio, A.F.: User preferences discovery using fuzzy models. Fuzzy Sets and Systems 161, 3044–3063 (2010)
13. Ziemba, P., Budziński, R.: Dobór kryteriów dla oceny serwisów informacyjnych w portalach internetowych. Studia i materiały Polskiego Stowarzyszenia Zarządzania Wiedzą 37, 368–378 (2011)
14. Ziemba, P., Piwowarski, M.: Procedure of Reducing Website Assessment Criteria and User Preference Analyses. Foundations of Computing and Decision Sciences 36(3-4), 315–325 (2011)
15. Jankowski, J.: Integration of collective knowledge in fuzzy models supporting web design process. In: Jędrzejowicz, P., Nguyen, N.T., Hoang, K. (eds.) ICCCI 2011, Part II. LNCS, vol. 6923, pp. 395–404. Springer, Heidelberg (2011)
16. Hsu, H.H., Hsieh, C.W., Lu, M.D.: Hybrid feature selection by combining filters and wrappers. Expert Systems with Applications 38, 8144–8150 (2011)
17. Kononenko, I., Hong, S.J.: Attribute Selection for Modelling. Future Generation Computer Systems 13, 181–195 (1997)
18. Ahmad, A., Dey, L.: A feature selection technique for classificatory analysis. Pattern Recognition Letters 26, 43–56 (2005)
19. Senthamarai Kannan, S., Ramaraj, N.: A novel hybrid feature selection via Symmetrical Uncertainty ranking based local memetic search algorithm. Knowledge-Based Systems 23, 580–585 (2010)
20. Perzyńska, J.: Zastosowanie metody Hellwiga do wyznaczania wag prognoz kombinowanych. Ekonometria 34, 292–302 (2011)
21. Rokach, L., Maimon, O.: Classification Trees. In: Maimon, O., Rokach, L. (eds.) Data Mining and Knowledge Discovery Handbook, 2nd edn., pp. 149–174. Springer, Heidelberg (2010)
22. Webb, G.I.: Association Rules. In: Ye, N. (ed.) The Handbook of Data Mining, pp. 25–40. Lawrence Erlbaum Associates (2003)
23. Rokach, L., Maimon, O.: Supervised Learning. In: Maimon, O., Rokach, L. (eds.) Data Mining and Knowledge Discovery Handbook, 2nd edn., pp. 133–148. Springer, Heidelberg (2010)
24. Jankowski, J.: Analysis of Multiplayer Platform Users Activity Based on the Virtual and Real Time Dimension. In: Datta, A., Shulman, S., Zheng, B., Lin, S.-D., Sun, A., Lim, E.-P. (eds.) SocInfo 2011. LNCS, vol. 6984, pp. 312–315. Springer, Heidelberg (2011)
25. Ben-David, A.: Comparison of classification accuracy using Cohen's Weighted Kappa. Expert Systems with Applications 34, 825–832 (2008)
26. Kuchenhoff, H., Augustin, T., Kunz, A.: Partially identified prevalence estimation under misclassification using the kappa coefficient. International Journal of Approximate Reasoning 53, 1168–1182 (2012)
27. Pham-Gia, T., Hung, T.L.: The Mean and Median Absolute Deviations. Mathematical and Computer Modeling 34, 921–936 (2001)
28. Sheskin, D.J.: Handbook of Parametric and Nonparametric Statistical Procedures, 3rd edn. CRC Press (2003)

# Latent Semantic Indexing
# for Web Service Retrieval

Adam Czyszczoń and Aleksander Zgrzywa

Wrocław University of Technology, Division of Computer Science and Management,
Institute of Informatics. Wybrzeże Wyspiańskiego 27, PL50370 Wrocław, Poland
{adam.czyszczon,aleksander.zgrzywa}@pwr.edu.pl
http://www.zsi.ii.pwr.edu.pl/

**Abstract.** This paper presents a novel approach for Web Service Retrieval that utilizes Latent Semantic Indexing method to index both SOAP and RESTful Web Services. Presented approach uses modified term-document matrix that allows to store scores for different service components separately. Service data is collected and extracted using web crawlers. To determine similarities between user query and services the cosine measure is used. Presented research results are compared to standard Latent Semantic Indexing method. We also introduce our Web Service test collection that can be used for many benchmarks and make research results comparable.

**Keywords:** Latent Semantic Indexing, Web Service Retrieval, Web Service.

## 1   Introduction

Web services are application components that are interoperable and platform independent. A large number of Web Services are being developed as building blocks to construct small or large-scale distributed Web Systems. There are two classes of Web Services – commonly referred to in the literature as SOAP Web Services and RESTful Web Services [1,2,3,4]. Services of the first class are used mainly in the industry. Their description is provided in WSDL (Web Service Description Language) documents and they exchange information using SOAP (Simple Object Access Protocol) protocol. RESTful Web Services are used mainly in Web applications. Their description is provided in HTML documents, can be and most frequently use JSON (JavaScript Object Notation) format to transfer data objects. Developers have access to a variety of services that are published on the Internet and some of the services provide similar functionality. Finding most suitable services from the vast collection available on the Web is still the key problem for service–oriented systems. Moreover, category-based Web Service Discovery methods were inefficient and ineffective as they relied on keyword matching.

In order to solve this problem, Platzer et al. [5] presented a Web Service Retrieval approach that used Information Retrieval (IR) methodology and utilized

D. Hwang et al. (Eds.): ICCCI 2014, LNAI 8733, pp. 694–702, 2014.

the Vector Space Model (VSM). This method, however, does not account for the order and association between terms [6]. To capture the higher-order association between terms and services, another IR approach can be applied – Latent Semantic Indexing (LSI). It also allows to find hidden semantic association between terms even though they do not appear together in any service. Such a higher-order association cannot be acquired by the VSM model or by keyword-based service discovery mechanism. It also allows to reduce the VSM term-document matrix by transforming it to matrices that represent the underlying structure.

In this paper we present a novel approach for Web Service Retrieval that utilizes LSI to index both SOAP and RESTful Web Services. Presented approach uses modified term-document matrix that allows to store scores for different service components separately. Service data is collected and extracted using web crawlers. Finally, we utilize the cosine measure to determine similarities between user query and services and to retrieve the corresponding relevant services sorted according to highest relevance. Presented research results concern retrieval effectiveness and are compared to standard LSI method. Additionally, we introduce and publish our Web Service test collection that can be used for many benchmarks and make research results comparable.

## 2  Related Work

Currently, two approaches to finding Web Services can be distinguished. The first one is Service Discovery [7] which concerned with matchmaking SOAP services assigned to prespecified categories in UDDI (Universal Description Discovery and Integration) repository. Second one, considered in this paper, is Web Service Retrieval which relies on Information Retrieval models and uses web crawlers to collect services that are publicly accessible on the Internet. In contrast to service discovery it can be applied to both SOAP and RESTful Web Services.

One of the earliest papers on LSI for Web Services was proposed in [6] where authors presented Service Discovery mechanism that retrieves services from UDDI registry based on concepts that were extracted from user query and retrieved from ontology framework. Descriptions of relevant services were later indexed into TF-IDF inverted index file (term-document matrix) and reduced using LSI method. Although the research presented the Service Discovery approach, its further drawback was indexing service descriptions only and concept categorization in UDDI. Despite its considerably high effectiveness, the ad-hoc indexing and operations concerned with ontology browsing put the performance of presented approach in question.

In [8] authors presented service retrieval solution using an approach that does not rely on ontology engineering. Presented approach utilized web crawling to collect WSDL documents, TF-IDF inverted index and LSI models. However, more information than just service description was indexed – authors also included operations, messages and data types. On the other hand, words occurring in selected service parameters were put together as one bag-of-words. Moreover, no effectiveness evaluation was presented. In further research [9] authors presented effectiveness evaluation but only for single term queries. The research also concerned only SOAP Web Services.

Another drawback of mentioned studies (and in Web Service Retrieval in general) is lack of common evaluation test collection of services for comparing different approaches. In many cases authors give only the information about destination URLs of public Web Service directories. Such a information becomes outdated as there appear new services, any of service providers makes changes to the service, or service becomes unavailable. Such a snapshot of a directory that is not published to others makes it impossible to compare the results on the same dataset.

## 3    Latent Semantic Indexing and Vector Space Model

Latent Semantic Indexing is a method for revealing hidden concepts in document data. This is achieved by finding the higher-order association between different keywords using the the Singular Value Decomposition (SVD). However, LSI is based on the VSM where documents are represented as a vectors. Every element in the document vector is calculated as weight that reflects the importance of a particular term that occurs in this document. In order to compute the weights, the *TF-IDF* (Term Frequency–Inverse Document Frequency) scheme is used. The file structure that allows to store such a data is called inverted index (hereinafter referred to as the index). This index is a term-document matrix where document vectors represent the columns and term vectors represent rows. User queries are also converted vectors in the same manner as documents. In order to calculate the similarity between a query and document the cosine value between this two vectors is calculated. Because vectors have different lengths they have to be normalized to the unit equal to one. The cosine value is referred to as the similarity degree and it is used for ranking documents relevance against user query. Based on the above we define VSM index as follows:

**Definition 1.** *The VSM index $A$ is a $m \times n$ matrix where $m$ denotes the number of terms in documents collection and $n$ represents collection's size. Every $a_{ij} \in A$ represents weight of $i - th$ term in $j - th$ document calculated using the TF-IDF scheme.*

The LSI is a variant of the VSM in which the original VSM matrix is replaced by a low-rank approximation matrix. Therefore, the original vector space is reduced to a sub-space as close as possible to the original matrix [9] using the Singular Value Decomposition. The result of the decomposition is following:

$$A = U \Sigma V^T \qquad (1)$$

where $A$ is the initial VSM term($t$)-document($d$) matrix, $U \in R^{t \times t}$ is the matrix which columns represent term vectors, $\Sigma$ is a diagonal matrix of size $R^{t \times d}$ containing singular values, and $V \in R^{d \times d}$ is the matrix which columns represent the document vectors. Both $U$ and $V$ are orthogonal [10]. The number of singular values in $\Sigma$ determines the dimensions of the vector space. As such, we can reduce matrix $\Sigma$ into $\Sigma_K$ as $K \times K$ matrix to containing only the K singular values. The projection of $\Sigma_K$ on term matrix $U$ and document matrix $V^T$ allows to reduce them into $S_K$ and $U_K$ with K columns and rows. In result, initial matrix A is now approximated by:

$$A_K = U_K \Sigma_K V_K^T \tag{2}$$

After reduction, the terms are represented by the row vectors of the $m \times K$ matrix $U_K \Sigma_K$, and documents are represented by the column vectors the of the $K \times n$ matrix $\Sigma_K V_K^T$.

## 4   Web Service Structure

The definition of Web Service structure was elaborated in our previous study [11]. We consider Web Service to be composed of quadruple of elements where the first three represent *parameters* which correspond to service *name*, *description* and *version*, and the fourth represents service *components* which are composed of six-tuple of following component elements: *name, input value, output value, description*. The biggest advantage of presented structure is that it conforms to both SOAP and RESTful Web Services. The definition is as follows:

**Definition 2.** $WS = \langle p_1, p_2, p_3, C \rangle$ *where* $p_1, p_2, p_3$ *are service parameters and parameter* $C$ *denotes service component set* $C = \{c_1, c_2, \ldots, c_n\}$ *where each component* $c_i \in C$ *is represented by following four-tuple* $c_i = \langle A_1, A_2, A_3, A_4 \rangle$.

## 5   LSI for Web Service Retrieval

Current LSI approaches treat all service components as single bag-of-words. Based on our research carried out in [11] Web Services are indexed as in the following manner:

**Definition 3.** *The extended index A is a* $m \times n$ *matrix where m denotes the number of terms in Web Service collection and n represents collection's size. Elements* $a_{ij}$ *represent service parameters a five-tuple* $< p_1, p_2, p_3, c >$ *where* $p_1$ *represents weights of service's name,* $p_2$ *service's description,* $p_3$ *service's version, and c represents weight of all service components. Weights are calculated using the modified TF-IDF scheme presented in Equation 3.*

For such a index, we model Web Services in vector space in following manner: each parameter and component is represented as a vector composed of weights where each weight corresponds to a term from the "bag of words" of particular service parameter or component. The bag-of-words content of all service components is merged into single content. Weights are calculated using modification of one of the best known combination of *TF-IDF*:

$$TFIDF = (1 + log(tf(t, \alpha))) \cdot log\frac{N_\alpha}{|\{\alpha \in WS : t \in \alpha\}|} \tag{3}$$

where $tf(t, \alpha)$ is term frequency of term $t$ in parameter or component $\alpha$, $N_\alpha$ is the total number of $\alpha$ parameter/component, and $|\{\alpha \in WS : t \in \alpha\}|$ is the number of parameters/components where the term $t$ appears.

Because parameter/component counts are stored separately there is substantial difference in weights between standard LSI indexing model and the extended one. For example: there are 2 services $A$ and $B$. In service $A$ term $X$ occurs 1 time in name and 2 times in components. In service $B$ term $X$ does not occur at all. In basic index the $TF = 3$, $DF = 1$, and thus $TF - IDF = 0.44$. However, in extended index the $TF$-$IDF$ is equal to 0.39 and $MWV = 0.01$, since for name the $TF = 1$, $DF = 1$, $TF - IDF = 0$, and for components $TF = 2$, $DF = 1$, $TF - IDF = 0.39$.

As we can see the extended VSM index is 4 times bigger than the standard one. In order to reduce it to its original size, we merge parameters/components using the $MWV$ method presented in [11] and computed as average weight of all service parameters. The final index structure is following:

**Definition 4.** *The MVW index $A'$ is a $m \times n$ matrix where $m$ denotes the number of terms in Web Service collection and $n$ represents collection's size. Elements $a_{ij} \in A'$ represent the MWV weight of five-tuple $b_{ij} = < p_1, p_2, p_3, c >$ where $b_{ij} \in A$.*

Afterwards, for such a index the LSI model is applied.

# 6  Test Collections

In many studies on Web Service Retrieval and Discovery authors do not share their evaluation test collections of services. Instead, authors give only the information about destination URLs of public Web Service directories. Such a information quickly becomes outdated as new services appear, get modified or become unavailable. In result, comparing research outcome of different approaches is impractical. In order to solve this problem we prepared two test collections of SOAP Web Services[1]: *SOAP_WS_12*, *SOAP_WS_14*.

First collection contains 267 Web Services collected from `xmethods.net` – a directory of publicly available Web Services, used by many researchers for service retrieval benchmarks. The crawl was performed in 2012 for the purpose of the research presented in [11]. Second collection contains 662 services collected from: `xmethods.net`, `service-repository.com`, `webservicex.net`, `venus.eas.asu.edu`, `visualwebservice.com` and `programmableweb.com` – popular and constantly updated directory of public Web Services.

---

[1] All test collections are available to download at:
   `http://www.ii.pwr.edu.pl/~czyszczon/WebServiceRetrieval`

At present, introduced in this paper test collections do not include any REST-ful Web Services because our methods of their identification on the Web are still being improved and currently developed collection is too small. This, however, does not influence experimental results presented in this paper.

In Table 1 we present summary data of the above datasets. Collection named *SOAP_WS_12* will be used in this paper for retrieval effectiveness analysis and the second one for indexing performance analysis.

**Table 1.** Test collections structure

|                | SOAP_WS_12 | SOAP_WS_14 |
|----------------|------------|------------|
| Services       | 267        | 662        |
| Parameters     | 432        | 886        |
| Components     | 5140       | 18353      |
| Total elements | 5572       | 19239      |

## 7    Evaluation

Based on the approach presented in this paper we implemented indexing system that allowed us to conduct evaluation experiments. The goal of the experiment was to measure the performance and effectiveness of proposed approach. Additionally, we compared our results to standard Latent Semantic Indexing and Vector Space Model indexing methods. In order to evaluate the effectiveness we used the *WS_SOAP_12* test collection and the following classical information retrieval measures: *Precision, Recall, F-measure* ($\beta = 1$) and *Mean Average Precision (MAP)*. In order to evaluate performance we checked the indexing time

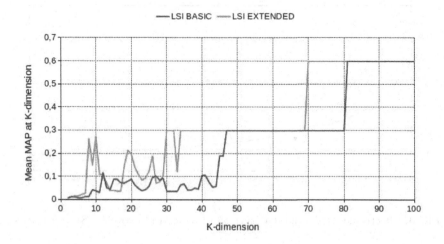

**Fig. 1.** Mean MAP at K-dimensions for basic and extended LSI

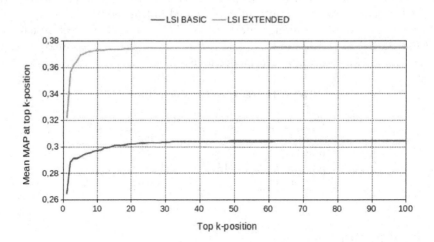

**Fig. 2.** Mean MAP at top $k$-positions for basic and extended LSI

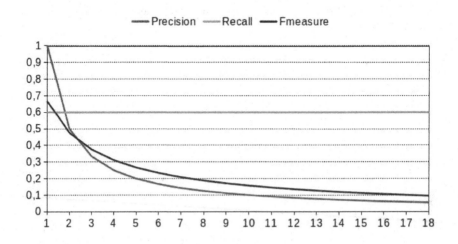

**Fig. 3.** Effectivenes evaluation for basic and extended LSI with K=81 dimensions

and memory consumption for *WS_SOAP_12*, *WS_SOAP_14*, *WS_SOAP_12-14* test collections of every index structure.

The experiment was carried out for following queries: *"temperature conversion"*, *"email validation"*, *"weather forecast"*, *"weather forecast service"*, *"currency exchange"*, *"demographics"*, *"new york"* and *"send sms"*, denoted as $q_1$, $q_2$, ..., $q_8$. The MAP is computed as the average precision for queries $q_{1..8}$.

## 7.1 Dimension Reduction Analysis

The first step was to define the threshold for dimension reduction of the LSI index. One of the common approaches is to reduce the space to $K$ largest singular values [6]. However, the best $K$ value needs to be found empirically [9]. To do that, for every index structure with $K = [2, 267]$, we calculated the Mean of MAP of every top $k$ positions, where $k = [1, 267]$. This allowed us to check which value of $K$ will give the best retrieval results on top of the results list. The dataset used was the *WS_SOAP_12* with 267 Web Services. The experiment was conveyed for basic and extended LSI matrices. The results are presented on Figure 1.

In basic index the Mean MAP achieved its highest value equal to 0.6 at $K = 81$. Additionally, if we look at the MAP results for LSI index with $K = 81$, we can see that the highest MAP was achieved at first top position (see Figure 3). In the case of extended index, the highest value of Mean MAP was was also achieved at first top position but it was reached at smaller index with $K = 70$. This assures the same effectiveness as in the basic size but with 13.6% smaller index.

In Figure 2 we illustrate Mean MAP of top $k$-positions, for every $K$. Illustration confirm that the extended LSI reaches maximal recall faster, but also shows that it returns more relevant services at higher top positions. Most of the relevant services were returned within top 10 results.

## 7.2 Effectiveness Evaluation

In Figure 3 we illustrate the retrieval effectivenes of both index structures using LSI index with $K = 81$ dimenstions at top 20 positions. Relevant services were found at the very top of the list. However, not all relevant services were returned (recall=0.6). The overall effectiveness of LSI indicate that proposed meyhod performs very well and results are satisfactory.

# 8 Conclusions and Future Work

The goal of presented research was to propose an alternative LSI approach for Web Service Retrieval. The modified LSI approach included modified matrix that allowed to count scores for different service components separately. In result, the overall retrieval effectiveness should be higher than in the basic LSI index.

The experimental results proved that proposed extended model of LSI is superior to the standard model in terms of indexing performance and retrieval effectiveness. The dimension reduction showed not only the optimal index size values for basic and extended methods, but also confirmed that extended index achieved its maximal effectiveness at 13.6% smaller index size. Additionally, the extended index returned more relevant services within smaller list of top results.

In further research we plan to check effectiveness of different indexing methods and *TF-IDF* variants, especially the *ltc.lnc* in SMART notation, since the calculation of *IDF* for queries requires additional computation. This, in result, requires to load additional data into computer memory. Some researchers on Web Service Retrieval also suggested that skipping length-normalization may bring better results. This is also a subject of our future work on Web Service Retrieval methods.

# References

1. Booth, D., Haas, H., McCabe, F., Newcomer, E., Champion, M., Ferris, C., Orchard, D.: Web services architecture. W3C Working Group Note World Wide Web Consortium (February 11, 2004), http://www.w3.org/TR/ws-arch (accessed: April 2014)
2. Erl, T., Bennett, S., Schneider, R., Gee, C., Laird, R., Carlyle, B., Manes, A.: Soa Governance: Governing Shared Services On-Premise and in the Cloud. In: The Prentice Hall Service-oriented Computing Series from Thomas Erl. Prentice Hall (2011)
3. Pautasso, C., Zimmermann, O., Leymann, F.: Restful web services vs. big web services: Making the right architectural decision. In: 17th International WWW Conference, Beijing (2008)
4. Richardson, L., Ruby, S.: RESTful Web Services: Web Services for the Real World. O'Reilly Media, Inc., Sebastopol (2007)
5. Platzer, C., Dustdar, S.: A vector space search engine for web services. In: Proceedings of the 3rd European IEEE Conference on Web Services (ECOWS 2005), pp. 14–16. IEEE Computer Society Press (2005)
6. Paliwal, A.V., Adam, N.R., Bornhovd, C.: Web service discovery: Adding semantics through service request expansion and latent semantic indexing. In: IEEE SCC, pp. 106–113. IEEE Computer Society (2007)
7. Mukhopadhyay, D., Chougule, A.: A survey on web service discovery approaches. CoRR abs/1206.5582 (2012)
8. Wu, C., Chang, E., Aitken, A.: An empirical approach for semantic web services discovery. In: Australian Software Engineering Conference, pp. 412–421. IEEE Computer Society (2008)
9. Wu, C., Potdar, V., Chang, E.: Latent semantic analysis – the dynamics of semantics web services discovery. In: Dillon, T.S., et al. (eds.) Advances in Web Semantics I, vol. 4891, pp. 346–373. Springer, Heidelberg (2009)
10. Hyung, Z., Lee, K., Lee, K.: Music recommendation using text analysis on song requests to radio stations. Expert Systems with Applications 41(5), 2608–2618 (2014)
11. Czyszczoń, A., Zgrzywa, A.: The concept of parametric index for ranked web service retrieval. In: Zgrzywa, A., Choroś, K., Siemiński, A. (eds.) Multimedia and Internet Systems: Theory and Practice. AISC, vol. 183, pp. 229–238. Springer, Heidelberg (2013)

# Author Index